Handbook of Research on Power and Energy System Optimization

Pawan Kumar
Thapar University, India

Surjit Singh
National Institute of Technology Kurukshetra, India

Ikbal Ali
Jamia Millia Islamia, India

Taha Selim Ustun
Carnegie Mellon University, USA

A volume in the Advances in Computer and Electrical Engineering (ACEE) Book Series

IGI Global
DISSEMINATOR OF KNOWLEDGE

Published in the United States of America by
IGI Global
Engineering Science Reference (an imprint of IGI Global)
701 E. Chocolate Avenue
Hershey PA, USA 17033
Tel: 717-533-8845
Fax: 717-533-8661
E-mail: cust@igi-global.com
Web site: http://www.igi-global.com

Library of Congress Cataloging-in-Publication Data

Names: Kumar, Pawan, 1977- editor. | Singh, Surjit, 1981- editor. | Ali, Ikbal, editor. | Ustun, Taha Selim, editor.
Title: Handbook of research on power and energy system optimization / Pawan Kumar, Surjit Singh, Ikbal Ali, and Taha Selim Ustun, editors.
Description: Hershey, PA : Engineering Science Reference, [2018] | Includes bibliographical references.
Identifiers: LCCN 2017024144| ISBN 9781522539353 (hardcover) | ISBN 9781522539360 (ebook)
Subjects: LCSH: Electric power systems--Management--Data processing--Handbooks, manuals, etc. | Electric power systems--Mathematical models--Handbooks, manuals, etc. | Mathematical optimization--Handbooks, manuals, etc.
Classification: LCC TK1005 .H29375 2018 | DDC 621.31--dc23 LC record available at https://lccn.loc.gov/2017024144

This book is published in the IGI Global book series Advances in Computer and Electrical Engineering (ACEE) (ISSN: 2327-039X; eISSN: 2327-0403)

British Cataloguing in Publication Data
A Cataloguing in Publication record for this book is available from the British Library.

All work contributed to this book is new, previously-unpublished material. The views expressed in this book are those of the authors, but not necessarily of the publisher.

For electronic access to this publication, please contact: eresources@igi-global.com.

Advances in Computer and Electrical Engineering (ACEE) Book Series

Srikanta Patnaik
SOA University, India

ISSN:2327-039X
EISSN:2327-0403

Mission

The fields of computer engineering and electrical engineering encompass a broad range of interdisciplinary topics allowing for expansive research developments across multiple fields. Research in these areas continues to develop and become increasingly important as computer and electrical systems have become an integral part of everyday life.

The **Advances in Computer and Electrical Engineering (ACEE) Book Series** aims to publish research on diverse topics pertaining to computer engineering and electrical engineering. **ACEE** encourages scholarly discourse on the latest applications, tools, and methodologies being implemented in the field for the design and development of computer and electrical systems.

Coverage

- VLSI Fabrication
- Applied Electromagnetics
- Computer Architecture
- Circuit Analysis
- Computer science
- Programming
- Algorithms
- Chip Design
- Computer Hardware
- Optical Electronics

IGI Global is currently accepting manuscripts for publication within this series. To submit a proposal for a volume in this series, please contact our Acquisition Editors at Acquisitions@igi-global.com or visit: http://www.igi-global.com/publish/.

The Advances in Computer and Electrical Engineering (ACEE) Book Series (ISSN 2327-039X) is published by IGI Global, 701 E. Chocolate Avenue, Hershey, PA 17033-1240, USA, www.igi-global.com. This series is composed of titles available for purchase individually; each title is edited to be contextually exclusive from any other title within the series. For pricing and ordering information please visit http://www.igi-global.com/book-series/advances-computer-electrical-engineering/73675. Postmaster: Send all address changes to above address.

Titles in this Series

For a list of additional titles in this series, please visit: www.igi-global.com/book-series

Big Data Analytics for Satellite Image Processing and Remote Sensing
P. Swarnalatha (VIT University, India) and Prabu Sevugan (VIT University, India)
Engineering Science Reference • copyright 2018 • 253pp • H/C (ISBN: 9781522536437) • US $215.00 (our price)

Modeling and Simulations for Metamaterials Emerging Research and Opportunities
Ammar Armghan (Aljouf University, Saudi Arabia) Xinguang Hu (HuangShan University, China) and Muhammad Younus Javed (HITEC University, Pakistan)
Engineering Science Reference • copyright 2018 • 171pp • H/C (ISBN: 9781522541806) • US $155.00 (our price)

Electromagnetic Compatibility for Space Systems Design
Christos D. Nikolopoulos (National Technical University of Athens, Greece)
Engineering Science Reference • copyright 2018 • 346pp • H/C (ISBN: 9781522554158) • US $225.00 (our price)

Soft-Computing-Based Nonlinear Control Systems Design
Uday Pratap Singh (Madhav Institute of Technology and Science, India) Akhilesh Tiwari (Madhav Institute of Technology and Science, India) and Rajeev Kumar Singh (Madhav Institute of Technology and Science, India)
Engineering Science Reference • copyright 2018 • 388pp • H/C (ISBN: 9781522535317) • US $245.00 (our price)

EHT Transmission Performance Evaluation Emerging Research and Opportunities
K. Srinivas (Transmission Corporation of Andhra Pradesh Limited, India) and R.V.S. Satyanarayana (Sri Venkateswara University College of Engineering, India)
Engineering Science Reference • copyright 2018 • 160pp • H/C (ISBN: 9781522549413) • US $145.00 (our price)

Fuzzy Logic Dynamics and Machine Prediction for Failure Analysis
Tawanda Mushiri (University of Johannesburg, South Africa) and Charles Mbowhwa (University of Johannesburg, South Africa)
Engineering Science Reference • copyright 2018 • 301pp • H/C (ISBN: 9781522532446) • US $225.00 (our price)

Creativity in Load-Balance Schemes for Multi/Many-Core Heterogeneous Graph Computing Emerging Research and Opportunities
Alberto Garcia-Robledo (Center for Research and Advanced Studies of the National Polytechnic Institute (Cinvestav-Tamaulipas), Mexico) Arturo Diaz-Perez (Center for Research and Advanced Studies of the National Polytechnic Institute (Cinvestav-Tamaulipas), Mexico) and Guillermo Morales-Luna (Center for Research and Advanced Studies of the National Polytechnic Institute (Cinvestav-IPN), Mexico)
Engineering Science Reference • copyright 2018 • 217pp • H/C (ISBN: 9781522537991) • US $155.00 (our price)

IGI Global
DISSEMINATOR OF KNOWLEDGE

701 East Chocolate Avenue, Hershey, PA 17033, USA
Tel: 717-533-8845 x100 • Fax: 717-533-8661
E-Mail: cust@igi-global.com • www.igi-global.com

Editorial Advisory Board

List of Contributors

Table of Contents

Section 2
Energy Management and Conversion

Detailed Table of Contents

show how they utilize the machine learning models trained from a large experimental database to perform precise prediction and optimization on a solar water heater (SWH) system. A new energy system optimization strategy based on a high-throughput screening (HTS) process is proposed. This chapter consists of: 1) comparative studies on varieties of machine learning models (artificial neural networks [ANNs], support vector machine [SVM], and extreme learning machine [ELM]) to predict the performances of SWHs; 2) development of an ANN-based software to assist the quick prediction; and 3) introduction of a computational HTS method to design a high-performance SWH system.

This chapter presents four effective evolutionary methods, namely grey wolf optimization (GWO), symbiotic organism search (SOS), JAYA, and teaching-learning-based optimization (TLBO), for solving automatic generation control (AGC) problem in power system. To show the effectiveness, two widely used interconnected power plants are examined. To extract maximum possible generation, distinct PID-controllers are designed employing ITAE-based fitness function. Further, to enhance the dynamic stability of concerned power systems, 2DOF-PID controllers are proposed in LFC area and optimally designed using aforesaid algorithms. To demonstrate the supremacy, obtained results are compared with some existing control algorithms. Moreover, robustness of the designed controller is believed under the action of random load perturbation (RLP). Finally, sensitivity analysis is carried out to show the stability of the designed system under loading and parametric disturbance conditions.

Wide-area measurement system (WAMS) is an important part of present power system structure as it provides real-time synchronized measurements of the system with the aid of phasor measurement units (PMUs). Due to economic considerations, PMUs should be installed at optimal locations. The optimal placement of PMUs (OPP) is a problem of optimally placing PMUs at strategic locations maintaining the full observability of the system. In this chapter, a novel binary whale optimization algorithm (BWOA) is applied to solve OPP problem. The maximization of measurement redundancy is considered in the objective function. The proposed algorithm is examined on five different test systems operating under normal operating conditions with or without inclusion of zero-injection buses (ZIBs) and compared with the reports available in literature. The results show the effectiveness of the proposed algorithm in solving OPPP.

The standalone solar photovoltaic (PV) powered water pumping system could have either two power conditioning units (PCUs) or a single PCU. The system with single PCU is advantageous over the prior because of low device count, smaller in size, low cost, low losses, and higher efficiency. This chapter includes four such single-stage PV water pumping systems. All the four systems are operated using the integrated control technique which assimilates maximum power point tracking (MPPT), pulse width modulation (PWM) technique, and motor control. But the PV systems differ in the inverter configuration used as well as the employed induction motor (conventional star connected and open-end winding induction motor [OEWIM]). The detailed description of mathematical modeling, design and analysis of all the four PV pumping systems along with the simulation results are presented in this chapter.

In this chapter, an interactive tool using graphic user interface (GUI) environment-based MATLAB is proposed to solve practical optimal power system planning and control. The main particularity of the proposed tool is to assist student and researchers understanding the mechanism search of new metaheuristic methods. The proposed tool allows users to interact dynamically with the program. The users (students or experts) can set parameters related to a specified metaheuristic method to clearly observe the effect of choosing parameters on the solution quality. In this chapter, a new global optimization method named grey wolf optimizer (GWO) and pattern search algorithm (PS) have been successfully applied within the interactive tool to solve the optimal power flow problem. The robustness of the two proposed metaheuristic methods is validated on many standard power system tests. The proposed interactive optimal power flow tool is expected to be a useful support for students and experts specialized in power system planning and control.

Today's power systems have been extended due to interconnections with the neighboring systems. Energy cost is increasing due to greater sensitivity of new transmission lines. These have demanded the search of new controllers. These objectives were met by reactive power control on transmission lines. Hence, FACTS controllers are becoming an important part of the modern power transmission systems. Different control techniques are in use with STATCOM to generate control signals. But the conventional controllers like P; PI provide a slow dynamic response during the fault condition. This chapter creates an

dynamic voltage restore (DVR), static compensator in distribution system (D-STATCOM), and unified power quality conditioner (UPQC). The third one is combination of DVR and D-STATCOM, which is known as UPQC. The UPQC can compensate all the PQ problems. Comprehensive knowledge of UPQC is presented here to understand the trade value of this device. This chapter is organized as follows: The evaluation of power electronic based devices is presented in the background part of this chapter where custom power devices (CPDs) are also discussed. And operating principle with the Matlab model of UPQC and results are presented in the section titled "Working Principle of Unified Power Quality Conditioner (UPQC)." The scope of the one of the functional power electronic based devices are presented in section titled "Scope of UPQC." At the end, this chapter is concluded.

Chapter 11
Performance of the Multicarrier Sinusoidal PWM-Based Multilevel Inverter With Reduced Power Loss and Fault Diagnosis Using Wavelets Transformation: Hybrid Multilevel Inverter.................... 353
Durga Prasad Garapati, Shri Vishnu Engineering College for Women, India
Jegathesan V., Karunya Institute of Technology and Sciences, India
Moorthy Veerasamy, Vishnu Institute of Technology, India

The performances of multilevel inverters (MLIs) are of high competence when compared to the conventional two-level inverters due to reduced harmonic distortions, lower electromagnetic interference, and higher dc link voltages. However, the increased number of components, complex PWM control, and voltage-balancing problem, component failure in the circuit are some of the disadvantages. The topology preferred in this chapter provides a dc voltage in the shape of a staircase which approximates the rectified shape of a commanded sinusoidal wave to the bridge inverter, which in turn alternates the polarity to produce an AC voltage with lesser total harmonic distortion. This topology requires fewer components and hence it leads to a reduction of overall cost and complexity particularly for higher output voltage levels. The component fault diagnostic procedure is developed using wavelets transform tool. Finally, the experimental prototype is developed and validated with the simulation results for different loading conditions.

Section 2
Energy Management and Conversion

Chapter 12
Physics, Modelling, and Optimization Studies of Photon-Enhanced Thermionic Emission-Based Hybrid Energy Conversion System: A Review... 399
S. C. Kaushik, Indian Institute of Technology Delhi, India
Ravita Lamba, Indian Institute of Technology Delhi, India
S. K. Tyagi, Indian Institute of Technology Delhi, India

The sustainable development of clean and efficient electricity generation techniques accelerated the research for invention of alternative electricity generation methods. In this chapter, the conceptual analysis of newly invented photon-enhanced thermionic emission (PETE) energy conversion process has been presented. It is a promising option for harvesting solar energy in terms of capturing photon as well as thermal energy simultaneously and converting solar energy into electrical energy based on photovoltaic and thermionic emission processes of energy conversion. Thus, the PETE process utilizes photons for PV conversion and heat of radiation for thermionic emission process. The main objective of this chapter is to review and analyze the performance of PETE converters including thermal modeling,

The growing demand for electrical energy calls for the assimilation of renewable energy sources to the main utility grid. Multiple renewable energy sources (RESs) like solar PV array, wind turbine, micro-hydro plant, etc. can be combined and controlled to form a microgrid. In spite of the availability of different microgrid topologies, DC microgrid largely facilitates the injection of DC power from various renewable energy sources into the stabilised DC power pool. The requirement for a minimal number of conversion stages, simple structure, economic operation, and numerous localised applications are driving factors for the DC microgrid technology. The mettle of the DC microgrid technology lies in choosing the appropriate microgrid participants for energy interchange and the suitable supervisory control to tap power from the microgrid partakers even after respecting their operating constraints. The use of high gain DC-DC converters is inevitable in DC microgrid due to the low terminal voltage levels of different RESs.

The foremost issues of the twenty-first century are the ever-increasing challenging demand of electrical energy and controlling the emission of greenhouse gases (GHG). Along with these issues and with limited energy resources, it is imperative to look for non-conventional methods of power generation like from renewable energy resources. Microgrid has emerged as a new field that can meet the energy demand with a special emphasis on good power quality, reliability, and security. A major concern with the use of renewable energy resources is their intermittent nature which makes their integration and operation a challengeable task. Energy storage devices like batteries can be used to overcome the problem of intermittent nature of renewable energy resources. This chapter focusses on different aspects of renewable energy resources in detail. It analyzes the effectiveness of the proposed topology of the microgrid for health clinic load profile with the help of PVSYST software.

In this chapter, the authors present a new scheduling algorithm that brings a reduction in dynamic power consumption by achieving components scheduling while holding the global latency of the application. The main idea of that algorithm is to augment the latency of some components without impacting the dependency constraint and degrading the global latency of the system. There exist many solutions that manage to increase component's latency; one of them is through decreasing the frequency of their corresponding clocks. Generally, such a method leads to an augmentation in global latency of a system. However, this algorithm manages to reduce the consumed power and hold the same global latency of the system. The presented algorithm has been tested and it provides a significant gain in power at both simulation and physical levels.

Foreword

I am delighted to write the foreword to this book, as its scope and content provide commercial and technical enterprises with the essential ingredients for implementing and managing power and energy systems techniques and management schemes.

The recent advances in the power and energy technology, especially in the optimization, renewable, energy management through analysis and simulation allowed a potentially enormous market for energy management and conversion. With advances in the latest field of electrical engineering including: energy efficiency, optimization and renewable energy resources have contributed to the solution of recent energy problems for any nation and the environmental issues.

However, now the major challenge is to maintain the high energy efficiency during operation and to explore sustainable energy resources. To deal with the technical challenges, two major areas in power and energy technologies are being identified:

1. Energy optimization and simulation
2. Energy management and conversion

This handbook provides answers to many challenging questions dealing power and energy system optimization. It addresses a variety of issues related to the energy management and the recently developed optimization techniques. This handbook comprised of 19 chapters divided into two parts as per the information providing according to the above areas. Part-A describes the various optimization approaches recently applied in the power and energy system optimization and the simulation results based upon the present problems faced by the research in energy optimization. However, Part-B describes the various energy management and conversion schemes for improving the energy efficiency in electrical utilities.

I recommend this handbook to researchers and practitioners in the field, and for scientists and engineers involved in power and energy systems optimization.

I really appreciate the efforts of all the Editors to compile this book. The managing editor Dr. Kumar and his team have meticulously collected the chapters, reviewed and place them in appropriate way for better in depth understanding.

I believe the readers of power and energy system will be benefited from the work presented in this book.

Srete Nikolovski
University of Osijek, Croatia

Srete Nikolovski *received his Ph.D. in Power system from Faculty of electrical engineering and computing Zagreb. His research fields of activities include active involvements in power and energy systems. Prof. Nikolovski was the coordinator of TEMPUS EU project EMSA "Electricity Market Simulation and Analysis Curricula for Engineering Education", EU ITEA 2 project ESNA "European Sensor Network Architecture". Prof. Nikolovski worked as Main researcher of scientific Croatian project MZOS Nr. 0165116 "Reliability of Electric Power System in deregulated electricity market and MZOS-a Nr.165-1651481-1482 "Quality and reliability of Croatian electric power system on regional electricity market. Dr. Nikolovski has authored seven books of the "Power System Protection" (2005, 2008), "The Basic of Reliability Analysis of Power System" (1995), Introduction to Electricity Market (2010), Electrical network I - solved problems" (1989, 2000), "Program Packages for Power System Analysis and Simulation" (1997, 2003). His main interests are power system protection, power system modeling, simulation and reliability. He has published 200 academic papers in journals and international conferences proceedings. He is IEEE Senior member of Reliability Society and PES Society, member of CRO CIGRE and CRO CIRED societies.*

Preface

In recent years, the world has witnessed a rapid growth in the demand for electrical energy and consequently the serious energy crises having direct impact on economics, society and development of any country. With the development of new technologies, the modern energy systems are changing rapidly and reveal increasingly multifaceted features. Nowadays, energy conversion and supply is not the only mission of power utilities rather energy stability, reliability, security and efficiency has become the major concern. This is to manage and operate the energy resources for a long-term future and to minimize their impacts on our ecology and environmental system. Energy efficiency in planning and operation of power systems allows maximizing the benefits of utilities and customers concurrently. Though, saving of single unit of energy is termed as energy resource so it contributes to the mitigation of carbon emissions for the globe's climate. Therefore, energy optimization also expected to secure their resources by loss reduction in operation.

In order to find the solution of the above-mentioned objectives, new technologies have continuously emerged with improved efficiency. With the implementation of modern technologies, it further helps to explore the renewable energy resources and their optimum usages. Although we achieved much, still there is a huge gap between present energy generation and the demand. To meet the load demand, we are dependent on the conventional resources since they are available continuously whereas, renewable energy sources such as solar and wind are available at the different time of the day. Therefore, for reliable operation dependence on conventional fossil fuels is still sturdy in most parts of the world.

On the other hand, predominant reliance on fossil fuel-based electrical energy use is a direct cause for a serious issue of the global warming. Therefore, promotion of the use of renewable energy sources, for meeting electrical energy demand, is an important strategy in order to enhance the energy security of any country. In this connection, wind and solar based electrical power generation have gained the attention of the researchers all over the world. This is enabling in high penetration of the renewable sources with the main electrical grid. Power generation from these renewable resources is intermittent because of its dependence on the environmental conditions. As a result, the power generation from solar and wind systems keep on fluctuating and has a direct impact on the voltage magnitude, supply frequency, and waveform, and hence on the quality and quantity of supplying power to the interconnected grid system.

The world energy requirement would be 70% higher in 2030 than today demand, even at the constant rate of growth. On the other hand, conventional resources are depleting day by day which raises the serious concern for energy utilization and optimization to improve operating efficiency. As a fact, this is the energy consumption per capita which describes the living standard in the country. Therefore, we need to explore more and more options for the energy resources, which are naturally available, and optimization techniques to improve energy efficiency.

The existing problems of modern energy systems indicate clearly that the available means of planning and operating the energy systems are far from perfect solution, and there is a large potential for the improvements. This book demonstrates the potential of energy systems engineering-based approach to systematically quantify different options at different levels of complexity (i.e. planning, operation, control and utilization) through state-of-the-art modeling, simulation, control, and optimization-based frameworks. The successful implementation of these approaches in a number of real-life case studies highlights further the significance of this integrated system-wide approach. This book presents the importance of fundamental and applied research in power and energy systems applications by developing mechanisms for the transfer of the new methodology, which is applicable to the real-time problem.

In recent years, power and energy is the major area for research. For optimal utilization of electrical power, we need to explore the several optimization techniques. The recent developments in the computational intelligence found to be a most effective tool for obtaining the optimal operation of power systems. In power system, loads are not of any specific type and they vary with the state of the economy of system operation. Therefore, it becomes very difficult to identify the optimal configuration of distribution system under the change in loading conditions. In this scenario, energy efficiency and auditing plays an important role in the energy management system. The automatic control of the power component improves the reliability and the bi-directional communication involves the smart control of system operation, which is termed as the smart grid.

DESCRIPTION OF THE BOOK

This handbook aims to be an essential reference source, building on the available literature in the field of power and energy system optimization, providing further research opportunities in this field. This specific text is expected to provide the primary and major resources necessary for researchers, academicians, students, faculties, and scientists, across the globe, to adopt and implement new inventions in thrust area of power generation from conventional and non-conventional resources and their utilization and energy management. Therefore, the handbook of research on power and energy system optimization is to provide a platform to share up-to-date scientific achievements in the core as well as related fields. There are three main objectives underlying this book:

1. Identifying and exploring the scope of different energy resources
2. Identifying the scope of energy management
3. Identifying the various optimization schemes/algorithm/approaches/techniques for intensification of energy efficiency in power generation, transmission, and utilization.

This handbook aims to be an essential reference source, building on the available literature in the field of power and energy system optimization, providing further research opportunities in this field. This specific text is expected to provide the primary and major resources necessary for researchers, academicians, students, faculties, and scientists, across the globe, to adopt and implement new inventions in thrust area of power generation from conventional and non-conventional resources and their utilization and energy management. According to the application domain and nature, the chapters in this book are categorized into two sections:

Section 1: Energy Optimization and Simulation
Section 2: Energy Management and Conversion

From recent trends, it can be observed that research is mainly focused on energy efficiency, renewable energy resource, and the computational intelligence. In the world of smart grid, application of artificial intelligence has become a trend. Three domains covered in this book will provide the vast knowledge to the researcher and it will also give common source of information to the power engineers.

Energy efficiency is the burning issue nowadays. Researchers are highly involved in this area and this is due to the fact that energy demand is exponentially growing whereas, energy resources are depleting day by day. This requires optimal use of electrical power. In order to find the optimal operation of the electrical system, we have to explore several optimization techniques. This book will be a source of motivation to the researcher and can exchange the knowledge of energy efficiency, optimization techniques and integration of DG for better utilization of the resources.

In order to meet the load demand and to improve the energy efficiency researchers are working in the area of integration of distributed generation, micro-grid, automation in power distribution, hybrid electric vehicle, and synchronized operation of solar power, battery and the grid supply for the development of smart power system for the home. They have developed several optimization techniques involving extensive search, analytical approach, and the computational or artificial intelligence. In recent years, the computational intelligence based on nature inspired meta-heuristic approaches are finding wider acceptance for optimization of discrete problems particularly for large systems with no fixed solution under different operating conditions.

In the first part, book chapters are describing the application of the different algorithm for optimization and simulation. A brief description of the chapters in Section 1 of this book is given as follows.

Chapter 1

This chapter presents a fuzzy logic approach for reactive power and voltage control in grid connected wind farms with different types of wind generator units to improve steady state voltage stability of power systems. The load buses voltage deviation is minimized by changing the reactive power controllers according to their sensitivity using fuzzy set theory. The proposed approach uses only a few controllers of high sensitivity to achieve the desired objectives. To prove the effectiveness of the proposed approach a comparative analysis is also carried out with the conventional linear programming based reactive power optimization technique. Results demonstrated that the proposed approach is more effective in improving the system performance as compared with the conventional existing techniques.

Chapter 2

This chapter presented the performance prediction and optimization of solar water heater via a knowledge-based machine learning method. Directly predicting their performance is challenging due to the complicated internal structures. Fortunately, a knowledge-based machine learning method can provide a promising prediction and optimization strategy for the performance of energy systems. In this Chapter, the authors will show how they utilize the machine learning models trained from a large experimental database to perform precise prediction and optimization on a solar water heater (SWH) system. A new energy system optimization strategy based on a high-throughput screening (HTS) process is proposed.

Chapter 3

This chapter presents recently developed four effective evolutionary methods, namely; grey wolf optimization (GWO), symbiotic organism search (SOS), JAYA, and teaching learning based optimization (TLBO) for solving automatic generation control (AGC) problem in power system. To show the effectiveness, two widely used interconnected power plants are examined. To extract maximum possible generation, distinct PID-controllers are designed employing ITAE based fitness function. Further, to enhance the dynamic stability of concerned power systems, 2DOF-PID controllers are proposed in LFC area and optimally designed using aforesaid algorithms. The robustness of the designed controller is believed under the action of random load perturbation (RLP). Moreover, sensitivity analysis is carried out to show the stability of the designed system under loading and parametric disturbance conditions.

Chapter 4

In this chapter, a novel binary whale optimization algorithm (BWOA) is applied to solve optimal placement of PMUs. The maximization of measurement redundancy is considered in the objective function. The proposed algorithm is examined on five different test systems operating under normal operating conditions with or without the inclusion of zero-injection buses (ZIBs) and compared with the reports available in the literature. The results show the effectiveness of the proposed algorithm.

Chapter 5

This chapter presented the modeling, design, and analysis of Solar PV powered standalone water pumping systems. The standalone solar Photo Voltaic (PV) powered water pumping system could have either two Power Conditioning Units (PCUs) or a single PCU. The system with single PCU is advantageous over the prior, because of low device count, smaller in size, low cost, low losses and higher efficiency. This chapter includes four such single-stage PV water pumping systems. All the four systems are operated using the integrated control technique which assimilates Maximum Power Point Tracking (MPPT), Pulse Width Modulation (PWM) technique and motor control. But the PV systems differ in the inverter configuration used as well as the employed induction motor. The detailed description of mathematical modeling, design, and analysis of all the four PV pumping systems along with the simulation results are presented in this chapter.

Chapter 6

In this chapter, an interactive tool using graphic user interface (GUI) environment based MATLAB is proposed to solve practical optimal power system planning and control. The main particularity of the proposed tool is to assist student and researchers understanding the mechanism search of new meta-heuristic methods. The proposed tool, allows users to interact dynamically with the program. The users (students or experts) can set parameters related to a specified meta-heuristic method to clearly observe the effect of choosing parameters on the solution quality. In this study a new global optimization method named Grey Wolf Optimizer (GWO) and pattern search algorithm (PS) have been successfully applied within the interactive tool to solve the optimal power flow problem. The robustness of the two proposed meta-heuristic methods is validated on standard power system tests. The proposed interactive optimal power flow tool is expected to be a useful support for students and experts specialized in power system planning and control.

Chapter 7

This chapter presents the controlling STATCOM by full order state feedback controller for stability improvement. Today's power systems have been extended due to interconnections with the neighboring systems. Energy cost is increasing due to the greater sensitivity of new transmission lines. These have demanded the search of new controllers. These objectives were met by reactive power control on transmission lines. Hence FACTS Controllers are becoming an important part of the modern power transmission systems. Different control techniques are in use with STATCOM to generate control signals. But the conventional controllers like P, PI provide a slow dynamic response during the fault condition. This chapter will be intended to create an awareness of control issues and has proposed observer based state feedback controller for STATCOM. The simulation results show that the observer based state feedback controller gives better performance than conventional PI controller during contingency condition.

Chapter 8

This chapter presents the possible scope of integration of solar power, battery and the grid supply for the energy management in a sample system under consideration. Integration of renewable resource is the need of the hour whereas there is a limited scope of power generation from these resources. In this chapter solar power is integrated with the grid through grid-tie-inverter. The solar irradiance and the temperature are varied differently to show the effect of variation of sun energy in a day time. Under these variations the system operation is studied and it has been observed that the power integration changes differently at different time of the day. As a result, it is challenging to find the accurate integration of the solar, grid and battery power in order to develop a smart power system for home.

Chapter 9

This chapter presents the design of synchronous controller to minimize response time and power. Since power dissipation and time constraint have become vital challenges during the creation of a digital circuit therefore, researchers and designers efforts have increased to figure out new ways of preserving power through the study of its sources and its impacts as well as through the decrease of response time to obtain faster treatments. The main objective of this chapter is to propose a design method that simultaneously builds a low power design and provides a faster circuit. For the achievement of that purpose, a controller based on a finite state machine (FSM) has been chosen as an example of synchronous system to prove that the new proposed design can optimize both parameters: time and power.

Chapter 10

This chapter introduces the custom power device to solve the various kinds of power quality (PQ) problems. It is generally known as flexible AC transmission system (FACTS) and D-FACTS (FACTs used for low voltage and low current rating in AC distribution system). The D-FACTS are various types like dynamic voltage restore (DVR), Static compensator in distribution system (D-STATCOM) and unified power quality conditioner (UPQC). The third one is combination of DVR and D-STATCOM which is known as UPQC. Comprehensive knowledge of UPQC is presented here to understand the trade value of this device.

Chapter 11

This chapter presented the performances of Multi-Level Inverters (MLIs). MLIs are found to be of high competence when compared to the conventional two level inverters due to reduced harmonic distortions, lower electromagnetic interference, and higher dc link voltages. The topology preferred in this chapter provides a dc voltage in the shape of a staircase which approximates the rectified shape of a commanded sinusoidal wave to the bridge inverter, which in turn alternates the polarity to produce an AC voltage with lesser total harmonic distortion. In the proposed topology fewer components are required which in turn reduces the cost and complexity particularly for higher output voltage levels.

In Section 2, book chapters are describing the energy management, energy conversion, energy trading, energy planning and its utilization. A brief description of the chapters in Section 2 of this book is given as follows:

Chapter 12

This chapter presented the conceptual analysis of newly invented photon enhanced thermionic emission (PETE) energy conversion process. The sustainable development of clean and efficient electricity generation techniques accelerated research for the invention of alternative electricity generation methods. The main objective of the study is to review and analyze the performance of PETE converters including thermal modeling, choice of materials and parametric optimization. The appropriate choice of material requirements for cathode and anode of PETE converters is necessary for the practical design of PETE converters. The PETE converter may be an efficient future option for electricity generation using solar energy.

Chapter 13

This chapter presented the significance of demand side management (DSM) for improvement in energy efficiency through demand response. Here, energy efficiency and demand response fulfilling the criteria of energy management which usually tries to take influence onto the energy consumption of a number of energy consumers are discussed. DSM objectives are peak clipping, valley filling, load shifting, load building, energy conservation and flexible load shape. A sample case study of an institution building is also presented to demonstrate the DSM approach and recommendations.

Chapter 14

This chapter presented the overview of market power in the deregulated power system. In today's competitive market, deregulation of the power industry is inevitable. The aim of deregulating the power markets is to bring competition into them and thereby make them more economically efficient. Though, the electricity wholesale market is not a perfect market and the potential for market power exploitation is an issue. The private companies collaborate with each other to get more profit, driving the prices to a higher level and thus acquiring a market power which is an anti-competitive practice. Thus market power is the capability of a seller or a group of sellers to profitably maintain the prices above a competitive level and control the total output for a noteworthy period of time.

Chapter 15

This chapter presented the dynamic energy management and integration of renewable resources in DC micro-grid system. The growing demand for electrical energy calls for the assimilation of renewable energy sources to the main utility grid. Multiple renewable energy sources (RES's) like solar PV array, wind turbine, micro-hydro plant etc., can be combined and controlled to form a micro grid. In spite of the availability of different micro grid topologies, DC micro grid largely facilitates the injection of DC power from various renewable energy sources into the stabilized DC power pool. The requirement for a minimal number of conversion stages, simple structure, economic operation, and numerous localized applications are few more driving factors for the DC micro grid technology. The use of high gain DC-DC converters is inevitable in DC micro grid due to the low terminal voltage levels of different RES's.

Chapter 16

This chapter focuses on different aspects of renewable energy resources in detail. It analyzes the effectiveness of the proposed topology of the micro-grid for health clinic load profile with the help of PV-SYST software. A major concern with the use of renewable energy resources is their intermittent nature which makes their integration and operation challengeable task. Energy storage devices like batteries can be used to overcome the problem of intermittent nature of renewable energy resources. Micro-grid has emerged as a new field which can meet the energy demand with a special emphasis on good power quality, reliability, and security.

Chapter 17

This paper presented a new scheduling algorithm that brings a reduction in dynamic power consumption by achieving components scheduling while holding the global latency of the application. The main idea of that algorithm is to augment the latency of some components without impacting the dependency constraint and degrading the global latency of the system. There exist many solutions that manage to increase component's latency; one of them is through decreasing the frequency of their corresponding clocks. Generally, such a method leads to an augmentation in global latency of a system. However, our algorithm manages to reduce the consumed power and hold the same global latency of the system. The presented algorithm has been tested and it has provided a significant gain in power at both: simulation and physical levels.

Chapter 18

This paper presented an optimal photovoltaic conversion system for future smart grids. Smart grid technology is the key for a reliable and efficient use of distributed energy resources. Amongst all the renewable sources solar power takes the prominent position due to its availability in abundance. In this chapter, the authors present smart grid infrastructure issues and integrating solar PV sourced electricity in the smart grid. The smart grid has many features, including reliability, flexibility on network topology, efficiency, sustainability, and market-enabling. The authors select a Photovoltaic Active Power Line Conditioner, as a case study. This line conditioner a device designed to extract the maximum power of a photovoltaic (PV) system and to compensate the nonlinear and unbalanced loads of the electrical power systems. The performance of the PV conditioner with the neuro-fuzzy control designed has been analyzed through a simulation platform.

Chapter 19

This chapter presented the system dynamics approach for energy policy. System dynamics approach was developed by Jay Forrester from MIT during 1950's to analyze especially the complex behavior in administration with computer simulation in social sciences. System dynamics is a form of the systems approach as a methodology to understand the dynamic behavior of complex systems. The basis of system dynamics is to understand how system structures cause system behavior and system events. System dynamics, policy analysis, and design are used for learning and decision making.

Handbook of research on power and energy system optimization contains 19 chapters of high-quality contributions from international leading researchers in the field of power and energy systems. The book chapters are divided into two sections depending upon the methodology and the approach described by the authors for the energy optimization and or management. First parts mainly consist of optimization techniques for voltage control, stability and reliability improvement in the electrical system's operation, control and planning. Recently, several heuristic and meta-heuristic optimization techniques are presented by the researchers in which genetic algorithm is the popular approach whereas other newly developed approaches such as binary whale optimization, grey wolf optimization, symbiotic organism search, JAYA (based on victory means Jaya), teaching learning based optimization and artificial neural network found to give significantly improved results in the power and energy optimization. However, simulation results have also appeared in some chapters showing the time response for dynamic operation. The second part consists of management and conversion techniques for improvement in energy efficiency and security. Since energy utilization is mainly by motors load and lighting. It is observed that the energy management and efficiency is the major concern in the present scenario when conventional resources are depleting day by day.

Acknowledgment

This handbook of research on power and energy system optimization is an outcome of the inspiration and encouragement given by many individuals for whom these words of thanks are only a token of our gratitude and appreciation for them.

Our sincere gratitude goes to the peoples who contributed their time and expertise to this book. We highly appreciate their efforts in achieving this project. The editors would like to acknowledge the help of all the people involved in this project and, more specifically, the editors would like to thank each one of the authors for their contributions and the editorial board/reviewers regarding the improvement of quality, coherence and the content presentation of this book.

Second, Editors would like to express their sincere thanks to Jordan Tepper, Mariah Gilbert, Jan Travers, Lindsay Wertman, and Courtney Tychinski of IGI Global for their continuous support and giving us an opportunity to edit this book.

The editors are thankful to our family members for their prayers, encouragement, and care shown towards us during the completion of this handbook of research on power and energy system optimization. Thank you all!!!

We also express our gratitude to the GOD for all the blessings!!!

Pawan Kumar
Thapar University, India

Ikbal Ali
Jamia Millia Islamia, India

Taha Selim Ustun
Carnegie Mellon University, USA

Surjit Singh
National Institute of Technology Kurukshetra, India

Section 1
Energy Optimization and Simulation

Chapter 1
Fuzzy-Logic-Based Reactive Power and Voltage Control in Grid-Connected Wind Farms to Improve Steady State Voltage Stability

Tukaram Moger
National Institute of Technology Karnataka, India

Thukaram Dhadbanjan
Indian Institute of Science Bangalore, India

ABSTRACT

This chapter presents a fuzzy logic approach for reactive power and voltage control in grid-connected wind farms with different types of wind generator units to improve steady state voltage stability of power systems. The load buses' voltage deviation is minimized by changing the reactive power controllers according to their sensitivity using fuzzy set theory. The proposed approach uses only a few high sensitivity controllers to achieve the desired objectives. A 297-bus-equivalent grid-connected wind system and a 417-bus-equivalent grid-connected wind system are considered to present the simulation results. To prove the effectiveness of the proposed approach, a comparative analysis is also carried out with the conventional linear-programming-based reactive power optimization technique. Results demonstrated that the proposed approach is more effective in improving the system performance as compared with the conventional existing techniques.

DOI: 10.4018/978-1-5225-3935-3.ch001

NOTATIONS

a: Transformer off-nominal tap
j: Complex operator
α: Angle of the primitive admittance of the line
y: Primitive admittance of the line
Y: Admittance matrix
B: Line charging susceptance
ΔQ: Incremental change in bus reactive power
ΔV: Incremental change in bus voltage magnitude
ΔT: Incremental change in transformer tap
θ: Angle of the bus admittance matrix
δ: Voltage phase angle
V^d_j: Desired voltage at load bus j
V^a_j: Actual voltage at load bus j
V^e: Error or difference voltage at load bus
V_{min}: Minimum or lower limit on bus voltage
V_{max}: Maximum or upper limit on bus voltage
X: Row vector of the voltage/reactive power controllers
$[C]$: Controlling ability of the controller matrix
m: Controller index
t_g: Number of grid OLTC transformers
t_{pcc}: Number of OLTC transformers at the point of common coupling of wind farm
g: Number of generators
svc: Number of switchable VAr compensators
vswg: Number of variable speed wind generators
r: Number of remaining buses
M: Voltage/reactive power controller step size
$[H]$: Voltage deviation sensitivity matrix
$[S]$: Linearized sensitivity matrix relating dependent and control variable
L-index: Static voltage stability index
n: Number of total buses
P_{loss}: System active power loss
Q_{loss}: System reactive power loss
L^{max}*-index*: Maximum value of voltage stability index

ABBREVIATIONS

AC: Alternating current
DFIG: Doubly fed induction generator
FERC: Federal energy regulatory commission
FLA: Fuzzy logic approach
FSWG: Fixed speed wind generator

GSC: Grid side converter
kV: Kilo volt
LP: Linear programming
MVAr: Mega volt ampere reactive
MVA: Mega volt ampere
MW: Mega watts
NL: Negative large
NLP: Non linear programming
NM: Negative medium
NS: Negative small
NVL: Negative very large
OLTC: On load tap changing transformer
OPF: Optimal power flow
PCC: Point of common coupling
P-V: Power-voltage
PR-FSWG: Pitch regulated fixed speed wind generator
pu: Per unit
pf: Power factor
PF: Power flow
PL: Positive large
PM: Positive medium
PS: Positive small
PVL: Positive very large
RSC: Rotor side converter
SCIG: Squirrel cage induction generator
SGFEC: Synchronous generator with front end converter
SR: Southern region
SR-FSWG: Stall regulated fixed speed wind generator
SSWG: Semi-variable speed wind generator
STDEV: Standard deviation
SVC: Switchable VAr compensator
VAr: Volt ampere reactive
V: Voltage
VSWG: Variable speed wind generator
UPF or UPF: Unity power factor
VS-WTGU: Variable speed wind turbine generating unit
VSC: Voltage source converter
WF: Wind farm
WGU: Wind generator unit
WTGU: Wind turbine generating unit
Z: Zero

INTRODUCTION

In recent years, the electric power industry around the world is changing continuously due to transformation from regulated market structure to deregulated market structure. The main aim of the transformation of electric supply industry under open access environment is to overcome the some of the limitations faced by a vertically integrated system. It is believed that this transformation will bring new technologies, and integration of other sources of energy or unconventional energy sources such as wind, solar, fuel cells, bio-gas, etc., which are self-sustainable and competitive. Moreover, it will also give better choice for the consumers. As a result, several new issues and challenges have emerged. One of the main issues in power systems is to support reactive power for maintaining the system voltage profile with an acceptable margin of security and reliability required for system operation.

In recent years, it has been observed that the voltage instability problem is the root cause for several major network blackouts in different countries such as France, Belgium, Sweden, Germany, Iran, Japan, USA and India (FERC, 2005; Srivastava, Velayutham, Agrawal, & Bakshi, 2012; Ajjarapu & Lee, 1998). Therefore, the system operator must make sure that there are enough reactive power reserve capacities available for the system to maintain the voltage profiles. Properly planned reactive power reserves minimize the risk of voltage collapse or low voltages as well as reduce transmission loss by keeping the appropriate voltage profiles (Elgerd, 1982; Taylor, 1994).

With concerns relating to climate change, energy prices, supply uncertainties and other factors, there is an increasing focus on renewable energy sources to satisfy energy requirements. Among the renewable energy resources, wind energy is gaining prominence due to the available technologies that allow for large scale power generation. Hence, the share of wind power in relation to the overall installed capacity has increased significantly. As wind energy installations are rapidly growing worldwide, the system operators are more concerned about the planning and operation of the system with wind farms.

The transmission grid code (Wu, Xu, & Østergaard, 2010; Tsili & Papathanassiou, 2009) in many countries requires wind farms to supply not only active power but also reactive power to the power grid. The grid code also specifies the reactive power ranges for wind generators as different types of wind generators will have different reactive power capability. The grid code regulations for reactive power/power factor requirements of wind farms in the system under steady state operating condition in some of the countries is summarized in Table 1.

Due to stochastic nature of wind, the wind farms adversely affect the voltage of the grid. It is necessary to control the load buses voltage and reactive power in the system so that the power quality of grid connected wind farms is maintained within the acceptable limits. Therefore, the reactive power and voltage control in grid connected wind farms is one of key issues that grid operators are concerned with, and received great attention from the researchers in recent years.

Presently, various types of WTGUs have been installed in the wind farms and can be broadly classified into two types, namely, fixed speed wind generators (FSWGs) which include both fixed and semi-variable speed type wind generators, and variable speed wind generators (VSWGs) which include doubly fed induction generator (DFIG) and synchronous generator with front end converter (SGFEC). The fixed speed wind generators are used in earlier wind farms installations. The lack of voltage or power factor regulation features in the FSWGs makes it difficult to operate these units with the grid as per the grid code requirements. To address these issues, researchers in the past have proposed several approaches/models to utilize the features of the FSWG in the system operation and make it compatible with the grid code requirements (Moger & Dhadbanjan, 2017). The fixed speed wind generators are

Table 1. Summary of existing grid code for reactive power/power factor requirement of wind farms

Country	Reactive Power/Power Factor Requirement		
	Max. Q Limit (Injection Limit)	**Min. Q Limit (Absorption Limit)**	**Remarks**
India (CERC, 2010)	0.95 lagging	0.95 leading	
UK (NGET, 2009)	0.95 lagging	0.95 leading	
US (FERC, 2005, 2012)	0.95 lagging	0.95 leading	
Alberta (Canada, 2004)	0.90 lagging	0.95 leading	
China (China, 2009)	0.97 lagging	0.97 leading	
Spain (Spain, 2005)	0.98 lagging	0.98 leading	Without any penalty
Ireland (EirGrid)	0.95 lagging	0.95 leading	
Australia (Australia, 2007)	1.0 UPF	0.95 leading	For 100% power
Germany (Germany, 2006)	0.95 lagging	0.95 leading	For capacity < 100 MW
	Voltage dependent reactive power limit for capacity > 100 MW		

Wu et al., 2010; Tsili & Papathanassiou, 2009.

of conventional asynchronous machine, which do not have capability to control the voltage, but rather absorbs large levels of reactive power during the normal operation. However, due to advancement in power electronics converter design, the variable speed wind generators have the ability to control the active and reactive power independently, and could be able to meet the grid code requirements. Hence, it is feasible to consider variable speed wind generators as possible sources of reactive power.

The growing importance of reactive power planning in the network has drawn the attention of system expansion planners to consider this important aspect of the resource planning. Reactive power controlling devices are being installed in transmission and distribution networks to enhance their operational efficiency in terms of better voltage profile, voltage control, increased power flow, and flexibility of power control and enhancing the stability of the system. These reactive power controls are scattered throughout the transmission network and are to be operated in coordination.

Judicious dispatch of reactive power is essential for the stable operation of the power system. It facilitates the flow of active power from generation sources to load centers (Kundur, Balu, & Lauby, 1994; Taylor, 1994) and maintains the bus voltages within prescribed limits (Abed, 1999). Stable operation of power systems requires the availability of sufficient reactive power generation in the system.

BACKGROUND

The integration of wind farms with power systems is taking place both at the transmission and distribution voltage levels. Several integration issues are also reported in the literature. The performance of grid connected wind turbine generating units could be quite different from the one that is operating in isolation (at nominal voltage and frequency). This is because the grid voltage and frequency fluctuate around the nominal value and hence, it is necessary to study the impact performance of the wind turbine generating units when connected to the grid to ensure the secure and reliable operation of the systems. These impacts can be positive or negative depending on the system operating characteristics and wind turbine generator itself (Trichakis, Taylor, Lyons, & Hair, 2008). Therefore, it is necessary to carry out an intensive research for reactive power and voltage control of wind farms, and review the related compensation facilities and strategies in order for the wind farms to be able to participate in grid voltage control and subsequently increase the grid stability.

The impact of wind generating units (WGUs) on various system aspects have been studied in the literature particularly the voltage stability of the system under steady state as well as transient conditions.

Vittal et al. (2010) investigated the impact of variable speed wind turbine generators on steady state voltage stability of the system. The voltage stability margin of both the distribution level and transmission level buses in the system can be increased through the proper implementation of voltage control strategies in the DFIG wind turbines. Freitas et al. (2005) investigated the long-term or small disturbance voltage stability of distribution systems with squirrel cage induction generators (SCIG) by using time-domain nonlinear dynamic simulations. The system stability margin is analyzed through the P-V curves. The system voltage stability margin may reduce due to the presence of SCIG. Usually, at the maximum loading point, if the system loading is increased even more, then the induction generator accelerates to a high speed and becomes unstable, which leading the system to a voltage collapse. Hossain et al. (2012) investigated the behavior of SCIG and DFIG on short-term voltage stability. Ullah and Thiringer (2007) proposed the improvements in grid voltage stability and transient stability with variable speed wind turbine generators using power electronics based converter with modified controller design. The voltage source converter (VSC) in modern variable speed wind turbine generators is utilized to achieve this enhancement. The suggested modifications can be incorporated in real installations through which the short-term voltage stability is improved as an additional feature of an existing VSC for the high voltage direct current (HVDC) installation.

As per the federal energy regulatory commission (FERC) order 661-A on interconnection requirements (FERC, 2005, 2012), the wind farm of capacity more than 25 MW should maintain the power factor range of 0.95 leading and lagging at the point of common coupling to the grid. Due to which, the reactive power capability of wind farms is drastically under-utilized. Fully utilizing the reactive power capability of DFIG wind farm in voltage control produces the significant improvements in the system performance and may prevent system voltage instability. Even though the FERC order 661-A gives general guidelines for interconnecting wind parks, but for specific parks employing DFIG units, the restriction on power factor can be lifted for enhancing the system voltage stability. In this connection, Konopinski et al. (2009) discussed the impact of utilizing reactive power capability curve of a doubly fed induction generator on steady state and dynamic power system operation.

Vittal et al. (2008) identified the ratio of DFIG wind turbine generators to fixed speed wind turbine generators required to maintain voltage and frequency stability during steady state and single line contingencies cases in the system. Further, authors discussed the effect of change in wind power generation and control strategies on system voltage profile (Vittal, O'Malley, & Keane, 2009) and voltage stability. P-V curve analysis was employed to demonstrate the improvement in voltage stability attained using terminal voltage control. In addition, other approaches are proposed to utilize the flexibility in reactive power generation of DFIG to reduce system losses (Meegahapola, Durairaj, Flynn, & Fox, 2011) and improve the reliability in static and dynamic system operation (Konopinski, Vijayan, & Ajjarapu, 2009). Chen et al. (2012) presented a coordinated reactive power and voltage control for wind farm gird integration to meet the operation requirements as described in the standards of state grid corporation of China. The control strategy of the proposed system takes into account the coordination between the voltage and power factor. The authors Kayikci and Milanovic (2007) discussed the different combinations of reactive power control of rotor side converter (RSC) and grid side converter (GSC) for voltage control purposes with the various network parameters and operation regimes. Chen et al. (2009) proposed a reactive/voltage control strategy of a wind farm to improve the voltage stability of the wind farm, but only one wind farm is considered. For a network integrating several large wind farms, how to improve the voltage stability by coordinating the reactive power among the wind farms is seldom reported in the literature. Aghatehrani et al. (2009) proposed a coordination method using both the rotor side and the grid side converters for voltage regulation and reactive power support considering the operational limits and network side voltage fluctuation. The central idea is to derive the reactive power reference signal in the rotor side converter (RSC) control loop from the terminal voltage controller in the grid side converter (GSC) control loop. Zhu et al. (2009) proposed a reactive power compensation control method considering voltage and reactive power command in integrated region. The voltage at the point of interconnection is chosen as controlled voltage and the related reactive power control is divided into three sections, namely, normal control, abnormal control and urgent control, which can be chosen according to detailed operation status of the system. Zheng et al. (2013) proposed a coordinated voltage control strategy using the sensitivity method. The coordinated reactive power and voltage control of wind farms is divided into two control hierarchies. In primary control, each wind farms regulate the voltage reference value at the point of interconnection of wind farm, and while in secondary control, the control center of wind farms give the values of voltage reference and reactive power adjustment. With the reactive power adjustment value of the wind farms based on sensitivity matrix, the voltage at the point of interconnection of the wind farms is adjusted to the reference value if it is exceeding the allowable range.

Most of the approaches reported in the literature (Keane, Ochoa, Vittal, Dent, & Harrison, 2011; Aghatehrani, Fan, & Kavasseri, 2009; Zhu, Chen, Wang, & Zhu, 2009) studied the impact of variations in reactive power output of variable speed wind generators (VSWGs) on voltage stability of the systems in isolation. Significant improvements in voltage stability can be attained if the reactive power output of the VSWGs is properly coordinated with other reactive power controllers in the grid. To achieve the reactive power requirement of the system in an optimal manner, grid operators may perform reactive power optimization within their own facilities. Some optimization based approaches have been reported in the literature (Meegahapola, Durairaj, Flynn, & Fox, 2011; Kumar, Reddy, & Dhadbanjan, 2014) for optimally dispatching reactive power to improve the voltage profile in grid connected wind farms.

De Almeida et al. (2006) proposed an optimized dispatch control strategy for active and reactive powers delivered by a doubly fed induction generator in a wind park. The control strategy used at the wind generator level exploits a combination of pitch control and control of the static converters to adjust the rotor speed for the required operating points. The optimization procedure is able to minimize the deviation between the total active and reactive power delivered by the wind farms as required by the system operator. Kumar et al. (2014) proposed a trust region framework for coordinating the reactive power output of VSWGs with other reactive power controllers in the grid for enhancement of system voltage stability and voltage profiles. However, the studies were carried only considering the VSWGs in the wind farms. In practical wind farms, the fixed speed wind generators are also present. The studies in the literature show that the fixed speed wind generating units have significant impact on system voltage performance (Moger & Dhadbanjan, 2017; Divya & Rao, 2006). This is due to the requirement of reactive power support from the grid especially during low voltage conditions. Hence, it must be considered while coordinating the reactive power output of VSWGs with other reactive power controllers in the grid.

In the last few decades, more attention was paid to reactive power and voltage control in power systems to improve the system voltage profile. The reactive power and voltage control is a subset of the optimal power flow (OPF) problem, which seeks to find an optimal set of parameters for steady state power system operation. It is a constrained, nonlinear problem of considerable complexity. Since the problem of reactive power optimization is non-linear in nature, non-linear programming (NLP) methods have been used to solve it. The NLP methods work quite well for small power systems but may develop convergence problems as system size increases. The studies performed on some IEEE standard test systems (Pudjianto, Ahmed, & Strbac, 2002) show that NLP based optimal power flow is comparatively less robust with respect to convergence under all the random starting points. A bad initial point, for instance near any operating limit of control and/or state variables, as well as a too narrow operating range of control or state variables may limit the permissible step length for those variables and therefore restrict the movement of the other variables. This would then cause non-convergence of the NLP based OPF. Linear programming technique with iterative scheme is certainly the most promising tools for solving these types of problems (Dhadbanjan, Parthasarathy, & Prior, 1984; Dhadbanjan & Yesuratnam, 2006).

In the context of real-time operation, the voltage stability analysis should be performed online. The conventional algorithms or approaches (Lof, Andersson, & Hill, 1993) such as linear programming (LP) require significant computation time as it uses all the controllers to achieve the desired objectives. For online applications, there is a need for tools which can quickly detect the potentially dangerous situations of voltage instability and provide guidance to the operators to steer the system away from a possible voltage collapse. Tinney et al. (1988) stated that there is no way to select a subset of the most important control actions from the total set of control actions in the conventional OPF solution since the actions are not ranked and importance of an action is not necessarily related to its magnitude. Therefore, conventional optimization problem uses all of the controllers to achieve the desired objectives. Researchers have explored other possibilities like fuzzy logic, expert systems and so on to address the issues of concern.

Recently, the efforts to improve the speed and ability to handle stressed power systems have led to the development of intelligent systems. Recent developments indicate that artificial intelligence techniques like fuzzy logic (Su & Lin, 1996; Bansal, 2003; El-Hawary, 1998), artificial neural networks (Scala, Trovato, & Torelli, 1996; El-Keib & Ma, 1995; Niebur, 1993) and expert systems (Dhadbanjan, Bansilal, & Parthasarathy, 1997; Dhadbanjan & Parthasarathy, 1995) and so on, may be appropriate for assisting the operators for real-time operation.

In recent years, many applications of fuzzy set theory to reactive power control problem have been reported in the literature (Su & Lin, 1996; Yokoyama, Niimura, & Nakanishi, 1993; Abdul-Rahman & Shahidehpour, 1993; Marques, Taranto, & Falc˜ao, 2005). In all these applications of fuzzy set theory to reactive power control problem, the objectives are to either minimize real power losses or improve the voltage profile of the given system.

Tomsovic (1992) proposed a fuzzy linear programming approach to reactive power and voltage control problem. Here, the objective and constraints are modeled using fuzzy sets and corresponding linear membership functions are defined and solved using LP technique. Udupa et al. (1999) presented an approach using fuzzy set theory for reactive power control with the purpose of improving the voltage stability of power systems. The voltage stability index (*L-index*) and controlling variables are translated into fuzzy set notations to formulate the relation between the voltage stability level and controlling ability of controlling devices using the linearized model. Then, fuzzy rule based system is formed to achieve the desired goal. Su and Lin (2001) presented an approach using fuzzy set theory for voltage and reactive power control of power systems. The purpose is to find a solution, which takes both voltage security enhancement and loss reduction into account for an electric power system. The proposed approach translates violation level of buses voltage and controlling ability of controlling devices into fuzzy set notations using linearized model. Nageswararao and Jeyasurya (1998) presented a fuzzy expert system approach to steady state voltage stability monitoring and control in power systems. In the proposed rule based expert system, a systematic procedure for voltage stability monitoring and control is developed. The key variables are the load bus voltage, generator MVAr reserve and generator terminal voltage. The three key variables (which have significant impact on the system performance) are selected based on the solution obtained by repeated load flow and modal analysis performed for various operating conditions. The information from the above analysis is stored in a knowledge base. A set of decision rules relating key system variables to voltage stability are established. The membership functions of the key variables and fuzzy rules may be defined based on the system requirements and operators experience. The designed expert system performs voltage stability monitoring and control through deductive reasoning. Tang et al. (1993) presented an expert system for voltage correction in interconnected power systems. As the control variables such as shunt capacitors and reactors, OLTC transformers are discrete, an algorithm for heuristic, one dimensional search in a discrete state space of control variables is proposed. The search is based on a heuristic function, which takes both the priority and the regulation effect of control variables into account. Shareeq et al. (2014) proposed a fuzzy rule based approach for addressing the reactive power and voltage control of grid connected wind farms. In the proposed approach the wind turbine generators are modeled as PV buses and therefore, the voltage of the wind turbine generators is considered as control variables in the reactive power optimization process. Further, Moger and Dhadbanjan (2016) proposed a new procedure in the fuzzy logic approach to address the issues associated with various controllers in grid connected wind farms. The FSWGs are also considered in the studies because of its impact on overall system voltage performance even though they do not support the system for voltage unlike the VSWGs (Moger & Dhadbanjan, 2017).

OBJECTIVES AND CONTRIBUTION

This chapter presents a fuzzy logic approach for reactive power coordination in grid connected wind farms with different types of wind generator units (wind farms with different wind generator types i.e., FSWGs and VSWGs) for the purpose of improving the steady state voltage stability of power systems. The load buses voltage deviation is minimized by changing the controllers according to their sensitivity. The control variables considered are switchable VAr compensators (SVCs), on load tap changing (OLTC) transformers, generator excitations and reactive power output of the VSWGs. The approach translates bus voltage violation and controlling ability of controlling devices (controllers) into fuzzy set notations using linearized model. Then, fuzzy rule based system is formed based on the operator's experience to select the important controllers to meet the desired objectives. The FSWGs are also considered in this study because of its impact on overall system voltage performance even though they do not support the system for voltage unlike the VSWGs. A 297-bus equivalent grid connected wind system and a 417-bus equivalent grid connected wind system are considered to present the simulation results. A comparative analysis is also carried out with the conventional LP based reactive power optimization technique to highlight the features of the proposed approach.

METHODOLOGY

The main objective of this work is to reduce the voltage deviation at the load buses by reactive power redistribution in the system. This can be achieved by adjusting the OLTC transformers tap, generator excitations, SVCs and reactive power output of the VSWGs. The problem can be stated in mathematical form as,

$$V^e = \sum_{j=g+1}^{n} \left[V_j^d - V_j^a\right]^2 \quad (1)$$

where,

V^d_j is the desired voltage at load bus j, which can be considered as nominal voltage (1.0 p.u) and
V^a_j is the actual voltage at load bus j.

For effective implementation of the proposed approach in grid connected wind farms, the total reactive power controllers in a particular types/groups (i.e., transformers or reactive power output of the elements) are separated into the subgroups depending on their power rating or capacity. In case of OLTC transformers, two separate subgroups are formed. In one subgroup only grid OLTC transformers are present, and another subgroup comprises of OLTC transformers at the point of common coupling (PCC) of wind farms since these have their power rating or capacity much lower than the grid OLTC transformers. Similarly, the SVCs are in one subgroup and VSWGs are in another subgroup.

Calculation of Controlling Ability of the Controllers

Consider a grid connected wind system comprising of n number of total buses. Let g be the number of grid generator buses, t_g be the number of grid OLTC transformers, t_{pcc} be the number of transformers at the PCC of wind farms, svc be the number of SVC buses, $vswg$ be the number of VSWG buses, and r be the remaining buses.

Let X be the row vector of the voltage/reactive power controllers in the system and C_{jm} be the controlling ability of the controller X_m on j_{th} load bus. The controlling ability of the controller C_{jm} can be defined as,

$$C_{jm} = H_{jm} \times M_m \quad (2)$$

where,

m is the controller index ($m = 1,\ldots t_g; t_g + 1,\ldots t_g + t_{pcc}; t_g + t_{pcc} + 1,\ldots t_g + t_{pcc} + g; t_g + t_{pcc} + g + 1, \ldots t_g + t_{pcc} + g + svc; t_g + t_{pcc} + g + svc + 1,\ldots t_g + t_{pcc} + g + svc + vswg$),
M_m is the margin available for controller X_m, and
H_{jm} is the sensitivity of the controller X_m with respect to the j_{th} load bus.

$$H_{jm} = \frac{\partial V_j^e}{\partial X_m} = 2\left(V_j^d - V_j^a\right)\left(-S_{jm}\right) \tag{3}$$

where,

$V^e_j = (V^d_j - V^a_j)$ is the voltage error to be corrected at j^{th} load bus, and
S_{jm} is the linearized sensitivity factor of load bus j with respect to the controller X_m.

Computation of Sensitivity Matrix ([*S*])

The sensitivity matrix [*S*] relating the dependent and control variables is evaluated (Dhadbanjan et al., 1984; Dhadbanjan & Yesuratnam, 2006) in the following manner. It is known that the reactive power injection at a bus does not change for a small change in the phase angle of bus voltage. Based on this assumption, the relationship between the net reactive power change at any node due to change in the transformer tap settings and voltage magnitudes can be written as,

$$\begin{bmatrix} \Delta Q_g \\ \Delta Q_{svc} \\ \Delta Q_{vswg} \\ \Delta Q_r \end{bmatrix} = \begin{bmatrix} \frac{\partial Q_g}{\partial T_{t_g}} & \frac{\partial Q_g}{\partial T_{t_{pcc}}} & \frac{\partial Q_g}{\partial V_g} & \frac{\partial Q_g}{\partial V_{svc}} & \frac{\partial Q_g}{\partial V_{vswg}} & \frac{\partial Q_g}{\partial V_r} \\ \frac{\partial Q_{svc}}{\partial T_{t_g}} & \frac{\partial Q_{svc}}{\partial T_{t_{pcc}}} & \frac{\partial Q_{svc}}{\partial V_g} & \frac{\partial Q_{svc}}{\partial V_{svc}} & \frac{\partial Q_{svc}}{\partial V_{vswg}} & \frac{\partial Q_{svc}}{\partial V_r} \\ \frac{\partial Q_{vswg}}{\partial T_{t_g}} & \frac{\partial Q_{vswg}}{\partial T_{t_{pcc}}} & \frac{\partial Q_{vswg}}{\partial V_g} & \frac{\partial Q_{vswg}}{\partial V_{svc}} & \frac{\partial Q_{vswg}}{\partial V_{vswg}} & \frac{\partial Q_{vswg}}{\partial V_r} \\ \frac{\partial Q_r}{\partial T_{t_g}} & \frac{\partial Q_r}{\partial T_{t_{pcc}}} & \frac{\partial Q_r}{\partial V_g} & \frac{\partial Q_r}{\partial V_{svc}} & \frac{\partial Q_r}{\partial V_{vswg}} & \frac{\partial Q_r}{\partial V_r} \end{bmatrix} \begin{bmatrix} \Delta T_{t_g} \\ \Delta T_{t_{pcc}} \\ \Delta V_g \\ \Delta V_{svc} \\ \Delta V_{vswg} \\ \Delta V_r \end{bmatrix} \tag{4}$$

The sub-matrices $\partial Q / \partial T$ and $\partial Q / \partial V$ are the partial derivatives and the values are calculated from the following equations;

$$\frac{\partial Q_k}{\partial T_{kl}} = \frac{2}{a^3} V_k^2 y_{kl} \sin(\alpha_{kl}) + \frac{1}{a^2} y_{kl} V_k V_l \sin\left(\delta_k - \delta_l - \alpha_{kl}\right) \tag{5}$$

$$\frac{\partial Q_l}{\partial T_{kl}} = \frac{1}{a^2} y_{kl} V_k V_l \sin\left(\delta_l - \delta_k - \alpha_{kl}\right) \tag{6}$$

where,

k and *l* are the nodes to which OLTC transformer is connected and *k* is the tap side bus.

$$\frac{\partial Q_r}{\partial T_{kl}} = 0, \quad r \neq k, l \tag{7}$$

$$\frac{\partial Q_k}{\partial V_k} = \frac{Q_k}{V_k} - B_{kk} V_k \tag{8}$$

$$\frac{\partial Q_k}{\partial V_l} = Y_{kl} V_k \sin\left(\delta_k - \delta_l - \theta_{kl}\right) \tag{9}$$

where, a is the transformer tap; y is the primitive admittance of the line; α is the angle of the primitive admittance of the line; Y is the admittance matrix; θ is the angle of the bus admittance matrix; δ is the voltage phase angle; B is the line charging susceptance.

Further, transferring all control variables to the right hand side and the dependent variables to the left hand side, and rearranging the equations, we can get,

$$\begin{bmatrix} \Delta Q_g \\ \Delta Q_{svc} \\ \Delta Q_{vswg} \\ \Delta V_r \end{bmatrix} = \left[S\right] \begin{bmatrix} \Delta T_{t_g} \\ \Delta T_{t_{pcc}} \\ \Delta V_g \\ \Delta V_{svc} \\ \Delta V_{vswg} \end{bmatrix} \tag{10}$$

where [S] is the linearized sensitivity matrix relating dependent and control variables.

For all critical load buses, the most effective controller from each group is obtained based on the controlling ability of the controller's value from (2). The number of suggested steps for each controller can be obtained using fuzzy logic approach as explained in the following section.

Fuzzy Logic Approach

The main objective of using fuzzy logic approach (FLA) is to incorporate the experience of a power system operator in the design of control strategy to improve the system voltage profile. From a set of linguistic rules which describe the operator's control strategy, a control algorithm is constructed where the words are defined as fuzzy sets. The block diagram of fuzzy logic system for the proposed approach is given in Figure 1. The main advantage of the fuzzy logic approach seems to be the possibility of implementing rule of the thumb experience, intuition, and heuristics. The control strategies employed by an operator are formulated as a set of rules that are simple to carry out manually but difficult to implement by using conventional algorithms (Zadeh, 1965; Yager & Zadeh, 2012; Zimmermann, 2011). This is because human beings use qualitative rather than quantitative terms when describing various decisions to be taken as a function of different states of the process.

Figure 1. Block diagram of fuzzy logic system

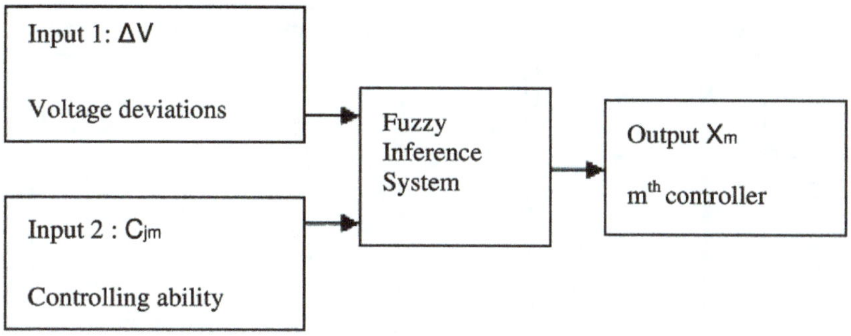

Input and Output Parameters

The fuzzy logic technique is formulated using membership functions for the input parameters such as voltage deviation (ΔV) at each of the load buses and the voltage deviation sensitivity to control variables that is, controlling ability of the controllers ($[C]$). The voltage deviation (ΔV) for a load bus j is defined as,

$$\Delta V_j = V_j^a - V_j^d \tag{11}$$

These input parameters are transformed into fuzzy set notations with the help of membership functions. The membership function for the voltage deviation and controlling ability of the controller with linguistic variables are shown in Figures 2 and 3 respectively.

Figure 2. Membership function for the voltage deviation

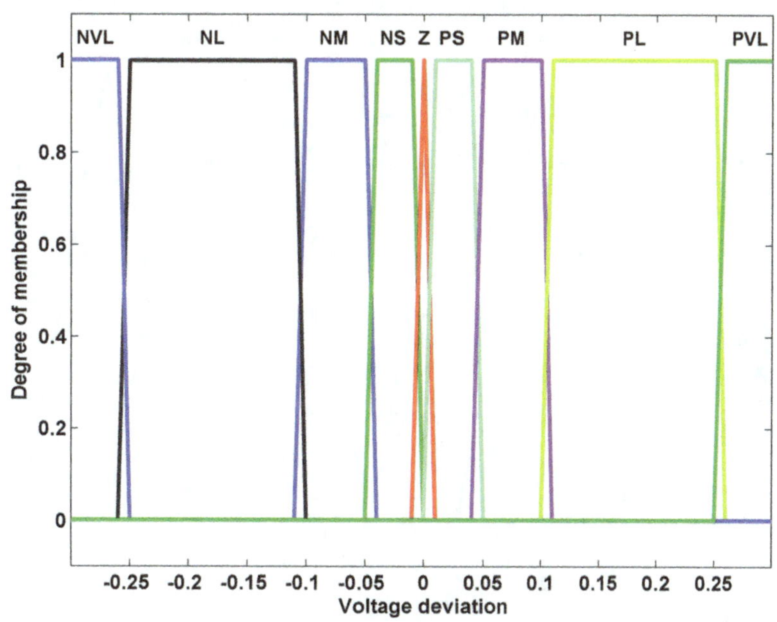

Figure 3. Membership function for the controlling ability of the controller

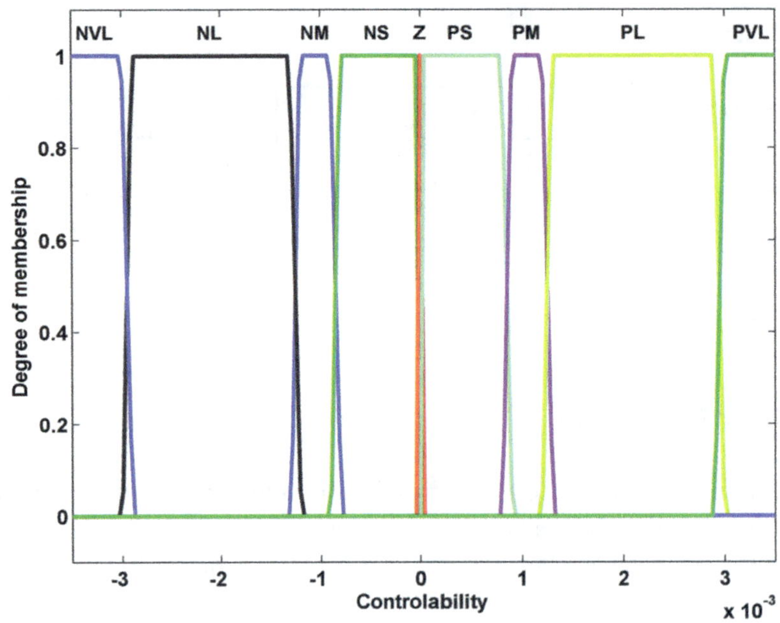

The output of fuzzy logic system is the new settings of the control variables such as grid OLTC transformers tap position, OLTC transformers tap position at the point of common coupling of the wind farms, generators excitation settings, switchable VAr compensator settings, and reactive power output of the VSWG settings. The membership function for the output with linguistic variables is shown in Figure 4.

Figure 4. Membership function for the controller settings

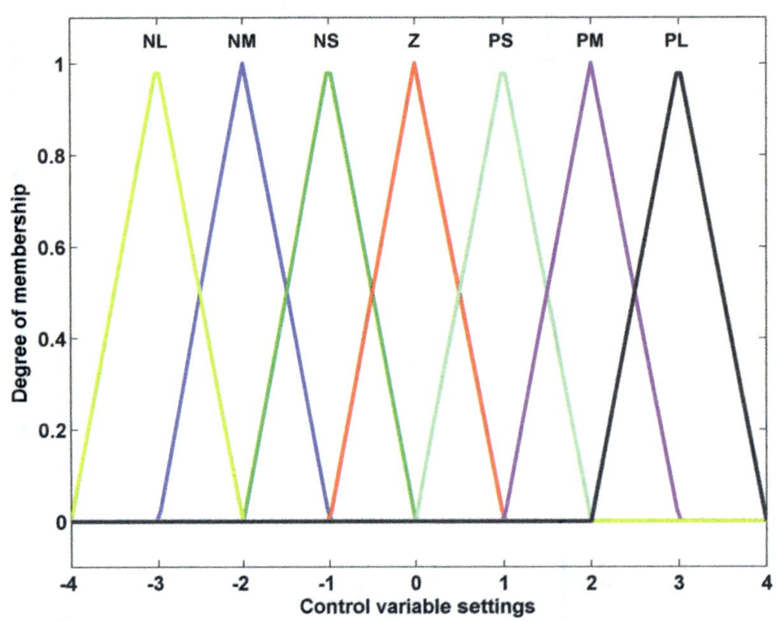

Definition of Fuzzy State Descriptions

The linguistic variables are used to describe the states of the systems. Nine linguistic variables are used for defining voltage deviation and controlling ability of the controller. The meaning of the linguistic variables are negative very large (NVL), negative large (NL), negative medium (NM), negative small (NS), zero (Z), positive small (PS), positive medium (PM), positive large (PL) and positive very large (PVL). Similarly, seven linguistic variables are used for defining the output, which are negative large, negative medium, negative small, zero, positive small, positive medium, and positive large.

Design of Fuzzy Rule

A rule based system consisting of 81 rules is developed to represent various system conditions, which is given in Table 2. Some of the control rules as fuzzy conditional statements of the type are described below. The number of steps in modifying the controller settings is limited to three. The implications or interpretations of the fuzzy output are given in Table 3.

- IF ΔV_j is PVL and C_{jm} is NVL THEN Xm is NL.
- IF ΔV_j is PL and C_{jm} is PM THEN X_m is PS.
- IF ΔV_j is PM and C_{jm} is NS THEN X_m is NS.
- IF ΔV_j is PS and C_{jm} is NL THEN X_m is Z.

where X_m is the output of the m_{th} controller.

Transformation From Suggested Actions to Crisp Control Actions

The controllers are given priorities according to their controlling ability towards the critical load buses. These will affect the final settings of the controllers. As it can be observed from Table 2 that if a controller is having highest sensitivity towards a load bus where voltage deviation is very large then FLA will suggest 3 steps change (PL or NL). But if a controller is having no significant influence on critical

Table 2. Fuzzy inference matrix or fuzzy rules

ΔV_j	C_{jm}								
	NVL	**NL**	**NM**	**NS**	**Z**	**PS**	**PM**	**PL**	**PVL**
NVL	PL	PL	PM	PS	Z	NS	NM	NL	NL
NL	PM	PM	PS	PS	Z	NS	NS	NM	NM
NM	PM	PS	PS	PS	Z	NS	NS	NS	NM
NS	PS	Z	Z	Z	Z	Z	Z	Z	NS
Z	Z	Z	Z	Z	Z	Z	Z	Z	Z
PS	NS	Z	Z	Z	Z	Z	Z	Z	PS
PM	NM	NS	NS	NS	Z	PS	PS	PS	PM
PL	NM	NM	NS	NS	Z	PS	PS	PM	PM

Table 3. Transformation of fuzzy output

Output Parameter	Implication
PL	3 positive steps
PM	2 positive steps
PS	1 positive step
Z	No change
NS	1 negative step
NM	2 negative steps
NL	3 negative steps

buses, then FLA will suggest Z. If the control action suggested by the FLA is Z, then no change is made in the controller setting. Thus, only significant controllers are changed to improve the voltage profile. In this way only few controllers of high sensitivity are used to achieve the desired objectives so that the reduction in complexity of the problem is ensured and also the voltage control operation is fast.

For an illustrative purpose, let us consider a sample system with three generators G1, G2 and G3, five OLTC transformers T1, T2,. . .T5, and four SVCs S1, S2,. . . S4. Totally, there are 12 controllers. Let us consider that five load buses L1, L2,. . . L5 are violating the voltage limit. For each of the critical load buses, a set of generator, OLTC, SVC controllers having highest sensitivity is obtained and a sample of that is given in Table 4.

It can be seen from Table 4 that generator G2, OLTC transformers T2, T4 and T5, and SVCs S2 and S4 are not in the list of most sensitive controllers for the critical load buses. Each column of controlling ability of the controllers $[C]$ matrix corresponds to one controller. The sensitivity values for the generators and SVCs are either positive or negative, which depends on the system operating conditions. The controller movement direction (steps) should be positive to improve the voltage. For a particular controller, the proposed fuzzy logic approach may suggest different control actions (steps).

For peak load condition in the sample system, generator G1 is the highest sensitive controller for load buses L1 and L2. Let the control action (steps) for G1 suggested by FLA be THREE positive steps for L1 and TWO positive steps for L2. In that case, the highest step (setting) is selected to decide the final control setting for that generator (controller) using the MAX operation. On the contrary, if the control action (steps) suggested by FLA is negative, MIN operation is used to determine the final control setting for the generator. Similarly, the final control settings for the SVCs are also generated.

Table 4. Most sensitive controllers for sample system

Load Bus	Generator	OLTC	SVC
	(G1,. . . G3)	(T1,. . . T5)	(S1,. . . S4)
L1	G1	T1	S1
L2	G1	T1	S3
L3	G3	T3	S3
L4	G3	T1	S3
L5	G3	T3	S1

However, the sensitivity value for OLTC transformers can be both negative as well as positive. In order to determine the correct controller movement direction (steps) for overall improvement in system voltage profile, the algebraic sum of the elements of the voltage deviation sensitivity matrix [*H*] (3) corresponding to the controller (transformer under consideration) is observed. A positive sign indicates that the reduction (negative controller movement direction) in tap improves the voltage, and a negative sign suggests that increase (positive controller movement direction) in tap will improve the voltage. So, if the suggested control action (positive/negative steps) by FLA is matching with this sign, then the suggested correction is incorporated in the final control settings of those controllers, otherwise no correction is made.

COMPUTATIONAL ALGORITHM

The algorithmic steps for the proposed approach are given below:

Step 1: For a given network topology, generation schedule and load data, run the state estimation/load flow program to get the initial operating condition of the given system such as voltage profile, real and reactive power losses, etc.
Step 2: Check for voltage violations. If there is any violations go to Step 3, otherwise go to Step 13.
Step 3: Advance fuzzy iteration count.
Step 4: Compute the voltage deviation (ΔV) at each of the load buses using (11).
Step 5: Compute the sensitivity matrix [*S*] relating the dependent and control variables using (10).
Step 6: Calculate voltage deviation sensitivity matrix [*H*] for all controllers using (3).
Step 7: Using the margin of the controllers, compute the controlling ability of the controllers' matrix [*C*] using (2).
Step 8: Design fuzzy logic approach as explained in the Subsection FLA.
Step 9: Feed the input parameters (voltage deviation (ΔV) and controlling ability of the controllers [*C*]) to fuzzy logic approach and produce fuzzy output.
Step 10: Transform fuzzy output to discrete controller settings using Table 3.
Step 11: Using the new control variable settings, again perform the load flow analysis.
Step 12: Check for voltage violations. If there is any violation go to Step 3. Otherwise, go to Step 13.
Step 13: Print the results and terminate the program.
Step 14: The above steps are repeated for different operating conditions in the system.

SYSTEM STUDIES AND DISCUSSIONS

The fixed speed wind generators are of conventional asynchronous machines, which do not have capability to control the voltage, but rather absorb large levels of reactive power during low voltage conditions. The studies in the literature show that the FSWGs have significant impact on the system voltage performance (Moger & Dhadbanjan, 2017; Divya & Rao, 2006). Since these units do show its contribution/impact on system voltage profile, the presence of FSWGs in the system is absolutely necessary while studying the steady state voltage stability of grid connected wind farms. The active power output of the wind generators are obtained directly from the power curve provided by the manufacturer. The reactive power output of the FSWG is computed by considering the power factor of 0.95 leading (inductive) as per the prevailing industry practice. However, the VSWGs can provide necessary voltage support because of its flexibility in control design. Therefore, the VSWGs can participate in reactive power optimization process.

The control variables are switchable VAr compensators, grid OLTC transformers, OLTC transformers at the point of the common coupling of wind farms, generator excitations, and reactive power output of the variable speed wind generators. For all studies, initially it is assumed that all control variables are to be at their nominal settings, that is, all OLTC transformers tap position and the excitation of all the generators at their nominal value (i.e., 1.0 per unit), and the reactive power output of SVCs and VSWGs are operated at nil.

A comparative analysis is carried out with conventional LP based optimization technique (Dhadbanjan & Yesuratnam, 2006) to highlight the effectiveness of the proposed approach. The detailed methodology of the LP based reactive power optimization technique is explained in Appendix.

297-Bus Equivalent Grid Connected Wind System

A 24-bus, 400 kV level equivalent system of Indian southern region power grid (SR 24-bus equivalent system) with wind integration is considered to test the proposed approach. The single-line diagram of the system is shown in Figure 5. The southern region power grid covers the electrical network of four south Indian states such as Karnataka, Andra Pradesh, Tamil Nadu and Kerala. The system data is taken from (Moger, 2016). The system comprises of four generators, 16 transmission lines which are of 400 kV level lines connecting the four south Indian states, and 12 transformers. The real and reactive loads are connected at eight locations. The shunt reactors are connected at few buses for transient over voltage protection. The system has an initial peak load of 2060 MW and 1040 MVAr.

Three practical wind farms (wind farms with different wind generator types, that is, FSWGs and VSWGs) are considered for the reactive power coordination in grid connected wind farms. The three wind farms are connected at buses 5, 6 and 8. In each wind farm, there are six groups and each group comprises of same type of wind generators either fixed speed wind generators or variable speed wind generators. Totally, there are 45 wind generators. With a pad mounted transformer at each wind generator and a collector system station, there are 91 buses in each wind farm. When wind farms are operated at rated power, the penetration of wind power in the system is around 10%.

The fixed speed wind generators were predominantly employed during the early installations of wind generators. Presently, the variable speed wind generators become more popular because of advancements in the power electronics controller design. Due to the presence of mixed types of wind generators in the wind farms, the reactive power coordination or reactive power and voltage control analysis is carried out for two different ratios of wind generator units, that is, FSWGs to VSWGs in the wind farm.

The two different ratios of wind generators units considered for the studies are:

Scenario-1: Wind generator units ratio of 26:74 (FSWG:VSWG)
Scenario-2: Wind generator units ratio of 5:95 (FSWG:VSWG)

The topology structure for two different ratios of wind generator units (FSWG to VSWG) in the wind farms is the same. However, the wind generator units in the respective groups are replaced with the same power rating of different type of wind generator units to make it to a particular ratio. For each scenario, studies are carried out for two operating points/conditions in the system:

Figure 5. Single-line diagram of SR 24-bus equivalent system with wind farms

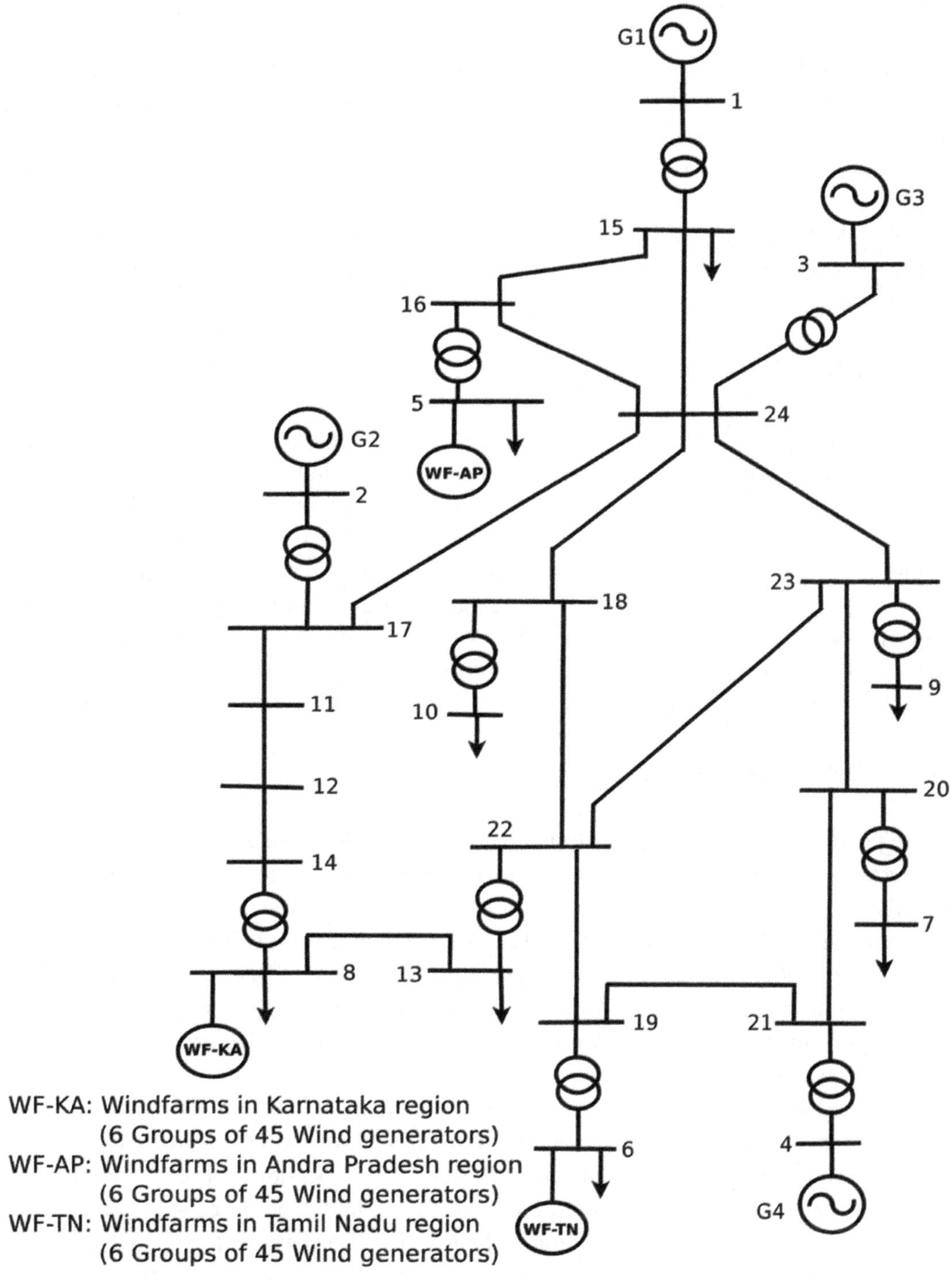

Case-1: Peak load condition.
Case-2: Single line outage/contingency.

As discussed in (Moger, 2016; Moger & Dhadbanjan, 2015), the line connected between the buses 24 and 23 is considered as one of the most severe contingency. This line is considered for the reactive power and voltage control analysis in the system.

For each operating point/condition, the system is evaluated with various performance indices or parameters, viz., system power loss (real and reactive power losses) and load bus voltage profile parameters (V_{min}, V_{max}, STDEV(V), $\Sigma(V^d - V^a)^2$ i.e., sum of the square of the voltage deviation from desired voltage of the buses, and voltage stability index at load buses (*L-index*)) to assess the effectiveness of the proposed approach.

Scenario-1: Wind Generator Units' Ratio of 26:74 (FSWG:VSWG)

In this scenario, Group-1 consists of five semi-variable speed wind generators (SSWGs) and each rated for 1.0 MW. Group-2 consists of five pitch regulated fixed speed wind generators (PR-FSWGs) and each rated for 0.5 MW. Group-3 consists of 10 VSWG-DFIGs and each rated for 1.0 MW. Group-4 consists of five SSWGs and each rated for 1.0 MW. Group-5 consists of 10 VSWG-SGFECs and each rated for 1.0 MW. Group-6 consists of 10 VSWG-DFIGs and each rated for 1.5 MW. In total, there are 45 wind generators out of that 15 are FSWGs, which supply the power of 12.5 MW. Out of the total wind power, there is about 26% of wind power supplied by the FSWGs (which always consume reactive power) and these units do not participate in the reactive power optimization process. The configuration of individual wind farms for wind generator units ratio of 26:74 (FSWG:VSWG) is shown Figures 6, 7 and 8.

Case-1: System Under Peak Load Condition

The reactive power or voltage control analysis is carried out for peak load condition in the system. The system performance parameters and voltage/reactive power controllers setting from the proposed approach and other existing technique (Dhadbanjan & Yesuratnam, 2006) are given in Tables 5 and 6, respectively. It can be seen from Table 5 that under peak load condition the minimum voltage of the system improves from 0.856 p.u. to 0.9654 p.u. using the proposed approach and to 0.9599 p.u. using the conventional LP based optimization technique. Similarly, the voltage stability index (*L-index*) is also improved from 0.5083 to 0.3940 using the proposed approach whereas it reduces to 0.3926 using the conventional LP based optimization technique.

In addition, the conventional LP based optimization technique reduces the real power loss of the system to 35.82 MW whereas proposed approach reduces it to 36.77 MW. The standard deviation (STDEV) of the load voltage is used to quantify the amount of variation or dispersion of a set of load voltages. The STDEV value of the load voltages also improved to 0.0157 using the proposed approach and to 0.0162 using the conventional LP based optimization technique. The improvement in system voltage profile and voltage stability index (*L-index*) at the load buses from both these techniques are shown in Figures 9 and 10, respectively.

Figure 6. Single-line diagram of wind farm at Karnataka: WGUs ratio of 26:74 (FSWG:VSWG)

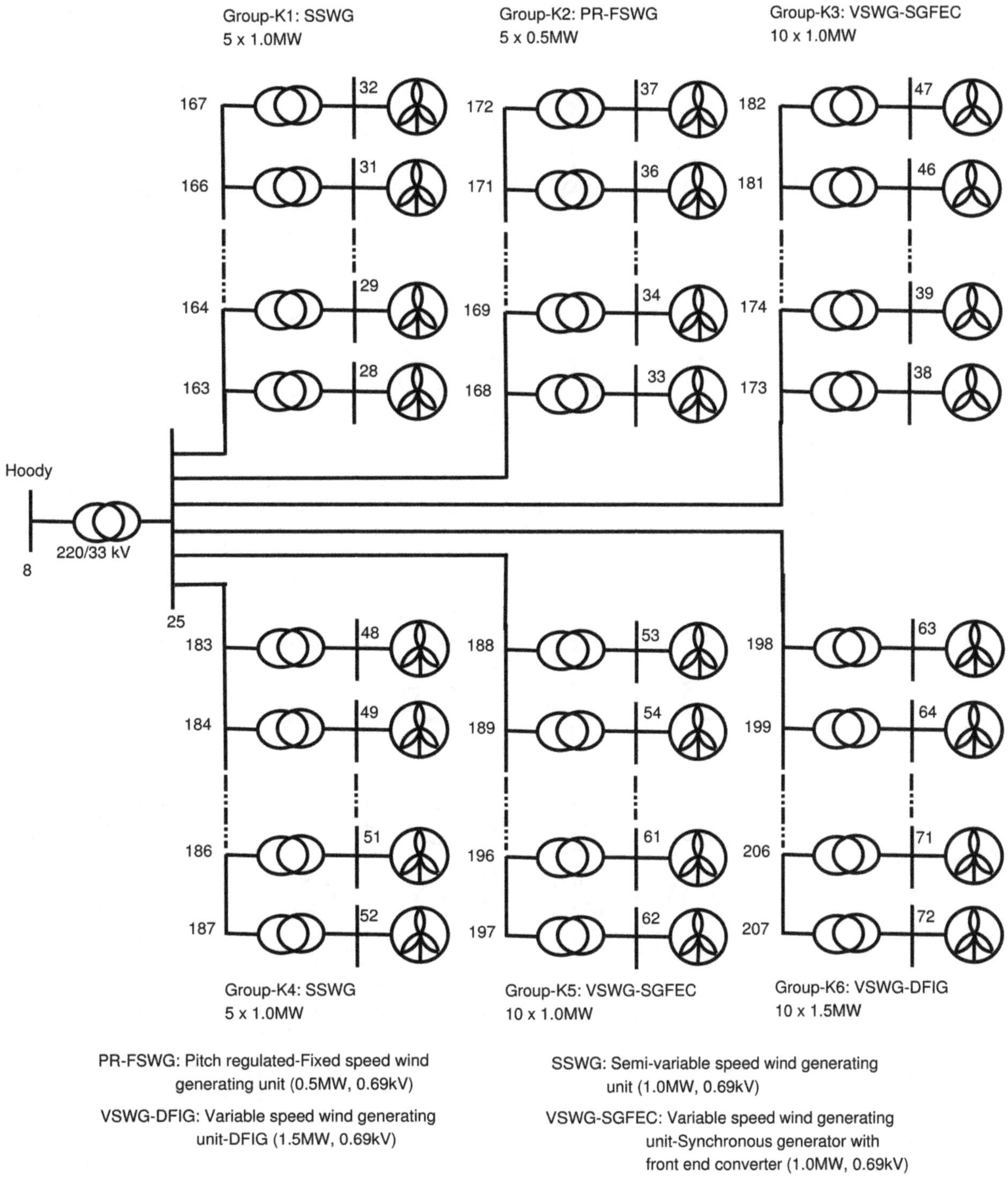

Figure 7. Single-line diagram of wind farm at Andra Pradesh: WGUs ratio of 26:74 (FSWG:VSWG)

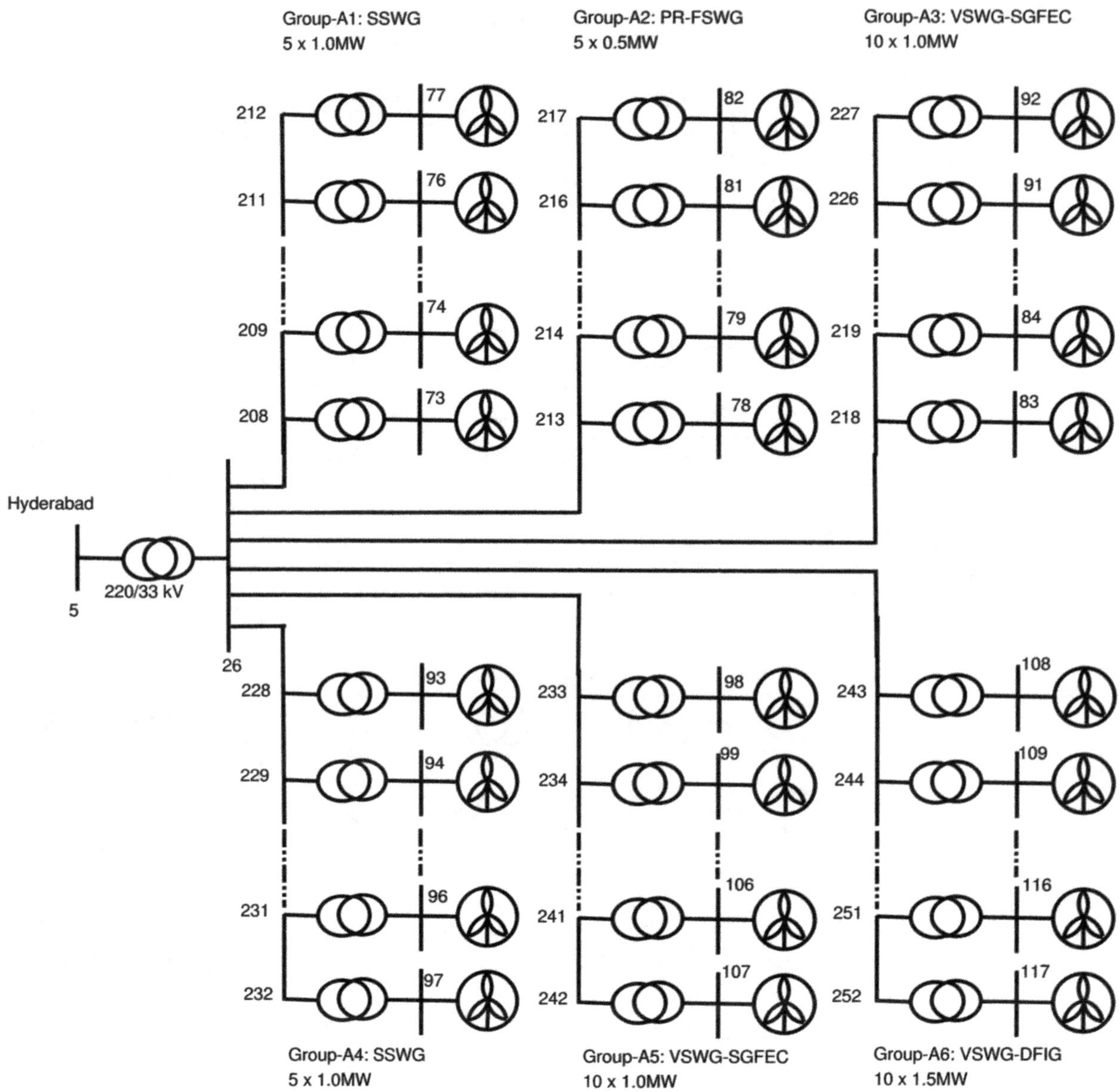

PR-FSWG: Pitch regulated-Fixed speed wind generating unit (0.5MW, 0.69kV)

VSWG-DFIG: Variable speed wind generating unit-DFIG (1.5MW, 0.69kV)

SSWG: Semi-variable speed wind generating unit (1.0MW, 0.69kV)

VSWG-SGFEC: Variable speed wind generating unit-Synchronous generator with front end converter (1.0MW, 0.69kV)

Figure 8. Single-line diagram of wind farm at Tamil Nadu: WGUs ratio of 26:74 (FSWG:VSWG)

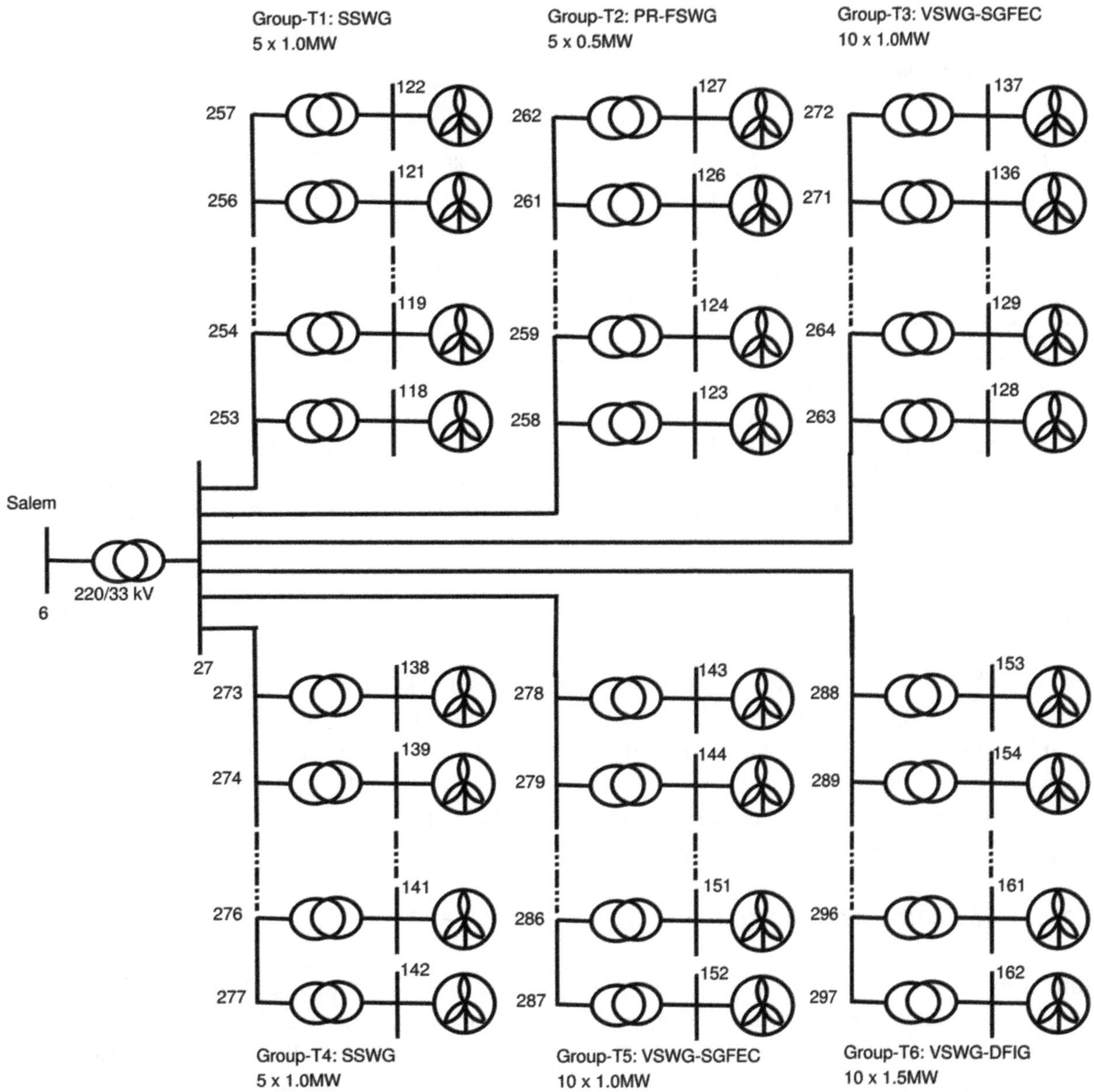

PR-FSWG: Pitch regulated-Fixed speed wind generating unit (0.5MW, 0.69kV)

VSWG-DFIG: Variable speed wind generating unit-DFIG (1.5MW, 0.69kV)

SSWG: Semi-variable speed wind generating unit (1.0MW, 0.69kV)

VSWG-SGFEC: Variable speed wind generating unit-Synchronous generator with front end converter (1.0MW, 0.69kV)

Table 5. System performance parameters of 297-bus equivalent grid connected wind system under different operating points/conditions: WGUs ratio of 26:74 (FSWG:VSWG)

Parameters	Case-1: Peak Load			Case-2: Outage		
	Initial	FLA	LP	Initial	FLA	LP
$P_{loss}(MW)$	46.25	36.77	35.82	38.46	28.9	28.2
$Q_{loss}(MVAr)$	674.86	503.9	487.58	518.99	362.38	348.82
$V_{max}(p.u.)$	1.0001	1.041	1.0476	0.9985	1.0255	1.0319
$V_{min}(p.u.)$	0.856	0.9654	0.9599	0.8573	0.9554	0.9514
STDEV(V)	0.0269	0.0157	0.0162	0.0257	0.0145	0.0159
$\Sigma(V^d - V^a)^2$	2.5076	0.0456	0.0992	1.9352	0.0337	0.1
L^{max}*-index*	0.5083	0.394	0.3926	0.4475	0.3458	0.3347

The number of controllers used during the first two iterations of the proposed fuzzy logic approach is given in Table 7. It can be seen from Table 7 that the proposed approach used 106 controllers during the first iteration and only 15 controllers during the second iteration. The number of controllers in subsequent iterations is further reduced.

Inference

It can be observed from the simulation results that the system performance produced by the proposed approach is closely in agreement with that produced from the conventional LP based reactive power optimization technique. During each iteration, the proposed fuzzy logic approach uses only few reactive power/voltage controllers of high sensitivity to achieve the desired objectives. Since the number of controllers in each iteration is reduced, the computational time for each iteration is also reduced. Hence, the overall computational effort is reduced.

Case-2: System Under Single Line Contingency Case

Similar to the analysis on peak load condition, the reactive power or voltage control analysis is carried out for single line contingency case. The system performance parameters and voltage/reactive power controllers setting from the proposed approach and other existing technique (Dhadbanjan & Yesuratnam, 2006) are also given in Tables 5 and 6, respectively. It can be seen from Table 5 that under single line contingency case the minimum voltage of the system improves from 0.8573 p.u. to 0.9554 p.u. using the proposed approach and to 0.9514 p.u. using the conventional LP based optimization technique. Similarly, the voltage stability index (*L-index*) is also improved from 0.4475 to 0.3458 using the proposed approach whereas it reduces to 0.3347 using the conventional LP based optimization technique.

In addition, the conventional LP based optimization technique reduces the real power loss of the system to 28.20 MW whereas proposed approach reduces it to 28.90 MW. Similarly, the STDEV value of the load voltages also improved to 0.0145 using the proposed approach and to 0.0159 using the conventional LP based optimization technique. The improvement in system voltage profile and voltage stability index (*L-index*) at the load buses from both these techniques are shown in Figures 11 and 12, respectively.

Table 6. Controllers settings of 297-bus equivalent grid connected wind system under different operating points/conditions: WGUs ratio of 26:74 (FSWG:VSWG)

Transformer (OLTC) Tap Settings						
Fbus-Tbus	Case-1: Peak Load			Case-2: Outage		
	Initial	FLA	LP	Initial	FLA	LP
16-5	1	0.9625	0.95	1	0.975	0.9875
19-6	1	0.975	0.975	1	0.9875	1.0125
20-7	1	0.9625	0.9625	1	0.9625	0.9625
14-8	1	1	0.95	1	1	0.95
23-9	1	0.9875	1	1	0.9625	0.975
18-10	1	0.9875	1	1	0.9875	1
22-13	1	0.9625	0.9875	1	0.975	1
8-25	1	0.975	1.025	1	0.975	1.025
5-26	1	0.975	1	1	0.975	1.0125
6-27	1	0.9875	1	1	0.9875	1.0125
Voltage Settings of the Generator Excitation (p.u.)						
Gen. Bus	**Case-1: Peak Load**			**Case-2: Outage**		
	Initial	**FLA**	**LP**	**Initial**	**FLA**	**LP**
1	1	1.05	1.05	1	1.025	1.025
2	1	1	1.025	1	1	1.025
3	1	1.0125	1.05	1	1	1.025
4	1	1.05	1.05	1	1.05	1.05
Reactive Power Output of SVCs (MVAr)						
SVC Bus	**Case-1: Peak Load**			**Case-2: Outage**		
	Initial	**FLA**	**LP**	**Initial**	**FLA**	**LP**
5	0	6	8	0	2	6
6	0	2	8	0	2	6
7	0	4	8	0	6	8
8	0	6	6	0	8	8
Total Reactive Power Output of VSWGs in the Wind Farms (MVAr)						
Wind Farms	**Case-1: Peak Load**			**Case-2: Outage**		
	Initial	**FLA**	**LP**	**Initial**	**FLA**	**LP**
WF-KAR	0	5.3	7	0	3.8	7
WF-AP	0	5.3	7	0	3.5	7
WF-TN	0	3.5	7	0	2	7

Table 7. Number of controllers used in fuzzy iterations

Total Number of Controllers = 108	
1st Iteration	2nd Iteration
106	15

Figure 9. Voltage profile of SR 24-bus equivalent system with wind integration ratio of 26:74 (FSWG:VSWG) under peak load condition

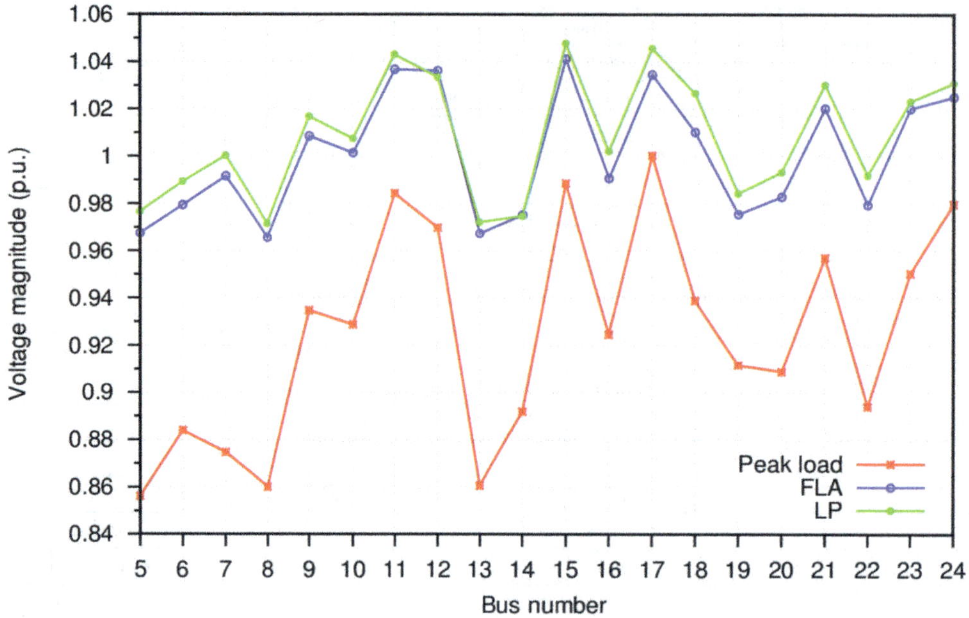

Figure 10. Voltage stability index (L-index) of SR 24-bus equivalent system with wind integration ratio of 26:74 (FSWG:VSWG) under peak load condition

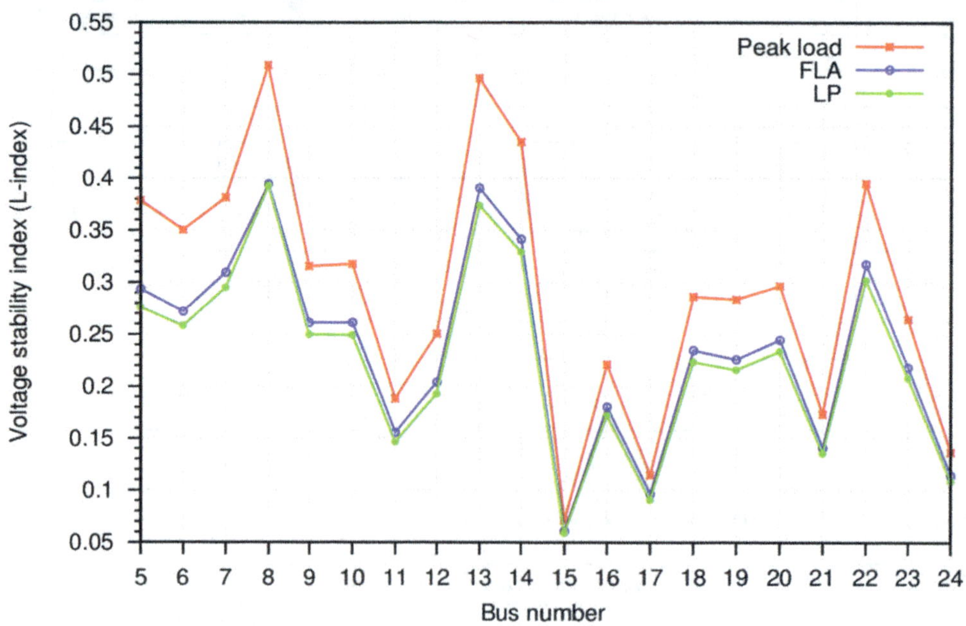

Figure 11. Voltage profile of SR 24-bus equivalent system with wind integration ratio of 26:74 (FSWG:VSWG) under outage/contingency condition

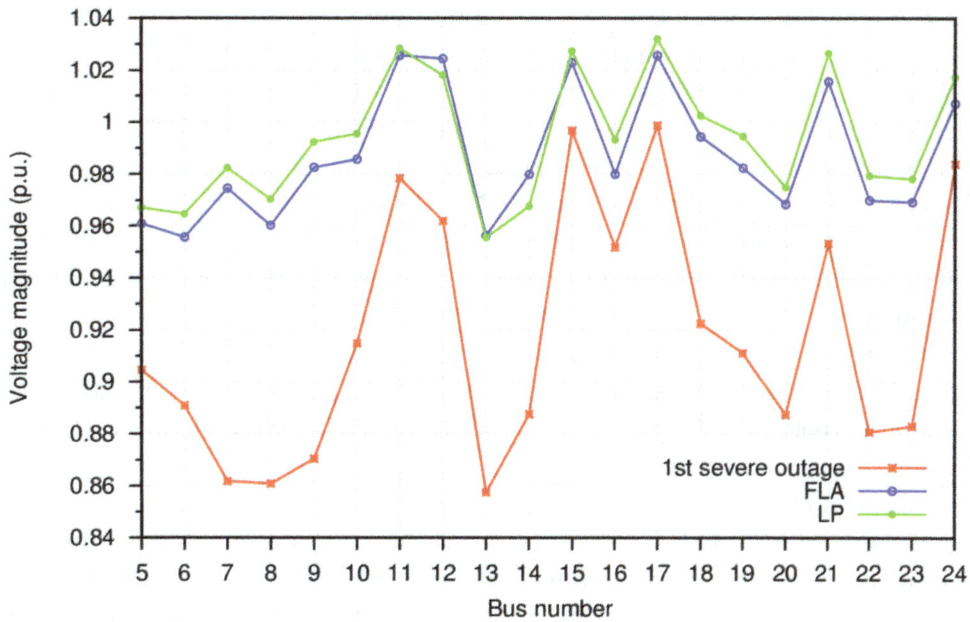

Figure 12. Voltage stability index (L-index) of SR 24-bus equivalent system with wind integration ratio of 26:74 (FSWG:VSWG) under outage/contingency condition

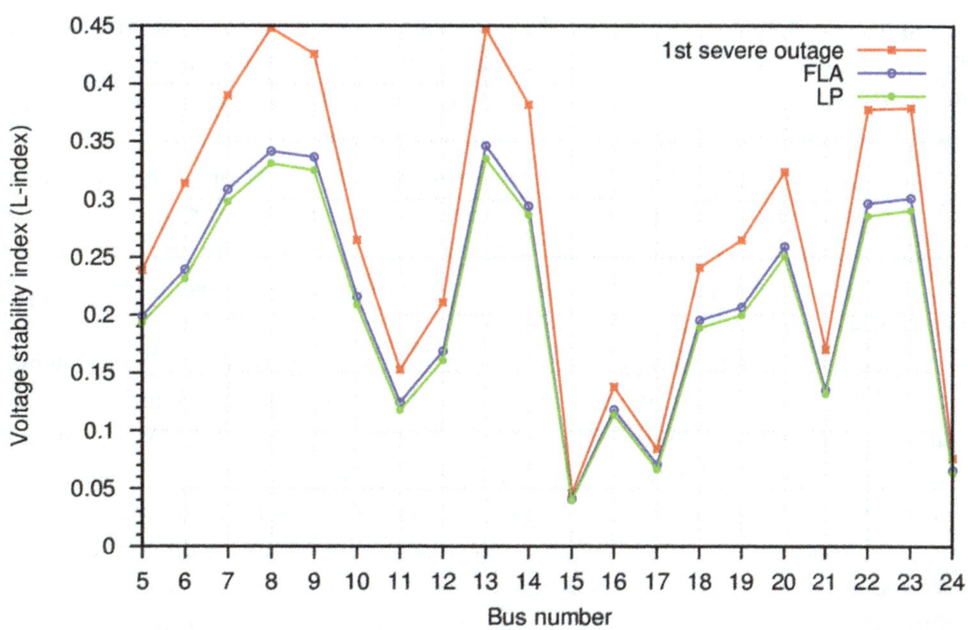

Scenario-2: Wind Generator Units' Ratio of 5:95 (FSWG:VSWG)

In this scenario, Group-1 consists of five VSWG-SGFECs and each rated for 1.0 MW. Group-2 consists of five pitch regulated fixed speed wind generators (PR-FSWGs) and each rated for 0.5 MW. Group-3 consists of 10 VSWG-SGFECs and each rated for 1.0 MW. Group-4 consists of five VSWG-SGFECs and each rated for 1.0 MW. Group-5 consists of 10 VSWG-SGFECs and each rated for 1.0 MW. Group-6 consists of 10 VSWG-DFIGs and each rated for 1.5 MW. In total, there are 45 wind generators out of that 5 are FSWGs, which supply the power of 2.5 MW. Out of the total wind power, there is about 5% of wind power supplied by the FSWGs (which always consume reactive power) and these units do not participate in the reactive power optimization process. The configurations of individual wind farms for wind generator units ratio of 5:95 (FSWG:VSWG) is shown in Figures 13, 14 and 15.

Similar to scenario-1, the reactive power or voltage control analysis is carried out for both peak load as well as single line contingency case. The system performance parameters and voltage/reactive power controllers setting for both the cases (Dhadbanjan & Yesuratnam, 2006) are given in Tables 8 and 9, respectively.

In addition, the improvement in system voltage profile for each of the cases is shown in Figures 16 and 17. Similarly, voltage stability index (*L-index*) at each of the load buses is shown in Figures 18 and 19.

In this scenario also it can be observed from the simulation results that the system performance produced by the proposed approach is closely in agreement with that produced from the conventional LP based reactive power optimization technique.

417-Bus Equivalent Grid Connected Wind System

The proposed approach is also tested on a 79-bus equivalent system with wind farms, which is a part of southern region power grid (SR 79-bus equivalent system). The single-line diagram of the system is shown in Figure 20. The southern region power grid covers the electrical network of four south Indian states such as Karnataka, Andra Pradesh, Tamil Nadu and Kerala.

The system data is taken from (Moger, 2016). The system is comprising of 10 conventional generators, 89 transmission lines which are of 400/200 kV level lines connecting Karnataka, Andra Pradesh, Tamil Nadu and Kerala states, and 12 transformers. The real and reactive loads are connected at 47 locations. The shunt reactors are connected at few buses for transient over voltage protection. The system has an initial peak load of 4113.5 MW and 1650.92 MVAr.

The system is connected with four practical wind farms (wind farms with different wind generator types, that is, FSWGs and VSWGs), which are connected at buses 36, 51, 43 and 71. Out of the total 164 wind generators, 56 units of PR-FSWG (FSWG) and each rated for 2.0 MW, 15 units of SGFEC (VSWG) and each rated for 2.0 MW, and 93 units of DFIG (VSWG) and each rated for 2.0 MW. Out of the total wind power, there is about 34% of wind power supplied by the FSWGs (which always consume reactive power) and these units do not participate in the reactive power optimization process. When wind farms are operated at rated power, the penetration of wind power in the system is around 10%. With a pad mounted transformer at each wind generator and the collector system stations, the total number of buses in all wind farms is 338. The configurations of individual wind farms are shown in Figures 21 and 22. Similar to 297-bus equivalent grid connected wind farm, system studies are carried out for two operating points/conditions in the system:

Figure 13. Single-line diagram of wind farm at Karnataka: WGUs ratio of 5:95 (FSWG:VSWG)

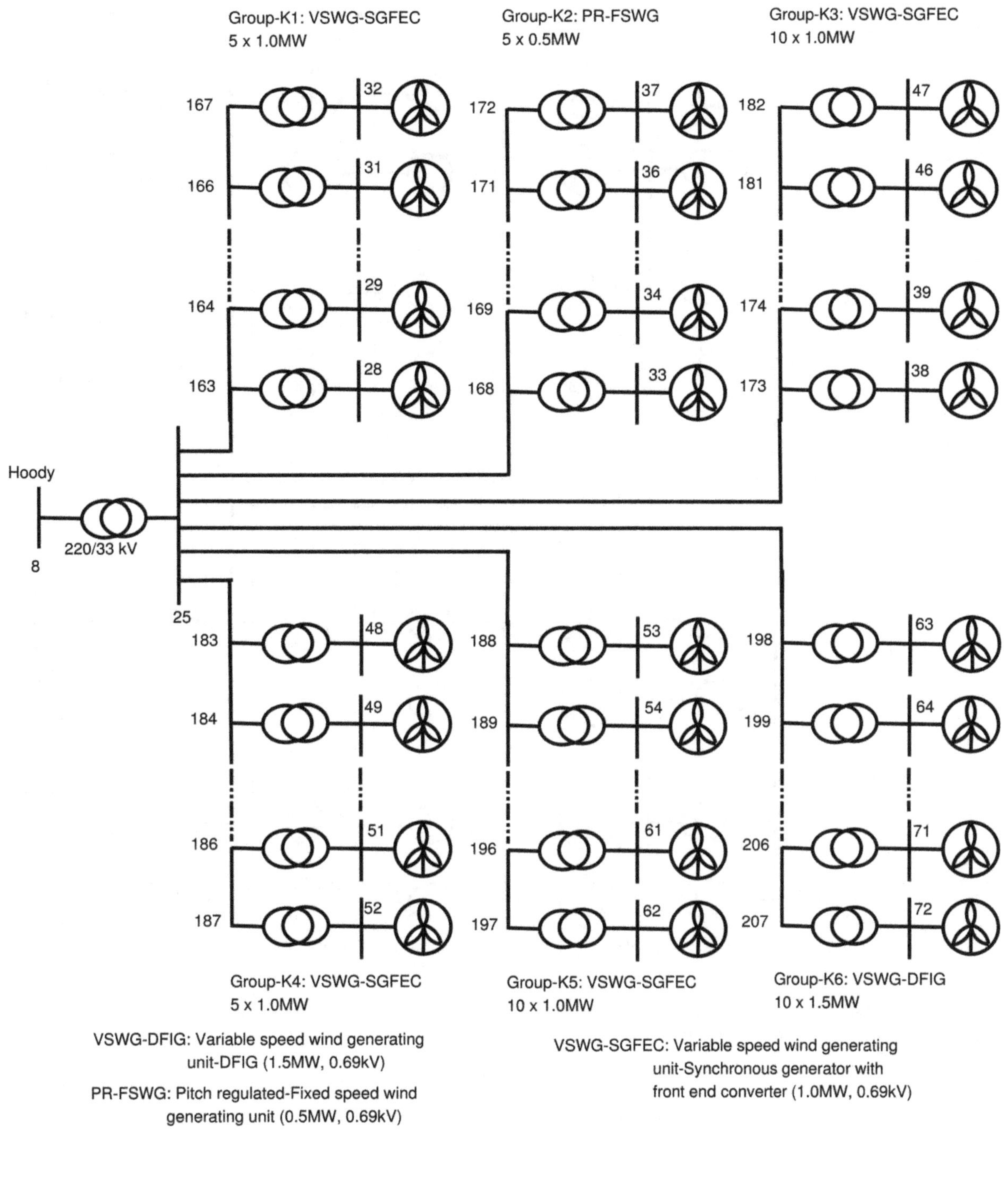

Figure 14. Single-line diagram of wind farm at Andra Pradesh: WGUs ratio of 5:95 (FSWG:VSWG)

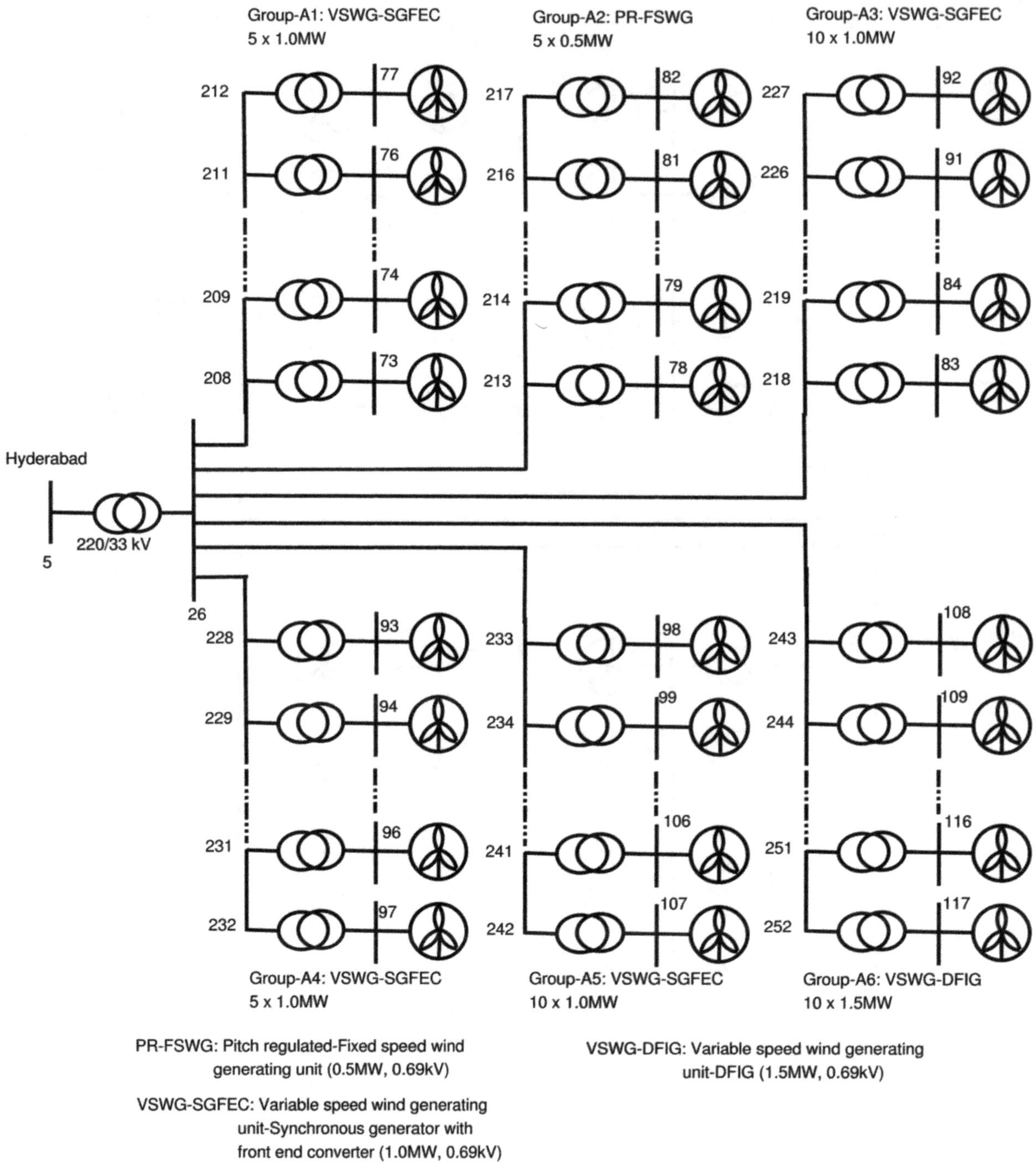

Figure 15. Single-line diagram of wind farm at Tamil Nadu: WGUs ratio of 5:95 (FSWG:VSWG)

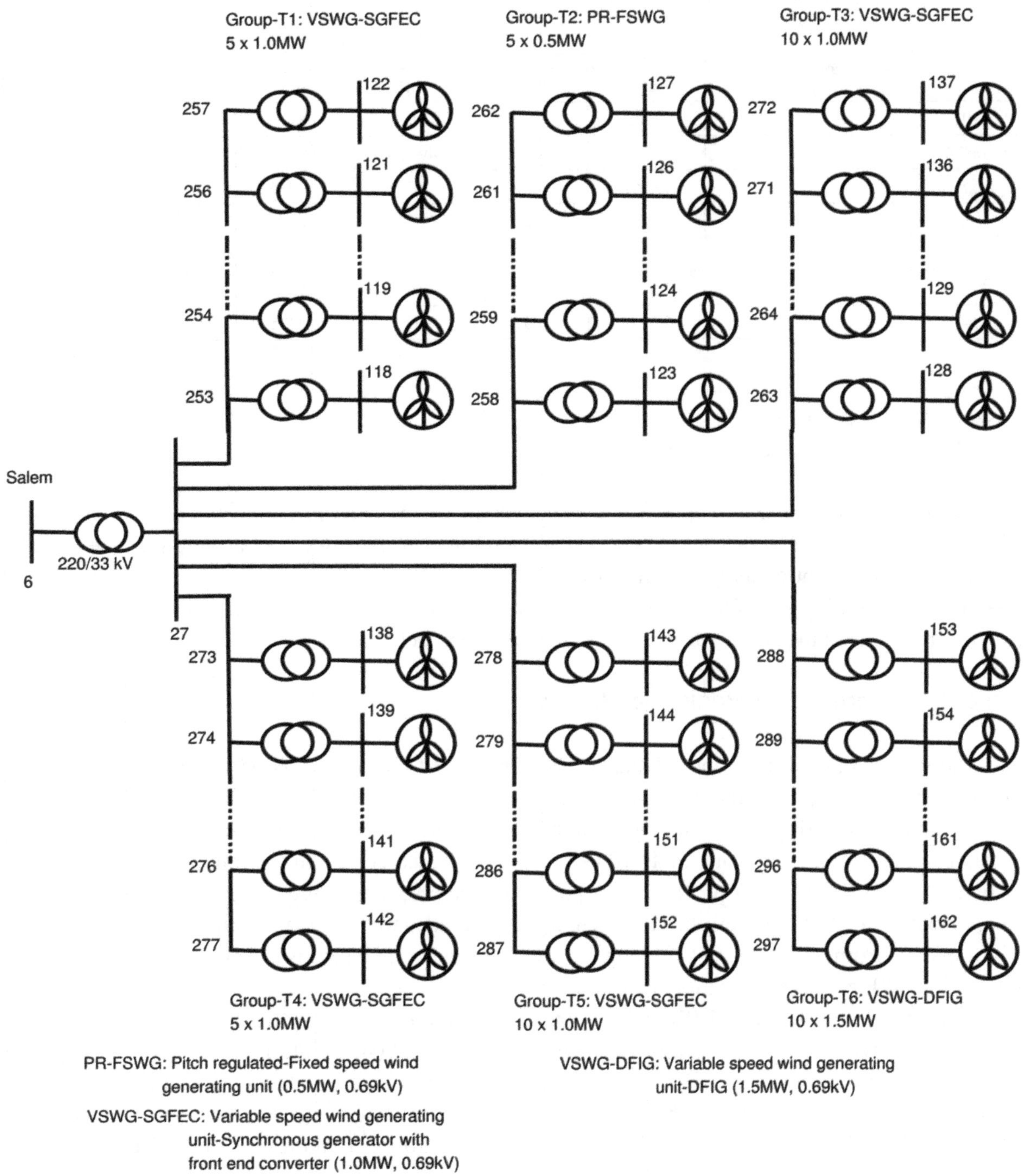

Table 8. System performance parameters of 297-bus equivalent grid connected wind system under different operating points/conditions: WGUs ratio of 5:95 (FSWG:VSWG)

Parameters	Case-1: Peak Load			Case-2: Outage		
	Initial	FLA	LP	Initial	FLA	LP
$P_{loss}(MW)$	46.25	36.34	36.6	38.46	28.35	27.8
$Q_{loss}(MVAr)$	674.86	499.07	497.38	518.99	355.7	344.23
$V_{max}(p.u.)$	1.0001	1.0349	1.0333	0.9985	1.0328	1.0427
$V_{min}(p.u.)$	0.856	0.9593	0.9511	0.8573	0.9621	0.9536
$STDEV(V)$	0.0269	0.015	0.0156	0.0257	0.0145	0.0157
$\Sigma(V^d - V^a)^2$	2.5076	0.059	0.0931	1.9352	0.0408	0.0986
L^{max}-*index*	0.5083	0.3913	0.3895	0.4475	0.3394	0.331

Case-1: Peak load condition.
Case-2: Single line outage/contingency.

As discussed in (Moger, 2016; Moger & Dhadbanjan, 2015), the line connected between the buses 20 and 23 is considered as one of the most severe contingency. This line is considered for the reactive power and voltage control analysis in the system.

For each operating point/condition, the system is evaluated with various performance indices or parameters to assess the effectiveness of the proposed approach.

Case-1: System Under Peak Load Condition

The reactive power or voltage control analysis is carried out for peak load condition in the system. The system performance parameters and voltage/reactive power controllers setting from the proposed approach and other existing technique (Dhadbanjan & Yesuratnam, 2006) are given in Tables 10 and 11, respectively. It can be seen from Table 10 that under peak load condition the minimum voltage of the system improves from 0.8377 p.u. to 0.9585 p.u. using the proposed approach and to 0.9508 p.u. using the conventional LP based optimization technique. Similarly, the voltage stability index (*L-index*) is also improved from 0.6973 to 0.5142 using the proposed approach whereas it reduces to 0.528 using the conventional LP based optimization technique.

In addition, the conventional LP based optimization technique reduces the real power loss of the system to 163.70 MW whereas proposed approach reduces it to 163.32 MW. Similarly, the STDEV value of the load voltages also improved to 0.0185 using the proposed approach and to 0.0188 using the conventional LP based optimization technique. The improvement in system voltage profile and voltage stability index (*L-index*) at the load buses using both these techniques are shown in Figures 23 and 24, respectively.

Case-2: System Under Single Line Contingency Case

Similar to the analysis on peak load condition, the reactive power or voltage control analysis is carried out for single line contingency case. The system performance parameters and voltage/reactive power controllers setting using the proposed approach and other existing technique (Dhadbanjan & Yesuratnam,

Table 9. Controllers settings of 297-bus equivalent grid connected wind system under different operating points/conditions: WGUs ratio of 5:95 (FSWG:VSWG)

Transformer (OLTC) Tap Settings						
Fbus-Tbus	Case-1: Peak Load			Case-2: Outage		
	Initial	FLA	LP	Initial	FLA	LP
16-5	1	0.9625	0.95	1	0.975	0.9875
19-6	1	0.975	1	1	0.9875	1.0125
20-7	1	0.9625	0.9625	1	0.9625	0.9625
14-8	1	1	0.95	1	1	0.975
23-9	1	0.9875	1	1	0.9625	0.975
18-10	1	0.9875	1	1	0.9875	1
22-13	1	0.975	0.9875	1	0.975	1
8-25	1	0.975	1.025	1	0.9875	1.025
5-26	1	0.975	1.025	1	0.975	1.025
6-27	1	0.9875	1.025	1	0.9875	1.025
Voltage Settings of the Generator Excitation (p.u.)						
Gen. Bus	**Case-1: Peak Load**			**Case-2: Outage**		
	Initial	**FLA**	**LP**	**Initial**	**FLA**	**LP**
1	1	1.05	1.025	1	1.025	1.025
2	1	1	1.0125	1	1	1.025
3	1	1.0125	1.025	1	1	1.025
4	1	1.05	1.05	1	1.05	1.05
Reactive Power Output of SVCs (MVAr)						
SVC Bus	**Case-1: Peak Load**			**Case-2: Outage**		
	Initial	**FLA**	**LP**	**Initial**	**FLA**	**LP**
5	0	6	6	0	4	6
6	0	4	8	0	2	6
7	0	4	8	0	8	8
8	0	8	8	0	10	6
Total Reactive Power Output of VSWGs in the Wind Farms (MVAr)						
Wind Farms	**Case-1: Peak Load**			**Case-2: Outage**		
	Initial	**FLA**	**LP**	**Initial**	**FLA**	**LP**
WF-KAR	0	4.8	9	0	5.1	9
WF-AP	0	4.95	9	0	4.65	9.2
WF-TN	0	4.65	9	0	3	9

Figure 16. Voltage profile of SR 24-bus equivalent system with wind integration ratio of 5:95 (FSWG:VSWG) under peak load condition

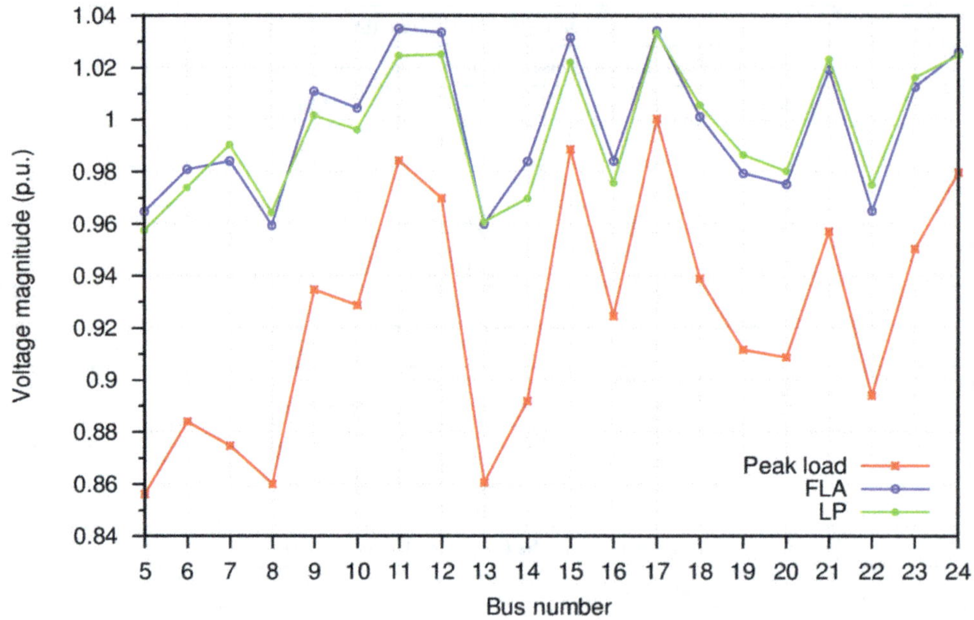

Figure 17. Voltage profile of SR 24-bus equivalent system with wind integration ratio of 5:95 (FSWG:VSWG) under outage/contingency condition

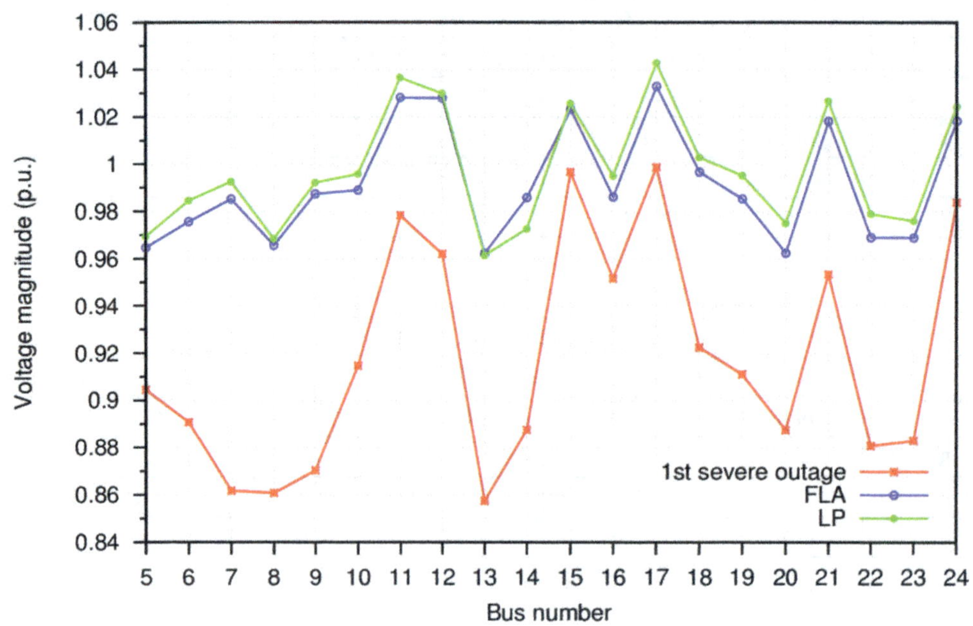

Figure 18. Voltage stability index (L-index) of SR 24-bus equivalent system with wind integration ratio of 5:95 (FSWG:VSWG) under peak load condition

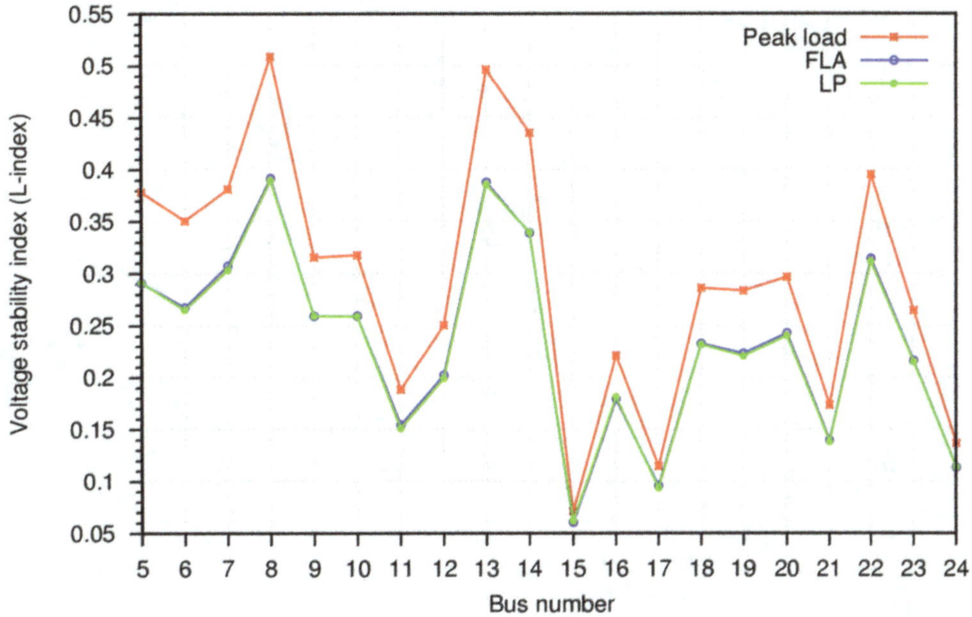

Figure 19. Voltage stability index (L-index) of SR 24-bus equivalent system with wind integration ratio of 5:95 (FSWG:VSWG) under outage/contingency condition

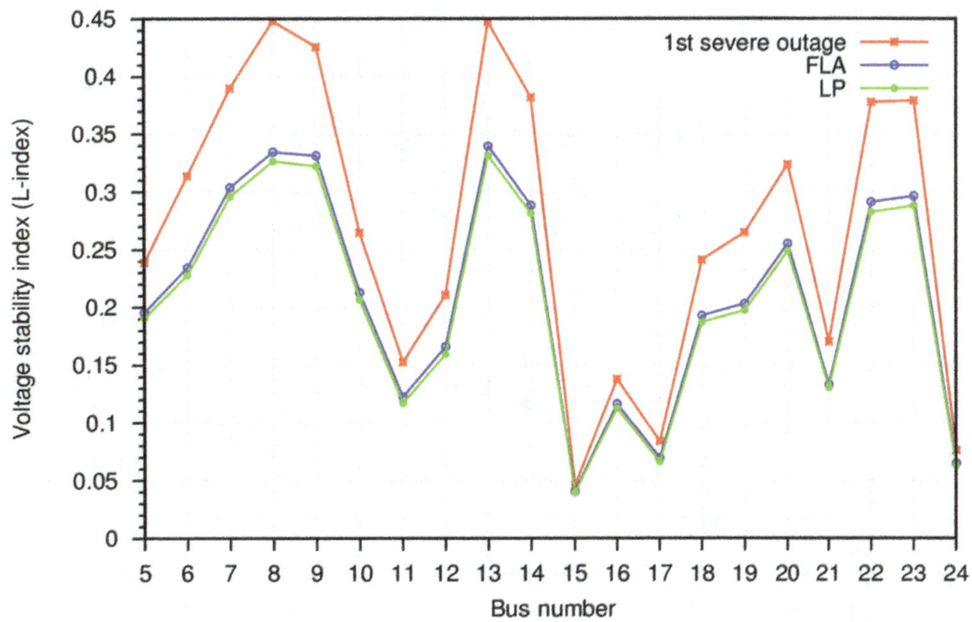

Figure 20. Single-line diagram of SR 79-bus equivalent system with wind farms

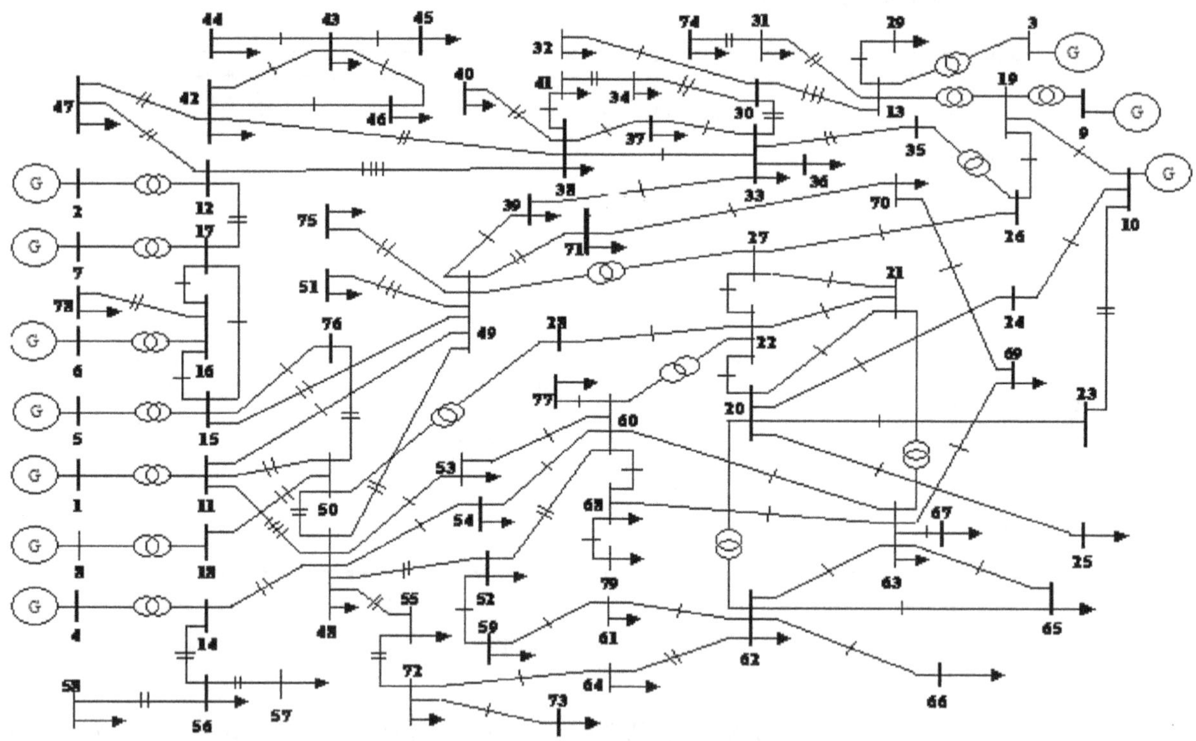

Figure 21. Single-line diagram of wind farm at Karnataka (Part-1)

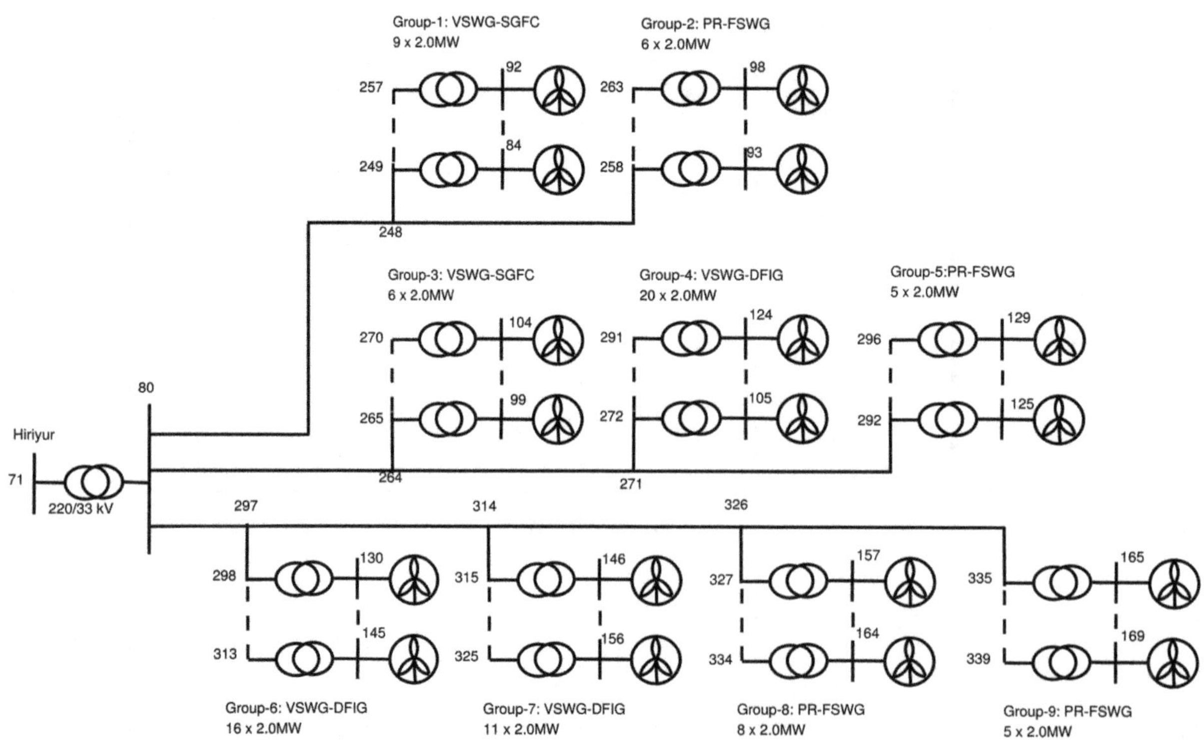

Figure 22. Single-line diagram of wind farm at Karnataka (Part-2)

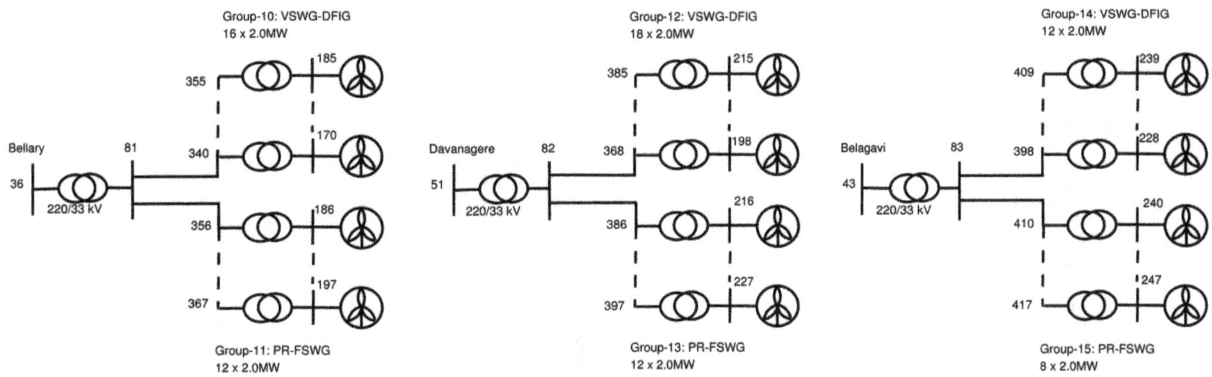

Table 10. System performance parameters of 417-bus equivalent grid connected wind system under different operating points/conditions

Parameters	Case-1: Peak Load			Case-2: Outage		
	Initial	FLA	LP	Initial	FLA	LP
$P_{loss}(MW)$	194.78	163.32	163.7	160.32	145.41	145.58
$Q_{loss}(MVAr)$	1549.85	1292.31	1282.69	1289.44	1148.33	1140.01
$V_{max}(p.u.)$	0.9823	1.0408	1.0381	0.988	1.037	1.0281
$V_{min}(p.u.)$	0.8377	0.9585	0.9508	0.8619	0.9621	0.9536
$STDEV(V)$	0.0266	0.0185	0.0188	0.0228	0.0172	0.0164
$\Sigma(V^d - V^a)^2$	1.923	0.1264	0.1868	1.1669	0.0808	0.1149
L^{max}-index	0.6973	0.5142	0.528	0.6684	0.5325	0.5327

2006) are also given in Tables 10 and 11, respectively. It can be seen from Table 10 that under single line contingency case the minimum voltage of the system improves from 0.8619 p.u. to 0.9621 p.u. using the proposed approach and to 0.9536 p.u. using the conventional LP based optimization technique. Similarly, the voltage stability index (*L-index*) is also improved from 0.6684 to 0.5325 using the proposed approach whereas it reduces to 0.5327 using the conventional LP based optimization technique.

In addition, the conventional LP based optimization technique reduces the system real power loss to 145.58 MW whereas proposed approach reduces it to 145.41 MW. Similarly, the STDEV value of the load voltages also improved to 0.0172 from proposed approach and to 0.0164 from the conventional LP based optimization technique. The improvement in system voltage profile and voltage stability index (*L-index*) at the load buses from both these techniques are shown in Figures 25 and 26, respectively.

Inference

Simulation results show that the system performance produced by the proposed approach is closely in agreement with that produced using the conventional LP based reactive power optimization technique. As observed in Table 7, the proposed fuzzy logic approach uses only few reactive power/voltage controllers

Table 11. Controllers settings of 417-bus equivalent grid connected wind system under different operating points/conditions

Transformer (OLTC) Tap Settings						
Fbus-Tbus	Case-1: Peak Load			Case-2: Outage		
	Initial	FLA	LP	Initial	FLA	LP
19-13	1	0.975	1	1	0.9875	0.975
26-35	1	0.975	1.025	1	0.9625	1
27-49	1	1.0125	0.9875	1	1	1.025
28-50	1	1.0125	1.025	1	1.025	1.025
22-60	1	0.925	0.9625	1	0.9625	1
20-62	1	0.925	0.9625	1	0.95	1
21-63	1	0.9875	0.9625	1	0.975	1
36-81	1	0.975	1.0375	1	0.9875	1.025
43-83	1	0.975	1.0375	1	0.9875	1.025
51-82	1	0.975	1.0375	1	0.9875	1.025
71-80	1	1.0125	1.0625	1	1	1.025
Voltage Settings of the Generator Excitation (p.u.)						
Gen. Bus	Case-1: Peak Load			Case-2: Outage		
	Initial	FLA	LP	Initial	FLA	LP
1	1	1.05	1.0375	1	1.0375	1.025
2	1	1.05	1.05	1	1.0375	1.025
3	1	1.05	1.05	1	1.05	1.025
4	1	1.0202	1.0077	1	1.0165	1.0165
5	1	1	1.0375	1	1	1.025
6	1	1	1.05	1	1	1.025
7	1	1	1.05	1	1	1.025
8	1	1	1.0375	1	1	1.025
9	1	1	1.05	1	1	1.025
10	1	1.028	1.003	1	1.025	1.025
Reactive Power Output of SVCs (MVAr)						
SVC Bus	Case-1: Peak Load			Case-2: Outage		
	Initial	FLA	LP	Initial	FLA	LP
41	0	25	25	0	15	15
44	0	15	25	0	10	15
57	0	30	25	0	10	15
59	0	15	20	0	15	15
70	0	15	20	0	5	15
72	0	25	20	0	15	15
74	0	25	25	0	20	15
77	0	25	20	0	15	15

continued on following page

Table 11. Continued

Wind Farms	Total Reactive Power Output of VSWGs in the Wind Farms (MVAr)					
	Case-1: Peak Load			Case-2: Outage		
	Initial	FLA	LP	Initial	FLA	LP
WF at 71bus	0	21.8	20.783	0	14.4	15.0852
WF at 36bus	0	3.2	10.5184	0	3.2	6.4
WF at 51bus	0	3.6	8.2332	0	3.6	7.2
WF at 43bus	0	2.6	7.8888	0	2.6	4.8

Figure 23. Voltage profile of SR 79-bus equivalent system with wind integration under peak load condition

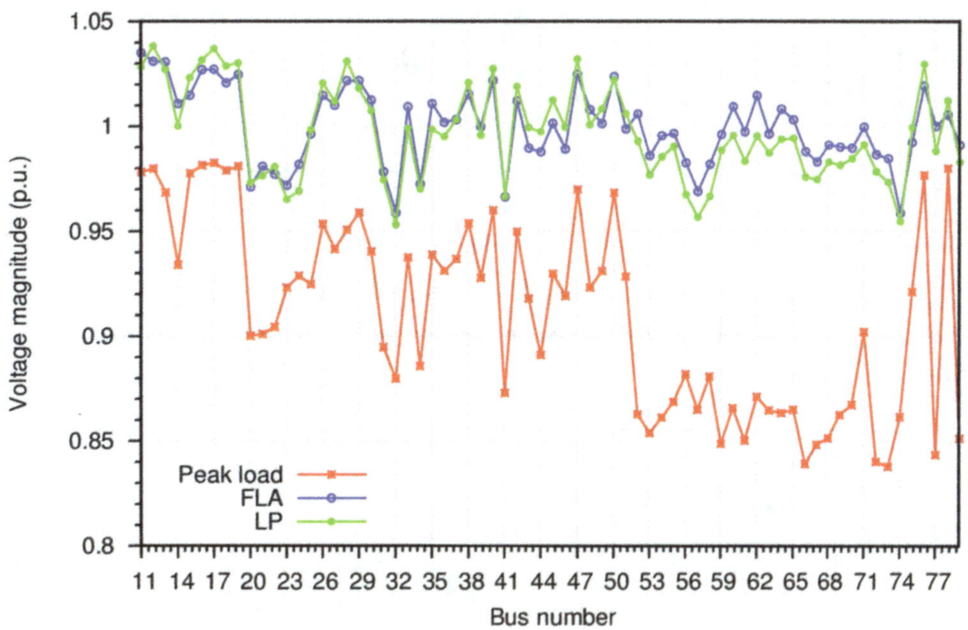

of high sensitivity to achieve the desired objectives so that the reduction in complexity of the problem is ensured and also the operation is fast.

FUTURE RESEARCH DIRECTIONS

The reactive power coordination in grid connected wind systems is carried out in accordance with the grid code compliance (Wu et al., 2010; Tsili & Papathanassiou, 2009). The mandatory requirement of power factor at the point of connection will restrict the full utilization of the reactive power capability of the variable speed wind generating units. However, the reactive power capability curve of the variable speed wind generators indicates that there is an additional supply of reactive power over the regulated power factor especially when the wind turbine operates below its rated power output. By utilizing the variable speed wind generators reactive power capability, the system operator will get more reactive

Figure 24. Voltage stability index (L-index) of SR 79-bus equivalent system with wind integration under peak load condition

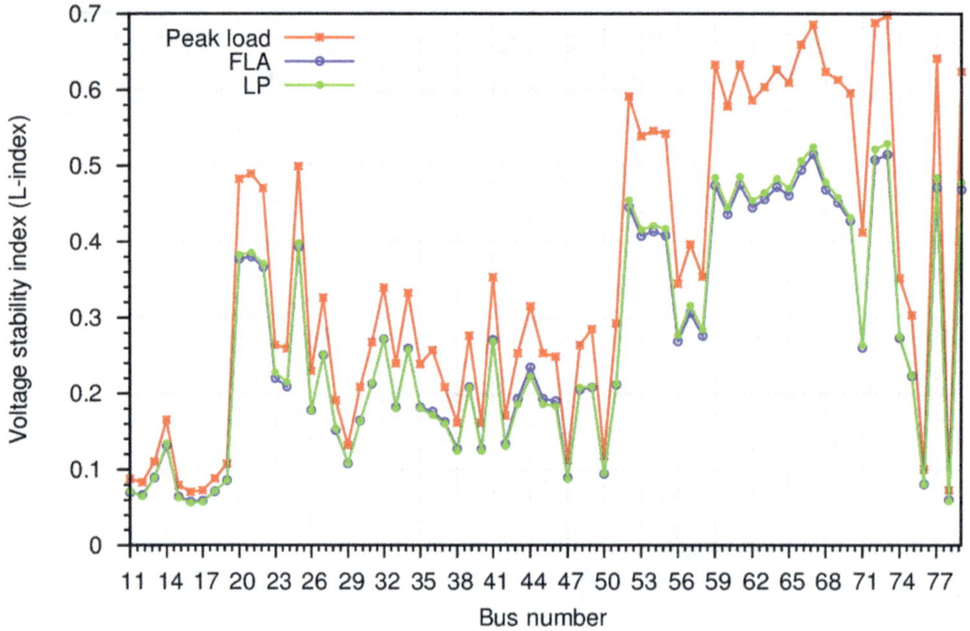

Figure 25. Voltage profile of SR 79-bus equivalent system with wind integration under outage/contingency condition

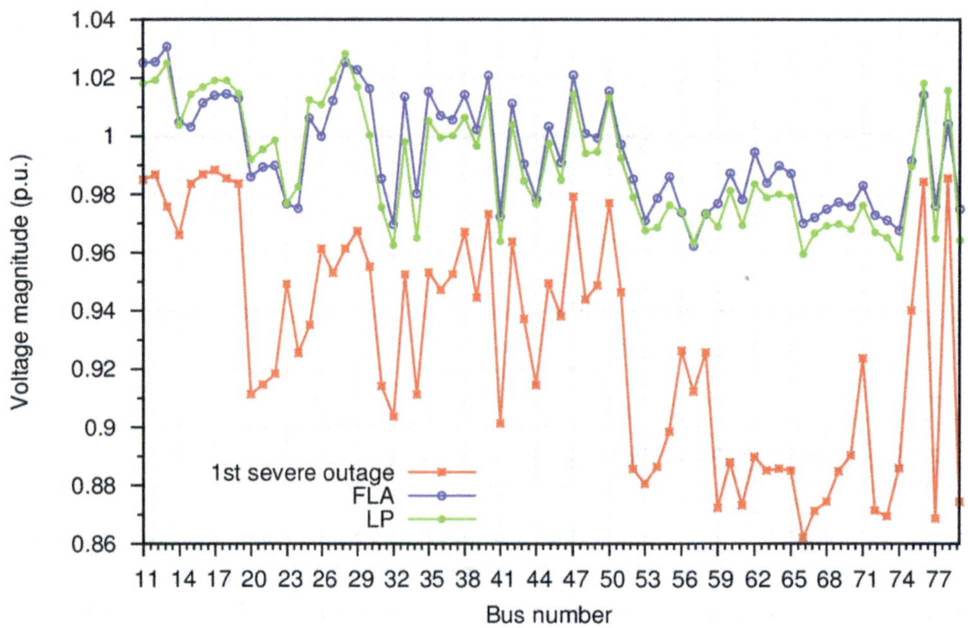

Figure 26. Voltage stability index (L-index) of SR 79-bus equivalent system with wind integration under outage/contingency condition

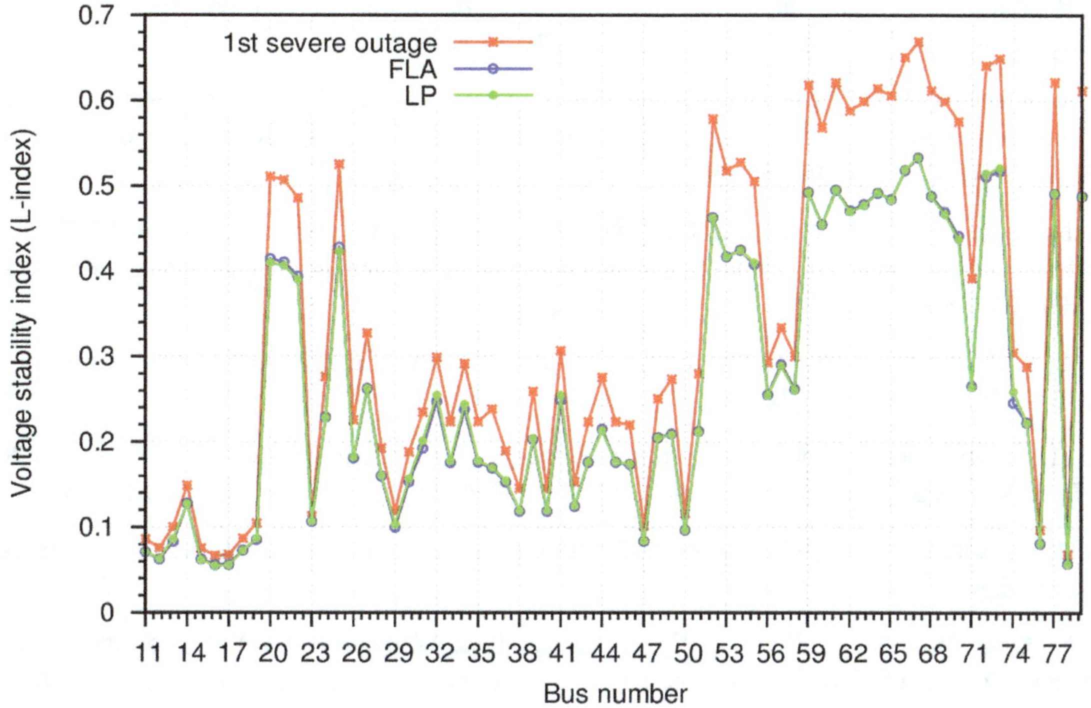

power from the wind farms that will be used along with other reactive power controllers from the grid for enhancing the system performance.

CONCLUSION

A fuzzy logic approach has been proposed for proper coordination of various reactive power controllers in grid connected with wind farms for voltage profile improvement. In addition, a new procedure has been incorporated in the fuzzy logic approach to address the issues associated with various controllers in grid connected wind farms. The FSWGs are also considered in the studies because of its impact on overall system voltage performance even though they do not support the system for voltage unlike the VSWGs. The simulation studies are carried out on 297-bus equivalent and 417-bus equivalent grid connected wind systems. Results demonstrated that the proposed approach is more effective in improving the system performance as compared with the conventional LP based reactive power optimization technique. The advantage of the proposed approach is that during each iteration, it uses only few reactive power/ voltage controllers of high sensitivity to achieve the desired objectives and hence, the overall computational effort is reduced. These are the desirable features for real-time implementation of the proposed approach for power system operation.

REFERENCES

Abdul-Rahman, K., & Shahidehpour, S. (1993). A fuzzy-based optimal reactive power control. *IEEE Transactions on Power Systems*, *8*(2), 662–670. doi:10.1109/59.260814

Abed, A. M. (1999). WSCC voltage stability criteria, undervoltage load shedding strategy, and reactive power reserve monitoring methodology. In *IEEE Power Engineering Society Summer Meeting* (Vol. 1, pp. 191-197). IEEE. 10.1109/PESS.1999.784345

Aghatehrani, R., Fan, L., & Kavasseri, R. (2009). Coordinated reactive power control of DFIG rotor and grid sides converters. In IEEE Power & Energy Society General Meeting (pp. 1-6). IEEE. doi:10.1109/PES.2009.5275235

Ajjarapu, V., & Lee, B. (1998). Bibliography on voltage stability. *IEEE Transactions on Power Systems*, *13*(1), 115–125. doi:10.1109/59.651622

Bansal, R. (2003). Bibliography on the fuzzy set theory applications in power systems (1994-2001). *IEEE Transactions on Power Systems*, *18*(4), 1291–1299. doi:10.1109/TPWRS.2003.818595

Central Electricity Regulatory Commission. (2010). *Indian Electricity Grid Code.* Retrieved from http://www.cercind.gov.in

Chen, N., Qian, M., Zhu, L., Yao, L., Wu, F., Chen, M., & Wang, N. (2012). A coordinated reactive power and voltage control system for wind farm grid integration. In *IEEE International Conference on Power System Technology* (pp. 1-6). IEEE.

Chen, N., Zhu, L. Z., & Wang, W. (2009). Strategy for reactive power control of wind farm for improving voltage stability in wind power integrated region. *Proceedings of the CSEE*, *10*, 019.

China Electric Power Research Institute. (2009). Retrieved from http://www.epri.ac.cn. China.

De Almeida, R. G., Castronuovo, E. D., & Peas Lopes, J. (2006). Optimum generation control in wind parks when carrying out system operator requests. *IEEE Transactions on Power Systems*, *21*(2), 718–725. doi:10.1109/TPWRS.2005.861996

Dhadbanjan, T. (1997). An expert system for alleviation of network overloads. *Electric Power Systems Research*, *40*(2), 143–153. doi:10.1016/S0378-7796(96)01146-7

Dhadbanjan, T., & Parthasarathy, K. (1995). An expert system for voltage control in a power system network. In *International Conference on Energy Management and Power Delivery* (Vol. 1, pp. 364-369). IEEE.

Dhadbanjan, T., Parthasarathy, K., & Prior, D. (1984). Improved algorithm for optimum reactive power allocation. *International Journal of Electrical Power & Energy Systems*, *6*(2), 72–74. doi:10.1016/0142-0615(84)90011-5

Dhadbanjan, T., & Yesuratnam, G. (2006). Comparison of optimum reactive power schedule with different objectives using LP technique. *International Journal of Emerging Electric Power Systems*, *7*(3), 1–31. doi:10.2202/1553-779X.1338

Divya, K., & Rao, P. N. (2006). Models for wind turbine generating systems and their application in load flow studies. *Electric Power Systems Research*, *76*(9), 844–856. doi:10.1016/j.epsr.2005.10.012

Eir Grid. (n.d.). *Eir Grid Plc*. Retrieved from http://www.EirGrid.com

El-Hawary, M. E. (1998). *Electric power applications of fuzzy systems*. Wiley-IEEE Press. doi:10.1109/9780470544457

El-Keib, A., & Ma, X. (1995). Application of artificial neural networks in voltage stability assessment. *IEEE Transactions on Power Systems*, *10*(4), 1890–1896. doi:10.1109/59.476054

Elgerd, O. I. (1982). *Electic energy systems theory-an introduction*. McGraw-Hill New York.

E.ON Netz Gmbh. (2016). *Grid Code-High and extra high voltage*. Retrieved from http://www.eon.com

Federal Energy Regulatory Commission. (2005). Principles for efficient and reliable reactive power supply and consumption staff report (Docket No. AD05-1-000). Author.

Federal Energy Regulatory Commission. (2005). Order no. 661 on interconnection for wind farms. Author.

Federal Energy Regulatory Commission. (2012). Order no. 661a (order on rehearing) on interconnection for wind farms. Author.

Freitas, W., Vieira, J. C., Da Suva, L. C. P., Affonso, C. M., & Morelato, A. (2005). Long-term voltage stability of distribution systems with induction generators. In *IEEE Power Engineering Society General Meeting* (pp. 2910-2913). IEEE.

Grid Code. (2004). *Wind power facility technical requirements - Revision 0*. Retrieved from http://www.aeso.ca

Hossain, M. J., Pota, H. R., Mahmud, M. A., & Ramos, R. A. (2012). Investigation of the impacts of large-scale wind power penetration on the angle and voltage stability of power systems. *IEEE Systems Journal*, *6*(1), 76–84. doi:10.1109/JSYST.2011.2162991

Kayikci, M., & Milanovic, J. V. (2007). Reactive power control strategies for dfig-based plants. *IEEE Transactions on Energy Conversion*, *22*(2), 389–396. doi:10.1109/TEC.2006.874215

Keane, A., Ochoa, L. F., Vittal, E., Dent, C. J., & Harrison, G. P. (2011). Enhanced utilization of voltage control resources with distributed generation. *IEEE Transactions on Power Systems*, *26*(1), 252–260. doi:10.1109/TPWRS.2009.2037635

Konopinski, R. J., Vijayan, P., & Ajjarapu, V. (2009). Extended reactive capability of dfig wind parks for enhanced system performance. *IEEE Transactions on Power Systems*, *24*(3), 1346–1355. doi:10.1109/TPWRS.2009.2023260

Kumar, V. S. S., Reddy, K. K., & Dhadbanjan, T. (2014). Coordination of reactive power in grid-connected wind farms for voltage stability enhancement. *IEEE Transactions on Power Systems*, *29*(5), 2381–2390. doi:10.1109/TPWRS.2014.2300157

Kundur, P., Balu, N. J., & Lauby, M. G. (1994). *Power system stability and control* (Vol. 7). McGraw-Hill New York.

Lof, P.-A., Andersson, G., & Hill, D. (1993). Voltage stability indices for stressed power systems. *IEEE Transactions on Power Systems*, *8*(1), 326–335. doi:10.1109/59.221224

Mamandur, K., & Chenoweth, R. (1981). Optimal control of reactive power flow for improvements in voltage profiles and for real power loss minimization. *IEEE Transactions on Power Apparatus and Systems*, (7): 3185–3194. doi:10.1109/TPAS.1981.316646

Marques, A. B., Taranto, G. N., & Falc˜ao, D. M. (2005). A knowledge-based system for supervision and control of regional voltage profile and security. *IEEE Transactions on Power Systems*, *20*(1), 400–407. doi:10.1109/TPWRS.2004.831672

Meegahapola, L., Durairaj, S., Flynn, D., & Fox, B. (2011). Coordinated utilisation of wind farm reactive power capability for system loss optimisation. *European Transactions on Electrical Power*, *21*(1), 40–51. doi:10.1002/etep.410

Moger, T. (2016). *Reactive power planning and operation of power systems with wind farms for voltage stability improvement* (Unpublished doctoral dissertation). Department of Electrical Engineering, Indian Institute of Science, Bangalore, India.

Moger, T., & Dhadbanjan, T. (2015). A novel index for identification of weak nodes for reactive compensation to improve voltage stability. *IET Generation, Transmission & Distribution*, *9*(14), 1826–1834. doi:10.1049/iet-gtd.2015.0054

Moger, T., & Dhadbanjan, T. (2016). Fuzzy logic approach for reactive power coordination in grid connected wind farms to improve steady state voltage stability. *IET Renewable Power Generation*, *11*(2), 351–361. doi:10.1049/iet-rpg.2016.0119

Moger, T., & Dhadbanjan, T. (2017). Impact of different pq models of wind turbine generating units (wtgus) on system voltage performance. *International Journal of Emerging Electric Power Systems*, *18*(4), 1–19. doi:10.1515/ijeeps-2017-0008

Nageswararao, B., & Jeyasurya, B. (1998). Fuzzy-expert system for voltage stability monitoring and control. *Electric Power Systems Research*, *47*(3), 215–222. doi:10.1016/S0378-7796(98)00068-6

National Electricity Amendment. (2007). *Technical Standards for wind and other generators connections rule.* Retrieved from http://www.aemc.gov.au

National Grid Electricity Transmission. (2009). *The Grid Code.* Retrieved from http://www.nationalgrid.com

Niebur, D. (1993). Artificial neural networks for power systems-a literature survey. *International Journal of Engineering Intelligent Systems for Electrical Engineering and Communications, 1*, 133–158.

Pudjianto, D., Ahmed, S., & Strbac, G. (2002). Allocation of VAR support using LP and NLP based optimal power flows. *IEE Proceedings. Generation, Transmission and Distribution, 149*(4), 377–383. doi:10.1049/ip-gtd:20020200

Scala, M. L., Trovato, M., & Torelli, F. (1996). A neural network-based method for voltage security monitoring. *IEEE Transactions on Power Systems, 11*(3), 1332–1341. doi:10.1109/59.535674

Shareeq, M. M., Moger, T., & Dhadbanjan, T. (2014). Fuzzy logic approach to curtail number of controllers for voltage control in grid connected wind farms. In *Eighteenth National Power Systems Conference (NPSC)* (pp. 1-6). IEEE. 10.1109/NPSC.2014.7103843

Spain Grid Code. (2005). *Grid code.* Retrieved from http://www.depeca.uah.es

Srivastava, S. C., Velayutham, A., Agrawal, K. K., & Bakshi, A. S. (2012). Report of the enquiry committee on grid disturbance in northern region on 30th July 2012 and in northern, eastern & north-eastern region on 31st July 2012. Delhi: Academic Press.

Su, C.-T., & Lin, C.-T. (1996). A new fuzzy control approach to voltage profile enhancement for power systems. *IEEE Transactions on Power Systems, 11*(3), 1654–1659. doi:10.1109/59.535711

Su, C.-T., & Lin, C.-T. (2001). Fuzzy-based voltage/reactive power scheduling for voltage security improvement and loss reduction. *IEEE Transactions on Power Delivery, 16*(2), 319–323. doi:10.1109/61.915501

Tang, G., Xu, H., & Chen, H. (1993). An Expert System for Voltage/VAR Control of Interconnected Power Systems. In *Proceedings of the Fourth Symposium on Expert Systems Application to Power Systems* (pp. 292-295). Academic Press.

Taylor, C. W. (1994). *Power system voltage stability.* McGraw-Hill Companies.

Tinney, W., Bright, J., Demaree, K., & Hughes, B. (1988). Some deficiencies in optimal power flow. *IEEE Transactions on Power Systems, 3*(2), 676–683. doi:10.1109/59.192922

Tomsovic, K. (1992). A fuzzy linear programming approach to the reactive power/voltage control problem. *IEEE Transactions on Power Systems, 7*(1), 287–293. doi:10.1109/59.141716

Trichakis, P., Taylor, P. C., Lyons, P. F., & Hair, R. (2008). Predicting the technical impacts of high levels of small-scale embedded generators on low-voltage networks. *IET Renewable Power Generation, 2*(4), 249–262. doi:10.1049/iet-rpg:20080012

Tsili, M., & Papathanassiou, S. (2009). A review of grid code technical requirements for wind farms. *IET Renewable Power Generation, 3*(3), 308–332. doi:10.1049/iet-rpg.2008.0070

Udupa, A. N., Dhadbanjan, T., & Parthasarathy, K. (1999). An expert fuzzy control approach to voltage stability enhancement. *International Journal of Electrical Power & Energy Systems*, *21*(4), 279–287. doi:10.1016/S0142-0615(98)00049-0

Ullah, N. R., & Thiringer, T. (2007). Variable speed wind turbines for power system stability enhancement. *IEEE Transactions on Energy Conversion*, *22*(1), 52–60. doi:10.1109/TEC.2006.889625

Visakha, K., Dhadbanjan, T., & Jenkins, L. (2004). Application of upfc for system security improvement under normal and network contingencies. *Electric Power Systems Research*, *70*(1), 46–55. doi:10.1016/j.epsr.2003.11.011

Vittal, E., Keane, A., & O'Malley, M. (2008). Varying penetration ratios of wind turbine technologies for voltage and frequency stability. In *IEEE Power and Energy Society General Meeting-Conversion and Delivery of Electrical Energy in the 21st Century* (pp. 1–6). IEEE. doi:10.1109/PES.2008.4596402

Vittal, E., O'Malley, M., & Keane, A. (2009). Impact of wind turbine control strategies on voltage performance. In IEEE Power & Energy Society General Meeting (pp. 1-7). IEEE. doi:10.1109/PES.2009.5275570

Vittal, E., O'Malley, M., & Keane, A. (2010). A steady-state voltage stability analysis of power systems with high penetrations of wind. *IEEE Transactions on Power Systems*, *25*(1), 433–442. doi:10.1109/TPWRS.2009.2031491

Wu, Q., Xu, Z., & Østergaard, J. (2010). Grid integration issues for large scale wind power plants (WPPs). In *IEEE Power and Energy Society General Meeting* (pp. 1-6). IEEE.

Yager, R. R., & Zadeh, L. A. (Eds.). (2012). *An introduction to fuzzy logic applications in intelligent systems* (Vol. 165). Springer Science & Business Media.

Yokoyama, R., Niimura, T., & Nakanishi, Y. (1993). A coordinated control of voltage and reactive power by heuristic modeling and approximate reasoning. *IEEE Transactions on Power Systems*, *8*(2), 636–645. doi:10.1109/59.260817

Zadeh, L. A. (1965). Fuzzy sets. *Information and Control*, *8*(3), 338–353. doi:10.1016/S0019-9958(65)90241-X

Zheng, T., Jiao, S., Ding, K., & Lin, L. (2013). A coordinated voltage control strategy of wind farms based on sensitivity method. In IEEE Grenoble PowerTech (POWERTECH) (pp. 1-5). IEEE.

Zhu, L. Z., Chen, N., Wang, W., & Zhu, X. D. (2009). Reactive control strategy for wind farm considering reactive power command. In *International Conference on Sustainable Power Generation and Supply* (pp. 1-5). IEEE. 10.1109/SUPERGEN.2009.5348206

Zimmermann, H. J. (2011). *Fuzzy set theory—and its applications*. Springer Science & Business Media.

KEY TERMS AND DEFINITIONS

Contingency: Contingency is an outage of a transmission line or transformer that may lead to overloads in other branches and/or sudden system voltage drop.

Electric Grid: An electric grid is a network of synchronized power providers and consumers that are connected by transmission and distribution lines and operated by one or more control centers.

Fuzzy System: It is a component of machine learning techniques that takes membership values within 0 to 1 unlike crisp sets.

Optimization: A mathematical method to find the solution of a problem towards achieving better performance either in form of minimum or maximum under one or more given constraints.

Reactive Power: In electric power transmission and distribution, volt ampere reactive (VAR) is a unit by which reactive power is expressed in an AC electric power system.

Standard Deviation (STDEV): The standard deviation of the load voltage is used to quantify the amount of variation or dispersion of a set of load voltages.

Voltage Stability: Voltage stability refers to the ability of a power system to maintain steady voltage at all buses in the system after being subjected to a disturbance from a given initial operating condition.

Wind Farm: An area of land with a group of energy producing windmills or wind turbines.

Wind Turbine: A wind turbine is a device that converts the wind's kinetic energy into electrical power.

APPENDIX: COMPARISON METHOD

Linear Programming Based Reactive Power Optimization Technique

The model selected for the system voltage profile improvement is to minimize the sum of the square of the voltage deviation from desired voltage of the load buses as an objective function. It can be expressed as,

$$v^e = \sum_{j=g+1}^{n} \left(V_j^d - V_j^a\right)^2 \tag{12}$$

The selected model for the reactive power optimization uses linearized sensitivity relationships to define the optimization problem.

For this objective function, the constraints are the linearized network performance equations relating the control and dependent variables, and their limits. To improve system voltage profile, redistribution of reactive power generations in power systems is necessary (Mamandur & Chenoweth, 1981). Reactive power distributions in the system can be controlled by suitably adjusting the following control variables:

- Transformer tap settings (ΔT)
- Generator excitation settings (ΔV)
- The Switchable VAr Compensator (SVC) settings (ΔQ)

The system voltage profile and thus the stability of the system are affected by any changes brought in these variables. The changes made in these variables affect the reactive power distribution in the system and also the reactive power output from generators. The variables which are affected due to the changes in control variables are the dependent variables.

The dependent variables are:

- The reactive power output of the generators (ΔQ)
- The voltage magnitude at the buses other than the generator buses (ΔV)

All these variables also have upper and lower limits.

In mathematical form, the problem can be expressed as,

Minimize

$$v^e = Cx \tag{13}$$

Subject to

$$b^{\min} \leq b = Sx \leq b^{\max} \tag{14}$$

and

$$x^{\min} \leq x \leq x^{\max} \tag{15}$$

where,

C is the row vector of the linearized objective function sensitivity coefficients,
S is the linearized sensitivity matrix relating the dependent and control variables,
b is the column vector of the linearized dependent variables,
x is the column vector of the linearised control variables,
b^{max} and b^{min} are the column vectors of the linearized upper and lower limits on the dependent variables, and
x^{max} and x^{min} are the column vectors of the linearized upper and lower limits on the control variables.

The linear programming technique is now applied to the above problem to determine the optimal settings of the control variables (Dhadbanjan & Yesuratnam, 2006).

The control vector in incremental variables is defined as

$$x = \left[\Delta T_1, \ldots, \Delta T_t, \Delta V_1, \ldots, \Delta V_g, \Delta Q_{g+1}, \ldots, \Delta Q_{g+s}\right]^t \tag{16}$$

The dependent vector in incremental variables is defined as

$$b = \left[\Delta Q_1, \ldots, \Delta Q_g, \Delta V_{g+1}, \ldots, \Delta V_{g+s}, \Delta V_{g+s+1}, \ldots, \Delta V_n\right]^t \tag{17}$$

The upper and lower limits on the both the control and dependent variables in linearized form are expressed as

$$x^{\max} = \left[\Delta T_1^{\max}, \ldots, \Delta T_t^{\max}, \Delta V_1^{\max}, \ldots, \Delta V_g^{\max}, \Delta Q_{g+1}^{\max}, \ldots, \Delta Q_{g+s}^{\max}\right]^t \tag{18}$$

$$x^{\min} = \left[\Delta T_1^{\min}, \ldots, \Delta T_t^{\min}, \Delta V_1^{\min}, \ldots, \Delta V_g^{\min}, \Delta Q_{g+1}^{\min}, \ldots, \Delta Q_{g+s}^{\min}\right]^t \tag{19}$$

$$b^{\max} = \left[\Delta Q_1^{\max}, \ldots, \Delta Q_g^{\max}, \Delta V_{g+1}^{\max}, \ldots, \Delta V_{g+s}^{\max}, \Delta V_{g+s+1}^{\max}, \ldots, \Delta V_n^{\max}\right]^t \tag{20}$$

$$b^{\min} = \left[\Delta Q_1^{\min}, \ldots, \Delta Q_g^{\min}, \Delta V_{g+1}^{\min}, \ldots, \Delta V_{g+s}^{\min}, \Delta V_{g+s+1}^{\min}, \ldots, \Delta V_n^{\min}\right]^t \tag{21}$$

where,

$$\Delta T^{\min} = T^{\min} - T^{actual}$$

$$\Delta T^{\max} = T^{\max} - T^{actual}$$

$$\Delta Q^{\min} = Q^{\min} - Q^{actual}$$

$$\Delta Q^{\max} = Q^{\max} - Q^{actual}$$

$$\Delta V^{\min} = V^{\min} - V^{actual}$$

$$\Delta V^{\max} = V^{\max} - V^{actual}$$

Computation of Sensitivity Matrix [*S*]

The sensitivity matrix [*S*] relating the dependent and control variables is evaluated (Dhadbanjan et al., 1984; Dhadbanjan & Yesuratnam, 2006) in the following manner. Considering the fact that the reactive power injection at a bus does not change for a small change in the phase angle of the bus voltage. Based on this assumption, the relationship between the net reactive power change at any node due to change in the transformer tap settings and voltage magnitudes can be written as,

$$\begin{bmatrix} \Delta Q_g \\ \Delta Q_s \\ \Delta Q_v \end{bmatrix} = \begin{bmatrix} \frac{\partial Q_g}{\partial T_t} & \frac{\partial Q_g}{\partial V_g} & \frac{\partial Q_g}{\partial V_s} & \frac{\partial Q_g}{\partial V_r} \\ \frac{\partial Q_s}{\partial T_t} & \frac{\partial Q_s}{\partial V_g} & \frac{\partial Q_s}{\partial V_s} & \frac{\partial Q_s}{\partial V_r} \\ \frac{\partial Q_r}{\partial T_t} & \frac{\partial Q_r}{\partial V_g} & \frac{\partial Q_r}{\partial V_s} & \frac{\partial Q_r}{\partial V_r} \end{bmatrix} \begin{bmatrix} \Delta T_t \\ \Delta V_g \\ \Delta V_s \\ \Delta V_r \end{bmatrix} \tag{22}$$

It can also written in this form

$$\begin{bmatrix} \Delta Q_g \\ \Delta Q_s \\ \Delta Q_v \end{bmatrix} = \begin{bmatrix} A_1 & A_2 & A_3 & A_4 \\ A_5 & A_6 & A_7 & A_8 \\ A_9 & A_{10} & A_{11} & A_{12} \end{bmatrix} \begin{bmatrix} \Delta T_t \\ \Delta V_g \\ \Delta V_s \\ \Delta V_r \end{bmatrix} \tag{23}$$

where,

$$\Delta Q_g = \left[\Delta Q_1, \ldots, \Delta Q_g\right]^t,$$

$$\Delta Q_s = \left[\Delta Q_{g+1},\ldots,\Delta Q_{g+s}\right]^t,$$

$$\Delta Q_r = \left[\Delta Q_{g+s+1},\ldots,\Delta Q_n\right]^t,$$

$$\Delta T_t = \left[\Delta T_1,\ldots,\Delta T_t\right]^t,$$

$$\Delta V_g = \left[\Delta V_1,\ldots,\Delta V_g\right]^t,$$

$$\Delta V_s = \left[\Delta V_{g+1},\ldots,\Delta V_{g+s}\right]^t,$$

and

$$\Delta V_r = \left[\Delta V_{g+s+1},\ldots,\Delta V_n\right]^t,$$

The sub-matrices A_1 to A_{12} are the corresponding terms of $\partial Q / \partial T$ and $\partial Q / \partial V$ and the values are calculated from the following equations; where,

$$\frac{\partial Q_k}{\partial T_{kl}} = \frac{2}{a^3} V_k^2 y_{kl} \sin(\alpha_{kl}) + \frac{1}{a^2} y_{kl} V_k V_l \sin\left(\delta_k - \delta_l - \theta_{kl}\right) \tag{24}$$

$$\frac{\partial Q_l}{\partial T_{kl}} = \frac{1}{a^2} y_{kl} V_k V_l \sin\left(\delta_l - \delta_k - \theta_{kl}\right) \tag{25}$$

where,

k and l are the index for the nodes and k is the tap-side bus

$$\frac{\partial Q_r}{\partial T_{kl}} = 0, \quad r \neq k,l \tag{26}$$

$$\frac{\partial Q_k}{\partial V_k} = \frac{Q_k}{V_k} - B_{kk} V_k \tag{27}$$

$$\frac{\partial Q_k}{\partial V_l} = Y_{kl} V_k \sin\left(\delta_k - \delta_l - \theta_{kl}\right) \tag{28}$$

Further, transferring all the control variables to the right hand side and the dependent variables to the left hand side, and rearranging the equations.

By properly arranging the above equations, we can get,

$$\begin{bmatrix} \Delta Q_g \\ \Delta V_s \\ \Delta V_r \end{bmatrix} = \begin{bmatrix} S_1 & S_2 & S_3 \\ S_4 & S_5 & S_6 \\ S_7 & S_8 & S_9 \end{bmatrix} \begin{bmatrix} \Delta T_t \\ \Delta V_g \\ \Delta Q_s \end{bmatrix} \tag{29}$$

where,

$$S_1 = (A_1 + A_3 S_4 + A_4 S_7) \tag{30}$$

$$S_2 = (A_2 + A_3 S_5 + A_4 S_8) \tag{31}$$

$$S_3 = (A_3 S_6 + A_4 S_9) \tag{32}$$

$$S_4 = \left(A_7 - A_8 \left(A_{12}\right)^{-1} A_{11}\right)^{-1} \left(A_8 \left(A_{12}\right)^{-1} A_9 - A_5\right) \tag{33}$$

$$S_5 = \left(A_7 - A_8 \left(A_{12}\right)^{-1} A_{11}\right)^{-1} \left(A_8 \left(A_{12}\right)^{-1} A_{10} - A_6\right) \tag{34}$$

$$S_6 = \left(A_7 - A_8 \left(A_{12}\right)^{-1} A_{11}\right) \tag{35}$$

$$S_7 = \left(-\left(A_{12}\right)^{-1} A_9 - \left(A_{12}\right)^{-1} A_{11} S_4\right) \tag{36}$$

$$S_8 = \left(-\left(A_{12}\right)^{-1} A_{10} - \left(A_{12}\right)^{-1} A_{11} S_5 \right) \tag{37}$$

$$S_9 = \left(-\left(A_{12}\right)^{-1} A_{11} S_6 \right) \tag{38}$$

Again (29), can be written in this form;

$$\begin{bmatrix} \Delta Q_g \\ \Delta V_s \\ \Delta V_r \end{bmatrix} = \left[S\right] \begin{bmatrix} \Delta T_t \\ \Delta V_g \\ \Delta Q_s \end{bmatrix} \tag{39}$$

where $[S]$ is the linearized sensitivity matrix relating dependent and control variables.

Computation of Objective Function ($v^e = C\Delta x$) Sensitivities With Respect to Control Variable

The row vector of objective function sensitivity coefficients (C) can be written as

$$C = \left[\frac{\partial v^e(x)}{\partial T_t}, \quad \frac{\partial v^e(x)}{\partial V_g}, \quad \frac{\partial v^e(x)}{\partial Q_s} \right] \tag{40}$$

Consider a system where, k = total number of control variables with k=1, 2...t be the number of OLTCs, t+1, t+2...t+g be the number of generator excitations and t+g+1....t+g+s be the number SVCs ($k = t+g+s$).

$$\left[\frac{\partial v^e(x)}{\partial T_m} \right] = \sum_{j=g+1}^{n} 2\left(V_j^d - V_j^a\right)\left(-S_{jm}\right) \tag{41}$$

where m = 1, 2....t for calculating the objective function sensitivities with respect to transformer taps, and S_{jm} is the corresponding sensitivity elements in (39).

$$\left[\frac{\partial v^e(x)}{\partial V_m} \right] = \sum_{j=g+1}^{n} 2\left(V_j^d - V_j^a\right)\left(-S_{jm}\right) \tag{42}$$

where $m = t+1, t+2, \ldots t+g$ for calculating the objective function sensitivities with respect to generator excitations, and S_{jm} is the corresponding sensitivity elements in (39).

$$\left[\frac{\partial v^e(x)}{\partial Q_m}\right] = \sum_{j=g+1}^{n} 2\left(V_j^d - V_j^a\right)\left(-S_{jm}\right) \tag{43}$$

where $m = t+g+1 \ldots k$ for calculating the objective function sensitivities with respect to SVCs, and S_{jm} is the corresponding sensitivity elements in (39).

Chapter 2
Performance Prediction and Optimization of Solar Water Heater via a Knowledge-Based Machine Learning Method

Hao Li
University of Texas at Austin, USA

Zhijian Liu
North China Electric Power University, China

ABSTRACT

Measuring the performance of solar energy and heat transfer systems requires a lot of time, economic cost, and manpower. Meanwhile, directly predicting their performance is challenging due to the complicated internal structures. Fortunately, a knowledge-based machine learning method can provide a promising prediction and optimization strategy for the performance of energy systems. In this chapter, the authors show how they utilize the machine learning models trained from a large experimental database to perform precise prediction and optimization on a solar water heater (SWH) system. A new energy system optimization strategy based on a high-throughput screening (HTS) process is proposed. This chapter consists of: 1) comparative studies on varieties of machine learning models (artificial neural networks [ANNs], support vector machine [SVM], and extreme learning machine [ELM]) to predict the performances of SWHs; 2) development of an ANN-based software to assist the quick prediction; and 3) introduction of a computational HTS method to design a high-performance SWH system.

INTRODUCTION

Predicting the thermal performance of solar energy systems is of huge challenge due to the complexity of the internal structures. In fact, the measurement of thermal performances of a typical solar energy system (*e.g.*, solar water heater (SWH)) requires a lot of time, economic costs and labors. Due to the complicated structures, conventional physical and mathematical models usually fail to estimate their

DOI: 10.4018/978-1-5225-3935-3.ch002

thermal performances. These problems not only dramatically hinder the acquisition of the thermal performances of solar energy systems, but also block the possibility of optimizing their thermal performance.

In the past decades, scientists have come up with a powerful prediction method to address these problems. People found that a knowledge-based machine learning model can help precisely predict the performances of some energy systems utilizing some simple independent variables as the computational inputs. With a proper machine learning algorithm, people only need to acquire a sufficient experimental database as well as perform the model training and testing, and then a predictive model can be acquired. During the training process, machine learning can "learn" the non-linear relationship between the independent and dependent variables via a "black box" fitting, and subsequently perform the predictions. The research group of Prof. S. Kalogirou, from Cyprus University of Technology, has conducted a majority of the pioneer application research on the prediction of thermal performances for energy systems (Kalogirou, 1999), leading to huge positive engineering impacts during the last two decades. Subsequently, relevant studies have become increasingly popular all over the world. Meanwhile, there are more and more new or revised machine learning algorithms developed. Among various machine learning approaches, there are some most widely used algorithms, such as artificial neural network (ANN) (Kalogirou, 1999), support vector machine (SVM) (Suykens & Vandewalle, 1999) and extreme learning machine (ELM) (Huang et al., 2006). ANN is the most prevalent algorithm due to its long history and powerful predictive capacity. The general schematic structure of an ANN model is presented in Figure 1.

So far, despite the great progress of machine learning, engineering and industrial requirements have been rising in recent years: how to cost-effectively design and optimize a solar energy system by utilizing machine learning? Now, machine learning is a proven powerful tool for varieties of numerical predictions, and people are trying to make full use of its predictive power, as well as provide a good optimization strategy in order to acquire higher performances. Nevertheless, to the best of the authors' knowledge, very few research reports have mainly focused on the relevant studies (Peng & Ling, 2008). Recently, the authors have found that with a sufficiently large experimental database, the machine learning models are

Figure 1. General schematic structure of a typical ANN model. The empty circles represent the neurons. All the neurons interconnect with other neurons in the adjacent layer(s). Each neuron in the input layer represents an independent variable. The neuron in the output layer represents the dependent variable.
Reproduced with permission from Reference (Li, Liu, Liu, & Zhang, 2017).

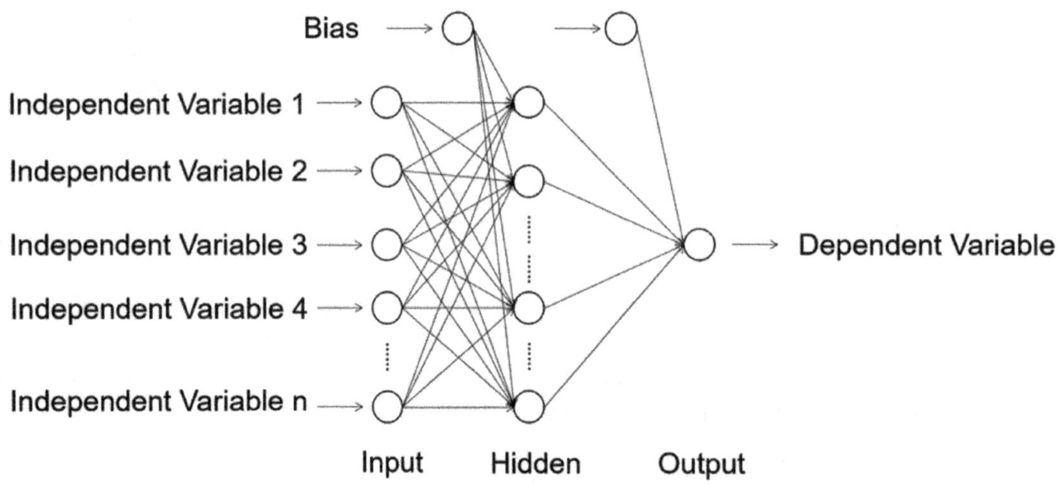

not only able to give excellent predictive performance for SWHs, but also assist an efficient and promising optimization of the thermal performances of SWHs, with the usage of a high-throughput screening (HTS) process (Zhijian Liu, Li, Liu, Yu, & Cheng, 2017). This is, so far, the first study of HTS on the optimization of energy system.

SWH is one of the most popular techniques where the solar collectors and concentrators are employed to gather, store, and utilize solar radiation to heat air or water (Mekhilef, Saidur, & Safari, 2011). Among several different types of stationary collectors, the evacuated tube solar collectors are featured by the edges of lowering the heat loss coefficient and lower economic cost than other conventional flat plate collectors. In lots of countries, the all-glass evacuated tubular solar water heaters are particularly popular due to the good thermal performance, easy installation, and good transportability. The annual production of evacuated solar tubes kept growing in lots of developing countries (*e.g.*, China) and the market share of them kept increasing at rapid speed during the past decades (Tang, Li, Zhong, & Lan, 2006).

In this Chapter, all the studies will focus on a typical SWH, the water-in-glass evacuated tube solar water heater (WGET-SWH, Figure 2). The reasons why WGET-SWH was selected as a case study are as below: i) WGET-SWH is one of the most common SWHs in developing countries; ii) experimental measurements of WGET-SWH's thermal performances is time-consuming and tedious; and iii) several most significant thermal properties (*e.g.*, heat collection rate and heat loss coefficient) of WGET-SWH have not been well-studied. In the authors' previous studies, the properties of 915 WGET-SWHs were experimentally measured according to a National Standard of China (GB/T 19141)(Zhijian Liu, Li, Zhang, Jin, & Cheng, 2015). The measured properties include heat collection rates (daily heat collection per square meter of a solar water system, MJ/m^2), heat loss coefficients (the average heat loss per unit, $W/(m^3K)$), tube length (mm), number of tubes, tube center distance (mm), tank volume (maximum mass of water in tank, kg), collector area (m^2), tilting angle (°) and final temperature (°C). Except the heat collection rate and heat loss coefficient, all other properties were measured by the "portable test instruments" (Table 1). Descriptive statistics of the measured data (maximum, minimum, range, average and standard deviation) are shown in Table 2.

According to the standard of measurements, measuring the thermal properties of a WGET-SWH requires around 15 days and a series of tedious setups. To provide a quick alternative for the measurement, the authors have developed a machine learning-based method to directly predict the heat collection rate and heat loss coefficient of a WGET-SWH with the inputs which can be measured from the "portable test instruments" (Zhijian Liu, Li, et al., 2015). In other words, once the machine learning model is properly trained with the use of the database, it will be able to predict quickly and precisely the heat collection rate and heat loss coefficient of a WGET-SWH. Such a novel measurement will help effectively reduce the measurement time from weeks to seconds, and thus will dramatically accelerate the measurements of SWHs in both industrial and commercial applications. More details will be introduced in the following Sections.

As a part of the handbook, here, the authors aim to present the predictive power of machine learning in the prediction and optimization of the thermal performances of SWHs, as well as pick the research of WGET-SWH as a case study. The content of this Chapter consists of three parts: i) Comparative studies on varieties of machine learning models (ANNs, SVM and ELM) to predict the thermal performances of SWHs; ii) Development of an ANN-based software to assist the quick and accurate prediction and iii) Introduction of a computational HTS strategy to design and optimize a high-performance SWH system.

Figure 2. Schematic picture of a representative WGET-SWH
Reproduced with permission from Reference (Zhijian Liu, Li, Liu, et al., 2017).

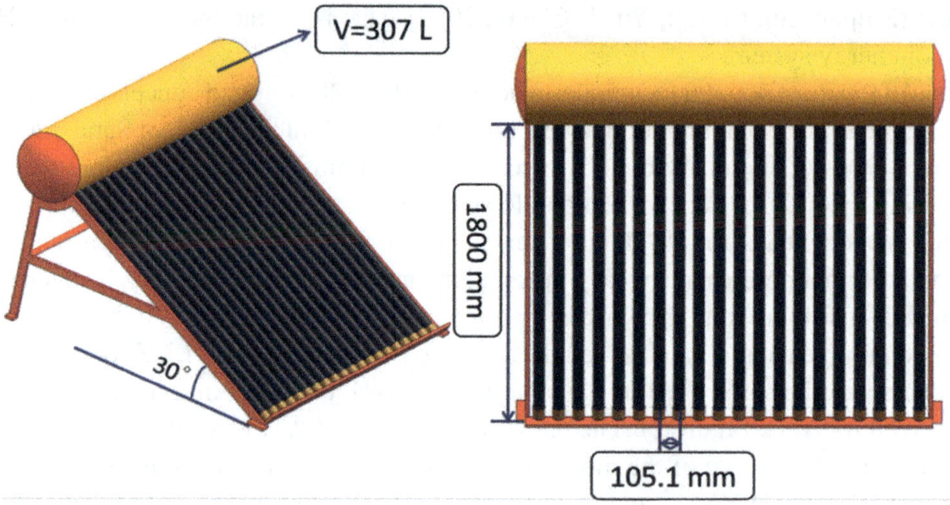

Table 1. "Portable test instruments" for measuring the properties of WGET-SWHs

Parameters	Portable Test Instruments	Accuracy
Final temperate of water	Digital thermoelectric thermometer	±0.5%
Tank Volume	Electric platform scale	±1.0%
Diameter, tube center distance, tube length, collector area	Taper ZSH-3	±0.5%

Reproduced with permission from Reference (Zhijian Liu, Li, et al., 2015).

Table 2. Descriptive statistics of the variables for 915 samples of WGET-SWHs

Items	Tube Length (mm)	Number of Tubes	TCD (mm)	Tank Volume (kg)	Collector Area (m^2)	Angle (°)	Final Temp. (°C)	HCR (MJ/m^2)	HLC (W/(m^3K))
Maximum	2200	64	151	403	8.24	85	62	11.3	13
Minimum	1600	5	60	70	1.27	30	46	6.7	8
Range	600	59	91	333	6.97	55	16	4.6	5
Average	1811	21	76.2	172	2.69	46	53	8.9	10
Standard deviation	87.8	5.8	5.11	47.0	0.73	3.89	2.0	0.48	0.77

Reproduced with permission from Reference (Zhijian Liu, Li, et al., 2015).

Abbreviations: TCD: tube center distance, final temp.: final temperature, HCR: heat collection rate, MJ/m^2, HLC: heat loss coefficient ($W/(m^3K)$). Tank volume was defined as the maximum mass of water in tank (kg).

BACKGROUND

Machine Learning Models

Machine learning is a powerful technique for numerical prediction, classification and pattern recognition, which has been widely used in chemical (Li et al., 2016; Li, Chen, Cheng, Zhao, & Yang, 2015), medical (Wernick, Yang, Brankov, Yourganov, & Strother, 2010), biological (Sommer & Gerlich, 2013), environmental (Zhijian Liu, Li, & Cao, 2017) and energy areas (Jordan & Mitchell, 2015; Sun, He, & Chang, 2015). In the authors' recent studies, various machine learning models were used to predict the heat collection rate and heat loss coefficient of WGET-SWH, such as ANN (Zhijian Liu, Liu, et al., 2015; Zhijian Liu, Li, et al., 2015), SVM (Zhijian Liu, Li, et al., 2015) and ELM (Z. Liu et al., 2016). In terms of ANN, several typical network algorithms were used, including general regression neural network (GRNN) (Specht, 1991) and multilayer feedforward neural network (MLFN) (Hornik, Stinchcombe, & White, 1989). Because this handbook is for readers with broad interests, details of their algorithms are not discussed in this Chapter. General principles of ANN, SVM and ELM can be referred to References (Huang, Zhou, Ding, & Zhang, 2012; Kalogirou, 1999; Suykens & Vandewalle, 1999).

Before developing a predictive machine learning model, the inputs and output(s) of the model should be rationally selected. The inputs of the models should be the independent variables that are related (or partially related) to the dependent variable(s), while the output(s) should be the dependent variable(s) that should be predicted. Since one of the basic missions of machine learning is to help people acquire the knowledge that are hard to be measured or observed, it is strongly recommended that the dependent variable(s) should be the properties that are experimentally hard to be detected (if the expected dependent variable(s) can be easily acquired from experiments or can be precisely predicted by multiple linear regression or any physical models, bring in machine learning again would be a waste of time).

Development of a machine learning model for numerical prediction consists of the training and testing of datasets. The training process is essentially the process of a non-linear fitting (also called a "black-box" fitting). A good training of the dataset means that the training is neither under- nor over-fitting. Usually, if the dataset for training is not sufficiently large, there will be a huge risk of over-fitting. To ensure a good training result, a high percentage of training set is usually necessary. The testing process is a process to validate if the trained model is good, utilizing the dataset that is not previously involved in the training process. By comparing the data in the testing set (also called the "actual values") with the data predicted by the training set (also called the "predicted values"), people can calculate the root mean square errors (RMS errors), prediction accuracies (with a given tolerance) and residual values. The RMS error, prediction accuracy and residual value can be calculated by Equations (1), (2) and (3), respectively:

$$RMS\ error = \sqrt{\frac{\Sigma_{i=1}^{n}\left(Z_i - O_i\right)^2}{n_{tot}}} \quad (1)$$

$$Prediction\ accuracy = \frac{n_{good}}{n_{tot}} \times 100\% \quad (2)$$

$$Residual\ value = Z_i - O_i \tag{3}$$

where Z_i and O_i are the predicted and actual values, respectively; n_{tot} is the number of tested samples; n_{good} is the number of tested samples with good predicted results under a given tolerance. Empirically, the tolerance is usually set as ±30%.

A well-trained model is usually accompanied by the testing result with low RMS error, high prediction accuracy and low absolute values of residual. To ensure a good target model, comparing the performances of models with different training and testing percentages is necessary. On one hand, if the percentage of training set is too high, the results given by the testing set would not be reliable. On the other hand, if the percentage of training set is too low, there is a risk of over-fitting. To further show the availability of the model, a cross-validation and/or sensitivity test should be performed (Browne, 2000). However, cross-validation usually requires extremely high computational cost. If the database is very large, a regular personal computer (PC) will no longer be effective. Fortunately, it is found that if both the training and testing datasets are sufficiently large, with a robust and stable ANN algorithm (*e.g.*, GRNN), a cross-validation process can be rationally skipped after a simplified sensitivity test.

To sum up, Table 3 shows a suggested model development process. Good training and testing processes guarantee that the machine learning model can be used for practical applications. For practical applications, people only need to acquire and input the independent variables into the model, and then the dependent variable(s) will be predicted and outputted automatically.

High-Throughput Screening (HTS)

HTS is generally defined as a method for experimentation previously used in medical and biological sciences (Hertzberg & Pope, 2000). With some state-of-the-art devices, algorithms and/or machines, an HTS process can help us quickly screen thousands or even millions of candidates (*e.g.*, chemical compounds (Hautier, Fischer, Jain, Mueller, & Ceder, 2010), materials (Greeley, Jaramillo, Bonde, Chorkendorff, & Nørskov, 2006), genes (Colbert, 2001) and biological designs (An & Tolliday, 2010; Wahler & Reymond, 2001)) with specific target performances for practical or scientific use. At first, the HTS was just used for drug discovery. But now it has been widely used in various areas, such as computer-assisted design of materials. The rough concept of a computational HTS process is pretty simple: the calculations of

Table 3. A recommended machine learning development process for numerical prediction

	Steps	Notes
Step 1	Preparation of experimental database	The database should be sufficiently large.
Step 2	Selection of independent and dependent variables	Independent variables should be easily-accessible.
Step 3	Model training using the training set	Different percentages of training and testing sets should be tried.
Step 4	Model testing using the testing set	
Step 5	Calculation of RMS error, prediction accuracy and residual values	If the testing results are not acceptable, the training should be performed again with different settings.
Step 6	Cross-validation and/or sensitivity test (if necessary)	NA
Step 7	Practical applications of the well-trained model	NA

all possible candidates in a short timescale (using fast algorithms) and the screening of candidates with acceptable target performances (Li et al., 2017). Previously, Greeley and his colleagues used density functional theory (DFT) calculations to screen and design thousands of bimetallic catalysts for chemical reactions via an HTS process (Greeley et al., 2006). The predicted performances of their target design were in excellent agreement with experimental validations. Ceder and his colleagues combined DFT calculations, machine learning and an HTS method to screen all the possible ternary oxide compounds in the periodic table (Hautier et al., 2010). A large database of the nature's missing ternary oxide compounds was then developed. This study indicated that a machine learning-assisted HTS process can potentially be a good method to discover human's unknown knowledge. However, though these studies successfully used the concept of HTS for new knowledge searching, in the following years, there are very few relevant studies in the engineering area, especially the field of energy system design and optimization.

In the previous Sections of this Chapter, it is shown that machine learning can potentially be a good method for thermal performance optimization of SWH. Meanwhile, it is known that the design and optimization of a high-performance SWH is particularly difficult due to its complicated internal structure. Thus, we can come up with a "bold" idea: can we use a machine learning-based HTS process to predict "infinite" possible designs of SWHs and screen the good candidates for practical use? The answer to this question will be given in the following Sections.

THERMAL PERFORMANCE PREDICTION AND OPTIMIZATION

Prediction of Thermal Performance

To find out a good machine learning model for predicting the thermal performances of WGET-SWHs, the authors have developed a series of models with the algorithms of GRNN, MLFN, SVM and ELM, respectively. The average RMSE errors and average prediction accuracies calculated from the testing results (after multiple training and testing processes) are shown in Table 4. The prediction accuracies were calculated with the tolerance of $\pm 30\%$. It should be noted that all the data used for model training and testing are in the same database, as shown in Table 2. In terms of the MLFN and ELM, only the results of the best network configurations (optimal numbers of hidden layer and hidden neuron) are shown here. Surprisingly, with the tolerance of $\pm 30\%$, all the four models have shown extremely good predictive performances for both heat collection rate and heat loss coefficient, with all the average prediction accuracies reach 100% (Table 4).

To show the training and testing results of the machine learning model, here, the authors pick one of the typical modeling results of an MLFN model, for the prediction of heat collection rate of WGET-SWHs (Figures 3 and 4, respectively). It can be clearly seen that, with a good ANN training process (Figure 3), the model can precisely predict the heat collection rates of the 137 data samples in a testing set (Figure 4), with relatively low absolute residual values. It should be noted that in the testing results, there are always some data points, which, more or less, deviate from the diagonal (Figure 4a). This is certainly acceptable—as gold cannot be pure and men cannot be perfect!

Although the performances of all the four models look quite similar (as shown in Table 4), there are still several criterions that can help us distinguish which model is the most practical one. First is the training time. As we may know, compared to GRNN and SVM, MLFN and ELM have the structures that are closer to a conventional ANN, containing undefined numbers of hidden layer and/or hidden neurons.

Table 4. Predictive performances and modeling information of GRNN, MLFN, SVM and ELM for the predictions of the heat collection rate and heat loss coefficient for WGET-SWHs

Model	Average RMS Errors (for Heat Collection Rate)	Average RMS Errors (for Heat Loss Coefficient)	Percentages of Training and Testing Sets	Average Prediction Accuracy for Heat Collection Rate	Average Prediction Accuracy for Heat Loss Coefficient
GRNN	0.33	0.71	Training set: 85%; Testing set: 15%;	100%	100%
MLFN	0.14	0.73	Training set: 85%; Testing set: 15%;	100%	100%
SVM	0.29	0.73	Training set: 85%; Testing set: 15%;	100%	100%
ELM	0.30	0.67	Training set: 85%; Testing set: 15%;	100%	100%

Data source: References from (Z. Liu et al., 2016; Zhijian Liu, Li, et al., 2015).

Figure 3. Training results of 778 samples using an MLFN model for the prediction of heat collection rate for WGET-SWHs. (a) Predicted values vs actual values; (b) residual values vs actual values; (c) residual values vs predicted values
Reproduced with permission from Reference (Zhijian Liu, Li, et al., 2015).

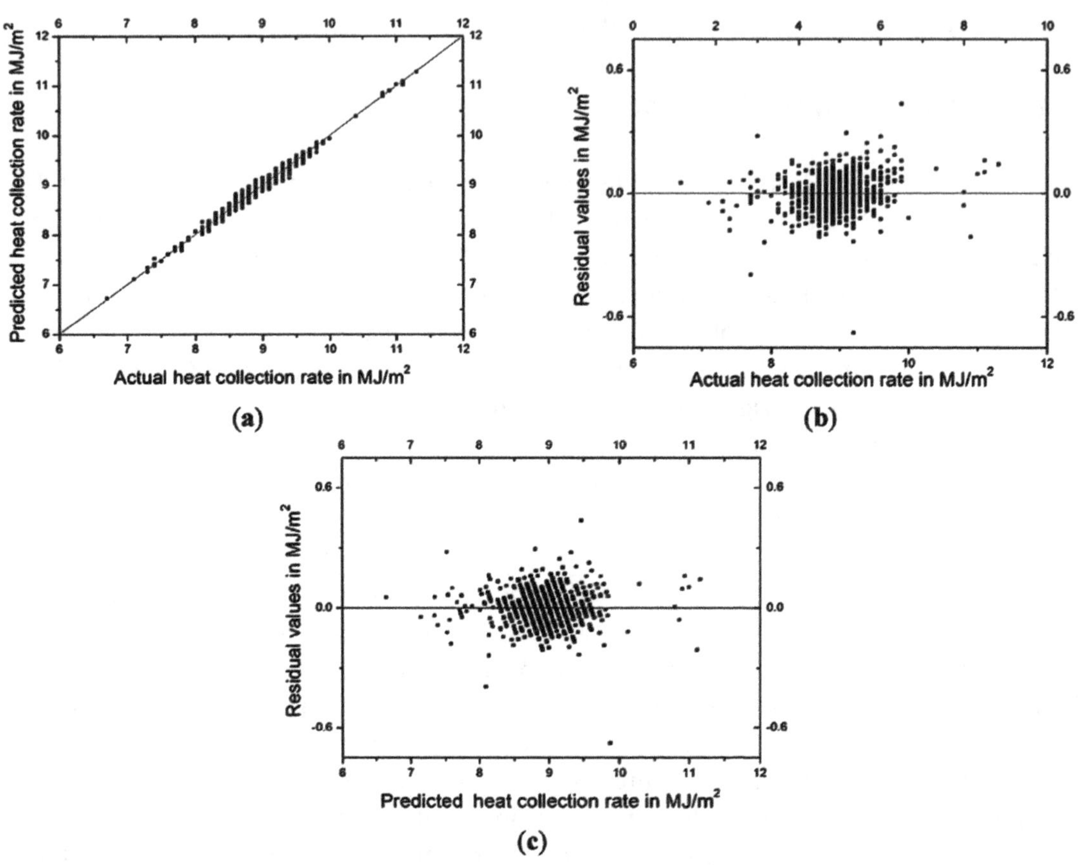

Figure 4. Testing results of 137 samples using an MLFN model for the prediction of heat collection rate for WGET-SWHs. (a) Predicted values vs actual values; (b) residual values vs actual values; (c) residual values vs predicted values
Reproduced with permission from Reference (Zhijian Liu, Li, et al., 2015).

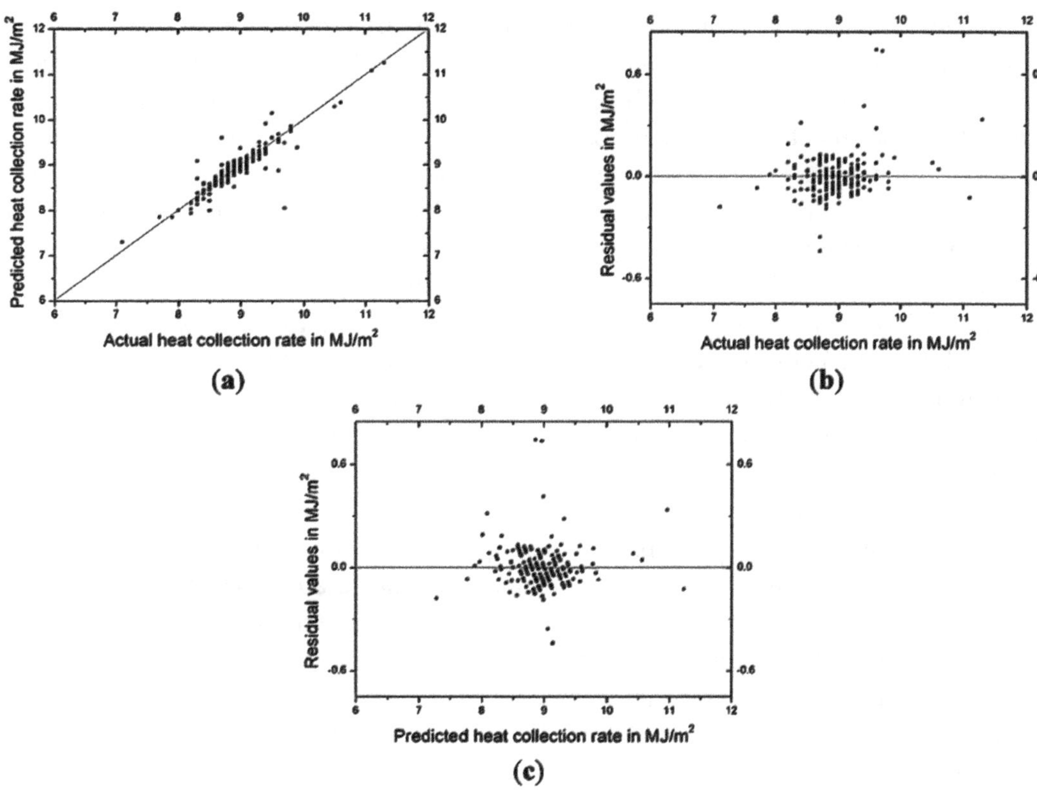

This means that people have to compare different network configurations (*e.g.*, different number of hidden layer and/or hidden neurons) in order to get the best algorithmic structure for prediction. Compared to GRNN and SVM (which only require the training once and for all), MLFN and ELM are clearly not the best options for this case study. In terms of the comparison between GRNN and SVM, both of them have the advantage of quick training and precise prediction. It is relatively hard to decide which one is the best since they have rather similar predictive performance in this study.

So far, the conclusion here is clear: machine learning is an effective method to predict the thermal performances of WGET-SWHs. Though it is very powerful, it is still far away from real applications: officers/workers in a company or industry usually don't want to spend a lot of money to learn and purchase a machine learning or Mathlab software— either a user-friendly machine learning package or Mathlab is rather expensive. So here is a new question: how to really use this method for practical application? In the next Section, a recently-developed ANN-based software will be introduced.

Developing an ANN-Based Software

To provide an effective support for practical applications, the authors developed an user-friendly package, *WaterHeater*, in both PC and Android platforms (Zhijian Liu, Liu, et al., 2015). The primary motivation of this study was to provide a software that could help people quickly acquire the ANN-predicted heat collection rate and heat loss coefficient of WGET-SWHs, with the simple inputs measured by the "portable test instruments" (Table 1). The reason why the authors also developed an Android-based version was that a mobile system is sometimes more user-friendly and applicable for industrial measurement (people do not always have their computers aside). Though in different platforms, there is no difference between the models inside the packages of the PC and the Android versions. With this software, people only need to use a computer or even a mobile phone to perform quick thermal measurements utilizing the "portable test instruments". The inner core of the package is a well-trained MLFN with a back-propagation algorithm and a Sigma function as the activation function. The inputs are the same as the machine learning models described in previous sections. The predictive performance of the packages was validated by the residual values calculated from the testing set (Figure 5). It should be noted that though this model has some slight deviations for predicting the heat loss coefficients when their actual values are very high or very low (Figure 5b), the general predictive performance of the network work is still acceptable for the prediction of heat loss coefficient (because most of the heat loss coefficient in the industry are around 10 W/(m^{3}K), which is neither too high nor too low). In this Section, the application of this software will be introduced. The details of software developments can be found in Reference (Zhijian Liu, Liu, et al., 2015).

Figures 6 and 7 show the main panels of the software in PC and Android platforms, respectively. Both these two platforms consist of three parts: i) input panels, ii) output panels and iii) buttons. As their name imply, the input panel is the place for inputting independent variables, while the output panel is the place to display the dependent variables. The buttons consist of the "Reset Parameters" button and the "Start to Predict" button. By clicking the "Start to Predict" button with the values inputted in the input panel, the predicted heat collection rate and heat loss coefficient will be instantly shown in the output panel. By simply clicking the "Reset Parameters" button, all the input and output data will be erased.

This software can provide a perfect solution for quick thermal performance estimation of WGET-SWHs. Here, a quick measurement flow chart is proposed, as shown in Figure 8. With simple measurements of independent variables using the "portable test instruments" (Table 1), fast prediction of the

Figure 5. Residual values vs. actual values in predicting (a) heat collection rates (in MJ/m^2) and (b) heat loss coefficient (in W/(m^{3}K)). All the actual values came from a testing set.
Reproduced with permission from Reference (Zhijian Liu, Liu, et al., 2015).

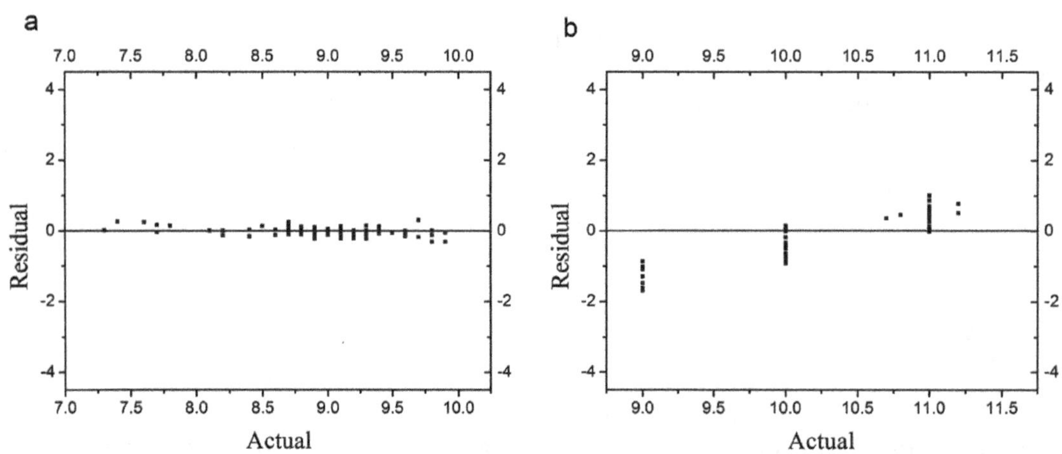

Figure 6. Overview panel of "WaterHeater" in a PC platform
Reproduced with permission from Reference (Zhijian Liu, Liu, et al., 2015).

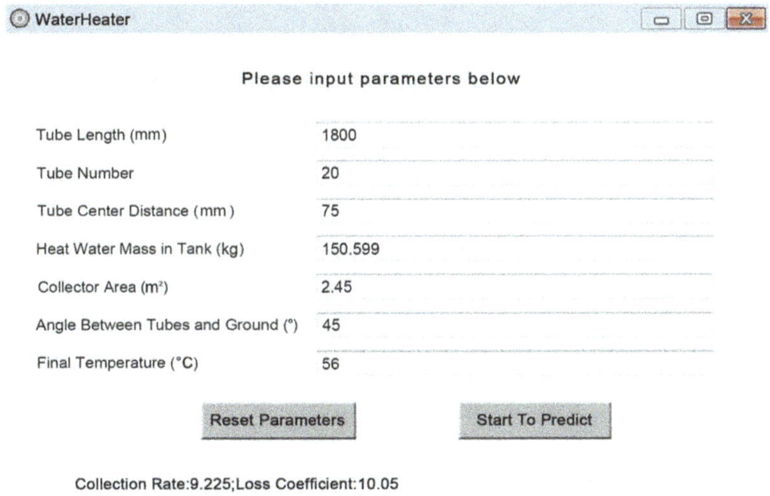

Figure 7. Overview panel of "WaterHeater" in an Android platform
Reproduced with permission from Reference (Zhijian Liu, Liu, et al., 2015).

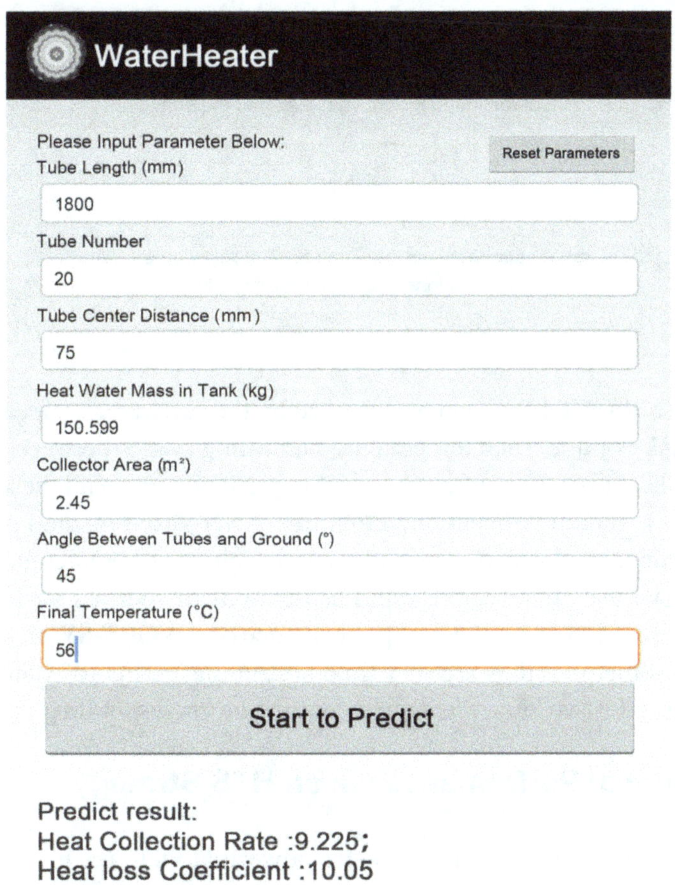

Figure 8. Flow chart of the novel method using the "portable test instruments" combined with the software "WaterHeater" for the predictions of heat collection rate and heat loss coefficient
Reproduced with permission from Reference (Zhijian Liu, Liu, et al., 2015).

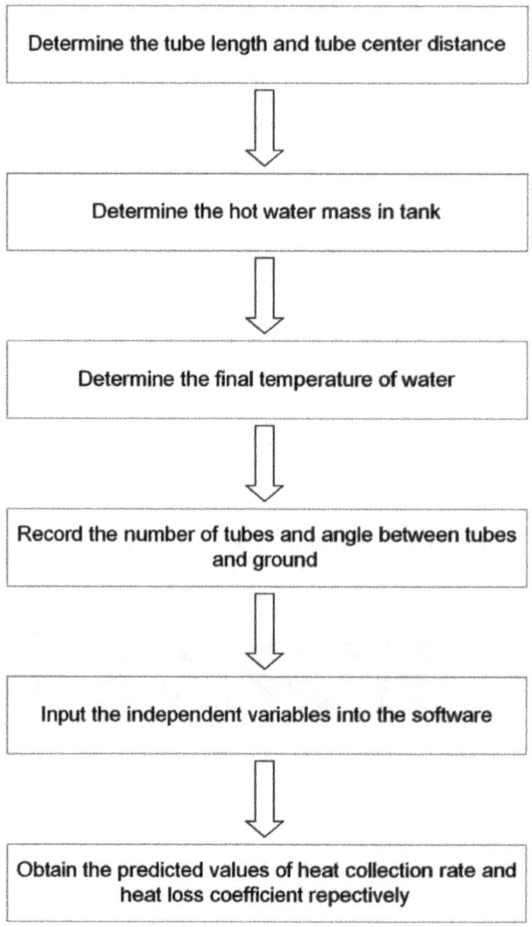

thermal performance of a WGET-SWH can be achieved by the software. By measuring the tube length, tube center distance, tank volume, final temperature and titling angle (between tubes and ground), the heat collection rate and heat loss efficient can be quickly predicted when all these variables are inputted into the software panel. It should be mentioned that this ANN-based method does not aim to replace the conventional measurements. Instead, it provides a quick choice to predict the thermal values for the industrial and commercial uses that require rapid estimations of thermal performances. It should be noted that, to more precisely assess the thermal performance of a WGET-SWH, standard measurements should be used (though sometimes they are very time-consuming). This software is available at http://t.cn/RLPKF08 and for the latest version, reader can contact the corresponding author of this Chapter.

Optimizing the Thermal Performance via an HTS Strategy

It has been shown that machine learning can be such a powerful tool to predict the thermal performance of WGET-SWHs, here comes a new question: can people use this technique to predict the thermal performance of a newly designed WGET-SWH without direct experiments? The answer is yes. And it can be done in an even crazier way: screening thousands or millions of design candidates by using a well-trained machine learning model, and then selecting the candidates with good target performances. This screening strategy, as mentioned in the previous sections, is called the HTS process.

Using the optimization of heat collection rates of WGET-SWHs as a case study, the authors recently found that an HTS process with a proper ANN can be used for this mission. Usually, a high-performance WGET-SWH should have the heat collection rate as high as possible. For the screening process, the first step was to generate a large number of independent variable combinations (around 3.5×10^8 possible design combinations) as the inputs of the previously-trained ANN. Since the final temperature is not a part of the SWH installation, all the integers of final temperature between 52 and 62 °C were selected as the input. With all these independent variables (except the final temperature), people can easily construct a completed WGET-SWH in industry. The heat collection rates of the WGET-SWHs with all these possible combinations were then numerically predicted and outputted. The designed WGET-SWHs with high predicted heat collection rates will be screened and collected. For validation, the authors also experimentally installed two selected candidates and measured their thermal performances. Here, the two selected designs are respectively called "Design A" and "Design B", with the design and predicted details shown in Table 5. The predicted heat collection rates of these two designs are relatively high, no matter with which final temperature between 52 and 62 °C. As a result, both Designs A and B showed high average heat collection rates after standard measurements, as shown in Table 6. It should be noted that the environmental conditions (*e.g.*, solar radiation intensity, ambient temperature, season and location) for measuring these two designs are very similar to all the WGET-SWHs in the authors' database. That is to say, these two designs are comparable with the previous 915 WGET-SWHs. Surprisingly, it was found that the two designs had the average heat collection rate higher than all the 915 WGET-SWHs in the previous database. In the following content, details about this screening process will be introduced.

Table 5. Predicted variables of two designed WGET-SWHs

	Tube Length (mm)	Number of Tubes	TCD (mm)	Tank Volume (kg)	Collector Area (m^2)	Angle (°)	Final Temp. (°C)
Design A	1800	18	105.5	163	1.27	30	52-62
Design B	1800	20	105.5	307	1.27	30	52-62

Reproduced with permission from Reference (Zhijian Liu, Li, Liu, et al., 2017).

Abbreviations: TCD: tube center distance, final temp.: final temperature. Tank volume was defined as the maximum mass of water in tank (kg).

Table 6. Measured heat collection rates (MJ/m^2) of the two new designs

	Day 1	Day 2	Day 3	Day 4	Average	Predicted	Error Rate
Design A	11.38	11.26	11.34	11.29	11.32	11.47	1.35%
Design B	11.47	11.43	11.42	11.45	11.44	11.66	1.90%

Reproduced with permission from Reference (Zhijian Liu, Li, Liu, et al., 2017).

Being similar to a previous computational HTS concept proposed by Aspuru-Guzik and his colleagues (Pyzer-Knapp, Suh, Gómez-Bombarelli, Aguilera-Iparraguirre, & Aspuru-Guzik, 2015), a modified HTS process for the optimization in this case has been proposed, as shown in Figure 9. Details about modeling and experimental contents can be found in Reference (Zhijian Liu, Li, Liu, et al., 2017). It can be clearly seen that though a large number of new designs are generated, they will be screened with the target criterions. Only a relatively small number of designs with the predicted results fulfill the criterions will be recorded in the database. Of course, sometimes it is hard to experimentally validate all these candidates after screening. It is recommended that picking at least two cases for experimental validation can ensure that the screening results are reliable.

A key step of this HTS process is the generation of inputs for the new designs. Without a rational criterion for input generation, there will be infinite possible combinations of inputs, which will lead to infinite computations. For an ANN model, an interesting way is to generate the inputs according to the final weights of the network: the independent variables with larger weight contributions will be assigned more possible input values. The basic assumption here is simple: the independent variables with higher weight will lead to more significant changes to the dependent variables. It should be noted that the numerical weights do not contain any physical meaning, and the weights are usually different under different repeated training of a given ANN structure, since the initial weights of the ANN are usually selected randomly by the weight optimization algorithm. Fortunately, if each repeated ANN is well-trained, the weight values would be relatively stable in multiple trainings. This method provides a quick decision of input generations, which does not require people to know the exact physical meanings of the variables. However, sometimes people have to artificially assign more values to some independent

Figure 9. An HTS process for solar energy system optimization. Each orange circle represents a possible design. The inset shows the schematic inter-connected structure of a general ANN algorithm.
Reproduced with permission from Reference (Li et al., 2017).

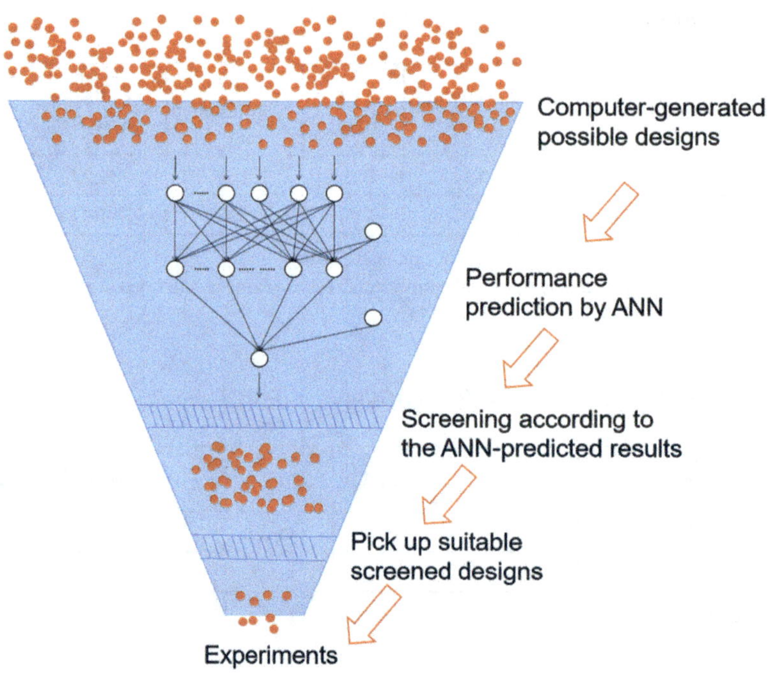

variables. Taking the final temperature in this case as an example, it is not a part of the WGET-SWH installation, but it highly correlates with the heat collection rate. Thus, the temperature effect should not be ignored. More final temperature should be assigned than it is expected from its weight contributions. The number of each selected value for generation is shown in Table 7. All these values randomly combine with each other, constructing around 3.5×10^8 possible design combinations as the inputs of the ANN. Also, if people want to use other machine learning algorithms that do not require weight calculations (*e.g.*, SVM), the variables' physical meanings should be considered: the independent variables with more significant physical influences to the dependent variable should be assigned more selected values.

A General HTS Process for Energy System Optimization

So far, all the essential processes for the design and optimization of a high-performance WGET-SWH have been introduced, which mainly include two parts: i) developing a predictive model and ii) screening possible candidates. Here a general HTS framework (that might be used for other energy systems) will be introduced and discussed.

The proposed HTS framework for the design and optimization of energy system is shown in Figure 10. When all the preconditions of the "cylinders" shown in Figure 10 are fulfilled, a completed machine learning-assisted HTS process can be achieved. Since the final target of this HTS process is to discover new designed candidates with optimized performance, there should be a database that record all the independent and dependent variables of the new candidates for future use. It should be noted that these predicted candidates will have the independent variables different (or partially different) from the experimental database. Also, the results from the validation experiments should be added to the previous experimental database. By combining the validation experiment results with the previous measurement database, people can reconstruct a new experimental database for future applications. People can refer to either the predicted candidate database or the new experimental data for industrial or commercial use. This framework can be achieved by a machine learning code combined with some additional simple coding. Here, the authors expect that this framework not only works for optimizing WGET-SWH, but also works for the optimization cases of other energy systems.

FUTURE RESEARCH DIRECTIONS

Though there is a great success on the prediction and optimization of SWHs using machine learning methods, there still remain some vital questions that should be addressed in the future: how to modify

Table 7. Number of selected values of different independent variables

	Tube Length (mm)	Number of Tubes	TCD (mm)	Tank Volume (kg)	Collector Area (m^2)	Angle (°)	Final Temp. (°C)
Number of Selected Values	5	30	5	111	50	5	17

Reproduced with permission from Reference (Zhijian Liu, Li, Liu, et al., 2017).

Abbreviations: TCD: tube center distance, final temp.: final temperature. Tank volume was defined as the maximum mass of water in tank (kg).

Figure 10. A proposed framework of machine learning-assisted HTS process for target performance optimization. "int" is the independent variable. "dep" is the dependent variable. {A_{in}} is the original experimental database. {B_{in}} is the generated independent variables as the inputs. {B_{in}(new)} is the generated independent variables and their predicted dependent variables. {C_{in}} is the new experimental database combining the original experimental database and the experimental validation results of the screened candidates.

Reproduced with permission from Reference (Li et al., 2017).

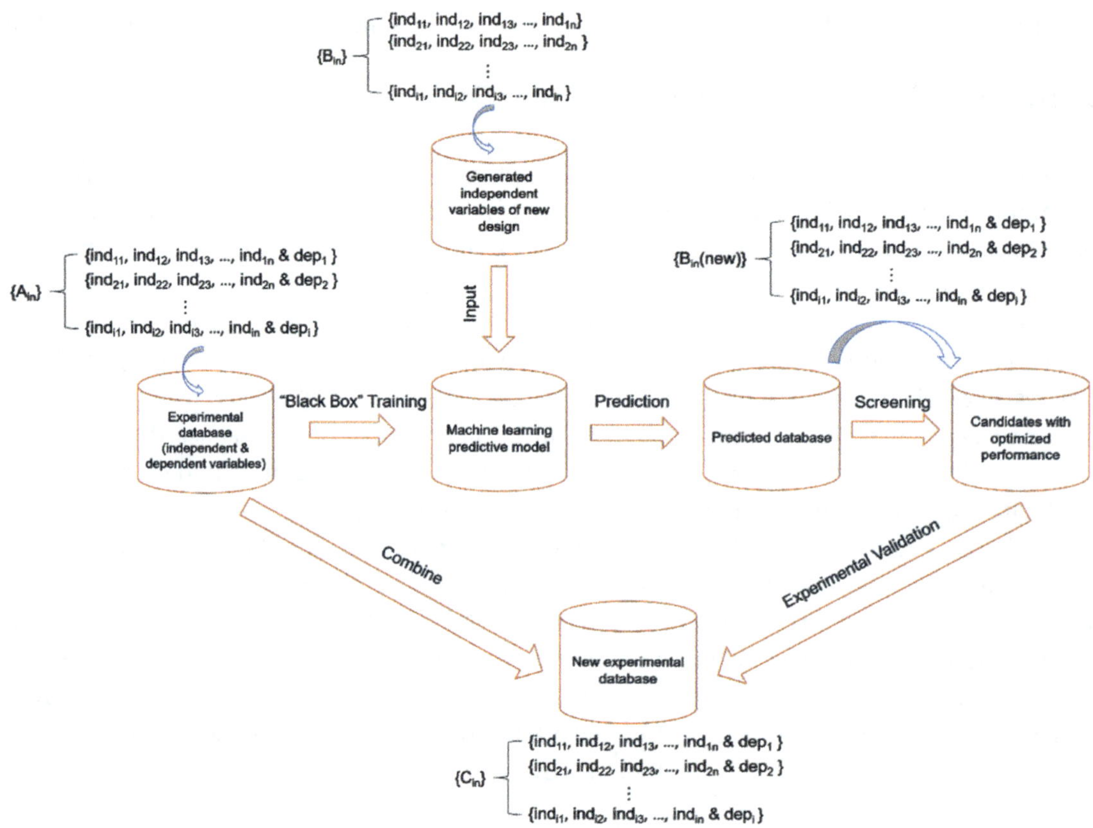

the HTS process for adopting different energy systems? Can people further simplify the HTS process and/or the generation rules of design inputs? Can people develop a user-friendly platform for the database of the designed candidates? Addressing these questions would be of great importance to make the HTS-based energy system optimization more applicable.

It should be noted that to achieve a successful HTS-assisted optimization of energy systems, both predictive model training and HTS process are necessary. Both of them require (more or less) some coding works, which meanwhile require some programming knowledge. If the thermal performance of an energy system can be easily optimized by the empirical knowledge, HTS process will no longer be recommended. Thus, the proposed HTS optimization process is only effective for the design of those energy systems with more complicated internal structures. This is the same as the basic mission of a machine learning model: to deal with those problems that are too complicated to be addressed by conventional methods. If things are simple, machine learning will no longer be cost-effective.

In the near future, it is expected to see that more commercial energy systems are optimized by a machine learning-based HTS process. With higher designed performances, it is expected that higher economic and environmental benefits can be achieved.

CONCLUSION

In this Chapter, the authors have shown that machine learning techniques are powerful tools for predicting and optimizing the thermal performance of SWHs. Picking WGET-SWH as a case study, the authors have developed a knowledge-based measurement of thermal performances using the simple inputs measured by the "portable test instruments". Various machine learning models (ANN, SVM and ELM) were subsequently compared for the prediction of the heat collection rate and heat loss coefficient of WGET-SWH. To provide a more user-friendly measurement for practical use, an ANN-based software was developed in both PC and Android platforms.

To optimize the heat collection rate of a WGET-SWH, the authors have developed an ANN-based HTS process to screen around 3.5×10^8 possible design combinations. Candidates with high predicted heat collection rates were screened and recorded into a database for future use. Validation experiments on two selected cases in the candidate database showed surprisingly high thermal performances. All these results show that machine learning not only provides a strong predictive power for thermal performance prediction, but also provide a brand new insight for the performance optimization of an energy system. The authors also expect that this new HTS-based optimization strategy can be more widely used in the near future in the area of energy engineering.

ACKNOWLEDGMENT

We are grateful to all the colleagues who involved in the data collection, programming and research discussions. This work was supported by the Major Basic Research Development and Transformation Program of Qinghai province (no. 2016-NN-141) and Natural Science Foundation of Hebei (no. E2017502051).

REFERENCES

An, W. F., & Tolliday, N. (2010). Cell-based assays for high-throughput screening. *Molecular Biotechnology*, *45*(2), 180–186. doi:10.100712033-010-9251-z PMID:20151227

Browne, M. W. (2000). Cross-Validation Methods. *Journal of Mathematical Psychology*, *44*(1), 108–132. doi:10.1006/jmps.1999.1279 PMID:10733860

Colbert, T. (2001). High-Throughput Screening for Induced Point Mutations. *Plant Physiology*, *126*(2), 480–484. doi:10.1104/pp.126.2.480 PMID:11402178

Greeley, J., Jaramillo, T. F., Bonde, J., Chorkendorff, I. B., & Nørskov, J. K. (2006). Computational high-throughput screening of electrocatalytic materials for hydrogen evolution. *Nature Materials*, *5*(11), 909–913. doi:10.1038/nmat1752 PMID:17041585

Hautier, G., Fischer, C. C., Jain, A., Mueller, T., & Ceder, G. (2010). Finding natures missing ternary oxide compounds using machine learning and density functional theory. *Chemistry of Materials*, *22*(12), 3762–3767. doi:10.1021/cm100795d

Hertzberg, R. P., & Pope, A. J. (2000). High-throughput screening: New technology for the 21st century. *Current Opinion in Chemical Biology*, *4*(4), 445–451. doi:10.1016/S1367-5931(00)00110-1 PMID:10959774

Hornik, K., Stinchcombe, M., & White, H. (1989). Multilayer feedforward networks are universal approximators. *Neural Networks*, *2*(5), 359–366. doi:10.1016/0893-6080(89)90020-8

Huang, G.-B., Zhou, H., Ding, X., & Zhang, R. (2012). Extreme learning machine for regression and multiclass classification. *IEEE Transactions on Systems, Man, and Cybernetics. Part B, Cybernetics*, *42*(2), 513–529. doi:10.1109/TSMCB.2011.2168604 PMID:21984515

Huang, G.-B., Zhu, Q., & Siew, C. (2006). Extreme learning machine: Theory and applications. *Neurocomputing*, *70*(1-3), 489–501. doi:10.1016/j.neucom.2005.12.126

Jordan, M. I., & Mitchell, T. M. (2015). Machine learning: Trends, perspectives, and prospects. *Science*, *349*(6245), 255–260. doi:10.1126cience.aaa8415 PMID:26185243

Kalogirou, S. (1999). Applications of artificial neural networks in energy systems. *Energy Conversion and Management*, *40*(10), 1073–1087. doi:10.1016/S0196-8904(99)00012-6

Li, H., Chen, F., Cheng, K., Zhao, Z., & Yang, D. (2015). Prediction of Zeta Potential of Decomposed Peat via Machine Learning. *Comparative Study of Support Vector Machine and Artificial Neural Networks*, *10*, 6044–6056.

Li, H., Liu, Z., Liu, K., & Zhang, Z. (2017). Predictive Power of Machine Learning for Optimizing Solar Water Heater Performance: The Potential Application of High-Throughput Screening. *International Journal of Photoenergy*, *2017*, 1–10. doi:10.1155/2017/4194251

Li, H., Tang, X., Wang, R., Lin, F., Liu, Z., & Cheng, K. (2016). Comparative Study on Theoretical and Machine Learning Methods for Acquiring Compressed Liquid Densities of 1,1,1,2,3,3,3-Heptafluoropropane (R227ea) via Song and Mason Equation, Support Vector Machine, and Artificial Neural Networks. *Applied Sciences*, *6*(1), 25. doi:10.3390/app6010025

Liu, Z., Li, H., & Cao, G. (2017). Quick Estimation Model for the Concentration of Indoor Airborne Culturable Bacteria: An Application of Machine Learning. *International Journal of Environmental Research and Public Health*, *14*(8), 857. doi:10.3390/ijerph14080857 PMID:28758941

Liu, Z., Li, H., Liu, K., Yu, H., & Cheng, K. (2017). Design of high-performance water-in-glass evacuated tube solar water heaters by a high-throughput screening based on machine learning: A combined modeling and experimental study. *Solar Energy*, *142*, 61–67. doi:10.1016/j.solener.2016.12.015

Liu, Z., Li, H., Tang, X., Zhang, X., Lin, F., & Cheng, K. (2016). Extreme learning machine: A new alternative for measuring heat collection rate and heat loss coefficient of water-in-glass evacuated tube solar water heaters. *SpringerPlus*. doi:10.118640064-016-2242-1 PMID:27330892

Liu, Z., Li, H., Zhang, X., Jin, G., & Cheng, K. (2015). Novel method for measuring the heat collection rate and heat loss coefficient of water-in-glass evacuated tube solar water heaters based on artificial neural networks and support vector machine. *Energies*, *8*(8), 8814–8834. doi:10.3390/en8088814

Liu, Z., Liu, K., Li, H., Zhang, X., Jin, G., & Cheng, K. (2015). Artificial Neural Networks-Based Software for Measuring Heat Collection Rate and Heat Loss Coefficient of Water-in-Glass Evacuated Tube Solar Water Heaters. *PLoS One*. doi:10.1371/journal.pone.0143624 PMID:26624613

Mekhilef, S., Saidur, R., & Safari, A. (2011). A review on solar energy use in industries. *Renewable & Sustainable Energy Reviews*, *15*(4), 1777–1790. doi:10.1016/j.rser.2010.12.018

Peng, H., & Ling, X. (2008). Optimal design approach for the plate-fin heat exchangers using neural networks cooperated with genetic algorithms. *Applied Thermal Engineering*, *28*(5-6), 642–650. doi:10.1016/j.applthermaleng.2007.03.032

Pyzer-Knapp, E. O., Suh, C., Gómez-Bombarelli, R., Aguilera-Iparraguirre, J., & Aspuru-Guzik, A. (2015). What Is High-Throughput Virtual Screening? A Perspective from Organic Materials Discovery. *Annual Review of Materials Research*, *45*(1), 195–216. doi:10.1146/annurev-matsci-070214-020823

Sommer, C., & Gerlich, D. W. (2013). Machine learning in cell biology – teaching computers to recognize phenotypes. *Journal of Cell Science*, *126*(24), 5529–5539. doi:10.1242/jcs.123604 PMID:24259662

Specht, D. F. (1991). A general regression neural network. *Neural Networks. IEEE Transactions on*, *2*, 568–576. PMID:18282872

Sun, W., He, Y., & Chang, H. (2015). Forecasting fossil fuel energy consumption for power generation using QHSA-based LSSVM model. *Energies*, *8*(2), 939–959. doi:10.3390/en8020939

Suykens, J. A., & Vandewalle, J. (1999). Least Squares Support Vector Machine Classifiers. *Neural Processing Letters*, *9*(3), 293–300. doi:10.1023/A:1018628609742

Tang, R., Li, Z., Zhong, H., & Lan, Q. (2006). Assessment of uncertainty in mean heat loss coefficient of all glass evacuated solar collector tube testing. *Energy Conversion and Management*, *47*(1), 60–67. doi:10.1016/j.enconman.2005.03.013

Wahler, D., & Reymond, J. L. (2001). High-throughput screening for biocatalysts. *Current Opinion in Biotechnology*, *12*(6), 535–544. doi:10.1016/S0958-1669(01)00260-9 PMID:11849935

Wernick, M., Yang, Y., Brankov, J., Yourganov, G., & Strother, S. (2010). Machine learning in medical imaging. *IEEE Signal Processing Magazine*, *27*(4), 25–38. doi:10.1109/MSP.2010.936730 PMID:25382956

KEY TERMS AND DEFINITIONS

Artificial Neural Network (ANN): A machine learning algorithm that performs "black box" non-linear fitting, similarly to the inter-connected neurons in human brain.

Heat Collection Rate (HCR): Daily heat collection per square meter of a solar water system, MJ/m^2.

Heat Loss Coefficient (HLC): The average heat loss per unit, $W/(m^3K)$.

High-Throughput Screening (HTS): The screening of thousands or millions of target candidates using state-of-the-art experimental or computational techniques.

Solar Water Heater (SWH): The heater that transfers solar energy (in the form of heat) to water.

Chapter 3
Robust Optimization Algorithms for Solving Automatic Generation Control of Multi–Constrained Power System:
Robustness Study of AGC Problem in Power System

Dipayan Guha
Dr. B. C. Roy Engineering College, India

Provas Kumar Roy
Kalyani Government Engineering College, India

Subrata Banerjee
National Institute of Technology Durgapur, India

ABSTRACT

This chapter presents four effective evolutionary methods, namely grey wolf optimization (GWO), symbiotic organism search (SOS), JAYA, and teaching-learning-based optimization (TLBO), for solving automatic generation control (AGC) problem in power system. To show the effectiveness, two widely used interconnected power plants are examined. To extract maximum possible generation, distinct PID-controllers are designed employing ITAE-based fitness function. Further, to enhance the dynamic stability of concerned power systems, 2DOF-PID controllers are proposed in LFC area and optimally designed using aforesaid algorithms. To demonstrate the supremacy, obtained results are compared with some existing control algorithms. Moreover, robustness of the designed controller is believed under the action of random load perturbation (RLP). Finally, sensitivity analysis is carried out to show the stability of the designed system under loading and parametric disturbance conditions.

DOI: 10.4018/978-1-5225-3935-3.ch003

1. INTRODUCTION

Balance between total power generation and load consumption is always an intricate task particularly at peak load. Hence this is assigned as serious issue to ensure stable and reliable operation of power system at dynamic period. Since the term frequency is directly correlated with system loading, thus any abrupt change in loading cause serious deviation of system frequency from its tolerance value. Owing to load variations, the oscillation in frequency and interchange tie-line power persist for long time. Primary controller served by speed governor would be unable to nullify these fluctuations after load perturbation. An ample variation of frequency may severely affects the equilibrium states of the system and causes high flow of magnetizing current in transformers, induction motors etc (Arya, 2017). Hence, continuous monitoring of power generation and loading demand is needed. This is accomplished by automatic generation controller (AGC). AGC continuously monitor the generator output to meet load demand so as to regulate the frequency error down to zero value. The AGC scheme is not only maintaining the frequency level, it offers steady flow of power through tie-line between the nearby control areas.

Sudden load fluctuation due to variation of load is the key source of disturbance in power system. In this connection, AGC play a significant role to ensure system stability under the action of load perturbation (Naidu et al., 2017). In the literature an extensive works have been undertaken for continuous improvement of AGC performance. The implementation and impact of AGC in power system has been documented by Chon (Chon, 1956). However, the success of optimal control theory in LFC has been shown by Elgerd and Fosha (Fosha & Elgerd, 1970). An extensive review of literature of LFC for conventional and deregulated power systems has been available in (Kumar and Kothari, 2005). Gradual expansion in dimension, change in structure and increasing load demand in power system necessitated establishment of new control strategies. The traditional control strategies may be inept to takeover such abrupt changes in AGC. Some of the recently addressed control strategies in the literature are dual mode linguistic hedge fuzzy logic (Ansari & Velusami, 2010), dual mode fuzzy logic (Kumar et al., 2015), feedback controller (Ahmadi & Aldeen, 2017), fractional order fuzzy proportional-integral-derivative (PID) controller (Arya, 2017), 3 degree-of-freedom integral-derivative (3DOF-ID) controller (Rahman et al., 2015). Load frequency control in deregulated environment via active disturbance rejection scheme is illustrated in (Tan et al., 2015). The infliction of PID controller or its variants in AGC analysis has been appeared in the literature because of its numerous practical amenities (Padhan et al., 2014, Shankar & Mukherjee, 2016, Dhillon et al., 2016, Guha et al., 2016a). Recently in (Guha et al., 2017) an optimized PID-controller is designed to resolve the load frequency control (LFC) issue of an interconnected power system.

The type and design of controller is an important aspect that influences the behavior and stability of system significantly. Classical methods such as Zeigler-Nichols (ZN), trial and error, steepest decent etc. used for design of controller shows sluggish responses and laborious. To revoke this issue, researchers are trying to implement different optimization techniques that amalgamate knowledge, advance theories from different sources to take care AGC problem efficiently (Arya, 2017). Some of the recently applied optimization techniques in AGC are backtracking search algorithm (BSA) (Guha et al., 2016b), ant-lion optimization (ALO) (Raju et al., 2016), quasi-oppositional harmony search algorithm (QOHSA) (Shiva and Mukherjee, 2015), cuckoo search algorithm (CSA) (Abdelaziz & Ali, 2015), improved particle swarm optimization (IPSO) (Zare et al., 2015), bat algorithm (Abd-Elazim & Ali, 2016) etc. The aforesaid optimization techniques show better results leaving behind few shortcomings. In the line of 'no-free-lunch (NFL)' theory, no optimization technique is well suited for all optimization problems.

Thus the introduction of novel control algorithm to solve AGC problem is always imperative as a good research work.

It is noted from existing literature that the use of FACTS controller is one of the best choice to increase the damping of power system oscillations. However less attention has been paid to the study of LFC problem in the presence of system nonlinearities. The utility of battery energy storage (BES) in LFC area is discussed in (Aditya & Das, 2000). Tripathy *et al.* have shown (Tripathy et al., 1992) the acceptability of superconducting magnetic energy storage (SMES) to improve the dynamic performance of power system. The use of redox flow batteries (RFB) in AGC of multi-area restructured power system was illustrated in (Arya & Kumar, 2016). The performance of the designed controller was compared with genetic algorithm based integral controller for the similar power system. The advantage of interline power flow controller (IPFC) and RFB in the performance of LFC was studied in (Gorripotu et al., 2015). The study was performed in the presence of time delay and GRC nonlinearities.

In view of foregoing discussion, following are the main objectives of the present chapter:

- To minimize the frequency fluctuations and supply better quality of power, a robust control algorithm has been briefly discussed in this chapter.
- To extract maximum benefit, distinct secondary controllers are designed employing grey wolf optimization, Jaya, symbiotic organism search and teaching learning based optimization.
- To revise the consequences of different inherent constraints of power system like generation rate constraint, governor dead band, boiler dynamics and time delay on the system dynamics.
- To affirm robustness of the designed controller in the presence of system nonlinearities.
- To improve transient responses, 2 degree-of-freedom proportional-integral-derivative controller is designed and incorporated in the system.
- To demonstrate the supremacy and efficacy of the proposed control algorithm, an extensive comparative study has been carried out.
- To enhance dynamic stability of power system, an optimized unified power flow controller has been designed.

2. MATHEMATICAL PROBLEM FORMULATION

2.1. System Under Study

The system under study comprises of three-area unequal reheat thermal power plant with capacity of area-1: 2000MW, area-2: 5000MW and area-3: 8000MW. Since the AGC analysis is associated with small signal modeling, hence it is relevant to consider linear approximated model of actual system for present analysis. The power system network is, thus, modeled in transfer function form for present analysis. The block diagram of the concerned power system is shown in Figure 1. The nominal values of system parameters are taken from (Raju et al., 2016) and displayed in Table 1. To add degree of nonlinearity, generation rate constraint (GRC) of 3%/min in each area is taken into consideration. GRC is mainly caused by the thermodynamics and mechanical constraints of the steam turbine. It is relevant to consider GRC in AGC because generator output is always changed at a maximum specified rate. If this constraint is not measured, power system may experiences an undesirable large momentary disturbances resulting wear and tear of controller. A step load increase of 1% ($0.01 p.u.$) in area-1 is applied to appraise the dynamic

Table 1. Nominal values of test system parameters

Parameter	Value	Parameter	Value	Parameter	Value	Parameter	Value
f	60 Hz	T_{sg}	0.08 sec	T_t	0.3 sec	T_r	10 sec
K_{ps}	120 Hz/pu MW	T_{ps}	20 sec	T_{12}	0.086 pu MW/rad	K_r	0.5
$B_1 = B_2 = B_3$	0.425 pu MW/Hz	$R_1 = R_2 = R_3$	2.4 pu Hz/MW	Loading	50%		

Figure 1. Transfer function model of three-area reheat thermal power plant

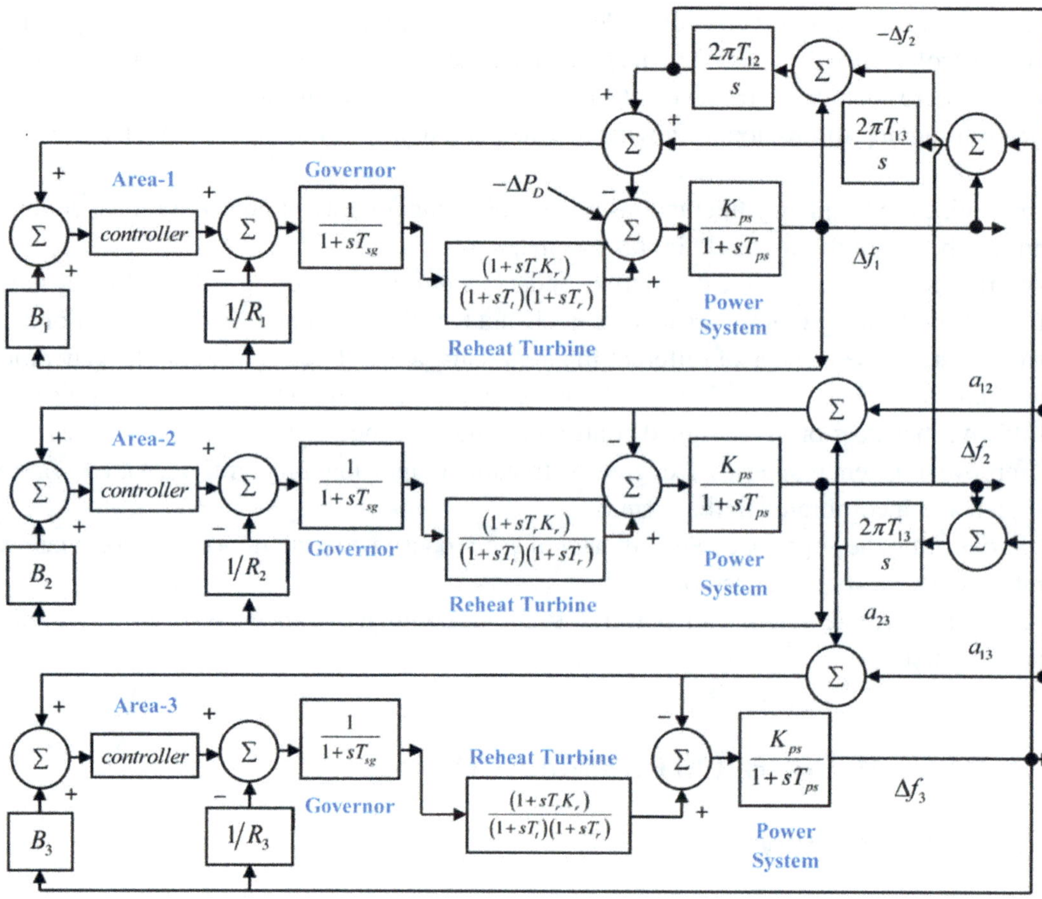

behavior of concerned test system. In Figure 1, R_i represents speed regulation parameter of governor, B_i is the frequency bias constant, T_{sg}, T_t, and T_r are the time constants of speed governor, turbine and reheater respectively, T_{ij} is the synchronizing time constant between i^{th} and j^{th} control areas, ΔP_D is the change of load in pu, Δf and ΔP_{tie} are the changes of frequency and tie-line power respectively, K_{ps} and T_{ps} are the gain and time constant of power system itself respectively.

2.2. Physical Constraints

Power system network is highly nonlinear and time-varying interconnected system. The inherent constraints of power system considered for present analysis are discussed in the following subheadings.

2.2.1. Governor Dead Band (GDB)

GDB is defined as a sustained speed change within which there is no resulting change in the valve position. Overlapping the values of hydraulic relays, effects of backlash and coulomb friction causes dead band. Typical backlash nonlinearity as shown in Figure 2 is used to demonstrate the GDB.

2.2.1.1. Mathematical Modeling of GDB

The linear approximated model of backlash is derived in this section employing describing function method. Describing function is an approximation technique for finding out the periodic orbits, amplitude and frequency of sustained oscillation. This gives the model of nonlinear system based on first harmonic approximation.

Let $y(t)$ be the non-sinusoidal periodic output of the nonlinear element subjected to sinusoidal input, $x(t) = X \sin(\omega t)$. Consequently, $y(t)$ can be expressed by Fourier series as

$$y(t) = y_0 + \sum_{n=1}^{\infty} A_n \cos(n\omega t) + B_n \sin(n\omega t) \tag{1}$$

where y_0 is the dc component of $y(t)$, A_n and B_n are the Fourier coefficients, ω is natural frequency of oscillation in *rad/sec*. Since backlash nonlinearity is 'odd-symmetric', hence $y_0 = 0$. Again, in the line of definition, higher order harmonic terms of Equation (1) can be discarded. Thus Equation (1) is simplified as in Equation (2).

$$y = A_1 \cos(\omega t) + B_1 \sin(\omega t) \tag{2}$$

The terms A_1 and B_1 in Equation (2) are calculated by using Equation (3).

$$\begin{cases} A_1 = \dfrac{1}{s\pi} \displaystyle\int_0^{2\pi} y(t) \sin(\omega t) d(\omega t) \\ B_1 = \dfrac{1}{s\pi} \displaystyle\int_0^{2\pi} y(t) \cos(\omega t) d(\omega t) \end{cases} \tag{3}$$

The describing function model of backlash nonlinearity is calculated as

Figure 2. Representation of GDB nonlinearity

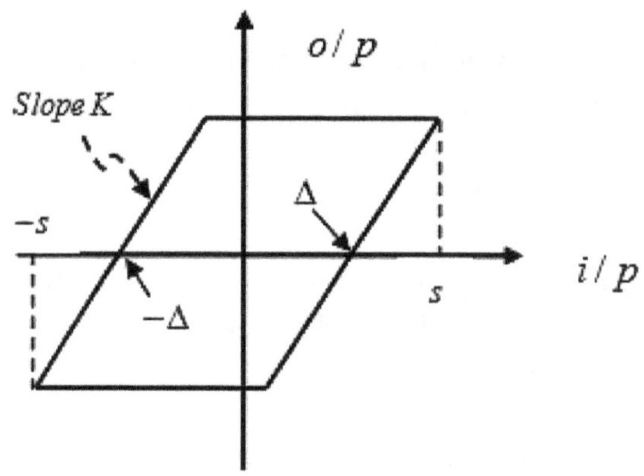

$$\begin{cases} A_1 = \frac{K}{\pi}\left\{\frac{\pi}{2} + \sin^{-2}\left(1 - \frac{2\Delta}{s}\right) + 2\left(1 - \frac{2\Delta}{s}\right)\left[\frac{\Delta}{s}K\left(1 - \frac{\Delta}{s}\right)\right]^{1/2}\right\}_{s \geq \Delta} \\ B_1 = -\frac{4K}{\pi}\frac{\Delta}{s}\left(1 - \frac{\Delta}{s}\right) \end{cases} \tag{4}$$

From Equation (2), we have

$$\begin{aligned} & y = A_1 \cos(\omega t) + B_1 \sin(\omega t) \\ & \Rightarrow y = N_1 x + \frac{N_2}{\omega}\frac{dx}{dt} \qquad \text{where } N_1 = \frac{B_1}{X} \text{ and } N_2 = \frac{A_1}{X} \end{aligned} \tag{5}$$

Laplace transform of Equation (5) yields

$$\begin{aligned} & Y(s) = \left(N_1 + \frac{N_2}{\omega}s\right)X(s) \\ & \Rightarrow \frac{Y(s)}{X(s)} = N_1 + \frac{N_2}{\omega}s \end{aligned} \tag{6}$$

The values of N_1 and N_2 as specified in (Tripathy et al., 1992) are selected as 0.8 and (– 0.2) respectively. Thus, Equation (6) is modified as

$$\frac{Y(s)}{X(s)} = \left(-\frac{0.2}{\pi}s + 0.8\right) \tag{7}$$

2.2.2. Transportation Lag or Time Delay

Transportation lag or time delay is becoming an important constraint in physical system because of restructuring, expanding of physical setups, and functionalities of different devices (Bevrani, 2014). In AGC, faster and time-varying components of frequency are almost unobservable due to the presence of various filters and delays involved in the system. The signal processing and data exchanging operation produces a considerable delay in AGC (Bevrani, 2014). Thus for getting better insight of dynamic responses of power system, this constraint must be included in the study. The Laplace transform of time delay of delay T sec is expressed by the following transcendental equation

$$G_{TD}(s) = e^{-sT} = 1 - sT + \frac{s^2T^2}{2} - \frac{s^3T^3}{6} + \ldots \tag{8}$$

2.2.2.1. Mathematical Modeling of Time Delay

In this section, Pade approximation is adopted to derive the rational model of time delay. The basic idea of Pade approximation is to match the initial few time moments of the actual transfer function to that of an approximated rational transfer function. Let, the actual transfer function of time delay is

$$G_{actual}(s) = \frac{a_{21} + a_{22}s + a_{23}s^2 + \ldots + a_{2,m+1}s^m}{1 + a_{12}s + a_{13}s^2 + \ldots + a_{1,n-1}s^n}, m \le n \tag{9}$$

where m and n are the number of zeros and poles available in the system, respectively. Second-order Pade approximation is used here to derive the model of time delay. The 2nd order approximated model of time delay is governed by Equation. (10).

$$G_{approx}(s) = \frac{a_{21} + a_{22}s + a_{23}s^2}{1 + a_{12}s + a_{13}s^2} \tag{10}$$

The initial time moments of the actual transfer function corresponds to Equation (8) are defined as

$$C_0 = 1, C_1 = -1, C_2 = \tfrac{1}{2}, C_3 = \tfrac{1}{6}, C_4 = \tfrac{1}{24}, \ldots \tag{11}$$

The Routh array with respect to Equation (10) is formed below.

$$\begin{vmatrix} 1 & a_{12} & a_{13} \\ a_{21} & a_{22} & a_{23} \\ a_{31} & a_{32} & \\ a_{41} & a_{42} & \\ a_{51} & a_{52} & \\ a_{61} & a_{62} & \\ \vdots & \vdots & \\ \vdots & \vdots & \end{vmatrix} \quad where \quad \begin{matrix} a_{31} = a_{21} & a_{12} = a_{22} \\ a_{32} = a_{21} & a_{13} = a_{23} \\ a_{41} = a_{31} & a_{12} = a_{32} \\ a_{42} = a_{31} & a_{13} \\ a_{51} = a_{41} & a_{12} = a_{42} \\ a_{52} = a_{41} & a_{13} \\ a_{61} = a_{51} & a_{12} = a_{52} \end{matrix} \tag{12}$$

The elements of the first column of Routh array give initial time moments with alternative signs. The initial time moments of Equation (11) is compared with the first column elements of Routh array to form the rational model of time delay. After simplification, transfer function model of time delay is defined as in Equation (13).

$$G_{TD}(s)\,|_{approximated} = \frac{1 - \frac{sT}{2} + \frac{s^2T^2}{12}}{1 + \frac{sT}{2} + \frac{s^2T^2}{12}} \tag{13}$$

In the present study a time delay of 50 ms (Sahu et al., 2015) is considered.

2.2.3. Boiler Dynamics (BD)

A boiler is a device meant for producing steam under pressure. It mainly has large time delay in the fuel system. Coal-fired well-tuned boiler dynamics as shown in Figure 3 is modeled in this section. The linearized model of BD is included in LFC model (Figure 1) for better study of system stability.

2.2.3.1. Mathematical Model of Boiler Dynamics

The Pade approximation and signal flow graph (SFG) methods help to derive the linear model of BD. The signal flow graph of coal-fired well-tuned boiler is shown in Figure 4. In Figure 4, G_{fs} and G_{pcu} are the transfer functions of fuel system and pressure control unit respectively. The transfer function form of G_{fs} and G_{pcu} are defined in Equations (14)-(15).

Figure 3. Block diagram of drum-type coal-fired boiler

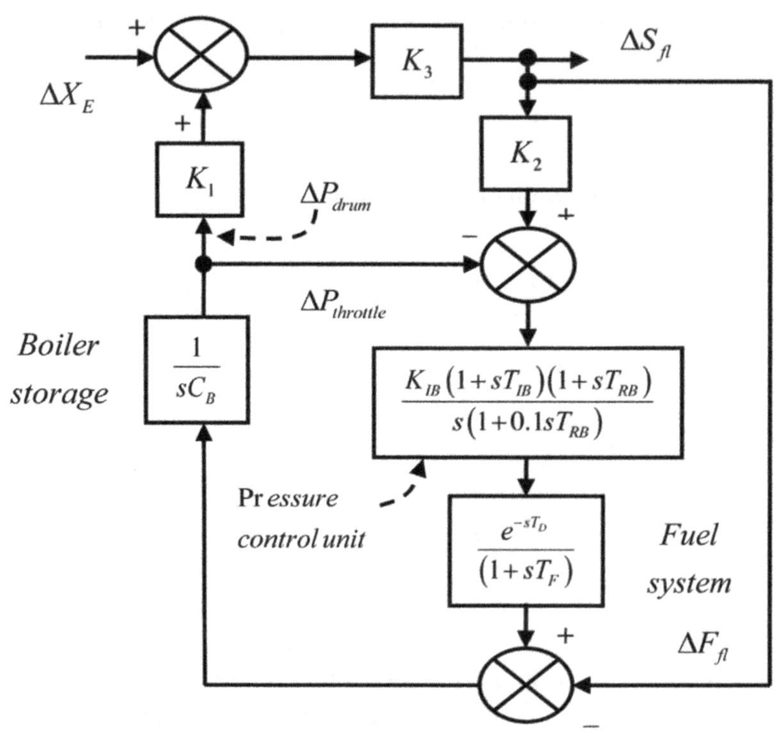

Figure 4. SFG of boiler dynamics

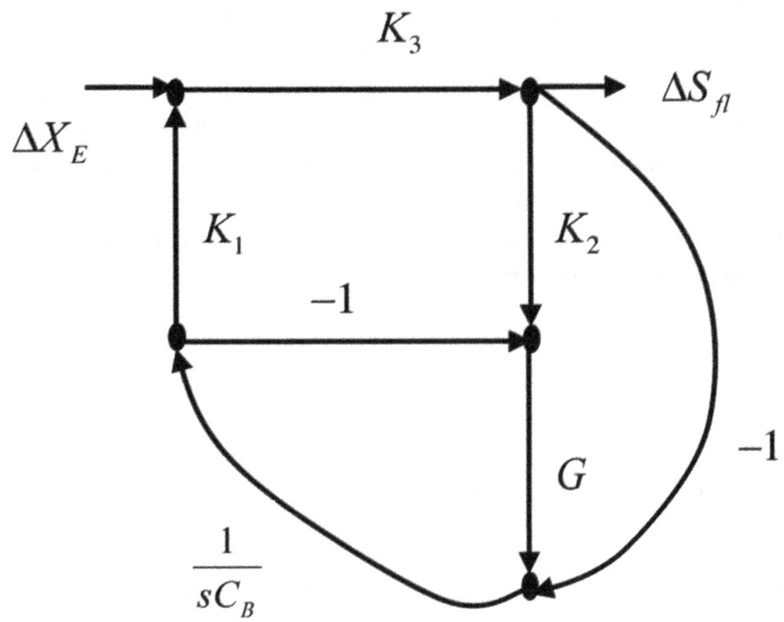

$$G_{fs} = \frac{1 - s\frac{T_d}{2} + s^2\frac{T_d^2}{12}}{1 + s\left(\frac{T_d}{2} + T_F\right) + s^2\left(\frac{T_d^2}{12} + \frac{T_d T_F}{2}\right) + s^3\frac{T_d^2 T_F}{12}} \quad (14)$$

$$G_{pcu} = \frac{K_{IB}\left(1 + s\left(T_{IB} + T_{RB}\right) + s^2 T_{RB} T_{IB}\right)}{s\left(1 + 0.1 s T_{RB}\right)} \quad (15)$$

$$G = G_{fs} \times G_{pcu} \quad (16)$$

Thus, the closed loop transfer function of BD is

$$\frac{\Delta S_{fl}}{\Delta X_E} = \frac{K_3\left(1 + \frac{G}{sC_B}\right)}{1 - \frac{K_1 K_2 K_3 G}{sC_B} + \frac{K_1 K_3}{sC_B}} \quad (17)$$

2.3. Controller Structure

In AGC study, the design of secondary controller is one of the major tasks to the designers. Despite of significant strides in the expansion of advanced control algorithms, PID-controller or its variants are remain popular because of its numerous practical advantages. These are: structural simplicity, ease realization, reliability, and favorable ration between cost and performance etc. It also offers simplified dynamic structure, lower user understanding, short terms error and provide highly responsive control system, gives desirable output in spite of the variations of the internal parameters etc. (Panda et al., 2013).

In this chapter, initially, PID-controller is optimally designed employing the proposed optimization techniques. To take out maximum possible generation from the power system, distinct PID-controllers are incorporated in the LFC loop. The errors input to the controllers are area control error (ACE) of the respective areas. ACE is mathematically described by Equation (18). The control input to the plant with PID-controller can be derived by Equation (19).

$$ACE_i = B_i \Delta f_i + \Delta P_{tie,i-j} \tag{18}$$

$$u_i = k_p(ACE_i) + k_i \int (ACE_i)dt + k_d \tfrac{d}{dt}(ACE_i) \tag{19}$$

In Equations (18) and (19), suffix i represents number of control areas (i.e., 1, 2, and 3) and $j = 2, 3$ (for $i \neq j$), k_p, k_i, k_d are the proportional, integral, and derivative gains of PID controller respectively.

Furthermore, to accelerate the speed of closed loop responses, two-degree-of-freedom PID (2DOF-PID) controller is optimally designed and included in AGC loops in lieu of PID-controller. The degree of freedom in control theory is defined as the number of closed loop transfer function that can be handled independently (Sahu et al., 2016). The controller produces controlled signal based on the difference between the reference and measured signals. It computes a weighted difference signals for each of the proportional, integral, and derivative control modes in accordance with the reference input. In control system design, constant set point and higher disturbance rejection facilities are highly desirable. However, due to variations of plants operating point, the control variable set pint may eventually changes. Specific criterion for command tracking includes rise time and settling time. Satisfying these two conditions simultaneously is not possible through conventional PID-controller, but 2DOF controller can do so. A 2DOF controller has two main objectives: one is set point tracking and another is disturbance rejection (Sahu et al., 2016).

Figure 5 illustrates the basic block diagram of 2DOF-PID controller. In Figure 5, $R(s)$, $Y(s)$, and $U(s)$ represents reference input, feedback from the measured devices and control output signals, respectively, PW and DW are the set pint weights of proportional and derivative controllers, respectively, N is derivative filter coefficient. A 2DOF-control system is shown in Figure 6. In Figure 6, $G_c(s)$ represents the feedback compensator and $G_{ff}(s)$ is the pre-filter on the reference signal or feed-forward compensator. The value of $G_{ff}(s)$ and $G_c(s)$ are calculated by using Equations (20) and (21), respectively.

$$\begin{aligned} G_{ff}(s) &= \tfrac{U(s)}{R(s)} = \left(\tfrac{sN}{s+N} k_d * DW\right) + \left(\tfrac{k_i}{S}\right) + PW * k_p \\ \Rightarrow G_{ff}(s) &= \frac{S^2(N*k_d*DW+PW*k_p)+s(k_i+N*PW*k_p)+k_iN}{s(s+N)} \end{aligned} \tag{20}$$

Figure 5. Block diagram of 2DOF-PID controller

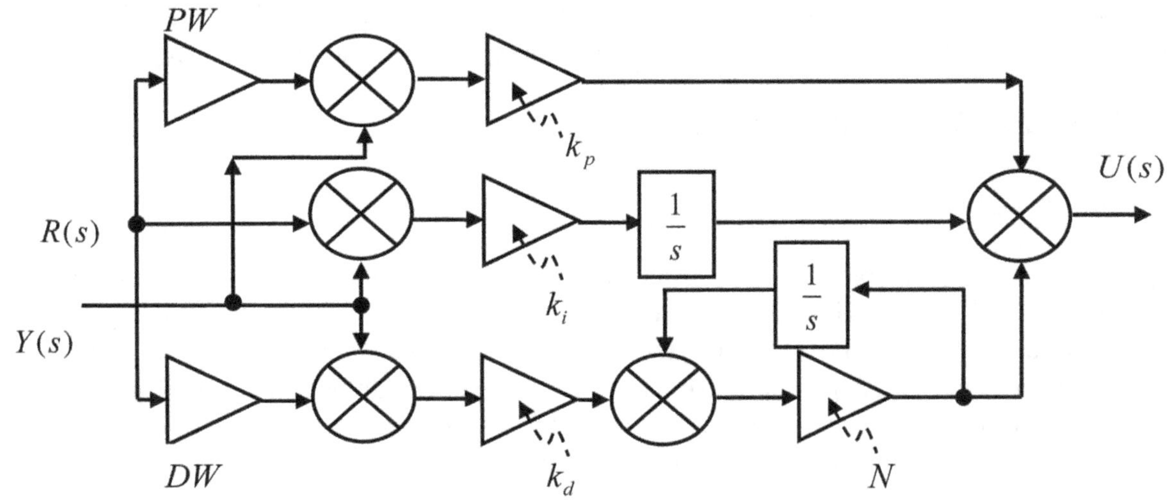

Figure 6. 2DOF control system

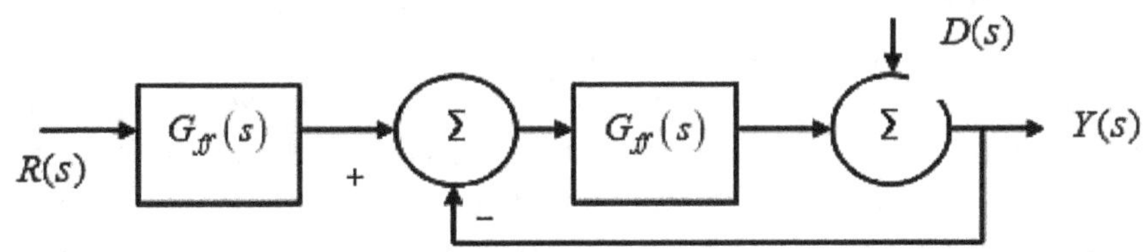

$$G_c(s) = \frac{U(s)}{R(s)} = \left(\frac{sN}{s+N} k_d\right) + \frac{k_i}{s} + k_p$$
$$\Rightarrow G_c(s) = \frac{S^2(k_d N + k_p) + s(k_i + k_p N) + k_i N}{s(s+N)} \tag{21}$$

3. OVERVIEW OF OPTIMIZATION

The word 'optimization' is originated from the Latin word '*optimum*' meaning '*the ultimate ideal*'. In today's scenario, computational challenges may exist in finding globally optimized solution from an immensely large solution space. Heuristic optimization techniques have therefore been forwarded that can find the candidate solution from large search area. Over the last few decades, nature-inspired optimization techniques have gained huge attention in the fields of science and technology. Following are the main advantages of population based optimization techniques over the conventional one:

1. Simplicity is the key advantage of population-based approaches.
2. Local optima entrapment is very high and making them suitable for real-time applications.
3. Unlike deterministic optimization methods, these algorithms considered the optimization problems as black boxes and don't require the derivative of the search space.
4. Flexibility is another key feature of the stochastic optimization method.

In this section, four parameter-free optimization algorithms, such as GWO, SOS, JAYA and TLBO are revisited, investigated, and applied to AGC for finding an effective and feasible solution. The inspirations and mathematical modeling of the proposed algorithms are discussed in the following sub-sections.

3.1. Grey Wolf Optimization (GWO)

3.1.1. Inspiration

Grey wolf, also known as timber wolf or western wolf belongs to Canidae family. Its scientific name is *Canis Lupus*. GWO algorithm is inspired by the social leadership and hunting behavior of grey wolves in nature (Mirjalili et al., 2014). Four types of grey wolves such as alpha (α), beta (β), delta (δ), and omega (Ω) are employed for simulating the leadership hierarchy in GWO. Additionally, the hunting mechanism in GWO involves three main strategies, viz. searching for prey, encircling prey, and attacking prey.

Grey wolves present at the top of the hierarchy are called '*alpha category wolves*' and they are the leader of the whole pack. This category of wolves may be male or female. Alpha wolves have decision-making power about hunting, sleeping place, time to wake etc. Their decisions are directed to the pack. However, some kind of democratic behavior is also observed in which they follow other wolves in the pack. In gatherings, the entire pack acknowledges alpha by holding their tails down. The alpha wolf is also known as *dominant wolf* since his/her direction should be followed by the entire pack.

In the second level of the hierarchy, the grey wolves are named as '*beta category wolves*' and they are the subordinate of alpha category wolves. They help alphas in the decision-making process and/or other pack activities. They are probably the best candidate and may transform into the alpha category wolves in case one of the alpha wolves passes away or become very old. The beta wolves respect alpha, but dominant other wolves in the pack. It plays the role of an advisor to alphas and discipliner in the pack.

The lowest stage of the hierarchy is occupied by the '*omega types of wolves*'. They are basically used as a scapegoat and always follow the decision made by the other dominant wolves. They are the worst category of wolves those are allowed to eat. It is noted that omega types of wolves are not so much important in the pack. But the whole pack may face internal fighting in case of losing the omegas. This is due to the venting of violence and frustration of all wolves by the omegas. Omegas are always maintaining the dominant structure in the hierarchy. In some cases the omega is also the babysitters in the pack.

The wolves which don't come under alpha, beta, and omega categories are grouped under '*delta or subordinate category*'. Delta types of wolves always follow the alphas and betas, but dominate omegas. In the hierarchy, delta wolves mainly perform five basic steps. These are: (i) scouts, (ii) sentinels, (iii) elders, (iv) hunters, and (v) caretakers. Scouts are responsible for watching the boundaries of the region and aware the pack in case of any hazard. Sentinels protect and promise the safety of the pack. Elders are the experienced wolves (alphas or betas) and their experiences are used to attack prey or any target elements. Hunters help alphas or betas when hunting prey and providing food for the pack. Finally, caretakers are responsible for caring the feeble, sick and injured wolves in the pack.

3.1.2. Mathematical Modeling

In this section, the social hierarchy of wolves, tracking, encircling, and attacking prey are discussed followed by the mathematical modeling of GWO algorithm.

3.1.2.1. Social Hierarchy

For modeling the social behavior of grey wolf, alpha is considered to be the fittest solution followed by beta and delta, respectively, and rest of the solutions is grouped under omega. In GWO, the hunting (optimization) process is guided by alpha, beta, and delta, whereas omega always follows these three wolves.

3.1.2.2. Encircling

During the optimization process, the positions of alpha, beta, and delta types of wolves with respect to prey position are updated. In (Mirjalili et al., 2014), Equations (22) and (23) are considered to model an encircling behavior of grey wolves around the prey.

$$\vec{D} = \left|\vec{C}.\vec{X}p(t) - \vec{X}(t)\right| \tag{22}$$

$$\vec{X}(t+1) = \vec{X}_p(t) - \vec{A} * \vec{D} \tag{23}$$

where t is the current iteration, $\vec{X}_p(t)$ denotes the current position of the prey. The coefficient vectors $\vec{A}$ and $\vec{C}$ are calculated by using Equations (24) and (25), respectively.

$$\vec{A} = \left(2\vec{a}\vec{r_1} - \vec{a}\right) \tag{24}$$

$$\vec{C} = 2\vec{r_2} \tag{25}$$

where $\vec{r_1}$ and $\vec{r_2}$ are two random vectors generated between (0, 1). The component $\vec{a}$ gives search direction matrix, which is linearly decreased from 2 to 0 over the course of iteration.

3.1.2.3. Hunting

In hunting phase, guided by the alpha wolves, the positions of the grey wolves are updated. Though alphas are the main agents in hunting phase, still occasionally betas and deltas also participate in the hunting process. So far we have the candidate solutions of grey wolves in terms of alphas, betas and deltas. But we don't know the exact or optimum position of prey. To find the optimum positions, three best solutions (obtained so far) in terms of alpha, beta and delta are saved and remaining solutions including omega are competed. The mathematical model proposed to adjust the position of omega wolves are defined in Equation (26).

$$\begin{cases} \overrightarrow{D_\alpha} = \left|\overrightarrow{C_1}\overrightarrow{X_\alpha} - \overrightarrow{X}\right| \\ \overrightarrow{D_\beta} = \left|\overrightarrow{C_2}\overrightarrow{X_\beta} - \overrightarrow{X}\right| \\ \overrightarrow{D_\delta} = \left|\overrightarrow{C_3}\overrightarrow{X_\delta} - \overrightarrow{X}\right| \end{cases} \tag{26}$$

where $\overrightarrow{X_\alpha}$, $\overrightarrow{X_\beta}$, and $\overrightarrow{X_\delta}$ are the positions of alpha, beta, and delta wolves, respectively, $\overrightarrow{C_1}$, $\overrightarrow{C_2}$, and $\overrightarrow{C_3}$ are the random vectors, $\overrightarrow{X}$ is the solution current population. Equation (26) calculate the approximate distance between the current solution and alpha, beta, and delta respectively. With this information, the final positions of wolves in the search space are updated by using Equations (27)-(28).

Figure 7. General flowchart of GWO algorithm

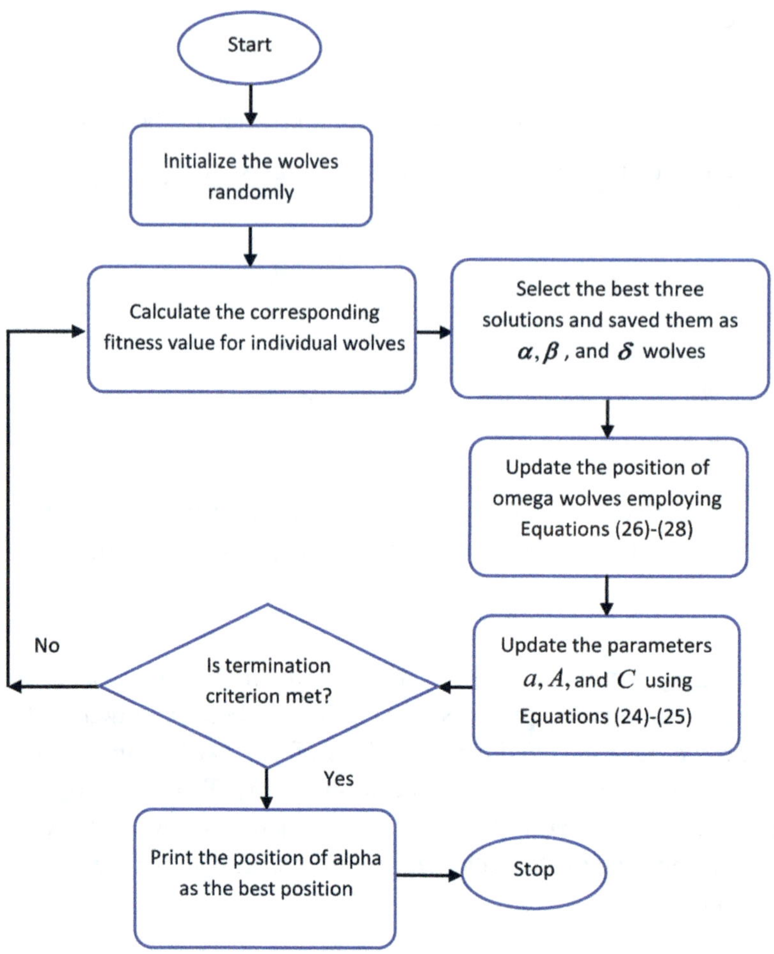

$$\begin{cases} \overrightarrow{X_1} = \overrightarrow{X_\alpha} - \overrightarrow{A_1}\left(\overrightarrow{D_\alpha}\right) \\ \overrightarrow{X_2} = \overrightarrow{X_\beta} - \overrightarrow{A_2}\left(\overrightarrow{D_\beta}\right) \\ \overrightarrow{X_3} = \overrightarrow{X_\delta} - \overrightarrow{A_3}\left(\overrightarrow{D_\delta}\right) \end{cases} \tag{27}$$

$$\overrightarrow{X}(t+1) = \frac{\overrightarrow{X_1}+\overrightarrow{X_2}+\overrightarrow{X_3}}{3} \tag{28}$$

where $\overrightarrow{A_1}$, $\overrightarrow{A_2}$, and $\overrightarrow{A_3}$ are three random vectors. It is viewed from Equation (26) that the step size of omega wolves is updated towards the position of alpha, beta, and delta types of wolves. Equations (27)-(28) are employed to update the final position of omega types of wolves w.r.t prey. It is also observed from the above equations that final position of the wolf is random in nature and it remains within the circle defined by the alpha, beta, and delta in the search space. The entire optimization process in GWO algorithm is guided by two vectors $\overrightarrow{A}$ and $\overrightarrow{C}$. Natural balance between exploration and exploitation phenomenon in GWO algorithm is accomplished by the vectors $\overrightarrow{A}$ and $\overrightarrow{C}$.

The exploration occurs when $\overrightarrow{A}$ is greater than 1 or less than – 1. The vector $\overrightarrow{C}$ also promotes the exploration when it is greater than 1. Unlike exploration, the exploitation is achieved with $|\overrightarrow{A}| < 1$ and $\overrightarrow{C} < 1$. In order to have a better exploitation in the search space, the vector $\overrightarrow{A}$ is linearly decreased with iteration (Mirjalili et al., 2014). On the other hand, the vector $\overrightarrow{C}$ is randomly generated throughout the course of a generation to exhibit better exploration and/or exploitation. These two vectors help optimization technique to avoid the local optimal solutions. The general flowchart of GWO algorithm is shown in Figure 7.

3.2. Symbiotic Organism Search (SOS)

3.2.1. Inspiration

SOS is a new population-based nature inspired meta-heuristic optimization technique developed by (Cheng and Prayogo, 2014). SOS mimics the symbiotic interaction techniques held by an organism to endure in the ecosystem. SOS imitates the symbiotic interactions within a paired organism relationship that are used to search the fittest organism in the ecosystem. The word '*symbiotic*' comes from the Greek word '*living together*' (Cheng and Prayogo, 2014). The symbiotic familiarity is either *obligate*, i.e. two organisms depend on each other for survival, or *facultative* means two organisms select to cohabitate in mutually beneficial but a nonessential relationship.

3.2.2. Mathematical Modeling

In SOS, new generations is governed by emulating the biological interaction between two organisms in an ecosystem. The common symbiotic relationships found in a real-world biological system are mutualism, commensalism, and parasitism. The symbiotic relationship may help the organism to improve the fitness value and survive for a long period. In the following subsections, different symbiotic relationships are briefly described.

3.2.2.1. Mutualism

Mutualism explains a symbiotic connection between two different species where both are profited. An example of mutualism is the relationship between bees and flowers (Cheng and Prayogo, 2014). Bees fly amongst flowers and gathering nectar to turn into honey. The same activity also helps flowers in the pollination process. The mutualism phase in SOS is mathematically modeled as follows:

Let X_i is an organism matched to the ith member of the ecosystem. Randomly select another organism X_j to interact with X_i. Both organisms are working in mutual basis to improve the fitness value and survive in the ecosystem. Modified solutions of X_i and X_j after mutualism can be found by using Equation (29).

$$\begin{cases} X_i^{new} = X_i^{old} + rand^*\left(X_{best} - \left(\frac{X_i+X_j}{2}\right) * round * (1+rand)\right) \\ X_j^{new} = X_j^{old} + rand^*\left(X_{best} - \left(\frac{X_i+X_j}{2}\right) * round * (1+rand)\right) \end{cases} \tag{29}$$

where *rand* is a uniformly distributed random number generated between 0 and 1, X_{best} is the best organism obtained so far, *round* is used to set the value either 2 or 1. In Equation (29) the part $X_{best} - \left(\frac{X_i+X_j}{2}\right) * round * (1+rand)$ imitating the mutualistic effort to achieve the goal of organism and survive in nature (Cheng and Prayogo, 2014).

3.2.2.2. Commensalism

In commensalism phase, between two different species, one gets benefits and other is unaffected. An example of this phase is the relationship between remora fish and sharks (Cheng and Prayogo, 2014). The remora fish is attached itself to the shark and eats food leftovers, thus receiving benefit. However, the shark is unaffected by the remora fish activities.

Similar to mutualism phase, an organism X_j is randomly selected from the ecosystem to interact with X_i. In this process X_i attempts to receive benefit from X_j, whereas X_j neither get benefit nor suffer from this relationship. The updated solution of Equation. (29) can be calculated by using Equation. (30).

$$X_i^{new} = X_i^{old} + rand^*(-1,1) * (X_{best} - X_j) \tag{30}$$

Figure 8. General flowchart of SOS algorithm

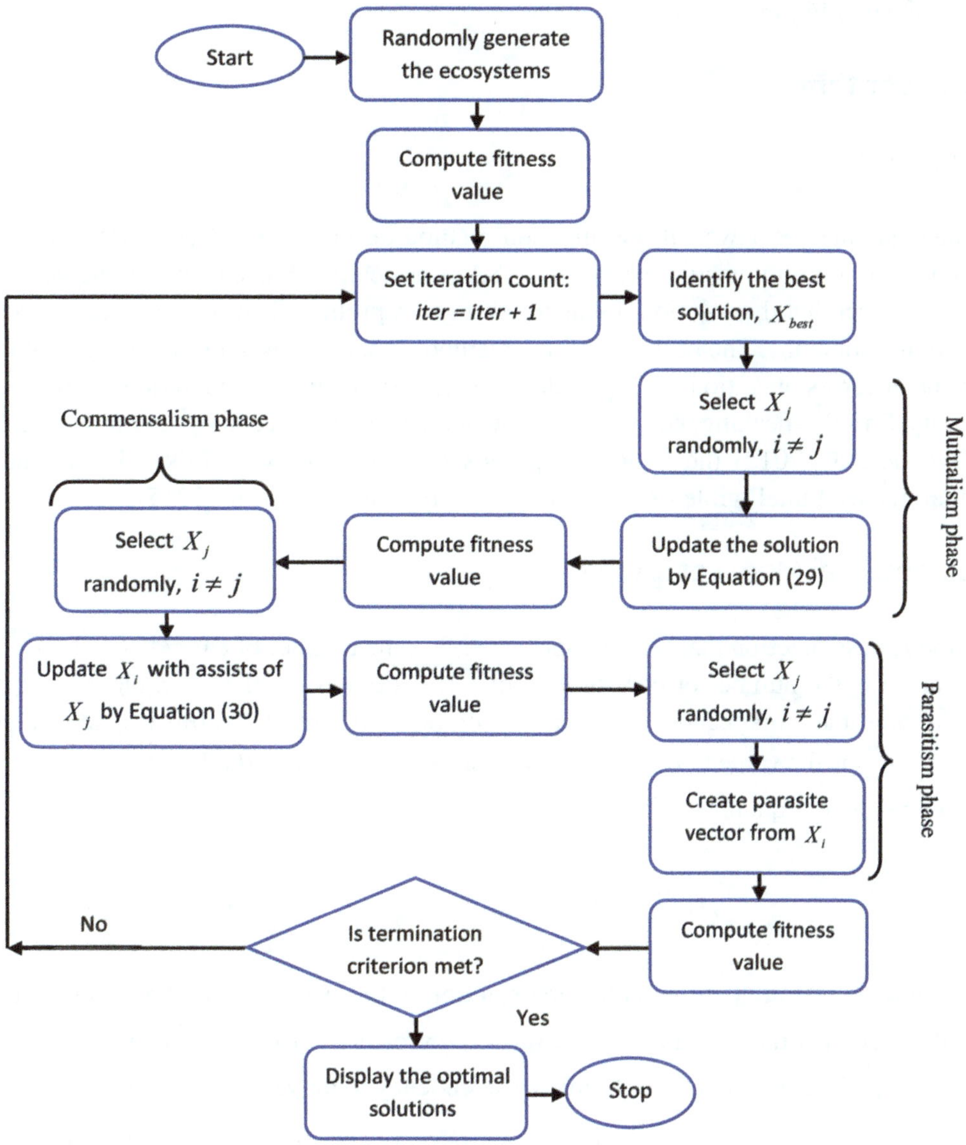

3.2.2.3. Parasitism

Parasitism is a symbiotic association between different organisms in which one get the benefit and other is actively injured with this association. Plasmodium parasite is a figure of parasitism phase. Plasmodium parasite uses this relationship with anopheles mosquito to pass between human hosts. In SOS, X_i shows the similar role of anopheles mosquito through the formation of an artificial parasite called '*parasite vector*'.

Parasite vector is the duplicating form of X_i. An organism X_j is randomly selected from the ecosystem and used as a host to parasite vector. Parasite vector attempts to modify X_j in the ecosystem. Both organisms are used then to measure the fitness value. If parasite vector has fittest fitness value, it will kill X_j in the ecosystem and adopt its position in the ecosystem, otherwise, X_j will have immunity from the parasite and the parasite-vector is no longer able to live in the system. Flowchart of proposed SOS algorithm is shown in Figure 8.

3.3. Jaya Algorithm

3.3.1. Inspiration

JAYA is a new, simple yet powerful optimization method developed by (Rao, 2015) for solving constraint and unconstraint optimization problem. JAYA is basically a Sanskrit word, meaning **victory**. It is inspired by the fact that the solution obtained for a given problem is move towards the best solution and discarded the worst one. The best part of this method is that it does not have any algorithm-specific control parameter for its operation, only population size and maximum generation count is defined for its functionality. Strictly speaking, keeping the view of the success of teaching learning-based optimization (TLBO) (Rao et al., 2012), the author was proposed JAYA. However, unlike TLBO, it has only one phase and that makes it intelligible to apply in optimization problem (Rao, 2015).

3.3.2. Mathematical Model of JAYA

Let, $f(x)$ be the fitness function that is to be optimized, k is the number of candidate solution (i.e. population size), and j is the number of control variables (i.e. dimension of the problem). Let, $f(x_{best})$ and $f(x_{worst})$ are the best and worst candidate solutions computed during the entire course of a generation, respectively. If $X^{i}_{j,k}$ is the value of the kth candidate of the jth variable at the ith generation, then the solution is updated by using Equation. (31).

$$X^{i+1}_{j,k} = X^{i}_{j,k} + r_1 * \left(X^{i}_{j,best} - \left|X^{i}_{j,k}\right|\right) - r_2 * \left(X^{i}_{j,worst} - \left|X^{i}_{j,k}\right|\right) \tag{31}$$

where $X^{i}_{j,best}$ and $X^{i}_{j,worst}$ are the best and worst solution of the jth variable at the ith generation, respectively, r_1 and r_2 are two random numbers those are uniformly selected between [0, 1]. The term " $r_1 * \left(X^{i}_{j,best} - \left|X^{i}_{j,k}\right|\right)$" defines that the solution has a tendency to move nearer to the fittest solution. The term " $r_2 * \left(X^{i}_{j,worst} - \left|X^{i}_{j,k}\right|\right)$" identifies that solution that avoids the worst value. The candidate solution will be selected or rejected based on the fitness value obtained at current iteration. The best solution will be considered for next iteration. Actually, the main philosophy of JAYA is to achieve victory and tries to avoid failure. That is why author named the algorithm as 'JAYA' (Rao, 2015). The flowchart of the proposed algorithm is available in Figure 9.

3.4. Teaching Learning Based Optimization (TLBO)

3.4.1. Inspiration

Teaching learning based optimization (TLBO) is a relatively new optimization method proposed by (Rao et al., 2012). It simply mimics the effect of influence of mentor on the knowledge of mentee in class. In

Figure 9. Flowchart of JAYA algorithm

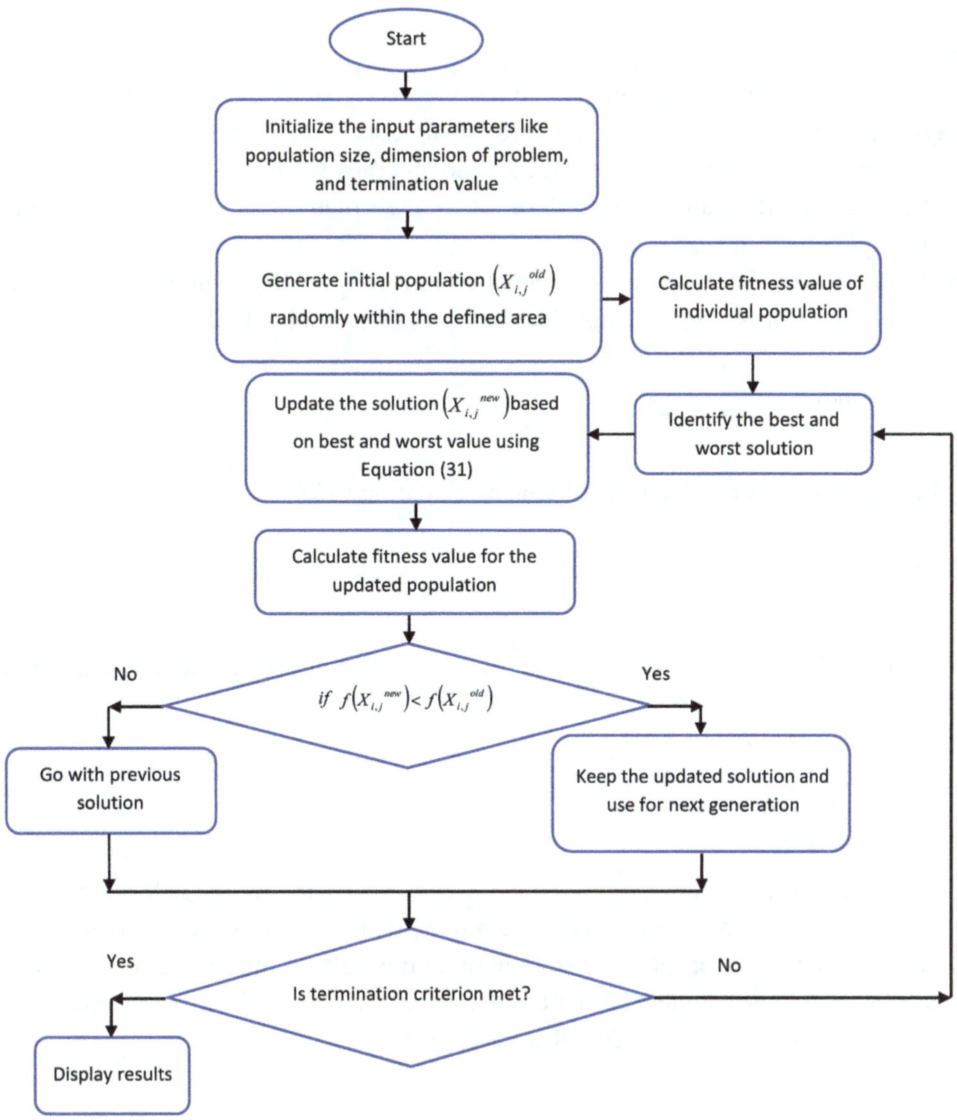

TLBO, students gather information or knowledge from lectures delivered by the teachers and also from the other students by the process of mutual interaction. In this algorithm, population consists of number of learners and control variables are identified as the number of subjects allotted to them. The fitness value of the optimization problem is defined as the knowledge of the students on a particular subject and best value is analogous to a teacher's knowledge obtained so far. The knowledge level of students is finally identified by the grade point in the class. The working philosophy of TLBO algorithm is divided into two phases, i.e., '*teaching phase*' and '*learning phase*'.

3.4.2. Mathematical Model of TLBO

3.4.2.1. Teacher Phase

In this phase, students acquire knowledge from the teachers. Depending upon his/her capability, a good teacher always tries to build up the knowledge level of students up to their own level. However, in actual practice it is somewhat difficult because the students are not only gathering knowledge from the lectures delivered by the teachers, it simultaneously depends on their own quality. So a teacher can only try to improve the average results in a class.

Let, μ_k be the mean and T_k be the teacher at any iteration k. T_k will try to move μ_k from its initial position to a new mean position say $\mu_{k,new}$ using Equation (32) (Rao et al., 2012).

$$Diff\_mean_k = rand(\mu_{k,new} - T_f \mu_k) \tag{32}$$

where T_f is the teaching factor and selected by using Equation (33).

$$T_f = round(1 + rand(0,1)) \tag{33}$$

Thus, the grade of the ith student of the jth subject at the kth iteration is calculated by using Equation (34).

$$X_{i,j}^{k+1} = X_{i,j}^{k} + Diff\_Mean_j^k \tag{34}$$

3.4.2.2. Learner Phase

In the course of the teaching-learning process, student increases their knowledge level either from teacher or through the interactions among themselves. The mode of interaction can be group discussion, presentation or formal conversion. In this phase, a student randomly selects another student for interaction and learn new things if the selected student has better knowledge than him. The learning phase is established by performing the following pseudo codes (Rao et al., 2012).

```
for i = 1:n_p
        Randomly select a student X_j, j ≠ i
        if f(X_j) > f(X_i)
                X_new,i = Xold,i + rand*(X_i - X_j);
        else
                X_new,i = Xold,i + rand*(X_j - X_i);
        end
end
```

3.5. Objective Function

A common problem encountered in the control system design and analysis is the identification of controller gains. The first constraint in the design of an optimal AGC is the choice of objective function. The selection of objective function is problem specific and should be selected accordingly. In the design of modern heuristic algorithms based optimal controller, the objective function is first defined based on system specifications and constraints. In AGC problem analysis, objective function is chosen by taking frequency and tie-line power deviation of each area of the concerned power system. In the literature, four types of performance indexes are defined for design of optimum controllers. These are integral square error (ISE), integral time square error (ITSE), integral absolute error (IAE), and integral time absolute error (ITAE). For any of the possible aforesaid criterions, best response corresponds to the minimum value of selected objective function and better system output in terms of transient specifications. It is stated in the literature that best results are obtained with ISE and ITAE based fitness functions. Hence in the current work, both ISE and ITAE based fitness functions are considered for fine tuning of controller gains. Selected fitness functions are defined in Equation (35).

$$\begin{cases} J_{ISE} = \int_{0}^{T} (ACE)^2\, dt \\ J_{ISE} = \int_{0}^{T} |\, ACE \,| * t * dt \end{cases} \tag{35}$$

In the design and analysis of AGC, the defined problem is viewed as constraint optimization problem bounded by the controller gains. It is described as

Minimize J_{ISE} (or J_{ITAE})

Subjected to:

$$\begin{cases} k_{p,\min} \le k_p \le k_{p,\max} \\ k_{i,\min} \le k_i \le k_{i,\max} \\ k_{d,\min} \le k_d \le k_{d,\max} \\ N_{\min} \le N \le N_{\max} \\ pw_{\min} \le pw \le pw_{\max} \\ dw_{\min} \le dw \le dw_{\max} \end{cases} \tag{36}$$

where $k_{pid,\min}$ and $k_{pid,\max}$ are the minimum and maximum values of PID-controller gains, respectively, $N_{\min}$ and $N_{\max}$ are the minimum and maximum values of derivative filter constants, respectively, $pw_{\min}$ and $pw_{\min}$ are the minimum and maximum values of set point weights of proportional block, respectively, and $dw_{\min}$ and $dw_{\max}$ represents the minimum and maximum values of set point weights of derivative block, respectively. As reported in (Raju et al., 2016), the minimum and maximum values of controller gains are chosen in the range of (–2.0, +2.0). The set point weights and filter coefficients for 2DOF-PID controller selected between (0,1) and (0,200) respectively.

4. ALGORITHMS APPLIED TO LFC PROBLEM

4.1. Algorithmic Steps of GWO Algorithm

A preview of applying GWO algorithm in AGC of power system is discussed below.

Step 1: Randomly initialize the controller variables, i.e. wolves in GWO, in the defined search space. Pseudo codes for randomly generated population is given below.

```
for i = 1:n_p
        for j = 1:dim
                X(i, j) = lb(j) + rand*(ub(j) - lb(j));
        end
end
```

Step 2: Calculate the fitness value, i.e. ACE, for the individual solution by Equation (35).
Step 3: Filtered out three best solutions and saved them as alpha, beta, and delta wolves.
Step 4: Update the positions of alpha, beta, and delta wolves by performing the following pseudo codes.

```
for i = 1:n_p
        if fitness < alpha
                alpha <--update-- fitness
        end
        if fitness > alpha && fitness < beta
                beta <--update-- fitness
        end
        if fitness > alpha && fitness < beta && fitness < delta
                delta <--update-- fitness
        end
end
```

Step 5: Defining two random numbers r_1, r_2 between [0, 1].

$$\begin{cases} r_1 = rand(0,1); \\ r_2 = rand(0,1); \end{cases}$$

The vector $\vec{a}$ is linearly decreased from 2 to 0 using $a = 2 - 1 * \left(\frac{2}{T}\right)$
l: current iteration, T: maximum number of iteration.

Step 6: Rest of the wolves i.e., omega are updated with respect to prey position employing Equations (26)-(28).

Step 7: Update the parameters $\vec{A}$ and $\vec{C}$ by Equations (24) and (25).
Step 8: Go to Step 2 until the termination criterion is met.
Step 9: Print the position of alpha wolf as an approximated optimal solution of control variables. Finally calculate the time domain specifications such as settling time, peak overshoot etc.

4.2. Algorithmic Steps of SOS Algorithm

The different steps of SOS algorithm for searching the global optimum solution are discussed hereunder.

Step 1: Initialize the input parameters of SOS algorithm, i.e. population size, number of control variables, and maximum iteration count.
Step 2: Randomly generate the initial population (controller gains), i.e. ecosystem in SOS, within upper and lower bounds by the following pseudo codes.

```
for i = 1:n_p
        eco(i,:) = rand(1,n).*(ub - lb) + lb
end
```

Step 3: Evaluate the fitness value with the generated population set using Equation (35).
Step 4: Select the best organism (elite solution) based on fitness value evaluated in Step 3.
Step 5: Randomly select an organism X_j where $j \neq i$ and update the solution by Equation (29).
Step 6: Evaluate fitness value of the newly created solution using Equation (35).
Step 7: Apply boundary condition to check whether the newly generated solutions are within the defined search space or not.
Step 8: Accept the new solutions based on their fitness value and update the initial populations by performing following steps:

```
if fitness_new1 < fitness(i)
      fitness(i) = fitness_new1
      eco(i,:) <--update--  econew1(X_i,new)
end
if fitness_new2 < fitness(j)
      fitness(j) = fitness_new2
      eco(j,:) <--update--  econew1(X_j,new)
end
```

Step 9: Randomly select an organism X_j where $j \neq i$.
Step 10: Calculate new solutions after commensalism phase using Equation (30). Apply boundary condition to check whether the newly generated solutions are within the search limit or not.
Step 11: Evaluate fitness function using Equation. (35) and accept the new solution based on lowest fitness value.

```
if fitness_new < fitness(j)
      fitness(j) = fitness_new
      eco(j,:) <--update-- econew1(X_{j,new})
end
```

Step 12: Randomly select an organism X_j where $j \neq i$.

Step 13: Determine the parasitic vector, i.e. the duplicate organism of X_i. Evaluate the fitness function with parasite vector by using Equation (35).

Step 14: Kill the organism j and replace it by the parasite vector.

```
if fitness_parasite < fitness(j)
      fitness(j) = fitness_parasite
      eco(j,:) <--update-- parasite_vector
end
```

Step 15: Update the best organism based on minimum fitness value.

Step 16: If the termination criterion is met, then stop iteration; otherwise go to Step 4 for next generation.

4.3. ALGORITHMIC STEPS OF JAYA ALGORITHM

The algorithmic steps of JAYA applied to AGC is discussed in this section. The steps are enumerated below.

Step 1: Initialize the population size (n_p), control variables, maximum iteration ($iter_{max}$), upper (up) and lower (lw) bounds of the control variables.

Step 2: Randomly initialize the initial population within the defined search space. Compute the fitness value, i.e. minimum ACE, of the individual population by using Equation (35).

```
for i = 1:n_p
        for j = 1:dim
                X(i,j) = lw(j) + rand*(up(j)-lw(j));
        end
end
```

Step 3: Identify the *best* and *worst* solutions from the generated population based on minimum fitness value.

Step 4: Update the current solution by performing the following pseudo codes.

```
for i = 1:n_p
        for j = 1:dim
                X(i,j)^new = X(i,j)^old + rand*(best-abs(X(i, j)))-rand*(worst-
abs(X(i, j)));
                if 0.1 > rand
```

```
                X(i,j) = lw(j) + rand*(up(j)-lw(j));
            end
        end
end
```

Step 5: Calculate the fitness value using Equation (35).

Step 6: If, $X(i,j)^{new}$ has fittest solution compared to $X(i,j)^{old}$, accept it for next generation. Otherwise go with the old solution.

Step 7: Go to Step 3, until the termination criterion ($iter_{max}$) is met.

Step 8: Display results and use the optimum solutions to find the system specifications like peak overshoot, undershoot and settling time.

4.4. Algorithmic Steps of TLBO Algorithm

The TLBO algorithm can be easily demonstrated by the following steps:

Step 1: Let n_p be the number of students and n_d be the number of subject allotted to individual students in the class. The initial population with this information can be randomly generated within the defined search space. After generation initial populations are stored in Equation (37).

$$X = \begin{bmatrix} x_{1,1} & \cdots & x_{1,j} & \cdots & x_{1,n_d} \\ \vdots & & \vdots & & \vdots \\ x_{j,1} & \cdots & x_{i,j} & \cdots & x_{i,n_d} \\ \vdots & & \vdots & & \vdots \\ x_{n_p,1} & \cdots & x_{n_p,j} & \cdots & x_{n_p,n_d} \end{bmatrix}_{n_p x n_d} \tag{37}$$

where $x_{i,j}$ is the initial grade of the ith student in j^{th} subject.

Step 2: Compute average grade point (μ_j), i.e. the fitness value, of individual students in each subject offered in the class. The mean grade of the jth subject is obtained as

$$\mu_j = mean(x_{1,j}, x_{2,j}, \ldots, x_{i,j}, \ldots, x_{n_p,j}) \tag{38}$$

Step 3: Based on the grade point, computed in Step 2, identify the best value and sort the population. Best fitness value will be treated as a teacher.

$$X_{teacher} = X \mid_{f(x)=\min} \tag{39}$$

Step 4: Improve the grade point of individual students by performing the following pseudo codes:

```
for i = 1:n_p
        for j = 1:n_d
                X_{i,j}^{k+1} = X_{i,j}^k + rand*[X_teacher-round(1+rand*)μ_j;
        end
end
```

Step 5: Individual population (students) is now updating their fitness value (grade point) of each control variable (subjects) by interacting with other students in the class. This process is included in learner phase of TLBO. Thus, the modification of grade point with this learner phase can be defined as

```
for i = 1:n_p
        a = randperm(n_p);
        l = a(l);
        X_i = X(i,:); X_l = X(l,:);
        if f(X_i) < f(X_l);
                for j = 1:n_d
                        X_{i,j}^{k+1} = X_{i,j}^k + rand*(X_{i,j}^k = X_{l,j}^k);
                end
        elseif f(X_i) < f(X_l)
                for j = 1:n_d
                        X_{i,j}^{k+1} = X_{i,j}^k + rand*(X_{l,j}^k = X_{i,j}^k);
                end
        end
end
```

5. RESULTS AND DISCUSSION

For any control process, quick response and higher degree of stability are enviable. It is remarkable that lower the performance indices, better the time domain specifications and higher the degree of stability. Under this heading, critical investigation of power system dynamics has been carried out and acceptability of the proposed controller is established by computing peak overshoot, undershoot, and settling time of system responses. The simulation model of test system is developed in MATLAB-SIMULINK environment, while the optimization codes are separately written in *.m-file*. The simulations were performed in an Intel core i3 processor of 2.4 GHz and 2GB RAM in MATLAB 7.10.0.499 (R2009a) background. In the entire analysis, the population size and maximum iterations are set to 50 and 100, respectively. Major observations of the current study are highlighted in the following headings and marked by **bold** faces. The obtained results are best in terms of lowest fitness value and minimum transient specifications.

5.1. Transient Analysis

5.1.1. With PID-Controller

At the first instant, three-area unequal thermal power plant equipped with PID-controllers has been investigated. To add some degree of nonlinearity, GRC is included in steam turbine. The proposed optimization techniques, i.e., GWO, SOS, JAYA, and TLBO, are separately applied to search the optimum gains of PID-controllers via the minimization of Equation (35). At the end of optimization, the controller gains are offered in Table 2. To show the supremacy, the output results are compared with ant-lion optimization (ALO) tuned PID-controllers (Raju et al., 2016) for the similar test system with identical controller structure. For fair comparison, the optimum PID-controllers settings as reported in (Raju et al., 2016) are also provided in Table 2. The critical review of Table 2 reveals that lowest minimum ITAE value (**0.1837**) is derived with GWO-tuned PID-controllers. Improved of 40% (SOS), 42.56% (TLBO), and 29.32% (JAYA) is achieved with GWO algorithm.

The comparative convergence characteristic of the proposed algorithms is painted in Figure 10(a). It is obtainable from Figure 10(a) that GWO algorithm approaches to global optimal solution at a faster

Table 2. Optimum gains of PID-controller with 1% SLP in area-1

Control algorithms	Area-1			Area-2			Area-3			Fitness value
	k_{p1}	k_{i1}	k_{d1}	k_{p2}	k_{i2}	k_{d2}	k_{p3}	k_{i3}	k_{d3}	
GWO	0.9996	0.9726	0.7840	0.8953	0.9394	0.7506	0.8294	0.9309	0.6120	**0.1837**
SOS	0.7944	0.5991	0.8255	0.9322	0.8648	0.7470	0.9800	0.9672	0.6560	0.3062
TLBO	0.3827	0.6249	0.7603	0.5585	0.9655	0.7952	0.9901	0.9066	0.4667	0.3198
JAYA	0.9209	0.6863	0.7463	0.9597	0.9059	0.8190	0.9489	0.9891	0.6749	0.2584
ALO	0.0273	0.4899	0.1862	0.1103	0.4031	0.0139	0.1039	0.3542	0.0203	NA

Bold face show the best results

Figure 10. Convergence characteristic of proposed algorithms (a) with PID-controller, (b) with 2DOF-PID controllers

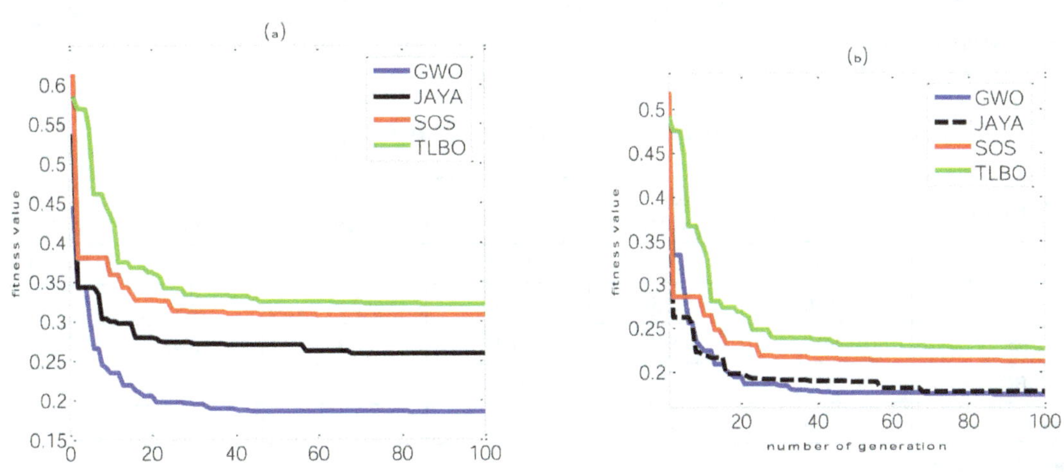

rate (takes 40-45 iterations) than JAYA, SOS, and TLBO. It is also confirmed that GWO algorithm offers minimum fitness value compared to others. It is to be noted that all the four optimization techniques approaches to global optimal point within 80 generations, which justifies the choice of maximum iterations be 100 for present study. Hence, from the aforesaid discussion it can be inferred that GWO algorithm has better tuning ability and provide more optimal controller settings compared to others for the concerned power system.

To assess the efficacy of designed GWO based PID-controllers, 1% SLP is applied to area-1 and corresponding closed loop responses are shown in Figure 11. The output results of the concerned test system is further derived with JAYA, SOS, and TLBO based PID-controllers and shown in Figure 11. To show the supremacy, the dynamic performance of the test system is compared with ALO based PID-controller (Raju et al., 2016). Typical transient specifications such as settling time and peak overshoot of Δf_1, Δf_2 and ΔP_{tie} are computed from Figure 11 and tabulated in Table 3 and 4, respectively. It is observed from Figure 11 that GWO-tuned PID-controllers provide highest damping to the system oscillations and force back to their steady value quickly, which is further verified from Table 4. The selling time of frequency and tie-line power oscillations is minimum compared to ALO and other proposed algorithms. The amount of improvement of settling time with GWO algorithm is highlighted in Table 4. Also compared to JAYA, SOS, TLBO and ALO based PID-controllers, GWO-based PID-controller provide minimum peak overshoot. Thus, it may be concluded that GWO outperformed over ALO and other EAs as listed in Table 4.

Table 3. Comparative study of settling time of test system using PID-controllers

Parameters	GWO	% of improvement	JAYA	% of improvement	SOS	% of improvement	TLBO	% of improvement	ALO
Δf_1	**11.34**	3.32	11.73	20.42	14.25	12.56	12.97	84.72	74.2
Δf_2	**18.94**	2.97	19.51	20.39	23.79	11.58	21.42	66.36	56.31
Δf_3	**21.43**	4.63	22.47	20.66	27.01	12.28	24.43	66.68	64.32
$\Delta P_{tie,12}$	**24.62**	2.37	25.22	15.37	29.09	7.69	26.67	68.88	79.11
$\Delta P_{tie,23}$	**23.01**	3	23.72	16.57	27.58	7.18	24.79	68.86	73.90
$\Delta P_{tie,31}$	**25.87**	1.82	26.35	14.19	30.15	7.67	28.02	64.34	72.54

Bold face show the best value

Table 4. Comparative study of peak overshoot of test system using PID-controllers

Control algorithms	Δf_1	Δf_2 (x10^{-4})	Δf_3 (x10^{-4})	$\Delta P_{tie,12}$ (x10^{-4})	$\Delta P_{tie,23}$ (x10^{-4})	$\Delta P_{tie,31}$ (x10^{-4})
GWO	0.0040	1.41	**0.448**	**2.38**	**5.46**	**4.19**
JAYA	0.0040	1.35	0.506	2.51	5.74	4.45
SOS	**0.0034**	**0.865**	0.629	2.88	6.45	5.17
TLBO	0.0055	2.76	0.748	3.54	7.53	6.08
ALO	0.0080	68	34	8	8	10

Bold face show the best value

Figure 11. Dynamic performance of concerned test system with PID-controllers, (a) change of frequency in area-1, (b) change of frequency in area-2, (c) change of tie-line power $\Delta P_{tie,12}$, (d) change of tie-line power $\Delta P_{tie,31}$

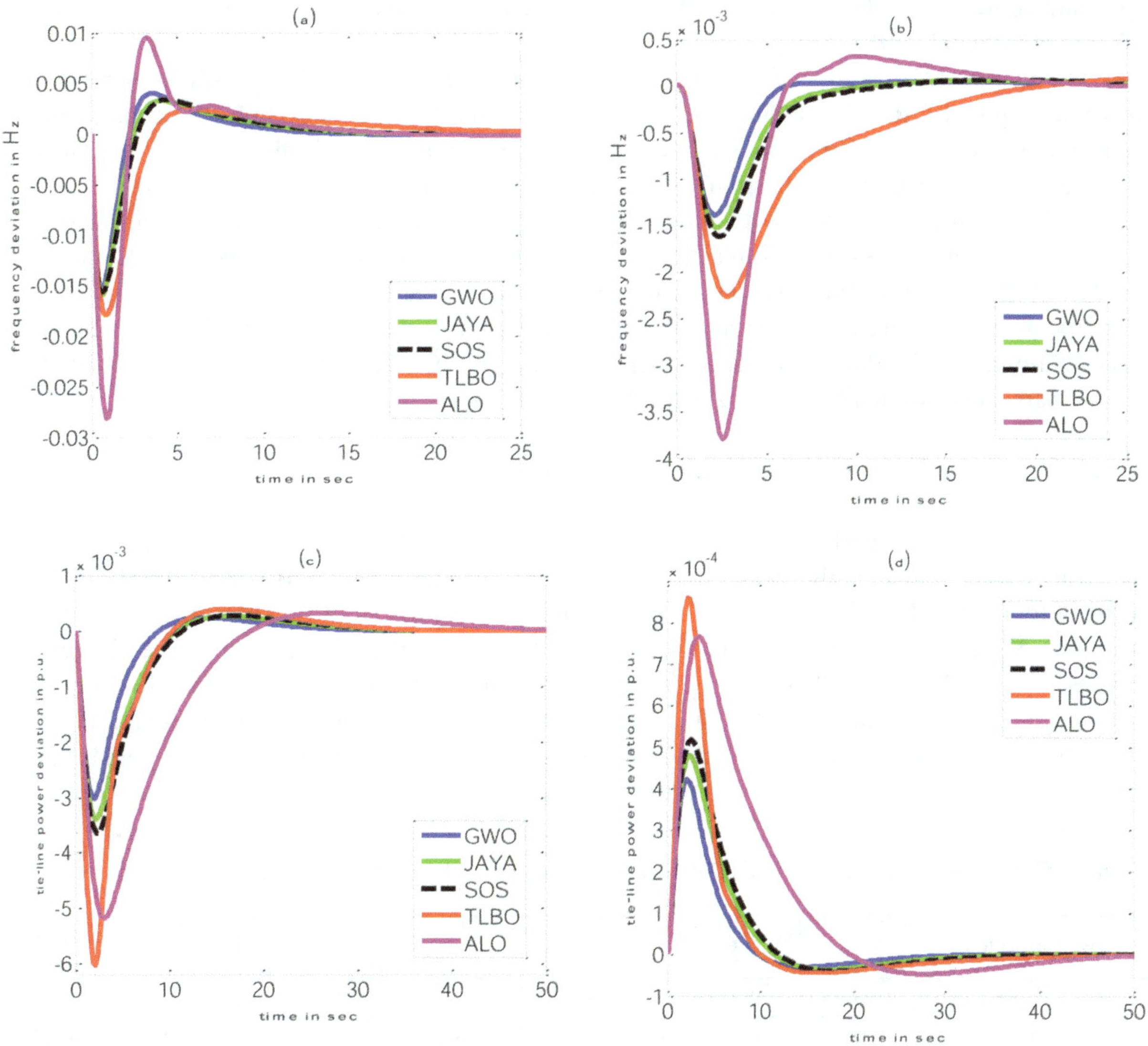

5.1.2. With 2DOF-PID Controller

To ennoble the dynamic stability of test system, 2DOF-PID controllers are optimally designed employing GWO, JAYA, TLBO, and SOS algorithms by minimizing Equation (35). Same procedure as discussed in Section 4 is followed to tune the 2DOF-PID controller gains by the aforesaid algorithms. The 2DOF-PID controller gains and minimum fitness value, offered by the proposed algorithms are shown in Table 5. Table 5 helps to infer that GWO-tuned 2DOF-PID controller's offers lowest minimum fitness value compared to JAYA, TLBO, and SOS algorithms. Improvement of 1.97% (JAYA), 17.79% (SOS), and 22.93% (TLBO) are noticed in fitness value with GWO algorithm. Hence, it may be stated at this stage that proposed GWO-tuned 2DOF-PID controllers gives more optimum controller settings compared to JAYA, SOS, and TLBO in terms of minimum fitness value. The convergence characteristic of proposed algorithms with 2DOF-PID controller is shown in Figure 10(b). It is evident from Figure 10(b) that GWO algorithm converges at a faster rate than JAYA, TLBO, and SOS that further justifies the computational efficacy of GWO algorithm.

The dynamic performance of concerned power system with GWO-tuned 2DOF-PID controllers are shown in Figure 12. Results obtained with JAYA, SOS, and TLBO based 2DIF-PID controllers are compared with GWO and plotted in the same figure. Time response specifications such as settling time and peak overshoot are calculated from Figure 12 and offered in Tables 6 and 7, respectively. Critical review of Figure 12 reveals that GWO-tuned 2DOF-PID controller's yields superior performance under the perturbed scenario. Results of Tables 6 and 7 further validate the efficacy of GWO over SOS, JAYA and TLBO. It is noteworthy that GWO-tuned 2DOF-PID controller's offer minimum settling time and lowest peak overshoot. Hence, examining the dynamic responses (Figures 11 and 12), it may conclude that GWO-tuned 2DOF-PID controller's outperformed over GWO-tuned PID-controller's and rest of the study is carried out with GWO-tuned 2DOF-PID controller's.

5.1.3. Transient Analysis With System Nonlinearities

In this section dynamic performance of concerned power system has been further investigated in the presence of more system nonlinearities, i.e. GDB, BD and time delay. At the first instant GDB is added in the system with GRC of steam turbine. The optimum settings of 2DOF-PID controller as calculated in Section 5.1.2 is considered to affirm robustness of the proposed GWO algorithm. The closed loop responses under the action of optimum 2DOF-PID controller are displayed in Figure 13. For better assessment of the performance of GWO algorithm, the output results obtained with SOS, JAYA and TLBO based PID controllers are also painted in Figure 13. Critical observation of the system behavior

Table 5. Optimum values of 2DOF-PID controller gains after 1% SLP in area-1

Control algorithm		k_p	k_i	k_d	N	pw	dw	Fitness value
GWO	Area-1	0.9631	0.9936	0.4845	101.06	0.0660	0.2670	**0.1741**
	Area-2	0.9951	0.9953	0.2124	68.35	0.1644	0.3408	
	Area-3	0.9421	0.9752	0.1197	54.87	0.0205	0.3477	
JAYA	Area-1	0.9803	0.9885	0.4649	145.83	0.0039	0.0013	0.1776
	Area-2	0.9705	0.9505	0.4463	164.79	0.0901	0.1165	
	Area-3	0.9819	0.9597	0.2599	29.07	0.1646	0.0621	
SOS	Area-1	0.8775	0.8309	0.4494	174.86	0.0417	0.2026	0.2118
	Area-2	0.8122	0.9497	0.3125	101.41	0.1802	0.6370	
	Area-3	0.9645	0.9370	0.5594	97.12	0.1323	0.2304	
TLBO	Area-1	0.8635	0.7930	0.6793	75.48	0.1776	0.1543	0.2259
	Area-2	0.9548	0.9951	0.4217	145.09	0.2067	0.2348	
	Area-3	0.5962	0.9309	0.5280	18.46	0.0398	0.6364	

Figure 12. Dynamic performance of concerned test system with 2DOF-PID controller, (a) change of frequency in area-1, (b) change of frequency in area-2, (c) change of tie-line power $\Delta P_{tie,12}$, (d) change of tie-line power $\Delta P_{tie,31}$

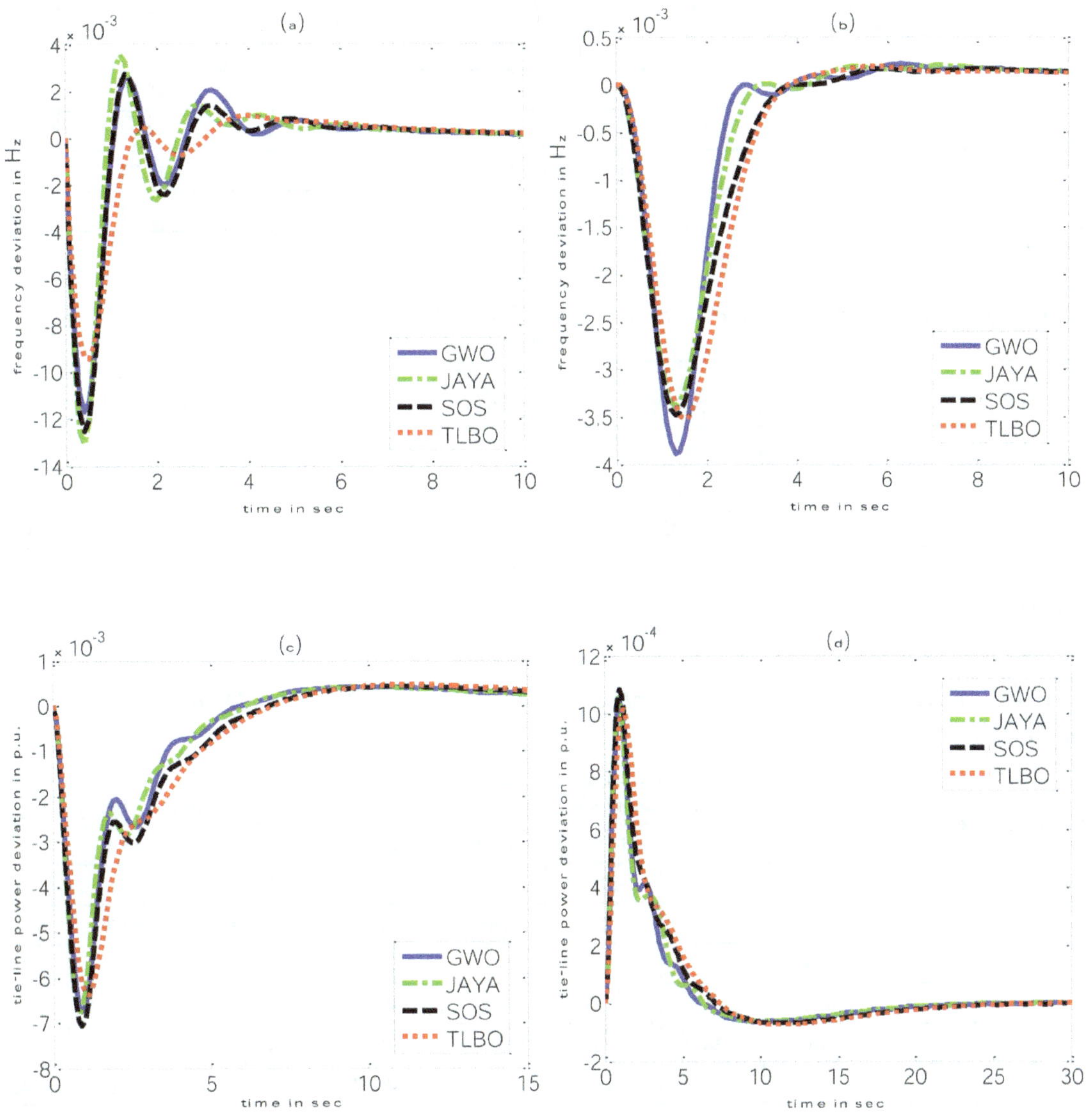

ravels that GWO-based 2DOF-PID controller shows relatively better outputs compared to other EAs as mentioned in Figure 13. The transient specifications of the system oscillations with GDB and GRC are displayed in Table 8. It is noteworthy from Figure 13 and Table 8 that the controllers designed at nominal condition can perform well in the presence of system nonlinearities and thus robustness of the controller is believed.

To add more degree of nonlinearities, BD and time delay of transmission system are considered in the model with GRC and GDB. The output results of the concerned power system with all said non-linearities are shown in Figure 14. For better analysis of GWO-tuned 2DOF-PID controller, the test

Table 6. Comparative study of settling time with 2DOF-PID-controller

Parameters	GWO	% of improvement	JAYA	% of improvement	SOS	% of improvement	TLBO
Δf_1	**7.99**	1.24	8.09	8.89	8.77	19.79	9.72
Δf_2	**11.39**	7.39	12.3	17.04	13.73	23.19	14.79
Δf_3	**13.01**	4.89	13.68	14.58	15.23	18.12	15.89
$\Delta P_{tie,12}$	**17.36**	2.96	17.89	10.14	19.32	14.23	20.24
$\Delta P_{tie,23}$	**15.76**	4.43	16.49	11.91	17.89	17.74	19.16
$\Delta P_{tie,31}$	**17.76**	3.43	18.39	10.84	19.92	13.91	20.63

Bold face show the best results

Table 7. Comparative study of peak overshoot with 2DOF-PID-controller

Control algorithms	Δf_1	Δf_2 (x10^{-4})	Δf_3 (x10^{-4})	$\Delta P_{tie,12}$ (x10^{-4})	$\Delta P_{tie,23}$ (x10^{-4})	$\Delta P_{tie,31}$ (x10^{-4})
GWO	0.0026	**2.21**	**2.135**	**4.103**	14	**10**
JAYA	0.0035	5.57	1.93	4.14	14	11
SOS	0.0027	2.32	1.7	4.30	14	11
TLBO	**9.34x10^{-4}**	3.14	1.86	4.50	**12**	**10**

Bold face show the best results

Table 8. Comparative study of peak overshoot and settling time with 2DOF-PID-controllers considering GRC and GDB

System parameters	Peak overshoot (×10^{-4})				Settling time (in sec)			
	GWO	SOS	JAYA	TLBO	GWO	SOS	JAYA	TLBO
Δf_1	251	**213**	228	237	**30.37**	31.71	31.65	30.82
Δf_2	86	69	73	**65**	**26.59**	27.01	26.97	30.74
Δf_3	45	**30**	34	35	**27.07**	27.63	27.55	27.15
$\Delta P_{tie,12}$	8.56	**7.83**	7.98	8.97	**40.44**	43.33	42.76	41.92
$\Delta P_{tie,23}$	**26**	**26**	**26**	**26**	**40.03**	43.05	42.29	41.18
$\Delta P_{tie,31}$	**25**	**25**	**25**	**25**	**40.57**	43.32	42.89	42.40

Bold face show the best results

system is simulated with SOS, JAYA and TLBO algorithms. The transient responses under the action of 2DOF-PID controllers are shown in Figure 14. Table 9 illustrates the comparative study among different algorithms in terms of peak overshoot and settling time of system oscillations. It is apparent from the presented results that settling time and peak overshoot of transient responses are increased with nonlinearities. However GWO-tune 2DOF-PID controller's gives satisfactory outputs compared to other control algorithms. Hence it may conclude that GWO algorithm provide more optimal solution of the designed controller and perform robustly in the presence of system nonlinearities.

Figure 13. Dynamic performance of concerned test system considering GRC and GDB, (a) change of frequency in area-1, (b) change of frequency in area-2, (c) change of tie-line power $\Delta P_{tie,12}$, (d) change of tie-line power $\Delta P_{tie,31}$

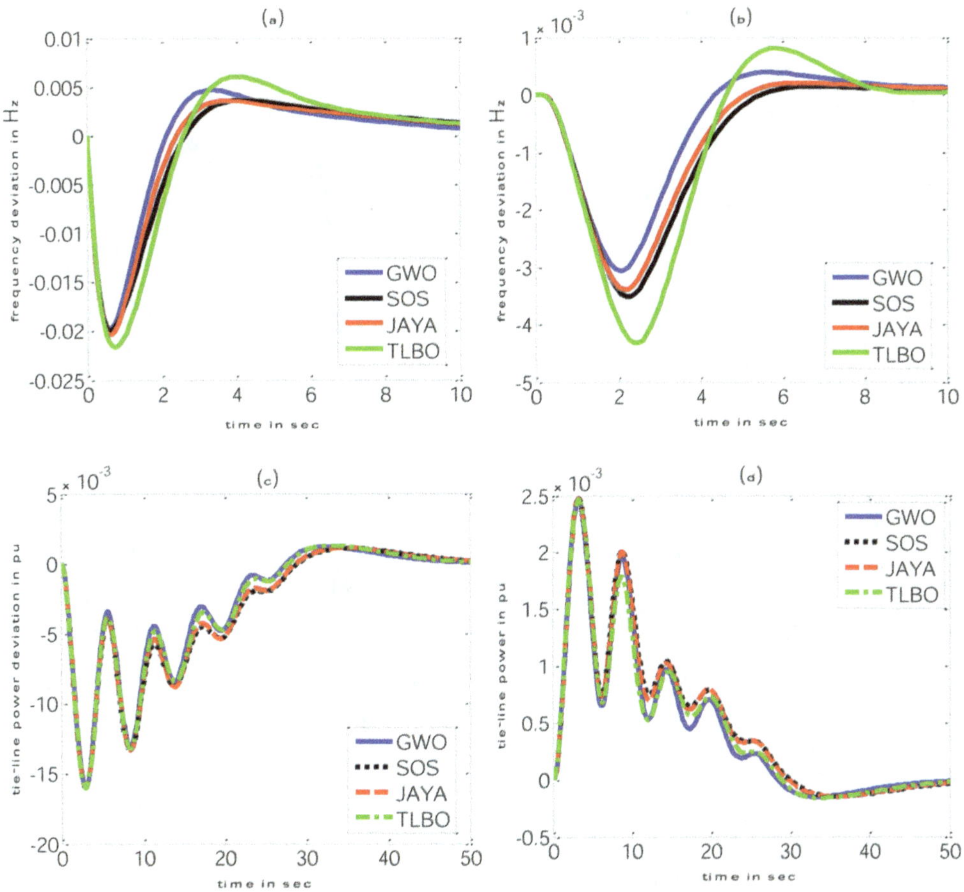

Table 9. Comparative study of peak overshoot and settling time with 2DOF-PID-controllers considering all system nonlinearities

System parameters	Peak overshoot (×10⁻⁴)				Settling time (in sec)			
	GWO	SOS	JAYA	TLBO	GWO	SOS	JAYA	TLBO
Δf_1	256	**217**	232	244	**30.85**	32.47	32.33	31.25
Δf_2	89	71	76	**67**	**31.58**	26.89	26.86	32.07
Δf_3	46	**31**	34	36	**26.92**	27.44	27.37	26.97
$\Delta P_{tie,12}$	12	**11**	11	13	**43.39**	47.16	46.34	44.87
$\Delta P_{tie,23}$	**26**	**26**	**26**	**26**	**42.49**	46.15	45.16	43.46
$\Delta P_{tie,31}$	**25**	**25**	**25**	**25**	**43.88**	47.78	46.92	45.68

Bold face show the best results

Figure 14. Dynamic performance of test system with all said nonlinearities, (a) change of frequency in area-1, (b) change of frequency in area-2, (c) change of tie-line power $\Delta P_{tie,12}$, (d) change of tie-line power $\Delta P_{tie,31}$

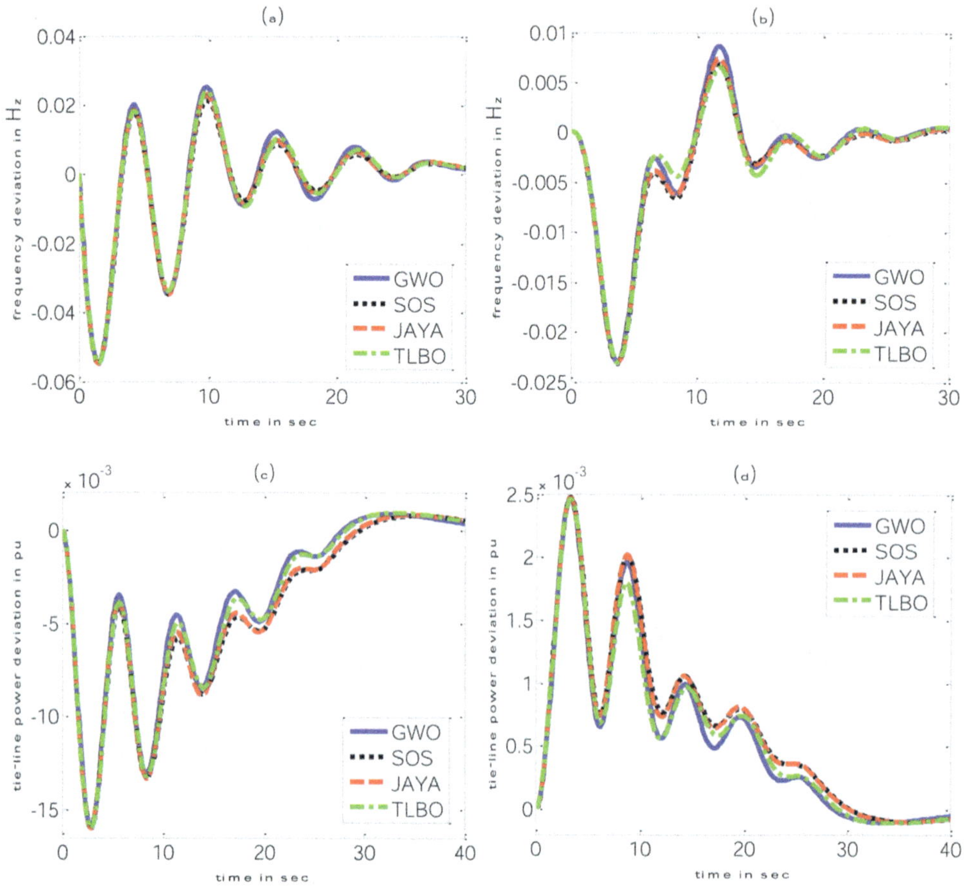

5.1.4. Transient Analysis With Facts Controller

Power system network is a large interconnected system which is often subjected to low frequency electromechanical oscillations. To ensure reliable power supply to customers, faster damping of these oscillations is highly demanded. The results presented in Section 5.1.3 clearly show that system oscillations and consequently settling time of frequency and tie-line power deviation are increased with nonlinearities. In addition, system oscillations are sustained with high peak overshoot and undershoot. To alleviate system oscillations, unified power flow controller (UPFC) has been designed and included in the AGC in coordination with 2DOF-PID controller.

UPFC is one of the powerful and advance FACTS controller used in power system to control both real and reactive power flow through tie-line. The UPFC is a combination of static synchronous compensator (STATCOM) and static synchronous series compensator (SSSC). These are connected via a common dc link to allow bidirectional power flow between STATCOM and SSSC (Hingorani & Gyugyi, 2000). Since AGC analysis is accomplished with small signal modelings, thus linear approximated model of

UPFC as shown in Figure 15 is considered to appraise its effects on system stability. The transfer function model of UPFC is defined in Equation (40).

$$G_{UPFC} = \underbrace{\left(\frac{K_{UPFC}}{1+sT_{UPFC}}\right)}_{UPFC \text{ gain \&time delay}} \underbrace{\left(\frac{sT_w}{1+sT_w}\right)}_{\text{Washout block}} \underbrace{\left(\frac{1+sT_1}{1+sT_2}\right)\left(\frac{1+sT_3}{1+sT_4}\right)}_{\text{2-stage lag-lead compensator}} \tag{40}$$

In Equation (40), washout block is basically representing a high pass filter. The washout block prevents the steady changes in speed with the steady change of frequency (Mondal et al., 2014). The value of T_w is taken as 10sec for present analysis.

The optimum values of coordinated 2DOF-PID and UPFC gains after optimization are tabulated in Table 10. Changes of frequency and tie-line power after 1% SLP in area-1 is shown in Figure 16. The maximum overshoot, undershoot and settling time of frequency and tie-line power deviations are specified in Table 10. It is noteworthy from the results of Figure 16 and Table 10 that coordinated 2DOF-PID and UPFC actively diminish the power system oscillations by contributing faster compensation. The value of peak overshoot and undershoot of transient responses are considerably reduced with the designed coordinated controller.

6. CONCLUSION

In this chapter, the dynamic and static performances of an interconnected power system are appraised in the presence of system nonlinearities. The nonlinearities which are mostly encountered in the power system are firstly defined with their approximated linear model and afterward, their effects on the system dynamics have been closely inspected. Describing function, signal flow graph and Pade approximation methods are used to derive the rational model of the aforesaid nonlinearities. GWO, SOS, JAYA and TLBO algorithms are applied for fine tuning of controller parameters. An extensive comparative study is carried out to show the feasibility and efficacy of proposed algorithms. It is noteworthy from the results that system responses deteriorate and take more time to settle down after the load perturbation. However designed GWO-tuned 2DOF-PID controller is competent to cope up with these nonlinearities and provide better results compared to others. Finally to ennoble system stability, an optimized UPFC is designed to work in coordination with 2DOF-PID controller. The performance of GWO optimized coordinated 2DOF-PID and UPFC is found better against load disturbance when compared with GWO-tuned PID-controller.

Figure 15. Model of UPFC as frequency stabilizer

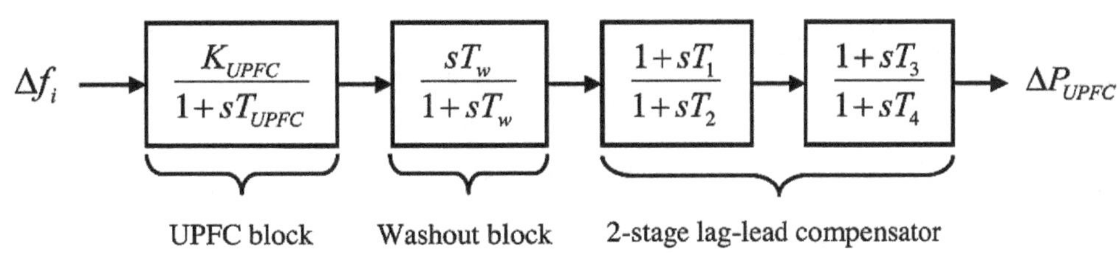

Figure 16. Dynamic performance of test system with all system nonlinearities and UPFC, (a) change of frequency in area-1, (b) change of frequency in area-2, (c) change of tie-line power $\Delta P_{tie,12}$, (d) change of tie-line power $\Delta P_{tie,31}$

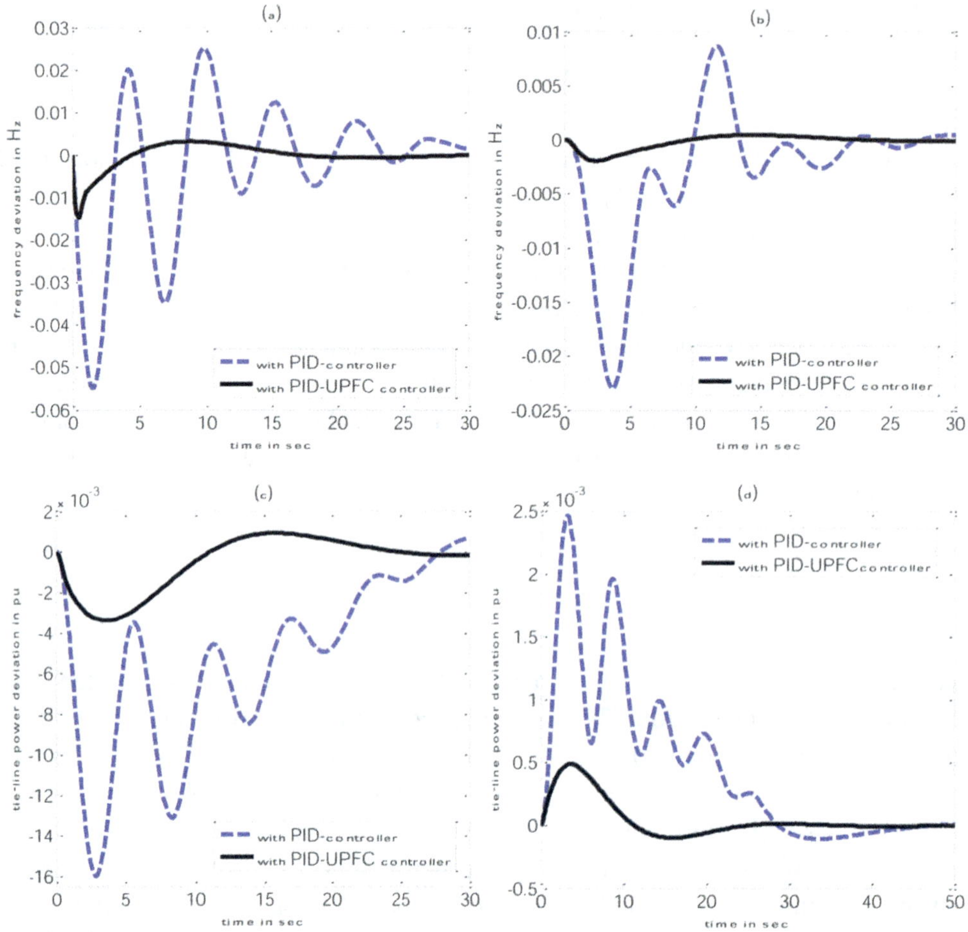

Table 10. Optimum settings of coordinated 2DOF-PID and UPFC controller

Controller gains	k_p	k_i	k_d	N	pw	dw	Fitness value
Area-1	0.9409	0.9713	0.9170	0.1068	0.4200	0.6577	0.1556
Area-2	0.4055	0.9209	0.4495	0.8677	0.6399	0.4186	
Area-3	0.8692	0.0887	0.2033	0.8289	0.7928	0.2008	
UPFC parameters	K_{UPFC}	T_{UPFC}	T_1	T_2	T_3	T_4	
	0.8352	0.4833	0.5630	0.2125	0.4079	0.6790	
Transient parameters		Δf_1	Δf_2	Δf_3	$\Delta P_{tie,12}$	$\Delta P_{tie,23}$	$\Delta P_{tie,31}$
Peak overshoot ($\times 10^{-4}$)		31	4.58	4.74	9.305	5.77	4.86
Undershoot ($\times 10^{-4}$)		149	19	16	34	2.18	0.960
Settling time (2% error band)		27.07	25.81	23.71	35.26	37.03	32.98

7. FUTURE SCOPE

Though the application of LFC is common in power system and a considerable amount of works are already available in the literature, still there are many areas open for research that may be considered as future extension of this present work. Some of them are listed below.

- The proposed techniques have the ability to search near the globally optimal solution. Nevertheless these algorithms may further be modified to produce probable global optimal solutions in reasonable time.
- In this chapter, single-objective evolutionary algorithms have been reviewed and implemented to enhance the power system performance. Study of AGC with multi-objective evolutionary algorithms will be the future extension of this current work.
- The future study of AGC may include more system nonlinearities. Advance FACTS devices like interline power flow controller (IPFC), thyristor-switched series reactor (TSSR) etc. can be implemented in the AGC for further improvement of existing results.

REFERENCES

Abd-Elazim, S. M., & Ali, E. S. (2016). Load frequency controller design via BAT algorithm for nonlinear interconnected power system. *International Journal of Electrical Power & Energy Systems*, *77*, 166–177. doi:10.1016/j.ijepes.2015.11.029

Abdelaziz, A. Y., & Ali, E. S. (2015). Cuckoo Search algorithm based load frequency controller design for nonlinear interconnected power system. *International Journal of Electrical Power & Energy Systems*, *73*, 632–643. doi:10.1016/j.ijepes.2015.05.050

Aditya, S. K., & Das, D. (2000). Load-frequency control of an interconnected hydrothermal power system with new area control error considering battery energy storage facility. *International Journal of Energy Research*, *24*(6), 525–538. doi:10.1002/(SICI)1099-114X(200005)24:6<525::AID-ER6033.0.CO;2-6

Ahmadi, A., & Aldeen, M. (2017). Robust overlapping load frequency output feedback control of multi-area interconnected power systems. *International Journal of Electrical Power & Energy Systems*, *89*, 156–172. doi:10.1016/j.ijepes.2016.12.015

Ansari, M. M. T., & Velusami, S. (2010). DMLHFLC (Dual mode linguistic hedge fuzzy logic controller) for an isolated wind diesel hybrid power system with BES (battery energy storage) unit. *Energy*, *35*(9), 3827–3837. doi:10.1016/j.energy.2010.05.037

Arya, Y. (2017). AGC performance enrichment of multi-source hydrothermal gas power systems using new optimized FOFPID controller and redox flow batteries. *Energy*, *127*, 704–715. doi:10.1016/j.energy.2017.03.129

Arya, Y., & Kumar, N. (2016). Optimal AGC with redox flow batteries in multi-area restructured power systems. *Eng Sci Tech. International Journal (Toronto, Ont.)*, *19*(3), 1145–1159.

Bevrani, H. (2014). *Robust Power System Frequency Control* (2nd ed.). Springer. doi:10.1007/978-3-319-07278-4

Cheng, M. Y., & Prayogo, D. (2014). Symbiotic Organisms Search: A new metaheuristic optimization algorithm. *Computers & Structures*, *139*, 98–112. doi:10.1016/j.compstruc.2014.03.007

Cohn, N. (1956). Some aspects of tie-line bias control on interconnected power systems. *Transactions of the American Institute of Electrical Engineers*, (Part 3), 75, 1415–1436.

Dhillon, S. S., Lather, J. S., & Marwaha, S. (2016). Multi objective load frequency control using hybrid bacterial foraging and particle swarm optimized PI controller. *International Journal of Electrical Power & Energy Systems*, *79*, 196–209. doi:10.1016/j.ijepes.2016.01.012

Elgerd, O. I., & Fosha, C. E. (1970). Optimum megawatt-frequency control of multiarea electric energy systems. *IEEE Transactions on Power Apparatus and Systems*, *PAS-89*(4), 556–563. doi:10.1109/TPAS.1970.292602

Gorripotu, T. S., Sahu, R. K., & Panda, S. (2015). AGC of a multi-area power system under deregulated environment using redox flow batteries and interline power flow controller. *Eng Sci Tech. International Journal (Toronto, Ont.)*, *18*, 555–578.

Guha, D., Roy, P.K., & Banerjee, S. (2016a) Quasi-oppositional differential search algorithm applied to load frequency control. *Engg Science and Technology, an Int. J*, *19*, 1635-54.

Guha, D., Roy, P. K., & Banerjee, S. (2016b). Application of backtracking search algorithm in load frequency control of multi-area interconnected power system. *Ain Shams Engg J.* doi:10.1016/j.asej.2016.01.004

Guha, D., Roy, P. K., & Banerjee, S. (2017). Study of differential search algorithm based automatic generation control of an interconnected thermal-thermal system with governor dead-band. *Applied Soft Computing*, *52*, 160–175. doi:10.1016/j.asoc.2016.12.012

Hingorani, N. G., & Gyugyi, H. (2000). *Understanding FACTS: concepts and technology of flexible AC transmission systems*. New York: IEEE Press, Inc.

Kumar, I. P., & Kothari, D. P. (2005). Recent philosophies of AGC strategies in power system. *IEEE Transactions on Power Systems*, *20*(1), 345–357.

Kumar, N. J. V., & Ansari, M. M. T. (2015). A new design of dual-mode type-II fuzzy logic load frequency controller for interconnected power systems with parallel AC/DC tie-lines and superconducting magnetic energy storage unit. *Energy*, *89*, 118–137. doi:10.1016/j.energy.2015.07.056

Mirjalili, S., Mirjalili, S. M., & Lewis, A. (2014). Grey wolf optimizer. *Advances in Engineering Software*, *69*, 46–61. doi:10.1016/j.advengsoft.2013.12.007

Mondal, D., Chakrobarty, A., & Sengupta, A. (2014). Power System small signals stability and Control. Academic Press.

Naidu, K., Mokhlis, H., Bakar, A. H. A., & Terzija, V. (2017). Performance investigation of ABC algorithm in multi-area power system with multiple interconnected generators. *Applied Soft Computing*, *57*, 436–451. doi:10.1016/j.asoc.2017.03.044

Padhan, S., Sahu, R. K., & Panda, S. (2014). Application of Firefly Algorithm for Load Frequency Control of Multi-area Interconnected Power System. *Electric Power Components and Systems*, *42*(13), 1419–1430. doi:10.1080/15325008.2014.933372

Panda, S., Mohanty, B., & Hota, P. K. (2013). Hybrid BFOA-PSO algorithm for automatic generation control of linear and nonlinear interconnected power systems. *Applied Soft Computing*, *13*(12), 4718–4730. doi:10.1016/j.asoc.2013.07.021

Rahman, A., Saikia, L. C., & Sinha, N. (2015). Load frequency control of a hydro-thermal system under deregulated environment using biogeography-based optimised three degree-of-freedom integral-derivative controller. *IET Generation, Transmission & Distribution*, *9*(15), 2284–2293. doi:10.1049/iet-gtd.2015.0317

Raju, M., Saikia, L. C., & Sinha, N. (2016). Automatic generation control of a multi-area system using ant lion optimizer algorithm based PID plus second order derivative controller. *International Journal of Electrical Power & Energy Systems*, *80*, 52–63. doi:10.1016/j.ijepes.2016.01.037

Rao, R. V. (2016). JAYA: A simple and new optimization algorithm for solving constrained and unconstrained optimization problems. *Int. J Industrial Eng. Compt.*, *7*, 1–16.

Rao, R. V., Savsani, V. J., & Vakharia, D. P. (2012). Teaching Learning Based Optimization: An optimization method for continuous non-linear large scale problems. *Information Science*, *183*(1), 1–15. doi:10.1016/j.ins.2011.08.006

Sahu, R. K., Panda, S., & Pradhan, P. C. (2015). Design and analysis of hybrid firefly algorithm-pattern search based fuzzy PID controller for LFC of multi area power systems. *International Journal of Electrical Power & Energy Systems*, *69*, 200–212. doi:10.1016/j.ijepes.2015.01.019

Sahu, R. K., Panda, S., Rout, R. K., & Sahu, D. K. (2016). Teaching learning based optimization algorithm for automatic generation control of linear and nonlinear interconnected power systems. *Applied Soft Computing*, *13*(12), 4718–4730.

Shankar, G., & Mukherjee, V. (2016). Load frequency control of an autonomous hybrid power system by quasi-oppositional harmony search algorithm. *International Journal of Electrical Power & Energy Systems*, *78*, 715–734. doi:10.1016/j.ijepes.2015.11.091

Shiva, C. K., & Mukherjee, V. (2015). A novel quasi-oppositional harmony search algorithm for automatic generation control of power system. *Applied Soft Computing*, *35*, 749–765. doi:10.1016/j.asoc.2015.05.054

Tan, W., Hao, Y., & Li, D. (2015). Load frequency control in deregulated environments via active disturbance rejection. *International Journal of Electrical Power & Energy Systems*, *66*, 166–177. doi:10.1016/j.ijepes.2014.10.036

Tripathy, S. C., Balasubramanian, R., & Nair, P. S. C. (1992). Effect of superconducting magnetic energy storage on automatic generation control considering governor dead band and boiler dynamics. *IEEE Transactions on Power Systems*, *7*(3), 1266–1273. doi:10.1109/59.207343

Zare, K., Hagh, M. T., & Morsali, J. (2015). Effective oscillation damping of an interconnected multi-source power system with automatic generation control and TCSC. *International Journal of Electrical Power & Energy Systems*, *65*, 220–230. doi:10.1016/j.ijepes.2014.10.009

KEY TERMS AND DEFINITIONS

Describing Function: Describing function approach was coined by M. F. Kryloy and N. Bogoliubov in the year 1930 and then forwarded by R. Kochenburgen. Describing function method is based on first harmonic approximation and only valid if the linear part of the separable system possesses low pass filter characteristic.

Dead-Band: Dead-band is defined as a total magnitude of sustained speed change within which there is no resulting change in the valve position. Dead-band possesses non-minimum phase behavior and thereby causes destabilizing effect to the system.

Load Frequency Control: Load frequency control (LFC) is a system to maintain reasonably uniform frequency, to divide the load between the generators, and to control the tie-line interchange schedules.

Optimization: The word optimization is originated from Latin word *optimus*, meaning "the best or most favorable point." Optimization is the process of making something better. In other words, optimization refers to the process of exploring the best possible solutions for a given problem.

Robustness: Robustness is the ability of a closed loop system to be insensitive while system parameters are varied over a wide range. Trade-offs between robustness and speed of response is a key issue for control system design.

Sensitivity: In control system application, the controller parameters are selected to match the process characteristics. However, in the due course of disturbance, process may changed. Therefore, controller parameters must be selected such that closed-loop response is less sensitive to these disturbances. Lower the sensitivity factor, higher the degree of relative stability of the system.

Time Moments: Time moments are defined as the coefficients of the series when the transfer function is expanded as series at the nearby of $s = 0$. If the time moments of two LTI-systems are close, their low frequencies are matched.

Two Degrees of Freedom: Degree of freedom (DOF) in control system is defined as number of closed loop transfer function that can be handled independently. It has two main objectives (i.e., set point tracking and disturbance rejection). 2DOF system mainly has two components, namely serial or main compensator and feed-forward controller.

Chapter 4
A Novel Binary Whale Optimization Algorithm-Based Optimal Placement of Phasor Measurement Units

Tapan Prakash
National Institute of Technology Raipur, India

Vinay Pratap Singh
National Institute of Technology Raipur, India

Soumya Ranjan Mohanty
Motilal Nehru National Institute of Technology Allahabad, India

ABSTRACT

Wide-area measurement system (WAMS) is an important part of present power system structure as it provides real-time synchronized measurements of the system with the aid of phasor measurement units (PMUs). Due to economic considerations, PMUs should be installed at optimal locations. The optimal placement of PMUs (OPP) is a problem of optimally placing PMUs at strategic locations maintaining the full observability of the system. In this chapter, a novel binary whale optimization algorithm (BWOA) is applied to solve OPP problem. The maximization of measurement redundancy is considered in the objective function. The proposed algorithm is examined on five different test systems operating under normal operating conditions with or without inclusion of zero-injection buses (ZIBs) and compared with the reports available in literature. The results show the effectiveness of the proposed algorithm in solving OPPP.

DOI: 10.4018/978-1-5225-3935-3.ch004

INTRODUCTION

There is tremendous advancement in communication technology in past decades. With these advancements, their applicability to various fields of sciences and engineering is increasing day-by-day. The power system is also under the influence of the advanced communication technologies. Presently, the power system is undergoing vast changes in terms of its expansion, interconnections of various areas and inclusion of advanced communication technologies. The primary reason behind this changing power system is fast growing power needs. Owing to these increasing power demands, the system experiences stressed conditions and the security and reliability of power supply is endangered (Kundur, Balu, & Lauby, 1994). Thus, the secure and reliable operation of the power system is turned into a major concern of electric utility. The security and reliability of power supply can be enhanced if it is possible to monitor the whole power system on real-time platform.

Real-time monitoring of the system in present scenario is very much possible with the aid of advanced communication infrastructure. Wide-area measurement system (WAMS) is one such system of advanced communication infrastructure which is used in monitoring of the power system. Phasor measurement units (PMUs) are vital part of WAMS. PMUs are the advanced measuring devices which take synchronized measurements of the system with the help of global positioning system (GPS) (Phadke & Thorp, 2008). They are installed at respective buses of the system to measure voltage phasors of the buses and the current phasors of the branches connected to those buses.

It is a well-known fact that when a PMU is installed at one of the buses of the system under consideration then, it will measure the synchronized voltage phasor of the bus with the synchronized measurement of current phasors of all the lines connected to the bus. As a result, the bus can be termed as directly observable in case stated above. However, the voltage phasor of the bus, connected with the line whose current phasor is known from other PMU measurement, can be computed easily. In this case, the bus is termed as indirectly observable. From the above discussion it can be concluded that a bus is observable if a PMU is installed on it or it is connected to a line whose current phasor and the voltage phasor of the bus present to the other side of the line are known from PMU measurements. This implies that the number of PMUs to be installed in a power system to make it fully observable can be reduced by choosing strategic locations for their placement. Thus, this problem of reducing the number of PMUs in a system provides a platform for a binary optimization problem and termed as optimal placement of PMU problem (OPPP). The main objective of OPPP is to reduce the number of PMUs in a system by placing them at strategic locations such that the system is fully observable.

Several attempts are reported in literature to solve OPPP. The solution methods adopted to solve OPPP can be categorized in two categories: conventional methods and meta-heuristic algorithm based methods. Conventional methods include linear programming, quadratic programming, mixed programming etc. However, the disadvantages associated with these methods like local minima trapping and slow convergence rate made a way for use of more efficient algorithms in solving OPPP. The evolution of meta-heuristic algorithms eradicated the disadvantages associated with conventional methods and prompted their application in solving OPPP. Some of the relevant reports are discussed in next section. However, there is always a scope of finding newer methods to solve the problem. In this work, a novel binary whale optimization algorithm (BWOA) is proposed and applied to solve OPPP. Whale optimization algorithm (WOA) is a recent population based optimization algorithm available in the literature and the application of the binary version of the algorithm is investigated in solving OPPP. The prime objectives of this work are outlined as below.

- Application of novel BWOA to solving OPPP
- Examining the effectiveness of BWOA by applying it on different standard test systems under different scenarios.
- Increasing the measurement redundancy of the whole system so as to maintain full observability
- Testing the superiority of BWOA by comparing it with results reported in literature.

To ensure full observability, the objective function is formulated considering minimum number of PMUs with increased measurement redundancy. Cases of inclusion of zero-injection buses (ZIBs) are included in solving OPPP.

BACKGROUND

During last two decades, the power system around the world is undergoing tremendous change due to fast growing power needs. Advanced communication technologies including WAMS are transforming the complete conventional communication used by electric utility in earlier stages. However, complete transformation of communication system with WAMS is not economically viable. Hence, WAMS are implemented in existing system in stage-by-stage manner. This prompted the need of installing minimum number of PMUs in the system so as to make it fully observable rather than installing PMUs at all the buses present in the system. The above stated problem is termed as OPPP. Different objectives can be considered in an OPPP but basic considerations taken are the minimization of number of PMUs while maintaining full observability of the system and increasing measurement redundancy.

The solution methods of OPPP include both conventional and heuristic methods. During initial years after the evolution of PMUs, OPPP is solved by many conventional methods. In (Gou, 2008b), an binary version of integer linear programming (ILP) is used to solve OPPP considering cases with and without flow measurements and ZIBs. The results are obtained for IEEE 14-bus test system. The method is proved to be computationally efficient in solving OPPP for a system with conventional measurements and ZIBs. A generalized ILP approach is adopted to solve OPPP in (Gou, 2008a). In this work, several cases are considered including redundant placement of PMUs, complete observability and incomplete observability of considered system. The system with and without conventional measurements and ZIBs are considered. Three different standard test systems (14-bus, 30-bus and 57-bus) are considered for testing the proposed method. The obtained results show that the generalized ILP solves OPPP in an effective manner. Optimal multistage scheduling of PMU placement with an ILP approach is reported in (Dua, Dambhare, Gajbhiye, & Soman, 2008). In this work, zero-injection constraints are converted to linear constraints in a framework based on ILP approach. Bus observability index (BOI) and system observability redundancy index (SORI) are proposed in this work which help in improving quality of OPPP. The proposed algorithm is tested on three different standard test systems. In (Chakrabarti, Kyriakides, & Eliades, 2009), integer quadratic programming (IQP) is used to minimize the number of PMUs and maximize the measurement redundancy of the power system. This work shows the solution of OPPP in presence of conventional measurements considering contingent scenarios of loss of transmission line or PMU outage while preserving the complete observability of the system. Five different test systems are simulated with proposed method. IQP is applied to maintain topological observability of the power system in (Chakrabarti, Kyriakides, & Albu, 2009). In this work, the uncertainties associated with state variables are estimated using three different approaches i.e. classical uncertainty propagation theory, the random fuzzy variables and the Monte Carlo method. The effective performance of the proposed method under different approaches is validated on IEEE 14-bus test system. Integer programming (IP) method is applied to analyse observability of the system and placement of PMUs in a mixed measurement scenario (Xu & Abur, 2004). The proposed method is tested on three different standard test systems to show distinctive properties of the system in presence of conventional flows and injections.

There are several works based on the use of conventional techniques available in literature but they are prone to trappings into local minima and are often unable to handle large set of constraints. To overcome these disadvantages, various modern heuristic and meta-heuristic optimization algorithms are developed and successfully applied to solve the problem. Some of the relevant works towards this contribution based on the application of these algorithms are described below.

The application of genetic algorithm (GA) in (Marin, Garcia-Lagos, Joya, & Sandoval, 2003) is proposed to solve OPPP. GA is a popular optimization algorithm based on the evolution theory. In this work, the minimum number of PMUs required to make system completely observable is determined while establishing the relationship of necessary measurements of current phasors for each PMU. Four different standard test systems are considered to show the performance of the proposed method.

Particle swarm optimization (PSO) algorithm is applied to the problem in (Ahmadi, Alinejad-Beromi, & Moradi, 2011; Chakrabarti, Venayagamoorthy, & Kyriakides, 2008; M Hajian, Ranjbar, Amraee, & Shirani, 2007). In (Ahmadi, et al., 2011), binary PSO is used to solve OPPP in presence of mixed measurement set. The complete observability and maximum measurement redundancy are ensured while solving the problem. Topology-based algorithm is adopted while considering several factors such as presence of conventional measurements, inclusion of ZIBs and maintaining system topology. The verification of the proposed method is done by testing it on four different standard test systems. Binary PSO is also used to solve OPPP in (Chakrabarti, et al., 2008) where the minimization of total number of PMUs and maximization of total measurement redundancy are considered as objectives. In (M Hajian, et al., 2007), topological rules are adopted for observability assessment while solving OPPP. A modified binary PSO algorithm is applied to achieve the desired results.

Non-dominated sorting genetic algorithm (NSGA) (Milosevic & Begovic, 2003) is used to optimize two conflicting objectives of minimization of total number of PMUs and maximization of measurement redundancy, simultaneously. The proposed method is applied to find out Pareto-optimal solutions. The algorithm is combined to graph-theoretic approach and simple GA to initially reduce the number of PMUs in the system. The method is tested on standard test systems of 39-bus and 118-bus to verify its effectiveness.

Simulated annealing (SA) algorithm is proposed to optimally place PMUs in the presence of conventional measurements in a power grid in (Gopakumar, Reddy, & Mohanta, 2015). A pragmatic multistage SA (PMSSA) is proposed in this work. The proposed algorithm is found to be much faster than the conventional SA. The efficacy of the proposed algorithm is tested on different standard test systems and Indian power systems.

Tabu search (TS) algorithm is used to solve OPPP in (Koutsoukis, Manousakis, Georgilakis, & Korres, 2013). In this work, recursive TS (RTS) algorithm is applied to numerical observability method for solving OPPP. RTS is proved efficient than TS in obtaining best results. The test of the algorithm is undertaken on several standard test systems ranging from small to large systems.

Differential evolution (DE) algorithm is a very popular meta-heuristic technique. Several reports can be found on the application of DE algorithm in solving OPPP (Al-Mohammed, Abido, & Mansour, 2011; Ketabi, Nosratabadi, & Sheibani, 2010; Peng, Sun, & Guo, 2010). In (Al-Mohammed, et al., 2011),

OPPP is solved using DE algorithm while maintaining complete observability of the system. The problem of installing new PMUs in the system along with retention or relocation of existing PMUs is considered in the objective function formulation. The effectiveness of the proposed method is proved by testing it on standard test systems. The work proposed in (Ketabi, et al., 2010) uses DE to solve OPPP by minimizing mean-square error (MSE) obtained from state estimation. The proposed method is applied on IEEE 14-bus test system considering the presence and absence of conventional measurements. In (Peng, et al., 2010), a multi-objective formulation for OPPP is adopted considering minimization of number of PMUs and maximization of $N-1$ measurement reliability. A non-dominated sorting DE is proposed to accomplish the task.

Artificial bee colony (ABC) algorithm is applied in (Kulanthaisamy, Vairamani, Karunamurthi, & Koodalsamy, 2014) to solve multi-objective PMU placement problem. A binary coded ABC is proposed to obtain the results. The objectives under consideration are minimization of total number of PMUs and maximization of measurement redundancy. Loss of line scenario is also considered in the problem. The proposed algorithm is tested on standard test systems.

Cuckoo search (CS) algorithm is presented in (Dalali & Karegar, 2016) to solve OPPP considering topological approach. A modified binary CS algorithm is proposed and applied to the problem. The algorithm is tested on standard test systems under different cases of normal operation and single loss of PMU and line outages. Two large test systems are also considered to test the efficiency of the proposed algorithm.

Bacteria foraging optimization algorithm (BFOA) is used to solve OPPP in (Mazlumi & Vahedi, 2010). While solving the problem, the complete observability of the system is maintained. Redundancy measurement analysis is also considered in the problem. The superiority of the proposed algorithm is demonstrated by applying it on IEEE 14-bus and Tehran 431-bus test system.

In (Basetti & Chandel, 2017), bat algorithm (BA) is presented to solve OPPP. This work proposed a novel Taguchi binary BA (TBBA) which is based on combination of reasoning ability of Taguchi method with conventional binary BA. The proposed algorithm is tested on different standard systems under different operating conditions. It is shown that the proposed algorithm is producing better results in comparison to earlier works.

The binary version of imperialistic competition algorithm (ICA) is presented in (Mahari & Seyedi, 2013) to solve OPPP. In this work, the complete observability of the system is maintained with minimum number of PMUs under diverse operating conditions. The maximization of measurement redundancy is also considered. Different contingent scenarios like single line outage and single PMU outage are considered in the problem. A new rule is framed to reduce the number of PMUs in order to maintain complete observability of the system.

A binary teaching learning based optimization (TLBO) algorithm is used to solve OPPP in (Recioui, Bentarzi, & Ouadi, 2015). The complete network observability is maintained with minimization of number of PMUs. Standard test systems are considered to verify the effectiveness of the proposed algorithm. An updated review on the application of these meta-heuristic techniques can be found in (Nazari-Heris & Mohammadi-Ivatloo, 2015).

Apart from the relevant works stated above regarding solution of OPPP, some other recent reports too appeared. In (Guo, Liao, & Chu, 2016), OPPP is formulated in a MILP framework with consideration of depth of observability propagation. A set of inequality constraints is formed by exchanging the information of interactions between PMUs to be placed and already placed PMUs. In this formulation, the inclusion of ZIBs, super ZIBs, flow measurements and power measurement injections are considered for both single stage and multi stage placement. An integrated model approach considering ZIBs and conventional measurements is proposed for OPPP in (Khajeh, Bashar, Rad, & Gharehpetian, 2017). The proposed approach studies the effect of limitations caused by ZIBs and conventional measurements on OPPP under normal and contingent operating conditions. In (Maji & Acharjee, 2017), multiple solutions of OPPP is presented with priority given to critical buses, generator buses and the buses which are likely to be used in future extension. An exponential binary PSO algorithm is used in this work for solving the formulated problem. In (Rahman & Zobaa, 2017), a modified version of BPSO is used to solve OPPP in order to attain complete observability and maximized measurement redundancy.

In most of the OPPP solution techniques stated above, the main objectives on which the solution are concentrated are the minimization of number of PMUs and maximization of measurement redundancy, both under different conditions of inclusion of ZIBs, inclusion of conventional measurements and outages of PMUs. Owing to the fast and continuous development of different optimization algorithms, the application of newer and better algorithms to solve OPPP must exist. In next section, the problem statement with its formulation is discussed.

PROBLEM STATEMENT

In this section, the basic OPP problem statement is discussed followed by the definition of measurement redundancy. The modified objective function is formulated with the inclusion of maximization of measurement redundancy in this section. Last part of the section deals with the definition of ZIBs and how they affect OPPP.

Basic Problem Statement

The prime objective of OPPP is to place PMUs at strategic locations in a given network such that the total counts of PMUs are minimum while the system remains in fully observable state. Considering the total number of buses in the system to be NB, the ith decision vector P_i is defined as

$$P_i = \begin{cases} 1 & \text{if } i\text{th bus is equipped with a PMU} \\ 0 & \text{otherwise} \end{cases} \tag{1}$$

Basically, the objective function of OPPP is represented as

$$\min \sum_{i}^{NB} P_i, i = 1, 2, \ldots, NB \tag{2}$$

such that

$$O(P) \geq b \tag{3}$$

where b is a unit vector of length equal to NB and is expressed as

$$b = \begin{bmatrix} 1 & 1 & \dots & 1 \end{bmatrix}^T \tag{4}$$

$O(P)$ is the observability constraint vector computed in the following manner

$$O(P) = \begin{cases} \text{non-zero} & \text{if the bus is in observable state w.r.t. decision vector} \\ 0 & \text{otherwise} \end{cases} \tag{5}$$

Thus, it can be concluded that full observability of the network is maintained by the constraint vector $O(P)$. The formation of constraint vector $O(P)$ depends upon a binary connectivity matrix C which contains the information regarding the connectivity of the buses in the network. It is computed as

$$C_{i,j} = \begin{cases} 1 & \text{if } i = j \\ 1 & \text{if bus } i \text{ is connected to bus } j \\ 0 & \text{otherwise} \end{cases} \tag{6}$$

The observability constraint $O(P)$ is expressed in terms of decision vector P and connectivity matrix C in the following manner

$$O(P) = CP \geq b \tag{7}$$

The corresponding entry of vector $O(P)$ at the ith bus of the network is calculated as

$$o_i = c_{i,1}P_1 + \dots + c_{i,i}P_i + \dots c_{i,NB}P_{NB} \tag{8}$$

where o_i is the ith entry of constraint vector $O(P)$.

From (8), it can be inferred that if the value of $c_{i,i}$ is equal to zero then, the output of $c_{i,i}P_i$ will be equal to zero and as a result, it will not appear in (8). However, if the value of P_i is non-zero with a non-zero value of $c_{i,i}$ then, the output of their product will be having a non-zero value. This implies that the respective bus is observable. Further, if all the values of o_i come out to be non-zero, then the network will be in a state of full observability.

Measurement Redundancy

Consideration of measurement redundancy is very important while solving an OPPP. Measurement redundancy M_R is defined as the total counts of each bus being observable by PMUs either directly or indirectly. From the above definition, it can be inferred that in order to ensure full observability of the network, the value of M_R should remain at least one. However, the higher values of M_R is always preferred as it leads the network to attain full observability. This necessitates the inclusion of maximization of M_R into objective function formulated in OPPP. However, for the ease of calculation, the maximization of M_R statement is converted into a minimization statement as suggested by (Enshaee, Hooshmand, & Fesharaki, 2012). The modified objective function after the inclusion of above suggestion in (2) is represented below.

$$\min\left(\sum_{i=1}^{NB} P_i\right) + w\left(\sum_{i=1}^{NB} M_{Ri}^{Ideal} - \sum_{i=1}^{NB} M_{Ri}^{Actual}\right) \tag{9}$$

where M_{Ri}^{Ideal} and M_{Ri}^{Actual} are the measurement redundancy of ith bus in ideal and actual conditions, respectively. w is termed as the scaling factor which is adjustably selected to make two parts of (9) comparable in magnitude. The first part of (9) is the basic objective function of an OPPP while the second part of (9) is calculating the difference of the measurement redundancy in ideal and actual conditions. This implies that if the difference of second part of (9) is minimized then, measurement redundancy in actual conditions should have to increase and, thus, the objective of maximization of measurement redundancy of whole network is fulfilled as it is maximized. The constraints associated with the modified objective function remain same as stated in (3), (5), (6), (7) and (8).

Zero-Injection Buses

ZIBs are the buses which do not have any generation or load. That implies that the summation of current injection at these buses is equal to zero. Their consideration in the system holds an important role in minimization of number of PMUs. It provides an added advantage for solving OPPP as the consideration of ZIBs can reduce the number of PMUs in the system while maintaining the full system observability. Consider a 4-bus network with bus 1 as ZIB and rest buses (i.e. bus 2, bus 3 and bus 4) are connected to it through lines. If KCL is applied at bus 1, the following equation is resulted.

$$I_{21} + I_{31} + I_{41} = 0 \tag{10}$$

where I_{21}, I_{31} and I_{41} are the current injections from bus 2, bus 3 and bus 4 into bus 1, respectively.

In this work, objective function defined by (9) is used throughout with consideration of the system with and without inclusion of ZIBs. Different test systems are considered to establish the effectiveness of the proposed algorithm. In the next section, the proposed algorithm is discussed.

BINARY WHALE OPTIMIZATION ALGORITHM

In this section, a brief discussion of basic WOA is outlined first followed by description of BWOA. In the end, the implementation of BWOA to OPPP is discussed.

Basic Whale Optimization Algorithm (WOA)

WOA is a population based meta-heuristic optimization algorithm based on the bubble-net feeding behaviour of humpback whales. It is a recent algorithm proposed in 2016 (Mirjalili & Lewis, 2016) and is applied to various optimization problems (Aljarah, Faris, & Mirjalili, 2016; Cherukuri & Rayapudi, 2016; Kaveh & Ghazaan, 2016; Prakash & Lakshminarayana, 2016; Touma, 2016). It involves three basic phases of encircling prey, spiral bubble-net feeding behaviour and searching for prey. The mathematical model of all the phases are described below.

Encircling Prey

It is considered that whales have the ability to locate and encircle the prey. The best candidate solution is regarded as the target prey and all whales try to update their positions towards the prey. Mathematically, this behaviour is represented in following manner

$$\vec{E} = \left|\vec{F} \cdot \vec{X}_n^{best} - \vec{X}_n\right| \tag{11}$$

$$\vec{X}_{n+1} = \vec{X}_n^{best} - \vec{B} \cdot \vec{E} \tag{12}$$

where $\vec{F}$ and $\vec{B}$ are coefficient vectors. $\vec{X}_n$ and $\vec{X}_{n+1}$ are the position vectors in $n\text{th}$ and $(n+1)\text{th}$ iterations, respectively. $\vec{X}_n^{best}$ is the best position vector in $n\text{th}$ iteration.

The vectors $\vec{B}$ and $\vec{F}$ are computed as

$$\vec{B} = 2\vec{b} \cdot \vec{\mu} - \vec{b} \tag{13}$$

$$\vec{E} = 2 \cdot \vec{\mu} \tag{14}$$

where $\vec{b}$ is linearly decreasing from 2 to 0 with each iteration and $\vec{\mu}$ is a vector of random numbers in the range $[0,1]$.

Bubble-Net Feeding Behaviour

This phase is based on two approaches namely, shrinking encircling mechanism and spiral updating position. In shrinking encircling mechanism, the feeding behaviour is attained by decreasing the value of $\vec{b}$ defined in (13). Since, $\vec{B}$ is dependent upon $\vec{b}$, so with the decrease in $\vec{b}$, $\vec{B}$ will decrease too. So, it can be said that $\vec{B}$ is a random value in the range $\left[-b,b\right]$. If the random values for $\vec{B}$ is set in the interval $\left[-1,1\right]$, then, the position of the new candidate solution can be defined in between the position of the current best solution and the original position.

In spiral updating positon phase, mimicry of helix-shaped movement of whales are modelled by a spiral mathematical equation and is represented in the following form

$$\vec{L} = \left|\vec{X}_n^{best} - \vec{X}_n\right| \tag{15}$$

$$\vec{X}_{n+1} = \vec{L} \cdot e^{al} \cdot \cos\left(2 \cdot \pi \cdot l\right) + \vec{X}_n^{best} \tag{16}$$

where $\vec{L}$ is the distance between the whale and the prey. a is a constant used for defining the shape of the spiral and l is any random number in the range $\left[-1,1\right]$.

Further, it is assumed that there is half probability of shrinking encircling mechanism and half of spiral updating position phase during the optimization process. If P is any random number between 0 and 1, then the updated position vector of the whales can be defined in the following manner.

$$\vec{X}_{n+1} = \begin{cases} \vec{X}_n^{best} - \vec{B} \cdot \vec{E} & \text{if } P < 0.5 \\ \vec{L} \cdot e^{al} \cdot \cos\left(2 \cdot \pi \cdot l\right) + \vec{X}_n^{best} & \text{if } P \geq 0.5 \end{cases} \tag{17}$$

The better solutions are moved to next phase.

Searching for Prey

In this phase, whales randomly search for the prey according to the position of each other. To implement this phase, the value of $\vec{B}$ is either taken greater than 1 or less than -1 so as to force the search agents to move farther within the search space. Thus, a global search is accomplished. It is represented in following manner

$$\vec{E} = \left|\vec{F} \cdot \vec{X}_n^{rand} - \vec{X}_n\right| \tag{18}$$

$$\vec{X}_{n+1} = \vec{X}_n^{rand} - \vec{B} \cdot \vec{E} \tag{19}$$

where $\vec{X}_n^{rand}$ is randomly chosen positon vector from the current population. The better solutions are selected and moved to take part in next iteration. Hence, the whole process is repeated till termination criterion is met.

The Proposed Algorithm

BWOA is the proposed binary version of basic WOA in which the updated position vector $\vec{X}_{n+1}$ is represented in binary form. The whole process of BWOA remains same as WOA except the binarization of initial and updated solution vector at the end of all phases of the algorithm. For any binary optimization problem, the initial population is generated randomly in the range $[0,1]$. The positon vector $\vec{X}$ is then converted into a binary vector in following manner.

$$\vec{X}_i = \begin{cases} 1 & \text{if } rand() \geq 0.5 \\ 0 & \text{otherwise} \end{cases} \tag{20}$$

where $\vec{X}_i$ is the $i\text{th}$ element of the respective position vector.

The updated position vector $\vec{X}_{n+1}$ obtained from (17) and (19) is converted into binary form by applying $\tanh$ transformation. It is represented as

$$\tanh\left(\left|\vec{X}_{n+1,i}\right|\right) = \frac{e^{\left(\left|2\vec{X}_{n+1,i}\right|\right)} - 1}{e^{\left(\left|2\vec{X}_{n+1,i}\right|\right)} + 1} \tag{21}$$

where $\vec{X}_{n+1,i}$ is the $i\text{th}$ element of the updated position vector.

The computed solution vector obtained using (21) is converted into binary form according to

$$\vec{X}_{n+1,i} = \begin{cases} 1 & \text{if } rand() < \tanh\left(\left|\vec{X}_{n,i}\right|\right) \\ 0 & \text{otherwise} \end{cases} \tag{22}$$

The better solutions are selected and the process is repeated until any termination criterion is met.

Implementation of BWOA to OPPP

The objective function for solving OPPP is defined in . Consider the number of solution vector to be N_{POP}. The total number of buses in the system is NB. The resultant size of the initial population would be $N_{POP} \times NB$. The involved steps of implementation of proposed BWOA to solution of OPPP are as follow.

Step 1: Randomly generate binarized initial population using (20).

Step 2: Evaluate objective function defined in (9) for each solution vector of the population.
Step 3: Identify best solution vector based on the evaluated objective function.
Step 4: Update each solution vector using (17).
Step 5: Convert the solution vector into binary form using (22).
Step 6: Select better solution from current and updated vectors.
Step 7: Update the solution vector using (19).
Step 8: Convert the solution vector into binary form using (22).
Step 9: Select better solution form current and updated vectors.
Step 10: Go to step 2 and repeat until any termination criterion is met.

RESULTS AND DISCUSSION

The proposed algorithm is applied to solve OPPP considering the minimum number of PMUs placement at strategic locations such that the full observability of the system is maintained and the measurement redundancy is maximized. To demonstrate the effectiveness of the proposed algorithm, it is applied to different standard test systems ranging from small power system to large power system. The standard test systems considered in this work are IEEE 14-bus, IEEE 30-bus, NE 39-bus, IEEE 57-bus and IEEE 118-bus test systems. The scenario of including ZIBs is also considered. A comparative analysis of proposed algorithm is accomplished with results reported in literature for the system operating under normal operating conditions with exclusion and inclusion of ZIBs. The relevant literature with which the proposed algorithm is compared are (Ahmadi, et al., 2011; Al-Mohammed, et al., 2011; Bei, Yoon, & Abur, 2005; Chakrabarti & Kyriakides, 2008; Chakrabarti, Kyriakides, & Eliades, 2009; Gou, 2008a; Mahdi Hajian, Ranjbar, Amraee, & Mozafari, 2011; Hurtgen & Maun, 2010; Mohammadi-Ivatloo, 2009; Roy, Sinha, & Pradhan, 2012). A total of 20 solution vectors are taken as initial population whereas total number of 100 iterations is considered.

Table 1 lists the specifications of the standard test systems considered in this work. In this table, the total number of transmission lines connecting the buses of the system with the total count of presence of ZIBs and their respective locations are presented. From the table, it can be observed that for smaller power system the number of transmission lines and ZIBs are less as compared to the larger power system. This implies that the complexity of OPPP increases with the increase in the size of the power system.

Table 1. Specifications of the considered test systems

Test System	Number of Lines	Number of ZIBs	Locations of ZIBs
IEEE 14-bus	20	1	7
IEEE 30-bus	41	6	6, 9, 22, 25, 27, 28
NE 39-bus	46	12	1, 2, 5, 6, 9, 10, 11, 13, 14, 17, 19, 22
IEEE 57-bus	80	15	4, 7, 11, 21, 22, 24, 26, 34, 36, 37, 39, 40, 45, 46, 48
IEEE 118-bus	186	10	5, 9, 30, 37, 38, 63, 64, 68, 71, 81

Table 2. Optimal PMUs placement excluding consideration of ZIBs for system under normal operating conditions

Test System	Optimal Number of PMUs	Locations of PMUs	Maximum Measurement Redundancy
IEEE 14-bus	4	2, 6, 7, 9	19
IEEE 30-bus	10	1, 5, 8, 9, 10, 12, 15, 20, 25, 29	52
NE 39-bus	13	2, 6, 9, 10, 12, 14, 17, 19, 22, 23, 25, 29, 34	52
IEEE 57-bus	16	1, 6, 9, 15, 19, 22, 25, 28, 32, 36, 38, 41, 47, 51, 53, 57	72
IEEE 118-bus	32	3, 5, 9, 12, 15, 17, 21, 26, 23, 28, 30, 36, 40, 44, 46, 51, 54, 57, 62, 64, 68, 71, 75, 80, 85, 86, 91, 94, 101, 105, 110, 114	184

Figure 1. Single line diagram of IEEE 14-bus system with optimally placed PMUs considering exclusion of ZIBs under normal operating conditions

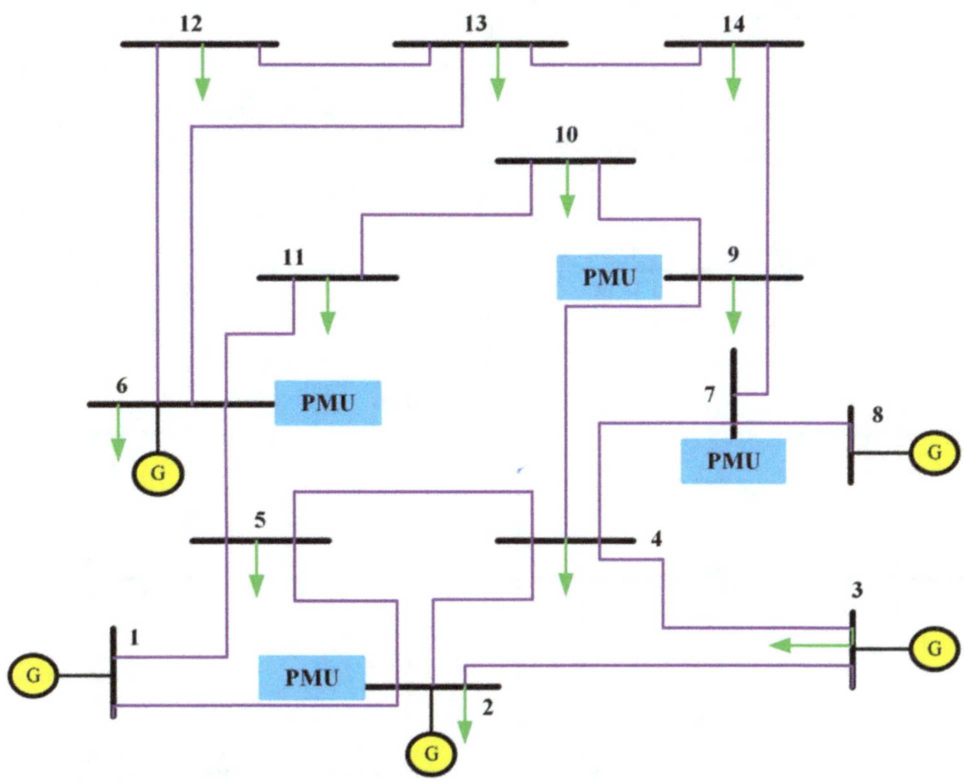

Table 2 presents the optimal number of PMUs required for maintaining full observability and maximized measurement redundancy for different test systems operating under normal operating conditions without including ZIBs obtained with proposed algorithm. The optimal number of PMUs required for IEEE 14-bus test system is equal to 4 and their respective locations are bus 2, bus 6, bus 7 and bus 9. The total measurement redundancy M_R is found to be 19. Figure 1 shows the single line diagram of

Figure 2. Single line diagram of IEEE 30-bus system with optimally placed PMUs considering exclusion of ZIBs under normal operating conditions

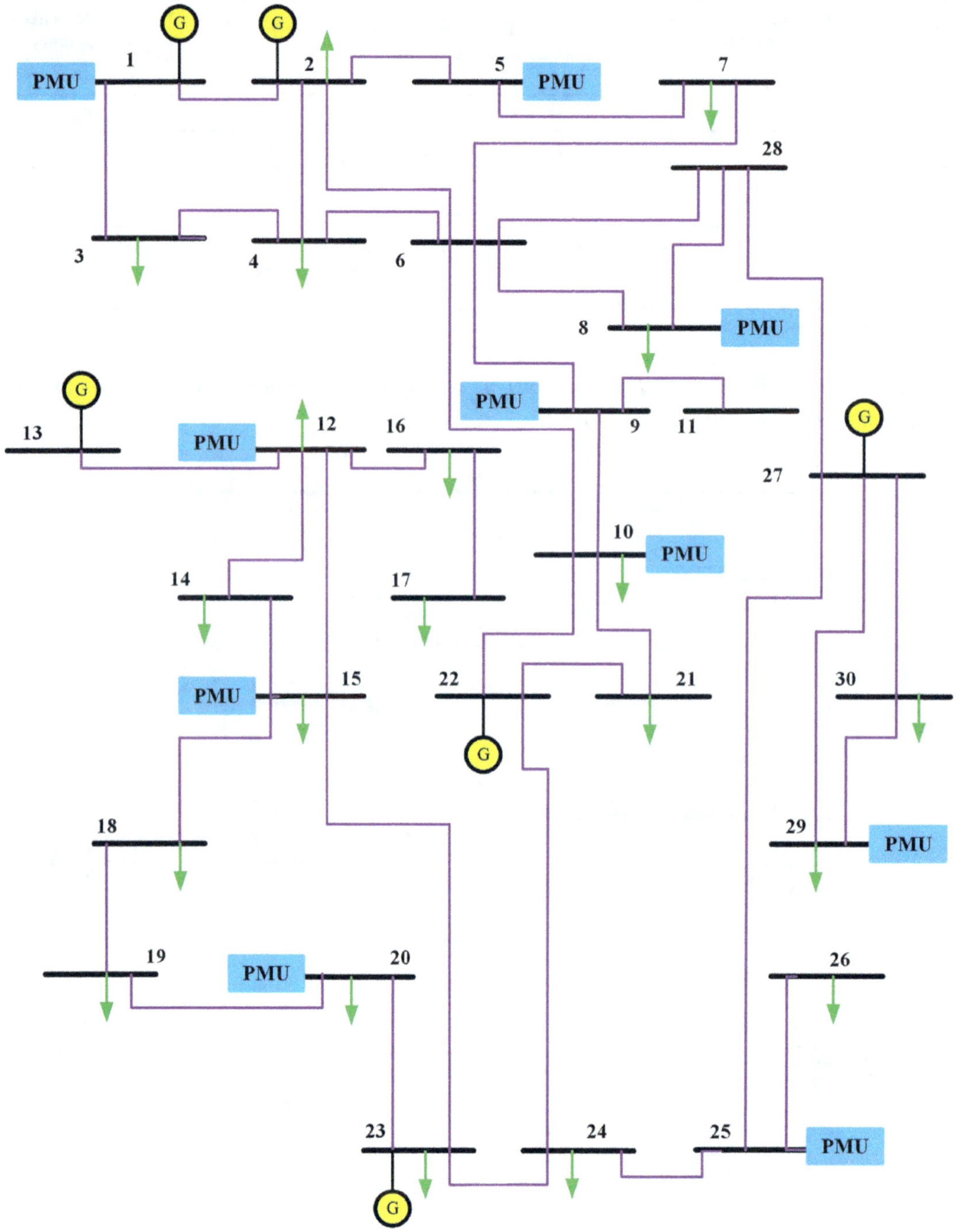

Figure 3. Single line diagram of NE 39-bus system with optimally placed PMUs considering exclusion of ZIBs under normal operating conditions

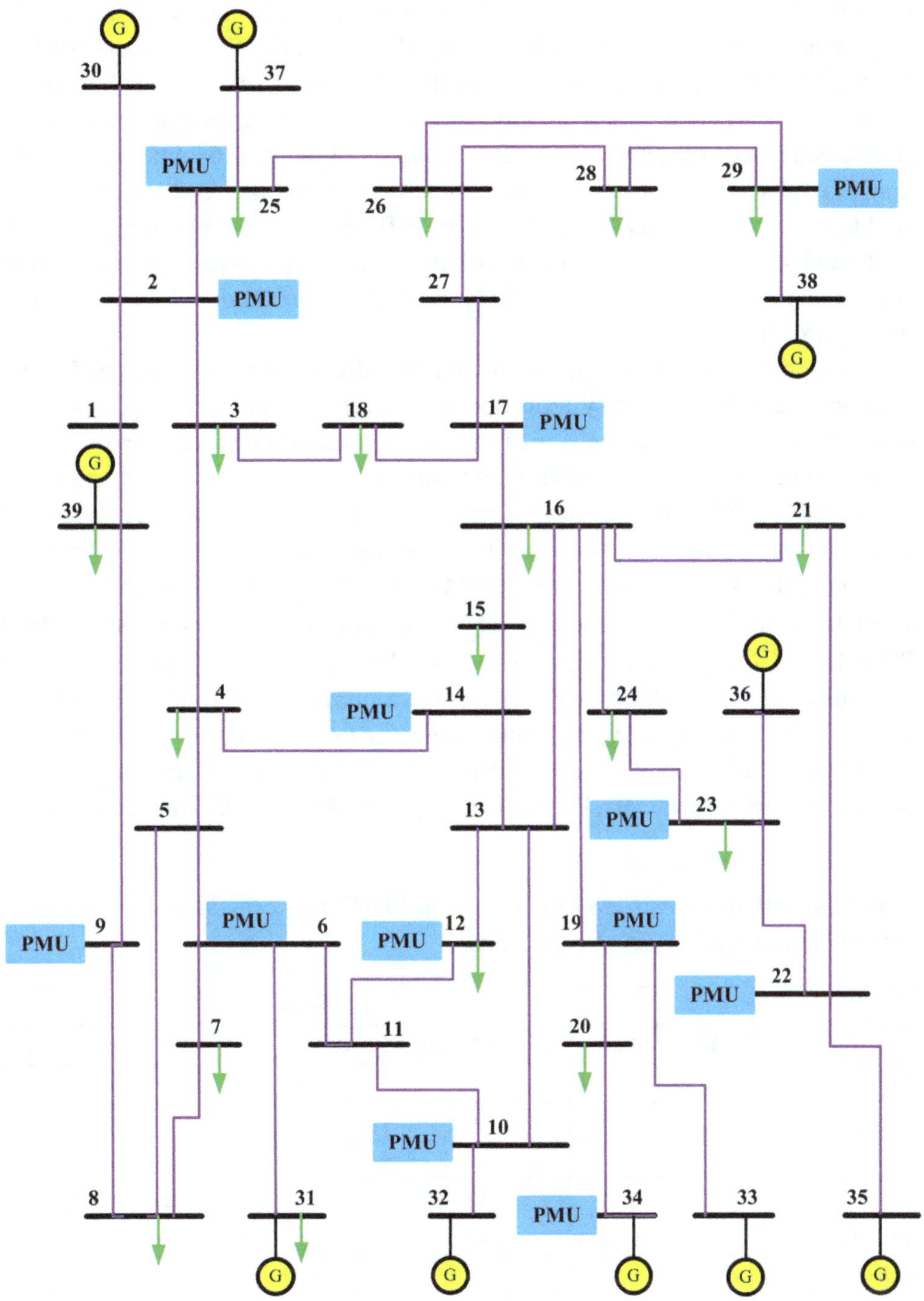

IEEE 14-bus test system with optimally placed PMUs for above case. For IEEE 30-bus, NE 39-bus, IEEE 57-bus and IEEE 118-bus test systems, the minimum number of PMUs required are 10, 13, 16 and 32, respectively. The respective locations of PMUs for the system can be seen in the Table 2. The maximum measurement redundancy for IEEE 30-bus, NE 39-bus, IEEE 57-bus and IEEE 118-bus are found to be 52, 52, 72, 184, respectively. Figure 2 and Figure 3 illustrate the single line diagram if IEEE 30-bus and NE 39-bus test systems with optimally placed PMUs without considering ZIBs under normal operating conditions, respectively. The comparative analysis of the results obtained with the proposed method is shown in Table 3. From the table, it can be observed that the required minimum number of PMUs obtained from all the methods for IEEE 14-bus, IEEE 30-bus, NE 39-bus and IEEE 118-bus are equal to 4, 10, 13 and 32, respectively. However, in case of IEEE 57-bus test system, the required number of PMUs is equal to 16 which is only obtained by proposed method and the method presented in (Mohammadi-Ivatloo, 2009).

Table 4 lists the optimal number of PMUs required for fully observable system and maximum measurement redundancy for different test systems operating under normal operating conditions including ZIBs obtained with proposed algorithm. As it is already discussed in earlier section that inclusion of ZIBs reduces the number of PMUs required for obtaining full system observability and the same can be validated from the Table 4. With the reduction of number of PMUs, there will be reduction in maximum measurement redundancy and it is proved from the results shown in Table 4. The optimal number of PMUs required for IEEE 14-bus test system is equal to 3 and their respective locations are bus 2, bus 6 and bus 9. The total measurement redundancy M_R is found to be 16. Figure 4 shows the single line diagram of IEEE 14-bus test system with optimally placed PMUs for above case. For IEEE 30-bus, NE 39-bus, IEEE 57-bus and IEEE 118-bus test systems, the minimum number of PMUs required are 7, 8, 11 and 28, respectively. The respective locations of PMUs for the system are listed in Table 4. The maximum measurement redundancy for IEEE 30-bus, NE 39-bus, IEEE 57-bus and IEEE 118-bus are found to be 41, 43, 59, 156, respectively. Figure 5 and Figure 6 illustrate the single line diagram of IEEE

Table 3. Comparison with reported literature for optimal PMUs placement excluding consideration of ZIBs for system under normal operating conditions

Method	Test System				
	IEEE 14-Bus	IEEE 30-Bus	NE 39-Bus	IEEE 57-Bus	IEEE 118-Bus
Ahmadi, et al., 2011	4	10	NA	17	32
Al-Mohammed, et al., 2011	4	10	13	17	NA
Bei, Yoon, & Abur, 2005	4	10	NA	17	32
Chakrabarti & Kyriakides, 2008	4	10	13	NA	NA
Hurtgen & Maun, 2010	4	NA	NA	17	32
Mohammadi-Ivatloo, 2009	4	10	13	16	32
Roy, Sinha, & Pradhan, 2012	4	10	13	17	32
Proposed method	4	10	13	16	32

'NA' indicates the data is not available.

Table 4. Optimal PMUs placement including consideration of ZIBs for system under normal operating conditions

Test System	Optimal Number of PMUs	Locations of PMUs	Maximum Measurement Redundancy
IEEE 14-bus	3	2, 6, 9	16
IEEE 30-bus	7	1, 7, 10, 12, 19, 24, 27	41
NE 39-bus	8	3, 8, 13, 16, 20, 23, 25, 29	43
IEEE 57-bus	11	1, 4, 13, 20, 25, 29, 32, 38, 51, 54, 56	59
IEEE 118-bus	28	2, 8, 11, 12, 17, 21, 25, 28, 33, 34, 40, 45, 49, 52, 56, 62, 72, 75, 77, 80, 85, 86, 90, 94, 101, 105, 110, 114	156

Figure 4. Single line diagram of IEEE 14-bus system with optimally placed PMUs considering inclusion of ZIBs under normal operating conditions

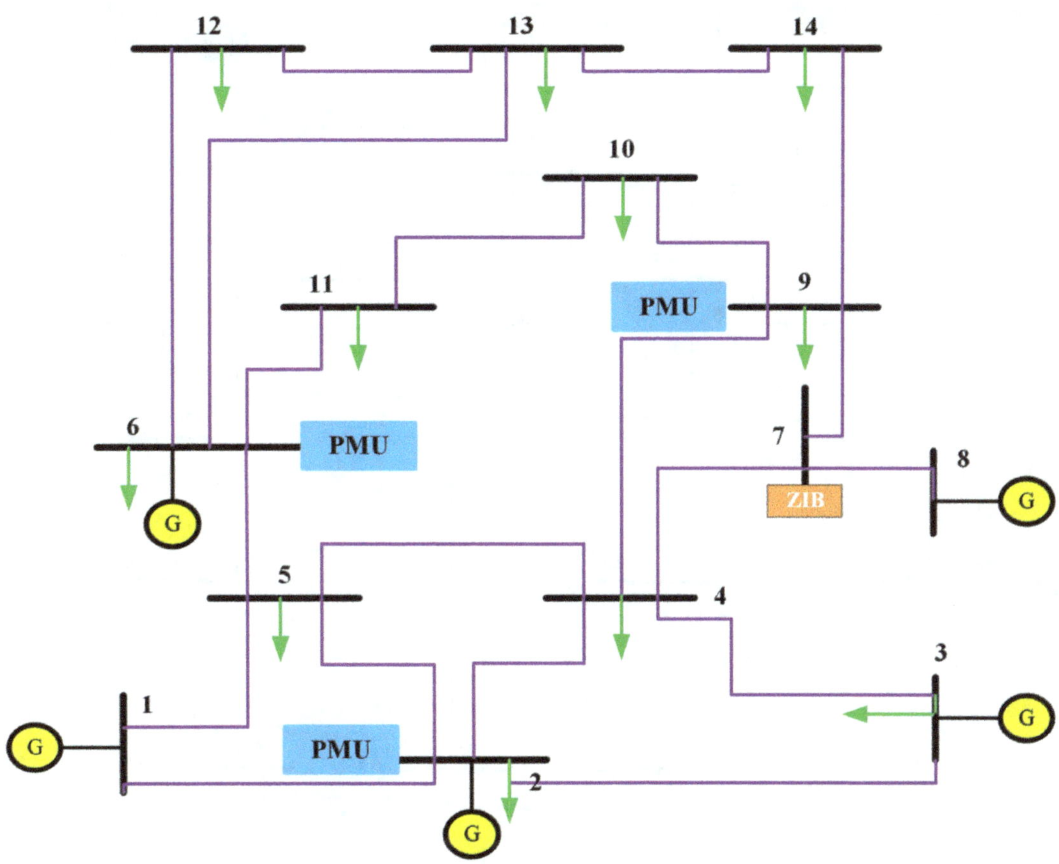

Figure 5. Single line diagram of IEEE 30-bus system with optimally placed PMUs considering inclusion of ZIBs under normal operating conditions

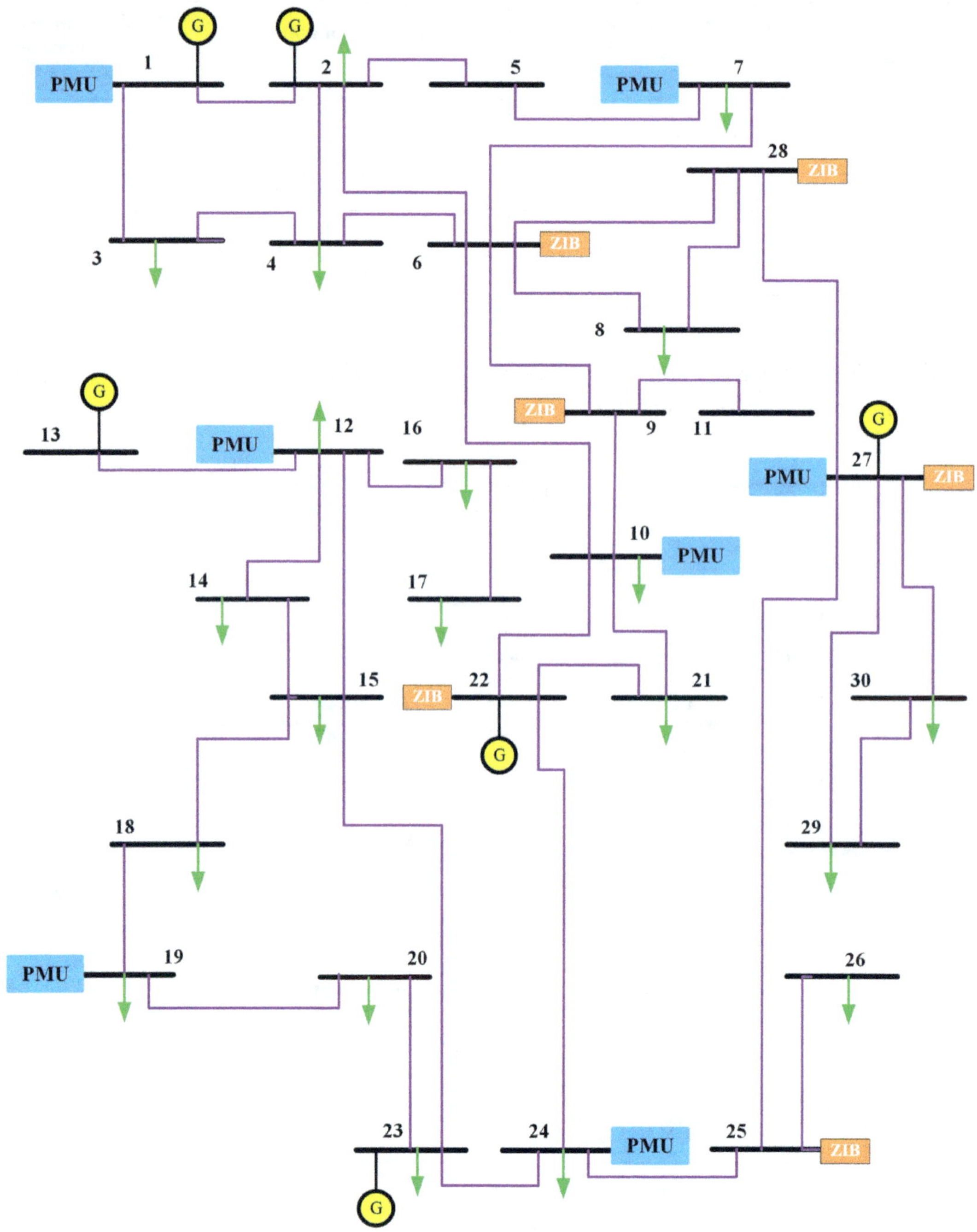

Figure 6. Single line diagram of NE 39-bus system with optimally placed PMUs considering inclusion of ZIBs under normal operating conditions

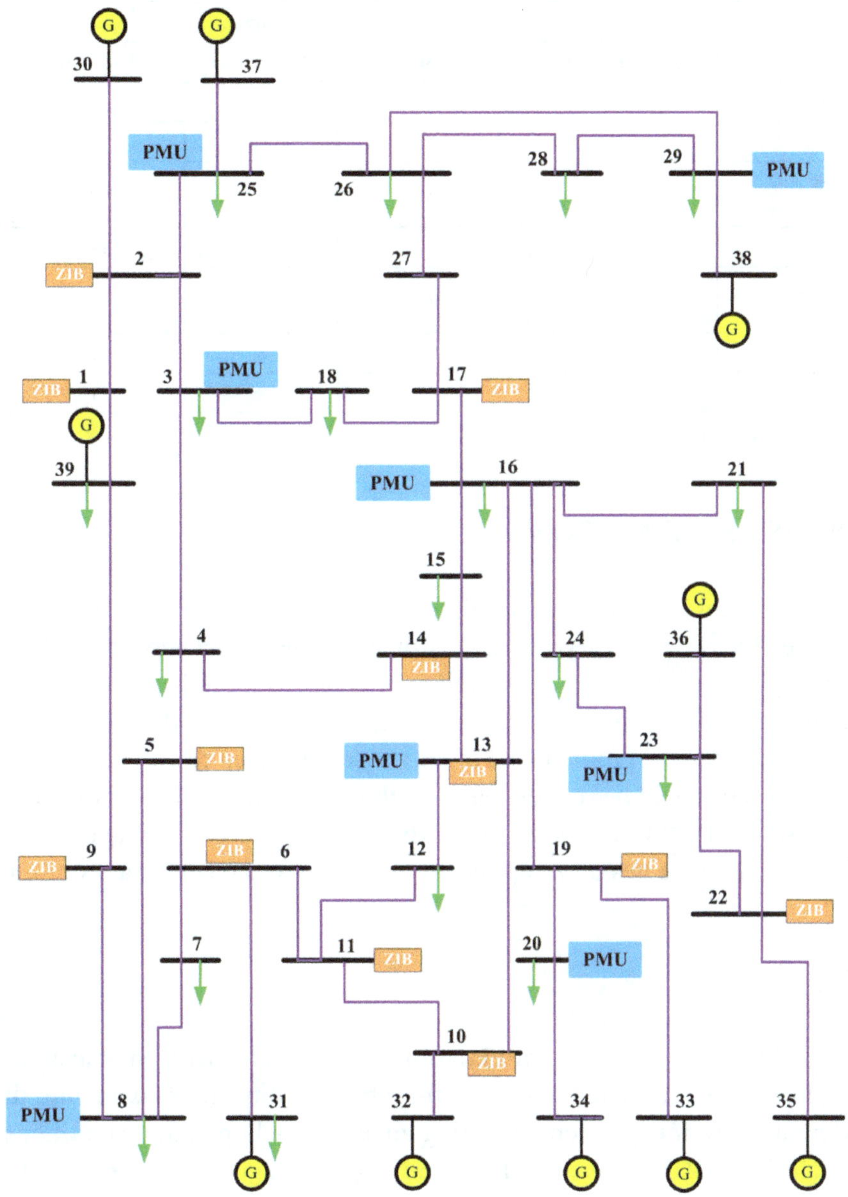

30-bus and NE 39-bus test systems with optimally placed PMUs considering ZIBs under normal operating conditions, respectively. The comparative analysis of the results obtained with the proposed method for the case where ZIBs are included in the system is shown in Table 5. From the table, it can be seen that the required minimum number of PMUs for IEEE 14-bus and IEEE 30-bus test systems are 3 and 7, respectively. However, for NE 39-bus, IEEE 57-bus and IEEE 118-bus test systems, the minimum number of PMUs required are equal to 8, 11 and 28, respectively. The proposed algorithm is found to be equally good in obtaining the minimum number of PMUs in above case.

Table 5. Comparison with reported literature for optimal PMUs placement including consideration of ZIBs for system under normal operating conditions

Method	Test System				
	IEEE 14-Bus	IEEE 30-Bus	NE 39-Bus	IEEE 57-Bus	IEEE 118-Bus
Ahmadi, et al., 2011	3	7	NA	13	29
Bei, Yoon, & Abur, 2005	3	7	NA	12	28
Chakrabarti, Kyriakides, & Eliades, 2009	3	7	8	NA	NA
Gou, 2008a	3	7	NA	11	NA
Mahdi Hajian, Ranjbar, Amraee, & Mozafari, 2011	3	7	8	11	28
Roy, Sinha, & Pradhan, 2012	3	7	8	11	28
Proposed method	3	7	8	11	28

'NA' indicates the data is not available.

FUTURE RESEARCH DIRECTIONS

In this work, OPPP problem is solved to find minimum number of PMUs in order to have full observability and maximized measurement redundancy with and without inclusion of ZIBs when the system is in normal operating conditions. Future research is possible with the inclusion of conventional flow measurements along with ZIBs. Further, contingency cases of loss of PMUs and outages of line can be considered in the PMU placement problem. All of the above suggested consideration will make an OPPP very complex problem and it will be interesting to analyse the performance of the proposed algorithm in such conditions. A novel BWOA is proposed in this work to solve OPPP and found to be an effective technique. The future scope lies in the application of BWOA to other binary optimization problem.

CONCLUSION

A novel application of BWOA to solve an OPPP problem is presented with an objective to minimize the number of PMUs in the network maintaining the full observability and maximizing the measurement redundancy. Different cases of the system operating under normal operating conditions are considered with or without inclusion of ZIBs in the problem. The proposed algorithm is examined on five different standard test systems to establish the effectiveness of the algorithm in solving the problem. The results obtained with the proposed algorithm suggested that it can be successfully applied to systems of different sizes and complexities. The comparative assessment with the reports available in the literature established that the proposed algorithm is equally capable or better in solving the problem. Further, it can be concluded that the proposed algorithm is very much competitive in nature in solving the problem and thus, can be attempted to be applied to solve other binary optimization problems.

REFERENCES

Ahmadi, A., Alinejad-Beromi, Y., & Moradi, M. (2011). Optimal PMU placement for power system observability using binary particle swarm optimization and considering measurement redundancy. *Expert Systems with Applications*, *38*(6), 7263–7269. doi:10.1016/j.eswa.2010.12.025

Al-Mohammed, A., Abido, M. A., & Mansour, M. (2011). *Optimal PMU placement for power system observability using differential evolution.* Paper presented at the Intelligent Systems Design and Applications (ISDA), 2011 11th International Conference on. 10.1109/ISDA.2011.6121668

Aljarah, I., Faris, H., & Mirjalili, S. (2016). Optimizing connection weights in neural networks using the whale optimization algorithm. *Soft Computing*, 1–15.

Basetti, V., & Chandel, A. K. (2017). Optimal PMU placement for power system observability using Taguchi binary bat algorithm. *Measurement*, *95*, 8–20. doi:10.1016/j.measurement.2016.09.031

Bei, X., Yoon, Y. J., & Abur, A. (2005). Optimal placement and utilization of phasor measurements for state estimation. *PSERC Publication*, 05-20.

Chakrabarti, S., & Kyriakides, E. (2008). Optimal placement of phasor measurement units for power system observability. *IEEE Transactions on Power Systems*, *23*(3), 1433–1440. doi:10.1109/TPWRS.2008.922621

Chakrabarti, S., Kyriakides, E., & Albu, M. (2009). Uncertainty in power system state variables obtained through synchronized measurements. *IEEE Transactions on Instrumentation and Measurement*, *58*(8), 2452–2458. doi:10.1109/TIM.2009.2016387

Chakrabarti, S., Kyriakides, E., & Eliades, D. G. (2009). Placement of synchronized measurements for power system observability. *IEEE Transactions on Power Delivery*, *24*(1), 12–19. doi:10.1109/TPWRD.2008.2008430

Chakrabarti, S., Venayagamoorthy, G. K., & Kyriakides, E. (2008). *PMU placement for power system observability using binary particle swarm optimization.* Paper presented at the Power Engineering Conference, 2008. AUPEC'08. Australasian Universities.

Cherukuri, S. K., & Rayapudi, S. R. (2016). A novel global MPP tracking of photovoltaic system based on whale optimization algorithm. *International Journal of Renewable Energy Development*, *5*(3), 225. doi:10.14710/ijred.5.3.225-232

Dalali, M., & Karegar, H. K. (2016). Optimal PMU placement for full observability of the power network with maximum redundancy using modified binary cuckoo optimisation algorithm. *IET Generation, Transmission & Distribution*, *10*(11), 2817–2824. doi:10.1049/iet-gtd.2016.0287

Dua, D., Dambhare, S., Gajbhiye, R. K., & Soman, S. (2008). Optimal multistage scheduling of PMU placement: An ILP approach. *IEEE Transactions on Power Delivery*, *23*(4), 1812–1820. doi:10.1109/TPWRD.2008.919046

Enshaee, A., Hooshmand, R. A., & Fesharaki, F. H. (2012). A new method for optimal placement of phasor measurement units to maintain full network observability under various contingencies. *Electric Power Systems Research*, *89*, 1–10. doi:10.1016/j.epsr.2012.01.020

Gopakumar, P., Reddy, M. J. B., & Mohanta, D. K. (2015). Pragmatic multi-stage simulated annealing for optimal placement of synchrophasor measurement units in smart power grids. *Frontiers in Energy*, *9*(2), 148–161. doi:10.100711708-015-0344-z

Gou, B. (2008a). Generalized integer linear programming formulation for optimal PMU placement. *IEEE Transactions on Power Systems*, *23*(3), 1099–1104. doi:10.1109/TPWRS.2008.926475

Gou, B. (2008b). Optimal placement of PMUs by integer linear programming. *IEEE Transactions on Power Systems*, *23*(3), 1525–1526. doi:10.1109/TPWRS.2008.926723

Guo, X.-C., Liao, C.-S., & Chu, C.-C. (2016). *Optimal PMU placements under propagation depth constraints by mixed integer linear programming*. Paper presented at the Smart Grid Communications (SmartGridComm), 2016 IEEE International Conference on. 10.1109/SmartGridComm.2016.7778836

Hajian, M., Ranjbar, A., Amraee, T., & Shirani, A. (2007). *Optimal placement of phasor measurement units: Particle swarm optimization approach*. Paper presented at the Intelligent Systems Applications to Power Systems, 2007. ISAP 2007. International Conference on. 10.1109/ISAP.2007.4441610

Hajian, M., Ranjbar, A. M., Amraee, T., & Mozafari, B. (2011). Optimal placement of PMUs to maintain network observability using a modified BPSO algorithm. *International Journal of Electrical Power & Energy Systems*, *33*(1), 28–34. doi:10.1016/j.ijepes.2010.08.007

Hurtgen, M., & Maun, J.-C. (2010). Optimal PMU placement using iterated local search. *International Journal of Electrical Power & Energy Systems*, *32*(8), 857–860. doi:10.1016/j.ijepes.2010.01.028

Kaveh, A., & Ghazaan, M. I. (2016). Enhanced whale optimization algorithm for sizing optimization of skeletal structures. *Mechanics Based Design of Structures and Machines*, 1–18.

Ketabi, A., Nosratabadi, S., & Sheibani, M. (2010). *Optimal PMU placement based on mean square error using differential evolution algorithm*. Paper presented at the Power Quality Conference (PQC), 2010 First.

Khajeh, K. G., Bashar, E., Rad, A. M., & Gharehpetian, G. B. (2017). Integrated model considering effects of zero injection buses and conventional measurements on optimal PMU placement. *IEEE Transactions on Smart Grid*, *8*(2), 1006–1013.

Koutsoukis, N. C., Manousakis, N. M., Georgilakis, P. S., & Korres, G. N. (2013). Numerical observability method for optimal phasor measurement units placement using recursive Tabu search method. *IET Generation, Transmission & Distribution*, *7*(4), 347–356. doi:10.1049/iet-gtd.2012.0377

Kulanthaisamy, A., Vairamani, R., Karunamurthi, N. K., & Koodalsamy, C. (2014). A multi-objective PMU placement method considering observability and measurement redundancy using ABC algorithm. *Advances in Electrical and Computer Engineering*, *14*(2), 117–128. doi:10.4316/AECE.2014.02020

Kundur, P., Balu, N. J., & Lauby, M. G. (1994). *Power system stability and control* (Vol. 7). McGraw-hill New York.

Mahari, A., & Seyedi, H. (2013). Optimal PMU placement for power system observability using BICA, considering measurement redundancy. *Electric Power Systems Research*, *103*, 78–85. doi:10.1016/j.epsr.2013.04.013

Maji, T. K., & Acharjee, P. (2017). Multiple Solutions of Optimal PMU Placement Using Exponential Binary PSO Algorithm for Smart Grid Applications. *IEEE Transactions on Industry Applications*, *53*(3), 2550–2559. doi:10.1109/TIA.2017.2666091

Marin, F., Garcia-Lagos, F., Joya, G., & Sandoval, F. (2003). Genetic algorithms for optimal placement of phasor measurement units in electrical networks. *Electronics Letters*, *39*(19), 1403–1405. doi:10.1049/el:20030919

Mazlumi, K., & Vahedi, H. (2010). *Optimal placement of PMUs in power systems based on bacterial foraging algorithm.* Paper presented at the Electrical Engineering (ICEE), 2010 18th Iranian Conference on. 10.1109/IRANIANCEE.2010.5506953

Milosevic, B., & Begovic, M. (2003). Nondominated sorting genetic algorithm for optimal phasor measurement placement. *IEEE Transactions on Power Systems*, *18*(1), 69–75. doi:10.1109/TPWRS.2002.807064

Mirjalili, S., & Lewis, A. (2016). The whale optimization algorithm. *Advances in Engineering Software*, *95*, 51–67. doi:10.1016/j.advengsoft.2016.01.008

Mohammadi-Ivatloo, B. (2009). Optimal placement of PMUs for power system observability using topology based formulated algorithms. *Journal of Applied Sciences (Faisalabad)*, *9*(13), 2463–2468. doi:10.3923/jas.2009.2463.2468

Nazari-Heris, M., & Mohammadi-Ivatloo, B. (2015). Application of heuristic algorithms to optimal PMU placement in electric power systems: An updated review. *Renewable & Sustainable Energy Reviews*, *50*, 214–228. doi:10.1016/j.rser.2015.04.152

Peng, C., Sun, H., & Guo, J. (2010). Multi-objective optimal PMU placement using a non-dominated sorting differential evolution algorithm. *International Journal of Electrical Power & Energy Systems*, *32*(8), 886–892. doi:10.1016/j.ijepes.2010.01.024

Phadke, A. G., & Thorp, J. S. (2008). *Synchronized phasor measurements and their applications*. Springer Science & Business Media. doi:10.1007/978-0-387-76537-2

Prakash, D., & Lakshminarayana, C. (2016). *Optimal siting of capacitors in radial distribution network using whale optimization algorithm*. Alexandria Engineering Journal.

Rahman, N. H. A., & Zobaa, A. F. (2017). Integrated mutation strategy with modified binary PSO algorithm for optimal PMU placement. *IEEE Transactions on Industrial Informatics*.

Recioui, A., Bentarzi, H., & Ouadi, A. (2015). *Application of a binary teaching learning-based algorithm to the optimal placement of phasor measurement units Progress in Clean Energy* (Vol. 1, pp. 817–830). Springer.

Roy, B. S., Sinha, A., & Pradhan, A. (2012). An optimal PMU placement technique for power system observability. *International Journal of Electrical Power & Energy Systems*, *42*(1), 71–77. doi:10.1016/j.ijepes.2012.03.011

Touma, H. J. (2016). Study of the economic dispatch problem on IEEE 30-bus system using whale optimization algorithm. *International Journal of Engineering Technology and Sciences*, *5*(1), 11–18.

Xu, B., & Abur, A. (2004). *Observability analysis and measurement placement for systems with PMUs.* Paper presented at the Power Systems Conference and Exposition, 2004. IEEE PES.

KEY TERMS AND DEFINITIONS

Binary Connectivity Matrix: A matrix containing the information of the connectivity of the buses in a network.

Binary Optimization Problem: An optimization problem in which the output is in binary form.

Bus Observability: A bus is said to be observable if its voltage and branch currents are measurable.

Measurement Redundancy: Total count of observability of a bus by different PMUs either directly or indirectly.

Observability Constraint Vector: A constraint vector having information about the observability of a network.

Optimal PMU Placement Problem: A problem of optimally placing PMUs in a network maintaining the full observability of the network.

Phasor Measurement Units: A modern measuring device which takes the synchronized measurements of voltage and current phasors.

Zero-Injection Bus: A bus with no generation and no load.

Chapter 5
Solar PV Powered Standalone Water Pumping Systems:
Modeling, Design, and Analysis

Sachin Jain
National Institute of Technology Warangal, India

Ramsha Karampuri
National Institute of Technology Warangal, India

Ramulu Chinthamalla
National Institute of Technology Warangal, India

ABSTRACT

The standalone solar photovoltaic (PV) powered water pumping system could have either two power conditioning units (PCUs) or a single PCU. The system with single PCU is advantageous over the prior because of low device count, smaller in size, low cost, low losses, and higher efficiency. This chapter includes four such single-stage PV water pumping systems. All the four systems are operated using the integrated control technique which assimilates maximum power point tracking (MPPT), pulse width modulation (PWM) technique, and motor control. But the PV systems differ in the inverter configuration used as well as the employed induction motor (conventional star connected and open-end winding induction motor [OEWIM]). The detailed description of mathematical modeling, design and analysis of all the four PV pumping systems along with the simulation results are presented in this chapter.

1. INTRODUCTION

In the present scenario of dwindling conventional energy resources, the role of renewable energy sources i.e., non-conventional energy resources need not to be over stated. Various renewable energy resources includes the solar PhotoVoltaic (PV), wind energy, bio-diesel, fuel cells etc. Amongst the available energy sources, the solar PV technology seems to be most promising proposition. The cost free nature of the primary energy source (i.e., irradiation from Sun), negligible maintenance and noise free operation

DOI: 10.4018/978-1-5225-3935-3.ch005

are some of the attractive features offered by the solar PV technology. Also, rapidly declining cost of the PV modules and installation expenses contribute to the proliferation of the solar PV systems.

2. BACKGROUND

The solar PV array directly converts the solar energy from the Sun rays into dc electric power. The generated dc power from the PV array is conditioned or transformed into the required form using Power Conditioning Unit (PCU). The PCU can be any inverter or converter circuit depending on the application of the PV system. With respect to the application, solar PV systems can be broadly divided into two categories: (i) Grid-connected PV system and (ii) Stand-alone PV system. In grid-connected PV systems, the solar PV power is fed into the grid in the form of high quality sinusoidal current using inverter. In contrast, in stand-alone PV systems, solar PV source with or without additional sources, maintains the defined voltage and current required by the stand-alone loads. Additional sources can be a battery bank, wind power, fuel cell, etc., and they are required due to the uncertainty in the PV power generation. However, stand-alone applications like solar water pump used in agriculture, multi-storey buildings or industries, etc., with water tank-storage may not require additional sources. Also, in such systems, PV source can always be operated near maximum power point (MPP), which gives the option of using generated PV power efficiently. Some good solutions presented in the literature for the PV fed water pump drive systems are briefed below.

Literature Survey

In the past, the initial solutions for PV pumping system were given in the literature which uses the dc motor. As PV source produces dc power, the use of dc motor offers the possibility of connecting the PV source directly without a PCU (Appelbaum J., 1986) as shown in Figure 1(a). However, the extraction of maximum available PV power may not be possible for all environmental conditions. Thus, solutions using a dc-dc converter (i.e., PCU) were also given as shown in Figure 1(b), which operates the PV source at an optimum voltage (Alghuwainem S. M., 1992, Elgendy M. A., 2010). However, the use of dc motor requires continuous maintenance and involves high cost. This restricts its application in PV pumping system. Hence, the research interest on PV pumping systems has been diverted towards the special motors like Brushless dc motor (Kumar R. & Singh B., 2016, Bhim Singh & Kumar R., 2016), permanent magnet motor (Sashidhar S. & Fernandes B. G., 2017) and switched reluctance motor (Bhim Singh, 2016), etc. Presently, the use of BLDC motor, permanent magnet motor and Switched Reluctance motor in a PV pumping system may not be a feasible solution because of its high cost and the requirement of complex control circuitry. Further, the requirement of continuous maintenance and high cost restricts the dc motors and special motors for their application in PV pumping system. Thus, there is a need of solution, which apart from using PV power effectively, uses a low cost, low maintenance, reliable and robust motor for pumping application. The most suitable candidate for above solution would be the usage of Induction Motor (IM).

There are not many literatures available on the usage of IM for PV powered water pump applications. One such proposal was given in (Bhat S. R., 1987). In the given proposal authors have used a two-stage system (i.e., system with two PCUs), where first-stage consists of a dc-dc converter and the second-stage composes of dc-ac inverter as shown in Figure 1(c). The purpose of first converter is to boost and/or regulate the dc output voltage as well as to track the maximum power available at PV source. The second stage is meant for the dc-ac inversion with the optimum V/*f* control required by the induction motor. Another good proposal given by (Yao Y., 1994) in which authors had suggested frequency control for a similar two-stage system. The proposed control strategy takes care of power mismatch between the generated PV power and motor output. This helps in minimizing the speed oscillations of the motor. The frequency of the inverter output voltage is controlled by sensing the PV power to improve the efficiency. Another elegant proposal given by (Mona N. Eskander, & Aziza M. Zaki, 1997) using two-stage systems. The authors have proposed two control schemes for the PV powered IM drive system. In the first scheme, PV source is operated at MPP using dc-dc converter and motor is operated at constant flux using V/*f* control. In second control strategy, authors operate the IM at optimum efficiency by matching the operating characteristics of the PV array with required voltage and current of the IM. Among two given control strategies, first one proved to be more efficient as it was able to pump water even during low insolation. Furthermore, authors have suggested to use closed loop speed control to improve the performance and efficiency of the system.

Some of such proposals with two PCUs were given by (Singh B. N., 1996, Montiê Alves Vitorino, & Maurício Beltrão de Rossiter Corrêa, 2011). In the given proposals the authors have presented the PV pumping system with closed loop speed control. The authors have used sensor-less control techniques to improve the efficiency of the system. Former PV system uses current control technique and the later uses complex control technique like IFOC and DTC for speed control. Also, the proposed control techniques suffer from the disadvantage of having more number of current sensors.

From the literature discussed till now, it can be concluded that the PV pumping system with two PCUs as shown in Figure 1(c), will result in poor performance and lesser efficiency. Thus, there is a need for the single-stage system with simple control as shown in Figure 1(d). One such good proposal was given by (Muljadi E., 1997), in which the author used a six-step quasi square wave inverter which can take care of dc-ac inversion as well as Maximum Power Point Tracking (MPPT). It is a better proposition since the author has given an integrated single-stage solution. But the drawback of this system is the six-step quasi-square nature of inverter output voltage, which will deteriorate the motor performance and hence the efficiency. It will also result in the requirement of filters with high value and increased size. Furthermore, the proposed system requires high voltage rating, input dc link capacitor which will increase the system cost.

To improve the performance of the motor and the efficiency of the PV pumping system (Betka A. & Moussi A., 2004) has used Pulse Width Modulation (PWM) technique where the frequency and amplitude of modulation index were controlled separately. In this paper, the authors have presented the flux weakening operation of induction motor which helps in optimal utilization of PV power even at lower insolation levels. It is advantageous over the scheme presented by (Yao Y., 1994), but the major drawback of this system is, it is applicable only for the light torque loads i.e., lower heads.

Recently, (Makhlouf M., 2011) has suggested the closed loop vector control for PV pumping system using a single PCU. It is a better suggestion since it is a single-stage system with better motor efficiency. But this system is more complex with more number of control variables and sensors involved in the control action. Most recently, (Tomás Perpétuo Corrêa, 2012) have published a paper with a stand-alone PV pumping system, it is also a single-stage system, wherein the authors have proposed MPPT and Minimum Losses Point Tracking (MLPT) methods of control. MPPT is for extracting maximum power from the PV source and MLPT is to minimize the motor losses. The proposed system is the best solution, however, still the proposed system suffers from a number of sensors, low bandwidth for PV operating

Figure 1. Block diagram of various PV powered water pumping systems

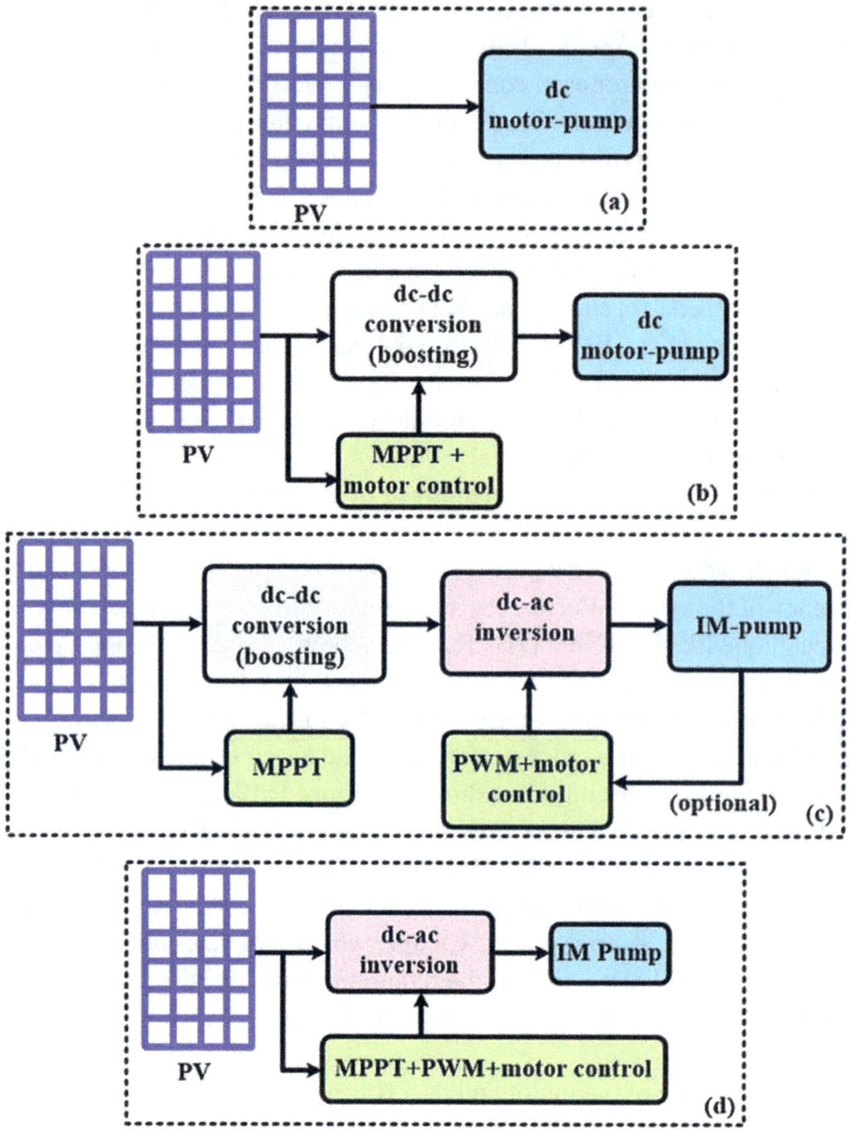

voltage, high voltage rating dc and two-level operation. Thus, from the above discussion, it is evident there is requirement of low cost single-stage PV pumping system with wide bandwidth of operating PV voltage. This chapter discusses the single-stage PV pumping system solutions with complete details which includes its simulation model, design and simulation results.

3. MAIN FOCUS OF THE CHAPTER

From the literature survey, it can be observed that the PV pumping system with single PCU and the use of induction motor is the better proposition (Sachin Jain, 2016). A tangible research could be performed to improvise the overall system efficiency and the effective utilization of all the components or equipment used in the single-stage PV pumping system. The research area could be in the use of various dc-ac inverter configurations in place of the PCU. In this chapter, the dc-ac inverter configurations used are:

1. Two-level three-phase H-bridge inverter.
2. Three-level cascaded inverter.
3. Three-level dual H-bridge inverter with common dc-bus.
4. Three/four-level dual H-bridge inverter with isolated dc-bus.

There are some issues and few problems related to various PV water pumping systems which uses different inverter configurations as mentioned in the literature review as well as the inverters listed above. Some of the issues and problems associated with various PV pumping systems are mentioned below.

Below are the issues and problems when two-level H-brigde inverter is used:

- The requirement of high PV array voltage,
- High voltage rating of dc-bus capacitors,
- High voltage rating of semiconductor switching devices,
- Low band-width of PV source operating voltage,
- Effect of shading on PV arrays,
- Less number of levels in the motor phase voltage.

To address the above issues the three-level cascaded inverter can be used in the single-stage PV pumping system. However, the first five problems pertains but the last one can be solved to some extent. All the above issues can be solved by introducing the Open-End Winding Induction Motor (OEWIM) which is connected to the dual-inverter either with single power source or with two electrically isolated power sources.

In addition to the above problems, the single-stage PV pumping system with any of the inverter configuration can encounter some other issues related to its control strategy. The concerns related to the control strategy of the PV pumping system are as follows:

- The control technique should be capable of performing MPPT and motor control together.
- This requires proper coordination or synchronization between MPPT and motor control along with the PWM technique to operate the inverter.
- The task of synchronization is very important because of the non-linear *I-V* characteristics of the PV source.
- The non-linear nature of the PV source should be matched with the parabolic characteristic of the centrifugal pump. This is very important to achieve MPPT.

All the above mentioned issues and concerns are reported in the next section.

4. SOLUTIONS AND RECOMMENDATIONS

This section give the details of PV powered water pumping systems which uses various single-stage PCUs. In addition, the conventional star connected and the OEWIM configurations coupled to the centrifugal pump are considered in this chapter. Below are the solutions which can be recommended to use as the standalone PV powered water pumping system.

1. Three-phase two-level H-bridge inverter connected to the star connected induction motor-pump.
2. Three-phase three-level cascaded inverter connected to the star connected induction motor-pump.
3. Three-phase three-level dual H-bridge inverter connected to the open-end winding induction motor-pump with common dc-link.
4. Three-phase three/four-level dual H-bridge inverter connected to the open-end winding induction motor-pump with electrically isolated dc-links.

The block diagram representation for all the four solutions is shown in Figure 2. In this section, the detailed explanation for the PV source model and the induction motor model is presented. The mathematical model for all the four inverter configurations is given separately in the consequent subsections. Also, the control technique integrates the MPPT algorithm, PWM strategy (with Common Mode Voltage (CMV) elimination or minimization) and the motor control technique. The MPPT technique extracts the maximum available electrical power from the PV source. The open-loop V/*f* control technique is used for the motor control. These two techniques are common for all the considered PV water pumping solutions. The control algorithm differs in the employed PWM technique which is used to operate various inverter configurations and is explained in detail in the consequent subsections. Initially, the mathematical model for PV source and the induction motor are presented.

PV Source Model

A PV array/source comprises the arrangement of several series (N_s) and parallel (N_p) connected PV modules. Each PV module is formed by connecting few PV cells in series. Considering only a single solar PV cell which can be modeled by using a current source, a diode and the combination of series and parallel resistors as shown in Figure 3 (Walker G., 2001). This equivalent circuit is known as a single diode model for solar PV cell. In Figure 3, the nomenclature of all the parameters is as given below:

I_λ: Photon current
I_d: Shockley temperature dependent diode current
I_p: PV cell leakage current
I_{pv}: PV cell output current
V_{pv}: PV cell voltage
V_d: Diode voltage
R_s: Series resistance usually 0.05 to 0.1Ω
R_p: Parallel resistance (around 300Ω) representing internal losses

The current generated by the PV module is given by the Equation (1).

$$I_{pv} = N_p I_{sc} - N_p I_s \left[e^{\frac{q(V_{pv} + I_{pv} R_s)}{N_s A k T}} - 1 \right] - \frac{\left[\frac{V_{pv}}{N_s} + I_{pv} R_s \right]}{R_p} \tag{1}$$

where

I_{sc}: PV cell short circuit current = photon current, I_λ (A)
I_s: Diode reverse saturation current (A)
N_p: Number of parallel connected PV cells
N_s: Number of series connected PV cells
A: Diode ideality factor
k: Boltzmann constant (J/K)
q: Electron charge (C)
T: Panel operating temperature (K)

The PV cell short circuit current is given by:

$$I_{sc} = \lambda I_{scref} \left(1 + k_i \left(T - T_{ref} \right) \right) \tag{2}$$

The diode reverse saturation current is given by:

$$I_s = I_{rs} \left(\frac{T}{T_{ref}} \right)^3 e^{\left(\frac{qeg}{Ak} \right) \left(\frac{1}{T_{ref}} - \frac{1}{T} \right)} \tag{3}$$

$$I_{rs} = \frac{I_{scref}}{e^{\frac{qV_{oc}}{N_s k A T}} - 1} \tag{4}$$

where

λ: Illumination intensity (kW/m^2) or (Suns)
I_{scref}: Short circuit current at STC (A)
T_{ref}: Temperature at STC (K)
k_i: Temperature coefficient of I_{sc} (/^{o}C)
e_g: Energy gap (eV)

The Equations (1) to (4) can be used to develop the mathematical model for the PV source. The model of PV source will give the output as PV current, I_{pv} as mentioned in Equation (1). However, the PV voltage can be obtained or calculated using the PV current and the converter current as shown in

Figure 2. Block diagram representation for various PV powered water pumping solutions

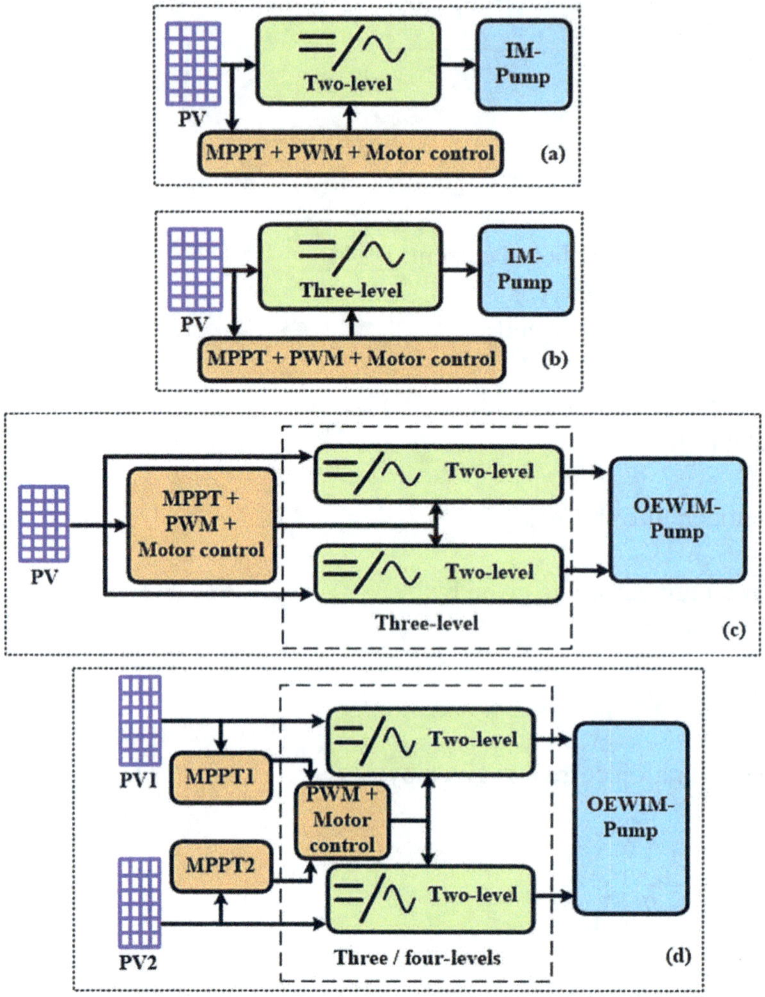

Figure 3. Equivalent circuit of solar PV cell

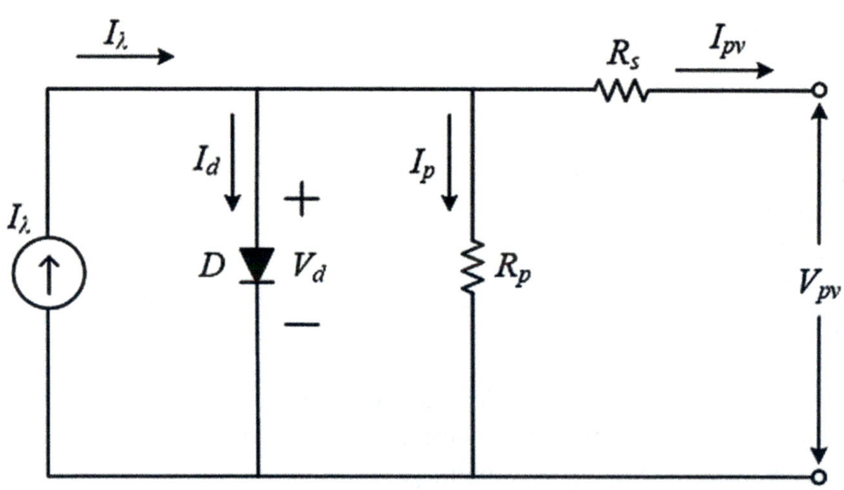

Figure 4. The converter current is different for different inverter configurations used in the PV pumping system (which is derived in the respective subsections).

The PV module used for the simulation is Solarex MSX60 (60 watt Multi-crystalline PV module) and the parameters are:

- Maximum power (P_{mpp}): 60W
- Voltage at P_{mpp} (V_{mpp}): 17.1V
- Current at P_{mpp} (I_{mpp}): 3.5A
- Short-circuit current (I_{sc}): 3.8A
- Open-circuit voltage (V_{oc}): 21.1V
- Temperature coefficient of I_{sc}: (0.065±0.015)%/°C
- Temperature coefficient of V_{oc}: -(80±10)mV/°C

The above parameters are used to simulate the PV module and the module characteristics are plotted in Figure 5. In Figure 5(a), the *power-voltage* characteristics are plotted for four different insolation (G) conditions at a fixed temperature (T) of 25°C. It can be observed from Figure 5(a), that the PV power increases with increase in the insolation and there is a slight increase in the PV voltage. In Figure 5(b), the *power-voltage* characteristics are plotted for four different temperature conditions at a fixed insolation of 1Sun. It can be observed from Figure 5(b), that the PV power and the voltage reduces with the increase in the temperature. In Figure 5(c), the PV *current-voltage* characteristics are plotted for the same insolation conditions as above at a fixed temperature of 25°C. It can be observed from Figure 5(c), that the PV current increases with increase in the insolation and there is a slight increase in the PV voltage. In Figure 5(d), the PV *current-voltage* characteristics are plotted for the same four temperature conditions at a fixed insolation of 1Sun. It can be observed from Figure 5(d), that the PV voltage reduces with the increase in the temperature, but there is a very slight increment in the PV current. In Figure 5(e), the PV current, voltage and the loci of the PV power at Maximum Power Point (MPP) are shown. These PV module characteristics are summarized in Table 1.

Induction Motor Model

The mathematical model of a three-phase induction motor can be described by the voltage equations in *q-d* frame of reference and is given below (Paul Krause, 2013).

Figure 4. MATLAB/Simulink model for PV source

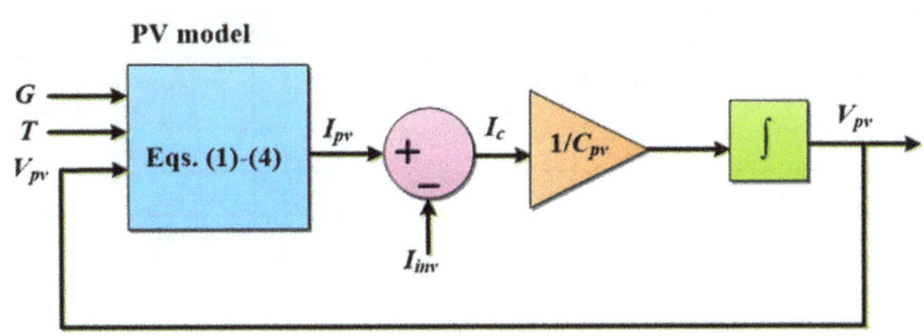

Figure 5. (a) Effect of insolation on P-V characteristics at 25°C (b) Effect of temperature on P-V characteristics at 1 Sun (c) Effect of insolation on I-V characteristics at 25°C (d) Effect of temperature on I-V characteristics at 1 Sun (e) Effect of insolation on I-V characteristics along with maximum power loci.

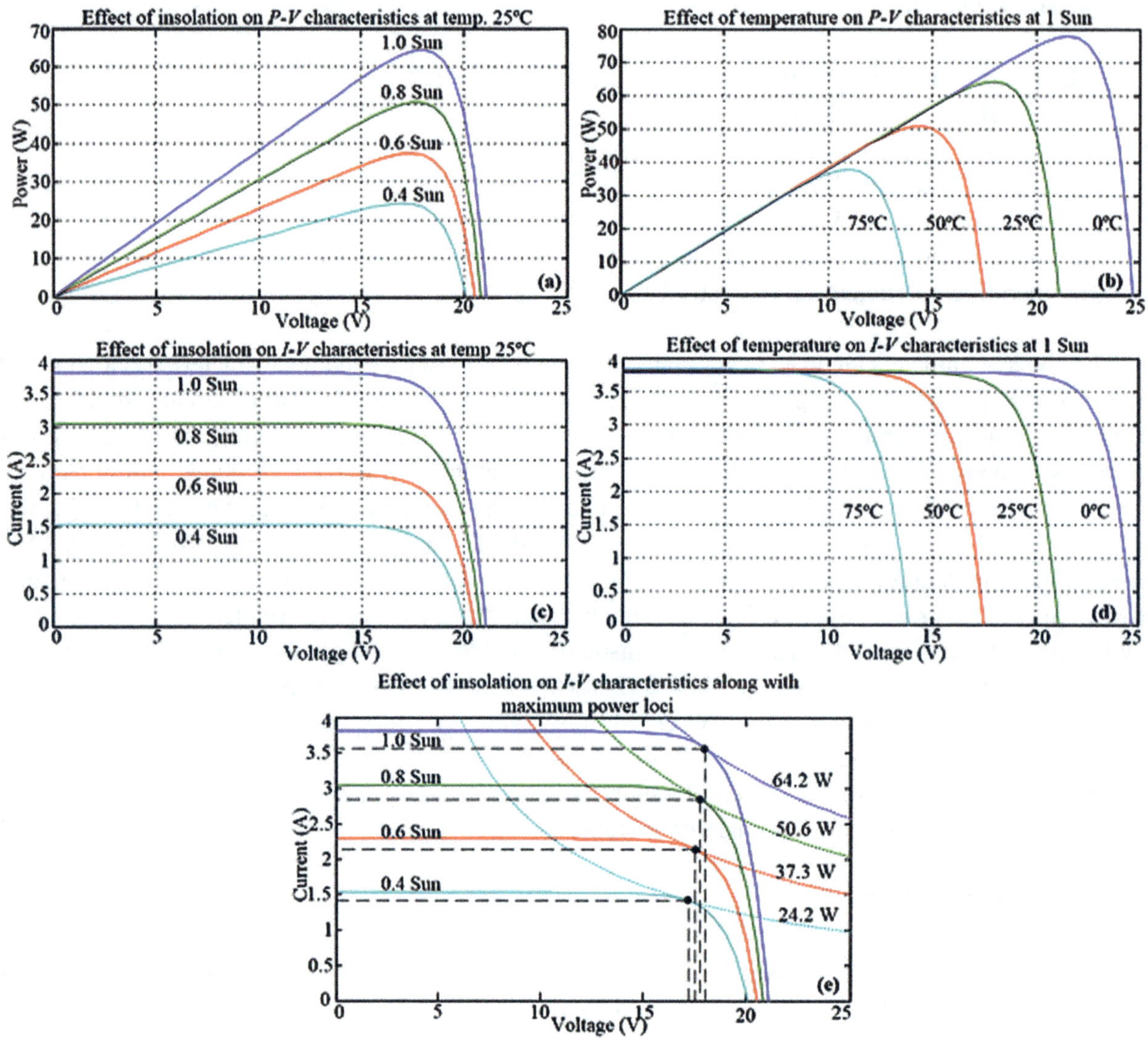

Table 1. Effect of change in environmental conditions on PV parameters

PV Parameters	Effect	
	Increase in Insolation at Constant Temperature	**Increase in Temperature at Constant Insolation**
Current	Increase	Slight increase (Almost constant)
Voltage	Slight increase	Decrease
Power	Increase	Decrease

$$v_{qs} = i_{qs} r_s + \omega \lambda_{ds} + \frac{d\lambda_{qs}}{dt} \tag{5}$$

$$v'_{qr} = i'_{qr} r'_r + \left(\omega - \omega_r\right) \lambda'_{dr} + \frac{d\lambda'_{qr}}{dt} \tag{6}$$

$$v_{ds} = i_{ds} r_s - \omega \lambda_{qs} + \frac{d\lambda_{ds}}{dt} \tag{7}$$

$$v'_{dr} = i'_{dr} r'_r - \left(\omega - \omega_r\right) \lambda'_{qr} + \frac{d\lambda'_{dr}}{dt} \tag{8}$$

where, 'q' is the quadrature-axis quantity, 'd' is the direct-axis quantity, " ' " represent rotor side quantities referred to stator, 'ω' is the synchronous speed, 'ω_r' is the rotor speed or instantaneous angular velocity of the motor shaft (rad/sec). Here 'v'_{qr}', 'v'_{dr}' are equal to zero, since a squirrel cage induction motor is used, in which rotor terminals are shorted.

Now, the equations for flux-linkage in q-d frame are given by,

$$\begin{bmatrix} \lambda_{qs} \\ \lambda_{ds} \\ \lambda'_{qr} \\ \lambda'_{dr} \end{bmatrix} = \begin{bmatrix} L_m + L_{ls} & 0 & L_m & 0 \\ 0 & L_m + L_{ls} & 0 & L_m \\ L_m & 0 & L_m + L'_{lr} & 0 \\ 0 & L_m & 0 & L_m + L'_{lr} \end{bmatrix} \begin{bmatrix} i_{qs} \\ i_{ds} \\ i'_{qr} \\ i'_{dr} \end{bmatrix} \tag{9}$$

where, 'L_m', 'L_{ls}' and 'L'_{lr}' are the magnetizing inductance, leakage inductance of stator winding and leakage inductance of rotor winding referred to stator. The above equations are obtained by transforming the phase quantities like voltage, current and the flux-linkage using the transformation matrix, T given in Equation (10).

$$T = \frac{2}{3} \begin{bmatrix} \cos\varphi & \cos\left(\varphi - \frac{2\pi}{3}\right) & \cos\left(\varphi + \frac{2\pi}{3}\right) \\ \sin\varphi & \sin\left(\varphi - \frac{2\pi}{3}\right) & \sin\left(\varphi + \frac{2\pi}{3}\right) \\ \frac{1}{2} & \frac{1}{2} & \frac{1}{2} \end{bmatrix} \tag{10}$$

where, the angle

$$\begin{bmatrix} v_{aa's} \\ v_{bb's} \\ v_{cc's} \end{bmatrix} = \begin{bmatrix} r_{aa's} & 0 & 0 \\ 0 & r_{bb's} & 0 \\ 0 & 0 & r_{cc's} \end{bmatrix} \begin{bmatrix} i_{aa's} \\ i_{bb's} \\ i_{cc's} \end{bmatrix} + p \begin{bmatrix} \lambda_{aa's} \\ \lambda_{bb's} \\ \lambda_{cc's} \end{bmatrix}$$

$\phi = \delta$ for stator transformation and 'γ' for rotor transformation. From Figure 6, 'δ' is the angle between stator '*as*' axis and q-axis, 'γ' is the angle between rotor '*ar*' axis and q-axis, or otherwise $\gamma = \delta - \delta_r$.

Conventional Star Connected Three-Phase Induction Motor

The winding diagram for the conventional star connected three-phase induction motor is shown in Figure 6. From Figure 6, the motor stator and rotor phase voltages can be expressed by the Equations (11) and (12) respectively.

$$\begin{bmatrix} v_{as} \\ v_{bs} \\ v_{cs} \end{bmatrix} = \begin{bmatrix} r_s & 0 & 0 \\ 0 & r_s & 0 \\ 0 & 0 & r_s \end{bmatrix} \begin{bmatrix} i_{as} \\ i_{bs} \\ i_{cs} \end{bmatrix} + p \begin{bmatrix} \lambda_{as} \\ \lambda_{bs} \\ \lambda_{cs} \end{bmatrix} \tag{11}$$

$$\begin{bmatrix} v_{ar} \\ v_{br} \\ v_{cr} \end{bmatrix} = \begin{bmatrix} r_r & 0 & 0 \\ 0 & r_r & 0 \\ 0 & 0 & r_r \end{bmatrix} \begin{bmatrix} i_{ar} \\ i_{br} \\ i_{cr} \end{bmatrix} + p \begin{bmatrix} \lambda_{ar} \\ \lambda_{br} \\ \lambda_{cr} \end{bmatrix} \tag{12}$$

where, (v_{as}, v_{bs}, v_{cs}), and (v_{ar}, v_{br}, v_{cr}) are the stator and rotor voltages respectively; (i_{as}, i_{bs}, i_{cs}), and (i_{ar}, i_{br}, i_{cr}) are the stator and rotor currents respectively; (λ_{as}, λ_{bs}, λ_{cs}), and (λ_{ar}, λ_{br}, λ_{cr}) are the stator and rotor flux-linkages respectively; r_s and r_r are the stator and rotor winding resistances respectively; p is the time derivative (d/dt).

Three-Phase Open-End Winding Induction Motor

An OEWIM can be realized physically by avoiding the shorting of stator windings (Stemmler H. & Guggenbach P., 1993). This facilitates to have six stator terminals namely '*as*', '*a's*', '*bs*', '*b's*', '*cs*' and '*c's*' as shown in Figure 7. The rotor winding is shorted and unaltered which is represented by '*ar*', '*br*' and '*cr*' axes. The phase voltages applied to the OEWIM stator winding can be expressed as,

$$\begin{bmatrix} v_{asa's} \\ v_{bsb's} \\ v_{csc's} \end{bmatrix} = \begin{bmatrix} r_s & 0 & 0 \\ 0 & r_s & 0 \\ 0 & 0 & r_s \end{bmatrix} \begin{bmatrix} i_{asa's} \\ i_{bsb's} \\ i_{csc's} \end{bmatrix} + p \begin{bmatrix} \lambda_{asa's} \\ \lambda_{bsb's} \\ \lambda_{csc's} \end{bmatrix} \tag{13}$$

And the rotor winding voltages can be expressed as,

Figure 6. Machine winding diagram for conventional star connected three-phase induction motor

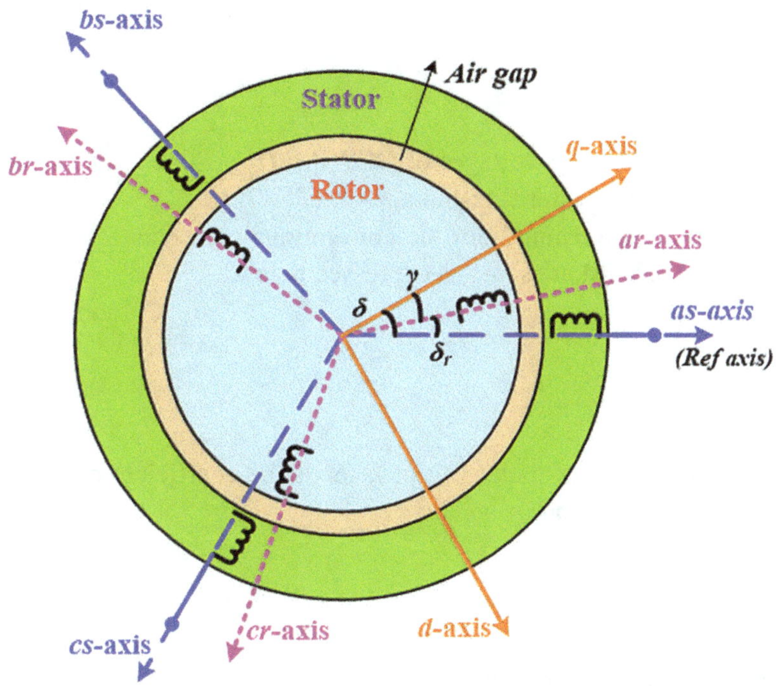

Figure 7. Machine model for three-phase OEWIM

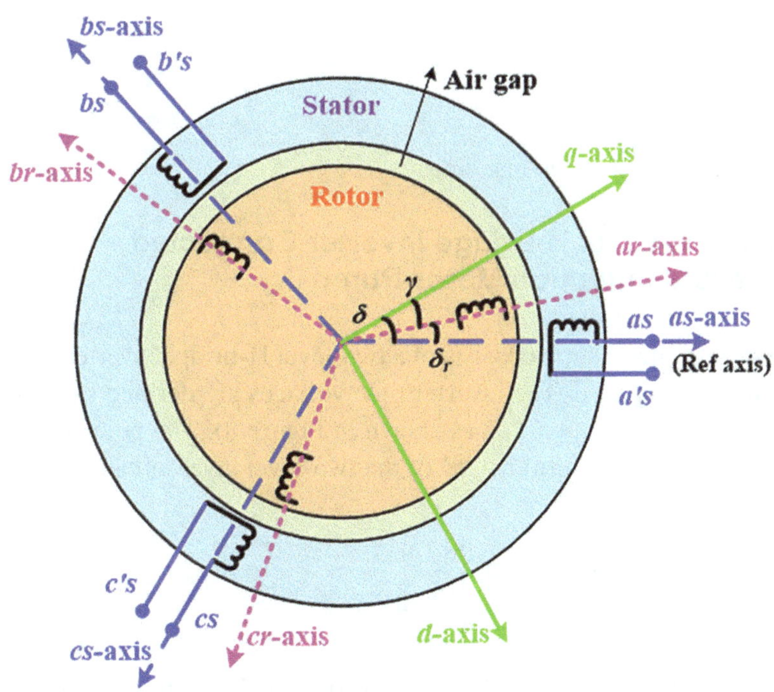

$$\begin{bmatrix} v_{ar} \\ v_{br} \\ v_{cr} \end{bmatrix} = \begin{bmatrix} r_r & 0 & 0 \\ 0 & r_r & 0 \\ 0 & 0 & r_r \end{bmatrix} \begin{bmatrix} i_{ar} \\ i_{br} \\ i_{cr} \end{bmatrix} + p \begin{bmatrix} \lambda_{ar} \\ \lambda_{br} \\ \lambda_{cr} \end{bmatrix} \tag{14}$$

where, ($v_{asa's}$, $v_{bsb's}$, $v_{csc's}$) are the stator voltages of OEWIM; ($i_{asa's}$, $i_{bsb's}$, $i_{csc's}$) are the stator currents of OEWIM; ($\lambda_{asa's}$, $\lambda_{bsb's}$, $\lambda_{csc's}$) are the stator flux-linkages in OEWIM.

The mechanical equations governing both the conventional star connected three-phase induction motor and the three-phase OEWIM are same as expressed below.

$$J\frac{d\omega_r}{dt} + B\omega_r + m_L = m_d \tag{15}$$

where 'J' is the moment of inertia of motor (kg-m^2), 'B' is the coefficient of centrifugal load torque, 'm_L' is the centrifugal pump load torque given by,

$$m_L = K_w \omega_r^2 \tag{16}$$

where 'K_w' is the centrifugal pump constant.

Also, the electromagnetic torque 'm_d' developed in the induction motor and OEWIM are same as expressed below.

$$m_d = \frac{3}{2}\frac{P}{2}\frac{L_m}{L_s}\left(\lambda_{qs} i'_{dr} - \lambda_{ds} i'_{qr}\right) \tag{17}$$

where 'P' is the number of poles of the machine.

4.1. Three-Phase Two-Level H-Bridge Inverter Connected to the Star Connected Induction Motor-Pump

The PV pumping system in this section consists of a two-level H-bridge inverter which is powered from the PV source as shown in Figure 8. The inverter pole voltages are defined from the respective three-phase poles to the dc ground point (say "o" as shown in Figure 8). The pole-voltages 'v_{ao}' at terminal 'a', 'v_{bo}' at terminal 'b', and 'v_{co}' at terminal 'c' of the two-level inverter can be expressed as,

$$v_{ao} = S_A\left(\frac{V_{pv}}{2}\right); v_{bo} = S_B\left(\frac{V_{pv}}{2}\right); \text{and } v_{co} = S_C\left(\frac{V_{pv}}{2}\right) \tag{18}$$

where, S_Y (where $Y \in \{A, B, C\}$) is the switching function defined for all the three poles (a, b, and c) of the two-level inverter. The switching function 'S_Y' hold the values 1 or -1, depending on top or bottom switch of corresponding pole of the inverter is turned ON; V_{pv} is the dc bus voltage applied from PV source.

Figure 8. Schematic circuit diagram of PV powered three-phase two-level H-bridge inverter connected to the induction motor-pump

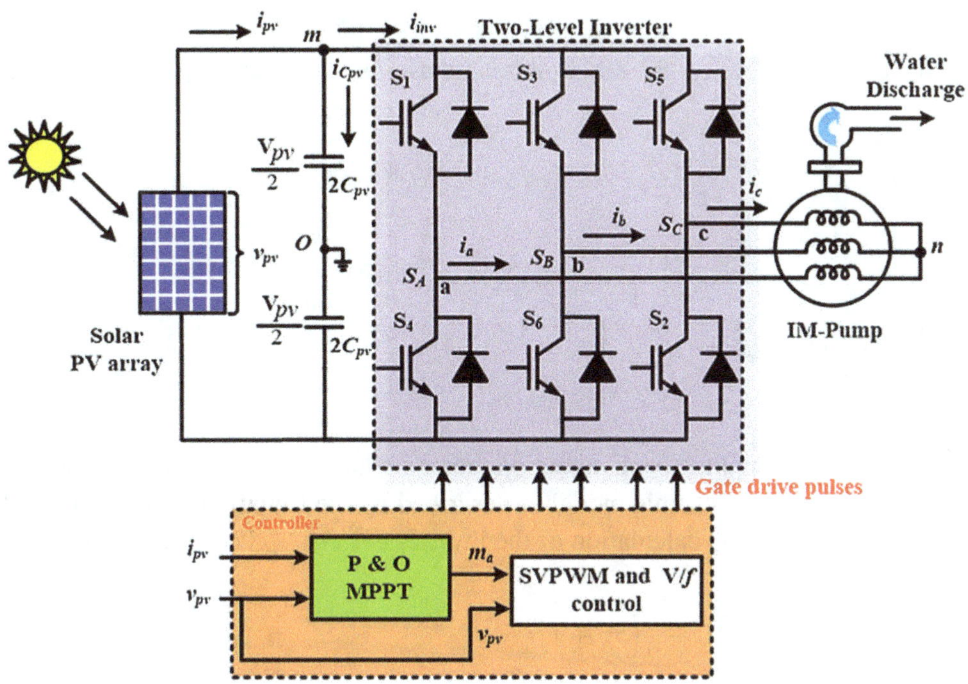

Once the pole-voltages are known, the common mode voltage of the inverter can be expressed as given in Equation (19).

$$v_{on} = \frac{1}{3}\left(v_{ao} + v_{bo} + v_{co}\right) \tag{19}$$

where, '*n*' represents the neutral point of the stator winding of induction motor.

Now, the motor phase voltage (phase-*an*) can be defined as:

$$v_{an} = v_{ao} - v_{on} \tag{20}$$

$$v_{an} = \left(S_A \frac{V_{pv}}{2}\right) - \frac{1}{3}\left[\left(S_A \frac{V_{pv}}{2}\right) + \left(S_B \frac{V_{pv}}{2}\right) + \left(S_C \frac{V_{pv}}{2}\right)\right] \tag{21}$$

Simplifying the Equation (21) results in,

$$v_{an} = \frac{V_{pv}}{2}\left\{\left(\frac{2}{3}\right)S_A - \frac{1}{3}\left(S_B + S_C\right)\right\} \quad (22)$$

Similarly, the phase voltages for the other phases (phase-*bn* and phase-*cn*) are given below,

$$v_{bn} = \frac{V_{pv}}{2}\left\{\left(\frac{2}{3}\right)S_B - \frac{1}{3}\left(S_C + S_A\right)\right\} \quad (23)$$

$$v_{cn} = \frac{V_{pv}}{2}\left\{\left(\frac{2}{3}\right)S_C - \frac{1}{3}\left(S_A + S_B\right)\right\} \quad (24)$$

Once the phase voltages applied to the induction motor are known, then the respective phase currents can be calculated using motor dynamic model as described by the Equations (11) to (12). The derived phase current can be used for the calculation of the inverter current 'i_{inv}' as given below.

$$i_{inv} = \frac{1}{2}\left(S_A + 1\right)i_a + \frac{1}{2}\left(S_B + 1\right)i_b + \frac{1}{2}\left(S_C + 1\right)i_c \quad (25)$$

The above derived inverter current can be used to model the PV source as explained above. The detailed design of the entire system is given below.

4.1.1. PV Sizing

The design of PV pumping system starts with the requirement of gallons of water/fluid to be transferred from one place to the other. Consider a case, where 'g' gallons of water/fluid need to be transferred to a dynamic head of 'H' in a minute. Then, the horse power 'P_m' required for pumping fluid/water is given by,

$$P_m(HP) = \frac{g \times H \times sg}{3960} \quad (26)$$

where, 'sg' is the specific gravity of water/fluid (sg for water = 1). The water/fluid horse power 'P_m' (in HP) also defines the output power of the motor pump-set to be chosen. Now, assuming that the PV array rating 'P_{PV_array}' required for the system is nearly equal to the rating of the motor i.e.,

$$P_{PV\_array} = K_c \times P_m \quad (27)$$

where, 'K_c' is the conversion constant (from horse power to watts).

This is done to simplify the design, since the objective is to design the PV array configuration for the proposed system. Once PV array rating is decided, the steps involved in the design of the PV configuration for the proposed system are described as follows:

1. Determine the total number of PV modules 'N_T'.
2. Determine the total number of PV modules in the string 'N_s' for the required dc-bus voltage.
3. Determine the PV array configuration based on 'N_T' and 'N_s' i.e., number of strings in parallel 'N_p'.

Now to determine above details, the specifications of the PV module used in the system is required. The PV module model no. Solarex MSX60 is considered with the below specifications at standard test conditions (STC).

V_{MPP_m}= 17.1 V; I_{MPP_m}= 3.5 A; I_{sc_m}= 3.8 A; V_{oc_m}= 21.1 V

where, 'V_{MPP_m}', and 'I_{MPP_m}' denote PV module operating voltage and current at MPP respectively, 'I_{sc_m}' and 'V_{oc_m}' denote PV module short-circuit current and open-circuit voltage respectively. Now, PV power of the module when operating at MPP under STC is given by,

$$P_{MPP\_m} = V_{MPP\_m} \times I_{MPP\_m} \quad (28)$$

Using 'P_{MPP_m}' and 'P_{PV_array}', the integer value 'N_T' can be determined as,

$$N_T = \frac{P_{PV\_array}}{P_{MPP\_m}} \quad (29)$$

Once the total number of modules 'N_T' is known, the next task is to determine the total number of modules to be used in the PV string that meets the total dc-bus voltage 'V_{dc_total}' requirements. This can be determined using,

$$N_s = \frac{V_{dc\_total}}{V_{MPP\_m}} \quad (30)$$

Using the information of 'N_s' and 'N_T' the number of strings which should be connected in parallel 'N_p' for the PV source is given by,

$$N_p = \frac{N_T}{N_s} \quad (31)$$

The arrangements or configurations of the two-level inverter connected PV pumping system can be calculated by considering the requirement of PV array with capacity 4 kW. Using the module of MSX60 the values of 'N_T' and 'N_s' results in, N_T= 67 and N_s = 33. The integer value of N_s is calculated by considering V_{dc_total} = 564V (since, it is the total required PV/dc-bus voltage to maintain the peak phase voltage of 325V in the linear modulation range for the SVPWM technique). Also, the required N_p is calculated using Equation (31), which results approximately equal to 2. Hence, the PV array configuration can be used as 33×2 (i.e., 33 modules in a string and 2 such strings in parallel). This result in the power of PV array nearly 3950W, the dc-bus voltage 564.3V. Once the PV array configuration is defined, then the task of designing or defining the rating of the components in the inverter can be done.

4.1.2. Design of DC Bus Capacitor And Inverter Switches

Inverter is the heart of the PV pumping system which converts the raw power generated at PV array into the required form to be used by the motor-pump set. The design of inverter mainly involves the design of the dc-bus capacitors and switches with respect to the voltage and the current being handled by them. The voltage ratings required for the components (switches and capacitor) is mainly governed by the maximum voltage appearing across the dc-bus of the inverter. The maximum dc voltage is actually the utmost open circuit voltage of the PV string. Thus, the voltage rating of the switches (V_{sw}) and the capacitors (V_{cap}) to be used is given by,

$$V_{sw} = V_{cap} = N_s \times V_{oc\_m} = 696.3V \tag{32}$$

The design value of dc-bus capacitor which acts as buffer between inverter and PV array can be decided based on the back up time to be provided or allowed ripple voltage at the dc-bus. Calculation of the capacitor value based on the ripple voltage assuming short circuit current flowing into it is given by,

$$C_{pv} = \frac{N_p \times I_{sc\_m}}{f_s \times \Delta V_{pv}} = \frac{2 \times 3.8}{554.2 \times 13.926} \approx 1000\mu F \tag{33}$$

where, 'f_s' is the minimum switching frequency of the semiconductor switches and is determined by using Equation (34), 'ΔV_{pv}' is the allowed ripple content in the PV source voltage which is assumed to be 2% of the maximum value of dc-bus voltage calculated using V_{oc_m}.

$$f_s = \frac{m_{a\_\min} \times sam \times f_{rated}}{m_{a\_linear}} \tag{34}$$

where, 'm_{a_min}' is the minimum value of the modulation index (considered m_{a_min} = 0.2), '*sam*' is the number of samples per fundamental cycle of the modulating waveform (considered *sam* = 48), f_{rated} =50 Hz and 'm_{a_linear}' is the maximum value of the modulation index in the linear range (considered m_{a_linear} = 0.866 for the SVPWM technique). Now, current rating for the switches in the inverter can be calculated from the below expression (assuming equal current is flowing through the switches of inverter),

$$i_{sw\_S_1}(t) = \frac{1}{2}\left(S_A + 1\right) i_a(t) \tag{35}$$

where, '$i_{sw\_S1}(t)$' and '$i_a(t)$' denotes the instantaneous value of the current flowing through the switch 'S_1' and motor phase-*a* respectively.

The current rating of the semiconductor switches used in the two-level H-bridge inverter could be decided by the maximum value of the instantaneous switch current 'i_{sw_S1}' (assuming continuous current flowing through the switch). In other words, the peak value of the induction motor phase current decides the current rating of the semiconductor switch.

4.1.3. Modulation Technique

The above designed PV pumping system requires a better control strategy for its good performance. This system employs an integrated control strategy which assimilates MPPT algorithm, V/*f* control and SVPWM technique as shown in Figure 9. In the given solution, simple and robust '*P & O*' algorithm is employed to extract maximum power from the solar PV array. Whereas, the open-loop V/*f* control is chosen to improve the motor performance and SVPWM technique is to operate the power semiconductor devices of the inverter with better dc-bus utilization.

MPPT Algorithm

The instantaneous values of PV voltage 'v_{pv}' and PV current 'i_{pv}' are sensed and given to MPPT algorithm. The instantaneous PV array power 'p_{pv}' is calculated. These values are averaged and defined as the present values '$V_{pv}(n)$' and '$P_{pv}(n)$'. The PV voltage and power are compared with the previously initiated values '$V_{pv}(n\text{-}1)$' and '$P_{pv}(n\text{-}1)$' to determine the operating region in the *p-v* curve as given in Equation (36) and Equation (37). In other words, the slope of *p-v* curve determine either increment or decrement in the value of modulation index.

$$if\ \frac{P_{pv}(n)-P_{pv}(n-1)}{V_{pv}(n)-V_{pv}(n-1)}<0\ \text{ then } m_a=m_a+\Delta m_a \tag{36}$$

$$if\ \frac{P_{pv}(n)-P_{pv}(n-1)}{V_{pv}(n)-V_{pv}(n-1)}>0\ \text{ then } m_a=m_a-\Delta m_a \tag{37}$$

Calculations for V/f Control

The updated value of modulation index 'm_a' is further processed with the operating voltage of the PV array for the calculation of resultant reference voltage vector '$\vec{V}_{sr}$' as given below,

$$\vec{V}_{sr}=m_aV_{pv} \tag{38}$$

The above equation determines the amplitude of the reference (modulating) voltage vector, whereas the frequency of the modulating vector can be determined using Equation (39).

$$f = \frac{\left|\vec{V}_{sr}\right| \times f_{rated}}{\left|\vec{V}_{sr(rated)}\right|} \tag{39}$$

where, '$\vec{V}_{sr(rated)}$' is the maximum output voltage across the motor phase winding operating at rated condition. For a peak motor phase voltage of 325V, $\vec{V}_{sr(rated)}$ is calculated as follows:

$$\vec{V}_{sr(rated)} = \left(\frac{\sqrt{3}}{2}\right) \times V_{dc} = \left(\frac{\sqrt{3}}{2}\right) \times 564 \approx 488 \text{ V} \tag{40}$$

where, V_{dc} = 564V is the total required dc-bus voltage to maintain the peak phase voltage of 325 V in the linear modulation range (0.866 for SVPWM technique).

In addition, the time period of fundamental cycle 'T' is calculated by inverting Equation (39) and the switching time 'T_s' is calculated by inverting Equation (34). The calculated values of 'T' and 'T_s' are used to generate the modulating and carrier signals respectively as shown in Figure 9. The fundamental time period 'T' is used to hold the sensed values of PV voltages and currents (see the blocks counter and S/H in Figure 9.). This is done for two reasons: one is to maintain the synchronization between both the modulating and carrier signals and the other is to maintain the fixed number of samples per fundamental cycle. This gives the opportunity to operate the system with variable switching frequency. The number of samples are calculated by the counter using the value of 'T' as shown in Figure 9.

Figure 9. Block diagram of the control technique for two-level H-bridge inverter connected induction motor-pump drive powered by PV source

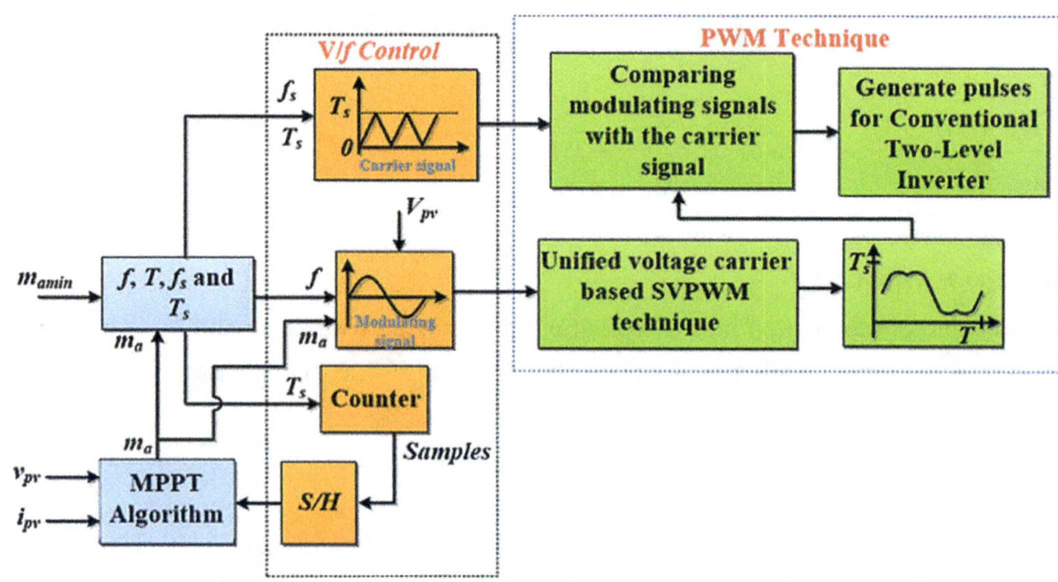

Unified Voltage Modulation SVPWM Technique

The generated modulating waveform can be processed using any conventional PWM technique like sine triangle PWM, Space Vector PWM (SVPWM), and so on. In this chapter, the SVPWM technique is chosen because of its better dc-bus utilization. However, the SVPWM technique can be implemented using two different approaches; either conventional look-up table based approach or the unified voltage modulation approach which is a carrier based technique (Dae-Woong Chung, 1998). In this chapter, the simplest unified voltage modulation SVPWM technique is used to generate the gating pulses required by the inverter switches.

4.1.4. Controller Requirements

The controller shown in Figure 8 can be any low cost device which can perform the given controller algorithm. For the system shown in Figure 8, the controller should have at least two Analog to Digital Converters (ADCs) to sense the PV voltage and the PV current from the PV source. This is done to achieve the MPPT operation. Another requirement is the controller should have at least three independent enhanced Pulse Width Modulation pins (ePWM pins) to drive the inverter switches. In addition, the controller should have sufficient flash memory and the Random Access Memory (RAM) where the code for the control algorithm can be encrypted. Furthermore, most importantly the chosen controller should have sufficient clock frequency, which could handle the calculation burden in a better way and could generate the switching signals with higher frequencies.

4.1.5. Results and Discussion

The PV pumping system shown in Figure 8 is simulated using MATLAB/Simulink software. The PV source is simulated by considering the parameters of a Solarex MSX60 panel as given in section 4.1.1. A 4kW, 400V, 1430 rpm three-phase star-connected induction motor is used. The PV array configuration used is 33×2 i.e., 33 modules are connected in series to form a string and 2 such strings are connected in parallel.

The simulation is performed for three different environmental conditions. Initially, the insolation and temperature values are considered to be 0.4Suns and 30°C respectively. Later, the insolation and temperature are increased to 0.7Suns and 40°C respectively. Finally, the insolation and temperature are decreased to 0.5Suns and 35°C respectively. These three variations in the environmental conditions are considered to show the working effectiveness of the PV system under dynamic weather conditions. The simulation results for these three conditions are shown in Figure 10. The system is started operating by applying an initial modulation index equal to 0.3 (this value can be varied depending on the initial environmental conditions). It can be observed from subplots (c) and (d) of Figure 10 that the value of PV current increases with the corresponding decrease in its voltage. This nearly confirms the PV array characteristics. Another observation is that by changing the insolation and temperature, the corresponding change in the PV variables like current, voltage and power of PV array can also be observed. Further, it can be observed from Figure 10(f), that modulation index 'm_a' of the conventional two-level inverter follows the PV power generated. Lower the value of PV power, lower is the value of modulation index 'm_a' at MPP, and vice versa. Thus, an increase or decrease in the PV power leads to the corresponding increase or decrease in the value of modulation index 'm_a'.

Another useful observation is that the operating voltage of PV array passes through optimum (MPP) voltage for every step increase in insolation and temperature. Also, the small oscillations in modulation index 'm_a' clearly indicates operation of the PV source near MPP. This can also be concluded with the matching value of peak power during transient tracking and steady-state near MPP as given in the PV power subplot (e) of Figure 10. Low ripple content in PV power and matching with peak PV power at transient and steady-state for PV array, thus confirms the operation near MPP for PV source. Further, it can be noted that PV voltage waveform shows a sudden rise or fall in its value during the step increase or decrease in insolation and temperature. This can be attributed to charging and discharging of PV capacitor 'C_{pv}' with excess or deficit PV power during the transient condition. In other words, in Figure 10, till 20 seconds the results were shown for increasing insolation and temperature conditions. Under these conditions, the PV voltage and current show inverse nature. But when the insolation and temperature are suddenly reduced at a time (say at 20s), then the PV source operation enters into the current source region. In current source region, the PV current is maintained constant at its value equal to short circuit current (for that particular insolation and temperature level) and the PV voltage falls below the operating voltage. For a system with improper control strategy, the PV voltage may fall to zero and settles at zero voltage. But, in this system the control algorithm will not allow the PV voltage dropping to zero value. This can be seen from Figure 10(d), the PV voltage is dropped nearly to 100V. However, the drop in the PV voltage can be minimized using a high value capacitor.

Figure 11 show the simulation results for motor and pump side parameters under different environmental conditions as given in subplots (a) and (b) of Figure 10. The variations in the modulation index 'm_a' with respect to PV power can also be seen here in the waveform of torque, speed of the motor and the flow rate of pump as shown in Figure 11. The constant head of 30m is considered for the calculation of flow rate of water. It can be observed that torque, speed, and flow rate follow 'm_a' or the PV power indirectly. Because of above reason after 20s a similar dip is observed in torque, speed, and flow rate. Additionally, the nature of motor phase voltage and the phase current at two different fundamental frequencies is shown in Figures 12 and 13. Figure 12 show the motor phase voltage (top) and the phase current (bottom) at lower fundamental frequency; whereas in Figure 13, the phase voltage and phase current are captured at higher fundamental frequency. Both the figures are plotted using equal time scale (i.e., time difference is 0.1 seconds), so as to depict the working of V/*f* control strategy.

4.2. Three-Phase Three-Level Cascaded Inverter Connected to the Star Connected Induction Motor-Pump

The cascaded inverter configuration consists of two 2-level inverter which are connected in a cascaded manner as shown in Figure 14 to give three-level output (Chinthamalla R., 2014 and 2016). The input dc-bus of inverter consists of two dc-link capacitors connected in series, which provides a three-level output in the pole voltages as $V_{Cpv1}+V_{Cpv2}$, V_{Cpv2}, and 0. The inverter can be modelled using the switching function analysis where a switching logic S_{XZ} (where $X \in \{a\ b, c\}$ and $Z \in \{1, 2\}$) is used. The switching logic S_{XZ} represents state '1' or '0' based on the state of the top switch whether turned-ON or turned-OFF respectively, corresponding to phase 'X' of inverter 'Z'. The pole voltage of the inverter can be calculated with respect to dc neutral point 'O', thus the cascaded inverter comprises of six poles ($a1, b1, c1, a2, b2$ and $c2$) and 12 switches (four switches per phase-leg). Now, the pole-voltage for '$a1$' phase of inverter-1 'v_{a1o}' can be calculated as,

Figure 10. Simulated PV source parameters

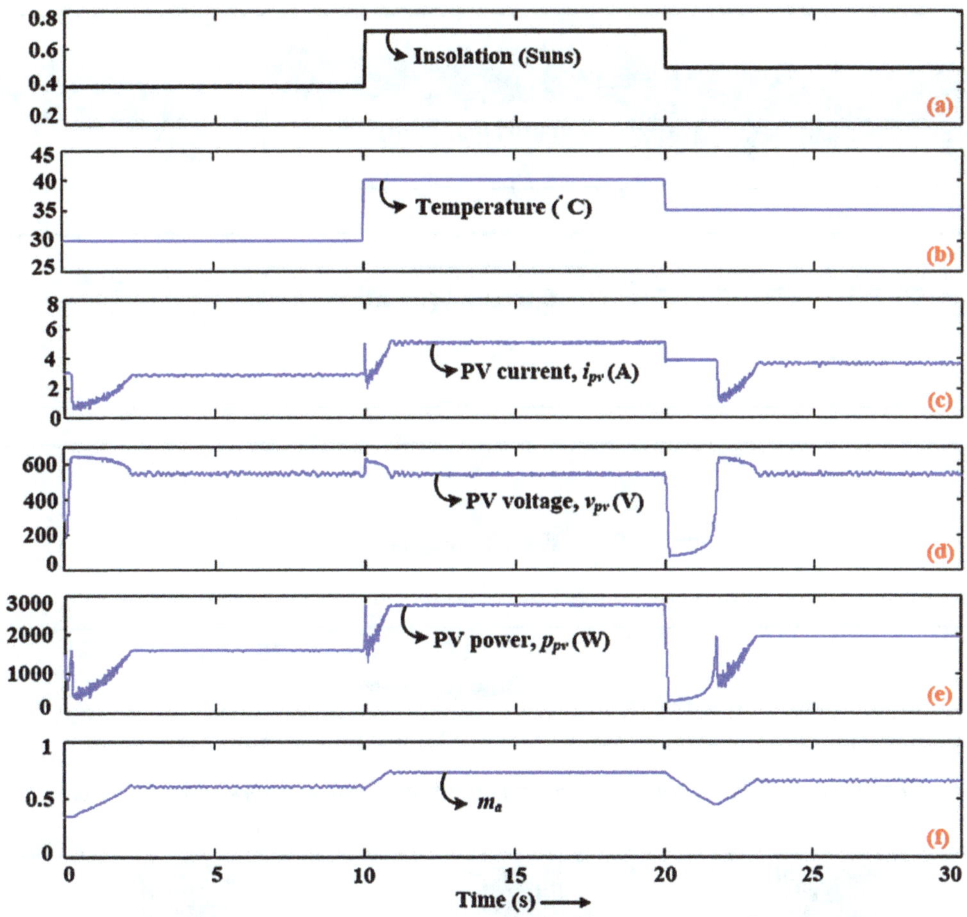

$$v_{a1o} = S_{a1} \times \left(V_{Cpv1} + V_{Cpv2}\right) + \bar{S}_{a1} \times V_{Cpv2} \tag{41}$$

where '–' represents complementary function i.e., if the switching state of top switch is 1, then the complementary switching state for bottom switch will be 0 and vice-versa. Similarly, the pole-voltages 'v_{b1o}' and 'v_{c1o}' of other phase's '*b1*' & '*c1*' respectively, for the other two legs of inverter -1 can be written as,

$$v_{b1o} = S_{b1} \times \left(V_{Cpv1} + V_{Cpv2}\right) + \bar{S}_{b1} \times V_{Cpv2} \tag{42}$$

$$v_{c1o} = S_{c1} \times \left(V_{Cpv1} + V_{Cpv2}\right) + \bar{S}_{c1} \times V_{Cpv2} \tag{43}$$

Using the same strategy, the pole-voltages 'v_{a2o}', 'v_{b2o}' and 'v_{c2o}' of '*a2*', '*b2*' and '*c2*' phases for inverter-2 can be expressed as,

Figure 11. Motor and pump parameters

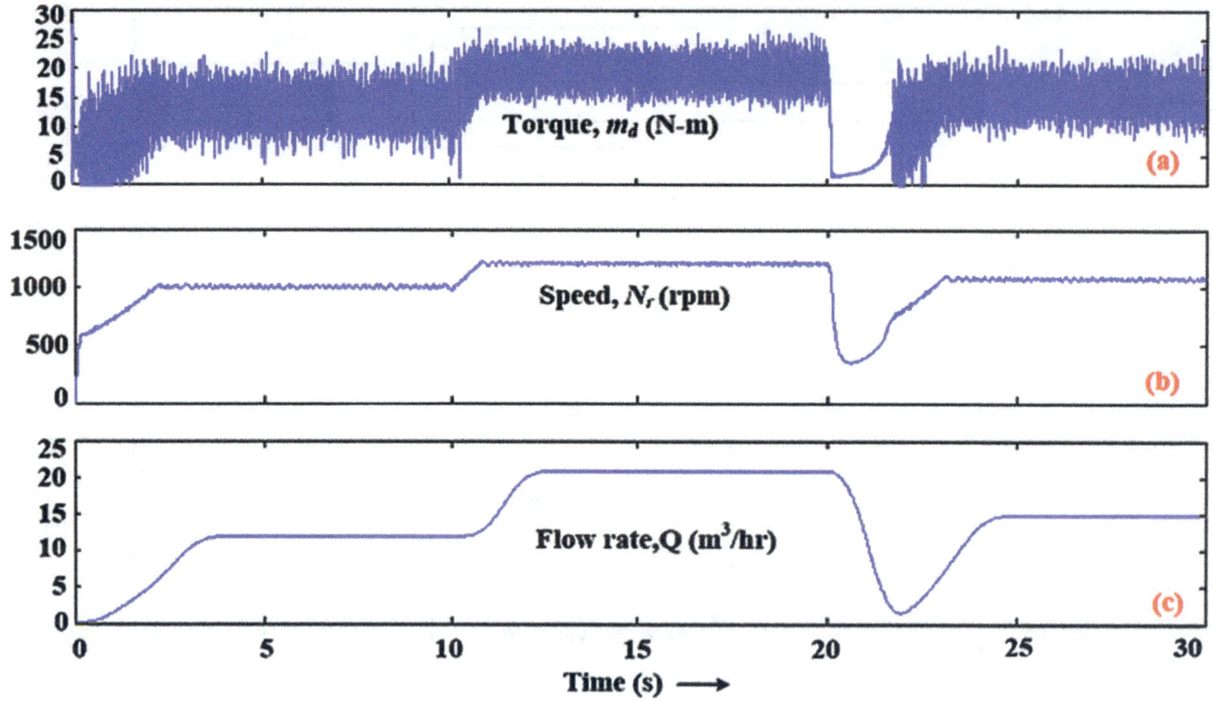

Figure 12. Induction motor phase voltage and phase current at lower modulation index

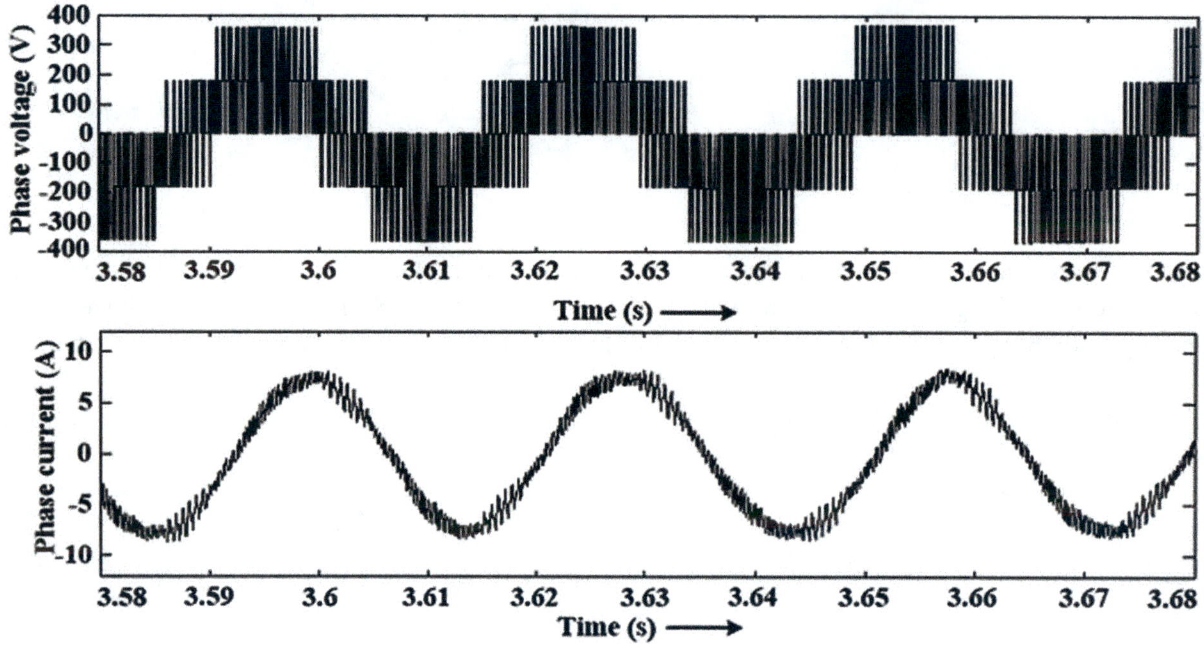

Figure 13. Induction motor phase voltage and phase current at higher modulation index

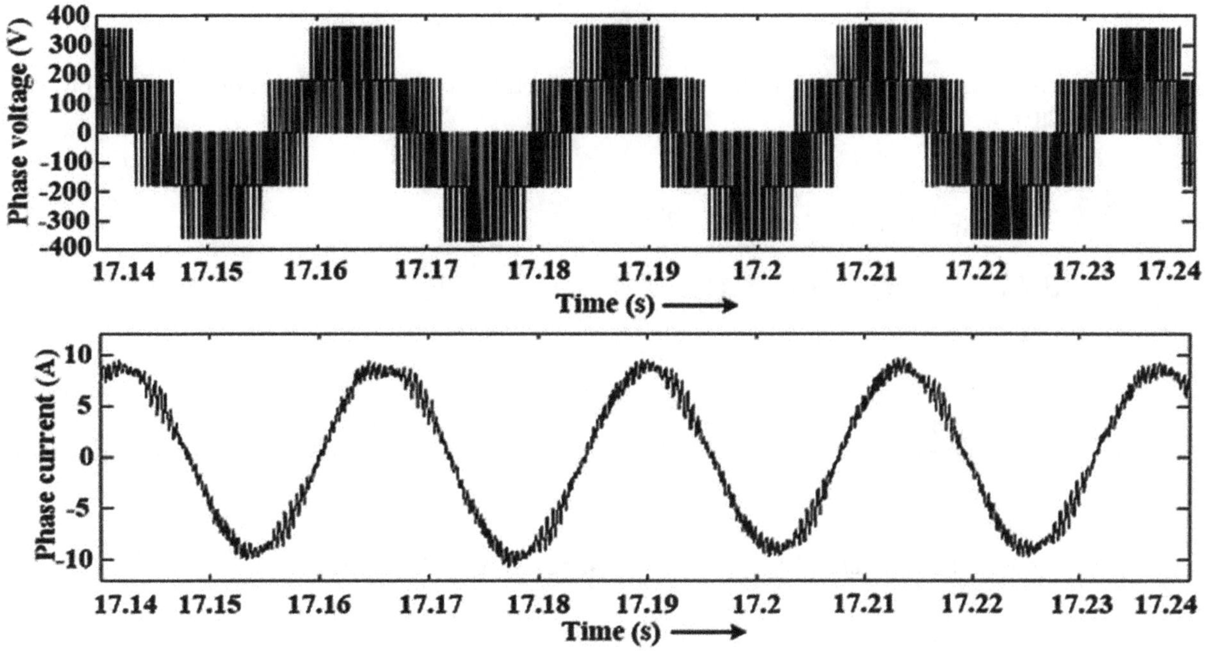

$$v_{a2o} = S_{a1}S_{a2} \times \left(V_{Cpv1} + V_{Cpv2}\right) + \bar{S}_{a1}S_{a2} \times V_{Cpv2} \tag{44}$$

$$v_{b2o} = S_{b1}S_{b2} \times \left(V_{Cpv1} + V_{Cpv2}\right) + \bar{S}_{b1}S_{b2} \times V_{Cpv2} \tag{45}$$

$$v_{c2o} = S_{c1}S_{c2} \times \left(V_{Cpv1} + V_{Cpv2}\right) + \bar{S}_{c1}S_{c2} \times V_{Cpv2} \tag{46}$$

Now, the common mode voltage 'v_{no}' can be determined as (from Figure 14),

$$v_{no} = \frac{1}{3}\left(v_{a2o} + v_{b2o} + v_{c2o}\right) \tag{47}$$

where, 'n' is the machine stator winding neutral point. Once the common mode voltage is known, the motor phase voltage can be calculated as,

$$v_{a2n} = v_{a2o} - v_{no} \tag{48}$$

where, 'v_{a2n}' is the motor a-phase voltage.

$$v_{a2n} = v_{a2o} - v_{no} = S_{a1}S_{a2} \times \left(V_{Cpv1} + V_{Cpv2}\right) + S_{a1}S_{a2}\left(V_{Cpv2}\right) - \frac{1}{3}\left(v_{a2o} + v_{b2o} + v_{c2o}\right) \tag{49}$$

The expression for motor phase voltage 'v_{a2n}' can be obtained by substituting Equations (44) to (46) in Equation (49).

$$\begin{aligned} v_{a2n} = \frac{2}{3}\left[S_{a1}S_{a2}\left(V_{Cpv1} + V_{Cpv2}\right) + \bar{S}_{a1}S_{a2}(V_{Cpv2})\right] - \\ \frac{1}{3}\left[S_{b1}S_{b2}(V_{Cpv1} + V_{Cpv2}) + \bar{S}_{b1}S_{b2}(V_{Cpv2}) + S_{c1}S_{c2}(V_{Cpv1} + V_{Cpv2}) + \bar{S}_{c1}S_{c2}(V_{Cpv2})\right] \end{aligned} \tag{50}$$

$$v_{a2n} = \left(V_{Cpv1} + V_{Cpv2}\right)\left[\frac{2}{3}S_{a1}S_{a2} - \frac{1}{3}S_{b1}S_{b2} - \frac{1}{3}S_{c1}S_{c2}\right] + V_{Cpv2}\left[\frac{2}{3}\bar{S}_{a1}S_{a2} - \frac{1}{3}\bar{S}_{b1}S_{b2} - \frac{1}{3}\bar{S}_{c1}S_{c2}\right] \tag{51}$$

Similarly, other phase voltages 'v_{b2n}' and 'v_{c2n}' can be expressed as follows.

$$v_{b2n} = \left(V_{Cpv1} + V_{Cpv2}\right)\left[\frac{2}{3}S_{b1}S_{b2} - \frac{1}{3}S_{c1}S_{c2} - \frac{1}{3}S_{a1}S_{a2}\right] + V_{Cpv2}\left[\frac{2}{3}\bar{S}_{b1}S_{b2} - \frac{1}{3}\bar{S}_{c1}S_{c2} - \frac{1}{3}\bar{S}_{a1}S_{a2}\right] \tag{52}$$

$$v_{c2n} = \left(V_{Cpv1} + V_{Cpv2}\right)\left[\frac{2}{3}S_{c1}S_{c2} - \frac{1}{3}S_{a1}S_{a2} - \frac{1}{3}S_{b1}S_{b2}\right] + V_{Cpv2}\left[\frac{2}{3}\bar{S}_{c1}S_{c2} - \frac{1}{3}\bar{S}_{a1}S_{a2} - \frac{1}{3}\bar{S}_{b1}S_{b2}\right] \tag{53}$$

Now, using phase voltage equations the respective phase currents can be calculated using IM model. The derived phase current can further be utilized in the calculation of current through inverter-1 'i_{inv1}' and inverter-2 'i_{inv2}' as given below.

$$i_{inv1} = S_{a1}S_{a2}i_a + S_{b1}S_{b2}i_b + S_{c1}S_{c2}i_c \tag{54}$$

$$i_{inv2} = \bar{S}_{a1}S_{a2}i_a + \bar{S}_{b1}S_{b2}i_b + \bar{S}_{c1}S_{c2}i_c \tag{55}$$

The above derived inverter current can be used to model the PV source as explained above. The detailed design of the entire system is given below.

4.2.1. PV Sizing

The basic requirements to design the PV pumping system are same as explained in section 4.1.1. The power rating of the PV source and the induction motor used in the PV system with three-level cascaded

Figure 14. Schematic circuit diagram of PV powered three-phase three-level cascaded inverter connected to the induction motor-pump

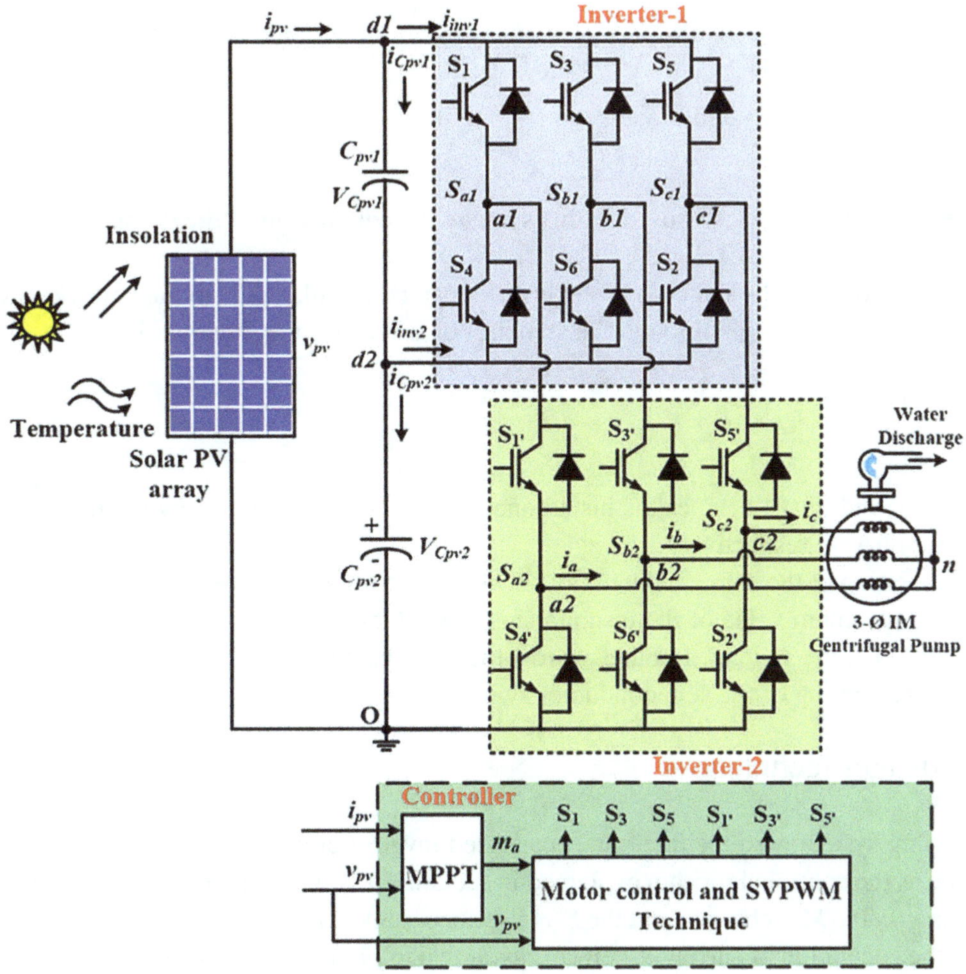

inverter is 4kW (same as in section 4.1.1). Hence, the required dc-bus voltage is $V_{dc_total} = 564$V. Also, the required N_p is calculated using Equation (31), which results approximately equal to 2. Hence, the PV array configuration can be used as 33×2 (i.e., 33 modules in a string and 2 such strings in parallel). This result in the power of PV array nearly 3950W, the dc-bus voltage 564.3V. Once the PV array configuration is defined, then the task of designing or defining the rating of the components in the inverter can be done.

4.2.2. Design of DC Bus Capacitor and Inverter Switches

The cascaded three-level inverter based PV pumping system shown in Figure 14, consists of 12 power semiconductor switching devices. In Figure 14, the bottom switches of Inverter-2 are to be rated for full PV bus voltage, and the remaining switches of Inveter-1 and Inverter-2 are rated for half the PV bus voltage. Thus, the voltage rating of the switches (V_{sw_inv1} and V_{sw_inv2} are the voltage ratings for switches to be used in Inverter-1 and Inverter-2 respectively) and the capacitor (V_{cap1} *and* V_{cap2} are voltage ratings for capacitors at dc-bus to be used in Inverter) to be employed in the proposed system is given by,

$$V_{sw\_inv1} = V_{sw\_inv2(top)} = \frac{1}{2} \times V_{oc\_m} \times 33 = 348.15 \text{ V}$$
$$V_{sw\_inv2(bottom)} = 33 \times V_{oc\_m} = 696.3 \text{ V} \quad (56)$$

Since the same PV array size is used in this system, the calculation of the dc-bus capacitor value is same as presented in section 4.1.2.

Now, current rating for the switches in the inverter can be calculated from the below expression (assuming equal current is flowing through the switches of Inverter-1 and Inverter-2),

$$i_{sw\_S_{11}}(t) = S_{A2}\left(S_{A1} + \bar{S}_{A1}\right) i_a(t) \quad (57)$$

where, '$i_{sw\_S11}(t)$' and '$i_a(t)$' denotes the instantaneous value of the current flowing through the switch 'S_{11}' and motor phase-*a* respectively.

The current rating of the semiconductor switches used in the three-level cascaded inverter could be decided by the maximum value of the instantaneous switch current 'i_{sw_S11}' (assuming continuous current flowing through the switch). In other words, the peak value of the induction motor phase current decides the current rating of the semiconductor switch.

4.2.3. Modulation Technique

The PV pumping system with the three-level cascaded inverter employs an integrated control strategy which is same as the control algorithm explained in section 4.1.3. The explanation for MPPT algorithm, unified voltage SVPWM technique and the S/H blocks is same as given in section 4.1.3. The only difference is in the generation of carrier waveforms as shown in Figure 15. To ensure the three-level output from the cascaded inverter, two carrier signals are generated: one with the magnitude varying from '0 to $T_s/2$' and the other is with the magnitude varying from '$T_s/2$ to T_s' (see Figure 15). Both the carrier signals are generated with a frequency 'f_s' as defined by Equation (34). The conditioned modulating signal is now compared with the two carrier signals to generate the required gating pulses for the power semiconductor switches of the cascaded inverter.

4.2.4. Controller Requirements

The controller shown in Figure 14 can also be any low cost device which can perform the given controller algorithm and it should have at least two Analog to Digital Converters (ADCs) to sense the PV voltage and the PV current from the PV source. This is done to achieve the MPPT operation. Another requirement is the controller should have at least six independent enhanced Pulse Width Modulation pins (ePWM pins) to drive the cascaded inverter switches. The other controller requirements stands same as explained in section 4.1.4.

Figure 15. Block diagram of the control technique for three-level cascaded inverter connected induction motor-pump drive powered by PV source

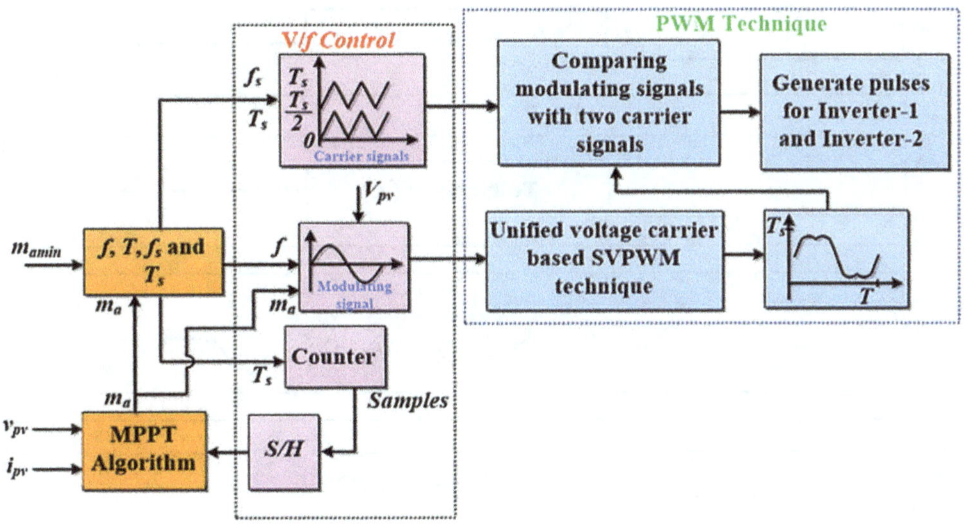

4.2.5. Results and Discussion

The PV pumping system shown in Figure 14 is simulated using MATLAB/Simulink software and considering the parameters as given in section 4.1.5. The simulation is performed for three different environmental conditions. Initially, the insolation and temperature values are considered to be 0.4Suns and 30°C respectively. Later, the insolation and temperature are increased to 0.7Suns and 40°C respectively. Finally, the insolation and temperature are decreased to 0.5Suns and 35°C respectively. These three variations in the environmental conditions are considered to show the working effectiveness of the PV system under dynamic weather conditions. The simulation results for these three conditions are shown in Figure 16. The system is started operating by applying an initial modulation index equal to 0.3 (this value can be varied depending on the initial environmental conditions). The PV parameters like current, voltage, and power shown in Figure 16 are nearly same as the PV current, voltage and power for two-level H-bridge inverter system (see Figure 10). The same is the case with the modulation index.

The cascaded inverter shown in Figure 14 is operated using single PV source, which could result in the neutral point voltage balancing problem. The other alternative is to use two electrically isolated PV sources, which could get rid of dc-bus capacitor voltage balancing. However, the control strategy explained in section 4.2.3 take care of the balancing in charging and discharging of the dc-bus capacitors. This can be supported by the voltage plots of both the dc-bus capacitors as shown in Figure 17. The ripple content in the voltage of dc-bus capacitors is about 2% of the total dc-bus voltage (as considered for the calculation of the value of dc-bus capacitors).

Figure 18 show the simulation results for motor and pump side parameters under different environmental conditions. The variations in the modulation index 'm_a' (see Figure 16(f)) with respect to PV power can also be seen here in the waveform of torque, speed of the motor and the flow rate of pump. The constant head of 30m (same as for two-level inverter connected PV pumping system) is considered for the calculation of flow rate of water. It can be observed that torque, speed, and flow rate follow 'm_a' or the PV power indirectly. The ripple content in the motor generated torque is less in case of cascaded

Figure 16. Simulated PV source parameters

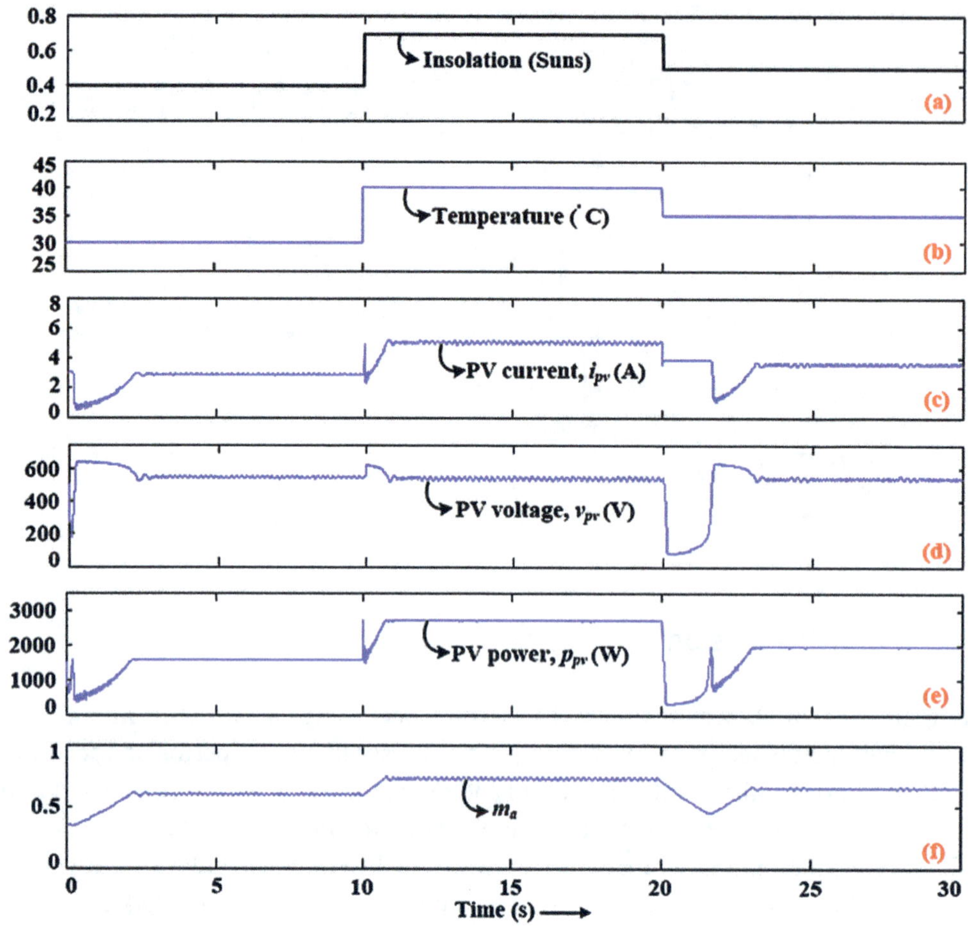

Figure 17. Voltage across PV/dc-bus capacitors, C_{pv1} and C_{pv2}

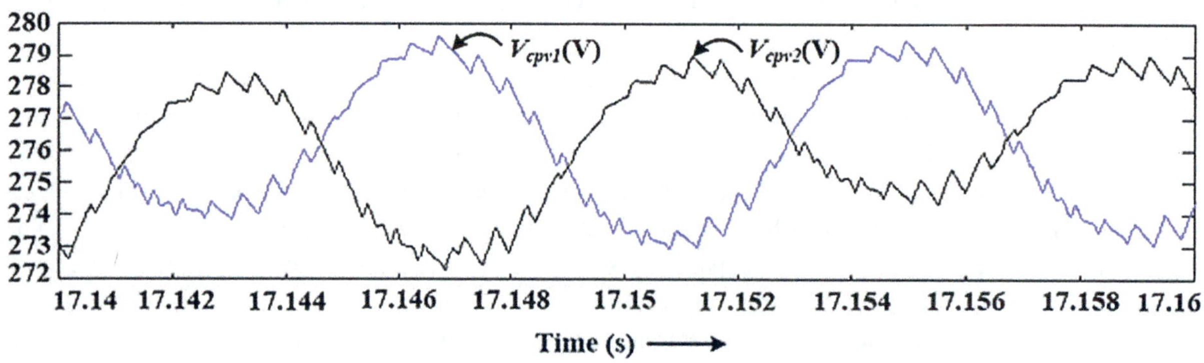

Figure 18. Motor and pump parameters

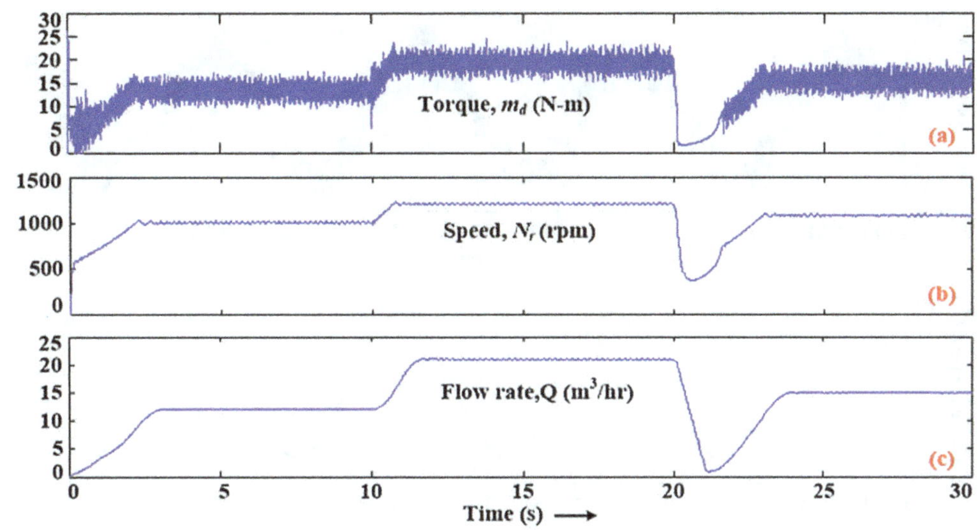

inverter connected PV pumping system when compared to the torque generated in case of two-level inverter connected PV pumping system. This is because of the higher levels in the motor phase voltage and more sinusoidal nature of the motor phase current. Additionally, the nature of motor phase voltage and the phase current at two different fundamental frequencies is shown in Figures 19 and 20. Figure 19 show the motor phase voltage (top) and the phase current (bottom) at lower fundamental frequency; whereas in Figure 20, the phase voltage and phase current are captured at higher fundamental frequency. Both the figures are plotted using equal time scale (i.e., time difference is 0.1 seconds), so as to depict the working of V/*f* control strategy.

Figure 19. Induction motor phase voltage and phase current at lower modulation index

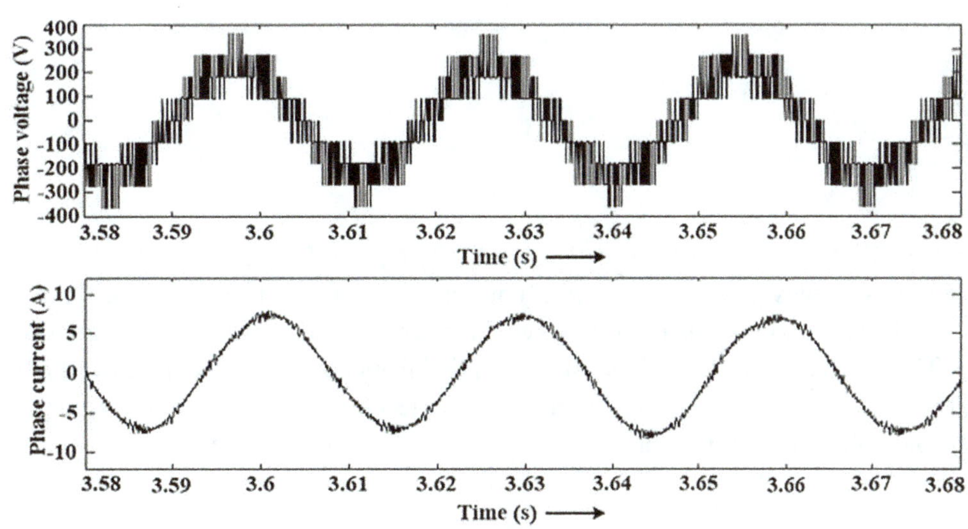

Figure 20. Induction motor phase voltage and phase current at higher modulation index

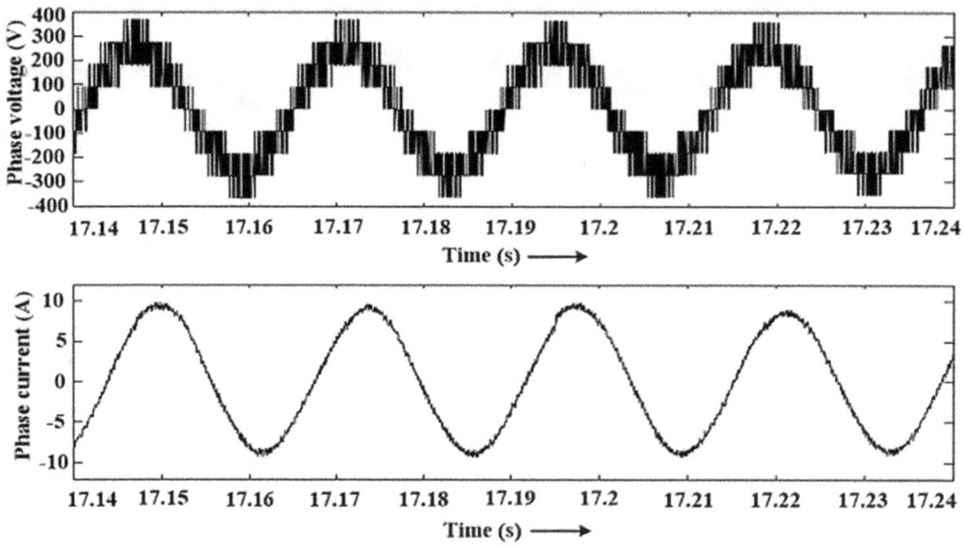

4.3. Three-Phase Three-Level Dual H-Bridge Inverter Connected to the Open-End Winding Induction Motor-Pump Powered From Single PV Source

The dual-inverter considered in this PV pumping system consists of two two-level H-bridge inverters (Inverter-1 and Inverter-2) which are powered from single PV source using a common dc-bus as shown in Figure 21 (Sachin Jain, 2015). The Inverter-1 and Inverter-2 pole voltages are defined from the respective three-phase poles to the dc ground points (say '*o*' as shown in Figure 21). The pole-voltages 'v_{ao}' at terminal '*a*' of Inverter-1 and '$v_{a'o}$' at terminal '*a'*' of Inverter-2 can be expressed as,

$$v_{ao} = S_A\left(\frac{V_{pv}}{4}\right) \tag{58}$$

$$v_{a'o} = S_{A'}\left(\frac{V_{pv}}{4}\right) \tag{59}$$

where, S_Y (where $Y\in \{A, B, C\}$ for Inverter-1 and $\{A', B', C'\}$ for Inverter-2) is the switching function defined for all the six poles (*a*, *b*, *c*, *a'*, *b'* and *c'*) of the dual-inverter. The switching function 'S_Y' hold the values 1 or -1, depending on top or bottom switch of leg '*Y*' in corresponding inverter is turned ON respectively. The voltage, V_{pv} is the dc-bus voltage applied from PV source. In this case, the PV bus voltage required is half of the voltage ($V_{pv}/2$) required for two-level inverter connected PV pumping system.

Similarly, the pole-voltages 'v_{Yo}' for Inverter-1 and Inverter-2 can be expressed as,

Figure 21. Schematic circuit diagram of three-phase three-level dual H-bridge inverter connected to the open-end winding induction motor-pump with common dc-bus

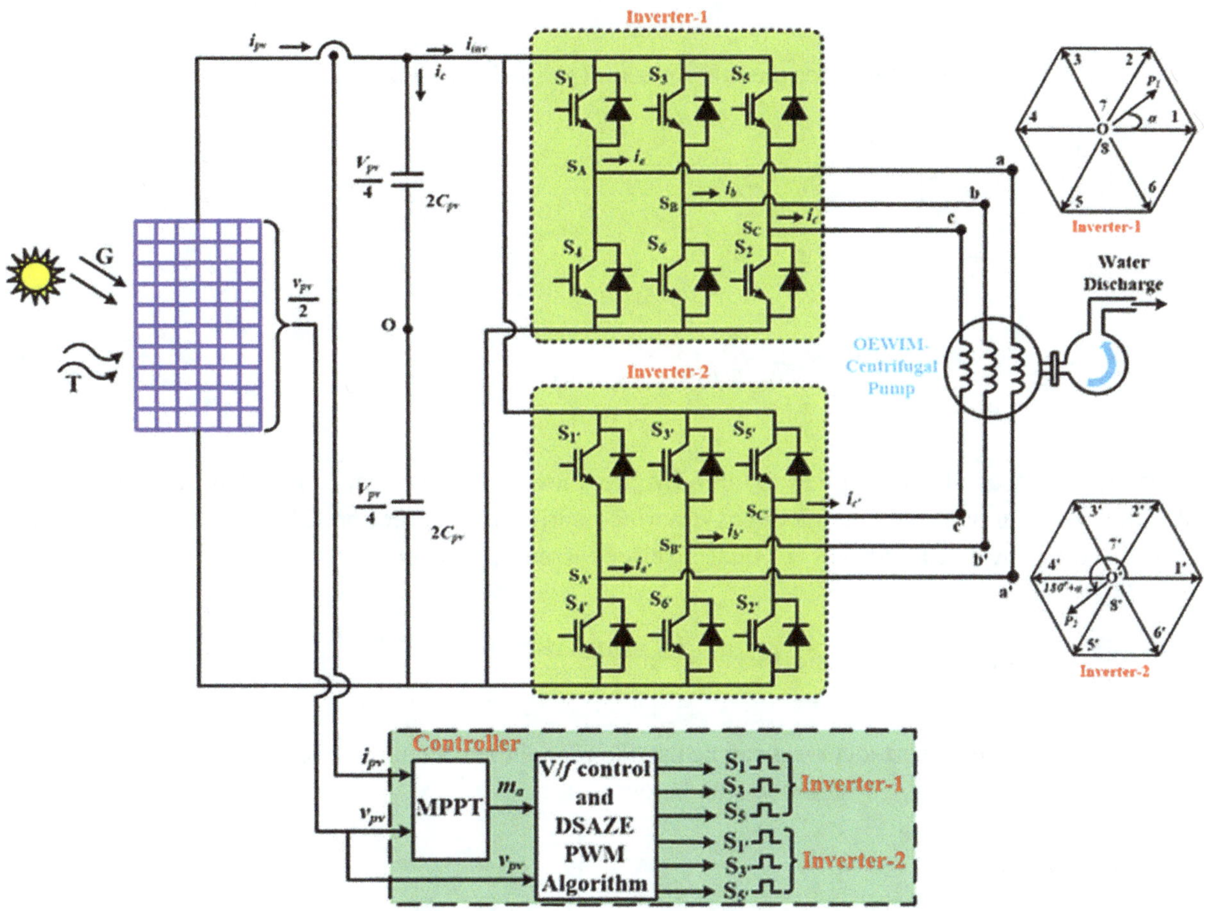

$$v_{bo} = S_{B}\left(\frac{V_{pv}}{4}\right) \text{ and } v_{b'o} = S_{B'}\left(\frac{V_{pv}}{4}\right) \text{for phase-bb'} \tag{60}$$

$$v_{co} = S_{C}\left(\frac{V_{pv}}{4}\right) \text{ and } v_{c'o} = S_{C'}\left(\frac{V_{pv}}{4}\right) \text{for phase-cc'} \tag{61}$$

Once the pole-voltages are known, the phase voltage (phase-*aa'*) for dual-inverter can be defined as:

$$v_{aa'} = v_{ao} - v_{a'o} \tag{62}$$

$$v_{aa'} = \left(S_A \frac{V_{pv}}{4} - S_{A'} \frac{V_{pv}}{4} \right) = \frac{V_{pv}}{4} \left(S_A - S_{A'} \right) \tag{63}$$

Similarly, the phase voltages for the other phases (phase-*bb'* and phase-*cc'*) are given below,

$$v_{bb'} = \left(S_B \frac{V_{pv}}{4} - S_{B'} \frac{V_{pv}}{4} \right) = \frac{V_{pv}}{4} \left(S_B - S_{B'} \right) \tag{64}$$

$$v_{cc'} = \left(S_C \frac{V_{pv}}{4} - S_{C'} \frac{V_{pv}}{4} \right) = \frac{V_{pv}}{4} \left(S_C - S_{C'} \right) \tag{65}$$

Once the phase voltages applied to OEWIM are known, then the respective phase currents can be calculated using motor dynamic model as described earlier in this section. The derived phase current can be used for the calculations of the dual-inverter current 'i_{inv}' as given below,

$$i_{inv} = \frac{1}{2}\left(S_A - S_{A'} \right) i_{aa'} + \frac{1}{2}\left(S_B - S_{B'} \right) i_{bb'} + \frac{1}{2}\left(S_C - S_{C'} \right) i_{cc'} \tag{66}$$

The above derived inverter current can be used to model the PV source as explained in Figure 4. The detailed design of the entire system is given below.

4.3.1. PV Sizing

The design considerations for the PV system shown in Figure 21 are same as the considerations presented in section 4.1.1. The PV array configuration for the dual-inverter connected PV system can be calculated by considering the requirement of PV array with capacity 4 kW. Using the solar PV module MSX60 the values of 'N_T' and 'N_s' can be calculated as,

$$N_T = \frac{P_{pv\_array}}{P_{MPP\_m}} = \frac{4000}{60} \approx 67 \tag{67}$$

$$N_s = \frac{V_{dc\_total}}{V_{MPP\_m}} = \frac{325}{17.1} \approx 19 \tag{68}$$

The integer value of N_s is calculated by considering V_{dc_total} = 325V. In case of two-level H-bridge inverter, the dc-bus voltage required to generate the peak phase voltage of 325V in the linear modulation region is V_{dc_total} = 564V. When the dual-inverter with common dc-bus is used, the dc-bus voltage required to generate the same amount of peak phase voltage is 564/2 i.e., 282V. But, the PWM strategy used in this section requires additional 15% of dc-bus voltage, which results in $V_{dc_total} = 325$V.

Now, the required N_p can be calculated as,

$$N_p = \frac{N_T}{N_s} = \frac{67}{19} \approx 3.5 \tag{69}$$

If the integer value of N_p is considered to be 4, then the PV array configuration can be formed as 19×4. This will result in the power rating of PV source is about 4.5kW (and $N_T = 76$), which is higher than the requirement. So, the integer value of N_p can be considered as 3, which results in the PV array configuration as 19×3 ($N_T = 57$) and the power is about 3.4kW. The first array configuration results in higher power rating, whereas the second configuration gives insufficient power. Hence, the integer value of N_p is fixed to 3 and the value of N_s is changed to 22 so as to meet the power rating of the PV panel. Hence, the PV array configuration can be used as 22×3 (i.e., 22 modules in a string and 3 such strings in parallel; $N_T = 66$). This result in the power of PV array nearly 3950W, the dc-bus voltage 376.2V. Once the PV array configuration is defined, then the task of designing or defining the rating of the components in the inverter can be done.

4.3.2. Design of DC Bus Capacitor and Inverter Switches

The voltage rating required for the components (switches and capacitor) is mainly governed by the maximum voltage appearing across the dc-bus of the dual-inverter. The maximum dc voltage is actually the utmost open circuit voltage of the PV string. Thus, the voltage rating of the switches (V_{sw}) and the capacitors (V_{cap}) to be used is given by,

$$V_{sw} = V_{cap} = N_s \times V_{oc\_m} = 22 \times 21.1 = 464.2V \tag{70}$$

The design value of dc-bus capacitor which acts as buffer between inverter and PV array can be decided based on the back up time to be provided or allowed ripple voltage at the dc-bus. Calculation of the capacitor value based on the ripple voltage assuming short circuit current flowing into it is given by,

$$C_{pv} = \frac{N_p \times I_{sc\_m}}{f_s \times \Delta V_{pv}} = \frac{3 \times 3.8}{554.2 \times 9.284} \approx 2200\mu F \tag{71}$$

where, 'f_s' is same as given in section 4.1.2, 'ΔV_{pv}' is the allowed ripple content in the PV source voltage which is assumed to be 2% of the maximum value of dc-bus voltage calculated using V_{oc_m}.

Now, current rating for the switches in the dual-inverter can be calculated from the below expression (assuming equal current is flowing through the switches of inverter),

$$i_{sw\_S11}(t) = \frac{1}{2}\left(S_A + 1\right) i_{aa'}(t) \tag{72}$$

where, '$i_{sw\_S11}(t)$' and '$i_{aa'}(t)$' denotes the instantaneous value of the current flowing through the switch 'S_1' and motor phase-*aa'* respectively.

The current rating of the semiconductor switches used in the dual-inverter could be decided by the maximum value of the instantaneous switch current 'i_{sw_S11}' (assuming continuous current flowing through the switch). In other words, the peak value of the OEWIM phase current decides the current rating of the semiconductor switch.

4.3.3. Modulation Technique

The control strategy for the PV pumping system which uses OEWIM and dual-inverter with single PV source is shown in Figure 22. The MPPT algorithm to extract maximum available power from PV source is same as that explained in section 4.1.3. The modified value of modulation index 'm_a' is further processed with the operating voltage of the PV array for the calculation of resultant reference voltage vector '$\vec{V}_{sr}$' as given below,

$$\vec{V}_{sr} = m_a \times \frac{V_{pv}}{2} \tag{73}$$

The modulating frequency can be calculated by using '$\vec{V}_{sr}$' as described in Equation (74).

$$f = \frac{\left|\vec{V}_{sr}\right| \times f_{rated}}{\left|\vec{V}_{sr(rated)}\right|} \tag{74}$$

where '$\vec{V}_{sr(rated)}$' is the maximum output voltage across the motor phase winding operating at rated condition. For a peak motor phase voltage of 325V, $\vec{V}_{sr(rated)}$ is calculated as follows:

$$\vec{V}_{sr(rated)} = m_a \times \frac{V_{pv}}{2} = 0.75 \times 325 \approx 244V \tag{75}$$

Here, the value of modulation index is considered as 0.75, since, it is the boundary of linear modulation region for the DSAZE PWM technique.

In addition, the time period of fundamental cycle 'T' is calculated by inverting Equation (74) and the switching time 'T_s' is calculated by inverting Equation (34). The calculated values of 'T' and 'T_s' are used to generate the modulating and carrier signals respectively as shown in Figure 22. The control algorithm generates two modulating signals of equal magnitude with 180° phase displacement. These modulating signals are individually processed using the unified voltage carrier based SVPWM technique. The only difference is in the use of offset time period, T_{offset} which is considered as,

Figure 22. Block diagram of the proposed dual-MPPT technique along with the motor control technique

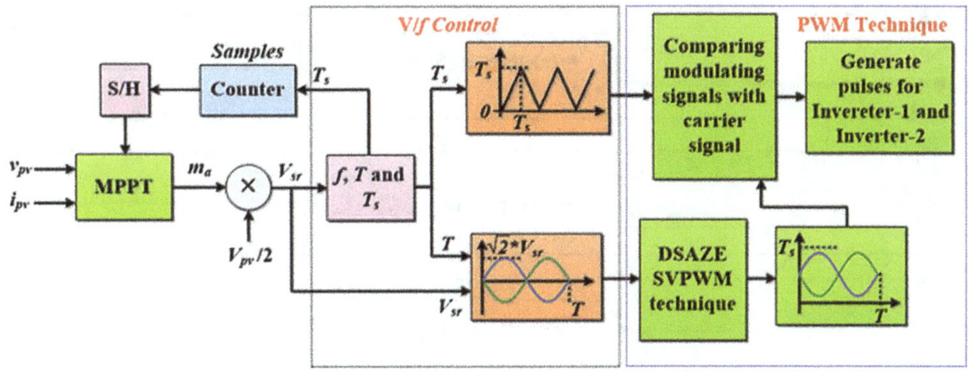

$$T_{offset} = \frac{T_s}{2} \tag{76}$$

The offset time is used for both the inverters so as to eliminate the common mode voltage and hence the common mode current which is because of the use of common dc-bus (Sachin Jain, 2015). The then obtained modulating signals are compared with the carrier signal to generate the gating pulses required by the dual-inverter switches.

4.3.4. Controller Requirements

For the system shown in Figure 21, the controller should have at least two Analog to Digital Converters (ADCs) to sense the PV voltage and the PV current from the PV source. This is done to achieve the MPPT operation. Another requirement is the controller should have six independent enhanced Pulse Width Modulation pins (ePWM pins) to drive the inverter switches. In addition, the controller should have sufficient flash memory and the Random Access Memory (RAM) where the code for the control algorithm can be encrypted. Furthermore, most importantly the chosen controller should have sufficient clock frequency, which could handle the calculation burden in a better way and could generate the switching signals with higher frequencies.

4.3.5. Results and Discussion

The PV pumping system shown in Figure 21 is again simulated using MATLAB/Simulink software. The PV source is simulated by considering its array configuration of 22x3. A 4kW, 400V, 1430 rpm three-phase OEWIM is used (same as the motor considered in section 4.1.5). For simulation, initially the insolation and temperature values for the PV source is considered as 0.4Suns and 30°C respectively. Then the environmental conditions are changed to 0.7Suns and 40°C, later are modified to 0.5Suns and 35°C. The simulation results for these three conditions are shown in Figure 23. The system is started operating by applying an initial modulation index equal to 0.3 (this value can be varied depending on the initial environmental conditions). During increasing environmental conditions the PV parameters like current, voltage, and power shown in Figure 23 are nearly same as the PV current, voltage and power for

Figure 23. Simulated PV source parameters

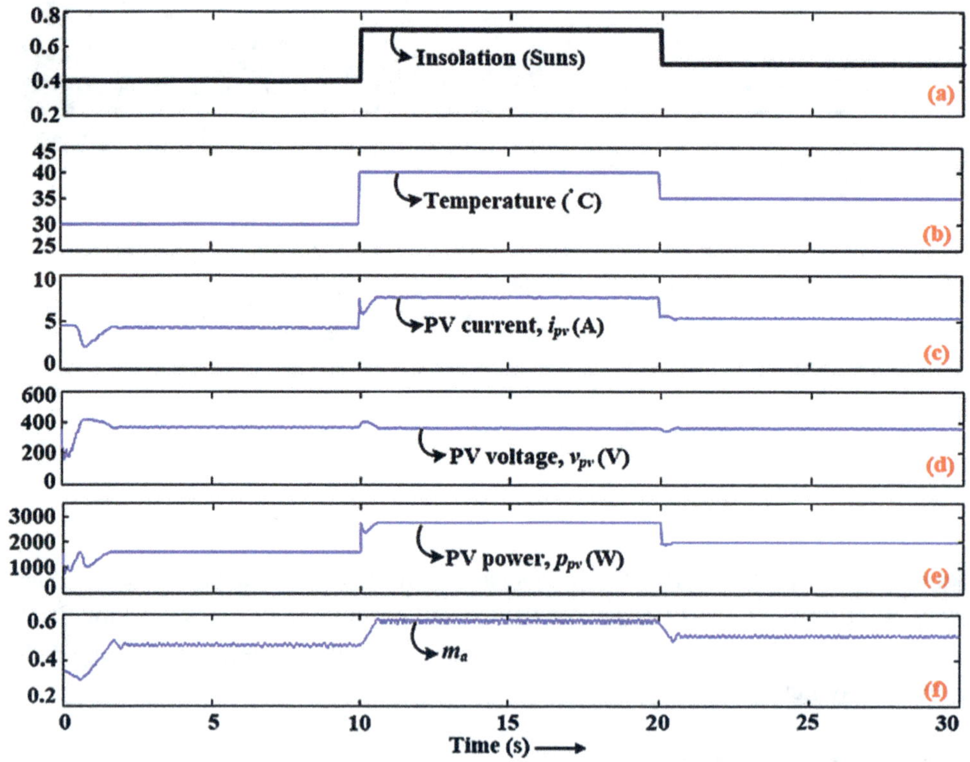

two-level H-bridge inverter system and three-level cascaded inverter system (see Figures 10 and 16). The same is the case with the modulation index. However, during the decreasing environmental conditions, the PV source is not entering into the current source region. This is different from both the PV systems with H-bridge inverter and the cascaded inverter. This is due to the value of dc-bus capacitor which could hold the PV/dc-bus voltage to the required value and hence the required current.

Figure 24 show the simulation results for motor and pump side parameters under different environmental conditions as mentioned above. The variations in the modulation index 'm_a' with respect to PV power can also be seen here in the waveform of motor developed torque, speed of the motor and the water flow rate of pump as shown in Figure 24. The constant head of 30m (same as for two-level inverter and three-level cascaded inverter connected PV pumping systems) is considered for the calculation of flow rate of water. It can be observed that torque, speed, and flow rate follow 'm_a' or the PV power indirectly. The ripple content in the motor generated torque is nearly same when compared to the torque generated in case of two-level inverter connected PV pumping system; and is more when compared to the cascaded inverter connected PV pumping system. Additionally, the nature of motor phase voltage and the phase current at two different fundamental frequencies (or two different modulation indices) is shown in Figures 25 and 26. Figure 25 show the motor phase voltage (top) and the phase current (bottom) at lower fundamental frequency; whereas in Figure 26, the phase voltage and phase current are captured at higher fundamental frequency. Both the figures are plotted using equal time scale (i.e., time difference is 0.1 seconds), so as to depict the working of V/*f* control strategy.

Figure 24. Motor and pump parameters

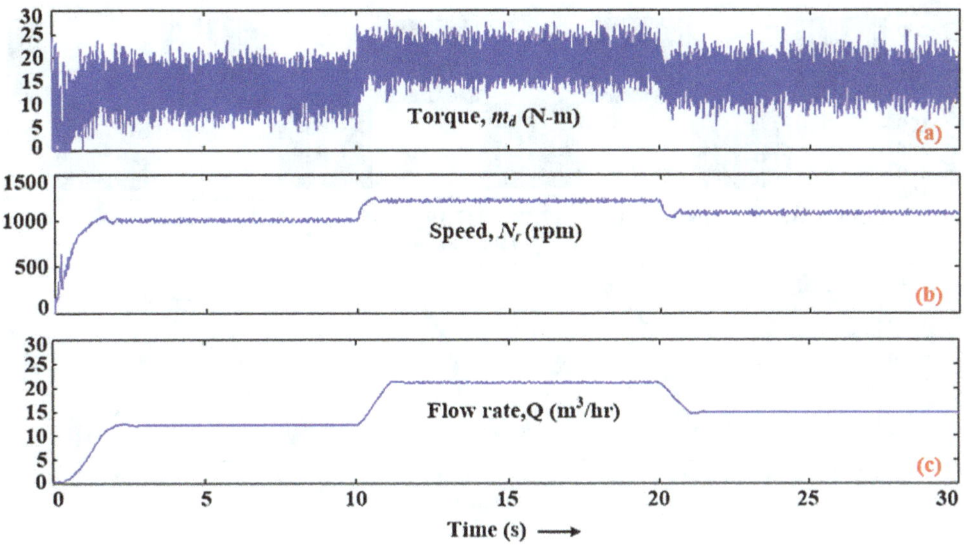

Figure 25. OEWIM phase voltage and phase current at lower modulation index

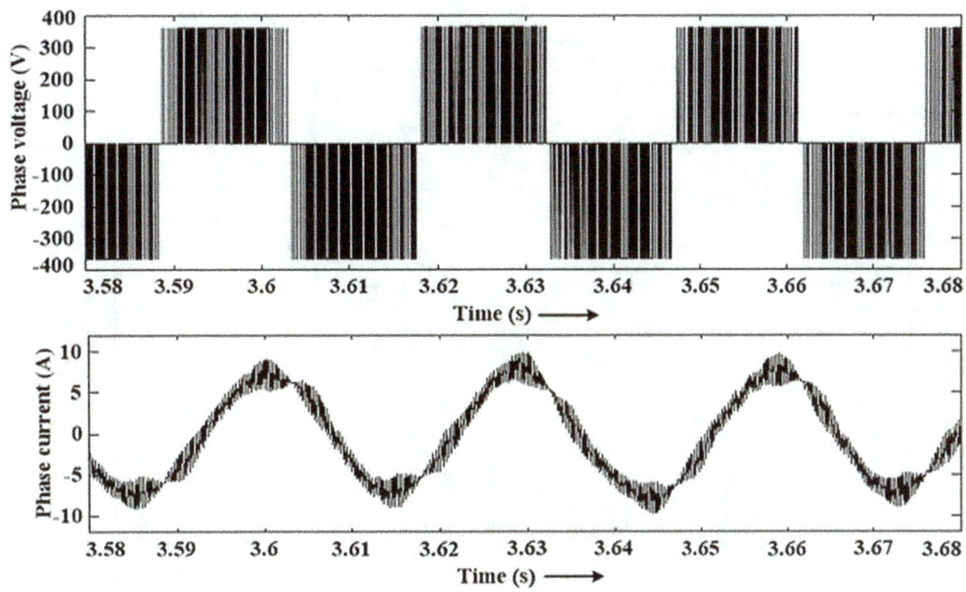

4.4. Three-Phase Three/Four-Level Dual H-Bridge Inverter Connected to the Open-End Winding Induction Motor-Pump With Electrically Isolated DC-Links

The dual-inverter considered in this PV pumping system consists of two two-level H-bridge inverters (Inverter-1 and Inverter-2) which are powered from two electrically isolated PV sources as shown in Figure 27 (Ramulu C., 2014 and Jain S. 2016). The Inverter-1 and Inverter-2 pole voltages are defined

Figure 26. OEWIM phase voltage and phase current at higher modulation index

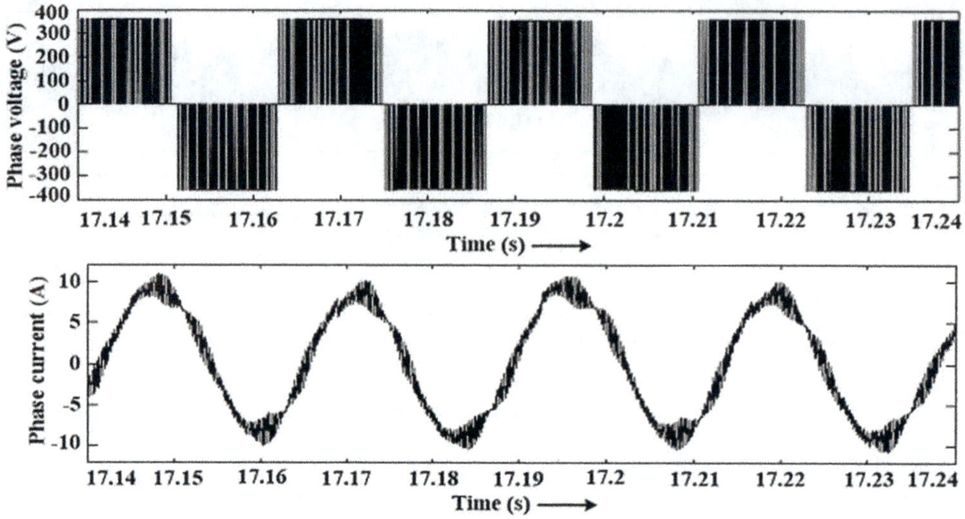

Figure 27. Schematic circuit diagram of three-phase three/four-level dual H-bridge inverter connected to the open-end winding induction motor-pump with electrically isolated dc-links

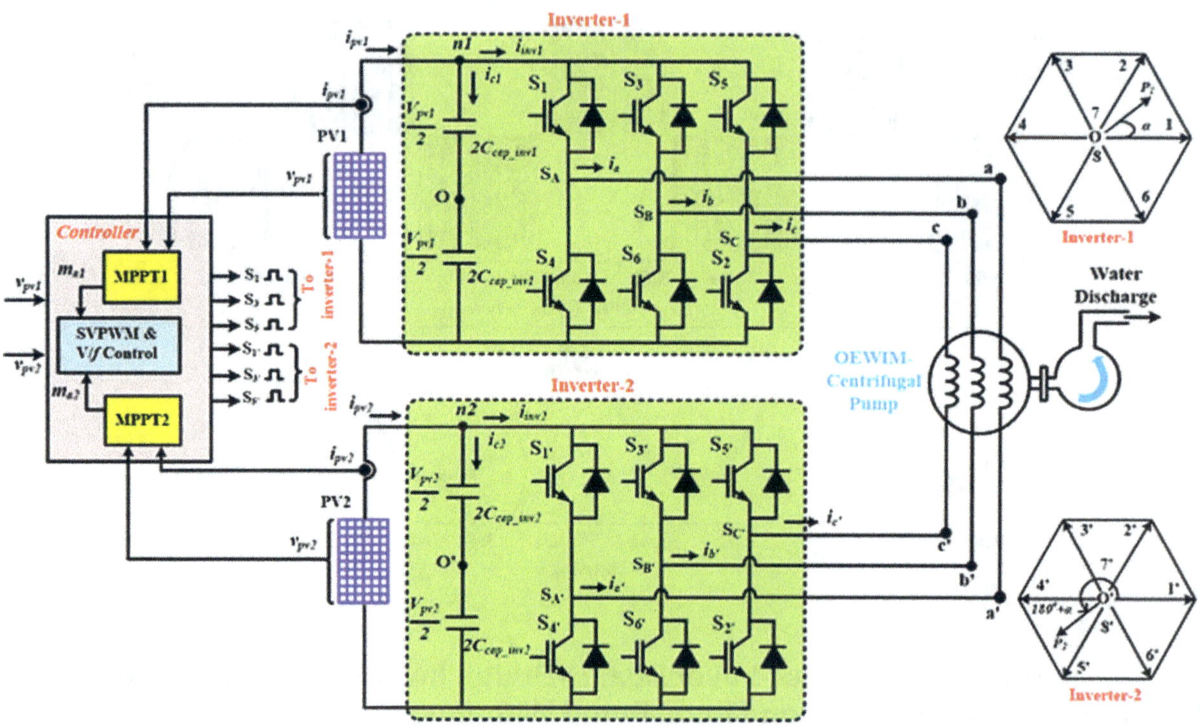

from the respective three-phase poles to the dc ground points (say *o* and *o'* as shown in Figure 27). The pole-voltages 'v_{ao}' at terminal '*a*' of Inverter-1 and '$v_{a'o'}$' at terminal '*a*'' of Inverter-2 can be expressed as,

$$v_{ao} = S_A \left(\frac{V_{pv1}}{2} \right) \tag{77}$$

$$v_{ao} = S_A \left(\frac{V_{pv1}}{2} \right) \tag{78}$$

where, S_Y (where $Y \in \{A, B, C\}$ for Inverter-1 and $\{A', B', C'\}$ for Inverter-2) is the switching function defined for all the six poles (*a*, *b*, *c*, *a'*, *b'* and *c'*) of the dual-inverter. The switching function 'S_Y' hold the values 1 or -1, depending on top or bottom switch of leg '*Y*' in corresponding inverter is turned ON respectively. The voltages V_{pv1} and V_{pv2} are the dc bus voltages applied from PV sources PV1 and PV2 respectively.

Similarly, the pole-voltages 'v_{Yo}' for Inverter-1 and '$v_{Yo'}$' for Inverter-2 can be expressed as,

$$v_{bo} = S_B \left(\frac{V_{pv1}}{2} \right) \text{and } v_{b'o'} = S_{B'} \left(\frac{V_{pv2}}{2} \right) \text{for phase-bb'} \tag{79}$$

$$v_{co} = S_C \left(\frac{V_{pv1}}{2} \right) \text{and } v_{c'o'} = S_{C'} \left(\frac{V_{pv2}}{2} \right) \text{for phase-cc'} \tag{80}$$

Once the pole-voltages are known, the common mode voltage of the Inverter-1 (v_{on}) and Inverter-2 (νo'n) are given by Equation (81) and Equation (82) respectively.

$$v_{on} = \frac{1}{3}\left(v_{ao} + v_{bo} + v_{co}\right) \tag{81}$$

$$v_{o'n} = \frac{1}{3}\left(v_{a'o'} + v_{b'o'} + v_{c'o'}\right) \tag{82}$$

where, '*n*' represents a hypothetical ground point of the OEWIM. The common mode voltage for dual-inverter is defined as,

$$v_{oo'} = v_{on} - v_{o'n} = \frac{1}{3}\left[\left(v_{ao} - v_{a'o'}\right) + \left(v_{bo} - v_{b'o'}\right) + \left(v_{co} - v_{c'o'}\right)\right] \tag{83}$$

Now, the phase voltage (phase-*aa'*) for dual-inverter can be defined as:

$$v_{aa'} = \left(v_{ao} - v_{a'o'}\right) - v_{oo'} \tag{84}$$

$$v_{aa'} = \left(S_A \frac{V_{pv1}}{2} - S_{A'} \frac{V_{pv2}}{2}\right) - \frac{1}{3}\left[\left(S_A \frac{V_{pv1}}{2} - S_{A'} \frac{V_{pv2}}{2}\right) + \left(S_B \frac{V_{pv1}}{2} - S_{B'} \frac{V_{pv2}}{2}\right) + \left(S_C \frac{V_{pv1}}{2} - S_{C'} \frac{V_{pv2}}{2}\right)\right] \tag{85}$$

Simplifying the Equation (85) results in,

$$v_{aa'} = \frac{V_{pv1}}{2}\left\{\left(\frac{2}{3}\right)S_A - \frac{1}{3}\left(S_B + S_C\right)\right\} - \frac{V_{pv2}}{2}\left\{\left(\frac{2}{3}\right)S_{A'} - \frac{1}{3}\left(S_{B'} + S_{C'}\right)\right\} \tag{86}$$

Similarly, the phase voltages for the other phases (phase-*bb'* and phase-*cc'*) are given below,

$$v_{bb'} = \frac{V_{pv1}}{2}\left\{\left(\frac{2}{3}\right)S_B - \frac{1}{3}\left(S_C + S_A\right)\right\} - \frac{V_{pv2}}{2}\left\{\left(\frac{2}{3}\right)S_{B'} - \frac{1}{3}\left(S_{C'} + S_{A'}\right)\right\} \tag{87}$$

$$v_{cc'} = \frac{V_{pv1}}{2}\left\{\left(\frac{2}{3}\right)S_C - \frac{1}{3}\left(S_A + S_B\right)\right\} - \frac{V_{pv2}}{2}\left\{\left(\frac{2}{3}\right)S_{C'} - \frac{1}{3}\left(S_{A'} + S_{B'}\right)\right\} \tag{88}$$

Once the phase voltages applied to OEWIM are known, then the respective phase currents can be calculated using motor dynamic model as described by Equations (5) to (10) and Equations (13) and (14). The derived phase current can be used for the calculations of the inverter current 'i_{inv1}' and 'i_{inv2}' of both Inverter-1 and Inverter-2 respectively, as given below,

$$i_{inv1} = \frac{1}{2}\left(S_A + 1\right)i_a + \frac{1}{2}\left(S_B + 1\right)i_b + \frac{1}{2}\left(S_C + 1\right)i_c \tag{89}$$

$$i_{inv2} = \frac{1}{2}\left(S_{A'} + 1\right)i_{a'} + \frac{1}{2}\left(S_{B'} + 1\right)i_{b'} + \frac{1}{2}\left(S_{C'} + 1\right)i_{c'} \tag{90}$$

The above derived inverter currents can be used to model the individual PV sources PV1 and PV2 as explained in PV source model.

The detailed design of the entire system is given below.

4.4.1. PV Sizing

The PV array configuration explained in section 4.3.1 is for a single PV source fed OEWIM-pump system, wherein the dual-inverter is connected with the common dc-link. There is no choice given to the user with respect to the arrangement and configuration of the PV array for single source system. However, for the system with two PV sources the user can define the arrangements or configurations of the two individual PV arrays. The arrangements or configurations of the PV array can be calculated as follows:

$$N_T = x \times y_1 + (N_s - x) \times y_2 \tag{91}$$

where 'x' is the variable representing number of PV modules used in the string of the PV array PV1, variables 'y_1' and 'y_2' represents number of strings connected in parallel in PV arrays PV1 and PV2 respectively.

Substituting different integer values for the variables 'y_1' and 'y_2', one can determine the various possible configurations for the PV system. Consider an example, the requirement of PV array with capacity 4 kW (same as the other PV pumping systems). Using the module of MSX60 the values of 'N_s' and 'N_T' results in, $N_T = 67$ and $N_s = 33$. The integer value of N_s is calculated by considering V_{dc_total} = 564V (since, it is the total required PV/dc-bus voltage to maintain the peak phase voltage of 325V in the linear modulation range for the proposed PWM technique).

Using Equation (91), one of the possible solutions for this PV system is given by,

For PV array PV1: $x = 17$ and $y_1 = 3$.

For PV array PV2: $(N_s\text{-}x) = 16$ and $y_2 = 1$.

This means, PV1 can be formed by connecting 17 modules as a string and 3 such strings are connected in parallel (3052W). Further, PV2 is formed by connecting 16 modules as a single string (958W). In this solution PV1 defines most of the system power, so it may not be the good proposal. A solution with near equal power rating by the two PV sources would be elegant. This can be done by simply adding or subtracting one module from the calculated 'N_T' i.e., $N_T = 66$ or 68. Thus, if one module is added into the system, then the two PV arrays can have identical arrangements (both 17×2 and 17×2 configurations, totaling to 68 modules). And if one module is eliminated then two strings of PV arrays with 16 and 17 modules can be arranged (one with 17×2 and the other with 16×2 configuration, totaling to 66 modules).

The above defined configurations result in nearly equal dc-bus voltage, and hence with this PV system one can achieve only three-levels in the motor phase voltage. This PV system also has the advantage of achieving four-levels in the motor phase voltage, by maintaining the ratio of dc-bus voltages to be 2:1 for both the inverters. So, to demonstrate the four-level operation and the effectiveness of the dual MPPT techniques, one more possible arrangement of PV arrays is considered as 22×2 and 11×2 (i.e., PV1 is configured with 22 modules in a string and 2 parallel strings; PV2 is with same current rating but the dc-bus voltage is halved). This result in the power of PV array PV1 is 2633.4W and the power contribution of PV2 is 1316.7W. The dc-bus voltages of PV1 and PV2 are 376.2V and 188.1V respectively, which sums up to 564V (V_{dc_total}). Once dual PV array configuration is defined, then the task of designing or defining the rating of the components in the inverter can be done.

4.4.2. Design of DC Bus Capacitor and Inverter Switches

The voltage rating of the components (switches and capacitor) is mainly governed by the maximum voltage appearing across the dc-bus of each inverter. The maximum dc voltage is actually the utmost open circuit voltage of the PV string. Thus, the voltage rating of the switches (V_{sw_inv1} and V_{sw_inv2} are the voltage ratings for switches to be used in Inverter-1 and Inverter-2 respectively) and the capacitors (V_{cap_inv1} and V_{cap_inv2} are voltage ratings for capacitors at dc-bus to be used in Inverter-1 and Inverter-2 respectively) to be used in the proposed system is given by,

$$\begin{aligned} V_{sw\_inv1} &= V_{cap\_inv1} = x \times V_{oc\_m} = 464.2 \text{ V} \\ \text{V}_{\text{sw\_inv2}} &= \text{V}_{\text{cap\_inv2}} = \left(Ns - x\right) \times V_{oc\_m} = 232.1 \text{ V} \end{aligned} \tag{92}$$

Now, the capacitor value based on the ripple voltage assuming short circuit current flowing into it is given by,

$$\begin{aligned} C_{cap\_inv1} &= \frac{y_1 \times I_{sc\_m}}{f_s \times \Delta V_{pv1}} = \frac{2 \times 3.8}{554.2 \times 9.284} \approx 1477 \text{ uF} \\ \text{C}_{\text{cap\_inv2}} &= \frac{y_2 \times I_{sc\_m}}{f_s \times \Delta V_{pv2}} = \frac{2 \times 3.8}{554.2 \times 4.642} \approx 2954 \text{ uF} \end{aligned} \tag{93}$$

where 'f_s' is the minimum switching frequency of the semiconductor switches and is determined in section 4.1.2. And 'ΔV_{pv1}' and 'ΔV_{pv2}' are the allowed ripple content in the PV source voltages 'V_{pv1}' and 'V_{pv2}' respectively which is assumed to be 2% of the maximum value of dc-bus voltages. Now current rating for the switches in the inverter can be calculated from the below expression (assuming equal current is flowing through the switches of Inverter-1 and Inverter-2),

$$i_{sw\_S_{11}}(t) = \frac{1}{2}\left(S_A + 1\right) i_{aa'}(t) \tag{94}$$

where '$i_{sw\_S11}(t)$' and '$i_{aa'}(t)$' denotes the instantaneous value of the current flowing through the switch 'S_{11}' and motor phase-*aa'* respectively.

The current rating of the semiconductor switches used in both the inverters could be decided by the maximum value of the instantaneous switch current 'i_{sw_S11}' (assuming continuous current flowing through the switch). In other words, the peak value of the OEWIM phase current decides the current rating of the semiconductor switch.

4.4.3. Modulation Technique

This system employs the similar integrated control strategy which is explained in section 4.1.3. The only difference is in the MPPT technique i.e., dual-MPPT algorithm. The control strategy is shown in Figure 28. The MPPT algorithms MPPT1 and MPPT2 are used separately to extract maximum available power from PV sources PV1 and PV2 respectively. The instantaneous values of PV1 voltage 'v_{pv1}' and PV1 current 'i_{pv1}' are sensed and given to MPPT1, whereas the instantaneous sensed values 'v_{pv2}' and 'i_{pv2}' of PV array PV2 are provided to MPPT2 algorithm. The instantaneous PV array powers 'p_{pv1}' and 'p_{pv2}' are calculated. These values are averaged and defined as the present values '$V_{pv1}(n)$', '$V_{pv2}(n)$' and '$P_{pv1}(n)$', '$P_{pv2}(n)$'. These PV voltages and powers are compared with the previously initiated values '$V_{pv1}(n\text{-}1)$', '$V_{pv2}(n\text{-}1)$' and '$P_{pv1}(n\text{-}1)$', '$P_{pv2}(n\text{-}1)$' to determine the operating region in the individual *p-v* curves of PV arrays PV1 and PV2 as given in Equation (95) and Equation (96). So, the dual-MPPT technique assures the extraction of maximum power from individual PV arrays by defining the corresponding operating points independently.

$$if\ \frac{P_{pvj}(n)-P_{pvj}(n-1)}{V_{pvj}(n)-V_{pvj}(n-1)}<0 \text{ then } m_{aj}=m_{aj}+\Delta m_{a} \tag{95}$$

$$if\ \frac{P_{pvj}(n)-P_{pvj}(n-1)}{V_{pvj}(n)-V_{pvj}(n-1)}<0 \text{ then } m_{aj}=m_{aj}+\Delta m_{a} \tag{96}$$

where, $j = 1$ for MPPT1, and 2 for MPPT2.

The modified values of modulation indices 'm_{a1}' and 'm_{a2}' for the individual inverters (see Equation (95) and Equation (96)) are further processed with the operating voltage of the PV array for the calculation of resultant reference voltage vector '$\vec{V}_{sr}$' as given below,

$$\vec{V}_{sr}=m_{a1}V_{pv1}+m_{a2}V_{pv2}=\left|\vec{V}_{sr1}\right|+\left|\vec{V}_{sr2}\right| \tag{97}$$

where, '$\vec{V}_{sr1}$' and '$\vec{V}_{sr2}$' are the reference voltage vectors of Inverter-1 and Inverter-2 respectively.

As both the inverters are operated using different reference voltage vectors, this motivates to define a common frequency of operation for the inverters. The problem of common modulating frequency can be resolved by using '$\vec{V}_{sr}$' for calculating the frequency '*f*' for both inverters as described in Equation (98).

$$f=\frac{\left|\vec{V}_{sr}\right|\times f_{rated}}{\left|\vec{V}_{sr(rated)}\right|} \tag{98}$$

where '$\vec{V}_{sr(rated)}$' is the maximum output voltage across the motor phase winding operating at rated condition. For a peak motor phase voltage of 325V, $\vec{V}_{sr(rated)}$ is calculated as follows:

$$\vec{V}_{sr(rated)} = \left(\frac{\sqrt{3}}{2}\right) \times V_{dc} = \left(\frac{\sqrt{3}}{2}\right) \times 564 \approx 488\text{V} \quad (99)$$

where, V_{dc} = 564V is the total required dc-bus voltage to maintain the peak phase voltage of 325 V in the linear modulation range.

In addition, the time period of fundamental cycle 'T' is calculated by inverting Equation (98) and the switching time 'T_s' is calculated by inverting Equation (34). The calculated values of 'T' and 'T_s' are used to generate the modulating and carrier signals respectively as shown in Figure 28. Furthermore, the fundamental time period 'T' is used to hold the sensed values of PV voltages and currents of both the PV sources. This is again done for two reasons: one is to maintain the synchronization between both the modulating and carrier signals and the other is to maintain the fixed number of samples per fundamental cycle. The control algorithm generates two modulating signals of equal magnitude with 180° phase displacement. This helps in improving the dc-bus utilization. The 180° phase displaced modulating signals are processed using the unified voltage carrier based SVPWM technique and it results in two camel-hump signals as shown in Figure 28. These signals are compared with the carrier signal and the gating pulses required by the inverter switches are generated.

4.4.4. Controller Requirements

The controller shown in Figure 27 can be any low cost device which can perform the given controller algorithm. For the system shown in Figure 27, the controller should have at least four Analog to Digital Converters (ADCs) to sense the PV voltage and the PV current from two isolated PV sources. This is done to achieve the MPPT operation. Another requirement is the controller should have six independent enhanced Pulse Width Modulation pins (ePWM pins) to drive the inverter switches. In addition, the controller should have sufficient flash memory and the Random Access Memory (RAM) where the code for the control algorithm can be encrypted. Furthermore, most importantly the chosen controller

Figure 28. Block diagram of the proposed dual-MPPT technique along with the motor control technique

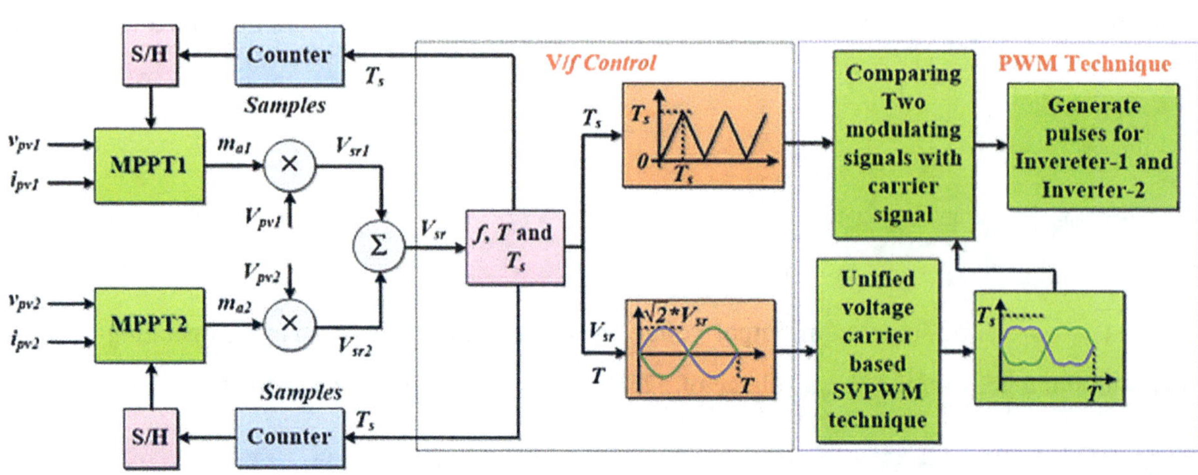

should have sufficient clock frequency, which could handle the calculation burden in a better way and could generate the switching signals with higher frequencies.

4.4.5. Results and Discussion

The PV pumping system shown in Figure 27 is simulated using MATLAB/Simulink software. The PV source is simulated by considering the parameters of a Solarex MSX60 panel as given above. A 4kW, 400V, 1430 rpm three-phase OEWIM is used (same as the motor considered in section 4.1.5). The PV array configurations used are 22×2 for PV1 and 11×2 for PV2. That means, the power contribution of PV1 is higher (twice) than the contribution of PV2. Also, the generated voltage of PV1 is twice that of PV2 voltage with equal current sharing. The equal current sharing is because of the strings connected in parallel i.e., 2.

For simulation, initially the insolation and temperature values for both the PV sources PV1 and PV2 are considered to be at two extreme conditions; then their roles are reversed and later are set to equal conditions as given in Table 2 and as shown in Figure 29(a) and (b). The simulation results for these three conditions are shown in Figure 29 and they are divided in to three regions *viz.* region-I, region-II and region-III. In region-I, the generated current (see Figure 29(c)) by both the PV sources is different (PV1 current is greater than PV2 current), this is because of the applied insolation conditions. Whereas, in region-II the PV1 current is reduced than the PV2 current, since the insolation condition is reversed. In region-III, both the PV sources contribute equal current generation, because of the equal insolation conditions. The generated PV voltage (see Figure 29(d)) by both the sources is nearly in the ratio 2:1, this is because of the PV array configuration. However, the slight variations in temperature conditions for both the PV sources will not affect the generated PV voltage in all the regions as shown in Figure 28(d). The power contributions of the sources PV1 and PV2 in these regions are furnished in Table 2. Since, the source PV1 is of larger size (22×2), the power contribution of PV1 is always higher compared to source PV2 irrespective of applied insolation and temperature conditions (see Figure 29(e)). The most important thing i.e., the beauty of the above explained control technique can be observed from the modulation index plot (Figure 29(f)). The modulation indices from MPPT1 and MPPT2 adjust themselves to track maximum power from both the sources.

As observed from the plot of modulation index, both the inverters are operated with different modulation indices. One of the way to operate the chosen drive system is by fixing the frequency of the inverter output voltage to its rated value and with different modulation indices (amplitude of reference voltage). But, this kind of control technique will not allow to implement the *V/f* control strategy. Hence, this is a challenging task to operate this type of system with single motor-pump load and with *V/f* control strategy. So to maintain the common frequency of operation for both the inverters, the frequency is defined using

Table 2. Environmental conditions and corresponding power contributions of the sources PV1 and PV2

Region	Insolation (Suns)		Temperature (°C)		Power (W)	
	PV1	PV2	PV1	PV2	PV1	PV2
I	0.7	0.4	40	30	1806	519
II	0.4	0.7	30	40	1039	903
III	0.5	0.5	35	35	1289	644

Figure 29. Simulated PV source parameters

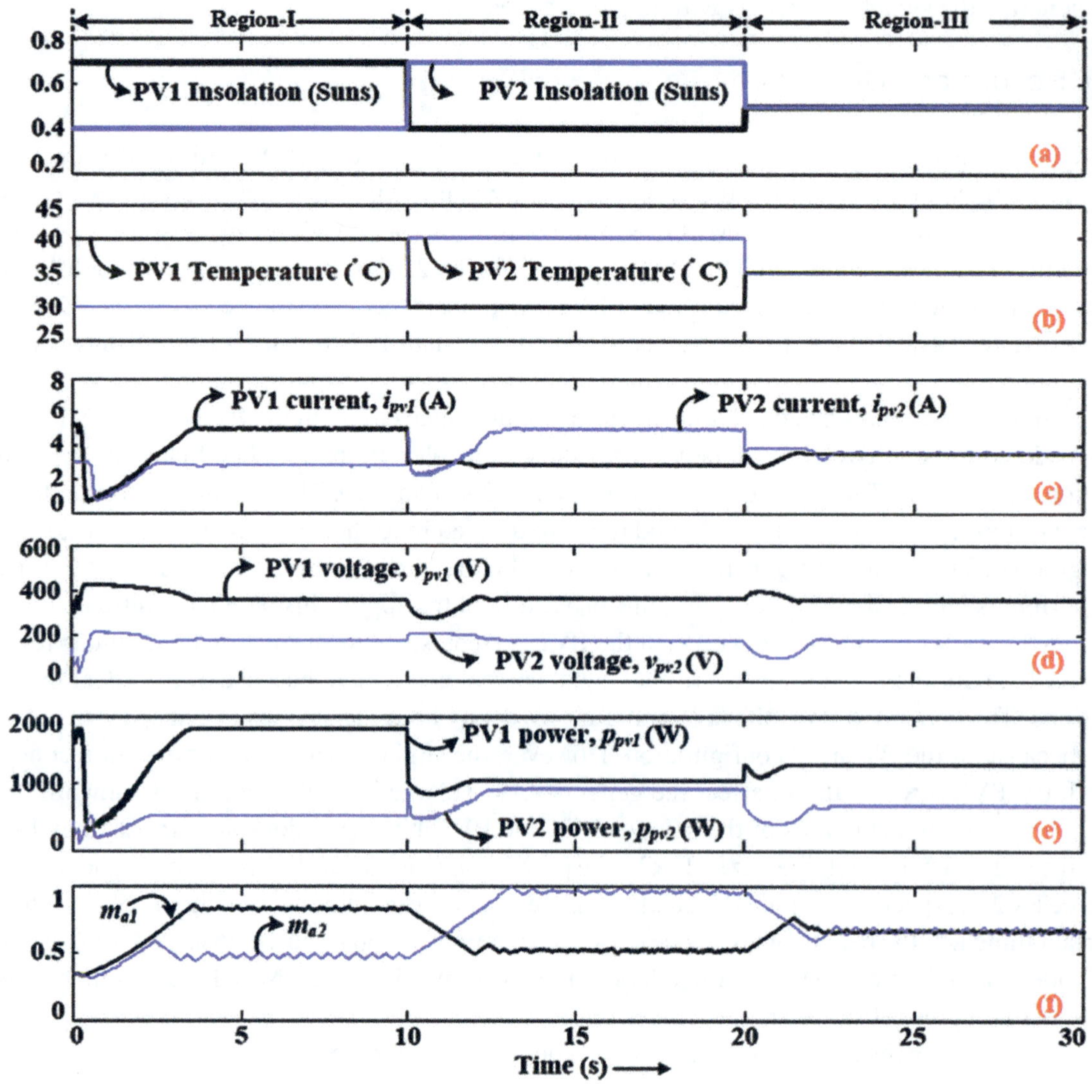

the reference voltage vectors as shown in Figure 30(a) and (b) (see Equations (97) and (98)). Furthermore, the operation of OEWIM-pump drive is depicted using the torque generated by the motor, speed of the motor and the obtained flow rate from the pump as shown in Figures 30(c) to (e) respectively. Additionally, the nature of motor phase voltage (four-levels) and the phase current (sinusoidal) is shown in Figure 31(a) and (b) respectively. The four-level nature in the motor phase voltage is because of the dc-bus voltage ratio i.e., 2:1, whereas if the same system is operated using the dc-bus voltages in the ratio 1:1 one can obtain three-levels in the motor phase voltage.

Figure 30. Motor and pump parameters

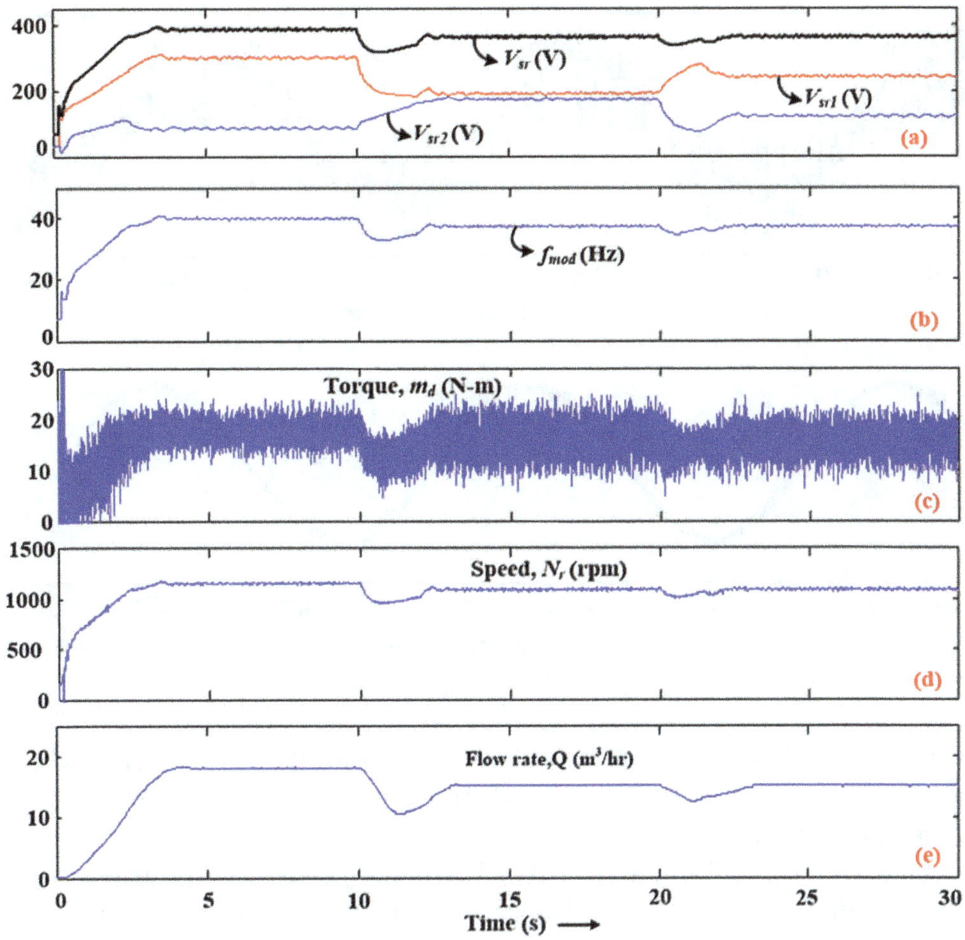

5. FUTURE RESEARCH DIRECTIONS

In this chapter, several aspects of various PV powered water pumping systems have been presented. Some suggestions or research directions for the future, in this field are presented as follows.

1. It would be an interesting study to use different PWM techniques, different MPPT techniques (for both unshaded and shaded conditions) and motor control techniques (like FOC, DTC, PTC, etc.) for the presented inverter configurations in the PV pumping system.
2. It would be an interesting study to extend the above research for PV powered drive systems for various applications (for example, Electric Vehicle).
3. It is also interesting to investigate the switching power loss for various inverter configurations employed in the PV pumping system to estimate the efficiency of the inverter.

Figure 31. OEWIM phase voltage and phase current

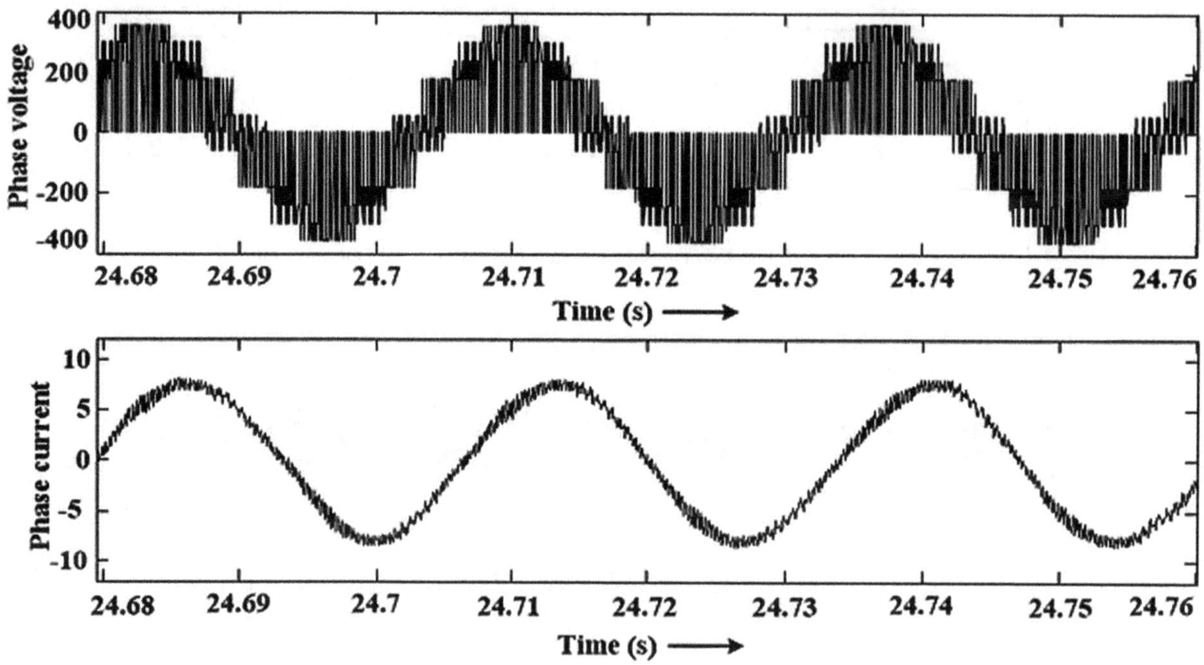

4. It would be very useful to explore the possibility of employing the two PV sources with dual-MPPT for one of the single-stage PV powered cascaded three-level inverter fed water pumping system for more better performance.
5. Furthermore, PV powered dual-inverter fed OEWIM configuration can be extended to have multiple PV sources with distributed MPPT technique. This may help in the generation of higher number of levels in the output voltage.

6. CONCLUSION

In this chapter, four different single-stage PV powered water pumping systems are presented. All the PV systems are operated using an integrated control algorithm which includes MPPT for extracting maximum power from PV source, PWM techniques to operate the inverters and the open-loop *V*/*f* control to control the speed of the motor and the flow rate of the centrifugal pump. Also, the detailed explanation for the mathematical model of PV source, induction motor, OEWIM and various inverter configurations are presented. In addition, the sizing of PV array configuration is presented in detail. Furthermore, the control technique used for all the four PV pumping systems is furnished along with the controller requirements. To support the analysis, all the PV pumping systems are simulated and the corresponding results are presented and discussed. Hence, the detailed study of all the presented PV pumping systems would help the reader to understand and gain knowledge in the field of standalone PV systems and can commission the indigenous standalone PV pumping system.

REFERENCES

Alghuwainem, S. M. (1992). Steady-state performance of DC motors supplied from photovoltaic generators with step-up converter. *IEEE Transactions on Energy Conversion*, *7*(2), 267–272. doi:10.1109/60.136220

Alves Vitorino, M., & Beltrão de Rossiter Corrêa, M. (2011). An effective induction motor control for photovoltaic pumping. *IEEE Transactions on Industrial Electronics*, *58*(4), 1162–1170. doi:10.1109/TIE.2010.2054053

Appelbaum, J. (1986). Starting and steady-state characteristics of dc-motors powered by solar cell generators. *IEEE Transactions on Energy Conversion*, *EC-1*(1), 17–25. doi:10.1109/TEC.1986.4765665

Betka, A., & Moussi, A. (2004). Performance optimization of a photovoltaic induction motor pumping system. *Elsevier Journal on Renewable Energy*, *29*(14), 2167–2181. doi:10.1016/j.renene.2004.03.016

Bhat, S. R., Pittet, A., & Sonde, B. S. (1987). Performance optimization of induction motor-pump system using photovoltaic energy source. *IEEE Transactions on Industry Applications*, *IA-23*(6), 995–1000. doi:10.1109/TIA.1987.4505020

Chinthamalla, Padmanaban, Karampuri, Jain, Ertas, & Fedak. (2016). A solar PV water pumping solution using a three-level cascaded inverter connected induction motor drive. *Engineering Science and Technology, an International Journal*, *19*(4), 1731-1741.

Chinthamalla, R., Ganesh, K. S., & Jain, S. (2014). *An optimal and efficient PV system using two 2-level cascaded 3-level inverter for Centrifugal pump*. Paper presented at IEEE International Conference on Power Electronics, Drives and Energy Systems (PEDES), Mumbai, India. 10.1109/PEDES.2014.7042060

Chung, D.-W., Kim, J.-S., & Sul, S.-K. (1998). Unified voltage modulation technique for real-time three-phase power conversion. *IEEE Transactions on Industry Applications*, *34*(2), 374–380. doi:10.1109/28.663482

Corrêa, T. P., Seleme, S. I. Jr, & Silva, S. R. (2012). Efficiency optimization in stand-alone photovoltaic pumping system. *Elsevier Journal on Renewable Energy*, *41*, 220–226. doi:10.1016/j.renene.2011.10.024

Elgendy, M. A., Zahawi, B., & Atkinson, D. J. (2010). Comparison of Directly Connected and Constant Voltage Controlled Photovoltaic Pumping Systems. *IEEE Transactions on Sustainable Energy*, *1*(3), 184–192. doi:10.1109/TSTE.2010.2052936

Eskander, M. N., & Zaki, A. M. (1997). A maximum efficiency photovoltaic induction motor pump system. *Elsevier Journal on Renewable Energy*, *10*(1), 53–60. doi:10.1016/0960-1481(96)00004-3

Jain, Karampuri, & Somasekhar. (2016). An integrated control algorithm for a single-stage PV pumping system using an open-end winding induction motor. *IEEE Transactions on Inductrial Electronics*, *63*(2), 956-965.

Jain, Ramulu, Sanjeevikumar, Ojo, & Ertas. (2016). Dual MPPT algorithm for dual PV source fed Open-End Winding Induction Motor Drive for pumping application. *Engineering Science and Technology, an International Journal*, *19*(4), 1771-1780.

Jain, S., Thopukara, A. K., Karampuri, R., & Somasekhar, V. T. (2015). A single-stage photovoltaic system for a dual-inverter-fed open-end winding induction motor drive for pumping applications. *IEEE Transactions on Power Electronics*, *30*(9), 4809–4818. doi:10.1109/TPEL.2014.2365516

Krause, P., Wasynczuk, O., Sudhoff, S., & Pekarek, S. (2013). *Analysis of Electric Machinery and Drive Systems*. John Willey & Sons. doi:10.1002/9781118524336

Kumar, R., & Singh, B. (2016). BLDC Motor-Driven Solar PV Array-Fed Water Pumping System Employing Zeta Converter. *IEEE Transactions on Industry Applications*, *52*(3), 2315–2322. doi:10.1109/TIA.2016.2522943

Kumar, R., & Singh, B. (2016). Solar photovoltaic array fed water pump driven by brushless DC motor using Landsman converter. *IET Renewable Power Generation*, *10*(4), 474–484. doi:10.1049/iet-rpg.2015.0295

Makhlouf, M., Messai, F., & Benalla, H. (2011). Vectorial command of induction motor pumping system supplied by a photovoltaic generator. *Journal of Electrical Engineering*, *62*(1), 3–10. doi:10.2478/v10187-011-0001-7

Muljadi, E. (1997). PV water pumping with a peak-power tracker using a simple six-step square-wave inverter. *IEEE Transactions on Industry Applications*, *33*(3), 714–721. doi:10.1109/28.585862

Ramulu, C., Praveen Kumar, T., & Jain, S. (2014). *Single stage PV source based Dual Inverter Fed Open-End Winding Induction Motor pump drive*. Paper presented at IEEE Students' Conference on Electrical, Electronics and Computer Science (SCEECS), Bhopal, India. 10.1109/SCEECS.2014.6804459

Sashidhar, S., & Fernandes, B. G. (2017). A Novel Ferrite SMDS Spoke-Type BLDC Motor for PV Bore-Well Submersible Water Pumps. *IEEE Transactions on Industrial Electronics*, *64*(1), 104–114. doi:10.1109/TIE.2016.2609841

Singh, Singh, Singh, Chandra, & Al-Haddad. (1996). *Optimized performance of solar powered variable speed induction motor drive*. Paper presented at International conference on Power Electronics, Drives and Energy systems for industrial growth, New Delhi, India.

Singh, B., Mishra, A. K., & Kumar, R. (2016). Solar Powered Water Pumping System Employing Switched Reluctance Motor Drive. *IEEE Transactions on Industry Applications*, *52*(5), 3949–3957. doi:10.1109/TIA.2016.2564945

Stemmler, H., & Guggenbach, P. (1993). *Configurations of high-power voltage source inverter drives*. Paper presented at European Power Electronic conference, Zurich Switzerland.

Walker, G. (2001). Evaluating MPPT converter topologies using a MATLAB PV model. *Journal of Electrical and Electronics Engineering (Oradea)*, *21*(1), 49–56.

Yao, Y., Bustamante, P., & Ramshaw, R. S. (1994). Improvement of induction motor drive systems supplied by photovoltaic arrays with frequency control. *IEEE Transactions on Energy Conversion*, *9*(2), 256–262. doi:10.1109/60.300150

Chapter 6
An Optimizer-Tool-Based Improved Metaheuristic Method for Solving Security Optimal Power Flow:
Interactive Power System Planning Tool

Belkacem Mahdad
University of Biskra, Algeria

ABSTRACT

In this chapter, an interactive tool using graphic user interface (GUI) environment-based MATLAB is proposed to solve practical optimal power system planning and control. The main particularity of the proposed tool is to assist student and researchers understanding the mechanism search of new meta-heuristic methods. The proposed tool allows users to interact dynamically with the program. The users (students or experts) can set parameters related to a specified metaheuristic method to clearly observe the effect of choosing parameters on the solution quality. In this chapter, a new global optimization method named grey wolf optimizer (GWO) and pattern search algorithm (PS) have been successfully applied within the interactive tool to solve the optimal power flow problem. The robustness of the two proposed metaheuristic methods is validated on many standard power system tests. The proposed interactive optimal power flow tool is expected to be a useful support for students and experts specialized in power system planning and control.

INTRODUCTION

This chapter aims to develop an interactive educational and research tool using graphic user interface (GUI) environment based MATLAB to solve practical optimal power system planning and control. The main contribution of this study compared to our previous works and other planning strategies developed is related to solve the security OPF considering two new indices: the critical reactive margin stability and

DOI: 10.4018/978-1-5225-3935-3.ch006

the global reactive margin stability. Among various global optimization methods only two algorithms GWA and PS have been successfully demonstrated in this study within the interactive tool to solve the optimal power flow problem. The feasibility and performances of the proposed optimizer tool in term of solution quality and simplicity is validated on many standard power system tests.

The proposed optimizer tool based metaheuristic methods open new research area for students, researchers and industrials in developing new competitive planning strategies for solving with accuracy many problems related to modern practical power system.

BACKGROUND

During the last two decades a large number of metaheuristic methods have been developed and applied to improve the solution of a variety of problems related to power system planning and control. The optimal power flow (OPF) is one of the most problems analyzed and discussed by researchers more than half a century. The main issue of OPF is to ensure delivering energy quality to customers by considering the cost and other objective functions such as, emission, voltage deviation, power loss, and voltage stability (Mahdad, 2014; Mahdad, 2015). As well indicated in many review papers, a large number of mathematical methods (Huneault, 1991; Momoh, 1993; Dommel, 1968; Alsac, 1974; Yan, 1999; Frank, 2012) have been proposed and applied to solve the standard OPF. Experience confirmed that the majority of the mathematical methods fail to achieve the near global solution when considering practical constraints related to generation units such as valve point effect and prohibited zones (Mahdad & Srairi, 2015). With the large integration of various types of FACTS devices and renewable sources characterized by their dynamic and intermittent constraints, the solution of OPF using conventional methods fail to achieve the optimal desired objective (Mahdad & Srairi, 2015). To overcome many drawbacks of the traditional optimization methods, a new category of optimization methods called metaheuristic techniques has been proposed to improve the solution quality of the OPF considering various practical operation constraints. Alsumait et al. (2010) proposed A hybrid GA–PS–SQP to solve the economic dispatch considering valve point effect, Bakirtzis et al. (2002) applied an enhanced genetic algorithm to solve the standard OPF problem, Lai et al. (1997) proposed an improved variant based genetic algorithm to solve the OPF problem under normal and contingent operation states, Attia et al. (2012) proposed an adaptive GA with adjusting population size to solve the standard OPF problem, Ghanizadeh et al. (2011) proposed a new metaheuristic method called imperialist competitive algorithm, Karaboga (2005) developed a new algorithm based on honey bee swarm for numerical optimization, Adaryani et al. (2013) proposed an artificial bee colony algorithm for solving multi-objective optimal power flow problem, Gnanambal et al. (2012) adapted and applied a hybrid differential evolution with particle swarm optimization to improve the maximum loadability limit of power system, Mahdad & Srairi (2012) proposed a hierarchical adaptive PSO for solving the multi-objective OPF considering emissions based shunt FACTS, Mahdad (2013) proposed various metaheuristic methods to solve the security OPF considering the integration of FACTS devices, Mahdad & Srairi (2013) proposed and applied the differential evolution algorithm to solve the multi objective OPF considering Wind and STATCOM, the state of the art of the majority of these methods proposed and applied to solve many problems related to power system planning and control is given in (Frank et al. (2012)).

Recently a large number of new metaheuristic methods have been proposed to improve the solution of many problems related to power system planning and control. Some of these methods have been adapted and applied with success to solve in particular the optimal power flow such as: hybrid chemical reaction optimization algorithm (Susanta et al. (2015)), a backtracking search algorithm (Mostafa Modiri & Nasrudin (2014)), a new self-adaptive particle swarm optimization technique (Mandal et al. (2015)), a novel modified hybrid PSOGSA based on fuzzy logic (Serhat et al. (2015)), new modified and hybrid modified imperialist competitive algorithms (Ghasemi et al. (2014)), adaptive biogeography based predator–prey optimization technique (Ananthi et al. (2014)), teaching-Learning-Based Optimization technique (Bouchekaraa et al. (2014)), an improved teaching–learning-based optimization algorithm using Lévy mutation strategy (Ghasemi et al. (2015)), hybrid fuzzy particle swarm (Joorabian & Afsan (2014)) [30], the grey wolf optimizer (Mirjalili et al. (2014); Mohd Herwan et al. (2015)),, adaptive group search optimization algorithm (Daryani et al. (2016)), a chaotic krill herd algorithm (Aparajita et al. (2015)), a new quantum inspired chaotic artificial bee colony algorithm (Xiaohui et al. (2015)), Gbest guided artificial bee colony algorithm (Roy et al. (2015)), improved group search optimization method (Tan et al. (2015)), a new interactive sine cosine algorithm (Mahdad & Srairi (2017)), and a non-dominated sorting hybrid cuckoo search algorithm (Balasubba reddy et al. (2015)).

In the literature, a variety of simulators under the user-friendly graphic user interface (GUI) based Matlab environments have been proposed to solve particular problems related to power system planning and control. Milano et al. (2008) proposed a robust tool named PSAT for solving various problems related to power system planning and control. In this simulator (PSAT), the practical objective functions such as valve point effect, prohibited zones are not considered.Woo Nam et al. (2011) proposed an educational simulator tool based PSO method to solve the economic dispatch problem considering valve point effect. Mahdad & Srairi (2011) proposed an interactive educational GUI Matlab package to understanding the effect of integration of FACTS devices to improve the power quality in practical power systems.

In this chapter an interactive tool using graphic user interface (GUI) based MATLAB is proposed to analysis and explains the concept and mechanism search of new metaheuristic optimization methods applied for solving the OPF problems. In this preliminary tool, the user can set the parameters of different optimization methods, to show their impact on the solution quality and convergence characteristics. The main contribution of this study compared to our previous works and other planning strategies developed is related to solving the security OPF considering two new indices: the critical reactive margin stability and the global reactive margin stability.

MAIN FOCUS OF THE CHAPTER

It is well known that there is no a generalized tool which is capable for solving various complex real-world optimization problems related to many research areas and in particular practical problems related to power system operation, planning and control. As a result, and during the last two decades a hundred of optimization methods have been developed. Among these methods, hybrid methods considered as an alternative and robust tool to solve complex problems related to power system planning, operation and control. The main focus of this chapter is to propose an initial version of a flexible power system planning strategy using two coordinated metaheuristic methods such as the Grey Wolf Optimizer considered as a global search mechanism and pattern search algorithm considered as a local search method to solve with interactivity the security OPF. The main particularity of the proposed optimizer tool is that offers an educational aspect which allows the user to well understand the mechanism search of these methods by dynamically adjusting their particular parameters and changing the original code. The proposed strategy allows to student, expert and industrials specialized in power system planning and control to examine and achieve the desired objectives with accuracy and credibility.

Power System Planning Strategy

Security optimal power flow is a fundamental sub-problem of power system planning, the main objective of this strategy is to find the optimal settings of control variables in terms of one or more objective function while satisfying operating equality and inequality constraints (Mahdad & Srairi (2014)). An interactive formulation of OPF is proposed to understand the basic principle of OPF problem. Figure 1 shows the basic structure strategy of security OPF.

Power Flow

Power flow (PF) is considered as the keystone in the structure of security OPF. The main role of power flow incorporated within the standard OPF is to determine the state variables and violation of constraints. The equality constraints are verified using PF algorithm. One of the most successful iterative methods used intensively by researchers and industrials is the Newton-Raphson method. This algorithm is preferred for its simplicity and fast convergence.

Figure 1. Basic structure of planning strategy for solving security multi objective OPF problems

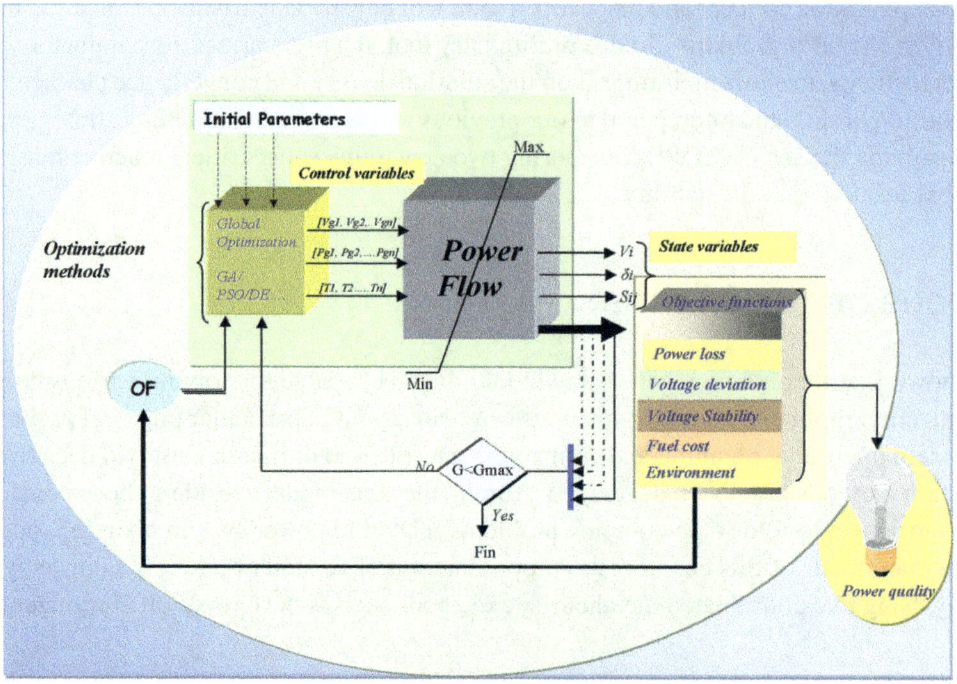

Economic Dispatch

Economic dispatch is an important sub problem of OPF. The main objective of ED is to distribute the active power of generating units to minimize the total fuel cost, while satisfying a specified equality and inequality constraints. The basic structure of economic dispatch is shown in Figure 2.

The following equations describe the basic mathematical formulation of the standard economic dispatch considering quadratic cost function.

$$\min C_T = \sum_{i=1}^{ng} C_i = \sum_{i=1}^{ng} \left(a_{i0} + a_{i1} Pg_i + a_{i2} Pg_i^2 \right) \quad (1)$$

Equality constraint: $\sum_{i=1}^{ng} Pg_i = P_D + Ploss$ (2)

Inequality constraint: $Pg_{i\min} \leq Pg_i \leq P_{i\max}; i = 1,..,ng$ (3)

Figure 2. The basic task of economic dispatch

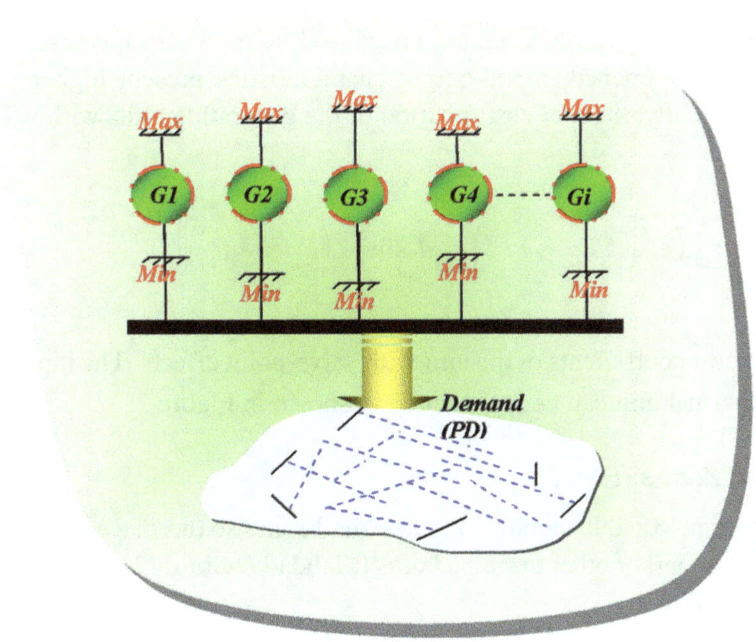

Objective Functions for Thermal Production Units

Fuel Cost

Minimization of fuel cost is an important task for experts and considered as an important index for power system planning and operation. In the literature, various objective functions are considered to represent the model of generating units.

Smooth Cost Function

The quadratic model is the simplest objective function used to minimize the total cost, mathematically formulated as follows:

$$J_1\left(x,u\right)=\min f_i=\min\left(\sum_{i=1}^{NG}\left(a_i+b_iP_{gi}+c_iP_{gi}^2\right)\right) \tag{4}$$

where, NG is the number of thermal units, P_{gi} is the active power generation at unit *i* and a_i, b_i and c_i are the cost coefficients of the *ith* generator.

Non-Smooth Cost Function

Effect of Valve Point (EVP)

In practice and due to valve-point loading effects caused by the sharp increase in losses when steam admission valves are first opened. Input–output characteristics present higher order discontinuities and nonlinearities. Typically, the fuel cost function of the generating units with valve-point loadings is represented as:

$$J_2\left(x,u\right)=\min f_i=\sum_{i=1}^{NG}\left[\left(a_i+b_iP_{gi}+c_iP_{gi}^2\right)+\left|d_i\sin\left(e_i\left(P_i^{\min}-P_i\right)\right)\right|\right] \tag{5}$$

d_i and e_i are the cost coefficients of the unit with valve-point effects. The input-output performance curve for a typical thermal unit can be represented as shown in Figure 3.

Effect of Prohibited Zones (EPOZ)

In practice, it is important to avoid operating in prohibited zones so thermal efficiency can be maintained during vibrations in the shaft or other machine faults (Mahdad & Srairi (2011)). This practical constraint can be formulated as follows:

Figure 3. Input-output performance curve for a typical thermal unit: With valve point effect

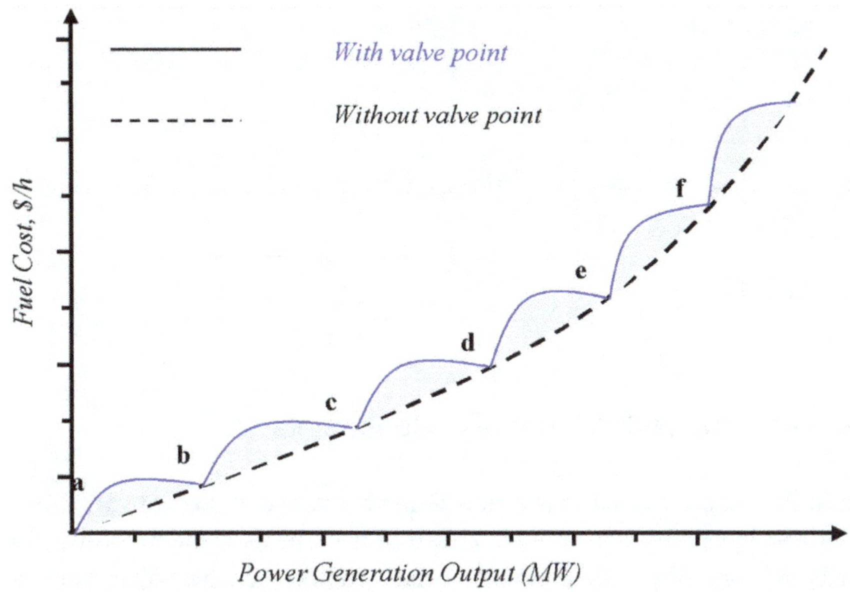

$$P_{Gi} = \begin{cases} P_{Gi}^{\min} \le P_{Gi} \le P_{Gi,1}^{L} \\ P_{Gi,k-1}^{U} \le P_{Gi} \le P_{Gi,k}^{L}, k = 2,3,.....ni \\ P_{Gi,ni}^{U} \le P_{Gi} \le P_{Gi}^{\max} \end{cases} \tag{6}$$

where, ni is the number of prohibited zones; k is the index of prohibited zones in unit i, and $P_{Gi,k}^{L}$, $P_{Gi,k}^{U}$ are the lower and upper limits, respectively, of the *kth* prohibited zone in the *ith* generator.

Optimal Power Flow

The problem of OPF can be formulated as a combination of the standard PF problem considering the associated inequality constraints and one or more specified objective functions (Mahdad & Srairi (2015)). Before giving the generalized mathematical formulation of OPF problem, we will introduce in this section a simple representation about the basic principle of power flow tool and its importance in solving power system planning problems.

Power Flow Solver Without Considering Inequality Constraints

This is the simple PF solver, it consists on determining all state and control variables while satisfying equality constraint. Figure 4 shows the basic principle of power flow solver without considering inequality constraints.

Figure 4. Power flow solver without considering inequality constraints

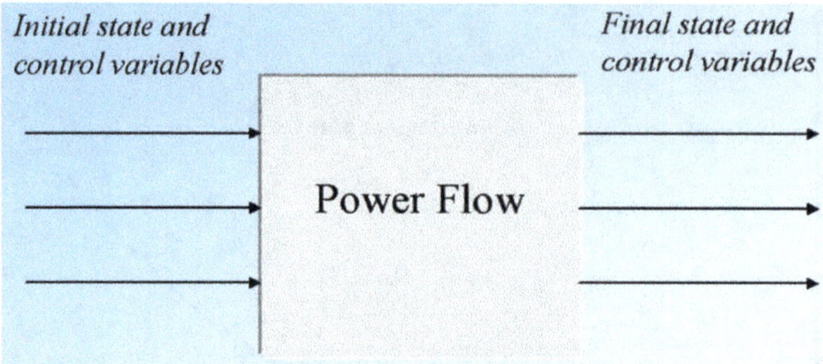

Power Flow Solver Considering Inequality Constraints

This is the standard PF solver, it consists in determining all state and control variables while satisfying equality constraint and inequality constraints. This tool becomes a standard solver for utilities to examine continuously the state of practical power system. The power flow solver considering inequality constraint is shown in Figure 5.

Power Flow Solver Considering Inequality Constraints Based on a Specified Objective Function

This solver is well known as the optimal power flow, it consists on determining all state and control variables considering a specified objective function, while satisfying equality and inequality constraints. The advanced power flow solver named optimal power flow based objective function and considering inequality constraints is well presented in Figure 6.

Figure 5. Power flow solver considering inequality constraint

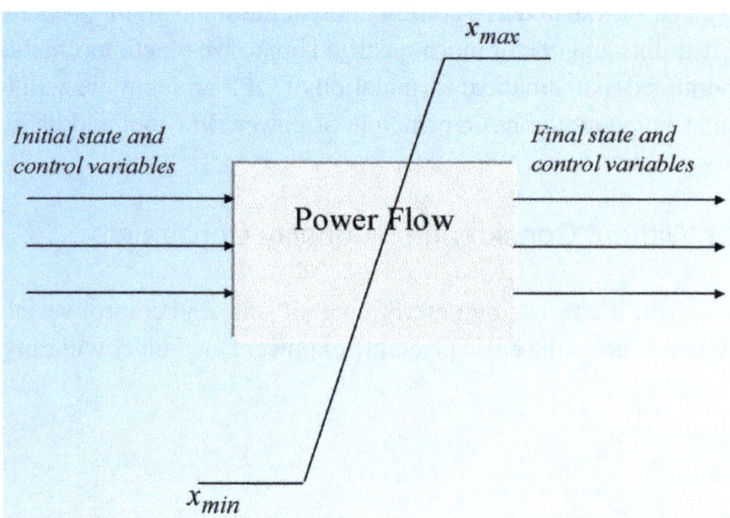

Figure 6. Structure of the standard power flow solver named optimal power flow

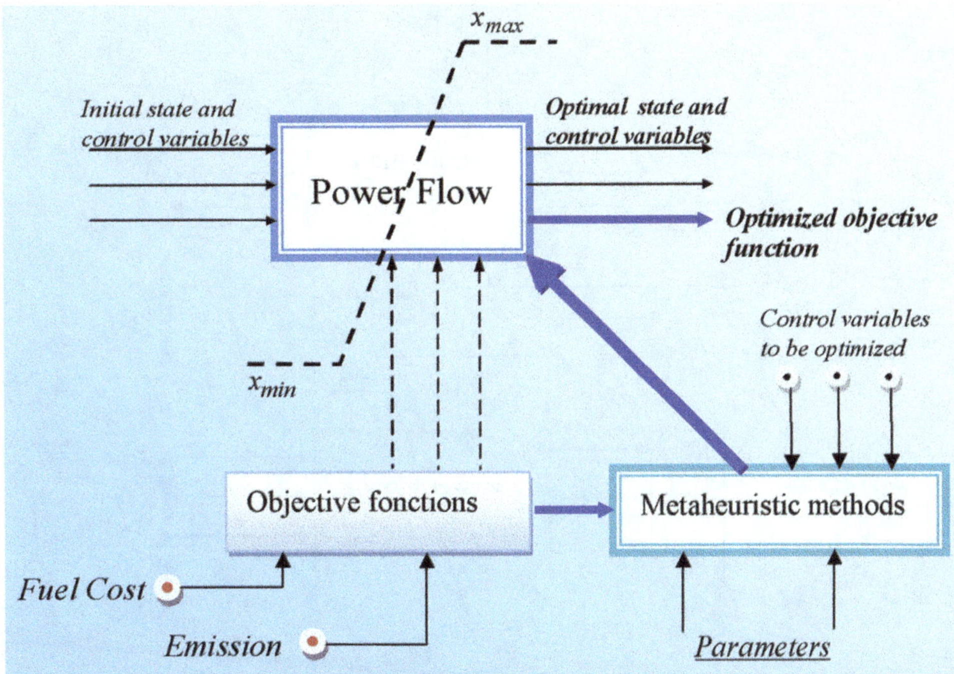

Reactive Power Planning

Reactive power planning is an important sub problem of power system planning and control. The main task of this tool is to optimize a specified objective function such as power loss, voltage deviation and voltage stability, while satisfying security and operation equality and inequality constraints (Mahdad, 2014; Mahdad, 2015). In most research papers, for solving reactive power dispatch problem, the vector control associated to generating units is fixed to a specified value. Dynamic control variables associated to different types of FACTS devices is used to improve the reliability and security of practical power system under sever loading conditions and considering contingency. Figure 7 shows the basic structure of reactive power planning.

GENERALIZED CONSTRAINTS

Equality Constraints

In general, the equality constraints $g(x)$ represent the balance between production and consummation. The real and reactive power balance equations are expressed by the two generalized equations:

$$P_{gi} - P_{di} - V_i \sum_{j=1}^{N} V_j \left(g_{ij} \cos \delta_{ij} + b_{ij} \sin \delta_{ij} \right) = 0 \quad (7)$$

Figure 7. Basic reactive power planning strategy

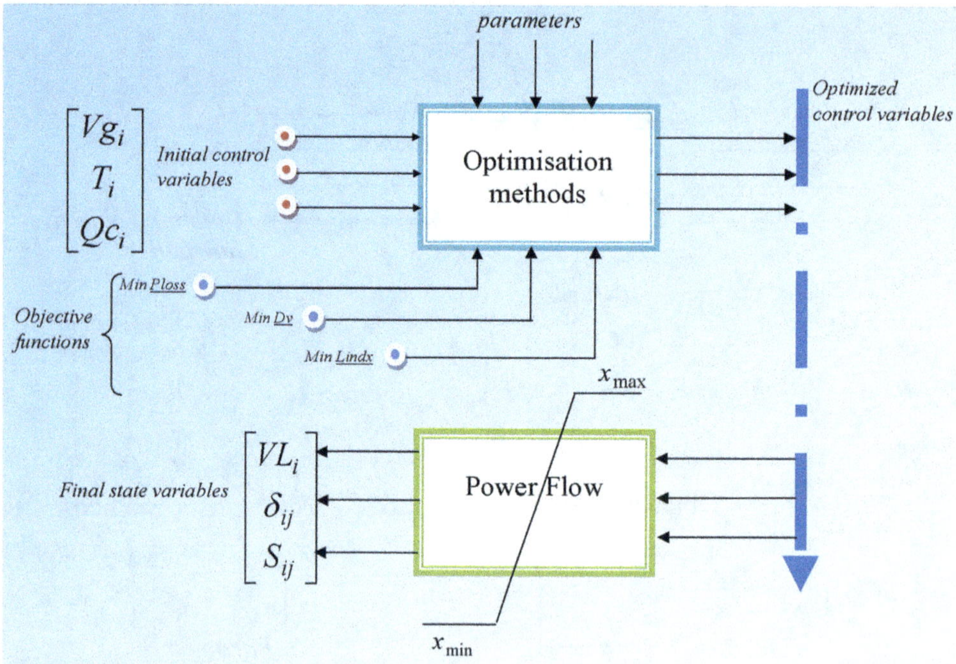

$$Q_{gi} - Q_{di} - V_i \sum_{j=1}^{N} V_j \left(g_{ij} \sin \delta_{ij} - b_{ij} \cos \delta_{ij} \right) = 0 \tag{8}$$

where, N is the number of buses, P_{gi}, Q_{gi} are the active and the reactive power generation at bus i; P_{di}, Q_{di} are the real and the reactive power demand at bus i; V_i, V_j, the voltage magnitude at bus i, j, respectively; δ_{ij} is the phase angle difference between buses i and j respectively, g_{ij} and b_{ij} are the real and imaginary part of the admittance (Y_{ij}).

Inequality Constraints

Generally, these security constraints are associated to reliable operation of all elements of power system, and represent all security limits associated to sate and control variables organized as follows:

$$V_{gi}^{\min} \leq V_{gi} \leq V_{gi}^{\max}, i = 1, 2, \ldots, Npv \tag{9}$$

$$P_{gi}^{\min} \leq P_{gi} \leq P_{gi}^{\max}, i = 1, 2, \ldots, Npv \tag{10}$$

$$Q_{gi}^{\min} \le Q_{gi} \le Q_{gi}^{\max}, i = 1,2,\ldots,Npv \tag{11}$$

$$t_i^{\min} \le t_i \le t_i^{\max}, i = 1,2,\ldots,Nt \tag{12}$$

$$B_{SVC}^{\min} \le B_{SVC} \le B_{SVC}^{\max} \tag{13}$$

$$V_{Li}^{\min} \le V_{Li} \le V_{Li}^{\max}, i = 1,2,\ldots,Npq \tag{14}$$

$$S_{li} \le S_{li}^{\max}, i = 1,2,\ldots,Nbr \tag{15}$$

where, $V_{gi}^{\min}$, $V_{gi}^{\max}$ are the limits on the generator bus voltage magnitude, $P_{gi}^{\min}$, $P_{gi}^{\max}$ are the limits on the output of active power generation, $Q_{gi}^{\min}$, $Q_{gi}^{\max}$ are the limits on the output of reactive power generation, $t_i^{\min}$, $t_i^{\max}$ are the limits on the tap ratio (t) of transformer, $B_{SVC}^{\min}$, $B_{SVC}^{\max}$ are upper and lower susceptance limits of shunt SVC Controllers, $V_{Li}^{\min}$, $V_{Li}^{\max}$ are the limits on voltage magnitude at loading buses (PQ buses) and $S_{li}^{\max}$ is the maximum transmission line loading.

OBJECTIVE FUNCTIONS RELATED TO POWER SYSTEM PLANNING

Minimization of Voltage Deviation

The objective function associated to the minimization of the total voltage deviation at all load buses is described by the following equation:

$$J_2(x,u) = \min\ VD = \min\left(\sum_{i \in NL} \left|V_i - V_{ref}\right|\right) \tag{16}$$

where, V_{ref} is the desired voltage at all load buses.

Minimization of Power Losses

The objective function associated to the total active power loss to be optimized is formulated by the following equation:

$$J_3\left(x,u\right)=\min P_{loss}=\min\left(\sum_{k=1}^{NL} g_k\left[\left(t_k V_i\right)^2+V_j^2-2t_k V_i V_j \cos\delta_{ij}\right]\right) \tag{17}$$

CONSTRAINTS MANAGEMENT

In order to take in consideration the violation of security constraints, a modified fitness function is adapted using the penalty function (Mahdad, 2014; Mahdad, 2015). The inequality constraints are immerged within the objective function. The mathematical formulation of the proposed augmented objective function can be expressed by the following expression:

$$J_{\mathrm{mod}}\left(x,u\right)=J\left(x,u\right)+Penalty \tag{18}$$

$$Penalty=\lambda_v\sum_{i=1}^{NL}\left(V_{Li}-V_{Li}^{\lim}\right)^2+\lambda_Q\sum_{i=1}^{NG}\left(Q_{gi}-Q_{gi}^{\lim}\right)^2+\lambda_P\left(P_{gs}-P_{gs}^{\lim}\right)^2+\lambda_{br}\sum_{i=1}^{NBR}\left(S_{bri}-S_{bri}^{\lim}\right)^2 \tag{19}$$

where; λ_v, λ_Q, λ_P and λ_{br} are penalty factor terms, if limits violation are occurred at dependent variables such as: load bus voltage magnitudes, real power generation at slack bus, reactive power generation output and transmission loading, they are fixed to their admissible limits (max or min). The following expressions illustrate the handling of constraints violations:

$$V_{Li}^{\lim}=\begin{cases}V_{Li}^{\min} & if\ V_{Li}<V_{Li}^{\min}\\ V_{Li}^{\max} & if\ V_{Li}>V_{Li}^{\max}\end{cases} \tag{20}$$

$$Q_{gi}^{\lim}=\begin{cases}Q_{gi}^{\min} & if\ Q_{gi}<Q_{gi}^{\min}\\ Q_{gi}^{\max} & if\ G_{gi}>Q_{gi}^{\max}\end{cases} \tag{21}$$

$$P_{gs}^{\lim}=\begin{cases}P_{gs}^{\min} & if\ P_{gs}<P_{gs}^{\min}\\ P_{gs}^{\max} & if\ P_{gs}>P_{gs}^{\max}\end{cases} \tag{22}$$

$$S_{bri}^{\lim}=S_{bri}^{\max}\quad if\ S_{bri}>S_{bri}^{\max} \tag{23}$$

SYSTEM COMPENSATION BASED SHUNT FACTS DEVICES

The Static VAr Compensator (SVC) (Mahdad, 2014; Mahdad, 2015) is a shunt connected VAR compensator. As well illustrated in Figure 8. a), the structure of SVC device is simple and consists of connected anti-parallel Thyristors to provide controllability. Air core reactors and high voltage AC capacitors are the reactive power elements used together with the Thyristor valves. The SVC device has the ability to exchange in real time reactive power (absorb or generate) with the network to control voltage at specified buses of the electric power system. Figure 8. b) shows the SVC device steady-state circuit representation. The principle of voltage control using shunt compensator based SVC device is shown in Figure 9.

The SVC model used in this study is based on representing the shunt Controller as a variable susceptance, the following equations describes the SVC model.

$$I_{SVC} = jB_{SVC}V \tag{24}$$

The reactive power Q_i^{SVC} exchanged with the bus i can be expressed as:

$$Q_i^{SVC} = B_i^{SVC} \cdot V_i^2 \tag{25}$$

METAHEURISTIC METHODS

In the last two decades a large number of meta-heuristic methods have been proposed to solve various industrial optimization problems and in particular the problems related to power system planning operation and control. As well shown in Figure 10, these optimization methods are classified on three large categories: mathematical methods, global optimization methods based on heuristic and stochastic aspect, and hybrid methods based on combination between different techniques (Mahdad, 2014; Mahdad, 2015). In this study, only two meta-heuristic methods have been proposed to solve the OPF considering different objective functions. The proposed methods are: GWO, and PS algorithm have been adapted and applied using an interactive GUI based Matlab for solving the OPF problems.

Figure 8. a) Basic circuit of Static VAr Compensator (SVC), b) SVC steady-state circuit representation

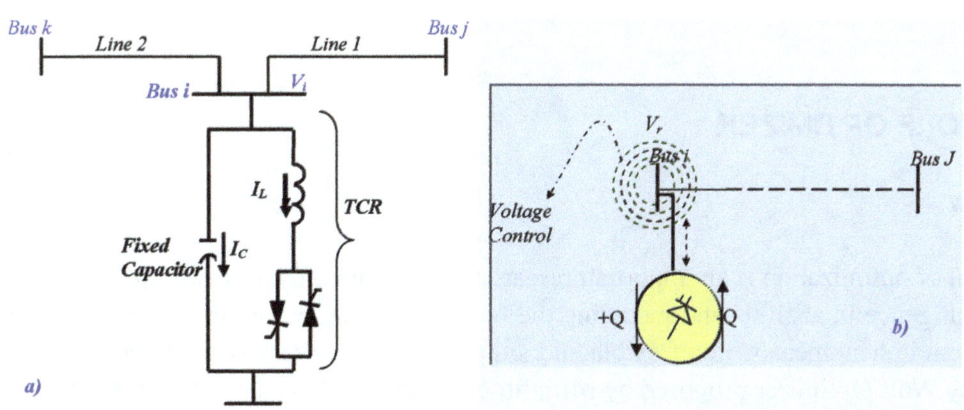

Figure 9. Principle of voltage control using shunt compensator based SVC device

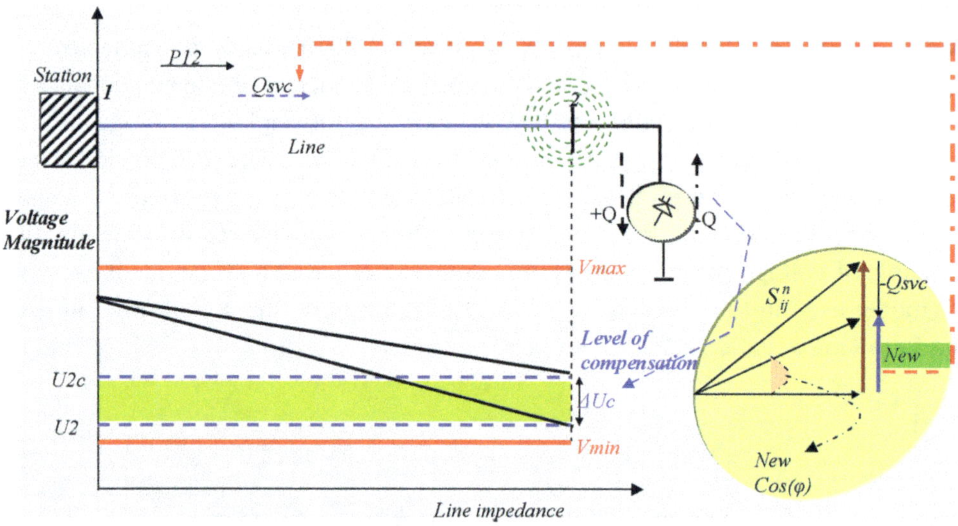

Figure 10. Classification of metaheuristic methods

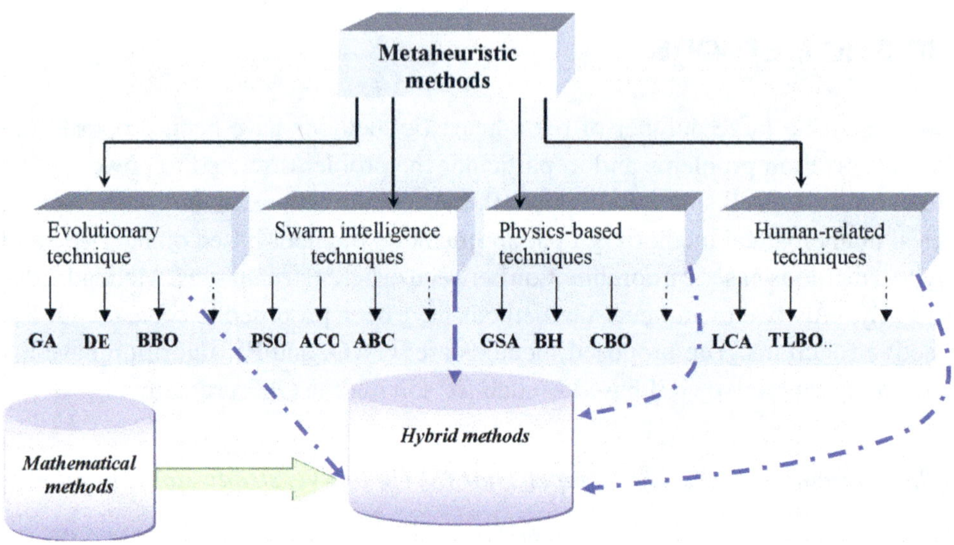

GREY WOLF OPTIMIZER

Overview

The domain of optimization is an important research area required by all research fields. It depends on human intelligence in analysis and modeling the behavior of nature, insects, animals and many kinds of creature and phenomena's, in a flexible and simple equations. Very recently, a new meta-heuristic named Grey Wolf Optimizer proposed by Mirjalili et al. (2014). This new algorithm inspired from the real life of an organized group of Grey Wolf by analyzing their behavior and communication for hunting in nature. The robustness of this proposed new metaheuristic algorithm tested and validated by authors using many standard benchmark functions.

Mechanism Search of GWO

This section is dedicated to give a description about the hierarchical structure of the grey wolves in nature. In general the group of grey wolves is formed between 5 and 12 individuals. The group of wolves is well organized and usually formed by the following coordinated sub groups (Mahdad & Srairi (2015)).

1. **Alphas Agent:** The group is guided by a special leader (male or female), called alphas. The main task of the alphas is to make the most powerful decisions for hunting, choosing location for sleeping, to safe the group.
2. **Beta Agents:** In nature and in many social lives, we can find democracy, in the hierarchy of grey wolves; a second category named betas works in coordination with the first group, agents from this sub group help the leader (alpha) in decision making to achieve the desired objectives of the pack. So the knowledge of leader is improved by the feedback transferred from associated agents of this second category. The hierarchy structure of grey wolf is shown in Figure 11.
3. **Delta Agents:** Called also subordinate, wolves within this category have to achieve many tasks such as: scouts, sentinels, elders and hunters, they have to communicate their ideas to alpha and beta to guarantee the safety of the pack.
4. **Omega Agents:** The omega is the lowest ranking grey wolf in the pack, they are the last wolves that are allowed to eat. The omega plays role of scapegoat, the wolves of this category always have to submit to all the other individuals from dominant category.

In unknown nature, hunting is one of the important social tasks of grey wolf group to survive within different constraints and critical situations. According to Mirjalili et al. (2014), the following descriptions summarize the main phases of grey wolf hunting:

- Tracking, chasing and approaching the prey.
- Pursuing, encircling, and harassing the prey until it stops moving.
- Attack towards the prey.

Mathematical Model of GWO Algorithm

In this subsection, we will introduce the mathematical models which interpret the hunting operation and social hierarchy of grey wolf proposed by Mirjalili et al. (2014). The global solution will be achieved principally by coordination between three best solutions, the fittest solution found by the leader alpha (α), consequently the second and the third solutions which named beta (β) and delta (δ). The rest of the candidate solutions found by the rest of the wolves named omega (ω).

Figure 11. Hierarchy structure of grey wolf optimizer

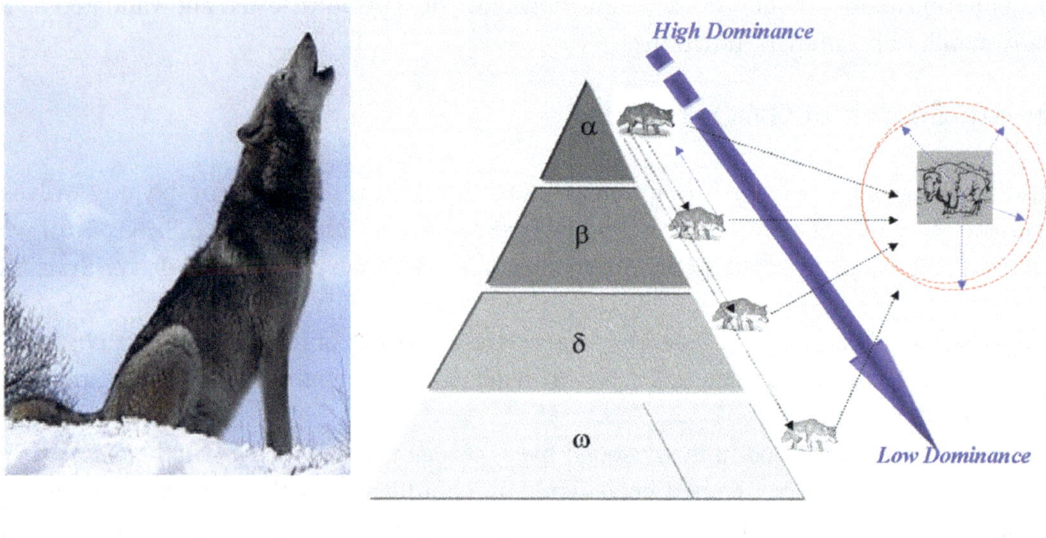

Encircling Prey

In this first step and during hunt the grey wolves encircle prey. The encircling behavior is modeled and described using the following equations:

$$\vec{D} = \left|\vec{C}.X_p(t) - \vec{X}(t)\right| \tag{26}$$

$$\vec{X}(t+1) = \vec{X}_p(t) - \vec{A}.\vec{D} \tag{27}$$

where t indicates the current iteration, $\vec{A}$ and $\vec{C}$ are coefficient vectors, $\vec{X}_p$ is the position vector of the prey, and $\vec{X}$ indicates the position vector of a grey wolf.

The two vectors $\vec{A}$ and $\vec{C}$ are calculated as follows:

$$\vec{A} = 2\vec{a}.\vec{r}1 - \vec{a}. \tag{28}$$

$$\vec{C} = 2.\vec{r}2 \tag{29}$$

where components of $\vec{a}$ are linearly decreased from 2 to 0 during iterations, and $r1$, $r2$ are random vectors in [0, 1]. The demonstration of the importance in choosing $\vec{a}$ is well discussed in details in (Mirjalili et al. (2014)).

Hunting

In order to simulate mathematically the hunting behavior of grey wolves, the location of prey is formulated by the combination of the best knowledge given by alpha and beta during the hunting process. The main idea introduced by authors in (Mirjalili et al. (2014)), is that the identification of the best location of the prey is based on the best solutions achieved during search process. The basic equations describing and guiding the hunting process are formulated as follows:

$$\vec{D}_\alpha = \left|\vec{C}_1.\vec{X}_\alpha - \vec{X}\right| \tag{30}$$

$$\vec{D}_\beta = \left|\vec{C}_2.\vec{X}_\beta - \vec{X}\right| \tag{31}$$

$$\vec{D}_\delta = \left|\vec{C}_3.\vec{X}_\delta - \vec{X}\right| \tag{32}$$

$$\vec{X}_1 = \vec{X}_\alpha - \vec{A}_1.\vec{D}_\alpha \tag{33}$$

$$\vec{X}_2 = \vec{X}_\beta - \vec{A}_2.\vec{D}_\beta \tag{34}$$

$$\vec{X}_3 = \vec{X}_\delta - \vec{A}_3.\vec{D}_\delta \tag{35}$$

$$\vec{X}\left(t+1\right) = \frac{\vec{X}_1 + \vec{X}_2 + \vec{X}_3}{3} \tag{36}$$

where; $\vec{X}\left(t+1\right)$ is the position vector of prey at iteration t+1, this position is updated based on the best information given by alpha, beta and delta agents during process search.

Attacking Prey (Exploitation)

The last phase in the hunting process is named attacking prey, this final situation is achieved when prey stops moving. In order to mathematically model approaching the prey, authors in (Mirjalili et al. (2014)) suggested the adjustment of particular parameters such as 'a' during process search. Based on simulation results demonstrated by authors in (Mirjalili et al. (2014)), this final stage is very important

Figure 12. Pseudo code of the GWO algorithm

```
Begin of algorithm……………………………………………………………………
Initialize the grey wolf population Xi: i=1,2,…………,n
Initialize parameters; a , A and C
Calculate the fitness of each search agent
X_α = the best search agent
X_β = the second best search agent
X_δ = the third best search agent
While t ≤ Max Gen
        For search space
        Update the position of the current pace agent by equation (*)
        End % for j loop
        Update a, A, and C
        Calculate the fitness of all search agents
        Update X_α, X_β and X_δ
        t=t+1
End % while loop ……………………………………………………………………
Return X_α
```

to escape stagnation in local solutions. Our main contribution is related to the enhancement of search mechanism of the standard algorithm to escape from local solutions by using communication process based PS algorithm to locate efficiently the position of the prey during the search process (Mahdad & Srairi (2015)). The standard pseudo code of the algorithm proposed by authors in (Mirjalili et al. (2014)) is summarized in Figure 12.

Pattern Search Algorithm (PS)

The Pattern Search (PS) is an evolutionary algorithm applied with success for solving a variety of complex optimization problems related to many research areas (Mahdad & Srairi (2015); Alsumait et al. (2010)). In this study the efficiency and particularity of the PS algorithm is exploited as local search mechanism in coordination with GWO algorithm. Figure 13 shows the basic flowchart related to the search mechanism of the PS algorithm. Detailed description of the algorithm can be retrieved from (Alsumait et al. (2010)).

Figure 13. a) Flow chart of the basic search mechanism of PS, b) Formation of meshes and pattern vectors

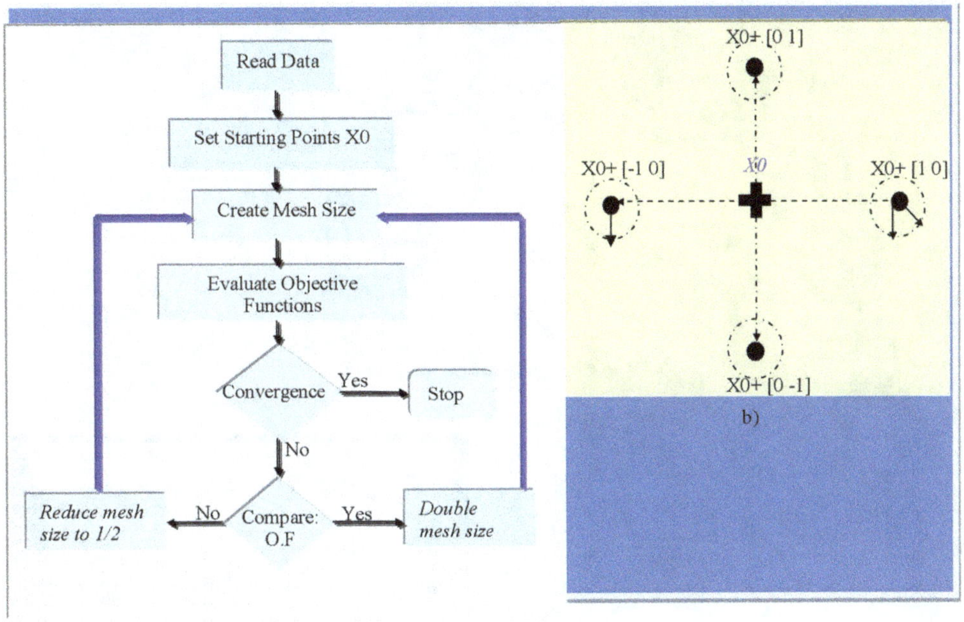

In this study, the standard parameters of PS algorithm such as mesh size, the mesh expansion and contraction factors, and the maximum number of generation are selected by user. By using PS as a local search mechanism in coordination with GWO, the vector of initial points X0, is dynamically updated based on communication process during search process.

Graphic User Interface Based Matlab

- The MATLAB graphical user interface development environment provides a set of tools for creating graphical user interfaces (GUIs) (Mahdad & Srairi (2011); Woo Nam et al. (2011)).

These tools greatly simplify the process of designing and building GUIs. We can use the GUIDE tools to:

- Lay out the GUI. Using the GUIDE Layout Editor, the user can lay out a GUI easily by clicking and dragging GUI components such as panels, buttons, text fields, sliders, menus, and so on into the layout area.
- GUIDE stores the GUI layout in a FIG-file.

GUIDE automatically generates an M-file that controls how the GUI operates. The M-file initializes the GUI and contains a framework for the most commonly used callbacks for each component the commands that execute when a user clicks a GUI component (Mahdad & Srairi (2011). Using the M-file editor, we can add code to the callbacks to perform the functions the user wants. Figure 14 shows the structure of the different Object contained in a GUI.

Figure 14. Proprieties of objects

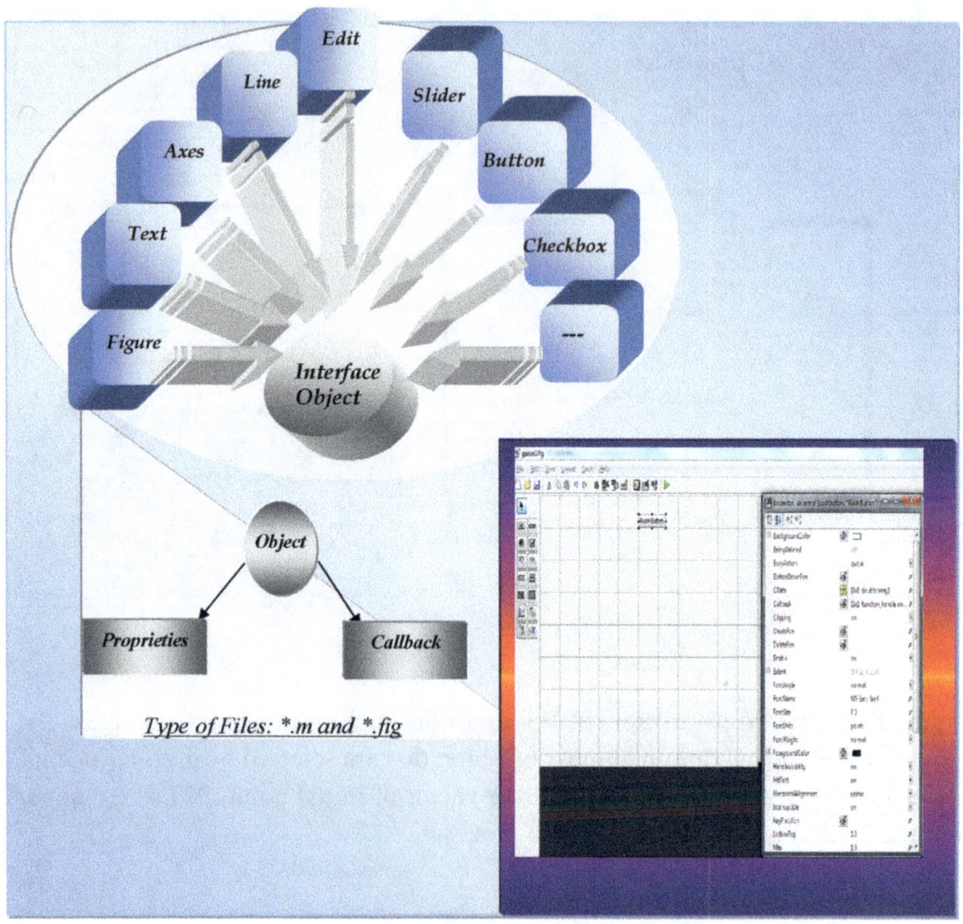

Structure of Simulator

The main task of the proposed optimizer tool is to facilitate and provide an efficient and interactive study for students and researchers to understand the mechanism search of various metaheuristic methods to improve the optimal power flow solution. The structure of the simulator is shown in Figures 15-16. The simulator consists of three parts:

1. Input data based on user setting of parameters of various metaheuristic methods as well as initial parameters of networks.
2. Output data based on optimal results obtained a well as the convergence characteristics of various objective functions.
3. Visualization tool to present the optimized results. It is important to note that, the main interaction between user and the simulator is performed through the GUI.

Figure 15. Structure of the developed metaheuristic simulator

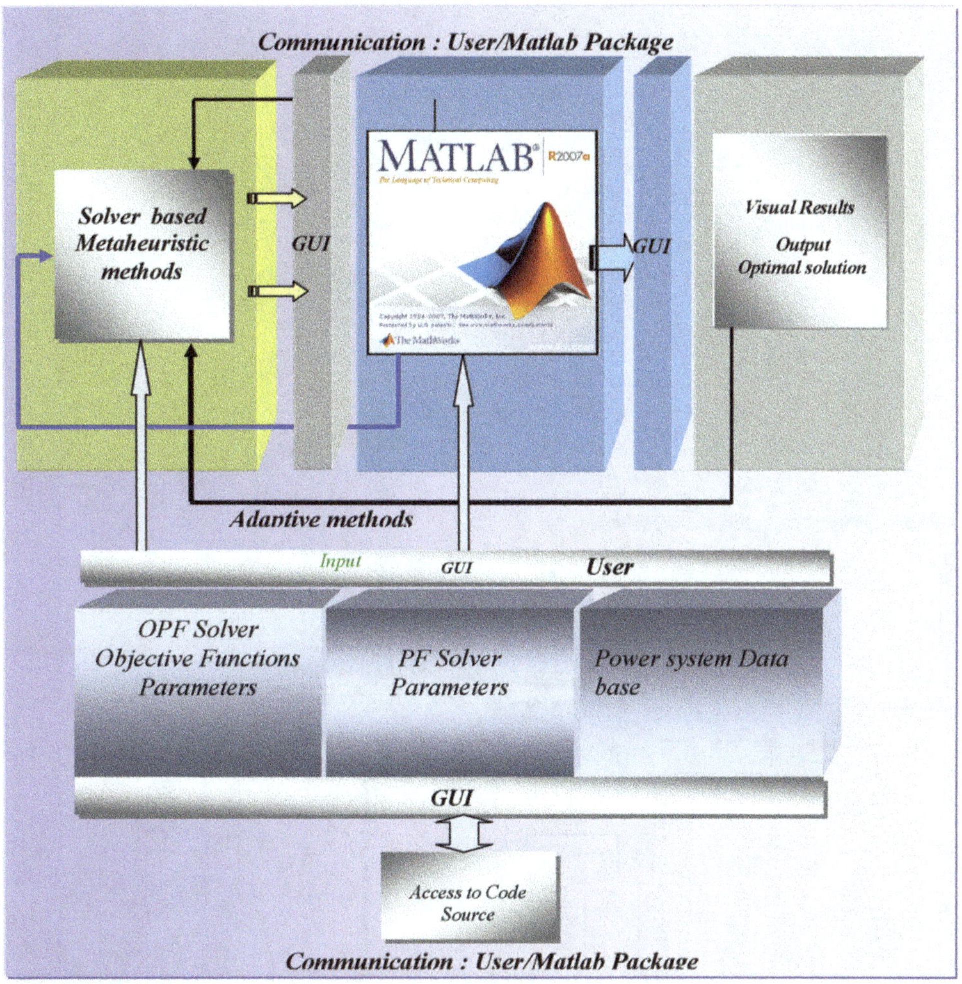

CASE STUDIES

Test System 1: IEEE 9-Bus

The first test system IEEE 9-Bus has three generators located at the buses 1, 2, and 3. The technical data such as bus and line data of the system and generator fuel cost parameters and limits are taken from (Zimmirrman et al. (2010)) and shown in Table 1. The total system demand to be satisfied is 315 MW for active power, and 115 MVAr for reactive power at 100 MVA base. The minimum and maximum voltage magnitudes of all buses were considered to be 0.95 and 1.1 in p.u, respectively. Figure 17 shows the various functionality of the optimizer tool. In this study few tests have been investigated to demonstrate the large functionality of the proposed optimizer tool. The one line IEEE 9-Bus test system is shown in Figure 18.

Figure 16. Structure of the proposed simulator based metaheuristic methods: GWO and PS

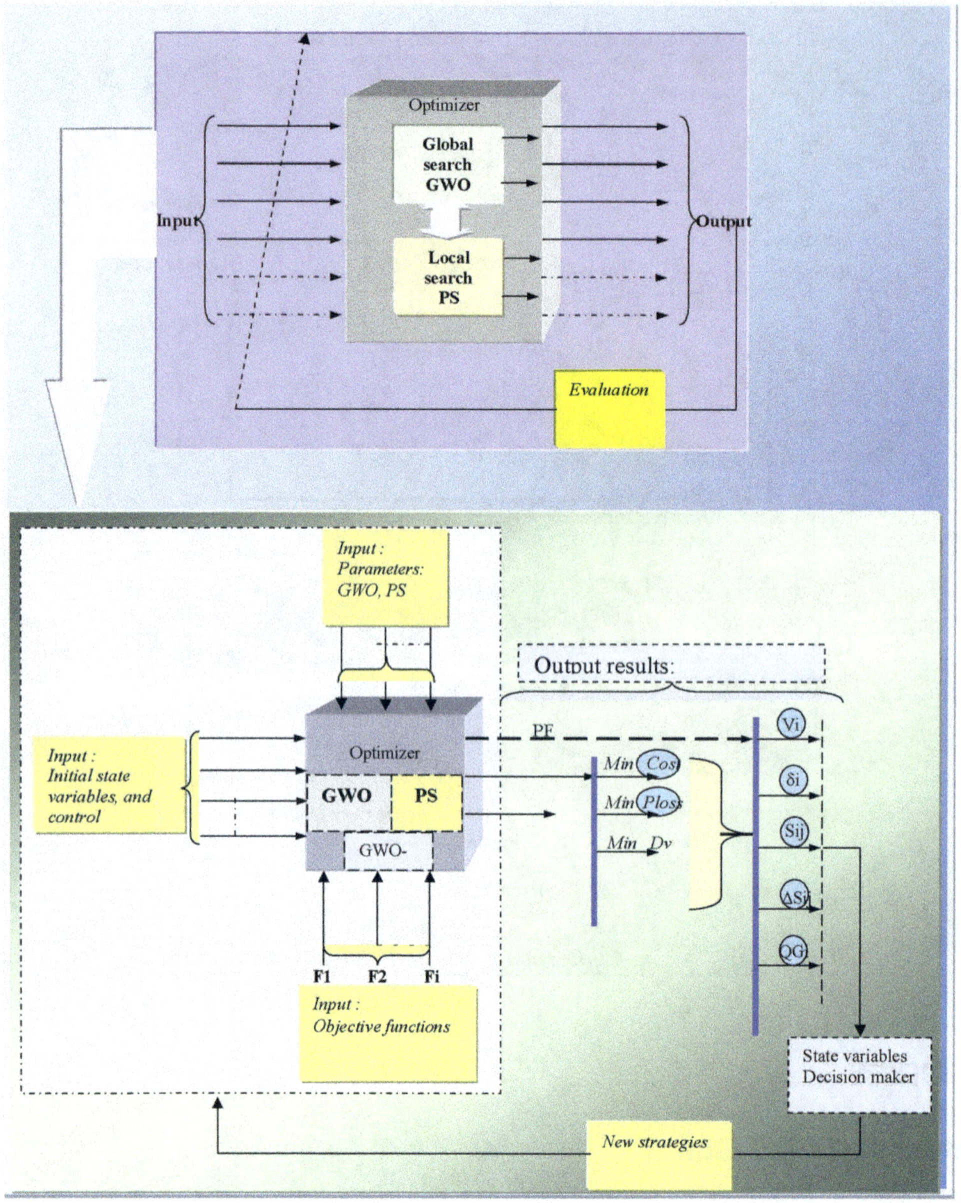

Case 1: Power Loss Minimization Without SVC Devices

In this case, the total power loss is minimized considering all security constraints. The optimal total power loss achieved is **2.4143** MW, the corresponding voltage deviations and fuel cost, are: 0.5472 p.u, 5691.10 $/h respectively. As well shown in Figure 19, voltages at all buses are within their admissible limits.

Figure 17. Various functionality of the optimizer tool

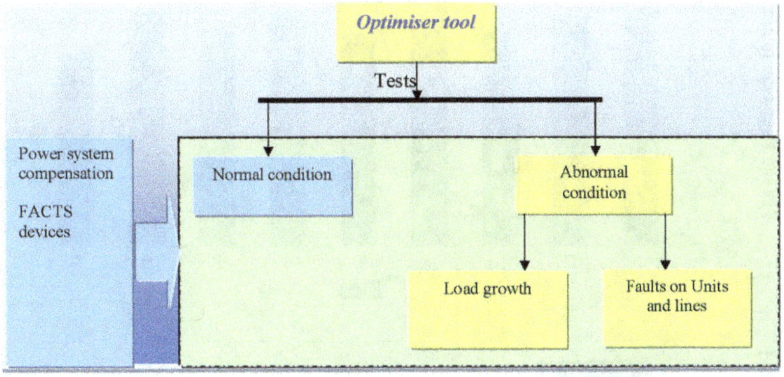

Table 1. Technical admissible generators and the fuel cost coefficientsparamaters of: IEEE 9-Bus

Unit Number	Pmin [MW]	Pmax [MW]	Qmin [MVAr]	Qmax [MVAr]	a [$/h]	b [$/MWh]	c [$/MW²h]
1 (slack)	10	250	-300	300	150	5	0.1100
2	10	300	-300	300	600	1.2	0.0850
3	10	270	-300	300	335	1	0.1225

Figure 18. One-line IEEE 9-Bus test system

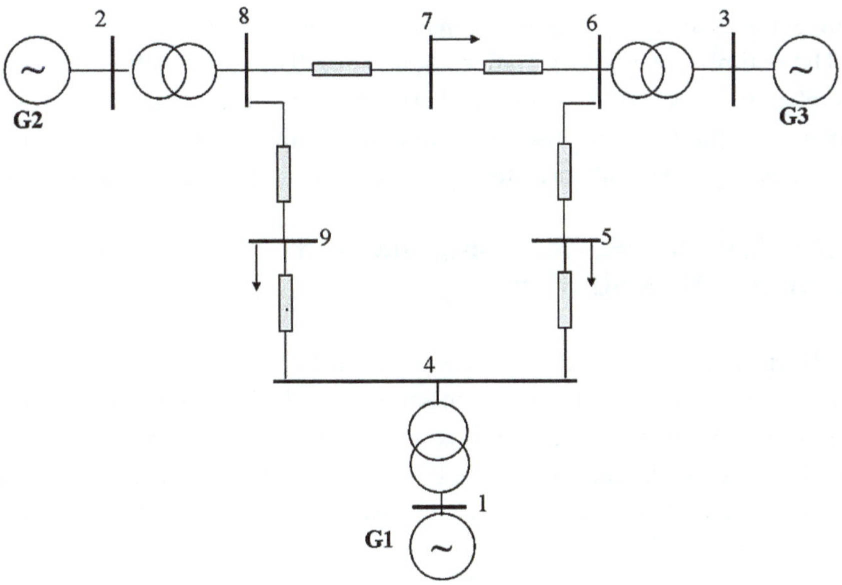

Figure 19. Optimal voltage magnitudes and active power generation: Case 1

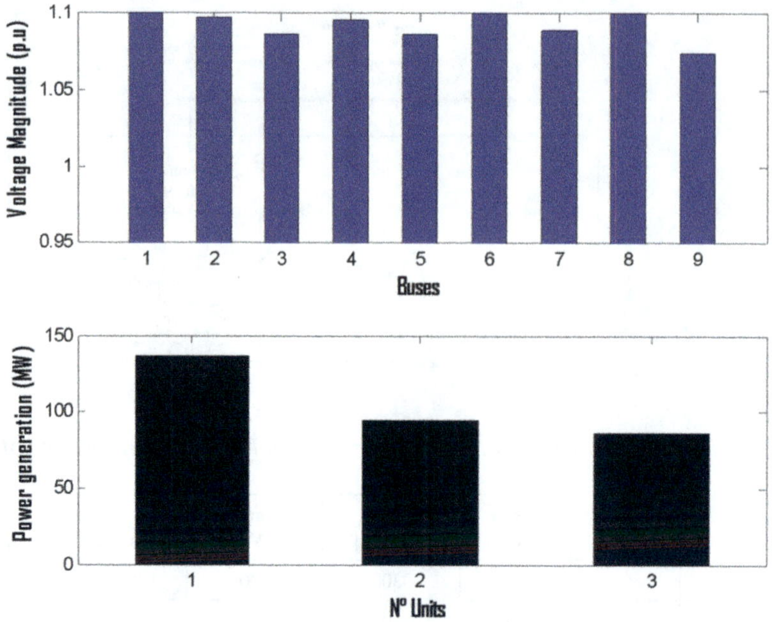

Case 2: Voltage Deviation Minimization Without SVC Devices

In this second case, the objective function selected was the total voltage deviation. All security constraints such as voltage magnitudes at all buses, power transit in branches and reactive power of generating units are considered. The optimized voltage deviation found is 0.1090 p.u, which is better than *case 1*. The corresponding total power loss and fuel cost are: 2.8056 MW, 5721.40 $/h, respectively. As well shown in Figure 20 voltage magnitudes at all buses are within their limits and based on Figure 21, the voltage profiles for this case is improved compared to *case 1* and is near the desired value 1 p.u.

Case 3: Effect of Reactive Power Compensation on Voltage Deviation Minimization

In this case the effect of reactive power compensation on the reduction of total power loss and total voltage deviation is investigated. In this study one shunt SVC device is installed at specified bus to exchange reactive power with the network. By injecting an optimized reactive power 22.46 MVAr at bus 9, the optimal total voltage deviation achieved is 0.0443 p.u. as well shown in Figure 22, this value is improved compared to the case without considering SVC device. Table 3 shows the optimal setting of control variables for two cases 3-4.

Case 4: Effect of Reactive Power Compensation on Power Loss Minimization

In this case the effect of reactive power compensation on the reduction of total power loss and total voltage deviation is investigated. In this study one shunt SVC device is installed at bus 9 to exchange reactive power with the network. The total power loss reduced to 2.3454 MW, this value is better than case 1.

Figure 20. Optimal voltage magnitudes and active power generation: Case 2

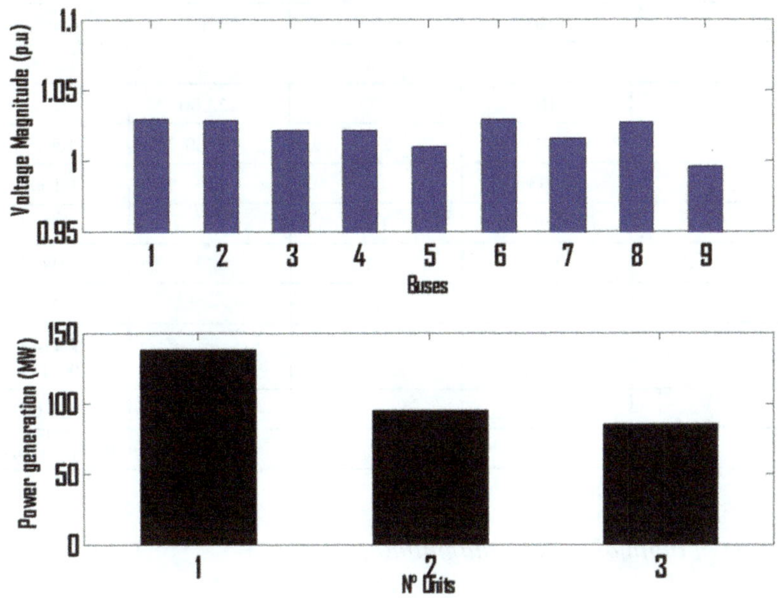

Figure 21. Comparison of optimal voltage magnitudes Case 1-2

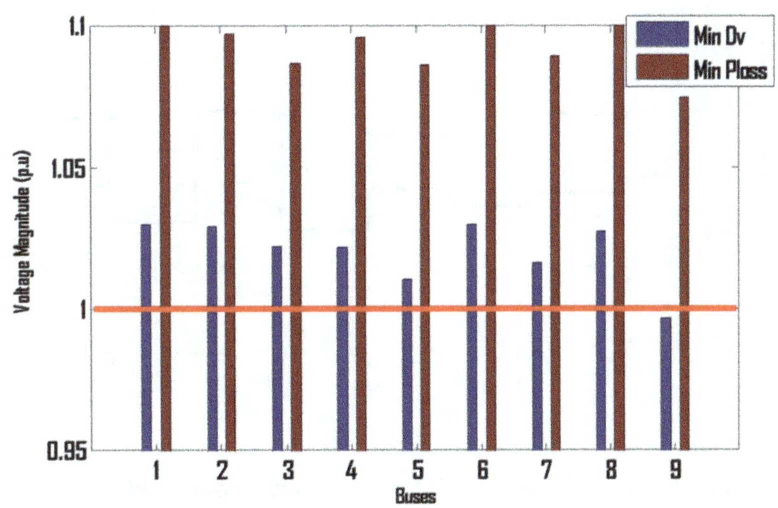

Case 5: Power Loss Minimization Under Contingency

The flexibility of the simulator allows the expert to optimize various objective functions under critical situations such as contingency. In this case the total power loss reduced with and without considering reactive power support of SVC device. The total power loss optimized without SVC device was 2.9251 MW, and by installing one SVC device at bus 9, the total power loss reduced to 2.6396 MW. It is important to note that all security constraints such as reactive power of generation units, power transit in branches and voltage magnitudes at all buses are within their security limits. The distribution of voltage

Table 2. Optimal setting of control variables: cases 1-2

Control Variables	Min	Max	PF	Case 1	Case 2
P_{g1} (Slack)	10	250	71.95	136.64	138.13
P_{g2}	10	300	163.00	95.80	94.39
P_{g3}	10	270	85.00	85.98	85.29
V_{g1}	0.95	1.1	1	1.1000	1.0299
V_{g2}	0.95	1.1	1	1.0973	1.0290
V_{g3}	0.95	1.1	1	1.0868	1.0221
Cost ($/h)	-	-	-	5691.10	5721.40
Power loss (MW) *Voltage deviation (p.u)*	- -	- -	4.955	**2.4143** **0.5472**	**2.8056** **0.1090**
Total load (MW)	-	-	315.00	315.00	315.00
Security constraints				*Within security limits*	

Figure 22. Comparison of optimal voltage magnitudes Case 2-3

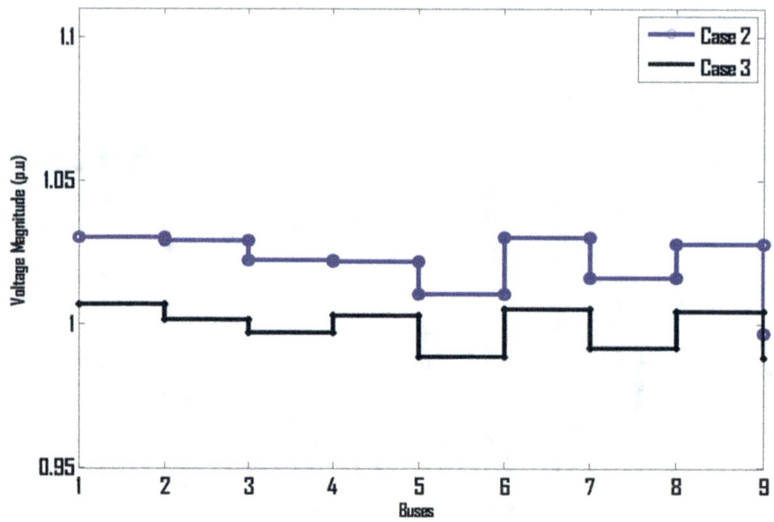

magnitudes at all buses with and without contribution of reactive power of SVC device installed at bus 9 are shown in Figure 23.

Case 6: Voltage Deviation Minimization Under Contingency

In this case, the total voltage deviation is also optimized considering the contribution of one shunt compensator installed at bus 9. The total voltage deviation optimized considering the SVC device was improved to 0.0831 p.u compared to the case without SVC device (0.2067 p.u). Table 4 shows the optimized control variables for cases 5-6. The distribution of voltage magnitudes at all buses with and without considering reactive power compensation are shown in Figure 24.

Table 3. Optimal setting of control variables: Cases 3-4

Control Variables	Min	Max	PF	Case 3	Case 4
P_{g1} *(Slack)*	10	250	71.95	139.30	137.06
P_{g2}	10	300	163.00	93.96	94.49
P_{g3}	10	270	85.00	84.58	85.79
V_{g1}	0.95	1.1	1	1.0068	1.0953
V_{g2}	0.95	1.1	1	1.0016	1.0909
V_{g3}	0.95	1.1	1	0.9971	1.0860
Cost ($/h)	-	-	-	5739.90	5696.5
Power loss (MW) *Voltage deviation (p.u)*	- -	- -	4.955	2.8315 **0.0443**	**2.3454** 0.5697
Total load (MW)	-	-	315.00	315.00	315.00
Qsvc (MVAR)			-	**22.49**	**22.49**
Security constraints				*Within security limits*	

Table 4. Optimal setting of control variables: Cases 5-6

Control Variables	Min	Max	PF	Case 5		Case 6	
				Without SVC	With SVC	Without SVC	With SVC
P_{g1} *(Slack)*	10	250	71.95	185.08	183.90	178.70	180.82
P_{g2}	10	300	163.00	59.83	60.29	62.88	61.89
P_{g3}	10	270	85.00	73.02	73.45	76.63	75.42
V_{g1}	0.95	1.1	1	1.1	1.1	1.0742	1.0363
V_{g2}	0.95	1.1	1	1.0996	1.0999	1.0467	1.0224
V_{g3}	0.95	1.1	1	1.0867	1.0882	1.0339	1.0100
Cost ($/h)	-	-	-	6880.60	6840.3	6698.80	6757.80
Power loss (MW) *Voltage deviation (p.u)*	- -	- -	4.955	**2.9251** 0.4615	**2.6396** 0.5343	3.2132 **0.2067**	3.1307 **0.0831**
Total load (MW)	-	-	315.00	315.00	315.00	315.00	315.00
Qsvc (MVAR)			-	**0**	**24.7390**	**0**	**25.8635**
Security constraints				*Within security limits*			

Test System 2: IEEE 30-Bus

Optimization Considering Critical Reactive Margin Security

The standard test system IEEE 30-Bus is proposed to demonstrate the efficiency of the proposed strategy, this test system consists of 6 generating units four transformers located at lines and nine shunt VAR compensation buses installed at buses 10, 12, 15, 17, 20, 21, 23, 24 and 29, the total load demand to satisfy is 283.4 MW. The generator data, the values of fuel cost coefficients, the minimum and maximum

Figure 23. Distribution of voltage magnitudes: case 5 with and without SVC device

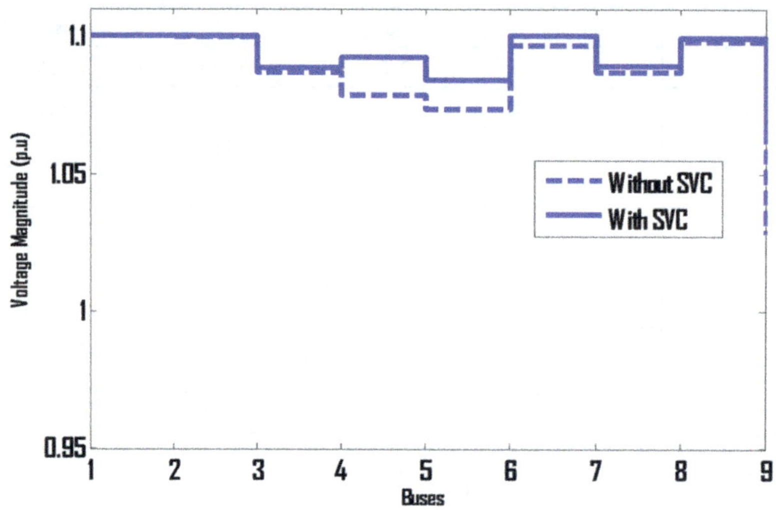

Figure 24. Distribution of voltage magnitudes: case 6 with and without SVC device

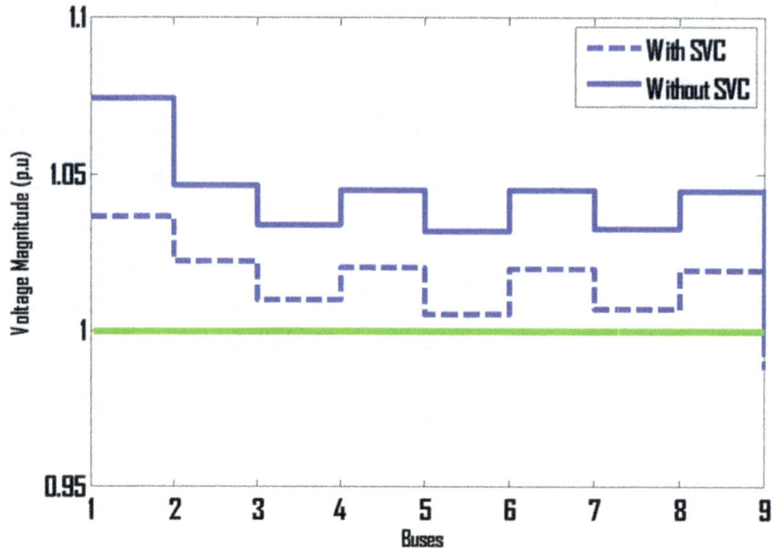

limits for active and reactive power of generating units are given in Table 1 and available in (Mahdad, 2014; Mahdad, 2015). The simulator allows user to solve with flexibility various problems related to power system operation and control such as:

- Fuel cost, power loss and voltage deviation minimization at normal condition
- Fuel cost, power loss and voltage deviation minimization at abnormal condition such as: load growth, and faults at generation units and important lines.
- Fuel cost, power loss and voltage deviation minimization at normal condition and critical situations considering reactive margin security.

- Fuel cost, power loss and voltage deviation minimization at normal condition and at critical situations considering FACTS devices.

In this study only one case is considered to validate the capability of the two metaheuristic methods in solving practical planning problems. In this study a new case study related to reactive margin security is introduced. The technical data of the test system IEEE 30-Bus are given in Table 5.

Critical Reactive Margin Security

At critical situations such as load growth and contingency, it is important to maintain reactive power of generating units at security levels. Optimal coordination of reactive power delivered by generating units and multi shunt FACTS devices is an important task to enhance the stability of power systems. In this study two indices are proposed to measure and evaluate the performance of practical power systems.

$$CRMS_i = \begin{cases} abs\left(\dfrac{QG_i}{QG_{\min}}\right) & if\ QG_i < 0 \\ \left(\dfrac{QG_i}{QG_{\max}}\right) & if\ QG_i > 0, i = 1....NG \end{cases} \tag{37}$$

Global Reactive Margin Security

The higher GRMS indicates the higher degree of security and stability of power system. This index must be maximized in coordination with other control variables to maintain the power system at reliable situation.

$$GRMS = \frac{1}{\sum_{i=1}^{NG} CRMS_i} \tag{38}$$

Table 5. Technical admissible parameters of generators and the fuel cost coefficients: IEEE 30-Bus

Unit Number	Pmin [MW]	Pmax [MW]	Qmin [MVAr]	Qmax [MVAr]	a [$/h]	b [$/MWh]	c [$/MW²h]
1	50	200	-20	200	0	2	0.00375
2	20	80	-20	100	0	1.75	0.01750
5	15	50	-15	80	0	1	0.06250
8	10	35	-15	60	0	3.25	0.00834
11	10	30	-10	50	0	3	0.02500
13	12	40	-15	60	0	3	0.02500

Minimization of Power Loss Considering Reactive Margin Security of Generating Units

In this first test, the total power loss is the main objective function to be minimized. Various tests performed to show the effect of the reactive margin security on power system stability. The total loss is optimized with and without considering the reactive power of SVC devices. Table 6 shows the optimal control variables achieved at the maximum margin security. The best GRMS found is 1.3806 p.u which is better compared to other levels, the total power loss maintained at 3.3669 MW. At this stage all units deliver their energy at safe and reliable condition. It is important to note that the maximum margin security was achieved by considering the reactive power of multi SVC devices installed at specified locations. The optimized CRMS related to three levels are shown in Figure 25, and the distribution of voltage magnitudes at all buses are shown in Figure 26, as well shown, the maximum CRMS is located at unit 13 which is lower than the CRMS corresponding to other levels.

FUTURE RESEARCH DIRECTIONS

Modern power systems characterized by large integration of multi type of FACTS devices and multi type of renewable sources become complex to manage using conventional tools. To achieve the desired technical and economical objectives with accuracy, it is mandatory to develop and adapt new interactive optimizer tools based coordination between various modern global optimization algorithms. This chapter introduces in brief the concept of interactive tools using combined metaheuristic based GUI to solve the security OPF problems. Extensive versions of this idea may be exploited to achieve the following important perspectives.

1. The proposed optimizer tool may be improved to include real models of FACTS devices and renewable sources.
2. Adapt and include new combined metaheuristic methods
3. In term of commercial aspect, it is possible to implement an extended a version of this optimizer tool for real time computations for large power systems based on parallel configuration using multi processors.
4. In term of educational aspect, a modified version of the proposed tool may be directed to an educational purpose by including flexible dialogue boxes with an interactive communication process between user and software to well understand the mechanism search of all integrated methods to achieve the best solutions.
5. The proposed tool may be generalized by including practical constraints related to generating units such as valve point effect, prohibited zones, multi fuels and transient stability criterions to solve practical multi objective functions.

CONCLUSION

Ensuring optimal management of modern power system is a challenge for experts and industrials. This fundamental task becomes more complex with the deregulation of electric industry throughout the world, and the large penetration of renewable energy sources and FACTS devices. On the other hand, the rapid

Table 6. Optimal setting of control variables at different level of RMS with SVC devices

Control Variables	Min	Max	Level 1	Level 2	Level 3
P_{g1} *(Slack)*	50	200	65.3201	65.4018	65.4549
P_{g2}	20	80	66.1212	66.2475	66.3141
P_{g5}	15	50	49.9998	49.9992	49.9991
P_{g8}	10	35	34.9999	34.9996	34.9996
P_{g11}	10	30	29.9999	29.9997	29.9997
P_{g13}	12	40	39.9999	39.9995	39.9995
V_{g1}	0.95	1.1	1.1000	1.1000	1.1000
V_{g2}	0.95	1.1	1.0953	1.0906	1.0852
V_{g5}	0.95	1.1	1.0741	1.0593	1.0487
V_{g8}	0.95	1.1	1.0766	1.0626	1.0527
V_{g11}	0.95	1.1	1.0999	1.0530	1.0600
V_{g13}	0.95	1.1	1.0993	1.0665	1.0744
T_{11}	0.90	1.1	1.0146	1.0368	1.0147
T_{12}	0.90	1.1	1.0054	1.0276	1.0055
T_{15}	0.90	1.1	0.9986	1.0208	0.9987
T_{36}	0.90	1.1	0.9916	1.0138	0.9917
Qc10	0	5	3.4976	3.4863	3.4786
Qc12	0	5	4.8975	4.8816	4.8708
Qc15	0	5	2.5743	2.5659	2.5602
Qc17	0	5	4.8985	4.8826	4.8718
Qc20	0	5	4.3113	4.2973	4.2878
Qc21	0	5	4.8914	4.8755	4.8648
Qc23	0	5	4.8730	4.8572	4.8464
Qc24	0	5	4.8985	4.8826	4.8718
Qc29	0	5	2.7684	2.7594	2.7533
Cost ($/h)	-	-	941.5768	942.2849	942.6875
Power loss (MW) *Voltage deviation (p.u)*	- -	- -	**3.0409** 1.7369	**3.2473** 0.9383	**3.3669** 1.1340
CRMSi			0.2495 **Bus 8**	0.1667 **Bus 13**	0.1250 **Bus 13**
GRMS	-	-	**0.9594**	**1.1616**	**1.3806**
Total load (MW)	-	-	283.4	283.4	283.4

development of communication technology and informatics has contributed to facilitate the treatment with accuracy of various conflict objectives. During the last two decades large number of optimization methods has been proposed for solving various practical optimization problems related to power system planning, operation and control. In this context, this chapter presents an interactive tool using graphic user interface (GUI) environment based various metaheuristic methods such as Grey wolf optimizer (GWO), and pattern search algorithm for solving OPF problems considering practical constraints. In this

Figure 25. Optimized critical reactive margin security

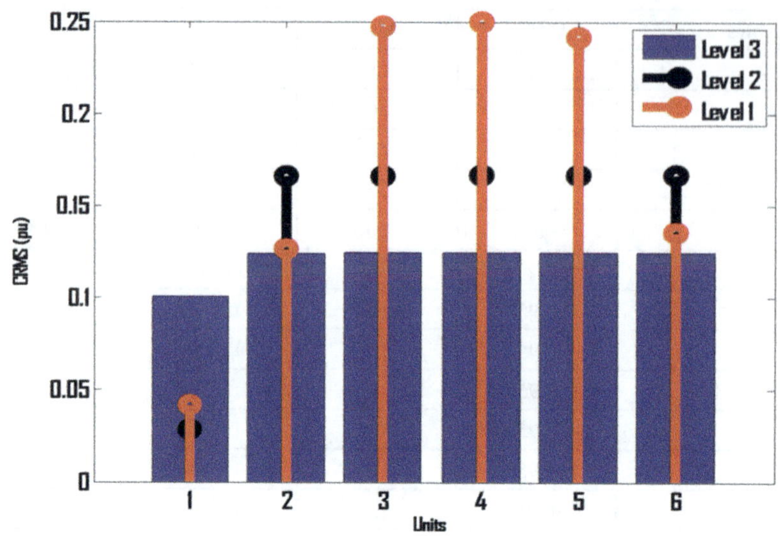

Figure 26. Distribution of voltage magnitudes: power loss minimization

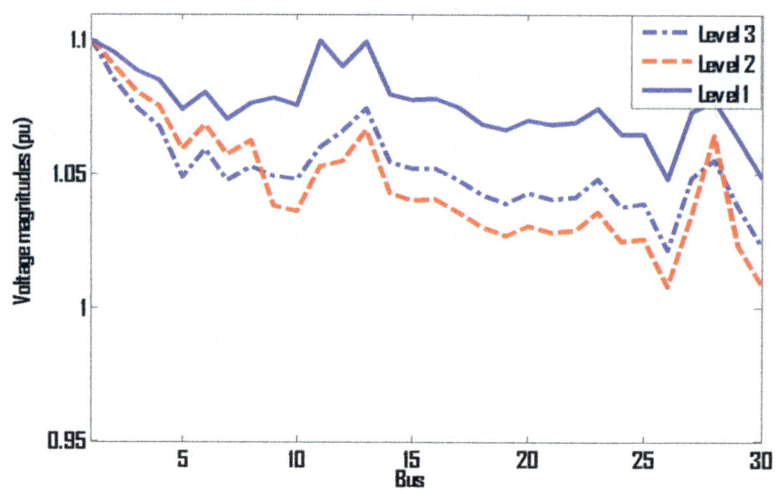

preliminary study, the proposed technical optimizer tool focuses on solving the security OPF considering new indices named critical and global reactive margin stability. The developed simulator will be useful as an educational tool for students as well for researchers to interact dynamically with the program to well understand the mechanism search of various metaheuristic methods. In this study only two metaheuristic methods have been studied, and successfully applied within the interactive tool to solve the optimal power flow problem considering various objective functions such as total cost, total power loss and total voltage deviation. The proposed interactive optimizer tool is expected to be a useful support for students and experts specialized in power system planning and control.

REFERENCES

Adaryani, M. R., & Adaryani, A. K. (2013). Artificial bee colony algorithm for solving multi-objective optimal power flow problem. Electrical Power and Energy Systems, 53, 219-230.

Alsac, O., Scott, B. (1974). Optimal load flow with steady state security. *IEEE Trans Power Appar Syst.*, 745–51.

Alsumait, J. S., Sykulski, J. K., & Al-Othman, A. K. (2010). A hybrid GA–PS–SQP method to solve power system valve-point economic dispatch problems. *Applied Energy*, *87*(5), 1773–1781. doi:10.1016/j.apenergy.2009.10.007

Ananthi, A. C., Ajay, D. P., & Vimal, R. (2014). Adaptive biogeography based predator–prey optimization technique for optimal power flow. *Electrical Power and Energy Systems (Elsevier)*, *62*, 344–352. doi:10.1016/j.ijepes.2014.04.054

Attia, A. F., Al-Turki, Y. A., & Abusorrah, A. M. (2012). Optimal power flow using adapted genetic algorithm with adjusting population size. *Electric Power Components and Systems*, *40*(11), 1285–1299. doi:10.1080/15325008.2012.689417

Bakirtzis, A. G., Biskas, P. N., Zournas, C. E., & Petridis, V. (2002). Optimal power flow by enhanced genetic algorithm. *IEEE Transactions on Power Systems*, *17*(2), 229–236. doi:10.1109/TPWRS.2002.1007886

Balasubbareddy, M., Sivanagaraju, S., & Chintalapudi, S. (2015). Multi-objective optimization in the presence of practical constraints using non-dominated sorting hybrid cuckoo search algorithm. *Engineering Science and Technology, an International Journal, 18*, 603-615.

Bouchekaraa, H. R. E. H., Abido, M. A., & Boucherma, M. (2014). Optimal power flow using Teaching-Learning-Based Optimization technique. *Electric Power Systems Research*, *114*, 49–59. doi:10.1016/j.epsr.2014.03.032

Dommel, H. W., & Tinney, T. F. (1968). Optimal power flow solutions. *IEEE Transactions on Power Apparatus and Systems*, *87*(5), 1866–1976. doi:10.1109/TPAS.1968.292150

Frank, S., Steponavice, I., & Rebennak, S. (2012). *Optimal power flow: a bibliographic survey I, formulations and deterministic methods*. Springer-Verlag.

Frank, S., Steponavice, I., & Rebennak, S. (2012). *Optimal power flow: a bibliographic survey II, nondeterministic and hybrid methods*. Springer-Verlag.

Ghanizadeh, A.J., Mokhtari, G., Abedi, M., Gharehpetian, G.B. (2011). Optimal power flow based on imperialist competitive algorithm. *Int. Rev. Electr. Eng, 6.*

Ghasemi, M., Ghavidel, S., Ghanbarian, M. M., Reza Massrur, H., & Mahboubi-Moghaddam, E. (2014). Multi-objective optimal power flow using new modified and hybrid modified imperialist competitive algorithms: A comparative study. *Information Sciences (Elsevier)*, *281*, 225–247. doi:10.1016/j.ins.2014.05.040

Gnanambal, K., & Babulal, C. K. (2012). Maximum loadability limit of power system using hybrid differential evolution with particle swarm optimization. *International Journal of Electrical Power & Energy Systems*, *43*(1), 150–155. doi:10.1016/j.ijepes.2012.04.033

Huneault, M., & Galiana, F. (1991). A survey of the optimal power flow literature. *Power Syst IEEE Trans*, *6*(2), 762–770. doi:10.1109/59.76723

Joorabian, M., & Afsan, E. (2014). Optimal power flow under both normal and contingent operation conditions using the hybrid fuzzy particle swarm optimization and nedler-mead algorithm (HFPSO-NM). *Applied Soft Computing*, *14*, 623–633. doi:10.1016/j.asoc.2013.09.015

Karaboga, D. (2005). *An idea based on honey bee swarm for numerical optimization.* Technical Report-TR06, Erciyes University of Engineering, Faculty of Computer Engineering Department.

Lai, L. L., Ma, J. T., Yokoyama, R., & Zhao, M. (1997). Improved genetic algorithms for optimal power flow under normal and contingent operation states. *International Journal of Electrical Power & Energy Systems*, *19*(5), 287–292. doi:10.1016/S0142-0615(96)00051-8

Mahdad, B. (2013). *Power Quality Management Based Security OPF Considering FACTS Using Metaheuristic Optimization Methods.* InTech Education and Publishing KG, Intechweb.org.

Mahdad, B., Srairi, K. (2011). *Understanding Power Quality based FACTS using Interactive Educational GUI Matlab Package.* InTech Education and Publishing KG, Intechweb.org.

Mahdad, B., & Srairi, K. (2012). *Hierarchical adaptive PSO for multi-objective OPF considering emissions based shunt FACTS.* IEEE.

Mahdad, B., & Srairi, K. (2013). Solving multi-objective optimal power flow problem considering wind-STATCOM using differential evolution. *Front. Energy*, *7*(1), 75–89. doi:10.100711708-012-0222-x

Mahdad, B., & Srairi, K. (2014). Multi objective large power system planning under sever loading condition using learning DE-APSO-PS strategy. *Energy Conversion and Mmanagement (Elsevier)*, *87*, 338–350. doi:10.1016/j.enconman.2014.06.090

Mahdad, B., & Srairi, K. (2015). Blackout risk prevention in a smart grid based flexible optimal strategy using Grey Wolf-pattern search algorithms. *Energy Conversion and Management*, *98*, 411–429. doi:10.1016/j.enconman.2015.04.005

Mahdad, B., & Srairi, K. (2015). Security optimal power flow considering loading margin stability using hybrid FFA–PS assisted with brainstorming rules. *Applied Soft Computing*, *35*, 291–309. doi:10.1016/j.asoc.2015.06.037

Mahdad, B., & Srairi, K. (2017). A new interactive sine cosine algorithm for loading margin stability improvement under contingency. *Electrical Engineering*, 1–21.

Mandal, K. K., Mandal, S., Bhattacharya, B., & Chakraborty, N. (2015). Non-convex emission constrained economic dispatch using a new self-adaptive particle swarm optimization technique. *Applied Soft Computing*, *28*, 188–195. doi:10.1016/j.asoc.2014.11.033

Milano, F., Vanfretti, L., & Morataya, J. C. (2008). An open source power system virtual laboratory: The PSAT case and experience. *IEEE Transactions on Education*, *51*(1), 17–23. doi:10.1109/TE.2007.893354

Mirjalili, S., Mirjalili, S. M., & Lewis, A. (2014). Grey Wolf Optimizer. *Advances in Engineering Software*, *69*, 46–61. doi:10.1016/j.advengsoft.2013.12.007

Modiri-Delshad, M., & Abd Rahim, N. (2014). Solving non-convex economic dispatch problem via backtracking search algorithm. *Energy*, *77*, 372–381. doi:10.1016/j.energy.2014.09.009

Mohd Herwan, S., Zuriani, M., Mohd Rusllim, M., & Aliman, O. (2015). Using the gray wolf optimizer for solving optimal reactive power dispatch problem. *Applied Soft Computing*, 286–292.

Mojtaba, G., Ghavidel, S., Gitizadeh, M., & Akbari, E. (2015). An improved teaching–learning-based optimization algorithm using Lévy mutation strategy for non-smooth optimal power flow. *International Journal of Electrical Power & Energy Systems*, *65*, 375–384. doi:10.1016/j.ijepes.2014.10.027

Momoh, J. A., El-Hawary, M. E., & Adapa, R. A. (1999). Review of selected optimal power flow literature to 1993. II. Newton, linear programming and interior point methods. *IEEE Transactions on Power Systems*, *14*(1), 105–111. doi:10.1109/59.744495

Mukherjee, A., & Mukherjee, V. (2015). Solution of optimal power flow using chaotic krill herd algorithm. *Chaos, Solitons, and Fractals*, *78*, 10–21. doi:10.1016/j.chaos.2015.06.020

Narges, D., Mehrdad, T. H., & Teimourzadeh, S. (2016). Adaptive group search optimization algorithm for multi-objective optimal power flow problem. *Applied Soft Computing*, *38*, 1012–1024. doi:10.1016/j.asoc.2015.10.057

Ranjit, R., & Jadhav, H. T. (2015). Optimal power flow solution of power system incorporating stochastic wind power using Gbest guided artificial bee colony algorithm. *International Journal of Electrical Power & Energy Systems*, *64*, 562–578. doi:10.1016/j.ijepes.2014.07.010

Serhat, D., Yorukeren, N., & Altas, I. H. (2015). A novel modified hybrid PSOGSA based on fuzzy logic for non-convex economic dispatch problem with valve-point effect. *Electrical Power and Energy Systems*, *64*, 121–135. doi:10.1016/j.ijepes.2014.07.031

Susanta, D., Roy, P. K., & Debashis, N. (2015). Optimal location of UPFC controller in transmission network using hybrid chemical reaction optimization algorithm. *International Journal of Electrical Power & Energy Systems*, *64*, 194–211. doi:10.1016/j.ijepes.2014.07.038

Woo, N. L., & Jong, B. P. (2011). Educational Simulator for Particle Swarm Optimization and Economic Dispatch Applications. In MATLAB - A Ubiquitous Tool for the Practical Engineer. InTech Education and Publishing KG.

Xiaohui, Y., Pengtao, W., Yanbin, Y., Yuehua, H., & Xiaopan, Z. (2015). A new quantum inspired chaotic artificial bee colony algorithm for optimal power flow problem. *Energy Conversion and Management*, *100*, 1–9. doi:10.1016/j.enconman.2015.04.051

Yan, X., & Quantana, V. H. (1999). Improving an interior point based OPF by dynamic adjustments of step sizes and tolerances. *IEEE Transactions on Power Systems*, *14*(2), 709–717. doi:10.1109/59.761902

Yi, T., Canbing, L., Yijia, C., Kwang, Y. L., Lijuan, L., Shengwei, T., & Lian, Z. (2015). Improved group search optimization method for optimal power flow problem considering valve-point loading effects. *Neurocomputing*, *148*, 229–239. doi:10.1016/j.neucom.2013.09.065

Zimmirrman, R. D., Murillo-Sanchez, C. E., & Gan, D. (n.d.). *MATPOWER: a Matlab power system simulation package*. Retrieved from http://www.pserc. cornell.edu/matpower/3.2

KEY TERMS AND DEFINITIONS

FACTS Devices: A new technology based on power electronics elements designed to ensure flexible control of modern power system in particular at critical conditions.

Global Optimization Methods: A new category of optimization methods inspired from nature based on interaction between exploration and exploitation, designed to improve the solution of complex optimization problems.

Graphic User Interface: An interactive platform based on visual programming designed to analyze data with simplicity and accuracy.

Hybrid Methods: A new category of global optimization methods, based on combining various methods, designed to improve the performances of the standard metaheuristic methods.

Reactive Power Planning: An important sub problem of power system planning. It consists of improving security optimal power flow by optimizing specified control variables.

Chapter 7
Controlling STATCOM by Full Order State Feedback Controller for Stability Improvement

Champa Nandi
Tripura University, India

Ajay Kumar Chakraborty
National Institute of Technology Agartala, India

ABSTRACT

Today's power systems have been extended due to interconnections with the neighboring systems. Energy cost is increasing due to greater sensitivity of new transmission lines. These have demanded the search of new controllers. These objectives were met by reactive power control on transmission lines. Hence, FACTS controllers are becoming an important part of the modern power transmission systems. Different control techniques are in use with STATCOM to generate control signals. But the conventional controllers like P; PI provide a slow dynamic response during the fault condition. This chapter creates an awareness of control issues and proposes an observer-based state feedback controller for STATCOM. The mathematical explanation of the observer-based state feedback controller for STATCOM is presented. Simulations have been carried out with one case study in MATLAB/SIMULINK environment. The simulation results show that the observer-based state feedback controller gives better performance than conventional PI controller during contingency condition.

INTRODUCTION

Voltage Instability or transient instability and voltage collapse can cause the major blackout in the power system. Installation of FACTS Devices in existing transmission line is a way to strengthen power and voltage stability. Voltage collapse typically occurs on power systems which are heavily loaded, faulted and has reactive power shortages. Voltage collapse is system instability in that it involves many power system components and their variables at once. Indeed, voltage collapse often involves an entire power system, although it usually has a relatively larger involvement in one particular area of the power system.

DOI: 10.4018/978-1-5225-3935-3.ch007

The present scenario of power stations of India demonstrates that the stress on transmission systems is increasing considerably because of the large population, modernization and vibrant economic growth. This growing demand for electricity and restriction on the set up of new transmission lines results in huge loss, even many existing transmission systems operate below their normal rating due to some limits. To tackle these problems now-a-days more advanced technologies have been developed for better control and reliable operation of power systems. In 1980s, the Electric Power Research Institute (EPRI) has introduced a new approach for increasing power transmission capability of existing power systems. The proposed concept is known as Flexible AC Transmission Systems (FACTS). Flexible AC Transmission System (FACTS) are Power Electronics based devices which have been developed for improving the utilization and capability of power transfer of the existing power systems. STATCOM is one of the most effective FACTS devices for providing dynamic compensation of ac transmission systems and also to improve the stability of the system. STATCOM can control various parameters of the transmission line .It has the capability to control both the real and reactive power flow through the line. Hence it brings new scopes in power electronics and power system design. During voltage instability reactive power support is so important. Now to control the amount of reactive power it requires a fast response and well designed controller. Controller of FACTS devices plays an important role to compensate the system in different system conditions. Different control techniques are in use with STATCOM to generate control signal. But the conventional controllers like P; PI provides slow dynamic response during the fault condition. This chapter presents with a new approach to design an Observer Based State Feedback Controller of VSI based STATCOM for improving transient stability of the system. Full order state feedback controller for VSI-STATCOM has been proposed. This chapter also presents detailed analysis to compare the performance of STATCOM with conventional PI and proposed Observer Based State Feedback Controller.

BACKGROUND

Past studies have explored the elaborate relationship that exist between insufficient reactive power support and voltage collapse (Hammed, 1998) .Some related study has done by Taylor (Taylor, 1993) & Johansson (Johansson, 1995).There are many conventional controllers such as transformer tap changers, phase shifters which are used for improving voltage stability, transient stability etc. But, these controllers are not so fast in response and suffer due to limitations. To avoid these drawbacks, FACTS devices are proposed (Padiyar, 2007). The FACTS controllers have the ability of enhancing transmission system control, reliability, and operation, and also improve the distribution-system power quality. In 1980s, the Electric Power Research Institute (EPRI) introduces a new approach to solve the problem of designing, controlling and operating power systems: the proposed concept is known as Flexible AC Transmission Systems (FACTS). Its main objectives are to increase power transmission capability, voltage control, voltage stability enhancement and power system stability improvement. FACTS controllers are based on voltage source converters such as Static Var Compensators (SVC), static Synchronous Compensators (STATCOM), Thyristor Controlled Series Compensators (TCSC), Static Series Synchronous Compensators (SSSC) and Unified Power Flow Controllers (UPFC) (Hingorani, 1998). Similar study has been analysed by Mathur (2000), Gama (1992), Taranto (1992), Gama (1992), Nowuh (2009) etc. During voltage instability reactive power support is important. Now to control the amount of reactive power it requires a fast response and well designed controller. Controller of FACTS devices plays an important role to compensate the system in different system conditions. Different control techniques are in use with STATCOM to generate control signal. But the conventional controllers like P; PI provides slow dynamic response during the fault condition. The author of Al-Mawsawi (2004), Voraphonpiput (2005), Norouzi (2005), Blazic (2006), Shen (2002) et al. discuss in their research about the effects of different FACTS controller performances. Hossain et al. (2014) presents in his research work that the design of pole placement controller for D-STATCOM and it shows that this type of controller is able to control the amount of injected current or voltage from the D-STATCOM inverters to mitigate the three-phase fault.

Spitsa et al. (2009) proposed one solution technique based on a zero set concept. It calculates a complete set of the admissible feedback gains that place closed-loop poles into a prespecified region in the complex plane under parametric uncertainties in the plant model. Conventional STATCOM control method is proportional-integral (PI) controllers. However, this work obtains the PI gains via a trial-and-error approach or extensive studies with a trade off of performance and applicability. Hence, control parameters for the optimal performance at a given operating point may not be effective at a different operating point. Xu (2014) in his research proposes a new control model based on adaptive PI control, which can self-adjust the control gains during a disturbance such that the performance always matches a desired response, regardless of the change of operating condition. Since the adjustment is automatic, this gives the plug-and-play capability for STATCOM operation. Rahim and Kandlawala (2003) in their work proposed a loop-shaping technique to select a suitable open-loop transfer function, from which the robust controller function is constructed. Here the damping is provided by the robust controller was compared with PID controller, whose gains were calculated through a pole-placement technique. The robust design demonstrated to provide extremely good damping characteristics over a range of operating conditions. Al-Mawsawi, M. R. Qader and G. M. Ali (2004) in their research work proposed a closed-loop STATCOM system with two types of controllers, PI with gain scheduling and fuzzy logic. The dynamic performance of the two controllers is tested and compared. It is found that, the fuzzy logic controller forces the system to settle to the steady state value faster than the PI controller with gain scheduling. Voraphonpiput et al (2005) presented a mathematical model of STATCOM based on space vector theory. The voltage equations are transformed into the synchronous rotating reference frame (dq-axis). In this paper the behaviour of STATCOM for voltage regulation was simulated with ATP-EMTP. The case study was made on a 230 kV radial system. Lu et al. (2009) presented fuzzy PI based direct output voltage control strategy for the STATCOM. The DOV control strategy not only reduces the active and reactive current control loops of the conventional double-loop control strategy but also achieves a decoupling control to regulate the dc-link voltage and maintain the voltages at the PCC. A scaled physical prototype rated at ±300 KVar was built to verify the validity of the STATCOM. Simulation and experimental results all demonstrate that this novel control strategy effectively improves the static and dynamic performances of the STATCOM system. Ajami and Younesi (2008) in their work proposed a state feedback controller based optimal linear quadratic regulator (LQR) for transient dynamic performance of current-source converter based STATCOM. Here the decoupled state-feedback control is formulated and applied to the VSC-based STATCOM. The performances of the STATCOM at steady-state and in response to step changes in the reference values of the system voltage and the dc-side voltage are evaluated using the simulation results from MATLAB package. Peyman Jafari et al (2010) proposed a closed-loop control technique for the D-STATCOM, aiming at load balancing as well as power factor correction. In this paper, a state-feedback-based current controller was proposed and the engaged pulse width modulation was enhanced by dynamically modification of the switching duty ratios to tackle the dc-side voltage oscillations effects. This paper also described that the current control technique provides faster response compared to the conventional approaches under unbalanced condition.

The whole method, including the control technique and the modulation compensation, was simulated by using PSCAD/EMTDC software to demonstrate the validity of the proposed approach. Farokhnia, Khoraminia and Gharehpetian (2010) described that according to nonlinear operation of STATCOM, nonlinear controller has a better performance in comparison with linear controller. Regulating the DC capacitor voltage in STATCOM is a common task and can improve the system dynamic. The nonlinear control is based on exact linearization via feedback. A PI controller exists in this control system to regulate the capacitor voltage. In conventional scheme, the trial and error method has been used to determine PI controller coefficients. In this paper, the effect of PI gains on responses of V_{dc}, I_d and Modulation Index (M) has presented. Homayounzade, Keshmiri, Danesh (2010) proposed an observer based state feedback controller for Rigid Link Electrically-Driven (RLED) manipulators actuated by a brushed DC (BDC) motors based on joint velocity estimation. The observer based controller is in fact a model based controller but it can be shown practically that the method is robust against parametric uncertainties. Pseudo-velocity proposed to estimate the joint's velocity, provided joint's position is measured. Using the measured and estimated states, a controller is designed through the direct Lyapunov method. It is shown that the controller is asymptotically stable in its region of attraction, both in position and velocity tracking. To show the validity of the proposed scheme, the algorithm was implemented on a two link planar manipulator actuated by BDC motors. Mark et al. (2011) analysed that the effects of Static Var Compensator (SVC) on voltage stability of a power system. The functional structure for SVC built with a Thyristor Controlled Reactor (TCR) and its model are described. The model represented the controller as variable impedance that changes with the firing angle of the TCR. A Power System Computer Aided Design Electromagnetic Transients including DC (PSCAD/EMTDC) is used to carry out simulations of the system. Ki Seok Jeong, Tae Young Jyung and Young Sik Baek (2011) proposed the theoretical analysis based on mathematical model of Static Synchronous Compensator (STATCOM). The STATCOM can rapidly supply dynamic reactive power required for short-tem voltage regulation. The performances of STATCOM under specified load variation were simulated with EMTP. Simulation results validate the ability of the STATCOM to improve short-term voltage regulation. With appropriate PI gains, the controller performance is observed reasonably well during capacitive and inductive modes of operation. The simulation results from EMTP/RV represented good dynamic performances of STATCOM during step changes of load variation, Sethy and Moharana (2012) proposed a linear model of STATCOM along with design of current and voltage controllers. The designed controllers with variation of DC-link voltage applied to the STATCOM and suitable DC-link voltage has been selected on basis of spike and over shoot of the responses. The values of the passive parameters decided on basis of selected DC-link voltage are used for the design and fabrication of a STATCOM. All responses are obtained through MATLAB Simulink tool box and presented here for clarity of the control strategy.S.Pradeepa, K.Uma Rao, Ravishankar Deekshit and Shantha (2013) proposed a new approach to the state feedback controller to enhance the performance of CSI based STATCOM. This paper describes that this type of controller can give fast dynamic response to the system. Another research work by Ponmurugavel and Mohamed Ghouse (2013) discussed a model of fuzzy and model of predictive controllers of STATCOM. This paper also described a comparative study between these two types of controller. Simulation results shows that the Predictive Controllers of STATCOM provides better control in transient stability improvement of power system. Yao at all (2010) design CSI –STATCOM with state feedback controller. According to their report, decoupled State feedback controller for CSI-STATCOM fulfils all the expected objectives

HISTORY OF POWER SYSTEM COLLAPSE

Insufficient reactive power leading to voltage collapse has been a causal factor in major blackouts in the worldwide. It is reported that the past massive collapse of India's power grid was the worst in the decade. Three out of the five regional power grids collapsed leaving about 670 million people powerless making July 2012 as the largest blackout month in history. 21 Indian states plunged into darkness as the northern, north-eastern and eastern grids collapsed (Anandane, 2014).In 2003 voltage collapse problem reported in North America (Beck G, 2005) While August 14, 2003, blackout in the United States and Canada was not due to a voltage collapse as that term has traditionally used by power system engineers, the task force final report said that" Insufficient reactive power was an issue in the blackout" and the report also "overestimation of dynamics reactive output of system generation" as common factor among major outages in the United States. Voltage collapse occurred in United States in the blackout of July 2, 1996, and August10, 1996 on the West Coast. Due to presence of electric loads which are predominantly induction motors the voltage recovery of the system follow severe disturbance due to lack of fast responding reactive power support (Krishnan V., 2006). The role of induction motors has also been highlighted in reference (Stefopoulos, 2006) where it is concluded that for the induction motor predominant load buses, the critical fault clearing time is important to ensure stable operating voltage.

VOLTAGE STABILITY

There are dozens of different definitions of stability, and several of these are presented by the author Dobson et al. (2003). One of the definitions is small disturbance stability of an operating point: An operating point of a power system is small disturbance stable if, following any small disturbance, the power system state returns to the identical or close to the pre-disturbance operating point. The stability of a system refers to the ability of a system to return back to its steady state when subjected to a disturbance. As mentioned before, power is generated by synchronous generators that operate in synchronism with the rest of the system. A generator is synchronized with a bus when both of them have same frequency, voltage and phase sequence. We can thus define the power system stability as the ability of the power system to return to steady state without losing synchronism. Usually power system stability is categorized into steady state, Transient and dynamic stability.

REACTIVE POWER SUPPORT

Practically reduce of reactive power will be improve system efficiency in power system. This is acceptable at some conditions. If system is purely inductive or capacitive it makes cause some difficulty in Electrical system. Alternating (AC) systems supply or consume two kind of power: real power and reactive power. Real power carry outs useful work while reactive power supports the voltage that must be controlled for system reliability. Reactive power has a profound effect on the security of power systems because it influences voltages throughout the system. Voltage control in an electrical power system is important for proper operation for electrical power equipment to prevent damage such as overheating of generators and motors, to reduce transmission losses and to maintain the ability of the system to resist and avoid voltage collapse. Decreasing of reactive power causes voltage to fall while increasing of reactive power it causes voltage to rise. A voltage collapse may be occurs when the system try to serve much more load than the voltage can support or any sort of fault occurs. When reactive power supply lower voltage, as voltage drops current must increase to maintain power supplied, causing system to consume more reactive power and the voltage drops further . If the current increase too much, transmission lines go off line, over loading other lines and potentially causing cascading failures. If the voltage drops too low, some generators will disconnect automatically to protect themselves. Voltage collapse occurs when an increase in load or less generation or transmission facilities causes dropping voltage, which causes a further reduction in reactive power from capacitor and line charging, and still there further voltage reductions. If voltage reduction continues, these will cause additional elements to trip, leading further reduction in voltage and loss of the load. The result in these entire progressive and uncontrollable declines in voltage is that the system unable to provide the reactive power required supplying the reactive power demands. Main limitations of reactive Power is that it does not travel very far. Reactive power support is important in different area of power system. The STATCOM is a solid-state shunt device that generates or absorbs reactive power and is one member of a family of devices known as flexible AC transmission system. The solid-state nature of the STATCOM controller can provide very fast and effective voltage control by supplying reactive power support.

STATCOM

The STATCOM is a shunt connected reactive power compensation device. It is a solid-state switching converter capable of generating or absorbing independently controllable real and reactive power at its output terminals as it is fed from an energy source or energy-storage device at its input terminals. Specifically, the STATCOM is a voltage-source inverter or current source inverter. VSI STATCOM produces a set of 3-phase ac-output voltages from dc input voltage, each phase is coupled to the corresponding ac system voltage through a relatively small reactance .The reactance is provided by either an interface reactor or the leakage inductance of a coupling transformer. The dc voltage is provided by an energy-storage capacitor. The STATCOM basically consists of a step-down transformer with a leakage reactance, a three-phase GTO or IGBT voltage source inverter (VSI), and a DC capacitor. The AC voltage difference across the leakage reactance produces reactive power exchange between the STATCOM and the power system, such that the AC voltage at the bus bar can be regulated to improve the voltage profile of the power system, which is the primary duty of the STATCOM. However, for instance, a secondary damping function can be added into the STATCOM for enhancing power system oscillation stability. The VSI generates a controllable AC voltage source behind the leakage reactance. This voltage is compared with the AC bus voltage system; when the AC bus voltage magnitude is above that of the VSI voltage magnitude. The AC system sees the STATCOM as an inductance connected to its terminals. Otherwise, if the VSI voltage magnitude is above that of the AC bus voltage magnitude, the AC system sees the STATCOM as a capacitance connected to its terminals. If the voltage magnitudes are equal, the reactive power exchange is zero (Hingorani, 2000), (Mathur, 2000).

OPERATING PRINCIPAL OF STATCOM

Figure 1 shows the phasor diagram of STATCOM. When the converter voltage (V_D) is lower than the bus voltage (V_B), the STATCOM acts like an inductance absorbing reactive power from the bus. When the converter voltage (V_D) is higher than the bus voltage (V_B), the STATCOM acts like a capacitor generating reactive power to the bus.

- **Normal (Floating) Excited Mode of Operation:** During No load mode, $V_B = V_D$, $I_D = 0$.The reactive power is zero and the STATCOM does not generate or absorb reactive power.
- **Under Excited Mode of Operation:** During Inductive mode of Operation, I_D appears to be lagging current, since the magnitude of the current can be controlled by adjusting V_D. The STATCOM will function as a reactor whose inductive reactance is continuously controllable. The current (I_D) flows through the transformer reactance from the AC system to STATCOM device. The device generates inductive reactive power.
- **Over Excited Mode of Operation:** During Capacitive mode of Operation I_D appears to be leading current, since the magnitude of the current can be controlled by adjusting V_D. The STATCOM will function as a capacitor whose capacitive reactance is continuously controllable. The current (I_D) flows through the transformer reactance from STATCOM device to the AC system. The device generates capacitive reactive power.

TRADITIONAL CONTROLLER FOR STATCOM

The static compensator (STATCOM), a popular device for reactive power control based on gate turn off (GTO) thyristors, has gained much interest in the last decade for improving power system stability. In many STATCOM models, the control logic is implemented with the PI controllers. The control parameters or gains play a key factor in STATCOM performance.

Figure 1. STATCOM with Phasor diagram

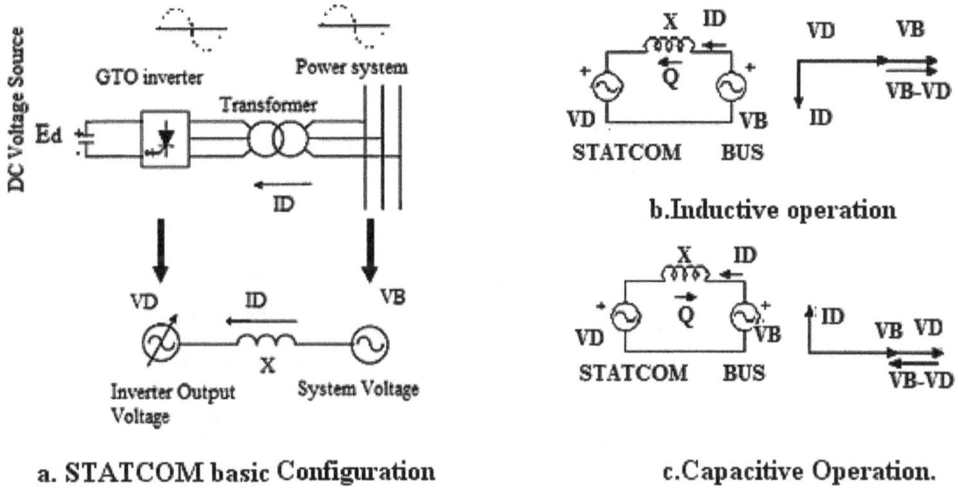

CONTROLLING STATCOM USING FULL ORDER STATE FEEDBACK CONTROLLER

This chapter deals with a new approach to design an Observer Based State Feedback Controller of VSI based STATCOM for improving transient stability of the system. Full order state feedback controller for VSI-STATCOM has been proposed. Full state feedback controller is developed using pole placement and LQR (Linear Quadratic Regulation) methods. Then considering the reality condition a full order state observer is designed. Finally, the state feedback controller and the state observer are summed up to give a pre compensator. Mathematical modelling is presented and simulations with their analysis have done using MATLAB Software .This book chapter also presents detailed analysis to compare the performance of STATCOM with conventional PI and proposed Observer Based State Feedback Controller

The whole results and discussions have been depicted in this chapter and detailed derivation is presented below.

DYNAMIC MATHEMATICAL MODEL OF STATCOM

Figure 2 shows the circuit diagram of a STATCOM. The dynamic equation of fig can be written as Mathur (2000):

$$\frac{di_a}{dt} = -\frac{R}{L}i_a + \frac{1}{L}V_a - \frac{1}{L}e_a \tag{1}$$

$$\frac{di_b}{dt} = -\frac{R}{L}i_b + \frac{1}{L}V_b - \frac{1}{L}e_b \tag{2}$$

$$\frac{di_c}{dt} = -\frac{R}{L}i_c + \frac{1}{L}V_c - \frac{1}{L}e_c \tag{3}$$

Figure 2. Circuit diagram of a STATCOM

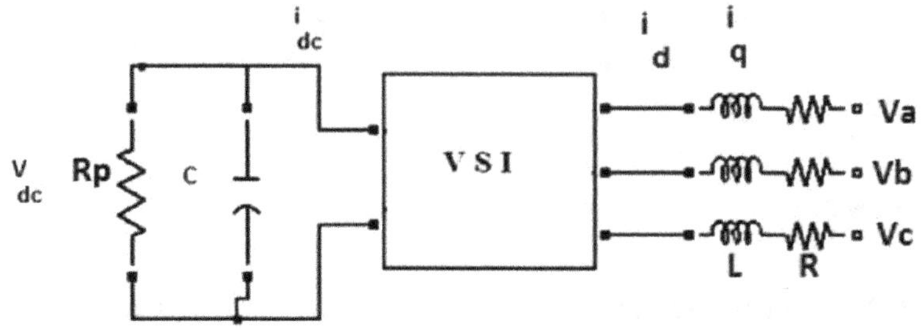

$$\frac{dV_{dc}}{dt} = -\frac{V_{dc}}{R_p C} + \frac{1}{C} i_{dc} \tag{4}$$

Now equations, (1), (2), and (3) can be transformed to synchronously rotating reference frame as:

$$\begin{bmatrix} i_\alpha \\ i_\beta \end{bmatrix} = \sqrt{\frac{2}{3}} \begin{bmatrix} 1 & -\frac{1}{2} & -\frac{1}{2} \\ 0 & \frac{\sqrt{3}}{2} & -\frac{\sqrt{3}}{2} \end{bmatrix} \times \begin{bmatrix} i_a \\ i_b \\ i_c \end{bmatrix}$$

$$\begin{bmatrix} i_d \\ i_q \end{bmatrix} = \begin{bmatrix} \cos\theta & \sin\theta \\ -\sin\theta & \cos\theta \end{bmatrix} \times \begin{bmatrix} i_\alpha \\ i_\beta \end{bmatrix}$$

$$\frac{di_d}{dt} = -\frac{R}{L} i_d + \frac{1}{L} V_d - \frac{1}{L} e_d + \omega i_q \tag{5}$$

$$\frac{di_q}{dt} = -\frac{R}{L} i_q + \frac{1}{L} V_q - \frac{1}{L} e_q + \omega i_d \tag{6}$$

The magnitude of converter voltage and dc voltage across capacitor is directly proportional to each other.

$$e_d = m V_{dc} \cos\delta \tag{7}$$

$$e_q = -m V_{dc} \sin\delta \tag{8}$$

where m is the co-efficient related to the type of the converter. Using equation (7) and (8) the equation (5) and (6) can be written as:

$$\frac{di_d}{dt} = -\frac{R}{L} i_d + \frac{1}{L} V_d - \frac{1}{L} m V_{dc} \cos\delta + \omega i_q \tag{9}$$

$$\frac{di_q}{dt} = -\frac{R}{L} i_q + \frac{1}{L} V_q - \frac{1}{L} m V_{dc} \sin\delta + \omega i_d \tag{10}$$

Now Power $P = V_{dc} \cdot I_{dc}$

$$= \frac{3}{2}\left(e_d i_d + e_q i_q\right)$$

$$= \frac{3}{2}\left(m V_{dc} \cos\delta i_d + m V_{dc} \sin\delta i_q\right) \tag{11}$$

Using this equation (11) i_{dc} can be written as,

$$i_{dc} = \frac{3}{2}\left(m \cos\delta i_d + m \sin\delta i_q\right) \tag{12}$$

Now the equation (9), (10), and (12) can be written as matrix form like:

$$\frac{d}{dt}\begin{bmatrix} i_d \\ i_q \\ V_{dc} \end{bmatrix} = \begin{bmatrix} \frac{-1}{T_1} & \omega & \frac{-m\cos\delta}{L} \\ -\omega & \frac{-1}{T_1} & \frac{m\sin\delta}{L} \\ \frac{3m\cos\delta}{2C} & \frac{-3m\sin\delta}{2C} & \frac{-1}{T_2} \end{bmatrix}\begin{bmatrix} i_d \\ i_q \\ V_{dc} \end{bmatrix} + \begin{bmatrix} \frac{1}{L} & 0 \\ 0 & \frac{1}{L} \end{bmatrix}\begin{bmatrix} V_d \\ V_q \end{bmatrix} \tag{13}$$

where $T_1 = \frac{1}{R}$ and $T_2 = R_p C$

Therefore, characteristic equation of STATCOM is

$$S^3 + S^2\left(\frac{2}{T_1} + \frac{1}{T_2}\right) + S\left(\frac{2}{T_2 T_1} + \frac{1}{T_1^2} + \frac{3m^2}{2LC} + \omega^2\right) + \left(\frac{3m^2}{2LCT_1} + \frac{1}{T_2 T_1^2} + \frac{\omega^2}{T_2}\right) = 0 \tag{14}$$

Linearization of this matrix equation No (13) around the operating firing angle δ_0 gives a set of linear equations as

$$\frac{d}{dt}\begin{bmatrix}\Delta i_d\\ \Delta i_q\\ \Delta V_{dc}\end{bmatrix}=\begin{bmatrix}\frac{-1}{T_1} & \omega & \frac{-m\cos\delta}{L}\\ -\omega & \frac{-1}{T_1} & \frac{m\sin\delta}{L}\\ \frac{3m\cos\delta}{2C} & \frac{-3m\sin\delta}{2C} & \frac{-1}{T_2}\end{bmatrix}\begin{bmatrix}\Delta i_d\\ \Delta i_q\\ \Delta V_{dc}\end{bmatrix}+\begin{bmatrix}\frac{1}{L} & 0 & \frac{mV_{dc0}\sin\delta_0}{L}\\ 0 & \frac{1}{L} & \frac{mV_{dc0}\cos\delta_0}{L}\\ 0 & 0 & -\frac{3m}{2C}\left(i_{d0}\sin\delta_0+i_{q0}\cos\delta_0\right)\end{bmatrix}\begin{bmatrix}\Delta V_d\\ \Delta V_q\\ \Delta\delta\end{bmatrix} \tag{15}$$

Comparing equation (15) with state equation, it can be writing

$$A=\begin{bmatrix}\frac{-1}{T_1} & \omega & \frac{-m\cos\delta}{L}\\ -\omega & \frac{-1}{T_1} & \frac{m\sin\delta}{L}\\ \frac{3m\cos\delta}{2C} & \frac{-3m\sin\delta}{2C} & \frac{-1}{T_2}\end{bmatrix},\ B=\begin{bmatrix}\frac{1}{L} & 0 & \frac{mV_{dc0}Sin\delta_0}{L}\\ 0 & \frac{1}{L} & \frac{mV_{dc0}Cos\delta_0}{L}\\ 0 & 0 & -\frac{3m}{2C}\left(i_{d0}Sin\delta_0+i_{q0}Cos\delta_0\right)\end{bmatrix}$$

CALCULATION OF STATE FEEDBACK GAIN MATRIX (K)

From the characteristic equation of STATCOM equation No (14) the phase-variable representation of the plant is,

$$A=\begin{bmatrix}0 & 1 & 0\\ 1 & 0 & 1\\ -\left(\frac{3m^2R}{2LC}+\frac{R^2}{R_pC}+\frac{\omega^2}{R_pC}\right) & -\left(\frac{2R}{R_pC}+R^2+\frac{3m^2}{2LC}+\omega\right) & -\left(2R+\frac{1}{R_pC}\right)\end{bmatrix}$$

$$B = \begin{bmatrix} \frac{1}{L} & 0 \\ 0 & \frac{1}{L} \\ 0 & 0 \end{bmatrix}; \; K = \begin{bmatrix} K_1 & K_2 & K_3 \end{bmatrix}$$

The characteristic equation of the state feedback system is

$$Det\left[SI - \left(A - BK\right)\right]$$

$$= \begin{bmatrix} \left(S + \frac{K_1}{L}\right) & 1 - \frac{K_2}{L} & 0 \\ -\frac{K_1}{L} & \left(S + \frac{K_2}{L}\right) & 1 \\ -\left(\frac{3m^2R}{2LC} + \frac{R^2}{R_pC} + \frac{\omega^2}{R_pC}\right) & -\left(\frac{2R}{R_pC} + R^2 + \frac{3m^2}{2LC} + \omega\right) & \left(S + 2R + \frac{1}{R_pC}\right) \end{bmatrix} = 0$$

$$= \left(S + \frac{K_1}{L}\right)\left\{\left(S + \frac{K_2}{L}\right)\left(S + 2R + \frac{1}{RpC}\right) + \left(\frac{2R}{RpC} + R^2 + \frac{3m2}{2LC} + \omega\right)\right\}$$

$$-\left(1 - \frac{K_2}{L}\right)\left\{-\frac{K_1}{L}(S + 2R + \frac{1}{R_pC} + \left(\frac{3m^2R}{2LC} + \frac{R^2}{RpC} + \frac{\omega^2}{RpC}\right)\right\}$$

$$= \left(S + \frac{K_1}{L}\right)\left(S^2 + 2RS + \frac{S}{RpC} + \frac{SK_2}{L} + \frac{2R}{L}K_2 + \frac{k2}{LRpC} + \frac{2R}{RpC} + R^2 + \frac{3m^2}{2LC} + \omega^2\right)$$

$$-\left(1 - \frac{K_2}{L}\right)\left(-K_1\frac{S}{L} - \frac{2RK_1}{L} - \frac{K_1}{LRpC} + \frac{3m^2R}{2LC} + \frac{R^2}{RpC} + \frac{\omega^2}{RpC}\right)$$

Let the desired characteristics equation is,

$$S^3 + d_1S^2 + d_2S + d_3 = 0$$

By comparing equations

$$d_1 = 2R + \frac{1}{RpC} + \left(k_1 + k_2\right)\frac{1}{L}$$

$$d_2 = k_2\left(\frac{2R}{L} + \frac{1}{LRpC}\right) + k_1\left(\frac{2R}{L} + \frac{1}{LRpC} + \frac{1}{L}\right) + \left(R^2 + \frac{3m^2}{2LC} + \omega^2\right)$$

$$d_3 = k_1\left(\frac{2R}{LRpC} + \frac{R^2}{L} + \frac{3m^2}{2L^2C} + \frac{\omega^2}{L} + \frac{2R}{L} + \frac{1}{LRpC}\right)$$

$$+k_2\left(\frac{3m^2R}{2L^2C} + \frac{R^2}{LRpC} + \frac{\omega^2}{LRpC}\right) + \left(\frac{3m^2R}{2LC} - \frac{R^2}{RpC} - \frac{\omega^2}{RpC}\right)$$

By simplifying these equations we get the feedback gains as:

$$K_1 = L\left\{d_2 - d_1\left(2R - \frac{1}{R_pC}\right) + \left(3R^2 + \frac{4R}{R_pC} + \frac{1}{R_p^2C^2} - \frac{3m^2}{2LC} - \omega^2\right)\right\}$$

$$K_2 = L\left\{\left(d_1 - 2R - \frac{1}{R_pC}\right) - d_2 - d_1\left(2R - \frac{1}{R_pC}\right) + \left(3R^2 + 4\frac{R}{R_pC} + \frac{1}{R_p^2C^2} - \frac{3m^2}{2LC} - \omega^2\right)\right\}$$

CALCULATIONS OF OBSERVER

$$Det\left[SI - A\right] = S^3 + S^2\left(2R + \frac{1}{RpC}\right)$$

$$+S\left(\frac{2R}{RpC} + R^2 + \frac{3m2}{2LC} + \omega\right) + \left(\frac{3m^2R}{2LC} + \frac{R^2}{RpC} + \frac{\omega^2}{RpC}\right) = 0$$

$$A_X = \begin{bmatrix} -\left(2R + \frac{1}{RpC}\right) & 1 & 0 \\ -\left(\frac{2R}{RpC} + R^2 + \frac{3m2}{2LC} + \omega\right) & 0 & 1 \\ -\left(\frac{3m^2R}{2LC} + \frac{R^2}{RpC} + \frac{\omega^2}{RpC}\right) & 0 & 0 \end{bmatrix}$$

$$L_X = \begin{bmatrix} l_1 \\ l_2 \\ l_3 \end{bmatrix}, C = \begin{bmatrix} 0 & 0 & 1 \end{bmatrix}$$

This is the observable canonical form. Here Lx is the observer gain matrix. The characteristic polynomial of the plant for observer design is

$$[A_X - L_X C] = \begin{bmatrix} -\left(2R + \frac{1}{RpC}\right) & 1 & 0 \\ -\left(\frac{2R}{RpC} + R^2 + \frac{3m2}{2LC} + \omega\right) & 0 & 1 \\ -\left(\frac{3m^2R}{2LC} + \frac{R^2}{RpC} + \frac{\omega^2}{RpC}\right) & 0 & 0 \end{bmatrix} - \begin{bmatrix} l_1 \\ l_2 \\ l_3 \end{bmatrix} \begin{bmatrix} 0 & 0 & 1 \end{bmatrix}$$

$$= \begin{bmatrix} -\left(2R + \frac{1}{RpC} + l1\right) & 1 & 0 \\ -\left(\frac{2R}{RpC} + R^2 + \frac{3m2}{2LC} + \omega + l2\right) & 0 & 1 \\ -\left(\frac{3m^2R}{2LC} + \frac{R^2}{RpC} + \frac{\omega^2}{RpC} + l3\right) & 0 & 0 \end{bmatrix}$$

$$[A_X - L_X C] = \begin{bmatrix} -\left(2R + \frac{1}{RpC}\right) & 1 & 0 \\ -\left(\frac{2R}{RpC} + R^2 + \frac{3m2}{2LC} + \omega\right) & 0 & 1 \\ -\left(\frac{3m^2R}{2LC} + \frac{R^2}{RpC} + \frac{\omega^2}{RpC}\right) & 0 & 0 \end{bmatrix} - \begin{bmatrix} l1 \\ l2 \\ l3 \end{bmatrix} \begin{bmatrix} 0 & 0 & 1 \end{bmatrix}$$

$$= \begin{bmatrix} -\left(2R + \frac{1}{RpC} + l1\right) & 1 & 0 \\ -\left(\frac{2R}{RpC} + R^2 + \frac{3m2}{2LC} + \omega + l2\right) & 0 & 1 \\ -\left(\frac{3m^2R}{2LC} + \frac{R^2}{RpC} + \frac{\omega^2}{RpC} + l3\right) & 0 & 0 \end{bmatrix}$$

Therefore,

$$Det[SI - (A_X - L_X C)] = \begin{bmatrix} S & 0 & 0 \\ 0 & S & 0 \\ 0 & 0 & S \end{bmatrix} - \begin{bmatrix} -\left(2R + \frac{1}{RpC} + l1\right) & 1 & 0 \\ -\left(\frac{2R}{RpC} + R^2 + \frac{3m2}{2LC} + \omega + l2\right) & 0 & 1 \\ -\left(\frac{3m^2R}{2LC} + \frac{R^2}{RpC} + \frac{\omega^2}{RpC} + l3\right) & 0 & 0 \end{bmatrix}$$

$$Det\left[SI-\left(A_X-L_XC\right)\right]=\begin{bmatrix} S & 0 & 0 \\ 0 & S & 0 \\ 0 & 0 & S \end{bmatrix}-\begin{bmatrix} -\left(2R+\dfrac{1}{RpC}+l1\right) & 1 & 0 \\ -\left(\dfrac{2R}{RpC}+R^2+\dfrac{3m2}{2LC}+\omega+l2\right) & 0 & 1 \\ -\left(\dfrac{3m^2R}{2LC}+\dfrac{R^2}{RpC}+\dfrac{\omega^2}{RpC}+l3\right) & 0 & 0 \end{bmatrix}$$

$$=\begin{bmatrix} \left(S+2R+\dfrac{1}{RpC}+l1\right) & -1 & 0 \\ \left(\dfrac{2R}{RpC}+R^2+\dfrac{3m2}{2LC}+\omega+l2\right) & S & -1 \\ \left(\dfrac{3m^2R}{2LC}+\dfrac{R^2}{RpC}+\dfrac{\omega^2}{RpC}+l3\right) & 0 & S \end{bmatrix}$$

$$=\left(S+2R+\frac{1}{RpC}+l_1\right)S^2+S\left(\frac{2R}{RpC}+R^2+\frac{3m2}{2LC}+\omega+l_2\right)+\left(\frac{3m^2R}{2LC}+\frac{R^2}{RpC}+\frac{\omega^2}{RpC}+l_3\right)$$

$$=S^3+\ S^2\left(2R+\frac{1}{RpC}+l_1\right)+S\left(\frac{2R}{RpC}+R^2+\frac{3m2}{2LC}+\omega+l_2\right)+\left(\frac{3m^2R}{2LC}+\frac{R^2}{RpC}+\frac{\omega^2}{RpC}+l_3\right)$$

Let the desired characteristics equation is

$$S^3+a_1S^2+a_2S+a_3=0$$

Now by comparing it is found,

$$l_1=a_1-\left(2R+\frac{1}{RpC}\right)$$

$$l_2=a_2-\left(\frac{2R}{RpC}+R^2+\frac{3m2}{2LC}+\omega\right)$$

$$l_3=a_3-\left(\frac{3m^2R}{2LC}+\frac{R^2}{RpC}+\frac{\omega^2}{RpC}\right)$$

OVERVIEW

One new controller for STATCOM has been proposed. The main objective of this work is to develop an observer based state feedback controller for STATCOM to investigate the performance of STATCOM during contingency. Generalized proposed observer based state feedback STATCOM controller model is shown in Figure 3.

OPEN LOOP STEP RESPONSE SYSTEM

It can be seen from Figure 4 that the system is unstable under step input.

Stability of the system:

The Eigen values of open loop system are:

1.0e+002 *[-2.1226 + 3.3021i, -2.1226 - 3.3021i, 2.5785]

There is one pole on the right hand plane at 257.85. This confirms that open loop system is unstable.

Figure 3. Generalized VSI-STATCOM model of observer based state feedback controller

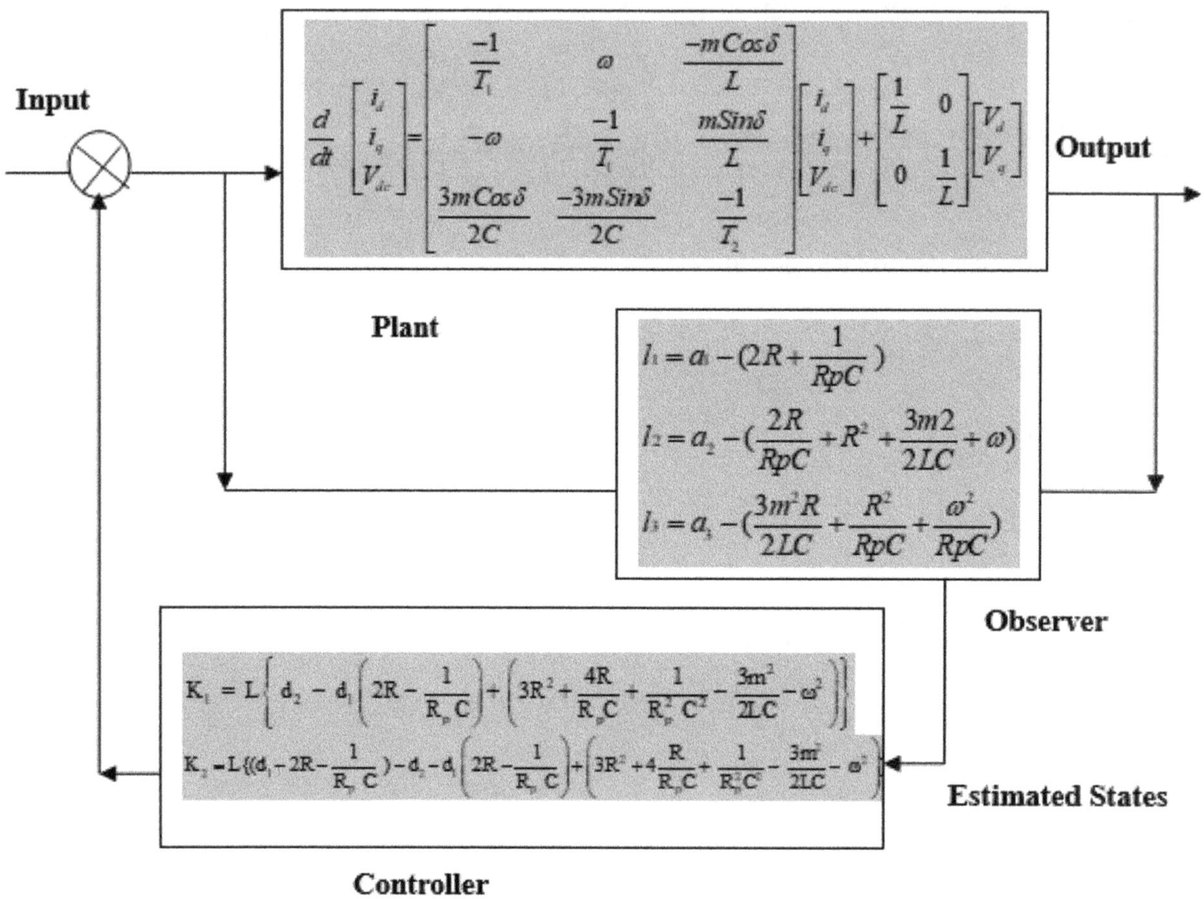

Figure 4. Open loop step response of STATCOM

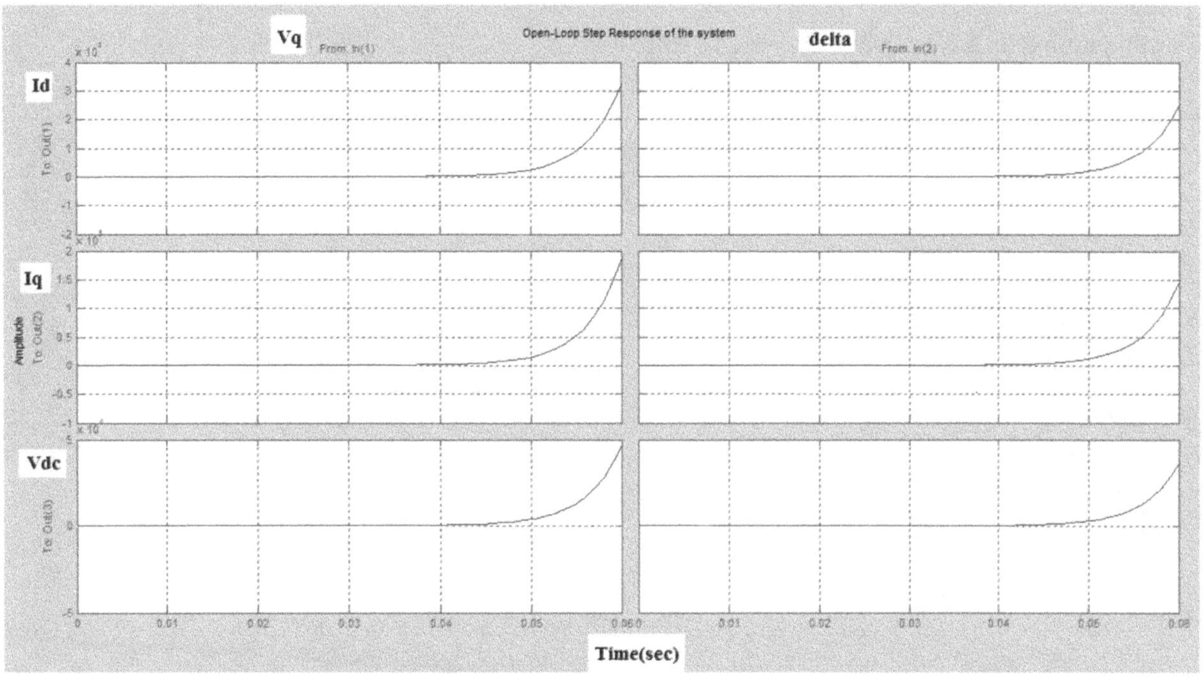

CONTROLLABILITY AND OBSERVABILITY OF THE SYSTEM

A system is controllable if there exists a control input u (t) that transfer any state of the system is zero to finite time. It can be shown that an LTI system is controllable if and only if its controllability matrix has full rank.Here; Controllability matrix has full rank, so the system is controllable.

A system is observable if all initial states are observable, the system is said to be completely observable otherwise the system is said unobservable. It can be shown that an LTI system is observable if and only if its Observability matrix has full rank. Here Observability matrix has full rank, so the system is observable

DESIGNING STATE FEEDBACK CONTROLLER

Designing a state feedback controller to stabilize the system by improving performance of the system using pole placement and Linear Quadratic Regulator (LQR) methods are presented here.

Here the main purpose is to design a controller so that when a step response is given to the system, the system gives stable response.

DATA FOR CONTROLLER DESIGN

For validation data has been consider are as follows

R = 0.5; L=20e^{-3}, C= 300e$^{-6;}$ R_p = 50; d = 0,

So

$$A = \begin{bmatrix} -25 & 314 & -38.98 \\ -314 & -25 & 0 \\ 3898.48 & 0 & -66.667 \end{bmatrix},$$
$$B = \begin{bmatrix} 50 & 0 \\ 0 & 50 \\ 0 & 0 \end{bmatrix},$$
$$C = \begin{bmatrix} 1 & 1 & 0 \end{bmatrix},$$
$$D = 0$$

$$B = \begin{bmatrix} 50 & 0 \\ 0 & 50 \\ 0 & 0 \end{bmatrix}$$

Eigen Values:

$$1.0e + 2 * \begin{bmatrix} .0203 \\ 1.1073 \\ 5.5279 \end{bmatrix}$$

$$K = \begin{bmatrix} -0.2333 & 6.2800 & .5083 \\ -6.200 & 0.4000 & 0 \end{bmatrix}$$

Rank of Controllability matrix: R =3

$$L = \begin{bmatrix} 203 \\ 11072 \\ 55279 \end{bmatrix}$$

Rank of Observability matrix: R = 3

So the system is Controllable and observable

USING POLE PLACEMENT METHODE

See Figures 5-8 and Table 1.

COMPARATIVE STUDY BETWEEN PI AND PROPOSED CONTROLLER IN TERMS OF MAXIMUM OVERSHOOT AND SETTLING TIME

See Figures 9-12 and Table 2.

COMPARATIVE SIMULATION STUDY

To validate this proposed controller it is implemented in a two bus power system network in MATLAB/SIMULINK environment. The test system comprises of a transmission line connecting an AC power source at the sending end and a receiving side connected to a RL load. The sending end voltage is stepped up with a transformer rating 100 MVA, 11/33 KV and transmitted over a distance of 150 km. After that the voltage is stepped down with a transformer rating 100 MVA, 33KV/11KV and connected to a three phase parallel RL load. A fault is applied on the load end side for 100 ms.The test system is first simulated without STATCOM, then with conventional PI controller based STATCOM, next with state

Figure 5. Step response using pole placement

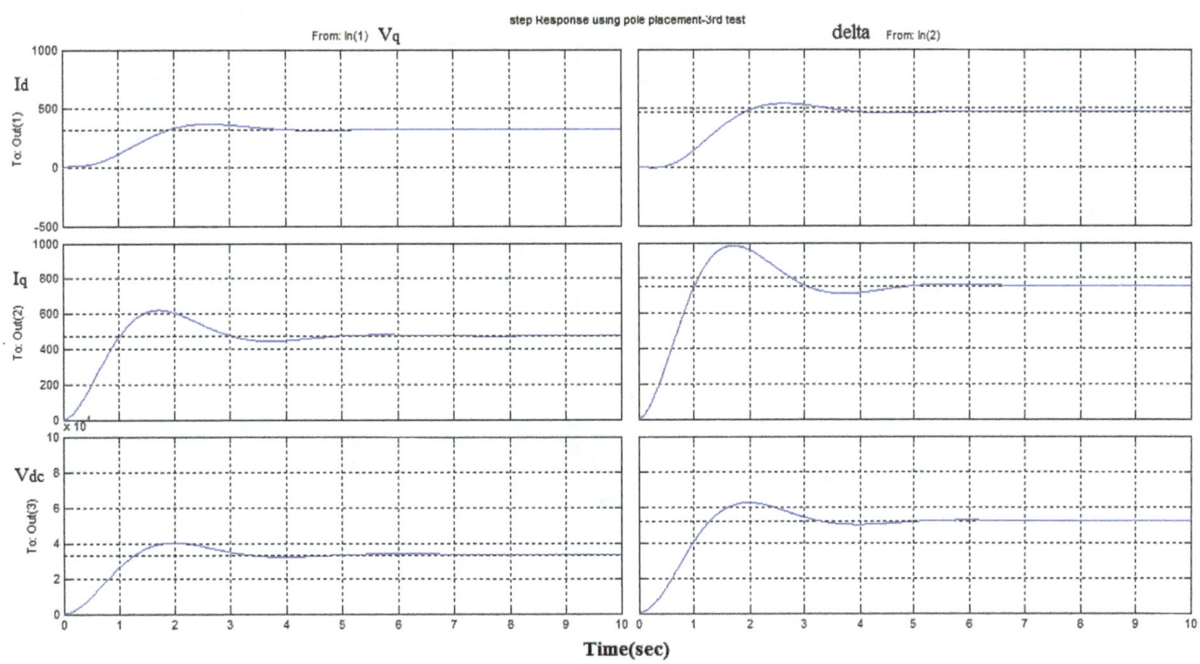

Figure 6. Step response using LQR technique

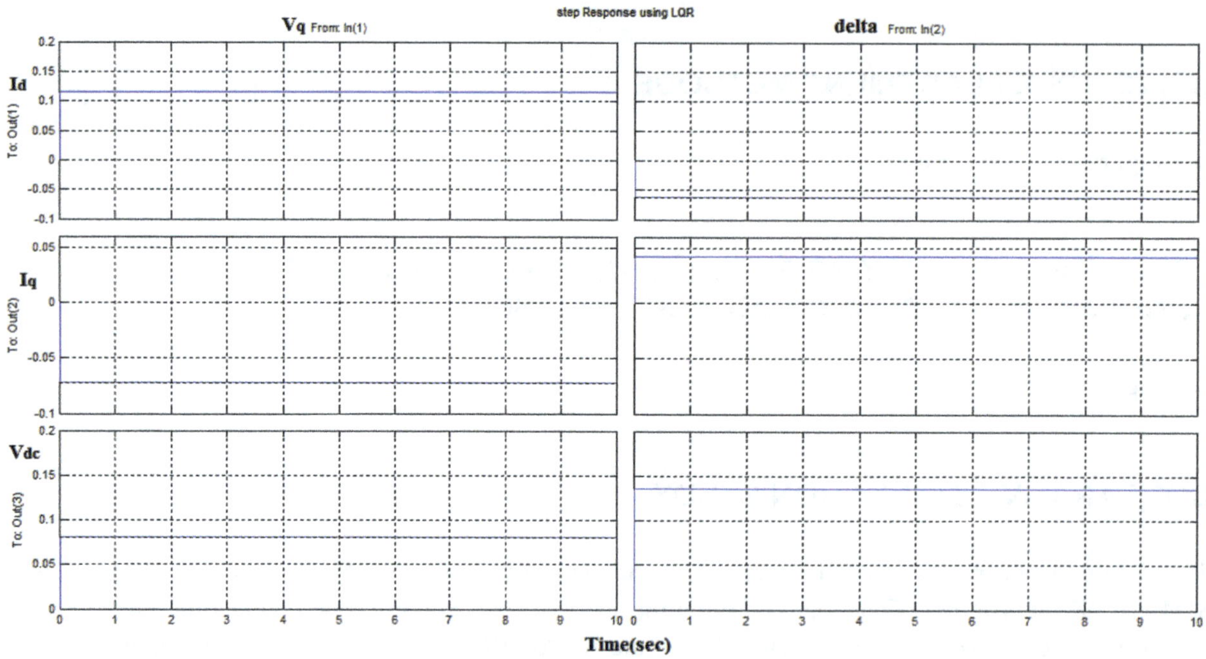

Figure 7. Step response using LQR technique (.02 sec span)

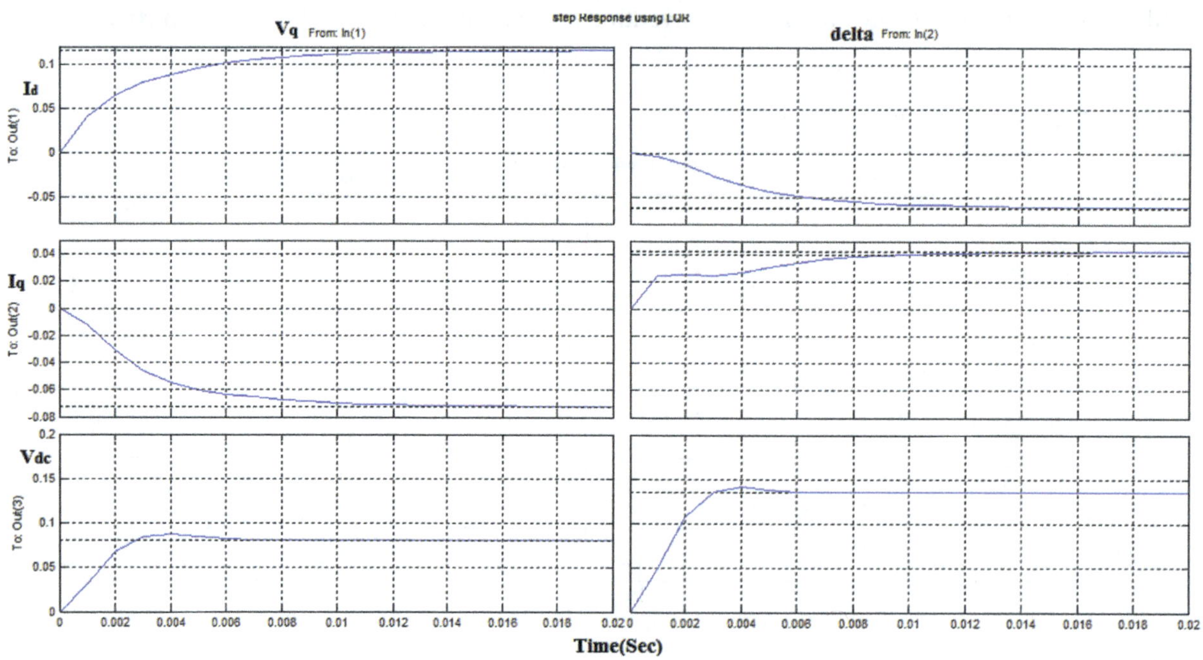

Figure 8. Step response using full order state feedback controller

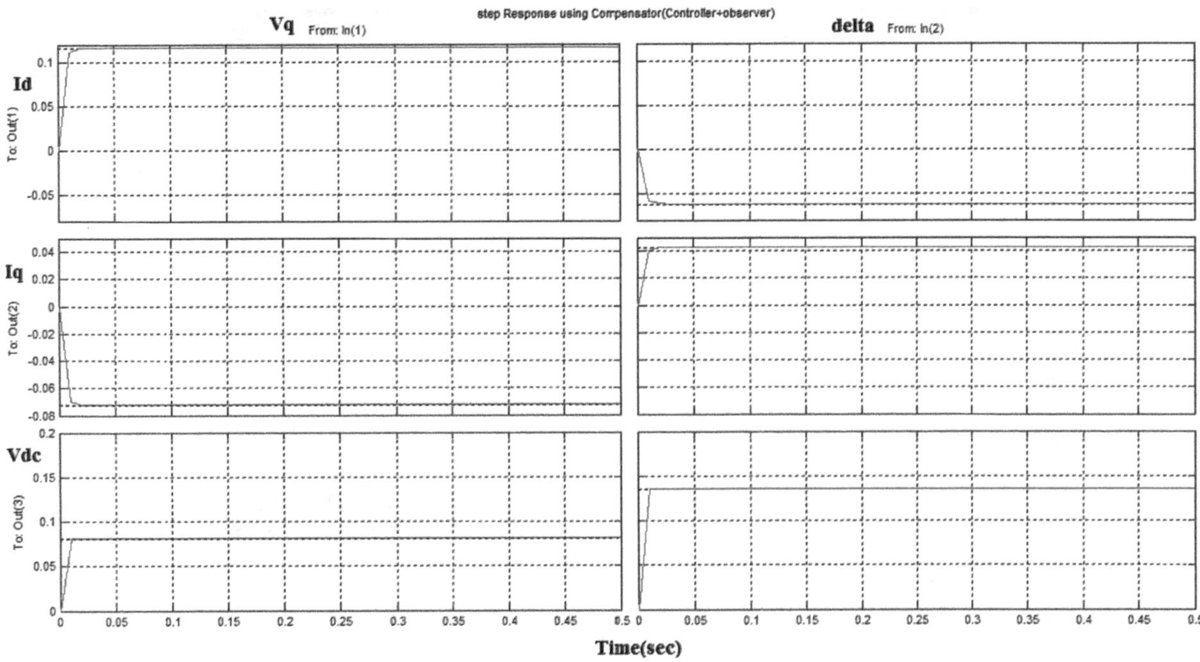

Table 1. Eigen values of STATCOM with and without state feedback controller

Sl No.	System	Eigen Values	Controller/Observer Gain Matrix
1	Open loop system	1.0e+002 *[-2.1226 + 3.3021i, -2.1226-3.3021i, + 2.5785]	NA
2	Pole Placement 1st test	-2.4330 -0.8110 + 1.5840i -0.8110 - 1.5840i	[-0.8980 1.7743 -0.4221 -1.7043 -1.1284 0.6565]
3	Pole placement 2nd test	-8.1100 -0.8110 + 1.5840i -0.8110 - 1.5840i	[-1.0555 1.5311 -0.4181 -1.5979 -0.9640 0.6539]
4	Pole placement 3rd test	-11.3540 -0.8110 + 1.5840i -0.8110 - 1.5840i	[-1.0700 1.5068 -0.4218 -1.5868 -0.9457 0.6562]
5	LQR method	1.0e+003 * [-1.7506 -3.1822 + 3.3226i -3.1822 - 3.3226i]	[1.4118 1.8630 2.6380 1.8630 6.2949 6.1848]
6	Observer with LQR method	1.0e+003 * [-1.7506 -3.1822 + 3.3226i -3.1822 - 3.3226i -3.4000 -3.2000 -3.0000]	1.0e+003 * [2.1667 1.3500 -0.3507 -1.3500 2.3667 0.5463 2.1045 3.2775 3.3740]

Figure 9. Error signal graph with PI Controllers ($K_p = 1$, $K_i = 25$)

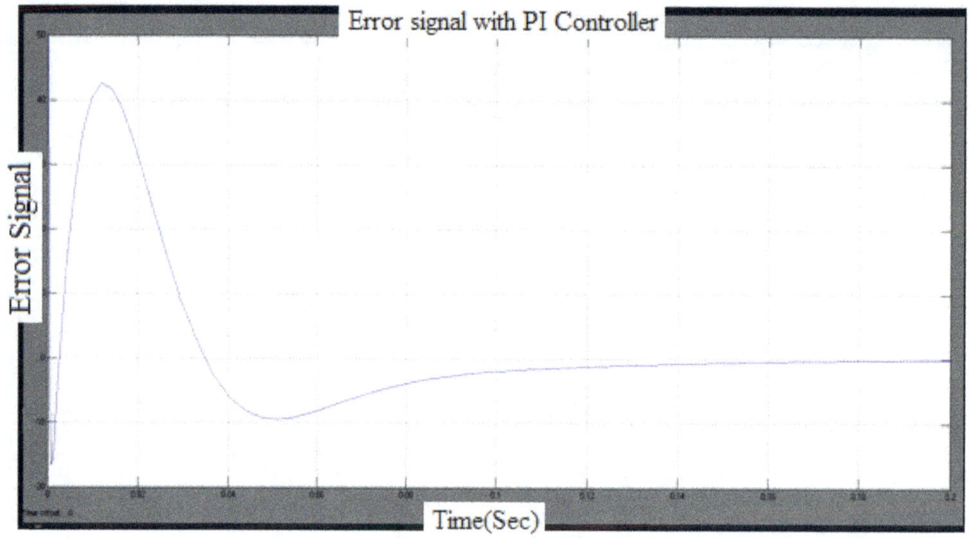

Figure 10. Error signal graph with PI Controllers ($K_p = 2.4$, $K_i = 1125$)

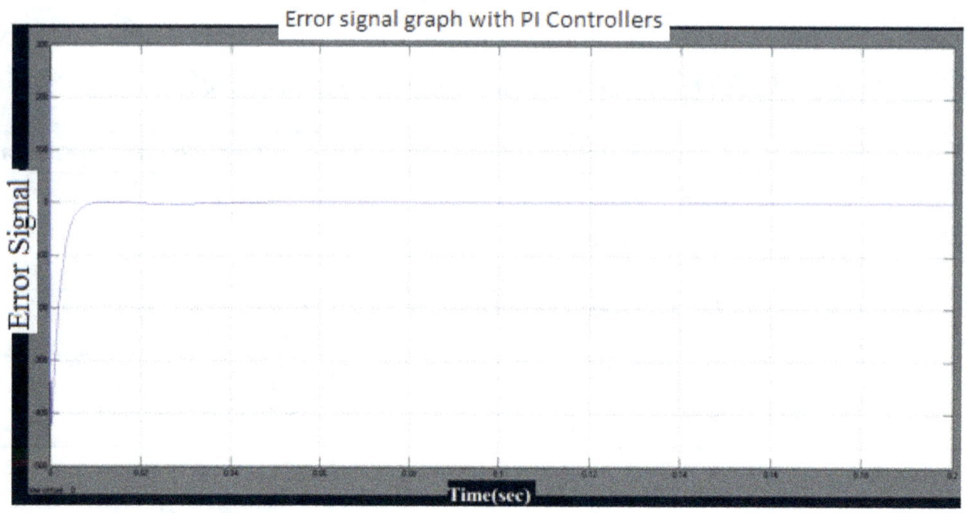

feedback CSI-STATCOM and finally with proposed observer based state feedback controller. Comparative analysis has been carried out in MATLAB Simulink Software.

Figure 13 shows the graph for rms value of receiving end voltage for all the three cases i.e. without STATCOM, with conventional PI controller based STATCOM and with proposed observer state feedback controller based STATCOM. From the above graph it has been observed that uncompensated system voltage is dropped to 0 P.U. from 1 P.U. during the fault. But in case of conventional PI controller based STATCOM, the voltage rises to 0.96 P.U and in case of proposed controller based STATCOM the voltage reaches to almost its rated value 0.985 P.U from 0 P.U during the fault condition. That means the proposed observer based state feedback controller can give better voltage support than conventional PI controller in three phases to ground fault condition.

Figure 11. Error signal graph with observer based state feedback controllers

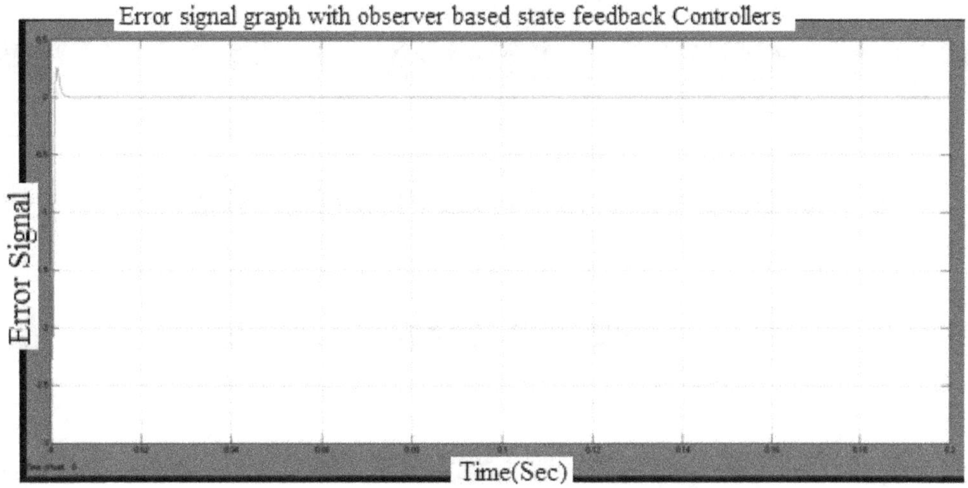

Table 2. Comparative study between PI and proposed controller

Sl. No.	Controller Type	Maximum Overshoot	Settling Time
1	PI (K_p=1,K_i=25)	M_p 42%, Negative shoot -17%	14msec
2	Tuning PI controller (using Zeigler-Nichols damped oscillation method) PI (K_p=2.4,K_i=1125)	M_p 0%, Negative shoot - 425%	4.9 msec
3	Observer based State Feedback Controller using LQR Technique	M_p 0.3%, Negative shoot -2.4%	0.5 msec

Figure 12. Graph for different states with observer based state feedback controller

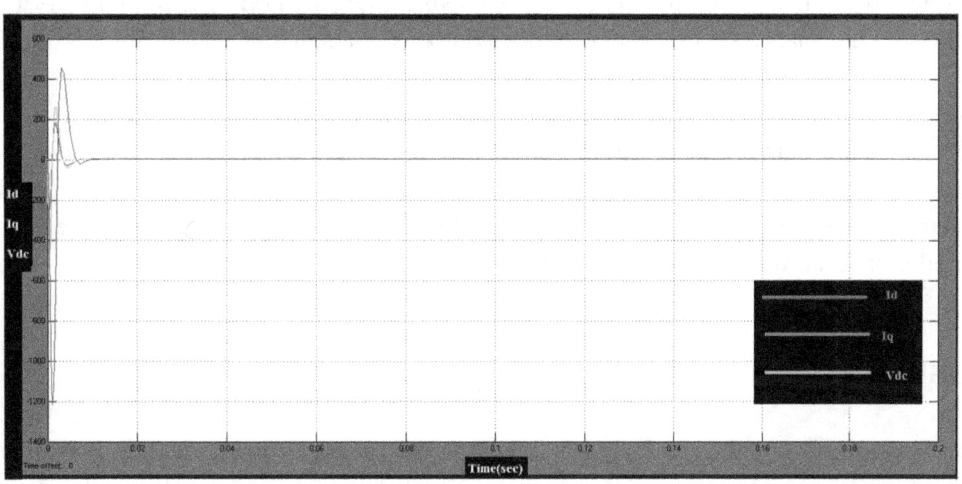

Figure 13. Simulation graph of rms voltage without STATCOM, with con-ventional PI Controller, with state feedback CSI-Controller and with proposed STATCOM observer based state feedback controller

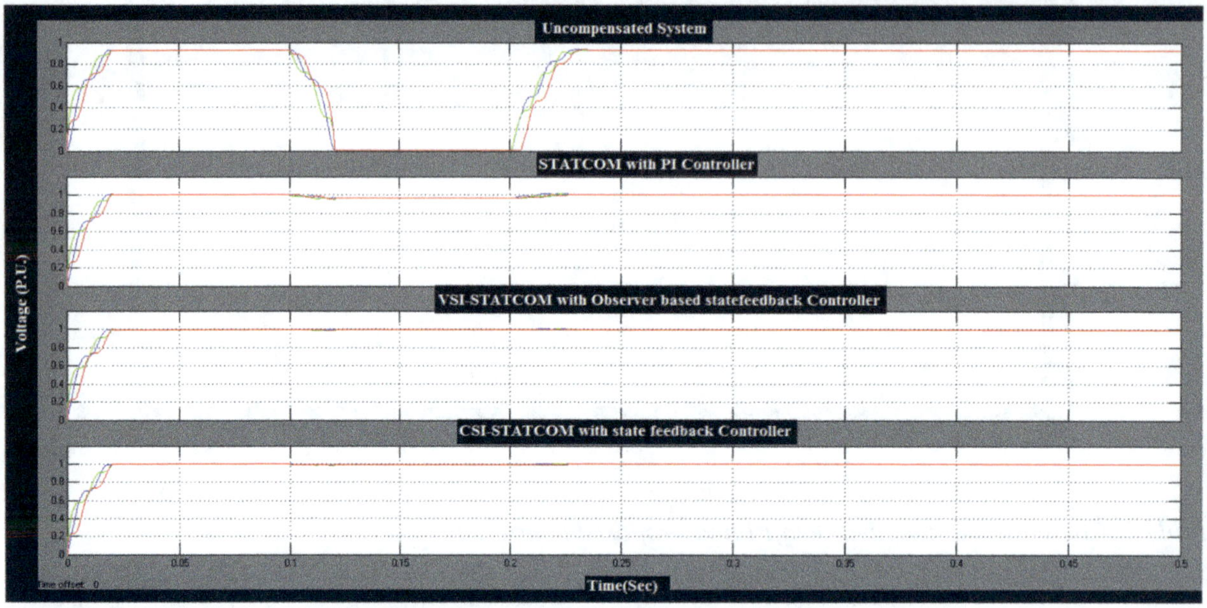

FUTURE RESEARCH DIRECTIONS

This chapter presents the observer-based state feedback controller for VSI-STATCOM. This chapter focuses on controller design of STATCOM. Here only full order state feedback controller for STATCOM has analyzed for load compensation, voltage regulation and voltage balancing in electric power distribution and transmission networks. The behavior of this controller is first validated and subsequently, the performance of this STATCOM controller is examined. State feedback controller can be designed in two ways: one is pole placement method and another one using LQR method. Big challenge is to study new control methodologies that can efficiently generate control signals, while accounting for the detection of the dynamics, changes in the process plans. By shifting the pole in pole placement method increase the control effort and consequently increase the cost of controller. In LQR method optimum solution can be possible as it gives optimum value of control effort and cost. Here cost analysis of proposed controller has not done for real world. Mathematical analysis has performed for pole placement method and LQR Method. A further research direction, cost of state feedback controller may be emphasized by designing hardware model of full order state feedback controller.

CONCLUSION

Today's power systems have been extended by applying interconnections to the neighboring systems in order to achieve technical and economical advantages and sharing huge load demands. Due to various restricted access such as Right of way (R.O.W) etc., related to lying of new transmission lines, growth in transmission networks fail to keep pace with the persistent increase in electricity demand. Stability of such stressed power systems is often endangered due to reduced stability margins which can be further worsen by power quality problem and voltage collapse problem. Installation of FACTS Devices in existing transmission line is a way to strengthen power and voltage stability. The research work also concerted on controller design of STATCOM. In this thesis new controller i.e. full ordered state feedback controller for VSI STATCOM is analyzed and found very effective. It is seen from the simulation result that new proposed controller gives better performance during contingency condition. It has been observed from the analysis that the proposed controller is efficient and can provide fast dynamic response to the system than conventional STATCOM PI controller.

REFERENCES

Ajami, A., & Younesi, M. (2008). Modeling and state feedback controller for current source inverter based STATCOM. *IEEE International Conference on Control, Automation and System*, 2418-2423.

Al-Mawsawi, S. A., Qader, M. R., & Ali, G. M. (2004). Dynamic Controllers Design for STATCOM. *Iranian Journal of Electrical and Computer Engineering*, *3*(1), 16–22.

Beck, G., & Povh, D. (2005). *Global Blackouts – Lessons Learned*. POWER-GEN Europe 2005, Milan, Italy.

Blazic, B., & Papic, I. (2006). Improved D-StatCom Control for Operation with Unbalanced Currents and Voltages. *IEEE Transactions On Power Delivery, 21*(1), 225-232.

Dayananda, B.R., Sekhar, V., & Babu, R.N. (2014). Implementation of Linear Controller of a STATCOM Design in DC Link Voltage. *International Journal of Emerging Technology and Advanced Engineering, 4*(9).

Dobson, I., Cutsem, T. V., Vourmas, C., Demarco, C.L., Bramanian, M.V., Ovenbye, T., & Canizares, C.A. (2003). Voltage Stability Assesment: Concepts Practices and Tools. *IEEE Power Engineering Society*.

Farokhnia, N., Khoraminia, R., & Gharehpetian, G. B. (2010). Optimization of PI Controller Gains in Nonlinear Controller of STATCOM Using PSO and GA. *International Conference on Renewable Energies and Power Quality (ICREPQ)*.

Gama, C. A., Ellery, E. H., Azevedo, D. C., & Poute, J. R. R. (1992). Static var Compensator (SVC) versus synchronous compensaters (SC) for inverter Stations. *Compensation- Technical and Economic Aspects in Electronorte studies CIGRE*, 14-103.

Gopal, M. (2014). *Digital Cntrol and State Variable Methods* (4th ed.). McGrawHill.

Hammed, A.E., & EI-Sadek, M.Z. (1989). Prevention of transient voltage instabilities due to induction motor loads by static Var Compensators. *IEEE Transactions on Power Systems*, (3).

Hingorani, N. G. (1988). High Power Electronics and Flexible AC Transmission System. *IEEE Power Engineering Review*, *8*(7), 3–4. doi:10.1109/MPER.1988.590799

Hingorani, N. G., & Gyugyi, L. (2000). *Understanding FACTS Concepts and Technology of Flexible AC Transmission Systems*. New York: IEEE Press.

Homayounzade, M., Keshmiri, M., & Danesh, M. (2010). An Observer-Based State Feedback Controller Design for Robot Manipulators Considering Actuators' Dynamic. *IEEE 2010 15th International Conference on Methods and Models in Automation and Robotics (MMAR)*, 240-248.

Hossain, S., Khan, A.R., & Noorm, J.E. (2014). Design of STATCOM for Power System Stability Improvement. *Global Journal of Researches in Engineering: Electrical and Electronics Engineering, 14.*

Jafarian, P., & Bina, M. T. (2010). State-Feedback Current Control of VSI-Based D-STATCOM for Load Compensation. *IEEE International Conference on Environment and Electrical Engineering*, 214-217. 10.1109/EEEIC.2010.5490431

Jeong, K. S., Jyung, T. Y., & Baek, Y. S. (2011). Modeling and Dynamic Analysis of STATCOM for Short-Term Voltage Stability Improvement using EMTP/RV. *Journal of International Council on Electrical Engineering*, *1*(2), 129–134. doi:10.5370/JICEE.2011.1.2.129

Krishnan, V., & McCalley, J. D. (2006). Role of Induction Motors in Voltage Instability and coordinated Reactive Power planning. Academic Press.

Lawan, A.U. (2014). State Feedback Approaches for Designing A Statcom Supplementary Controller for Oscillations Damping. *The International Journal Of Engineering And Science, 3*(7), 27-37.

Lu, A., Tang, C., Shuai, Z., Tang, J., Xu, X. Y., & Chen, D. (2009). Fuzzy PI based direct output voltage control for STATCOM. *IEEE Transactions on Industrial Electronics, 56*(7), 2401-2412.

Mathur, R. M., & Varma, R. K. (2000). *Thyrustor_based FACTS Controllers for Electrical Transmission Systems*. Piscataway, NJ: IEEE press.

Mukhopadhyay, A. K., Kothari, D. P., Chakrabarty, A., & De, A. (2010). *An Introduction to Reactive Power Control and Voltage Stability in Power Transmission System*. PHI.

Norouzi, A. H., & Sharaf, A. M. (2005). Two control schemes to enhance the dynamic performance of the STATCOM and SSSC. *IEEE Transactions on Power Delivery*, *20*(1), 435–442. doi:10.1109/TPWRD.2004.839725

Nwouh, N. M. (2009). Voltage Stability Improvement using Static Var Compensator in Power Systems. *Leonardo Journal of Sciences*, (14), 167-172.

Padiyar K. R. (2007). *Facts Controllers in Power Transmission and Distribution*. New Age International (P) Limited.

Ponmurugavel, P. S., & Ghouse, S. M. (2013). Design and Modelling of Fuzzy and Model Predictive Controllers for STATCOM to enhance Transient Stability of Power System. *IACSIT International Journal of Engineering and Technology*, *5*(3), 2609–2619. doi:10.1109/ICPEC.2013.6527635

Pradeepa, S., Rao, K. U., Deekshit, R., & Shantha, M. S. (2013). State feedback Control of a Voltage Source Inverter-based STATCOM. *IEEE International Conference on Power, Energy and Control (ICPEC)*, 120-123.

Rahim, A. H. M. A., & Kandlawala, M. F. (2004). *Robust STATCOM voltage controller design using loop-shaping technique*. Elsevier Electric Power Systems Research.

Safari, A., Mawsawi, A., Qader, M. R., & Ali, G. M. (n.d.). Dynamic Controllers Design for STATCOM. *Iranian Journal Of Electrical And Computer Engineering, 3*(1), 16-22.

Sethy, S.K., & Moharana, J.K. (n.d.). Design, Analysis and Simulation of Linear Model of a STATCOM for Reactive Power Compensation with Variation of DC link Voltage. *International Journal of Engineering and Innovative Technology, 2*(5).

Shen, D., & Lehn, P. W. (2002, January). Modeling, Analysis, and Control of a Current Source Inverter-Based STATCOM. *IEEE Transactions on Power Delivery, 17*(1), 248–253.

Somsai, K., Kulworawanichpong, T., & Voraphonpiput, N. (2012). Design of DC Voltage Control for D-STATCOM. *International Scholarly and Scientific Research and Innovation, 6*, 917–923.

Song, Y. H., & Johns, A. T. E. (1999). Flexible AC Transmission Systems (FACTS). IEEE Press.

Spitsa, V., Alexandrovitz, A., & Zeheb, E. (2009). Design of a Robust State Feedback Controller for a STATCOM Using a Zero Set Concept. *IEEE Transactions on Power Delivery, 25*(1), 456 – 467.

Stefan & Fredrik. (1995). *Voltage collapse in power systems* (Licentiate thesis). Chalmers University of Technology.

Stefopoulos, G. K., & Meliopoulos, A. P. (2006). Induction Motor Load Dynamics: Impact on Voltage Recovery Phenomena. *Proceedings of IEEE Transmission and Distribution Conference and Exposition*, 752 – 759. 10.1109/TDC.2006.1668591

Taranto, G. N., Pinto, L. M. V. G., & Pereira, M. V. F. (1992). Representation of FACTS Devices in Power System Economic Dispatch. *IEEE Transactions on Power Systems, 7*(2), 572–576. doi:10.1109/59.141761

Taylor Carson, W. (1993). *Modelling of Voltage collapse Including Dynamic Phenomena.* CIGRE Publication.

Taylor Carson, W. (1995). *Criteria and countermeasures for voltage collapse.* CIGRE Publication.

Voraphonpiput, N., & Chatratana, S. (2005). STATCOM Analysis and Controller Design for Power System Voltage Regulation. IEEE Transmission and Distribution Conference and Exhibition, 1-6.

Wenjin, D., & Xiangjie, C. (2009). Transient and Steady-state Analysis for STATCOM Mathematical Model. *2nd International Conference on Power Electronics and Intelligent Transportation System*, 271-275. doi:10.1109/TDC.2005.1546873

Xu, Y. (2014). Adaptive PI control of STATCOM for Voltage Regulation. *IEEE Transaction on Power Delivery, 29*(3), 1002-1011.

Yanushkevich, A., Muller, Z., Svec, J., Tlusty, J., & Valouch, V. (2014). Power Quality Enhancement using STATCOM with Energy Storage. *International Conference on Renewable Energies and Power Quality*, 12. 10.24084/repq12.336

KEY TERMS AND DEFINITIONS

Contingency: A future event or circumstance which is possible but cannot be predicted with certainty.

FACTS Devices: Flexible alternating current transmission system (FACTS) is static equipment used for the AC transmission of electrical energy. It is meant to enhance controllability and increase power transfer capability. It is generally a power electronics-based device.

Instability: Lack of stability.

LQR (Linear Quadratic Regulation): The theory of optimal control is concerned with operating a dynamic system at minimum cost.

PI Controller: A proportional-integral controller (PI controller) is a control loop feedback mechanism widely used in industrial control systems and a variety of other applications requiring continuously modulated control.

Pole Placement Method: Pole placement method is a controller design method in which you determine the places of the closed loop system poles on the complex plane by setting a controller gain.

Voltage Collapse: Voltage collapse is a system instability involving many power system components. In fact, a voltage collapse may involve an entire power system. Voltage collapse is typically associated with reactive power demand of load not being met due to shortage in reactive power production and transmission.

VSI (Voltage Source Inverter): The simplest dc voltage source for a VSI may be a battery bank, which may consist of several cells in series-parallel combination.

Chapter 8
Synchronized Operation of Grid Power, Solar Power, and Battery for Smart Energy Management

Pawan Kumar
Thapar University, India

Dip V. Thanki
Lovely Professional University, India

ABSTRACT

This chapter gives details of solar photovoltaic, starting from its general pros and cons. It covers the basics of site evaluation when installing a solar powered plant and various ways to overcome the uncertainties in the predicted output of the solar arrays. The efficiency of the plant can be improved with the help of maximum power point tracker (MPPT), which works on algorithms based on perturb and observe, incremental conductance, constant voltage, etc. The output of the solar PV arrays can be utilized more effectively by integrating it with grid to supply ac loads. This integration requires a power conditioning system (PCS), enabling smooth operation. Continuity of supply can be maintained by having a battery backup, for the time when both grid and solar array fail to meet the load demand. Such a system can have wide range of applications and has the potential to meet the energy demand.

INTRODUCTION

The sun is a noteworthy wellspring of energy. The energy supplied by the sun to the earth for 60 minutes can meet the worldwide energy requirements for one year. However, it is just 0.001 percent of that energy. The earth gets 1336 watts for every square meter of direct sun oriented radiation and with the assistance of Photovoltaic (PV) cells it can be changed over straight forwardly into electrical energy. PV is intended to absorb the warm radiations of the sun and change over it to power. However, various arrangement and parallel associated PVs are required to get an adequate measure of energy since the output of PV is small which is due to its small size and technical limitations. Such a gathering of series and parallel associated PV cells is known as a solar array. The output from solar panel majorly depends

DOI: 10.4018/978-1-5225-3935-3.ch008

upon the temperature and the solar irradiance. So it is challenging to track the maximum power at different temperature and irradiance using different methodology.

BACKGROUND

Kumar et al. (2015) have developed a simulation model of the synchronized operation of the grid power, solar power and battery for a smart home. In this model authors have shown that under different temperature and irradiance the output power varies and the operating time of the grid and solar power changes accordingly. Abdulkadir et al. (2014), presented an improved particle swarm optimization (IPSO) based MPPT technique for photovoltaic system operating under different environmental conditions. Chandwani and Kothari (2016) comprise of detailed study in the field of MPPT in a simulative environment using MATLAB and the same results are reproduced in hardware using Perturb and Observe (P&O) algorithm. However, a method for the reduction of the steady state oscillation (to practically zero) once the maximum power point (MPP) is located is described by the Ishaque et al. (2012) and Shukla et al. (2015) studied the effect of variation in temperature and irradiance and the results are observed from simulation in MATLAB environment. Here, authors have developed program which allows the prediction of PV module behavior under different temperature and irradiance.

A detailed analysis and modeling of Solar and Fuel cells using Cadence SPICE, and to investigate dynamic interactions between the modules and power conversion circuits is presented in the thesis of Krishnamurthy (2009). The results are simulated using equivalent electronic static and dynamic models for Solar and Fuel Cells. Kumar et al. (2011, 2013, 2016 and 2017) presented a detailed analysis of the different load models which are voltage dependent. In these studies author have observed that the system voltage profile not only affect the power loss rather it majorly changes the system loadability and load profile. During peak load conditions the voltage profile usually found to be low and hence the operating efficiency of the system decreases. In order to improve the operating efficiency the integration of the solar power may help in relieving the overload and simultaneously improve the voltage and the load profile at respective nodes. On the other hand, in the ever growing demand scenario a comprehensive stability analysis, reactive power injection, loss minimization under different loading pattern for distribution system are appears in the work presented by Kumar and Singh (2014a, 2014b, 2014c)

Kumar et al. (2015) observed that PV generation depends on solar irradiance, site location and environmental conditions including temperature and wind. As a results, PV output keep on varying and it is found challenging to extract maximum power under these conditions. Kuo et al. (2013) has developed a self-synchronization error dynamics formulation based controller for MPPT in Micro-Grid Systems. Here also authors observed that the output of PV system mainly depends upon the temperature and irradiance and its efficiency is low. Smida and Sakley (2015) presented genetic algorithm based approach for MPPT under partial shaded conditions and uses two-stage conversion system which composed of a dc-dc boost converter and a dc-ac inverter whereas, Venugopalan et al. (2013) uses PSO techniques for the design of MPPT. Hill et al. (2012) presented the challenges in integration of solar power system with distribution system whereas, Jinn *et al.* (2014) presented the same for small capacity grid connected solar power generation system.

Qinghui et al. (2001) have proposal of different methods in order to improve the efficiency solar power generation system. Liu et al. (2008) compared and the performance of P & O and hill climbing method through theoretical analysis and digital simulation. Lee et al. (2006) proposes an advanced Incremental conductance MPPT algorithm with a variable step size whereas, Sung-Yeul et al. (2010) developed a flexible control scheme for a three-phase GTI system for MPPT.

Guo and Wu (2010) presented the synchronized operation three phase grid connected voltage source inverter and Chiang et al., (2010) have developed a smart system where uninterruptible power supply is available in hybrid power system. Zhigang et al. (2011) presented a PV Module-Integrated DC/DC Converter for PV Energy Harvest. Zhilei et al. (2010) developed an integrated system in order to provide power to the emergency load during system outages, which operate in grid-tied and off-grid modes. Dasgupta et al., (2011) proposed a technique for active and reactive power control from a renewable energy source which is feeding a microgrid system through a single-phase parallel-connected inverter. However, Bojoi et al. (2010) proposed a control strategy for single-phase inverters to improve power quality in distributed generation systems. Yang and Smedley (2008) developed a inverter for single power stage that converts dc power to grid-connected ac power by injecting three in phase sinusoidal currents into grids for the reduction in loss and circuit complexity. Hussein et al. (1995) and Masoum et al. (2002) presented the theoretical and experimental analyses of MPPT using voltage and current based approaches whereas, Yishu et al. (2009) has developed circuit topology for single-phase grid connected system for improving the energy efficiency. Blaabjerg et al. (2006) presented the detailed analysis of the synchronization of solar power with the grid system whereas, Svensson (2001) has initially developed four different methods for the synchronization of the voltage source inverter with grid system.

Sun powered energy is one of the cleanest and greenest innovations. All over the world, solar energy is driven by solar warmth or thermal, it is normal that solar PV will turn out to be the single biggest resource of energy in the circumstances to come. It is thus expected that solar energy is and will keep on playing an overwhelming part in power scenario because of different advantages it offers over other inexhaustible innovations. In PV power generation, ideal use of the accessible sun based energy must be guaranteed because of the high cost of modules. Likewise, PV framework requires uncommon plan contemplations because of the varying nature of the sun oriented power resulting from unpredictable and sudden changes in climate conditions, which change the radiation level and also the cell working temperature. Sunlight based photovoltaic frameworks execution relies upon a few ecological parameters like sun oriented insolation, temperature, wind speed and shading. The execution of such framework requires exact information of the I-V and P-V qualities bend.

ADVANTAGES OF SOLAR POWER

1. The fuel is clearly free.
2. The operation of photovoltaic frameworks does not deliver any poisonous gas emanations, alleged green-house gasses, or clamor. For correlation, the ordinary creation of power from coal discharges 1.5 pounds of carbon dioxide for every kilowatt-hour.
3. No moving parts are required unless a sun based tracker is introduced.
4. The utilization of sun oriented energy decreases our reliance on outside oil. This decreases the measure of cash streaming to radical and totalitarian administrations of some oil exporters.
5. Solar innovation is flexible as it can be utilized as a part of wide scope of utilizations from little adding machines to generation plants.
6. Capability to generate power in remote areas that are not connected to a grid, and also in space.

7. Grid-tie frameworks can bring down electricity bills. PV has a capacity to give reinforcement control when the grid is in failure. Times of most extreme daylight are generally corresponded with the times of peak energy utilization.

DISADVANTAGES OF SOLAR POWER

1. High capital cost,
2. Solar energy requires substantial surface territories for valuable measures of power era. Contingent upon the board productivity, it may require from 6 to 12 square meters per kilowatt of pinnacle limit in non-focused light. Under normal insolation of 4.4 (say) sun-hours can generate aggregate of 0.39 to 0.77 kW-hours per square-meter amid a whole day. This would be under 17 to 33 watt for each square-meter found the middle value of over a 24-hour time frame.
3. PWM inverters infuse high-frequency into house-wiring, which goes about as a huge radio wire and acts as an antenna. As per a few examinations, this electromagnetic radiation can cause negative impacts in the surrounding.
4. Because the power of daylight differs with the season of day and climate conditions, an off- network solar frame-work requires either substantial capacity batteries or alternative supplemental energy resources.
5. Roof-top establishments introduce fire peril. Likewise the generation of PV modules utilizes some perilous materials, (for example, cadmium and lead). This can show wellbeing and security perils, if legitimate safeguards are not taken amid assembling, establishment and reusing.

Taking everything into account, sunlight-based controlled power obviously is naturally more benevolent than power delivered from coal and other non-renewable energy sources. It additionally lessens our reliance on oil imports. The daylight energy is utilized in power generation while oil in transportation. On a negative side, non-concentrated photovoltaic generators are not exactly fetched in competition due to their cost. The discounts and motivation can frequently balance the greater part of the establishment cost for the consumers. Recently, concentrated sun based power and additionally the thermal framework, among daylight applications, is more monetarily practical arrangements than photovoltaic-based power.

SITE EVALUATION FOR SOLAR PLANT

This area plots the comprehensive declaration expected to assess particular locales for sunlight based energy establishments. To begin with, the accompanying is required:

1. A solar screening program, which is a distinctive state, preparatory investigation used to decide a site's reasonable suitability.
2. A solar feasible inspection, which is a more thorough engineering and monetary examination to characterize particular framework outline contemplations for use in demands for recommendations and additionally extent of work advancement.

Solar Screening Program

A solar-based screening program ought to give structural or mechanical contemplations and monetary contemplations, as given beneath:

Structural or mechanical contemplations encompass:

- Roof condition and period of rooftop in case of housetop framework.
- Shading examination i.e. distinguishing proof of obstacles that may shade the exhibit area.
- Available area for a solar system.
- Preliminary gauge of the framework's size.
- Structural matter when framework is to be mounted on a building then elevation contemplations are taken into consideration.
- Historic building matter when framework will be on a building that could be a noteworthy property or is situated in a historically significant area.

Financial contemplations encompass:

- Cost of vitality at a site, in addition to any points of interest of rate plans that could support or condemn solar.
- Economic investigation of undertaking e.g., straight forward payback, internal rate of return, net present value, life cycle cost, anticipated reserve funds.
- Estimated yearly energy generation.
- Hot water or space-warming interest.
- Incentives (government, state, neighborhood, utility, RECs) and their chance sensitivities.

Solar Feasibility Study

A solar feasibility study must include the mechanical and monetary consideration as under,

Structural or mechanical consideration includes:

- Capacity of the neighborhood business to supply and keep up such frameworks
- Utility interconnection issues (if arranging an electric undertaking, it's critical to know whether the utility has exceptional equipment or authoritative necessities)
- Electrical and mechanical head margin issues (e.g., space for accessories, substitute area, sufficiency limits, access amongst framework and hardware room).

Monetary consideration includes:

- Recommended framework capacity.
- Site load necessities.
- Analysis of 15-minute load information for maximum demand.
- Estimated month to month maximum demand.
- Yearly operations and upkeep.
- Magnitude and timing of the electrical and heating loads at site.

- Size, condition, and effectiveness of existing heating frameworks.

MAXIMUM POWER POINT TRACKER (MPPT)

If we simply associate a sun oriented cluster and abandon it at that, we will not get the best or most extreme output from it. Output enhances in the event that we steadily associate voltage and current. A MPPT empowers us to do as such. A MPPT, or most extreme power point tracker is an electronic DC to DC converter that advances the match between the sun powered cluster (PV boards), and the battery bank or utility framework. Basically, they change the level from a higher voltage DC output from solar down to the lower voltage expected to charge batteries.

Usually, MPPT is an electronic tracking process. In this process a charge controller compares the battery voltage with the panel's output. On comparing, controller figures out what is the best power that the board panel can develop to charge the battery. It takes this and converts it to best voltage to get maximum amperes into the battery. At present MPPT is highly productive in the transformation.

The most extreme power point is not quite the same as the STC (Standard Test Conditions) rating under all circumstances. Several literatures reveal that MPPT is best under these test conditions:

- **Cold Climate, Overcast or Hazy Days:** Normally, PV module works better at lower temperatures and MPPT is used to extract most extreme power accessible from them.
- **When Battery Is Profoundly Discharged:** MPPT can extricate more current and charge the battery if the condition of charge in the battery is brought down.

The electrical execution of any solar array is represented by their current-voltage (I-V) characteristics. For solar system I-V characteristics speaks to the unbounded working points under various working state of solar-based radiation and the temperature. In numerical plans, the regularly created terminologies are described by Kumar et al. (2015), and they are as under,

The generalized characteristic equation of a solar cell, which relates solar cell parameters to the output current and voltage, is given by:

$$I = I_L - I_0\left(e^{\frac{q(V+IR_s)}{nKT}} - 1\right) - \frac{V + IR_s}{R_{sh}} \tag{1}$$

Figure 1. Equivalent circuit of solar cell

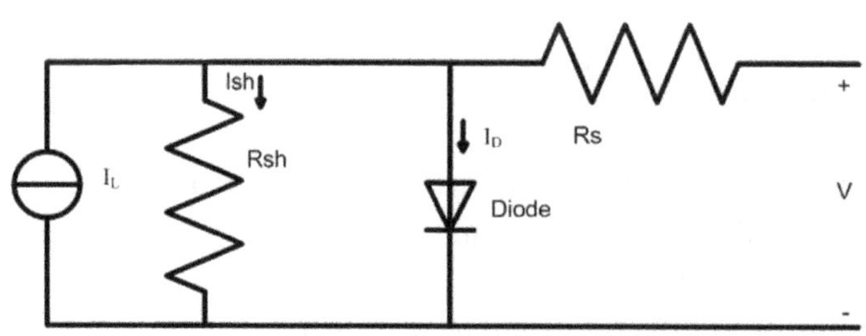

Here,

I = current following through load terminals,
I_L = photo-generated current,
I_0 = reverse saturated current,
V = load terminal voltage,
R_S = series resistance,
R_{sh} = shunt resistance,
η = diode ideality factor, which is 1 for an ideal diode,
q = elementary charge,
k = Boltzmann's constant,
T = absolute temperature

1. **Open Circuit Voltage (V_{OC}):** This is the peak DC voltage created over the receiving-end terminals when no load is associated in the framework. It is autonomous of the region of the solar-cell and is the pointer of the most extreme voltage restrain for solar-panel board.
2. **Short Circuit Current (Isc):** This is the peak current that can move through the solar-panel board when load terminals are short-circuited. By and by, it is straightforwardly relative to the solar-based irradiance. However, solar cells are naturally current restricted in this way; precaution is to be taken if they are shorted for testing reason with proper shorting gadget.
3. **Maximum Power Point (P_{MP}):** Under specific working states of solar radiation and the temperature, the most extreme value of the result of current and voltage, in I-V qualities, show the greatest conceivable power that can be drawn from the solar system. The relating point in I-V curve is known as the maximum power point.
4. **Maximum Power Voltage (V_{MP}):** The value of voltage corresponding to maximum power point is termed as the maximum power voltage. The estimation of voltage comparing to most extreme power point is named as the maximum power voltage.
5. **Maximum Power Current (I_{MP}):** The value of current corresponding to maximum power point is termed as the maximum power current. Alternatively, it can be computed as the proportion of the maximum power (P_{MP}) to the maximum power voltage (V_{MP}).

The MPPT works out the maximum power point and gives changing signs to the PWM framework in the inverter which switches by balancing the pulse width control and the power contribution to the load. However, an ordinary power generator controls the output by controlling the field excitation and energy input. In case of PV cell focus is on the maximum energy it can deliver. The control is with respect to the load and MPPTs are said to depict the load. In a voltage-reference-based MPPT, referred as VMPPT, the reference is made V_{oc} of the cell. This course of action is appropriate for resistive burdens. Commonly, it has expanded the power output by 20%.

On the other hand, in current-based MPPT (i.e. CMPPT), the reference is made to I_{sc}. This course of action is reasonable for a pump load. In this case, the losses are marginally higher and the announced increment in power output remains at 10%. The general effectiveness of a MPPT-based control framework is genuinely more than 97%. The expansion in output is not decreased by the losses in the MPPT. However, VMPPT is more reasonable for low-voltage, high-current loads, for example, battery chargers and low resistance levels whereas, CMPPT is more reasonable for high-voltage, low-current loads, for example, DC motors on agricultural pumps.

Solar Charge Controller for MPPT

Solar charge controller (SCC) is the charge controller embedded with MPPT algorithm in order to maximize the amount of current going into the battery from PV module. Solar charge controllers are more useful when supply system is isolated from the grid power such as stand-alone system, home system and water pump system, etc. Main features of MPPT-SCC are described as under:

1. During application when PV module worked as energy source, MPPT-SCC is utilized to redress the recognized varieties in the present voltage characteristics of solar cell as appeared by the I-V curve.
2. It is necessary since for any solar power framework needs to extricate maximum power from PV module and it powers PV module to work at voltage close to maximum power point to draw maximum accessible power.
3. MPPT-SCC enables clients to utilize PV module with a higher voltage output than operating voltage of battery framework.
4. MPPT-SCC reduces complexity of framework while output of framework has high efficiency.
5. It can be connected to use with more energy sources, since PV output control is utilized to control DC-DC converter specifically.
6. MPPT-SCC can be applied to other renewable energy sources such as small water turbines, wind power turbines, etc.

MPPT sunlight based charge controllers are valuable for off-lattice sun oriented power frameworks, for example, remain solitary sun based power framework, sun oriented home framework and sun based water pump framework, and so on.

MPPT Implementation

When a load is straight forwardly associated with solar panel, the tracking of the operating point of the solar panel at peak power is difficult. In this case, the impedance seen by the panel will affect the operating point of the solar panel. Therefore, on changing the impedance operating point can be controlled and moved towards the peak power point. The panel systems are DC devices, so DC-DC converters must be utilized to transform the impedance of one circuit (source) to the next circuit (load). At a specific impedance (or duty proportion) the operating point will be at the peak power exchange point. The I-V curve of the solar power system can shift impressively with change in radiance and the temperature. In this manner it is not attainable to settle the duty proportion with such progressively varying working conditions.

MPPT executions use algorithms that frequently sample the panel voltages and currents, in order to change the obligation proportion as required. In this scenario, microcontrollers are utilized to actualize the calculations.

MPPT Classification

Controllers utilize the different methodologies to optimize the power output of an array. Therefore, MPPT may execute diverse algorithms and switch between them in light of the working states of the array.

Perturb and Observe

In this strategy the controller modifies the voltage by a little sum from the array and measures power. In case power increases, it facilitate changes toward that path until control never again increments. This is known as the Perturb and Observe (P & O) technique and it is commonly used, in spite of the fact that this strategy can bring about oscillations of energy output. It is similar to a hill climbing technique, since it relies upon the ascent of the curve of energy against voltage underneath the maximum power point and the fall over that point. P & O is the most regularly utilized MPPT strategy because of its simplicity of execution. P & O technique may bring about best level effectiveness, gave that a legitimate predictive and versatile hill climbing procedure is embraced.

Incremental Conductance

In the incremental conductance technique, the controller measures the incremental changes in PV array's current and voltage to anticipate the impact of a voltage change. This strategy requires more calculation in the controller, however can track changing conditions more quickly than the P & O. However, it can also create oscillations in power output like P&O algorithm. This strategy uses the incremental conductance ($\partial I / \partial V$) of the photovoltaic array to figure the indication of the adjustment in control as for voltage ($\partial P / \partial V$).

The incremental conductance strategy obtains the maximum power point by calculation of the incremental conductance ($I\delta / V\delta$) to the array conductance (I / V). At the point when these two are the same (i.e. $I / V = I\delta / V\delta$), it indicates that the output voltage is the MPP voltage. The controller keeps up this voltage until irradiation changes and the procedure is rehashed.

The incremental conductance method is an observation techniques in which $\partial P / \partial V = 0$, and that P = IV is to be satisfied for MPP. Equation (2) represents the current of the array as a function of the voltage:

$$P = I(V) * V \tag{2}$$

Therefore,

$$\frac{\partial P}{\partial V} = \frac{\partial (VI)}{\partial V} \tag{3}$$

Setting this equal to zero yields:

$$\frac{\partial P}{\partial V} = -\frac{I}{V} \tag{4}$$

Therefore, the maximum power point is achieved when the incremental conductance is equal to the negative of the instantaneous conductance.

- **Current Sweep:** This technique utilizes a breadth waveform for the PV array current with the end goal that the I-V normal for the PV array is obtained and updated at settled time interims. The MPP voltage would then be able to be obtained from the characteristic curve at similar interims.
- **Constant Voltage:** This technique is utilized by different researchers in different ways. In one case, the output voltage is managed to a steady value under all conditions whereas; in other case the output voltage is controlled in view of a steady proportion to the deliberate Voc. Later case is alluded to conversely as the "open voltage" technique. In the event that the output voltage is held consistent, there is no endeavor to track the maximum power point, so it is not a most extreme power point following system in a strict sense, however it has a few preferences in situations when the MPPT has a tendency to come up short, and in this manner it is infrequently used to supplement a MPPT technique in those cases.

In the "constant voltage" MPPT strategy usually referred as "open voltage technique", the power conveyed to the load is quickly hindered and the V_{oc} with zero current is measured. The controller at that point resumes operation with the voltage controlled at a settled proportion. The working purpose of the PV array is along these lines kept close to the MPP by directing the array voltage and coordinating it to the settled reference voltage

$$V_{ref} = KV_{oc} \tag{5}$$

The calculation of 'V_{ref}' gives ideal execution w.r.t. different factors and the MPP. But, in this strategy 'V_{ref}' is resolved as a proportion to V_{OC}. However, one of the intrinsic approximations to the "constant voltage" proportion technique is that the proportion of the MPP voltage to V_{OC} is just around consistent, so it leaves space for further possible enhancement.

COMPARISON OF METHODS

The P&O and incremental conductance are the special case of hill climbing techniques that can locate the local maximum of the power curve for the working condition of the PV array and gives a genuine maximum power point. However, P&O strategy requires oscillating power output around the maximum power point even under steady state irradiance.

The incremental conductance strategy has the preferred standpoint over the P&O technique that it can decide the MPP without oscillating. It can perform MPPT under quickly fluctuating irradiation conditions with higher precision than the P&O strategy. However, the incremental conductance strategy can create oscillations and can perform unpredictably under quickly changing environmental conditions. The sampling frequency is diminished because of the higher many-sided quality of the algorithm contrasted with the P&O technique.

In the consistent voltage proportion or simply open voltage strategy, the current from the photovoltaic array must be set to zero momentarily to measure the open circuit voltage and afterward a short time it is set to a predetermined level of the measured voltage. In this case, energy is wasted amid the time the current is set to zero. Also, the estimation of MPP/V_{oc} proportion is not really precise. Further, basic and minimal effort to execute the interferences diminishes array efficiency and don't guarantee finding the genuine maximum power point.

MPPT Placement

Traditional solar inverters perform MPPT for the whole PV array. Under such framework current directed by the inverter moves through all modules in the string. Since various modules have diverse I-V curves and distinctive MPPs (because of assembling tolerance, fractional shading and so on.) this engineering implies a few modules will be performing beneath their MPP, bringing about lower efficiency. A few organizations are presently setting maximum power point tracker into singular modules, enabling each to work at crest efficiency in spite of uneven shading and electrical mismatch.

Operation With Batteries

Off-grid PV system may utilize batteries to supply loads. In spite of the fact that the completely charged battery pack voltage might be near the PV board's maximum power point voltage. This is probably not going to be valid when the battery has been somewhat discharged. Charging may start at a voltage extensively beneath the PV board maximum power point voltage and a MPPT can resolve this mismatch.

At the point when the batteries, in an off-grid framework, are completely charged and PV production is greater than nearby loads a MPPT can no longer operate the solar system at its maximum power point as the abundance power has no load to absorb it. The MPPT should then move the PV system working point far from the peak power point until the point when generation precisely coordinates demand. In a grid-network associated photovoltaic framework, all conveyed power from sun powered modules will be sent to the grid. In this manner, the MPPT in a grid associated PV system will dependably try to work the PV modules at its maximum power point.

INTEGRATING SOLAR POWER WITH GRID

Network associated PV frameworks are intended to work in parallel and interconnected mode with the electric utility grid. The essential segment is the inverter or power conditioning unit (PCU). The inverter changes the DC power created by the PV array into AC power which is reliable with the voltage and power quality required by the utility framework. The inverter naturally quits providing energy to the framework when the utility lattice is de-energized. A bi-directional interface is made between the PV framework AC output circuits and the electric utility system, commonly at on-line distribution systems. This permits the power delivered by the PV framework to either supply nearby electrical loads, or to back feed the grid when the PV system output exceeds the on-line load demand. On the other hand, if electrical demand exceeds the PV system output the additional energy required is delivered by the electric utility. This security factor is required in all grid associated PV system which is additionally guarantees that the PV system won't proceed to work and feedback onto the utility grid when the grid is down for service or repair.

The incorporation of solar system with the current grid-network is the immense test in outlining and usage of the upcoming generation smart grid. However, with a specific end goal to help incorporation of the sun based power into the grid-network an extra battery storage system should be introduced in parallel. This is to enhance the power use, and it is fit for absorbing and conveying both real and reactive power with sub-second response times. It additionally helps in moderating the ramp rate, frequency, and voltage issues in the sunlight based power generation. Likewise, mix of the energy market makes the sun powered asset more practical and realistic.

INCORPORATING SOLAR POWER WITH GRID

Typical grid-tied solar systems are based on the following additional components:

- Charge Controller
- Battery Bank
- DC Disconnect (additional)
- Battery-Based GTI
- Power Meter

Figure 2. Integrating solar panels with grid and battery bank

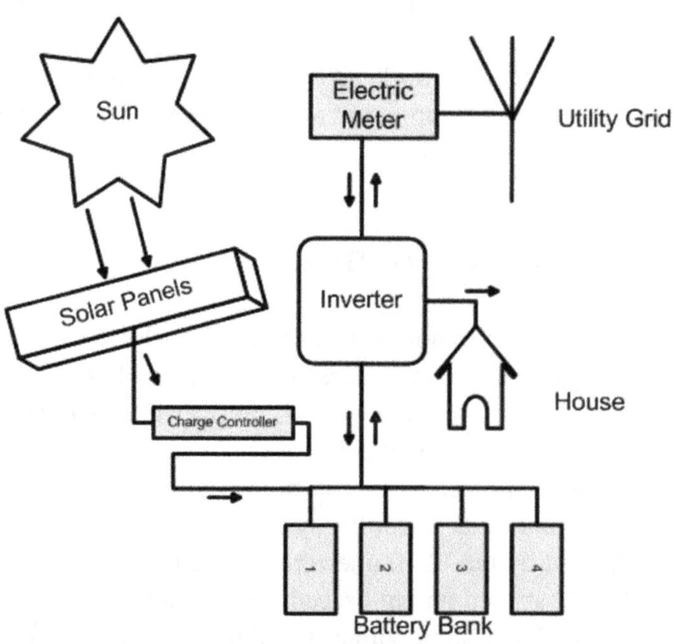

POWER CONDITIONING SYSTEM (PCS)

The DC output of a PV system has restricted direct uses, including private system for water warming and straight forwardly coupled agricultural pumps in an independent system for which the driving motors are rated at DC voltages. In private buildings, PV systems are utilized to substitute for network-grid energy when the sun's energy is accessible. PV supply parameters must be identical with the network-grid supply parameters for synchronization. In this situation, a utility may utilize a PV system possessed by it to lessen the power peak demand via ventilation systems amid the twelve and evening times of hot summer months. The utility may likewise need to utilize the PV system to help its distribution at far away locations, in terms of both active and reactive powers. All these require an interface or a power conditioning system (PCS) that will coordinate and tie-up with the grid system in terms of voltage, frequency, transient conduct, etc. The PCS transfers over DC into AC with parameters required at the utility connection point. The PCS has three fundamental capacities to perform:

1. It needs to screen and modify to get maximum power from the PV source and match its output with a MPPT. For this, the load trademark represented by the PCS needs to suit the output characteristic for the PV–MPPT framework so that there is ideal power under all states of insolation.
2. It must change DC into AC and clutch the quality of power on the AC side.
3. It must screen the utility framework and keep up the steady state similarity, and in addition react appropriately to utility-side elements.

BASIC REQUIREMENTS OF PCS

- It must have unwavering quality and work ideally under practical situation.
- It must have a high effectiveness.
- It should not add general expenses of a PV-vitality framework.
- It should deal with harmonic distortion and enhance power factors where feasible and important.
- The high-frequency switching of the inverter, particularly at higher power handling, tends to make electromagnetic impedance on the output side. This must be checked and prepared for.
- Safety to utility work force and hardware must be the highest need.

FUNCTIONAL UNITS OF PCS

1. **Solar Charger:** The solar charger converts the solar energy collected by the solar PV panel to charge the battery. There are two fundamental sorts of sun oriented chargers-PWM sort and MPPT sort:
   a. In the PWM sort sun oriented charger, the solar panel is directly connected with the battery through electronic switch. The solar panel board is a current source which delivers charging current at a voltage equal to thc battery voltage plus the drop across the switching element. However, this is a very effective and low cost charger whereas; the drawback of this method is that the battery voltage controls the panel voltage which does not allow it to operate at the peak power point of the panel. Recently, different technologies are used for switching the panel

to the battery where a combination of MOSFET and diode is commonly used. In a zero drop type switching the losses is very negligible which results very high conversion efficiency.

b. In a MPPT type solar charger, an electronic conversion unit changes the voltage of the solar board panel to the battery level. Though, the board panel voltage is not reliant on the battery voltage level so solar system can operate at its peak power point. This allows the maximum utilization of the panel. This scheme is costly and it is mainly used only for larger systems.

2. **Inverter:** The inverter is the heart of the solar PCS as it converts the DC voltage from the battery to AC power to the output. Different technologies exist for inverter. In the transformer based inverter, the DC to AC inversion occurs at low voltage of the battery and then this is stepped up to 220V ac using a transformer. This technology is being replaced world over by high efficiency Switched Mode Power Supply (SMPS) based inverter. Here, the DC voltage is first transformed to high voltage DC using a high frequency SMPS. The high voltage DC is then inverted.
3. **Grid Charger:** The grid charger is an auxiliary charger in a solar PCS. It is used to charge the battery from the grid supply when solar charger is found to be inadequate. Different technologies using SMPS type off- grid charger or the SCR based phase controlled charger, or the transformer based charger are available in the market. Some inverters itself are bi-directional and have the ability to charge the battery in the reverse direction from the AC grid. This design reduces cost but suffers with the drawback as it takes considerable amount of time to reverse the power flow. As a result, in case of grid failure there may be considerable delay for the power flow to reverse direction. The delay causes the connected appliance to stop and re-start which further create operational disturbances. However, the delay can be up to few seconds as it depends upon the system design. Also, any attempt to reduce this changeover time can have high reliability issues.
4. **Output Selector Mechanism:** Under the operation of solar power the AC output from the PCS originate from the inverter. But, during operation if solar energy is inadequate the appliances connects with the grid supply. To achieve this is action we generally use of a relay.
5. **Battery Bank:** The battery bank is used to store the sunlight based charge. Throughout the day the incident sun powered energy continuously varying depending upon the cloud development, shadow and time of day, etc. The AC machines connected with the inverter likewise have its own

Figure 3. Schematic diagram of PCS

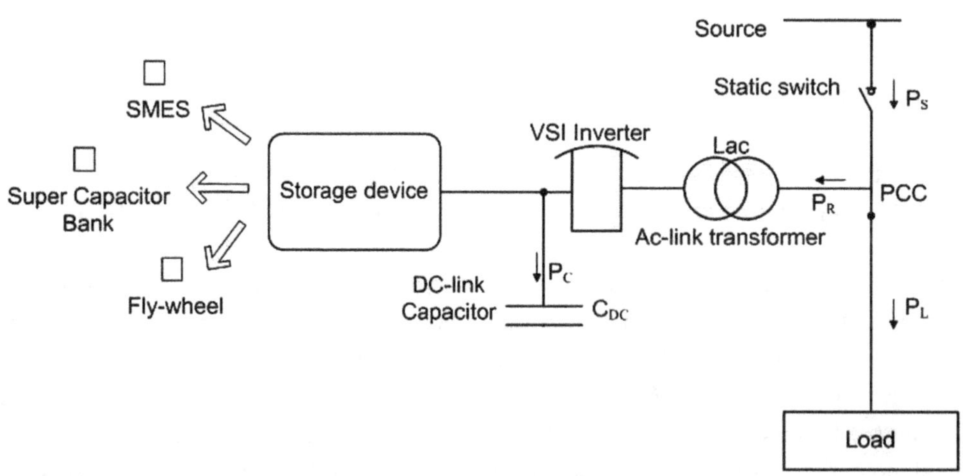

variety in loading pattern. Therefore, it is basic to have a battery bank which has a buffer and a storage component for smooth and reliable operation.

6. **Control Algorithm:** The control algorithm distinguishes a solar PCS from an ordinary inverter and solar charger. It has ability to allot priority and select the source of charging either from source of grid or both for optimal operation. Therefore, it can also select the source of AC output to be either from the inverter or the grid. It can operate with different polarity modes like Solar-Battery-Grid or as Solar-Grid-Battery.

In the Solar-Grid-Battery method of operation, the sunlight based charger initially charges the battery and furthermore underpins the inverter to control the apparatuses. Once the sun oriented energy quits coming, the apparatuses are moved to the network-grid utility by the output selector. The battery is not discharged in this case till there is a grid failure. This type of operation is suitable for installations where grid supply is irregular and the user want to use the grid power as long as it is available. The battery is kept reserved for use when the grid fails. At the same time, in daytime, the loads will work in solar power.

FEATURES OF NEW TYPE OF PCS

New PCS use two-level pulse width modulation (PWM) and achieve small size and high efficiency through features that include a transformer-less configuration and high output voltage (400 V). In general, three-level PWM inverters have only been used in applications which require the suppression of harmonics or a higher converter voltage, such as in railway rolling stock or large motor drives. However, with insulated-gate bipolar transistor (IGBT) modules for three-level inverters are commercially available in recent years. The expected benefits of using three-level PWM inverters includes the following,

1. Low switching loss
2. Use smaller and lighter filter reactors with lower losses
3. Less high-frequency electrical noise
4. Quieter operation.

The lower losses make a major contribution to increasing annual power output.

GRID-TIE INVERTER

A grid-tie inverter (GTI) changes direct current (DC) into an alternating current (AC) appropriate for injecting into an electrical power grid. GTI are utilized between neighborhood electrical power generators and the grid. In order to inject electrical power efficiently and securely into the grid, GTIs should precisely coordinate the voltage and period of the grid sine wave AC waveform and keep up the output voltage marginally higher than the grid voltage at any moment. A high-quality GTI has a settled unity power factor which implies its output voltage and current are impeccably arranged. The inverter has an on-board PC which detects the present AC grid waveform, and yields a voltage to compare with the grid. However, providing reactive energy to the framework may be important to keep the voltage in the neighborhood grid inside permitted impediments. Otherwise, in a grid section with considerable power from sustainable sources, voltage levels may rise excessively on occasion of high generation.

Depending upon voltage and input voltage levels, GTIs circuits regularly have from one to three stages. Figure 4 shows the operation of a three stage GTI. Such a topology can be helpful for low voltage inputs in grounded frameworks.

In this circuitry, the boost converter framed with inductor L_1, MOSFET Q_1, diode D_1 and capacitor C_2 raise the input voltage. If a PV array is appraised for more than 50V, generally one of the input coordinate current busses must be grounded. In this illustration, isolation is given by a high frequency transformer in the second transformation stage.

This stage is essentially a pulse-width modulated balanced DC-DC converter. The schematic (Figure 4) demonstrates a full extension (also known as H-bridge) isolating converter. It contains Q_2-Q_5, T_1, D_2- D_5, L_2, and C_3. For power levels under 1000 watt it could likewise be a half-bridge or a forward converter. A low-frequency (LF) transformer can be utilized in the output stage instead of a high frequency one in the DC-DC segment. The hardware with a LF transformer has a fundamentally bigger weight and size however it won't infuse a DC part into the load. Here is a lesser known detail: UL 1741 allows transformer-less inverters and exempts them from dielectric voltage withstand test amongst input and output. In this manner the separating stage can be wiped out.

The transformer-less inverter apparently bring down the weight and cost. However, in view of the absence of the isolation, these models display potential electrical risks. In this setup if a man touches a terminal of the PV board or the battery he will show up under AC line potential. The transformer-less frameworks require extra security device and unique cautioning names set wherever energized circuits might be uncovered amid benefit.

In order to step up the input voltage T_1 can be a so-called step-up type and it can omit the first stage (boost converter). But high turns ratio leads to large leakage inductance which in turn causes voltage spikes on the FETs and rectifiers as well as other undesirable effects. The converter maintains a DC-link to the output AC inverter. In this case DC must be higher than the peak of the utility AC voltage. For example, for 140V AC service the V_{dc} should be $>140*\sqrt{2}=198V$.

The third transformation stage organize transforms DC into AC by utilizing another full bridge converter. It comprises of IGBT Q_6-Q_9 and LC-channel L_3, C_4. T_1 can be a supposed stride up sort to open up the info voltage. With a stage up T_1, the primary stage might be excluded. Notwithstanding, high turns proportion prompts extensive spillage inductance. This thus causes voltage spikes on the FETs and rectifiers and additionally other undesirable impacts.

The IGBTs Q_6 - Q_9 fill in as electronic switches that work in PWM mode. This topology requires anti- parallel freewheeling diodes to give a substitute way for the current when the switches are off. These diodes are either included inside IGBTs or included remotely. By controlling distinctive switches in the H-bridge, a positive, negative, or zero potential can be connected crosswise over inductor L_3. The output LC channel at that point decreases high frequency sounds to create a sine wave. Any grid-tie-power source needs to synchronize its frequency, stage and adequacy with the utility and encourage a sine wave current into the load. Note that if inverter output (V_{out}) is higher than utility voltage, the GTI will be over-loaded. If it is lower, the GTI may sink current rather than source it. In order to allow a limited current flow into the loads as well as back into the line, and because the grid acts as a source with a very low impedance, a typical GTE is designed to work as a current controlled source, rather than a voltage source. Ordinarily there is an extra coupling inductor (L_{grid}) amongst GTE and the mains that goes about as a shim that "retains" the additional AC voltage. It additionally decreases the current harmonics

Figure 4. Circuit connections of a GTI

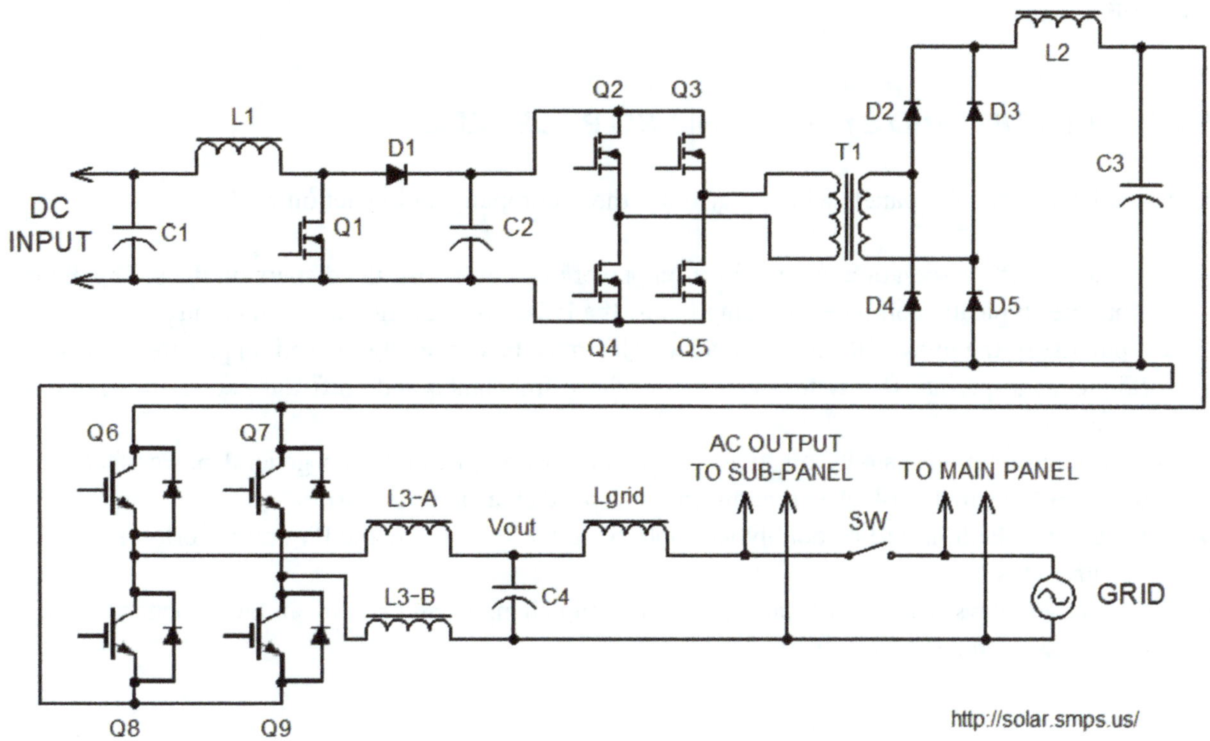

Figure 5. PWM to synthesize sinusoidal output

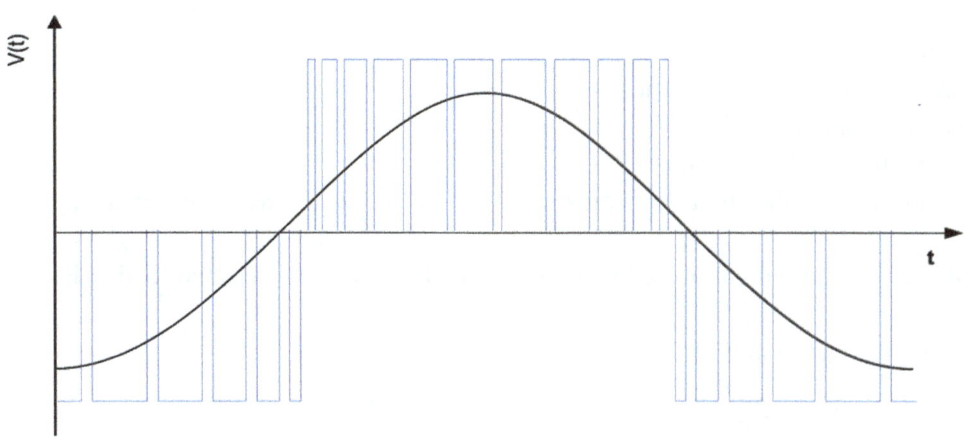

created by the PWM. A disadvantage of L_{grid} is it introduces additional posts in the control loop, which conceivably may prompt the framework instability.

A GTI likewise needs to give anti-islanding assurance. At the point when mains fails or when its voltage level or frequency goes outside of satisfactory limits, the programmed switch ought to rapidly detach the framework output from the line. The clearing time relies upon the mains conditions. In the more terrible cases, when utility voltage dips under 0.5 of nominal, or its frequency aviates off by +0.5

or - 0.7 Hz from the evaluated value, GTI should stop to send out power back to the network in under 100 milliseconds.

PV PANEL AND GRID SYNCHRONIZING PROBLEMS

Synchronization of solar panels with the grid has the accompanying impact on grid:

- An ad-hoc PV association into a utility framework can over-load the dispersion lines. The dispersion lines typically continue including loads. If a PV output additionally flows along these lines to a purchaser, the lines will get over-loaded. This must be especially viewed on private PV system.
- Single-stage private PV system can cause phase imbalance and shift the neutral to hazardous esteems.
- Line commutation, usually on low-voltage frameworks, debases the general power factor and damages the supply voltages. It additionally expands harmonic distortion.
- If there is a fault, it will be doubly fed, both by the utility and by the PV frameworks. This could be hazardous.
- Protective relays, voltage regulation with load tap changes, and so on can be influenced by a recently associated PV framework.

SOLAR PV SYSTEM DESIGN

A solar PV system design can be done in four steps:

- Load estimation
- Estimation of number of PV panels
- Estimation of battery bank
- Cost estimation of the system.
- Base condition: 2 Tube-lights (36 watts each), 2 fans (65 watts each) for 8hrs a day.

The total energy requirement of the system (total load) i.e. Total connected load to PV panel system

= No. of units × rating of equipment

$=2 \times 36 + 2 \times 65 = 202$ watts

Total watt-hours rating of the system

= Total connected load (watts) × Operating hours

$= 202 \times 8 = 1616$ watt-hours

Actual power output of a PV panel

= Peak power rating × operating factor= 40 × 0.75 = 30 watt

The power used at the end use is less (due to lower combined efficiency of the system

= Actual power output of a panel × combined efficiency

= 30 × 0.81 = 24.3 watts (VA)

= 24.3 watts

Energy produced by one 40 Wp panel in a day

= Actual power output × 10 hours/day (peak equivalent)

= 24.3 × 10 = 243 watts-hour

Number of solar panels required to satisfy given estimated daily load:

= (Total watt-hour rating (daily load)/(Daily energy produced by a panel)

=1616/243

= 6.65≈ 7

Inverter size is to be calculated as:

- Total connected load to PV panel system = 202 watts
- Inverter are available with rating of 100, 200, 500 VA, etc.
- Therefore, the choice of the inverter should be 200 VA.

COST ESTIMATION OF A PV SYSTEM

Cost of arrays = No. of PV modules × Cost/Module

= 7 × 8000 (for a 40 Wp panel @ Rs.200/Wp) = Rs.56000

Cost of batteries = No. of Batteries × Cost/Module

=1 × 7500= Rs.7500

Cost of Inverter = No. of inverters × Cost/Inverter = 1 × 5000

$= 1 \times 5000 =$ Rs.5000

Total cost of the system = A + B + C = 56000 + 7500 + 5000 = Rs.68500

[Additional cost of auxiliary items may be taken as 5% of total system cost]

ASSUMPTIONS TAKEN FOR DESIGN

- Inverter converts DC into AC power with efficiency of about 90%.
- Battery voltage used for operation = 12 volts
- The combined efficiency of inverter and battery will be calculated as:

Combined efficiency = inverter efficiency × battery efficiency = $0.9 \times 0.9 = 0.81 = 81\%$

- Sunlight available in a day = 8 hours/day (equivalent of peak radiation.
- Operation of lights and fan = 8 hours/day of PV panels.
- PV panel power rating = 40 Wp (Wp, meaning, watt (peak), gives only peak power output of a PV panel)
- The actual output from a PV module is usually estimated using operating factor.

The operating factor between 0.60 and 0.90 (implying the output power is 60 to 90% lower than rated output power) in normal operating conditions, depending on temperature, dust on module, etc.

PV GRID-PARITY

Electrical energy from private and business PV system is at present 30% more costly than that from utility-scale. Grid parity occurs when a solar system or wind system (or may be any renewable power generating system) can generate power at a levelized cost of electricity that is less than or equal to the price of the purchasing power from the conventional electricity grid. For grid parity there are several factor need to be considered such as transmission fees, taxes etc. In order to include the various possibilities, Japan's NEDO defines the grid parity in three phases as under,

- **1st Phase Grid Parity:** Residential grid-connected PV systems
- **2nd Phase Grid Parity:** Industrial/transport/commercial sectors
- **3rd Phase Grid Parity:** General power generation

Grid parity under solar power generation is more complex than other sources due to a side-effect of one of its main advantages. As compared to other systems PV power generation can be designed as a single panel for small power whereas it can scale successfully to systems as large as millions. A small system can be installed at individual customer's house. Therefore, solar power generation competes against the retail price of grid power. However, retail prices are generally higher than wholesale prices, so grid parity may not have been reached for the very same system installed on the supply-side of the grid.

DESIGN FOR SMART POWER SYSTEM

Figure 6 demonstrates the block diagram of the smart power system. Three diverse power sources are synchronized with a specific end goal to take care of the load demand at client's premises. The different sources are sunlight based board, batteries and the grid supply.

The sun oriented power and the battery yields are interfaced and the resultant power can likewise be sustained to the DC loads. Since the framework control is AC consequently, inverter circuit is utilized to change over DC into AC of required voltage level at grid-network frequency. Static changes are utilized to change the info supply to the AC load according to the accessibility. In day time, the sun based power is encouraged to the load and in night time, the primary power is changed to the grid-network supply.

However, the battery is constantly connected to charge through battery charger as showed in Figure 6. Steady charging of the capacity battery permits the usage of sun based power in maximum load hours. Interface permits the association between sun powered yield and the battery with the end goal that power can stream one way as it were. Further, battery charger is a converter circuit used to converter back ac in to dc for charging at wanted level of voltage.

Practically, particular devices are utilized to decide the working point on the I-V bend. The load on sun oriented cell can be a battery or AC load associated through inverter hardware. In this way, the working point on I-V bend is observed to be load subordinate (dependent). Likewise, the Maximum Power Point Tracking (MPTT) circuits are indispensable (integral) to the charge controllers and interactive inverters.

Figure 6. Block diagram of the smart power system

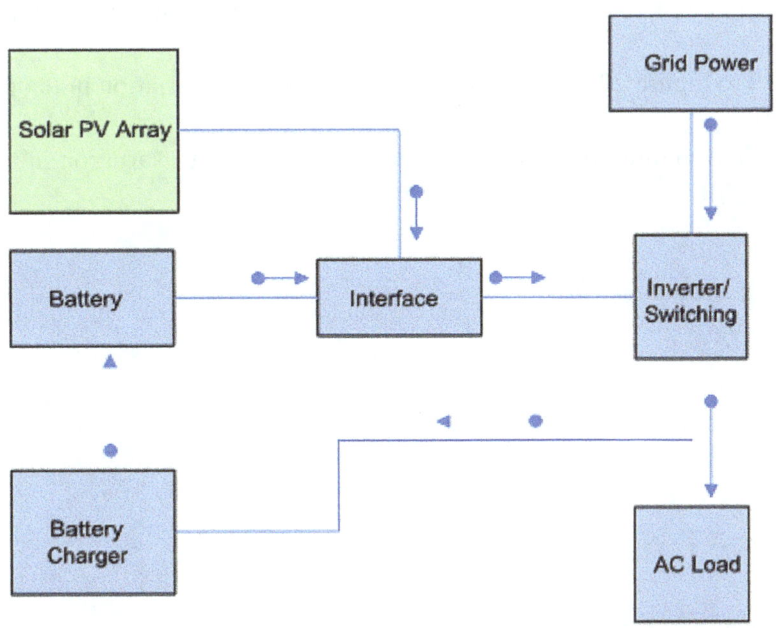

Figure 7. Simulation diagram of solar, grid and battery power in MATLAB environment

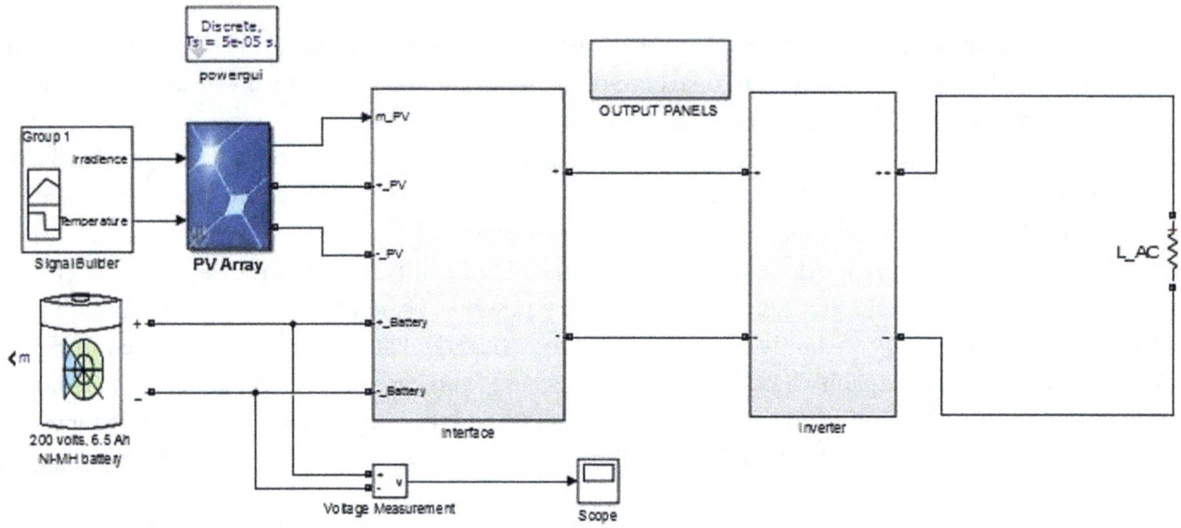

SIMULATION RESULTS

In order to show the effect of variation of temperature and irradiance three different cases are considered and the performance of the proposed model as shown in Figure 7 is evaluated as under,

CASE- I: Figure 8 to Figure 18 shows the simulation results with variation in temperature and irradiance in Case-I.

CASE- II: Figure 19 to Figure 29 show the simulation results with variation in temperature and irradiance in Case-II.

CASE-III: Figure 30 to Figure 40 shows the simulation results with variation in temperature and irradiance in Case-III.

Figure 8. Variation in irradiance Case-I

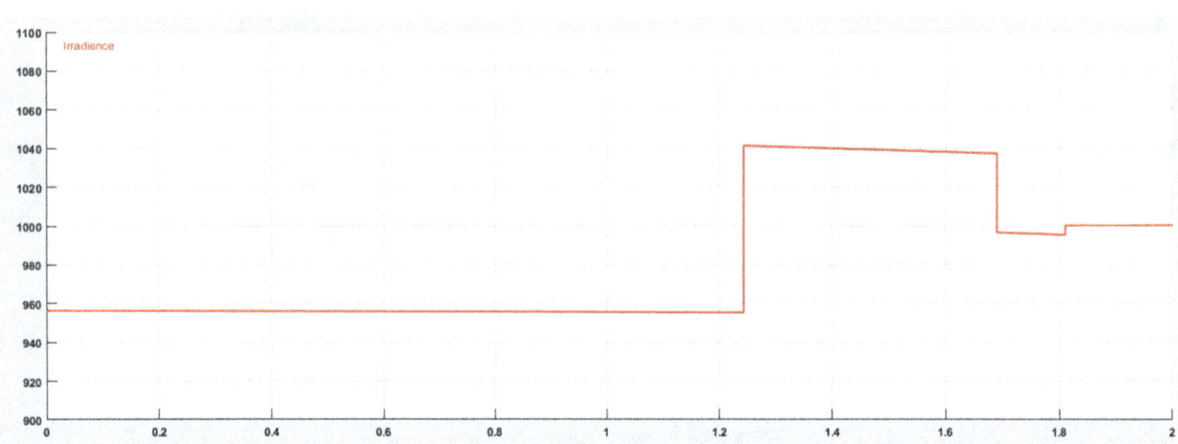

Figure 9. Variation in temperature in Case-I

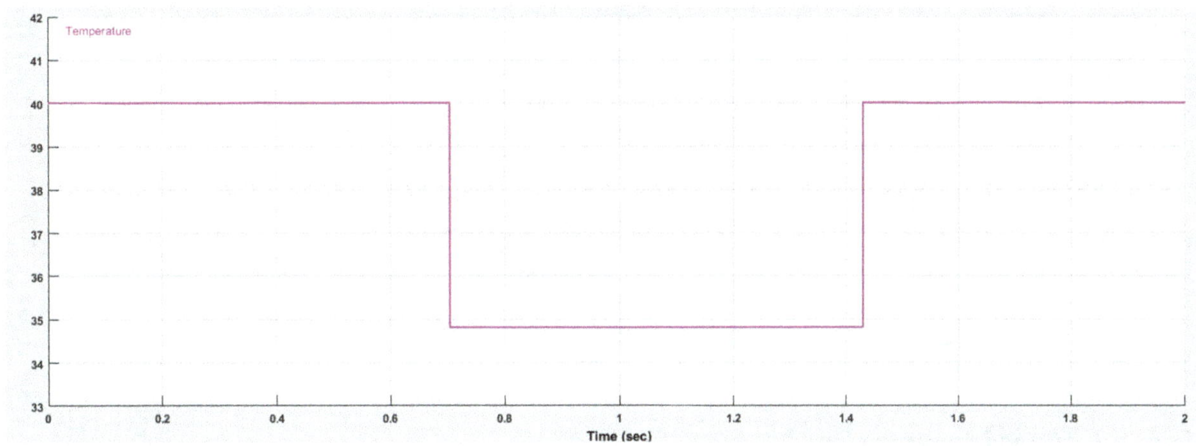

Figure 10. Voltage of battery in Case-I

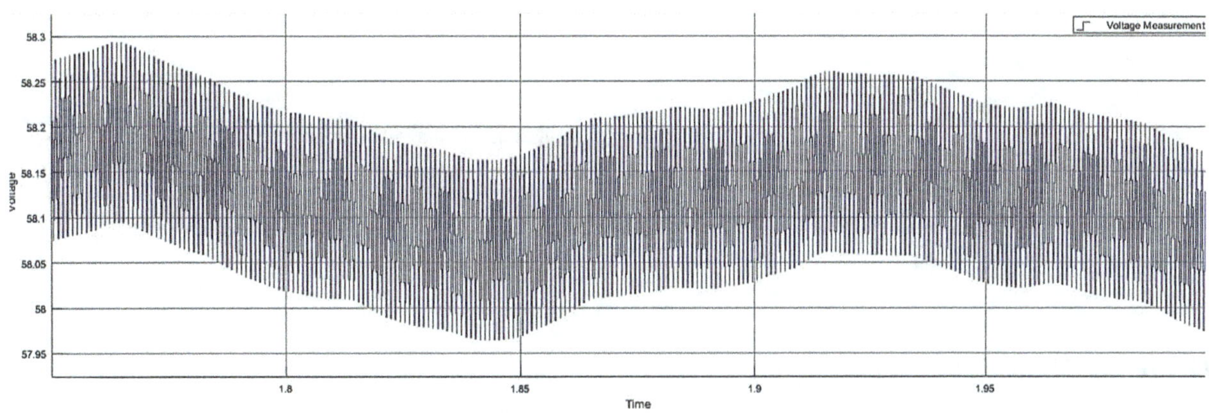

Figure 11. AC load current in Case-I

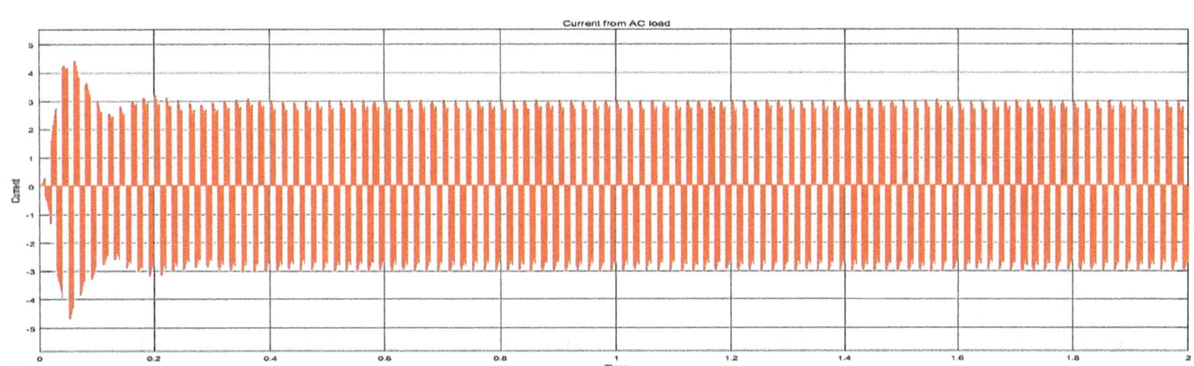

Figure 12. AC load voltage in Case-I

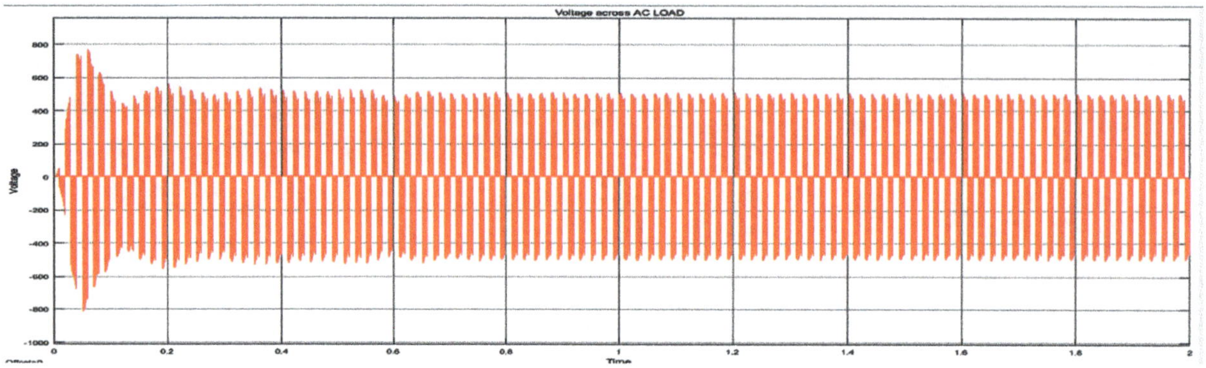

Figure 13. AC load power in Case-I

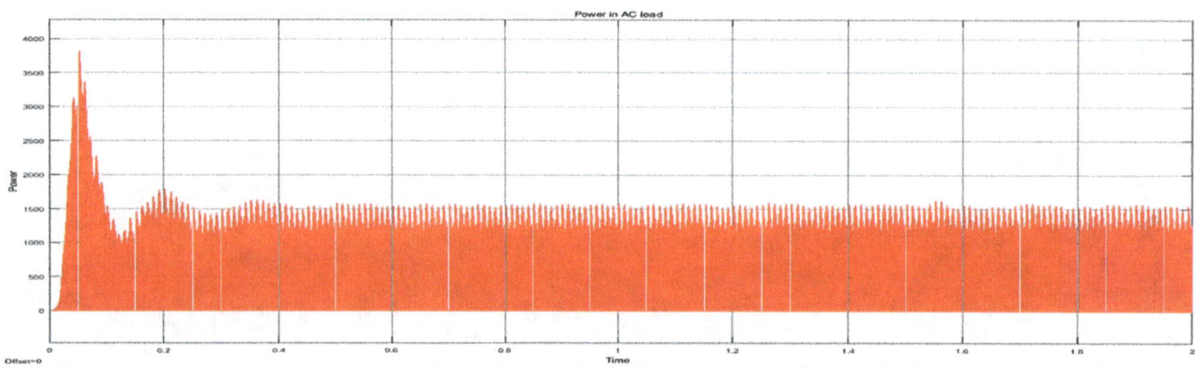

Figure 14. DC load current in Case-I

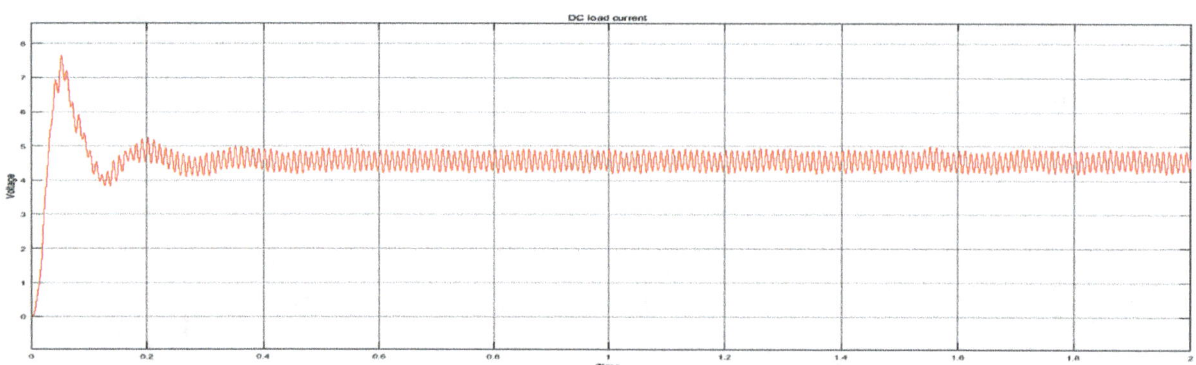

Figures 10, 21, and 32 show the battery voltage under three cases. The variation in temperature and irradiance can be clearly observed in these cases of study. Similarly, the variation in DC current, power and the AC current and power at the inverter input and the load terminals found to be differently. However, in this chapter the power at load end is the sum of power supplied by the grid supply and the solar power when battery is charging whereas it sum of grid power and the battery power when solar

Figure 15. DC total load current in Case-I

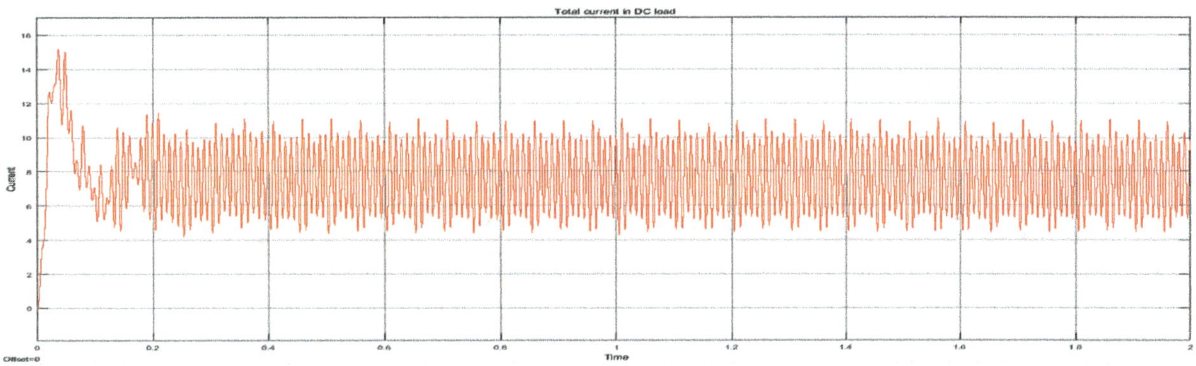

Figure 16. DC load Voltage in Case-I

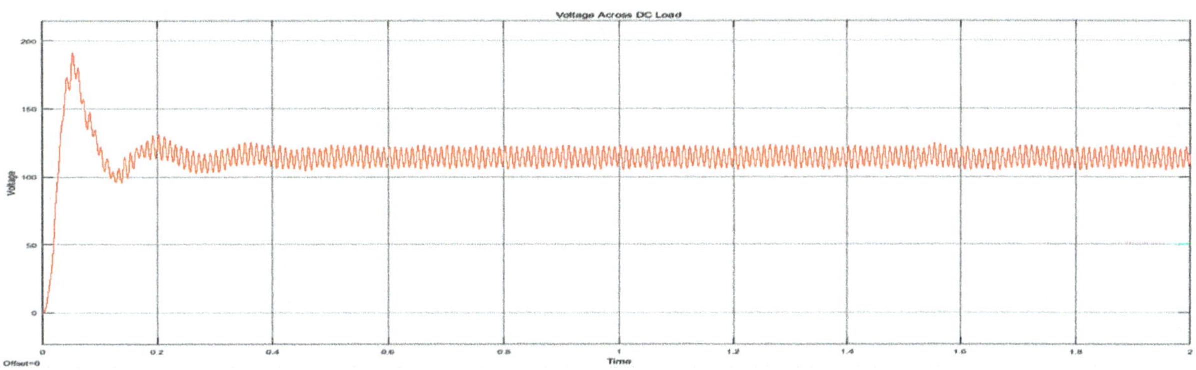

Figure 17. DC load power in Case-I

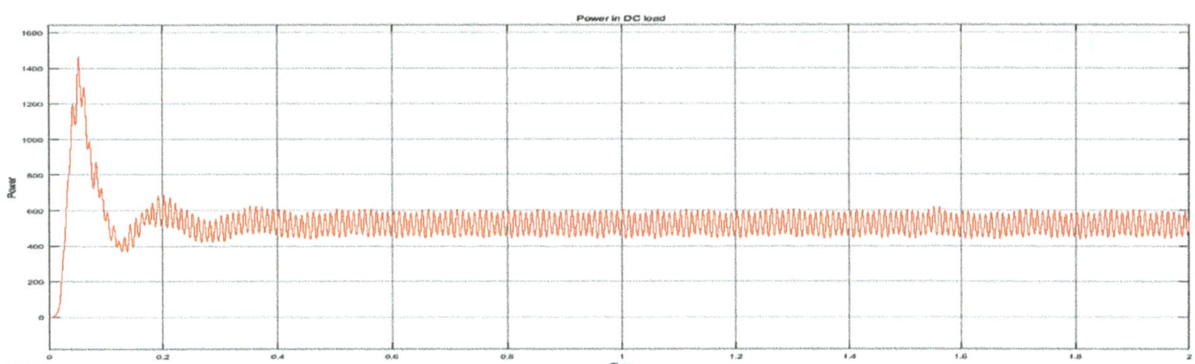

Figure 18. Total DC load Power in Case-I

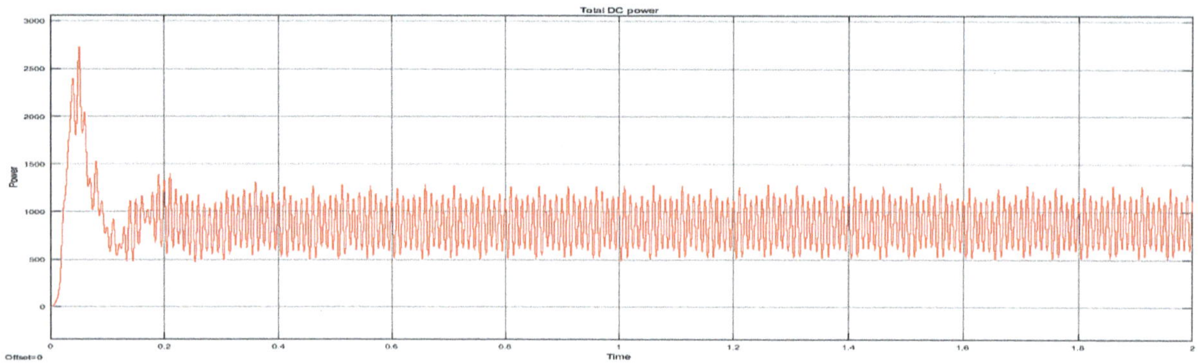

Figure 19. Variation in irradiance in Case-II

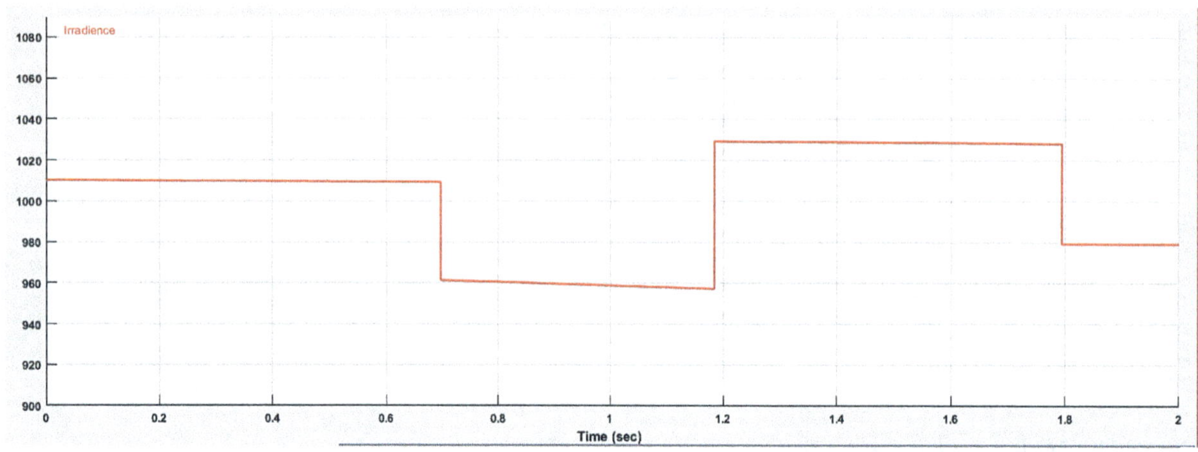

Figure 20. Variation in temperature in case-II

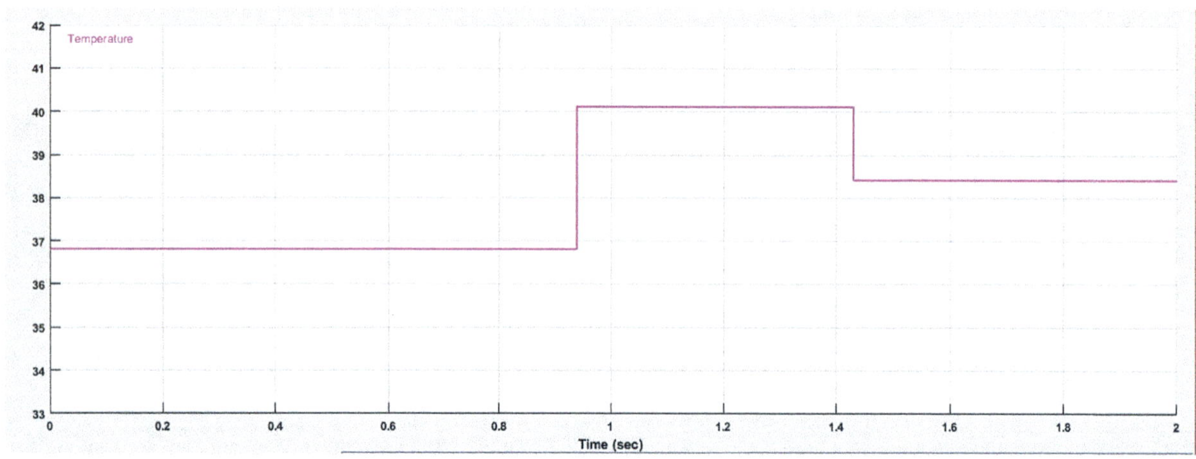

Figure 21. Voltage of battery in case-II

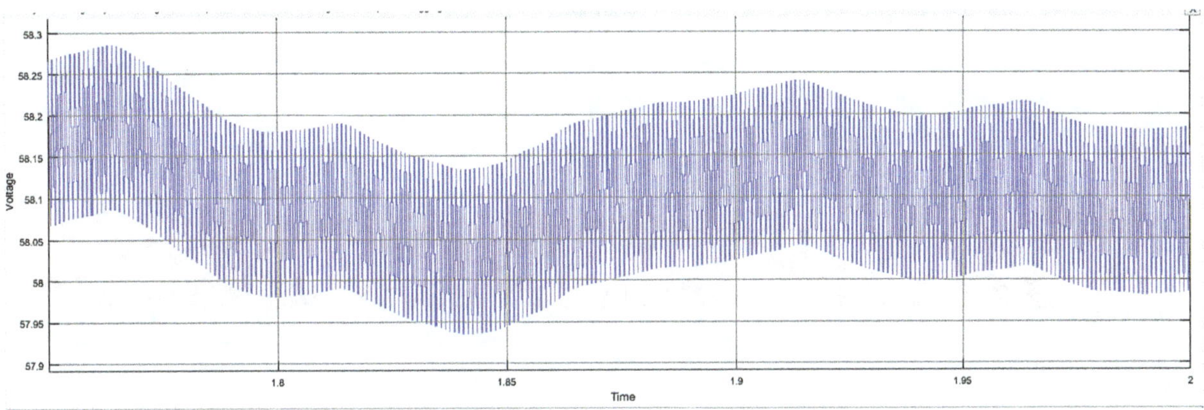

Figure 22. AC load current in case-II

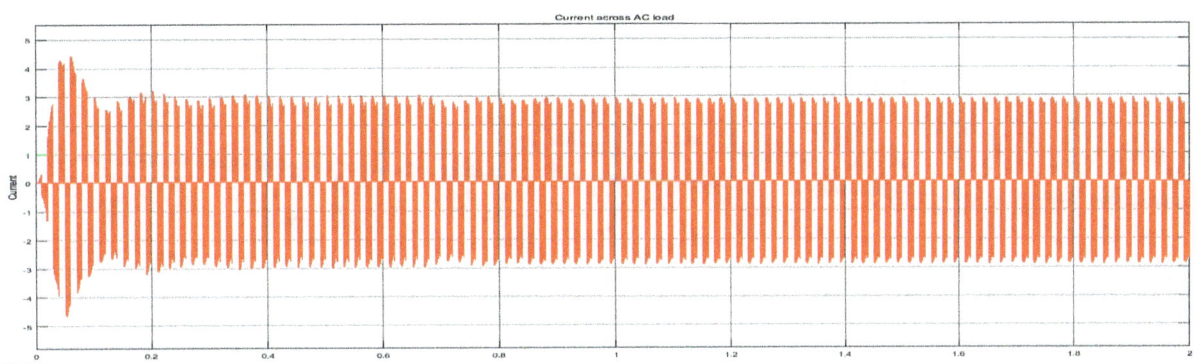

Figure 23. AC load voltage in case-II

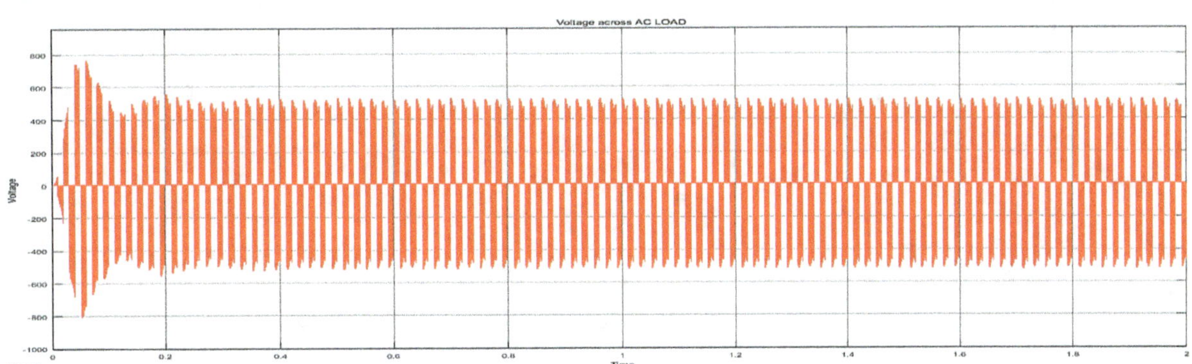

Figure 24. AC load power in case-II

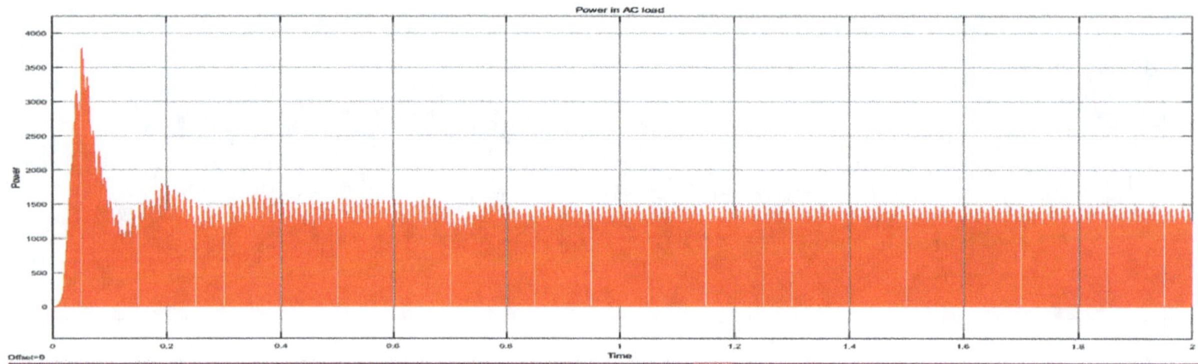

Figure 25. DC load current in case-II

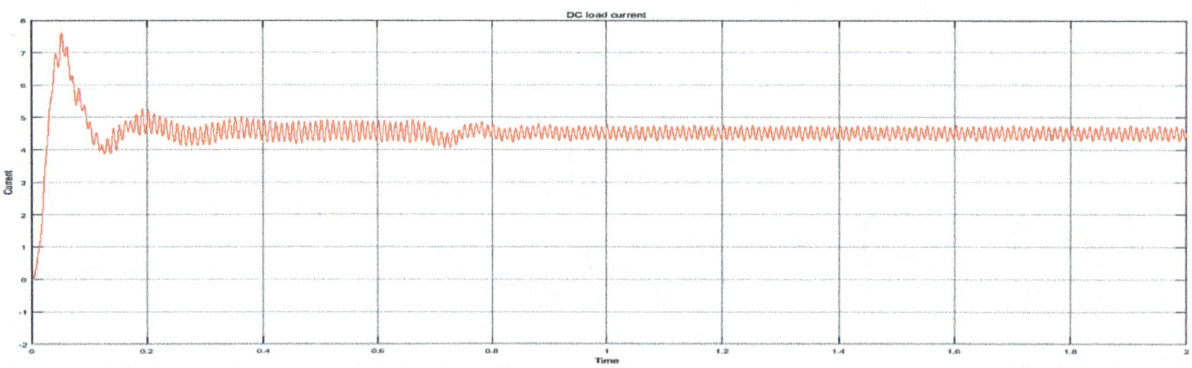

Figure 26. DC total load current in case-II

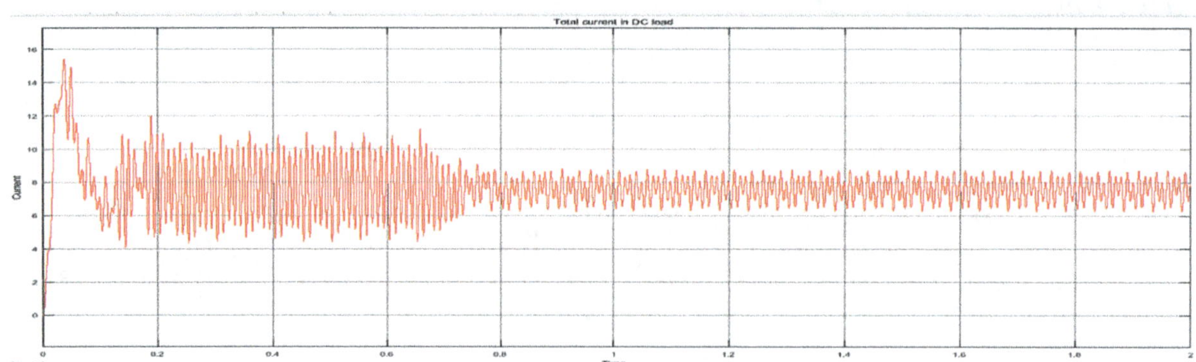

Figure 27. DC load Voltage in case-II

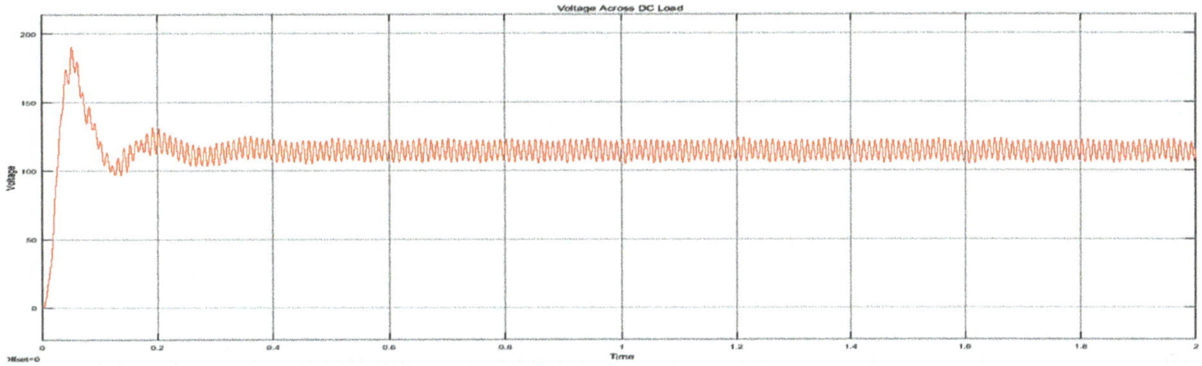

Figure 28. DC load power in case-II

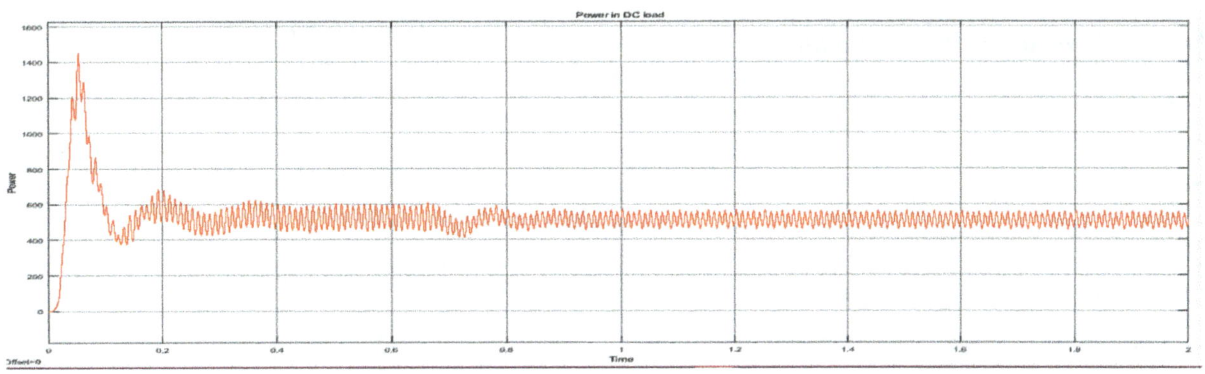

Figure 29. Total DC load Power in case-II

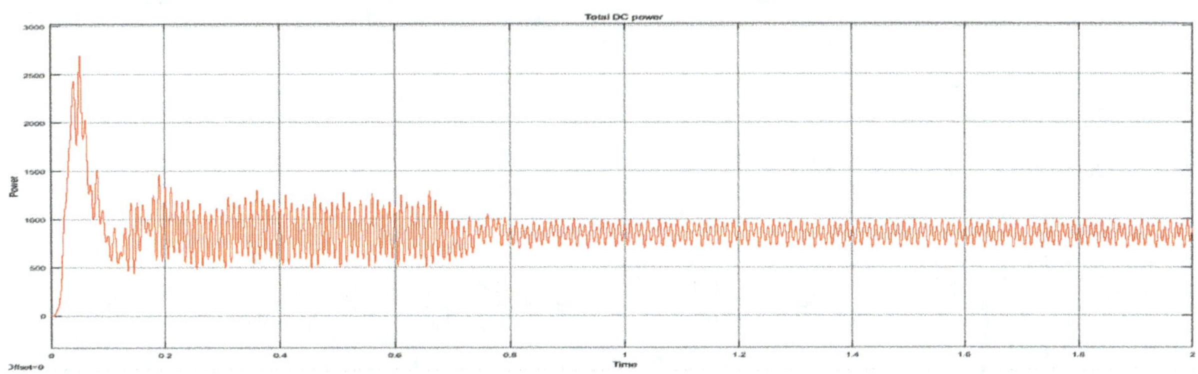

Figure 30. Variation in irradiance in case-III

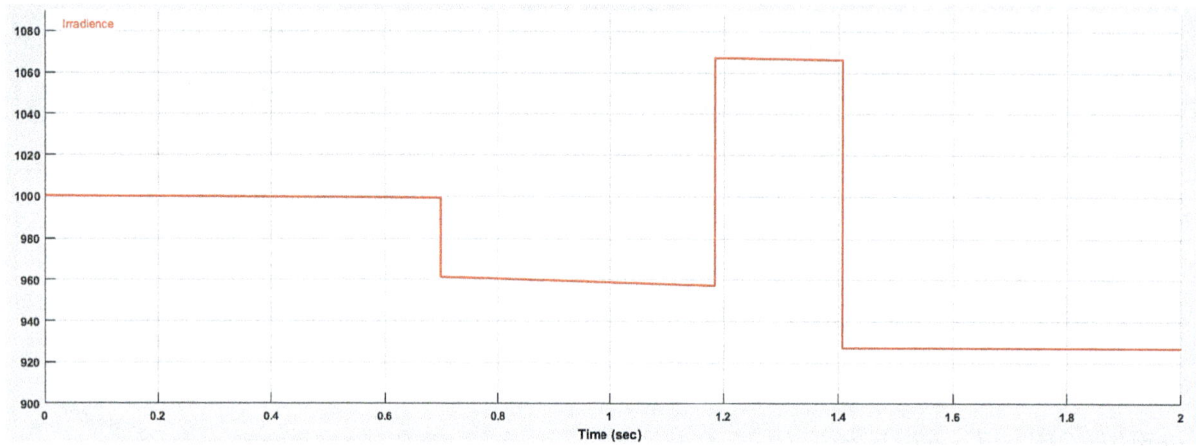

Figure 31. Variation in temperature in case-III

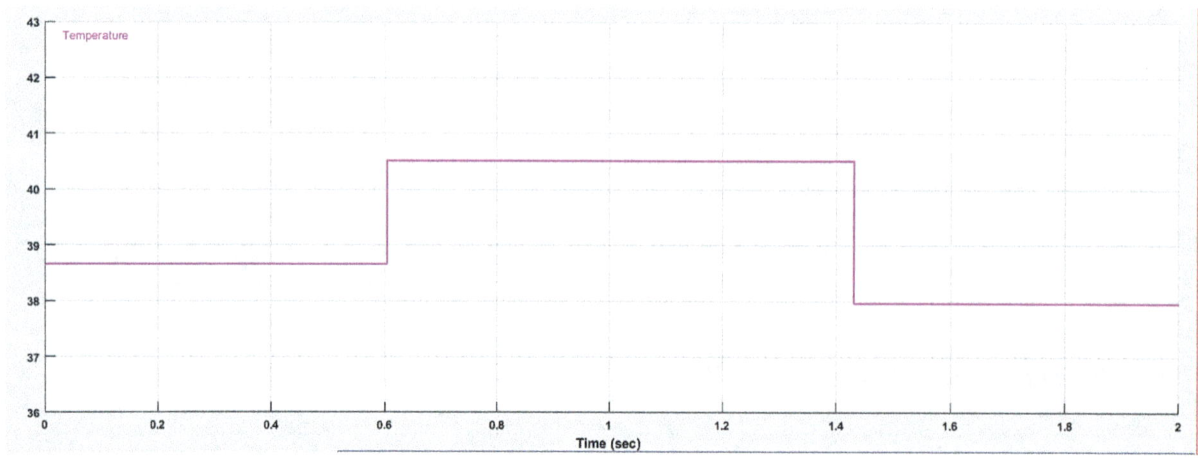

Figure 32. Voltage of battery in Case-III

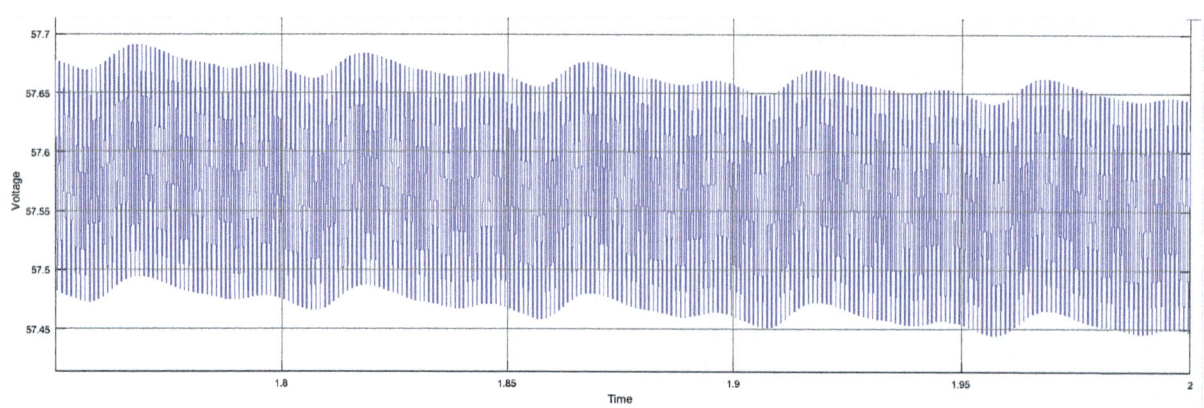

Figure 33. AC load current in Case-III

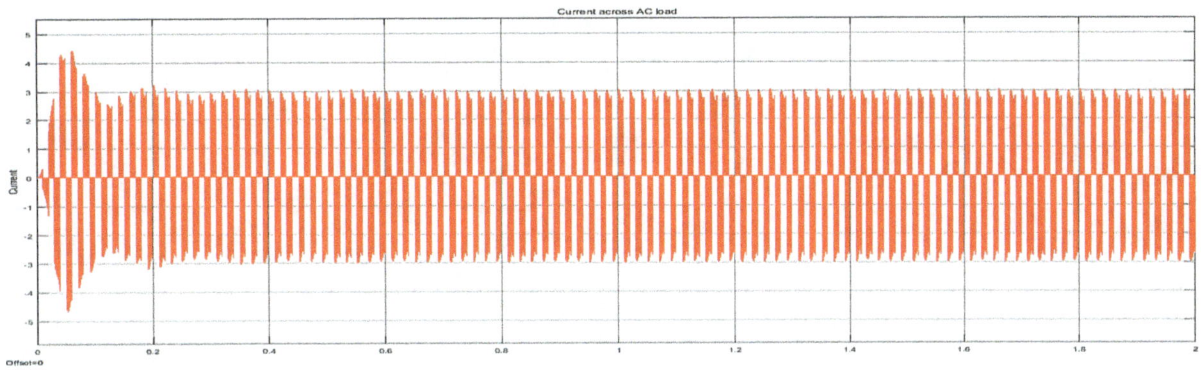

Figure 34. AC load voltage in Case-III

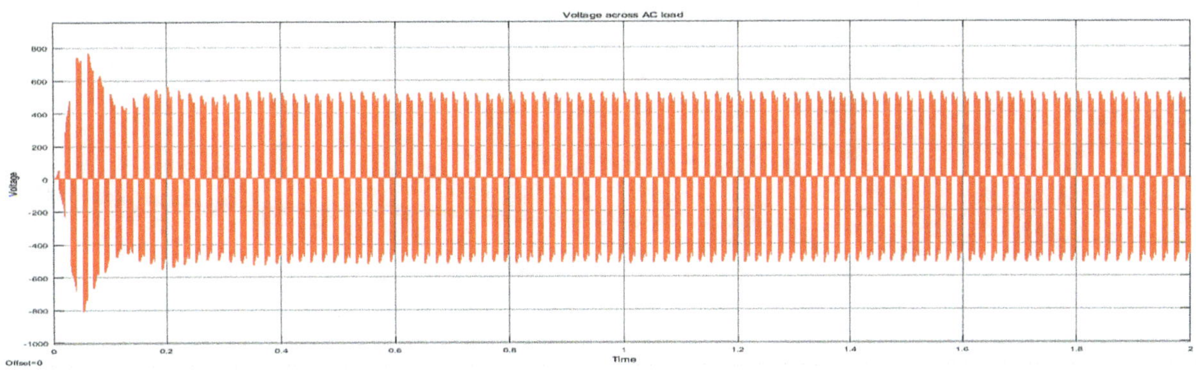

Figure 35. AC load power in Case-III

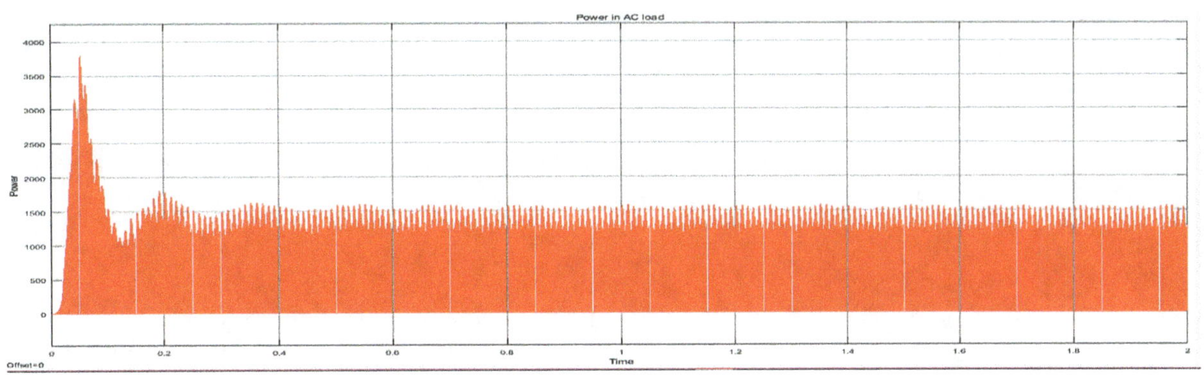

Figure 36. DC load current in Case-III

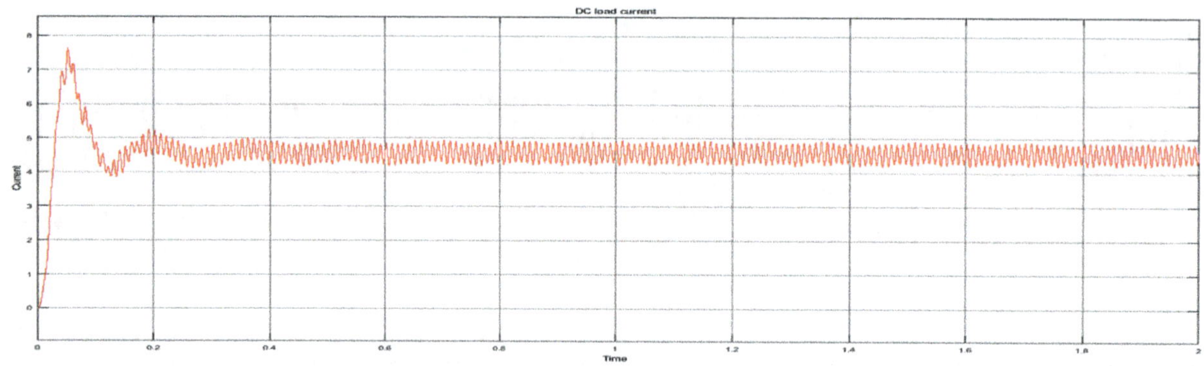

Figure 37. DC total load current in Case-III

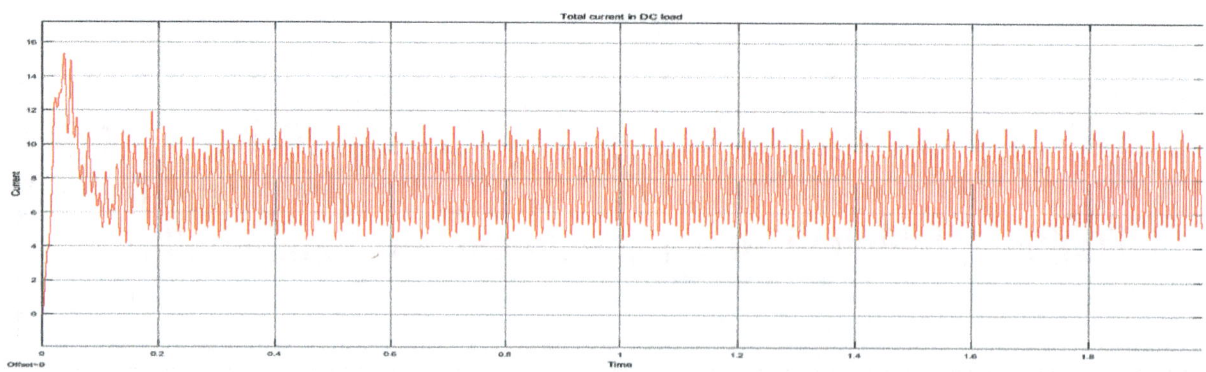

Figure 38. DC load voltage in Case-III

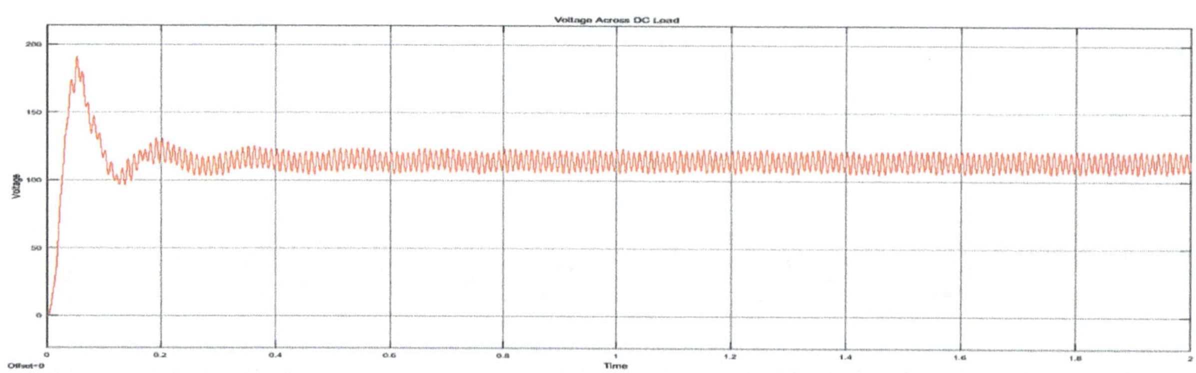

Figure 39. DC load power in Case-III

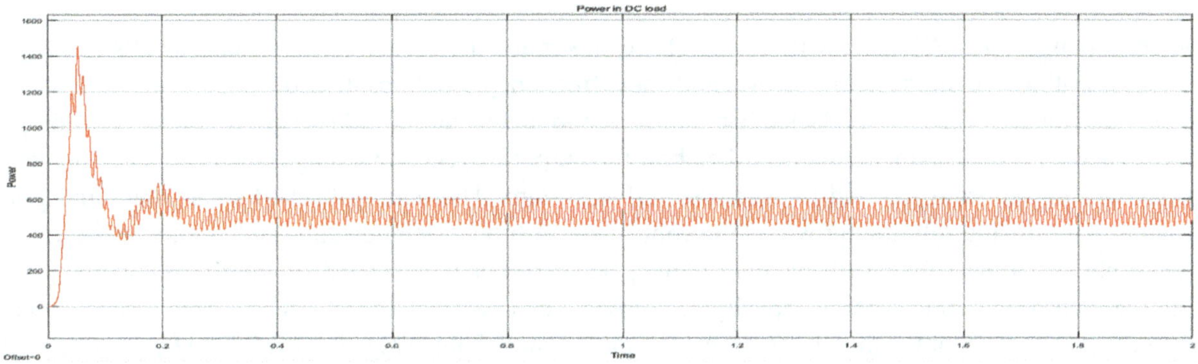

Figure 40. Total DC load power in Case-III

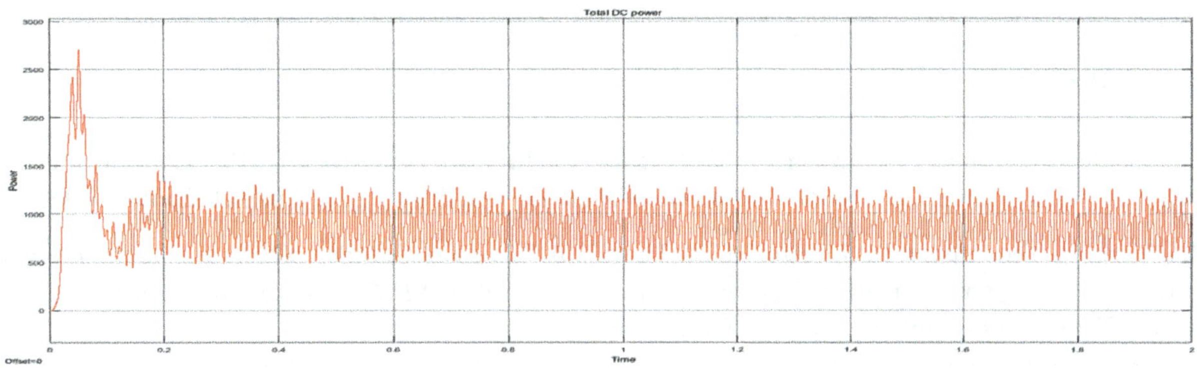

power is not available. Therefore, under the conditions when variation in temperature and irradiance is not uniform it is challenging to have synchronized operation of solar power, battery and grid power for the smart power system for local grid system.

FUTURE RESEARCH DIRECTION

Form the simulation result it has been observed that the variation in solar irradiance and temperature can affect the solar power output and the voltage generated by the solar panel. This further causes the serious issues in synchronization of solar power with grid power. Therefore, it is challenging to have accurate integration under these conditions and it further need to be address in the extension of this work for practical results under real time loading scenario.

CONCLUSION

Synchronizing a solar PV panel with grid can have power quality issues such as voltage swells and sags, harmonic pollution of the system, and over and under frequency problems. In addition, PV systems cannot be operated ideally when there is lack of irradiance or completely no sunlight on the systems. Therefore, it is critical to comprehend the fundamental on how the output current and voltage fluctuates with changes in daylight power from sun and temperature in PV system. The battery backed solar integrated grid provides a promising solution, provided the control algorithm is designed and implemented effectively. The combined output of solar PV panels and the grid should be in accordance with the connected load, with due consideration to reserve capacity or contingency operation of the system. New technological advancements have enabled highly efficient Maximum Power Point Trackers (MPPT) to be used, as well as the design of effective Power Conditioning Systems (PCS). Thus, the overall advantages of integrating solar panels with grid are far more prominent than the corresponding disadvantages. Solar energy is, and will remain a widely accepted renewable energy source.

REFERENCES

Abdulkadir, M., Yatim, A. H. M., & Yusuf, S. T. (2014). An Improved PSO-Based MPPT Control Strategy for Photovoltaic Systems. *International Journal of Photoenergy*, *2014*, 1–11. doi:10.1155/2014/818232

Blaabjerg, F., Teodorescu, R., Liserre, M., & Timbus, A. V. (2006, October). Overview of control and grid synchronization for distributed power generation systems. *IEEE Transactions on Industrial Electronics*, *53*(5), 1398–1409. doi:10.1109/TIE.2006.881997

Bojoi, R., Limongi, L. R., Roiu, D., & Tenconi, A. (2010). Enhanced power quality control strategy for single-phase inverters in distributed generation systems. *Industrial Electronics (ISIE), 2010 IEEE International Symposium on*, 2727-2732.

Chandwani, A., & Kothari, A. (2016). Design, simulation and implementation of Maximum Power Point Tracking (MPPT) for solar based renewable system. *2016 International Conference on Electrical Power and Energy Systems (ICEPES)*, 539-544. 10.1109/ICEPES.2016.7915987

Chiang, H. C., Ma, T. T., Cheng, Y. H., Chang, J. M., & Chang, W. N. (2010). Design and implementation of a hybrid regenerative power system combining grid-tie and uninterruptible power supply functions. *Renewable Power Generation, IET*, *4*(1), 85–99. doi:10.1049/iet-rpg.2009.0033

Dasgupta, S., Sahoo, S. K., & Panda, S. K. (2011). Single-Phase Inverter Control Techniques for Interfacing Renewable Energy Sources With Microgrid-Part I: Parallel-Connected Inverter Topology With Active and Reactive Power Flow Control Along With Grid Current Shaping. *IEEE Transactions on Power Electronics, 26*(3), 717-731.

Guo, X. Q., & Wu, W. Y. (2010). Improved current regulation of three-phase grid-connected voltage-source inverters for distributed generation systems. *Renewable Power Generation, IET*, *4*(2), 101–115. doi:10.1049/iet-rpg.2009.0044

Harakawa, T., & Tujimoto, T. (2001). A proposal of efficiency improvement with solar power generation system. *Proc. of 27th Annual Conference of IEEE Industrial Electronics Society, 1*, 523-528.

Hill, C. A., Such, M. C., Chen, D., Gonzalez, J., & Grady, W. M. (2012). Battery Energy Storage for Enabling Integration of Distributed Solar Power Generation. *IEEE Transactions on Smart Grid, 3*(2), 850–857. doi:10.1109/TSG.2012.2190113

Ho, B. M. T., Chung, H. S. H., & Lo, W. L. (2004, March). Use of system oscillation to locate the MPP of PV panels. *IEEE Power Electronics Letters, 2*(1), 1–5. doi:10.1109/LPEL.2004.828443

Ishaque, K., Salam, Z., Amjad, M., & Mekhilef, S. (2012, August). An Improved Particle Swarm Optimization (PSO)–Based MPPT for PV With Reduced Steady-State Oscillation. *IEEE Transactions on Power Electronics, 27*(8), 3627–3638. doi:10.1109/TPEL.2012.2185713

Jinn, C. W., Kuen, D. W., Hurng, L. J., & Sheng, K. C. (2014). Small capacity Grid connected solar power generation system. *IET Power Electronics, 7*(11), 2717–2725. doi:10.1049/iet-pel.2014.0015

Krishnamurthy, S. (2009). *Solar and Fuel Cell Circuit Modeling*. Analysis and Integrations with Power Conversion Circuits for Distributed Generation, Electronic Theses and Dissertations.

Kumar, P., Ali, I., & Thomas, M. S. (2011). Optimal capacitor placement in smart distribution systems to improve its loadability and energy efficiency. *International Journal of Engineering Science and Technology, 3*(8), 271–284.

Kumar, P., Ali, I., & Thomas, M. S. (2013). Distributed resource planning for improved voltage stability of radial distribution system. *International Journal of Advanced Research in Electrical, Electronics and Instrumentation Engineering, 2*(1), 95-101.

Kumar, P., Ali, I., & Thomas, M. S. (2015). Synchronizing solar cell, battery and grid supply for development of smart power system for home. *Proc. IEEE International Conference INDICON.*

Kumar, P., Ali, I., & Thomas, M. S. (2016). Energy efficiency analysis of reconfigured distribution system for practical loads. *Elsevier Journal of Perspectives in Science, 8*, 498–501. doi:10.1016/j.pisc.2016.05.007

Kumar, P., Ali, I., Thomas, M. S., & Singh, S. (2017, March). Imposing voltage security and network radiality for reconfiguration of distribution system using efficient heuristic and meta-heuristic approaches. *IET Generation, Transmission & Distribution, 11*(10), 2457–2467. doi:10.1049/iet-gtd.2016.0935

Kumar, P., Jain, G., & Palwalia, D. K. (2015). Genetic Algorithm Based Maximum Power Tracking in Solar Power Generation. *International Conference on Power And Advanced Control engineering (ICPACE)*, 1-6.

Kumar, P., & Singh, S. (2014a). Comprehensive stability analysis of radial distribution system with load growth. *Proc. IEEE International Conference IICPE-2014.*

Kumar, P., & Singh, S. (2014b). Energy efficiency performance of radial distribution system with reactive power injection. *Proc. IEEE International Conference IICPE-2014.*

Kumar, P., & Singh, S. (2014c). Reconfiguration of radial distribution system with static load models for loss minimization. *Proc. IEEE International Conference PEDES-2014.*

Kuo, C. L., Lin, C. H., Yau, H. T., & Chen, J. L. (2013). Using Self-Synchronization Error Dynamics Formulation Based Controller for Maximum Photovoltaic Power Tracking in Micro- Grid Systems. *IEEE Journal on Emerging and Selected Topics in Circuits and Systems*, *3*(3), 459–467. doi:10.1109/JETCAS.2013.2272839

Lee, J., Han, B., & Choi, K. (2011). High-efficiency grid-tied power conditioning system for fuel cell power generation. *8th International Conference on Power Electronics - ECCE Asia(ICPE & ECCE)*, 1492-1497.

Lee, J. H., Bae, H., & Cho, B. H. (2006). *Advanced Incremental Conductance MPPT Algorithm with a Variable Step Size*. IEEE Power Electronics and Motion Control. doi:10.1109/EPEPEMC.2006.283227

Liu, F., Zhang, Y., & Duan, S. (2008). Comparison of P&O and hill climbing MPPT methods for grid-connected PV converter. IEEE Industrial Electronics and Applications, 804-807.

Masoum, M. A. S., Dehbonei, H., & Fuchs, E. F. (2002). Theoretical and Experimental Analyses of Photovoltaic System with Voltage and Current-Based Maximum Power-Paint Tracking. *IEEE Transactions on Energy Conversion*, *17*(4), 514–522. doi:10.1109/TEC.2002.805205

Qinghui, D., & Jun, C. (2009). Improving the efficiency of solar photovoltaic power generation in several important ways. *IET International Technology and Innovation Conference*, 13-15.

Shukla, A., Khare, M., & Shukla, K. N. (2015). Modeling and Simulation of Solar PV Module on MATLAB/Simulink, International Journal of Innovative Research in Science. *Engineering and Technology*, *4*(1), 18516–18527.

Smida, M. B., & Sakley, A. (2015). Genetic based algorithm for maximum power point tracking (MPPT) for grid connected PV systems operating under partial shaded conditions. *7th International Conference on Modelling, Identification and Control (ICMIC)*,18-20.

Sung-Yeul, P., Jih-Sheng, L., & Woo-Cheol, L. (2010). An easy, simple, and flexible control scheme for a three-phase grid-tie inverter system. In *Energy Conversion Congress and Exposition (ECCE)* (pp. 599–603). IEEE.

Svensson, J. (2001). Synchronization methods for grid connected voltage source converters. *IEE Proceedings Generations, Transmission and Distribution, 148*(3), 229-235.

Venugopalan, R., Krishnakumar, N., Sarjila, R., & Rajasekar, N. (2013). Application of Particle Swarm Optimization Technique for the Design of Maximum Power Point Tracking. *Advanced Materials Research*, *768*, 47–56. doi:10.4028/www.scientific.net/AMR.768.47

Yadav, R. G., Roy, A., Kharparde, S. A., & Pentayya, P. (2005). India's fast growing power sector from regional development to the growth of a national grid. *IEEE Power & Energy Magazine*, *3*, 39–48. doi:10.1109/MPAE.2005.1458229

Yang, C., & Smedley, K. (2008, September). Three-Phase Boost-Type Grid-Connected Inverter. *IEEE Transactions on Power Electronics*, *23*(5), 2301–230. doi:10.1109/TPEL.2008.2003025

Yishu, Z. (2009). The circuit topology for single-phase Grid connected system and the control technology on converters. *International Conference on Sustainable Power Generation and Supply, SUPERGEN '09*, 1-5. 10.1109/SUPERGEN.2009.5348130

Zhigang, L., Rong, G., Jun, L., & Huang, A. Q. (2011, March). A High-Efficiency PV Module-Integrated DC/DC Converter for PV Energy Harvest in FREEDM Systems. *IEEE Transactions on Power Electronics, 26*(3), 897–909. doi:10.1109/TPEL.2011.2107581

Zhilei, Y., Lan, X., & Yangguang, Y. (2010). Seamless Transfer of Single-Phase Grid-Interactive Inverters Between Grid-Connected and Stand-Alone Modes. *IEEE Transactions on Power Electronics, 25*(6), 1597–1603. doi:10.1109/TPEL.2009.2039357

KEY TERMS AND DEFINITIONS

AC and DC Load: The electrical components which operate at AC and DC supply is termed as the AC and DC loads, respectively.

Grid Power: The power received from electrical grid is known as grid power.

Grid Tie Inverter (GTI): It is a device that converts DC supply into AC supply suitable for injecting into an electrical power grid at desired voltage and frequency level.

Insolation: The solar power density incident on a surface of stated area and orientation, usually expressed as watts per square meter or Btu per square foot per hour.

Irradiance: The direct, diffuse, and reflected solar radiation that strikes a surface. It is usually expressed in kilowatts per square meter. Irradiance multiplied by time equals insolation.

Maximum Power Point Tracking (MPPT): It is an algorithm that included in charge controllers used for extracting maximum available power from PV module under certain operating conditions.

Operating Point: The current and voltage that a photovoltaic module or array produces when connected to a load. The operating point is dependent on the load or the batteries connected to the output terminals of the array.

Photovoltaic (PV) Conversion Efficiency: The ratio of the electric power produced by a photovoltaic device to the power of the sunlight incident on the device.

Solar Constant: The average amount of solar radiation that reaches the earth's upper atmosphere on a surface perpendicular to the sun's rays; equal to 1353 watts per square meter or 492 Btu per square foot.

Storage Battery: It is a battery (or cell) used for storing electrical energy.

Chapter 9
Typical Design of Synchronous Controller to Minimize Response Time and Power:
Power and Response Time Optimization

Siwar Ben Haj Hassine
National Engineering School of Sousse, Tunisia

Bouraoui Ouni
National Engineering School of Sousse, Tunisia

ABSTRACT

As power dissipation and time constraint have become vital challenges during the creation of a digital circuit, researchers' and designers' efforts have increased to figure out new ways of preserving power through the study of its sources and its impacts as well as through the decrease of response time to obtain faster treatments. However, it is widely acknowledged that these two parameters are antagonistic in synchronous systems. In fact, current technologies have managed to further decrease the response time to have a faster circuit at the cost of a considerable simultaneous augmentation in its power or vice versa, which leaves no option for designers but to choose from these two important parameters. Hence, the main objective of this chapter is to propose a design method that simultaneously builds a low power design and provides a faster circuit. For the achievement of that purpose, a controller based on a finite state machine (FSM) has been chosen as an example of synchronous system to prove that the new proposed design can optimize both parameters: time and power.

INTRODUCTION

Power consumption has become one of the most important parameter while designing a digital system. In fact, according to the International Energy Agency, electricity consumption is expected to increase by up to 75% between 2007 and 2030. The average annual increase in global electricity consumption will be about 2.4%. This critical growth is now the overriding responsibility for more than 40% of greenhouse

DOI: 10.4018/978-1-5225-3935-3.ch009

gas projections. Thus, a universal awareness in all sectors is observed in order to implement new solutions to solve such problem. In this context, the designers of electronic systems adhere to the efforts of the universal community to develop appropriate methodologies and techniques in order to obtain less consuming systems.

The power consumption affects many sectors and levels such as: the socio-economic level and the technological one. Concerning the socio-economic level, it is observed that with the increase of electricity prices, the power consumption of several informatics, electrical, microelectronic and domestic equipments has become a major concern. Consequently, media mobilization is necessary in order to evolve behaviors and mentalities.

In addition, social and political action plans are being implemented and focused on optimizing the average per capita energy consumption rate. This rate differs from one country to another and reaches its peak in countries with strong industrialization such as Europe and America.

On the social level, high electricity prices push people to take into account the energy consumption criterion. For example, according to the magazine science and life the use of an ecological light bulb offers an annual gain of electricity that reaches a level of more than 100 Euros compared to a standard bulb.

While choosing an electronic system, it is important to consult its technical information, specifically the estimation of its power consumption. Indeed, several magazines that deal with high technology make comparisons between different products in terms of power dissipation. These products are evaluated in terms of average electricity consumption per year in order to make the user aware about the importance of the power constraint. For instance, the magazine named the "Internaute High Tech" presented a comparative study between a Plasma TV and an LCD television in terms of electrical power. Another study concerning the use of computers has been shown. It is estimated that a family computer turned on 24 hours a day can increase the bill by 300 euros per year. Concerning mobile systems, their energy consumption has a greater effect. These systems are limited by the lifetime of their batteries. Thus, reduced power consumption leads to a certain increase in the operating time of these systems.

Reducing power consumption increases battery lifecycle and reduces its production and its storage. Such solution would produce less threatening mobile systems and be also imperative to emphasize the danger of batteries on the environment.

On the environmental side, ecological associations are worried about the influence and the effect of new information and communication technologies (ICTs) on greenhouse gas emissions. Unfortunately, the emergence of the Internet has turned mobility into a big enemy of the environment. Practically speaking, huge computing centers have been developed to store applications and data. These centers, which are mainly located in China and the United States, are powered by electricity produced from coal. According to the World Green Organization study "GreenPeace", the impressive development of the Internet is the main responsible for the doubling of greenhouse gas emissions between 2007 and 2010.

Finally, Smart phones and laptops that have shaken the world of business and information influence the overall energy balance because of their power consumption. The reduction in the energy consumption of these systems will therefore make it possible to control existing primary energy and preserve the environment.

Concerning the second level, which is the technological one, power consumption has not only an impact on the environment, but also on the industry. In fact, the power consumption of electronic systems has an influence on their final cost and reliability.

The final cost of an electronic system is particularly dependent on its energy consumption. In fact, a high consumption leads to an increase in the thermal dissipation of the circuit, which causes an augmentation of the cost of cooling and packaging. This increase accelerates the mechanism of circuit aging that produces the rupture of the junctions, the breakdown of the oxide of the grid or even the deterioration of the encapsulation box.

The electrical behavior of the microelectronic circuits can also be altered by their energy consumption. An important current carried by the supply network can be responsible for potential drop in the supply lines. In certain parts of the electronic circuit, the supply voltage may be varied and during the design phases the expected nominal voltages will no longer be reached. In addition, desired current in a portion of the electronic circuit will cause different noises in the power supply network. In order to limit their effects, the use of capacity may be indispensable. This solution causes an increase of the circuit area. Besides, the sudden variations of the current demand will involve changes in voltage which may influence the operation of other part of the circuit. This phenomenon is known as "feed bounce".

These electrical and physical phenomena mentioned above are amplified with new submicron technologies. This amplification results from various factors. As geometry models become smaller, the amount of integrated transistors increases, followed by an increase in operating frequency. From here, it come the importance of reducing power consumption.

In this context many techniques have been proposed. These techniques present at different abstraction levels that are going to be described in the following section all along with an overview of the existing works and proposed methods in the literature at each level. The problematic will be discussed after, followed by the proposed new design. That design has been tested at both levels: simulation and physical one. Finally, the chapter will end up with a conclusion that briefs the present findings and the future research in this theme.

BACKGROUND: SYSTEM ON CHIP POWER CONSUMPTION

While designing a system on chip, the power consumption is considered as one of the most critical and important constraint. The increase of the integration density coupled with the increase of the frequency is explained by a large augmentation in power dissipation. This increase is due to several factors such as the improvement of technologies used, architectural evolutions and improvement of design tools.

The increase of integration density has provided the possibility of architectural innovations due to the large number of integrated transistors.

The reduction of transistors dimensions has offered the possibility of increasing the density of implemented logic. Consequently, the complexity and number of functionalities as well as the frequency of operation of the circuits may increase.

Table 1 presents the roadmap provided by Semiconductor Industry Association SIA.

Table 1 shows that although the size and voltage supply (V_{DD}) of the circuits continue to decrease, the execution frequency and power dissipation increase.

The power consumed in CMOS circuits is related to two main factors (Chandrakasan et al., 1992; Gaillardon et al., 2015):

Table 1. International roadmap for semi-conductors technology

Year	2001	2004	2007	2010	2013	2016
Size (nm)	130	90	65	45	32	22
V_{DD} (V)	1.1-1.2	1-1.2	0.7-1.1	0.6-1	0.5-0.9	0.4-0.9
Frequency (MHz)	1684	3990	6739	11511	19348	28751
Maximum Power (W)	130	160	190	218	251	288

- The dynamic power which depends on the system activity (Bellaouar et al., 2012). It varies according to the commutations observed in the logic elements of the circuit.
- The static power that is consumed when the circuit is inoperative (Yeap, 2012). This power depends on the leakage currents circulating in the circuit when it is on the standby mode.

Equation (1) presents the expression of total power:

$$P = P_{dynamic} + P_{static} \quad (1)$$

The dynamic power is divided into two parts:

- The power dissipated due to the switching activity during charging and discharging of the load capacities.
- The short-circuit power caused by a short-circuit current that occurs during changes in the states of the logic gates. This current depends on the rise time and the dimensions of the transistor. For correctly designed circuits, this current represents 15% of the dynamic consumption (Kao et al., 2002).

In older CMOS technologies (Chandrakasan et al.,1992), dynamic power has represented the dominant part of the total power of a CMOS gate. With the miniaturization of transistors and technological progress, an increase in the static part is observed that reached 50% of the total consumed power (Kao et al., 2002).

Dynamic Power

Dynamic Switching Power

As mentioned above, the dynamic power during the switching of circuit activities constitutes the primary source of the total dynamic power consumption. Equation (2) presents the expression of this power:

$$P_{dynamic} = A \times C \times V_{DD}^{2} \times F \quad (2)$$

where A is the switching activity that depends on the probability of an output transition, F refers to the execution frequency, V_{DD} is the supply voltage and C is the total capacity load. According to equation

(2), it can be seen that the power dissipated is proportional to the execution frequency and the activity A. After applying a transition, the output of a gate may have several transitions before reaching a steady state. From these transitions, only one is the essential transition and all others are unnecessary that are called "glitches". Therefore, the dynamic power is composed of two parts: a logical switching power which is provided by the signal transitions and the power dissipated by glitches.

Short-Circuit Power

While the mutation of logical state of a logic gate input, the PMOS and NMOS transistors can be simultaneously conductive. This transition phase is responsible for the occurrence of a short-circuit current between the ground and the power supply. This phenomenon is implied when the variation of the input is slow compared to the variations of the output. The PMOS transistor slightly conducts the current when the two transistors are on. However, a short-term leakage current is observed between the supply and the ground. This current mainly depends on the input transition. The faster it is, the longer the transistors are on.

Static Power

The static power is due to leakage currents from reversed polarized PN junctions and conduction currents under thresholds. It is proportional to the surface and to the temperature. The leakage component depends on the threshold voltages of the device. For systems that have a high ratio between standby and active operation, static power may be the main factor affecting the overall battery life.

Theoretically, a MOS transistor should not dissipate power in a stable state. However, leakage currents occur even when the transistor is in its blocked state. Generally, these currents are neglected during the design phase. Unfortunately, such negligence can be dangerous in some cases, such as with equipment that operates with low supply voltages or equipment that remains in standby mode for a long time. In these cases, leakage currents must be taken into consideration to ensure correct functionality and good performance. The leakage current varies according to the physical dimensions of the channel, the doping, the depth of the drain / source junctions, the supply voltage V_{DD}, the threshold voltage and the temperature.

Estimated Power Consumption of ASIC vs. an FPGA

At this stage, it is imperative to note that the power consumption differs from one technology to another. For example, the power consumption of a digital circuit designed with an ASIC will be different from the consumption of the same circuit implemented with an FPGA. Similarly, the efficiency of the power reduction techniques consumed varies from one technological target to another (Olivier et al., 2005; Kuon et al.,2007).

The Power Consumed by ASICs

Concerning the ASIC technologies, Synopsys has proposed a power rating tool called PrimePower. This tool is incorporated into the logical synthesis chain "Design Compiler" which is devoted to ASIC technology targets. The power estimation flow of the ASIC technology is shown in Figure 1.

Figure 1. Power estimation flow of ASICs

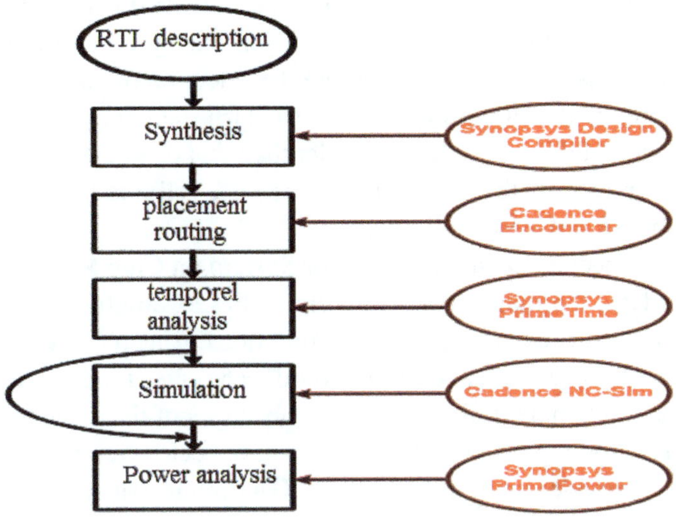

Generally, dynamic power estimation is performed by a Value Change Dump file with. VCD extension. This file, which is generated at the functional simulation stage of the circuit, encompasses the activity of the various internal commutations with respect to the stimuli employed.

After the power estimation phase, the results obtained can be reused by the synthesis tools to improve the circuit implantation and reduce the power consumption. In order to achieve the circuit improvement, the synthesis tool studies the links with products that have greater activity and the connections between the functional instances, more particularly the target instances.

Despite the remarkable and important development of the CAD tools; particularly the power estimation tools; the accuracy of estimation values proposed by these tools remain rather low. It is clear that the use of these tools is done precisely to estimate the amount of power consumed rather than to have an accurate evaluation of the power dissipated (Nathalie, 2006).

The Power Consumed by FPGAs

The disadvantage of an FPGA is that it consumes more power than an ASIC. In fact, FPGA circuits lose a significant amount of power on the standby mode. This type of circuit therefore has large static power dissipation compared to the ASIC circuits. As generally reckoned, there are two types of FPGA circuits: volatile FPGAs and nonvolatile FPGAs. Like the ASIC circuits, the volatile FPGAs have a dynamic component and a static one. For this kind of circuit generally based on SRAP, the designers must necessarily take into account two other phenomena which are the consumption of the circuit functions and the consumption during the loading of the system configuration. This additional consumption is important before stabilizing the supply voltage. In addition, the volatile FPGAs must be initiated by an existing program in an external circuit, usually a PROM, which causes the increase of power consumption as well as the start-up time of the system.

These consumptions that are linked to system startup can be important if a common power supply is used for several boards or if several of these types of FPGAs are integrated on a single board. Therefore, designers must take into consideration power consumption during configuration and startup time to predict the amount of the necessary power or facilitate the selection of appropriate batteries.

On the other side, there are the FPGAs based on real flash memory. This kind of FPGAs is called non-volatile. The integration of this technology is easier than that of SRAM-based FPGAs and its power consumption is much lower. Despite these advantages, their use is limited due to the low integration that they have compared to the SRAM-based components.

Power consumption has become a vital concern for System on Chip SoC designers, especially SoC based on FPGAs. Nowadays, the major manufacturers of FPGA circuits have proposed CAD design methodologies that take into consideration power consumption during the design and manufacture phase of circuits. These tools evolve over time and offer several techniques for estimating power dissipated at different levels of abstraction to help designers during the stages of the design cycle.

It has been noticed that various power estimation tools have been developed. These tools are generally offered by companies that provide synthesis tools for FPGA circuits (such as Xilinx and Altera). In fact, the estimation processes are nevertheless very similar from one manufacturer to another.

As well, Xilinx has proposed XPower tool (the one used in this chapter book), which is integrated into the ISE Funding and dedicated to FPGA targets designed by Xilinx. Similarly, there is the Power Play produced by Altera which targets the FPGAs designed by that company. This software is integrated in the Quartus logical synthesis tool.

LITERATURE REVIEW

The exponential rise of complexity of embedded systems, the significant increase in integration capacity, the remarkable space of research space and the reduction of design time have been largely noticed nowadays. These findings imply a vital necessity of using a design methodology. Such a methodology makes it possible to explore the architectural space with a rewarding way, to limit design errors, to facilitate repetitive and laborious tasks by using CAD tools, to offer independence of the technology as long as possible and to take into account material constraints as soon as possible. For these reasons, the use of a design methodology has become a vital necessity for designers of digital circuits. Since then, several techniques have been proposed in order to have a less consuming and more reliable system. These techniques are located on several design flow levels and impact the final consumption of the circuit with a different ways. There are several levels of abstraction where the designer can propose new techniques to optimize the total consumption of the system. Based on the "bottom up" approach, there are: technological, circuit, architectural, algorithmic and system levels. Each level is specialized by many known methodologies. For instance, the libraries specification is one of the most used techniques at the technological level. The routing is a well used method at the circuit level. The architectural level contains many optimization techniques such as pipeline, parallelism, clock gating and scheduling algorithm. Among the power optimization methods at the algorithmic level, there exist the algorithm complexity, regularity and allocation...etc. Concerning the system level, the hardware / software partitioning presents the major used technique.

Rabaey et al. (2012) as well as (cadence document) have demonstrated that different power optimization techniques can be considered at different levels of abstraction. At the system level, a good partitioning can offer a significant reduction in the power dissipated (Havinga & Smit, 2012; Unsal & Koren,2003).

At the algorithmic level, the optimization of algorithms complexity (Unsal & Koren,2003) reduces consequently the power dissipated. Similarly, the use of optimized arithmetic resources positively affects the total power consumed. Among the techniques commonly used are Booth or Booth-modified algorithms (Prakash & Sanju, 2014; Seo & Kim, 2010; Chong et al.,2007; Chang et al.,2005; Hwang et al.,1999), redundant addition algorithms (Huang et al.,2005; Dumonteix & Mehrez,2000; Phatak et al. 1998; Banerjee et al.,2005), Wallace's reduction methods (Li et al.,2003) and boolean optimizations.

At the architectural level, several approaches are available. Among the most widely used methods, there are the famous clock gating technique (Bhattacharjee et al., 2017; Bhattacharjee & Majumder, 2016; Shashidar et al., 2017; Gustafsson et al.2005), the parallelism and pipeline techniques (Havinga & Smit, 2000; Unsal & Koren, 2003), the voltage scaling (Lukefahr et al.,2014; Bortolotti et al., 2014; Unsal & Koren, 2003 ; Butzen et al.,2007), the encoding states of a finite state machine (Chandrakasan et al.1994), frequency scaling, and scheduling algorithms.

At the circuit level, the isolation of unused combinatorial paths is a frequently used method. Also, the elimination of false transitions (glitches) that occurs through the balancing of data paths also reduces the dynamic power consumption (Qu & Potkonjak, 2002). The placement and routing phase as well, plays an important role in the determination of system consumption (Phatak et al.,1998; Banerjee et al.,2005; Verma & Ienne,2007). In addition, the optimization of associated libraries and routing algorithms offers the reduction of the power dissipated by a digital circuit (Banerjee et al., 2005).

Regarding the technological level, which is the lowest level in the design flow, the possible optimizations generally concerns the scalability of the technology and the adjustment of the voltage thresholds (Homayoun et al.2008; Kim et al., 2007).

According to the SoC driver's power roadmap presented by the latest ITRS system drivers (Smith, 2014), the power exceeds a budget of 5W in 2016 and will become about 10W in 2028. Therefore, engineers must absolutely apply the technological innovations in design to limit the power to 5W. Each innovation affects static and dynamic power with different ways. It is assumed that three of these innovations, adaptive body polarization, power tightening, and dynamic voltage / frequency scaling will affect the power of a SoC during the five, seven and ten years respectively. Innovations in devices also reduce power and are classified in the ITIS chapter (ITRS, 2013). The improvements brought by the innovations at the device level are integrated with the power model of a SoC. Typically; a combination of these innovations will be required to limit power in 5W.

Power Optimization at the Technological Level

At the technological level, capacities can be reduced by finer geometries, smaller junction capacities and optimized interconnections. From one technological generation to another one, significant gains can be made. The reduction of the technologies geometrical parameters realizes a gain of scale: the increase in integration densities allows the realization of SoC by integrating circuits previously assembled in discrete components. The total number of chips in the final product is reduced. Hence, its energy consumption is reduced as well.

At the layout level, placement and routing are adjusted to reduce interconnection wire capacity and to balance wire delays. At this level, several techniques target the reduction of static power and consequently the leakage current (Kim & Kim, 2007). Among the techniques described in the literature, there are methods based on unconventional polarizations of the source and the substrate, which leads to a modification of the current below the threshold. In this context, they quote the inverse polarization of the voltage VGS between the gate and the source of a transistor in idle mode (Itoh et al., 1999).

Power Optimization at the Circuit Level

Reducing the activity of clocks makes it possible to considerably reduce the total power of the circuit. In fact, the synchronization of digital systems requires one or more signals to coordinate the operations. Most often, a clock tree is constructed to provide the clock signal to the various modules of the system. The power dissipated in a CMOS circuit is always dependent on the clock that manages its activity. The construction of a clock tree is therefore a delicate point since it results from a compromise between different objectives. It's largely known that the clock of an element keeps consuming power even when its corresponding element is in an idle state.

Using the clock-gating technique (Choi et al.,2005), these modules can be disconnected via specific cells in order to obtain dynamic power optimization. This method is very effective and widespread in most low-power systems. Banerjee et al. (2005) have proved that a 70% reduction in dynamic power has been achieved in an MPEG4 video decoder by massively using this technique (90% of the flip-flops are isolated by a Clock-Gating block

Similarly, it is possible that an element of the data stream can only be accessed periodically. This access depends on an "enable" signal. Thus, if this signal is inactive, the input of the combinatorial path can be forced to a constant value in order to reduce its dynamic power (Anagnostopoulos & Economakos,2005). The isolation technique was used (Verma & Ienne, 2006) to reduce the activity of a multiplier (array-multiplier) and showed reduction rates ranging between 20% and 80% in terms of power consumption.

The elimination of glitches or parasitic transitions makes it possible to effectively reduce the dynamic power and in particular the activity of the system. The elimination of the glitch is done by balancing the combinatorial data paths. Glitches reduction is achieved (Qu & Potkonjak, 2002) by replacing conventional logic gates with specific logic gates to control glitch propagation.

At the circuit level, a well known technique has been proposed which is the MSV (or Multiple Scaling Voltage) method. Its principle is to operate different parts of the design with different supply voltages. Designers use many voltages for different parts of the circuit depending on the performance requirements. The MSV technology allows a significant reduction in power consumption since the supply voltage has a quadratic effect on the dynamic consumption of the circuit. In processors, this technique is massively used with the system-level partitioning technique (Kapoor, 2015; Bezerra et al., 2015).

The Multi-Vth technique consists on using logic gates with different threshold voltage values to optimize the power and the occupied surface. Most library constructors provide libraries that have elements operating at different threshold voltages. The synthesis tools enable the generation of low power consumption circuits and are capable of mixing the elements with different threshold voltage to reach a surface or an execution frequency with minimal power dissipation.

Dynamic Voltage and Frequency Scaling (DVFS) (Dosluoglu & Gelman, 2016; Choi et al.,2004; Nathanael et al.,2010; Arroba et al., 2017) consists on applying dynamically the techniques of Voltage Scaling and Frequency Scaling. The DVFS technique consists on finding solutions to reduce the supply voltage and the execution frequency based on a specified performance of the circuit. This technique offers considerable gains in terms of power, since it reduces both: the execution frequency and the supply voltage, which are two dominant parameters.

Power Optimization at the Architectural Level

At the architectural level, the optimization of the power also requires the reduction of the effective load. This can be achieved by increasing the locality of the data (Buyuktosunoglu et al., 2001). Architectural choices affect the power consumption of the system. In fact, a carry propagation adder can be replaced by a faster structure such as a carry forward adder. The latter has a larger surface area which results an increase in physical capacity. However, this is inhibited by the fact that the fastest adder can work at a lower supply voltage for equivalent performance. Several other methods have been proposed. Some research groups have opted for the Dynamic Voltage Scaling method DVS (Fan et al.2005; Shin et al.2001; Hsu & Feng, 2004), while others have focused on the clock gating technique (siwar et al., 2014).

It has been proven that the clock is responsible for 70% of the consumed dynamic power. For that, designers and researchers have gathered their efforts to decrease such percentage. The dilemma is that the clock keeps functioning all the time even when its corresponding block is in an idle state. One may wonder why the clock keeps functioning in such circumstances. Undeniably, it will be more convenient if the clock will be enabled from functioning when its corresponding element is on the standby mode. For that, some designers have proposed to use an AND logic gate, an OR logic gate, a NOR logic gate or a flip flop in order to enable the clock from functioning. However, authors have presented the inconveniences of these methods and proven its inefficiency since they affect the global system performance and functionality (Sulaiman, 2008).

Power Optimization at the Algorithmic Level

At the algorithmic level, power optimizations involve transforming the algorithms to reduce their complexity. Thus, certain transformations on the format of the data make it possible reducing the number of operations to be carried out and, by the same, to minimize the power consumption. The reduction in operations complexity consists on a replacement of the basic operations with more adequate ones. A typical example of this type of optimization is to replace a multiplication operation by a sequence of additions and shifts (Fujino & Moshnyaga, 2003).

Several other transformations can be used to minimize power consumption. Thus, it has been proved (Zheng & Albicki, 1995) that an appropriate decomposition of a CIC filter associated with a transformation of the first stage reduces significantly its power consumption. These gains can reach 80% of the total filter consumption. The algebraic transformations associated with the data format named Signed Magnitude have made it possible (Aboushady et al.,2000) to reduce the power dissipation of the Multiply and Accumulate (MAC) by influencing the activity of its modules. Several structures for efficiently implementing multiplication operations are proposed and compared in terms of power consumption (Verma & Ienne, 2006; Fujino & Moshnyaga,2003; Huang & Ercegovac,2005; Dumonteix & Mehrez, 2000; Fonseca et al., 2005; Chong et al., 2007; Chang et al.,2005; Hwang et al.,1999). Thus, it has been proved (Fan et al.,2005; Shin et al.,2001; Hsu et al.,2003; Hsu & Feng, 2004; siwar et al.,2014; Sulaiman, 2008; Zhen & Albicki, 1995; Aboushady et al., 2000) that a tree multiplication structure is more efficient in terms of power consumption than the equivalent "array" structure. Aboushady et al. (2000) have shown that a dynamic transformation of inputs of an accumulation multiplication unit can allow a power gain of up to 50% compared to a conventional architecture.

Among the algorithmic transformations used in the literature, to reduce the power consumption, the choice of the format of data representation is given (Huang & Ercegovac, 2005; Fonseca et al.,2005). Indeed, the representation of the data can have a notable effect on the number of operations required for the implementation of an arithmetic operation. The number 7, for example, can be represented by (2^3-2^0) or ($2^2 + 2^1 + 2^0$). The most suitable representation remains the one that reduces the number of elementary operations required for the implementation of a given function. The representation of numbers greatly affects the design of a multiplier as well as its power consumption. The use of the complement by 2 representation which is well suited to signed numbers is commonly used to encode inputs/outputs of a multiplication operation. Inside the multiplier other data formats may be used to reduce the complexity of the step of generating partial products. Thus, it has been proved (Lin et al., 2008) that Signed Digit format consumes more silicon area and power than the complement-by-2 approach.

A redundant representation of some operands can also effectively reduce the power dissipated (Dumonteix & Mehrez, 2000). It has been proved in this work that redundant operators can offer better performance compared to classical operators while minimizing power consumption. It has been proved (Fonseca et al., 2005; Chong et al., 2007) that hybrid operators can reduce power consumption compared to redundant and conventional operators.

Some work particularly target the multiplication operation. The majority of works aim to optimize the step of generating the partial products in order to speed up the phase of adding it. Among the algorithms used, there exist the algorithms of Booth and Booth-modified. Several techniques for optimizing the generation of partial products exist in literature (Zheng & Albicki, 1995; Aboushady et al., 2000; Huang & Ercegovac,2005; Dumonteix & Mehrez, 2000; Fonseca et al.,2005; Chong et al.,2007; Chang et al.,2005). Thus, the application of the modified Booth algorithm, radix-2, radix-4 and mixed-radix was evaluated (Dumonteix & Mehrez, 2000).

Digital filters have required special attention since they are included in most digital communication channels. Thus, several transformations of these filters are proposed in the literature to reduce their power consumption. Most of works aim to transform the filter architecture to eliminate multiplication operations which are very costly in terms of area and power consumption. Thus, Stan et al. (1995) have proposed that the use of specific MAC units in order to reduce the consumption of multiplication operations. The approaches propose architectures based on distributed arithmetic available in FPGA-type circuits in order to eliminate the multiplication functions. Other authors (Wolf, 2003) propose to act from the step of quantifying the coefficients in order to eliminate the multiplication operators. The canonical signed digit representation is evaluated all along the common elements sharing method to replace the multiplication operations by an addition and shift sequence.

The reduction of system activity at the same level can be achieved using the pipeline and parallelism technique. In fact, in order to reduce parasitic transitions that cause unnecessary activities, pipeline stages can be added to stabilize signals at the input of certain logic gates. It has also been demonstrated (Hsu et al. 2003) that parallelism can be very effective to reduce the dissipated power for unconstrained systems in terms of area.

Power Optimization at the System Level

It has been proved in several works presented in the literature that a good partitioning can induce a significant reduction in power consumption for a given system. Thus, good partitioning of tasks optimizes the power consumed. It has also been proved that partitioning a processor allows the processors to be disconnected when they are unused (using the power gating method). Moreover, the impact of state encoding has been shown (Buyuktosunoglu et al., 2001) when synthesizing a state machine. In this case, effective state partitioning coupled with efficient encoding of the state-associated value significantly reduces power consumption. System partitioning is also recommended. It has been proved that this method uses the DVS (Dynamic Voltage Scaling) method especially for processors and more generally for real-time systems.

The effectiveness of Hardware/Software partitioning is not limited to processing systems. Good memory allocation also generates significant gains in power consumption. Indeed, the use of the cache memories as well as the registers makes it possible to reduce the power consumption of a given circuit by reducing the access to the external storage resources. This increases the locality of the data, reduces the effective load of the system and consequently reduces the power consumption.

The persistent quest for high level of performance has contributed to the need of creating new less consuming systems and faster one. For that, researchers have gathered their efforts and come up with several approaches at different levels of abstraction. Many techniques have been proposed such as the most elegant and used one: the "clock gating" technique. Based on the theory that the clock is the major consumer of power in synchronous systems, many works have opted for enabling it but none of them have proposed to replace it. Motivated by the fact, the following section presents a new method that replaces the standard known clock by a new clock generator which aims not only to reduce the global consumed power but also to offer a faster treatment.

PROBLEM DEFINITION

As digital systems have become the center of present and future technologies, the search for higher levels of performance have been persisting in the last several years. In fact, the rising quest for higher performance and efficient system has contributed to the creation of more effective types of digital circuits. As generally reckoned, there are two types of digital systems which are the combinatory system and the sequential one. The combinatory system is an unorganized circuit made up of a plurality of similar elements; the macro behavior of the system, as a unit, derives from the combination of the analogous behaviors of its similar elements, according to a feedback relation between micro and macro behaviors. The sequential systems use current input variables and previous input variables in order to store the information and put it back into the circuit on the next clock cycle. There are two types of sequential circuits, synchronous and asynchronous. The synchronous type uses a clock input and various pulsed inputs to drive the circuit. However, the asynchronous only uses the pulsed inputs to drive the circuit. When integrated in a big and sophisticated circuit, the synchronous system; which is the most commonly used, turns into a sequential block such as in image and video processing. As mentioned previously, with the synchronous system there is a need for two important conditions to get a response which are the input condition and the rising edge of the clock. In fact, if the condition is valid the system waits for the rising edge to give the command. This wasted time is called the waiting time $\mathbf{T}_{wait}$.

Time is a very important constraint in the process of digital system creation. A system is considered efficient when it manages to bring its response time to its lowest possible rate. However, a decrease in the response time can only be attained by decreasing the period of the clock which unfortunately requires a corresponding augmentation in frequency value which leads to an augmentation in consumption of power. So, designers have to choose between saving power and having a faster circuit. For that reason, we came up with a new approach that optimizes both parameters: time constraint and power. This chapter book has presented a new design that provides a faster circuit with a less consumption of power. The new design has been applied and tested on particular synchronous systems, which is a finite state machine-based controller.

The Response Time of Synchronous Systems

In digital system, the response time (**T_{res}**) is defined as the time elapsed from the moment the condition is valid until the system responds. Hence the response time can be written as follows:

$$T_{res} = (T_{output} - T_{input}) \tag{3}$$

In addition, the response time varies in an interval of time that its maximum can be obtained as follows:

$$\max (T_{res}) = \max (T_{res} = (T_{output} - T_{input})) \tag{4}$$

The synchronous system responds only if its input is valid and the rising edge (or falling edge) of its clock is active as shows Figure 3. Hence:

$$T_{output} = T_{rising\_edge} \tag{5}$$

where T_{rising_edge} is the time when the rising-edge takes place. The latter refers to the first rising-edge that takes place as soon as the condition is valid.

Based on equation (5), the response time can be written as follows:

$$T_{res} = T_{rising\_edge} - T_{input} \tag{6}$$

It is obvious that the rising edge presents after one period T, so the maximum time to obtain it is T. From here we deduce that in general:

$$\max (T_{res}) = \max (T_{res} = (T_{output} - T_{input})) = T \tag{7}$$

In reality, some designers use HDL description (like VHDL languages) to conceive the synchronous systems. Knowing that, the semantic of HDL language requires a delay of one clock period, thus:

$$\max(T_{res}) = \max(T_{res}) + T = T + T = 2T \tag{8}$$

To better clarify the above equations (5, 6, 7, and 8); let's consider the synchronous finite state machine shown in Figure 2.

Figure 2. FSM example

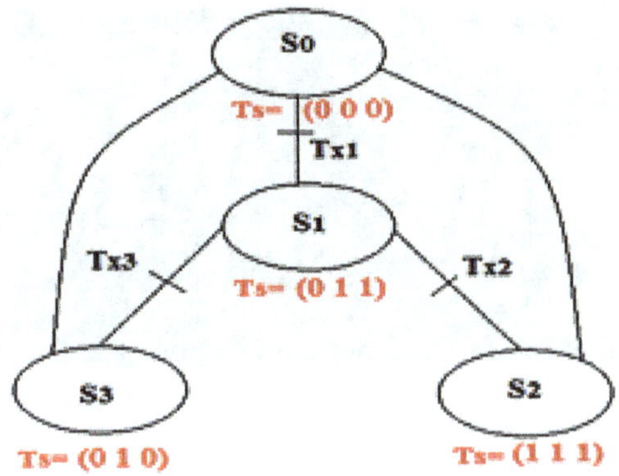

Designs Without HDL

Using this kind of designs, even if the input condition is true the system has to wait till the next rising edge (or falling edge) of the clock in order to provide the desired response.

Following to Figure 3, in the case when the condition tx1 is true at t=31ns, the desired response is obtained at t=35ns.

As a result:

$$T_{res} = T_{rising\_edge} - T_{input} = 35-31 = 4ns$$

Since the arrival time of tx1 is arbitrary, thus

$$\max(T_{res}) = \max\left(T_{res} = (T_{output} - T_{input})\right) = 5ns = T$$

Designs With HDL

Using the HDL design, it is obvious that the system has to wait an additional time equal to one period of the clock. It has been proved that the HDL caused a delay of one period. The simulation result shown in Figure 4 confirms that theory. In that case, the period of the clock is 4 ns. At t =4ns the condition Tx1 is

Figure 3. Illustrative simulation

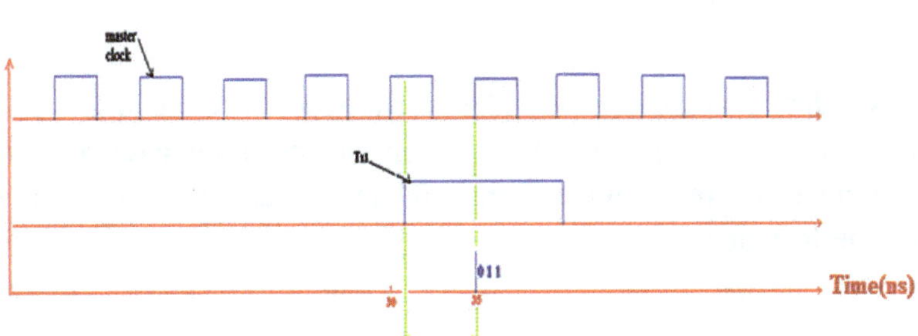

Figure 4. Simulation result

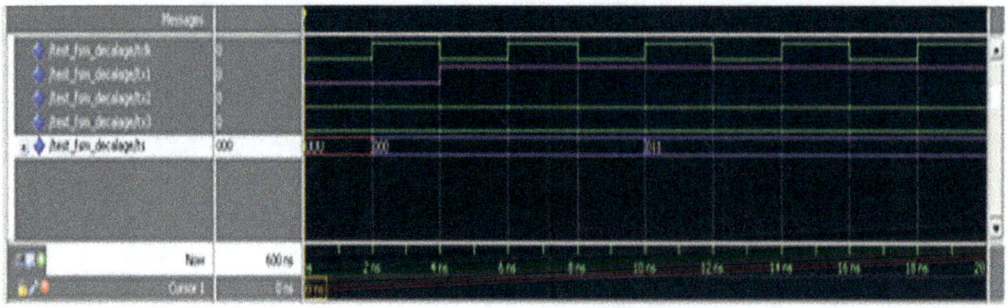

true; normally the desired response would be obtained when the first triggering edge occurs which is at t=6ns. However, the desired response provided by the FSM coded with VHDL is presented at t= 10ns which is the time when the second triggering edge occurs. As conclusion we have:

$T_{res} = T_{rising\_edge} - T_{input} = 10\text{-}4 = 6\text{ns}$

$\max(T_{res}) = \max(T_{res} = (T_{output} - T_{input})) = 2T = 8\text{ ns}$

As generally reckoned, the dynamic power $P_{dynamic}$ depends on the logic used by the design and the frequency and it can be written as follows:

$$P_{dynamic} = A \times C \times V^2 \times F \quad (2)$$

where A refers to the percentage of active logic gates; which are charged dynamically; C equals the total capacitance load, V represents the supply voltage and F is the execution frequency, as previously explained.

As

$$F = \frac{1}{T} = \frac{1}{\max\left(\text{Tres}\right)} \quad (8)$$

The equation of the dynamic power equals:

$$P_{dynamic} = A \times C \times V^2 \times \frac{1}{\max\left(\text{Tres}\right)} \quad (9)$$

The equation 9 proves that parameters, time and power, are proportional. In fact, a decrease in response time will lead to an augmentation in power consumption. Furthermore, a decrease in power consumption results in a simultaneous decrease in frequency and an increase in response time which ultimately leads to inaccuracy of performance.

PROPOSED APPROACH

Typically, the synthesized design consists of two parts: a control part (controller) and an operative part. The control part is intended to process the data coming from environment to control the operative part. However, the operative part of a system is the subset that receives orders from the control part and executes them. Often, the operative part performs the physical actions (movement, light emission ...), measures physical quantities (temperature, humidity, light ...) and provides reports to the control part.

In this chapter, authors focus on the control part which is formally described by finite-state machine FSM that generates control signals to activate the different operations in specific clock cycles. This latter is a model of computation that is one of the most widely used to design both sequential logic circuits and computer programs. It is a model of behavior using states and transitions. A transition is a change from current state to the next one when initiated by a triggering event or condition. The set of states should be finite. Subsequently, the set of transitions is finite as well. Therefore, FSM is defined by a list of finite states and the triggering condition for each transition. There are two basic ways to design clocked sequential circuits: the Mealy Machine and the Moore Machine. The difference between them is that in a Mealy machine the outputs are determined by both its current state and the values of its inputs. However, in a Moore machine the outputs depend solely on its current state.

The proposed method in this chapter has kept the traditional blocks (sequential block and combinatory block) of Finite state machine. However, a new block called "clock generator" has been added (Figure 5). Thus, the clock becomes auto generated by the FSM which leads to minimize both response time and power, as explained later, while guaranteeing the correct functionality of the system.

As mentioned previously, the new design is based on adding a new block to the classical FSM design, this block is called clock generator which contains a multiplexer.

A multiplexer is a device that selects one digital input and forwards the selected input to the output. Generally, a multiplexer has "n" address inputs, 2^n data inputs and one output. According to our approach n is equal to the number of bits used to encode states.

The address inputs are mapped to the outputs of the sequential block. The data inputs are mapped to the input conditions. For example, a data input (i) of the multiplexer is mapped to the condition related to the binary code (i). The output "mux_out" of the multiplexer is mapped to the clock of the sequential block.

Figure 5. Our top level design

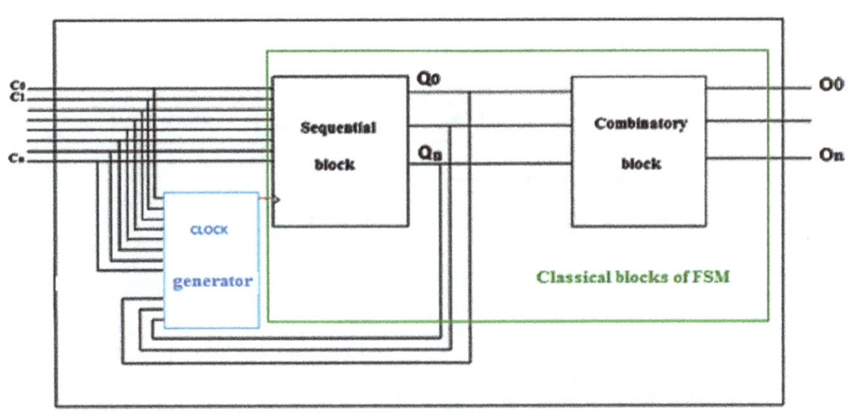

Illustrative Example

The proposed approach can be applied to any FSM regardless of its complexity and sophistication. For a simple and practical illustration of the approach's concept, it has been applied on the FSM in Figure 6.

As previously explained, this new block contains a multiplexer. The number of bits used to encode states equals two. Hence n=2, the number of the address inputs is 2; the number of data inputs equals $2^2=4$ and one output called "mux_out".

The address inputs are mapped to the outputs of the sequential block; knowing that the latter produces always the binary code of the current state. The input data (i) of the multiplexer is mapped to the condition related to the binary code (i). For example, whenever the C0 is true, a transition from state coded by "00" to state coded by "01" will automatically take place. Therefore, the data input (0) is mapped to C0. Besides, the transition from state coded by "01" to state coded by "10" is produced if (C1) is true and therefore the data input (1) is mapped to (C1).The same mechanism can be applied to the remaining cases, as a result data input (2) is mapped to (C2), data input (3) is mapped to (C0 OR C3) as shows Figure 7.

The Figure 8 presents the obtained simulation results.

Through an analysis of the Figure 8, it is obvious that the proposed approach provides accurate solution and correct functionality of the target circuit.

Figure 6. FSM

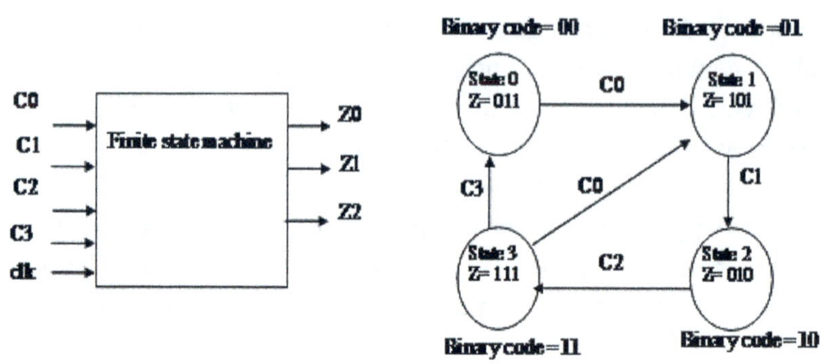

Figure 7. Our FSM architecture

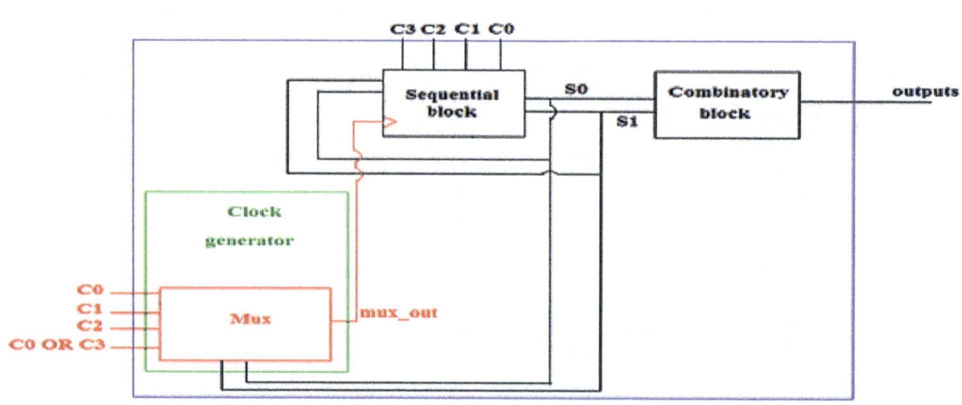

Figure 8. Simulation results

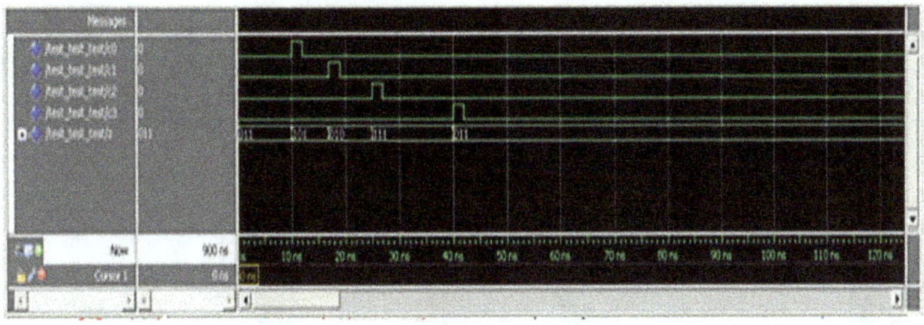

EXPERIMENT

In this part, a comparative study between the new proposed approach and the traditional one has been introduced concerning the following fields which are: power and response time. This study is meant to illustrate the amount of gain in power consumption using the new approach compared to the traditional one and to prove that authors' new approach provides a faster circuit. To put that into practice, different tools have been used consisting of a processor Intel core i2 running at 2,49 GHz, windows 7 as an operative system, Xilinx ISE (version 12.3) as synthesis tool, ModelSim (version 6.5) as simulation tool and nexys 2 based on Spartan 3E. An application has been built that aims to compare two images and indicate if they are similar or not. The target system contains four sequential blocks: the first one is responsible for registering the captured image. The second block is designed to calculate the histograms. The third block is meant to calculate the difference between the two histograms of two images (the histogram of the captured image and the image to be compared). The last block is presumed to make the comparison to the threshold value and take a decision if the two images are similar or not. To further clarify the functionality of the application, it has been described in the Figure 9.

To conceive the system, it has been decomposed into two parts which are the control part (controller) and the operative part as shows Figure 10. In fact, when TSt1 is true the treatment begins its execution and the output Ts is obtained when the treatment is over.

Figure 9. The description of our application

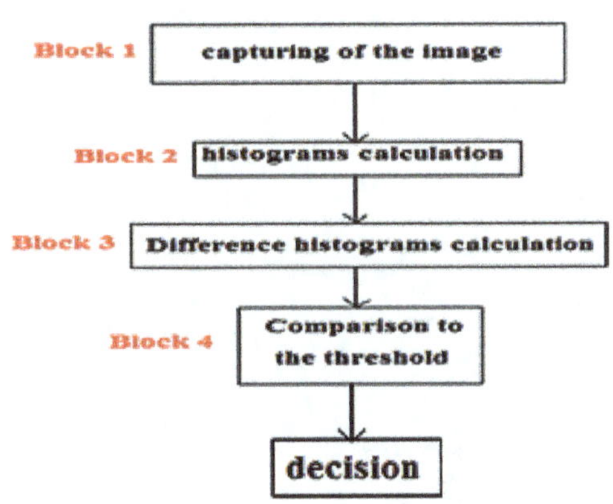

Figure 10. Proposed design

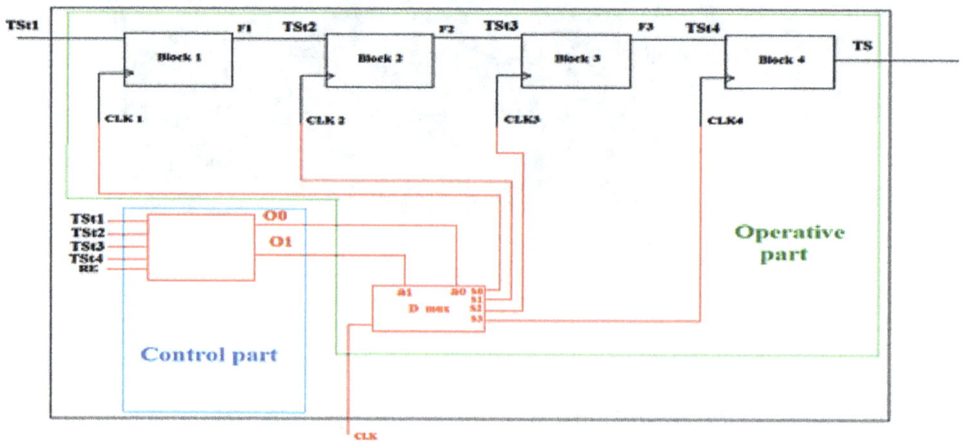

In this chapter book, authors have focused on the control part which is formally described by a finite-state machine. Since the blocks of the target application are sequential; the controller can be modeled by the Moore machine (Figure 11).

In the rest of the section, authors are going to focus on implementing the FSM of figure 11, using the new design and the classical one then introducing a comparative study between the two approaches in terms of power and response time.

The term "classical design" means the classical design of a Moore machine whose outputs depend only on the values of the current state.

Concerning the proposed design, it is clear that the binary code to encode the current state and the output of the FSM are similar. Thus, the combinatory block, shown figure 5, can be eliminated and the sequential block can be considered as a counter. Hence, the new design of the FSM is as shows Figure 12.

Simulation Result by Classical Approach

The calculation of the response time and the dynamic energy dissipated by the circuit when designing the FSM with classical approach will be the concern of the next phase.

Figure 11. Moore FSM

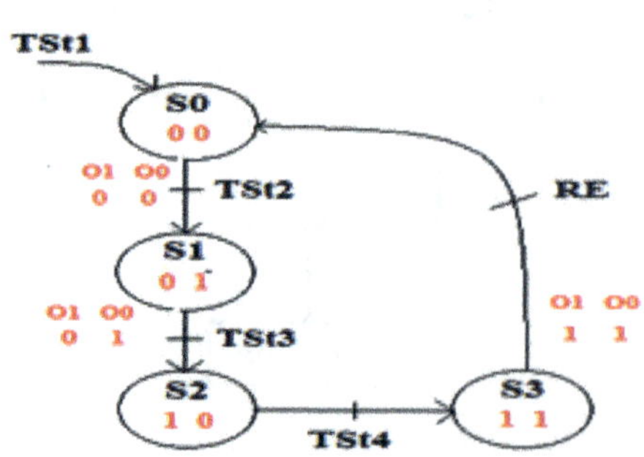

Figure 12. Our FSM design

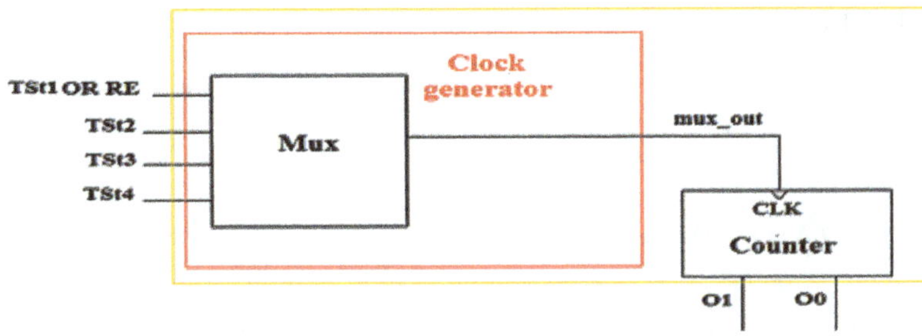

Figure 13. Simulation results provided by classical approach

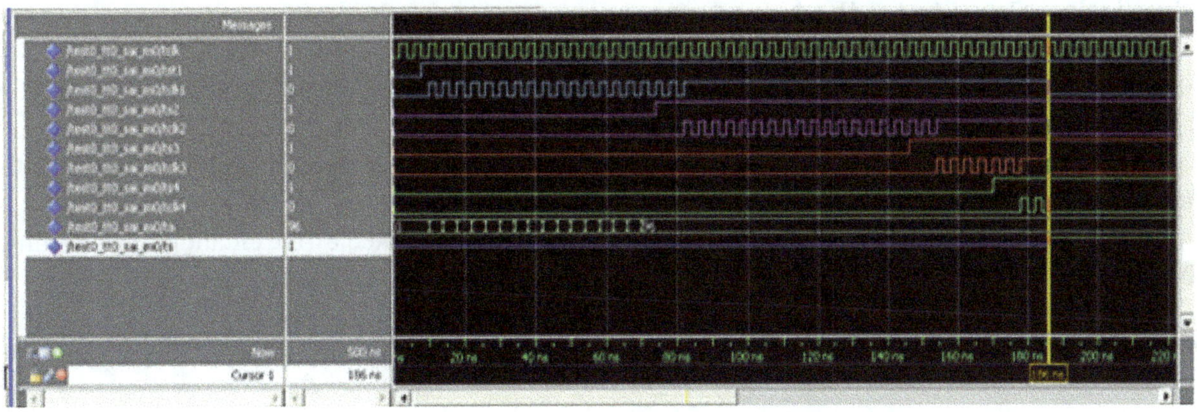

Response Time

As mentioned previously: $T_{res}= (T_{output}- T_{input})$. In that case, the T_{input} is the time when the first condition is valid. T_{output} is the time when the system provides the response.

According to figure 13, the system starts the treatment at TSt1=8 ns and provides the response at Ts= 186 ns. Therefore:

$$T_{resp}= 186\text{-}8=178 \text{ ns} \tag{10}$$

Dynamic Energy

Knowing that: Dynamic energy= Dynamic power × Time

$$E_d=P_{dynamic}\times T= T\,(P_{d1}+P_{d2}+P_{d3}+P_{d4}) \tag{11}$$

P_{d1}: dynamic power dissipated in the interval [10 ns, 81 ns]
P_{d2}: dynamic power dissipated in the interval [81 ns, 155 ns]

P_{d3}: dynamic power dissipated in the interval [155 ns, 179 ns]
P_{d4}: dynamic power dissipated in the interval [179 ns, 186 ns]

$$E_d = E_{d1} + E_{d2} + E_{d3} + E_{d4}$$

$$= (A\times C \times V^2 \times F) \times 71 + (A\times C \times V^2 \times F) \times 74$$

$$+ (A\times C \times V^2 \times F) \times 24 + (A\times C \times V^2 \times F) \times 7$$

$$E_d = 176\,(A\times C \times V^2 \times F) \quad (12)$$

Simulation Results by the New Approach

Figure 14 represents the simulation result using the new proposed approach. In fact, researchers can definitely deduce through an analysis of the above Figures (13 and 14) that the new approach as well as the traditional one provide the same simulation results. Thus, the proposed approach has kept the functionality of the target circuit.

By the same way, the response time and the dynamic energy dissipated by the circuit when designing the FSM with new approach will be calculated.

Response Time

T_{res} is the time elapsed between the time of the first triggered condition and the given time response. Following to Figure 14, authors deduce that the system starts the treatment at TSt1= 8 ns and produces a response at Ts= 162 ns. Therefore, the response time is

$$T_{res} = 162\text{-}8\text{=}154 \text{ ns} \quad (13)$$

Based on equations (10) and (13), researchers can conclude that the new approach gives an interesting gain in response time which is 13.48%.

Figure 14. Simulation results of the new approach

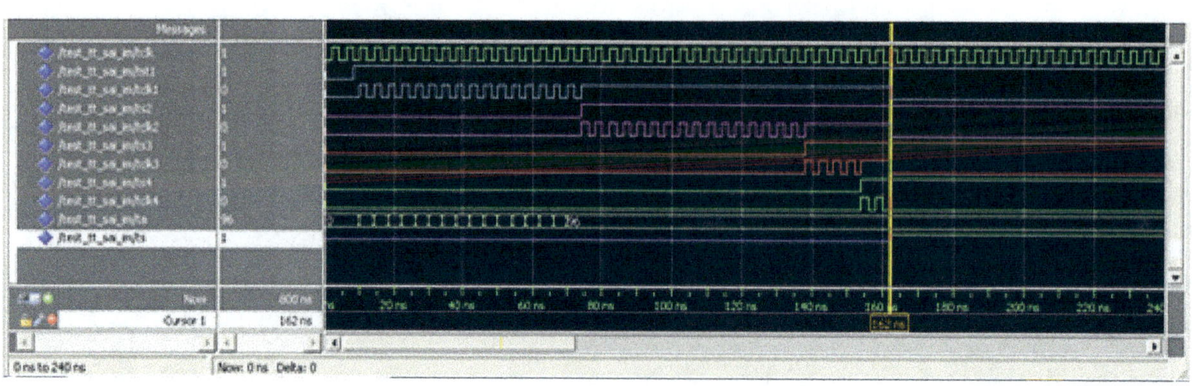

Dynamic Energy Dissipated by Our Approach Is

$E_d = E_{d1} + E_{d2} + E_{d3} + E_{d4}$

$= (A\times C \times V^2 \times F)\ 64 + (A\times C \times V^2 \times F)\ 65$

$+ (A\times C \times V^2 \times F)\ 16 + (A\times C \times V^2 \times F)\ 7$

$$E_d = 152\ (A\times C \times V^2 \times F) \tag{14}$$

Based on equations 12 and 14, authors deduce that the proposed approach manages to bring a decrease in power consumption that attains a level of 13.64% compared to the classical approach

Physical Results

To further clarify the proposed approach, Figure 15 has been proposed to physically calculate the response time which is provided by the counter referring to the time elapsed between the moment of the first valid condition and the time when the system responds.

Classical Approach

In this part, authors physically calculate the value of response time and the value of the dynamic energy dissipated by the circuit when designing the FSM with classical approach.

Figure 15. Proposed design

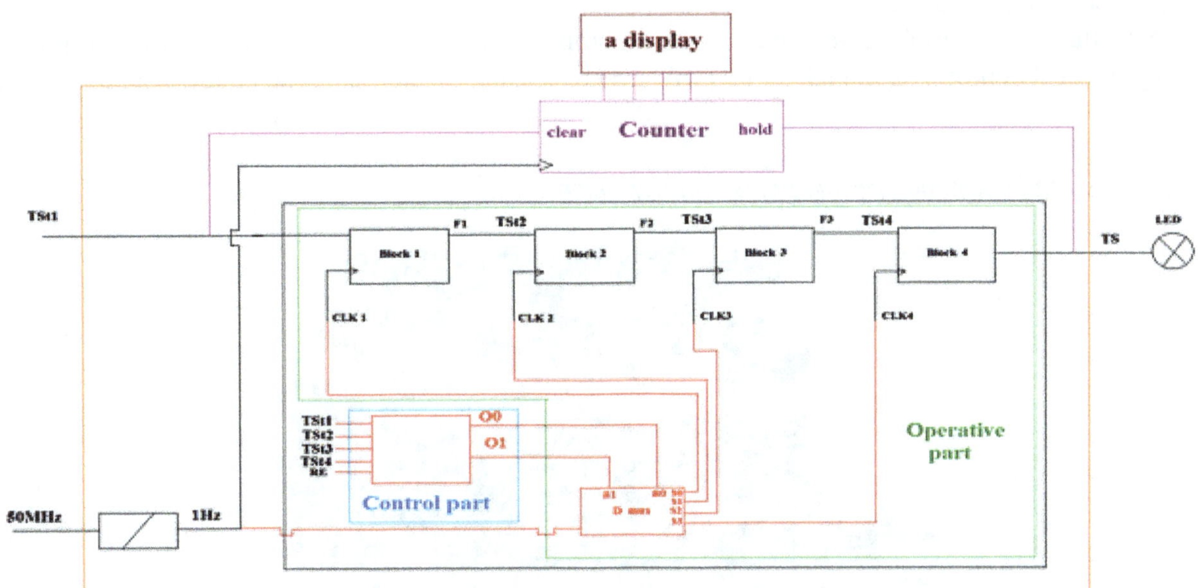

As mentioned previously, T_{res}= (T_{output}- T_{input}) where T_{input} is the time when the clear of the counter is deactivated. T_{input} is the time when the hold of the counter is active.

Following to Figure 16, authors figure out that the physical result using classical approach provides a response time equals to 42 s.

To physically calculate the dynamic power dissipated by the circuit; Xpower tool has been used. With a circuit running at 50 MHz, the following values have been obtained:

$P_{dynamic}$= 1.29 mW

E_d=1, 29 × 42= 54.18 mJ

The Proposed Approach

In this part, the value of response time and the value of the dynamic energy dissipated by the circuit using the proposed approach are physically calculated.

The same experience conditions have been kept. For that, T_{res}= (T_{output}- T_{input}) where T_{input} is the time when clear of the counter is deactivated. T_{input} is the time when the hold of the counter is active.

Following to Figure 17, authors figured out that the response time using the new approach is 39s.

By the same manner, to calculate the power dissipated Xpower tool has been used. With a circuit running at 50 MHz. the following values have been obtained.

$P_{dynamic}$= 0, 98 mW

E_d=0, 98×39=38.22 mJ

To summarize, a physical comparative study between the new approach and the traditional design is illustrated in the Table 2.

A further indicator of the effectiveness of the proposed approach is the fact that dissipated power using classical approach is as high as 1.29 mW whereas using the new approach there is an important

Figure 16. Implementation result of classical approach

Figure 17. Implementation result of proposed approach

Table 2. Design results

	Classical Approach	Our Approach	Gain (%)
Power (mw)	1.29	0.98	24.03
Energy (mJ)	54.18	38.22	29.46
Response time (s)	42	39	7.14

decrease of power consumption reaching a minimum level of only 0.98 mW. In other words, the new approach gives an interesting gain in dynamic power which equals 24.03%. Additionally, the new design provides a gain in response time which equals 7.14%. As for energy, the new approach offers a gain of 29.46% of energy compared to the traditional design.

From here, researchers deduce that the physical results confirm the simulation results. Therefore, designers looking to meet less power consumption and faster circuit have to opt for a new efficient approach.

FUTURE RESEARCH DIRECTIONS

As reckoned, a global system is composed of many synchronous blocks. Certainly, at a given instant, there are some inoperative synchronous blocks. The dilemma is that these blocks which are in an idle state still consume power as long as their corresponding clocks are running. For that, many researchers have opted for the clock gating technique that aims to prevent the clock from operating when its corresponding block is inoperative. However, there exists another situation that must be taken into consideration which is the "useless" transition. In fact, in the FSM when the condition is valid a transition from one state to another takes place. In some cases a transition to one state to the same current state can be produced to generate the same output at each triggering edge of the clock. One may nevertheless wonder, why the block re-function to provide the same output especially that such a situation will ultimately generate an inevitable waste of energy and decrease the speed of the circuit. Undeniably, it will be more convenient if we consider these blocks as inoperative blocks and prevent their clocks from functioning as well. This technique will be the objective of future research in the area of clock gating methods.

CONCLUSION

The clock signal is the most basic signal for any digital circuit. Recently, it has been realized that the clock signal consumes the greatest amount of total dynamic power since it consumes power even when its corresponding block is in an idle state. Up to 70% or more of dynamic power can be spent at clock buffers.

This result is intuitive because these buffers have the highest rate of failover in the system. They are numerous and they often have a high driving force to minimize the clock delay. Moreover, the components receiving the clock dissipate a certain dynamic power even if there is absolutely no change of input and output. The clock is particularly useless for blocks that must maintain the same logic values over many clock cycles. From here comes the idea of stopping the clock from functioning if its corresponding block is in an idle state (clock gating).

Another principle is proposed in this chapter book; authors have opted for eliminating the master clock of synchronous circuits and replaced it by another "clock generator" block that aims to offer a less consuming system and a faster one at the same time. In fact, as development of tools that support a low-power design methodology has become a critical issue while developing digital systems and an area of active research, and time constraint has become high-priority while creating digital circuit, efforts have increased in the last several years to obtain more efficient circuit. For that reason, this chapter book comes up with a new approach that helps to reduce the power consumption when designing a finite state machine-based controller while guaranteeing a correct and faster treatment. In fact, the new design has been tested and compared to classical design to prove its effectiveness in terms of energy and response time.

REFERENCES

Aboushady, H., Dumonteix, Y., Louerat, M. M., & Mehrez, H. (2000). Efficient polyphase decomposition of Comb decimation filters in/spl Sigma//spl Delta/analog-to-digital converters. In *Circuits and Systems, 2000. Proceedings of the 43rd IEEE Midwest Symposium on* (*Vol. 1*, pp. 432-435). IEEE.

Anagnostopoulos, K., & Economakos, G. (2005, December). Lowpower design of multipliers using a full-adder isolation technique. In *Electronics, Circuits and Systems, 2005. ICECS 2005. 12th IEEE International Conference on* (pp. 1-4). IEEE. 10.1109/ICECS.2005.4633542

Arroba, P., Moya, J. M., Ayala, J. L., & Buyya, R. (2017). Dynamic Voltage and Frequency Scaling-aware dynamic consolidation of virtual machines for energy efficient cloud data centers. *Concurrency and Computation*, *29*(10), e4067. doi:10.1002/cpe.4067

Banerjee, N., Raychowdhury, A., Bhunia, S., Mahmoodi, H., & Roy, K. (2005, October). Novel low-overhead operand isolation techniques for low-power datapath synthesis. In *Computer Design: VLSI in Computers and Processors, 2005. ICCD 2005. Proceedings. 2005 IEEE International Conference on* (pp. 206-211). IEEE. 10.1109/ICCD.2005.80

Bellaouar, A., & Elmasry, M. (2012). *Low-power digital VLSI design: circuits and systems*. Springer Science & Business Media.

Bezerra, P. A., Krismer, F., Andersen, T. M., Kolar, J. W., Sridhar, A., Brunschwiler, T., . . . Wang, N. (2015, November). Modeling and multi-objective optimization of 2.5 D inductor-based Fully Integrated Voltage Regulators for microprocessor applications. In *Power Electronics Conference and 1st Southern Power Electronics Conference (COBEP/SPEC), 2015 IEEE 13th Brazilian* (pp. 1-6). IEEE.

Bhattacharjee, P., & Majumder, A. (2016, December). LECTOR Based Gated Clock Approach to Design Low Power FSM for Serial Adder. In *Nanoelectronic and Information Systems (iNIS), 2016 IEEE International Symposium on* (pp. 250-254). IEEE. 10.1109/iNIS.2016.064

Bhattacharjee, P., Nath, B., & Majumder, A. (2017). LECTOR Based Clock Gating for Low Power Multi-Stage Flip Flop Applications. 대한전자공학회 학술대회, 106-109.

Bortolotti, D., Mamaghanian, H., Bartolini, A., Ashouei, M., Stuijt, J., Atienza, D., ... Benini, L. (2014, August). Approximate compressed sensing: ultra-low power biosignal processing via aggressive voltage scaling on a hybrid memory multi-core processor. In *Proceedings of the 2014 international symposium on Low power electronics and design* (pp. 45-50). ACM. 10.1145/2627369.2627629

Butzen, P., Reis, A., Kim, C., & Ribas, R. (2007). Subthreshold leakage modeling and estimation of general CMOS complex gates. *Integrated Circuit and System Design. Power and Timing Modeling, Optimization and Simulation*, 474-484.

Buyuktosunoglu, A., Schuster, S., Brooks, D., Bose, P., Cook, P., & Albonesi, D. (2001). An adaptive issue queue for reduced power at high performance. *Power-Aware Computer Systems*, 25-39.

Cadence document. (n.d.). *"A Practical Guide for Low-power Design": User experience with CPF.* Retrieved from www.cadence.com

Chandrakasan, A. P., Sheng, S., & Brodersen, R. W. (1992). Low-power CMOS digital design. *IEICE Transactions on Electronics*, *75*(4), 371–382.

Chang, C. H., Satzoda, R. K., & Sekar, S. (2005, May). A novel multiplexer based truncated array multiplier. In *Circuits and Systems, 2005. ISCAS 2005. IEEE International Symposium on* (pp. 85-88). IEEE.

Choi, K., Cheng, W. C., & Pedram, M. (2005). Frame-based dynamic voltage and frequency scaling for an MPEG player. *Journal of Low Power Electronics*, *1*(1), 27–43. doi:10.1166/jolpe.2005.005

Choi, K., Lee, W., Soma, R., & Pedram, M. (2004, November). Dynamic voltage and frequency scaling under a precise energy model considering variable and fixed components of the system power dissipation. In *Proceedings of the 2004 IEEE/ACM International conference on Computer-aided design* (pp. 29-34). IEEE Computer Society.

Chong, K. S., Gwee, B. H., & Chang, J. S. (2007). Low energy 16-bit Booth leapfrog array multiplier using dynamic adders. *IET Circuits, Devices & Systems*, *1*(2), 170–174. doi:10.1049/iet-cds:20060109

Dosluoglu, T., & Gelman, A. (2016). *U.S. Patent No. 9,369,040.* Washington, DC: U.S. Patent and Trademark Office.

Dumonteix, Y., & Mehrez, H. (2000). A family of redundant multipliers dedicated to fast computation for signal processing. In *Circuits and Systems, 2000. Proceedings. ISCAS 2000 Geneva. The 2000 IEEE International Symposium on* (Vol. 5, pp. 325-328). IEEE. 10.1109/ISCAS.2000.857430

Fan, X., Ellis, C., & Lebeck, A. (2005). The synergy between power-aware memory systems and processor voltage scaling. *Power-Aware Computer Systems*, 151-166.

Fonseca, M., da Costa, E., Bampi, S., & Monteiro, J. (2005, September). Design of a radix-2 m hybrid array multiplier using carry save adder format. In *Proceedings of the 18th annual symposium on Integrated circuits and system design* (pp. 172-177). ACM. 10.1145/1081081.1081126

Fujino, M., & Moshnyaga, V. G. (2003, May). Dynamic operand transformation for low-power multiplier-accumulator design. In *Circuits and Systems, 2003. ISCAS'03. Proceedings of the 2003 International Symposium on* (*Vol. 5*, pp. V-V). IEEE. 10.1109/ISCAS.2003.1206276

Gaillardon, P. E., Beigne, E., Lesecq, S., & Micheli, G. D. (2015). A survey on low-power techniques with emerging technologies: From devices to systems. *ACM Journal on Emerging Technologies in Computing Systems*, *12*(2), 12. doi:10.1145/2714566

Gustafsson, O., & Ohlsson, H. (2005, May). A low power decimation filter architecture for high-speed single-bit sigma-delta modulation. In *Circuits and Systems, 2005. ISCAS 2005. IEEE International Symposium on* (pp. 1453-1456). IEEE. 10.1109/ISCAS.2005.1464872

Havinga, P. J., & Smit, G. J. (2000). Design techniques for low-power systems. *Journal of Systems Architecture*, *46*(1), 1–21. doi:10.1016/S1383-7621(98)00057-5

Homayoun, H., Veidenbaum, A., & Gaudiot, J. L. (2008, October). Adaptive techniques for leakage power management in L2 cache peripheral circuits. In *Computer Design, 2008. ICCD 2008. IEEE International Conference on* (pp. 563-569). IEEE. 10.1109/ICCD.2008.4751917

Hsu, C. H., & Feng, W. C. (2004, December). Effective dynamic voltage scaling through CPU-boundedness detection. In PACS (pp. 135-149). Academic Press.

Hsu, C. H., & Kremer, U. (2003). The design, implementation, and evaluation of a compiler algorithm for CPU energy reduction. *ACM SIGPLAN Notices*, *38*(5), 38–48. doi:10.1145/780822.781137

Huang, Z., & Ercegovac, M. D. (2005). High-performance low-power left-to-right array multiplier design. *IEEE Transactions on Computers*, *54*(3), 272–283. doi:10.1109/TC.2005.51

Hwang, E., Vahid, F., & Hsu, Y. C. (1999, January). FSMD functional partitioning for low power. In *Proceedings of the conference on Design, automation and test in Europe* (p. 7). ACM.

Itoh, K., Kimura, S. I., & Sakata, T. (1999, September). VLSI memory technology: Current status and future trends. In *Solid-State Circuits Conference, 1999. ESSCIRC'99. Proceedings of the 25th European* (pp. 3-10). IEEE.

ITRS. (2013). *Process Integration, Devices, & Structures (PIDS) Chapter*. Retrieved from http://www.itrs.net/Links/2013ITRS/Home2013.htm

Kao, J., Narendra, S., & Chandrakasan, A. (2002, November). Subthreshold leakage modeling and reduction techniques. In *Proceedings of the 2002 IEEE/ACM international conference on Computer-aided design* (pp. 141-148). ACM.

Kapoor, A. (2015). *U.S. Patent Application No. 14/973,616*. Washington, DC: US Patent Office.

Kim, K. K., & Kim, Y. B. (2007, May). Optimal body biasing for minimum leakage power in standby mode. In *Circuits and Systems, 2007. ISCAS 2007. IEEE International Symposium on* (pp. 1161-1164). IEEE. 10.1109/ISCAS.2007.378256

Kuon, I., & Rose, J. (2007). Measuring the gap between FPGAs and ASICs. *IEEE Transactions on Computer-Aided Design of Integrated Circuits and Systems*, *26*(2), 203–215. doi:10.1109/TCAD.2006.884574

Li, H., Bhunia, S., Chen, Y., Vijaykumar, T. N., & Roy, K. (2003, February). Deterministic clock gating for microprocessor power reduction. In *High-Performance Computer Architecture, 2003. HPCA-9 2003. Proceedings. The Ninth International Symposium on* (pp. 113-122). IEEE.

Lin, T. J., Chen, S. K., Kuo, Y. T., Liu, C. W., & Hsiao, P. C. (2008). Design and implementation of a high-performance and complexity-effective VLIW DSP for multimedia applications. *Journal of Signal Processing Systems for Signal, Image, and Video Technology*, *51*(3), 209–223. doi:10.100711265-007-0061-x

Lukefahr, A., Padmanabha, S., Das, R., Dreslinski, R. Jr, Wenisch, T. F., & Mahlke, S. (2014, August). Heterogeneous microarchitectures trump voltage scaling for low-power cores. In *Proceedings of the 23rd international conference on Parallel architectures and compilation* (pp. 237-250). ACM. 10.1145/2628071.2628078

Nathalie Julien. (2006). *Caractérisation et modélisation de la consommation sur FPGA, Groupe Lester France*. Nice: Ecofac.

Nathanael, R., Pott, V., Kam, H., Jeon, J., Alon, E., & Liu, T. J. K. (2010). Four-terminal-relay body-biasing schemes for complementary logic circuits. *IEEE Electron Device Letters*, *31*(8), 890–892. doi:10.1109/LED.2010.2050133

Olivier Rochecouste. (2005). *Architecture et bits significatifs*. Thèse en informatique, Université de Rennes 1, Rennes 1.

Osmulski. (2000). A probabilistic power prediction tool for Xilinx FPGA. *Int. Wksp. on Computer Embedded/Distributed HPC Systems and Applications*, 776 - 783.

Phatak, D. S., Kahle, S., Kim, H., & Lue, J. (1998, June). Hybrid signed digit representation for low power arithmetic circuits. *Proceedings of the Lower Power Workshop in Conjunction with ISCA.*

Prakash, K. R., & Sanju, A. V. (2014). Design and Implementation of Signed, Rounded and Truncated Multipliers using Modified Booth Algorithm for Dsp Systems. *International Journal (Toronto, Ont.)*, *2*(1).

Qu, G., & Potkonjak, M. (2002). Techniques for energy-efficient communication pipeline design. IEEE Transactions on Very Large Scale Integration (VLSI). *Systems*, *10*(5), 542–549.

Rabaey, J. M., & Pedram, M. (Eds.). (2012). *Low power design methodologies* (Vol. 336). Springer Science & Business Media.

Seo, Y. H., & Kim, D. W. (2010). A new VLSI architecture of parallel multiplier–accumulator based on Radix-2 modified Booth algorithm. *IEEE Transactions on very large scale integration (vlsi) systems, 18*(2), 201-208.

Shashidar, R., Kumar, R. S., MahalingaSwamy, A. M., & Roopa, M. (2017). FPGA Implementation of Low Power Pipelined 32-Bit RISC Processor Using Clock Gating. In *Proceedings of the International Conference on Data Engineering and Communication Technology* (pp. 769-777). Springer Singapore. 10.1007/978-981-10-1678-3_74

Shin, D., Kim, J., & Lee, S. (2001, June). Low-energy intra-task voltage scheduling using static timing analysis. In *Proceedings of the 38th annual Design Automation Conference* (pp. 438-443). ACM. 10.1145/378239.378551

Siwar, B.H.H, Bouraoui, O., & Hatem, S. (2014). Efficient design of Finite State Machine to Minimize Power and Response time. *Wulfenia Journal, 21*(12), 364-383.

Smith, G. (2014, October). Updates of the ITRS design cost and power models. In *Computer Design (ICCD), 2014 32nd IEEE International Conference on* (pp. 161-165). IEEE. 10.1109/ICCD.2014.6974676

Stan, M. R., & Burleson, W. P. (1995). Bus-invert coding for low-power I/O. *IEEE Transactions on very large scale integration (VLSI) systems, 3*(1), 49-58.

Sulaiman, D. R. (2008, May). Using clock gating technique for energy reduction in portable computers. In *Computer and Communication Engineering, 2008. ICCCE 2008. International Conference on* (pp. 839-842). IEEE. 10.1109/ICCCE.2008.4580723

Unsal, O. S., & Koren, I. (2003). System-level power-aware design techniques in real-time systems. *Proceedings of the IEEE, 91*(7), 1055–1069. doi:10.1109/JPROC.2003.814617

Verma, A. K., & Ienne, P. (2006, July). Towards the automatic exploration of arithmetic-circuit architectures. In *Proceedings of the 43rd annual Design Automation Conference* (pp. 445-450). ACM. 10.1145/1146909.1147027

Verma, A. K., & Ienne, P. (2007, April). Automatic synthesis of compressor trees: reevaluating large counters. In *Proceedings of the conference on Design, automation and test in Europe* (pp. 443-448). EDA Consortium. 10.1109/DATE.2007.364632

Wolf, W. (2003). A decade of hardware/software codesign. *Computer, 36*(4), 38–43. doi:10.1109/MC.2003.1193227

Yeap, G. K. (2012). *Practical low power digital VLSI design*. Springer Science & Business Media.

Zheng, M., & Albicki, A. (1995, October). Low power and high speed multiplication design through mixed number representations. In *Computer Design: VLSI in Computers and Processors, 1995. ICCD'95. Proceedings., 1995 IEEE International Conference on* (pp. 566-570). IEEE.

KEY TERMS AND DEFINITIONS

ASIC: Application specific integrated circuit is an integrated circuit (IC) manufactured for a particular use, rather than intended for general purpose use.

Clock: Is a signal used in electronics and especially in synchronous digital systems. Generally, it is used to coordinate actions of electronic circuits.

Clock Gating: Is a well-known technique used in many synchronous circuits to reduce dynamic power dissipation. It is based on preventing the clock from functioning when it is required.

Dynamic Power: Is the power dissipated due to the switching activity during charging and discharging of the load capacities and due to the short-circuit power caused by a short-circuit current that occurs during changes in the states of the logic gates.

FPGA: Field programmable gate array is an integrated circuit designed to be programmed by the user after manufacturing.

RTL (Register Transfer Level): Is a high level description which models a synchronous circuit in terms of the flow of digital signals between hardware registers and the logical operations performed on those signals.

Synchronous Circuit: Is an electronic circuit in which the changes in the state of memory elements are synchronized by a clock signal.

Chapter 10
Application of Power Electronics in Power Systems

Mashhood Hasan
Galgotias Engineering College, India

ABSTRACT

This chapter introduces the custom power device to solve the various kinds of power quality (PQ) problems. It is generally known as flexible AC transmission system (FACTS) and D-FACTS (FACTs used for low voltage and low current rating in AC distribution system). The D-FACTS are various types like dynamic voltage restore (DVR), static compensator in distribution system (D-STATCOM), and unified power quality conditioner (UPQC). The third one is combination of DVR and D-STATCOM, which is known as UPQC. The UPQC can compensate all the PQ problems. Comprehensive knowledge of UPQC is presented here to understand the trade value of this device. This chapter is organized as follows: The evaluation of power electronic based devices is presented in the background part of this chapter where custom power devices (CPDs) are also discussed. And operating principle with the Matlab model of UPQC and results are presented in the section titled "Working Principle of Unified Power Quality Conditioner (UPQC)." The scope of the one of the functional power electronic based devices are presented in section titled "Scope of UPQC." At the end, this chapter is concluded.

INTRODUCTION

The application of power electronics in power system is presented to understand the trade value of power electronics evaluation. The area of power electronics is wide in power system such as High Voltage DC Transmission (HVDC), Flexible AC Transmission System (FACTS) controlling devices, which is generally applicable for stability of power system and reliability of power systems in transmission system. Moreover, D-FACTS is used in AC distribution system to enhance the power quality. In this chapter, the main focus is that how to enhance the power quality of AC distribution system using the power electronics switches. The chapter is organized as follows, background of the power electronics, principle of operation of unified power quality conditioners (UPQCs), scope of the UPQC, future research directions and conclusion of the presented chapter.

DOI: 10.4018/978-1-5225-3935-3.ch010

BACKGROUND

The power electronics based loads supports to generate harmonics in power system. It has some serious disadvantage in terms of power quality of the AC distribution system. The AC mains has to generate extra reactive power because of harmonics polluted equipment. The latest issue needs a technical solution for the betterment of the power system. The concept of custom power devices (CPD) has been exposed in the late 1980s. This is an application of power electronics, which includes the controller to improve the power quality and controllability of the power system (Hingorani, 1995; Hasan, et al., 2013; Fujita & Akagi, 1998; Singh, et al., 1999). There are two type of methodology behind the power electronics equipment, first one is conventional thyristor switched capacitors (TSC) or thyristor switched reactors (TSR) and second one is self-commutated switching converters. These two methods could control real and reactive powers efficiently but the self-commutated devices only could be used which eliminate currents harmonics and voltages harmonics. Moreover, compensation flexibility and response time in the self-commutated switching converters have been explored better than conventional thyristor switched converters. The static VAR compensators (SVCs) are used to improve the quality and control of voltage by supplying or absorbing reactive power using inductors and capacitors. The essentials parts of a SVC are thyristors in anti-parallel and passive elements (capacitor and inductor). A SVC contains a switched capacitor as a passive element is called as TSC or a controlled reactor is called TCR. The effective value of the passive elements is changed regularly by controlling the angle of gate pulse of the thyristors. One of the main problem of SVC is handling of the reactive power. The size of passive elements is limited by the reactive power demand. On the other hand, custom power devices overcome the limitation of SVC. Most adaptable custom power device is STATCOM. It has a voltage source converter (VSC) with fast response. It could be implemented to support stability and voltage profile in the power system. The same voltage and current profile in the distribution system can be enhance by D-STATCOM. The structure of D-STATCOM is similar to the STATCOM.

The static synchronous series compensator (SSSC) injects the voltage in series with the AC mains in power system and behaves as a voltage controller of the transmission system. SSSC is capable to maintain desired output voltage. SSSC contains a voltage source converter with suitable controller if it also contains DC link, the SSSC is able to eliminate harmonics in the voltage, to compensate reactive power, regulate voltage sag/swell, voltage unbalance and voltage flicker. The same profile can be enhanced in the distribution system using dynamic voltage restore (DVR). The most promising custom device DVR, D-STATCOM and unified power quality conditioners (UPQCs) are used in the distributions system to enhance the power quality. The UPQC is most promising device to enhance the power quality in the distribution system which offers advantages over DVR and D-STATCOM. In a UPQC), two converters are connected back to back with a common DC link (Hasan et al., 2015; Khadkikar, 2012). The series converter mitigates the voltage related issue whereas shunt converter compensates current related issue. This is the most common and popular converter topology which is shown in Figure 1.

WORKING PRINCIPLE OF UNIFIED POWER QUALITY CONDITIONER (UPQC)

The main objective of this chapter is the classification and configuration of the unified power quality conditioners (UPQCs. This chapter is started with the detail of working principle of the UPQC. This is also described the separate description of series converter and shunt converter of the UPQC. This

Figure 1. Schematic diagram of UPQC

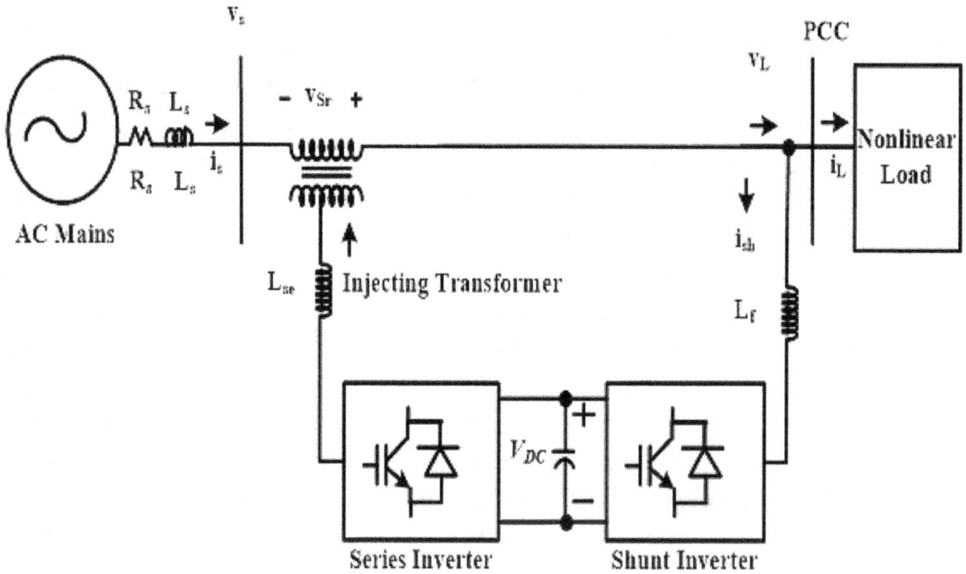

is incorporated a broad overview on the classification and configuration of the UPQC for 1P2W and 3P3W and 3P4W, the classification on the basis of physical structure and the mode of the series voltage injection (series voltage injection angle based). It is observed that many researchers have been used unique function such as application, topology and task based UPQC. Figure 2 shows a combination of series voltage converter and shunt converter of a UPQC. It can be easily understood, the principle of operation by describing the shunt converter and series converter, separately of the UPQC with the help of 3P3W configuration. The literature on shunt converter and series converter as an active power filter has been reported massively.

Principle Operation of Shunt Converter

The shunt converter, when it realized as shunt part of the UPQC is shown in Figure 3 This works as an active filter which is widely used to compensate current harmonics, load current unbalanced, reactive power and neutral current (for 3-phase 4-wire system) (Elnady et al., 2001; Ghosh et al., 2004; Khoor et al., 2005; Kazemi et al., 2006; Kolhatkar & Das, 2007; Akagi, 1996). Many researchers have improved power quality of the distribution system. The shunt converter of the UPQC is connected across the load (parallel to the load) to compensate the power quality problems which are created by the load. The shunt converter is also called as shunt active filter and parallel active filter. The shunt coupling inductance, L_f and shunt coupling resistance, R_f provide smoothing and isolation for high frequency components. The numerical value of R_f is very small as compared to shunt interfacing inductor.

The quality of currents can be obtained by using the proper control of converter switches such as insulated gate bipolar transistors (IGBTs). The behavior of current wave-shape depends on switching frequency and driving voltage across the coupling inductance. The maximum driving voltage across the coupling inductance (Ldi/dt) is so important because higher harmonics components may be cancelled by high value of (Ldi/dt). Thus, the size of coupling inductor is a trade-off involvement in AC power system.

Figure 2. Three phase three wire system configuration of UPQC

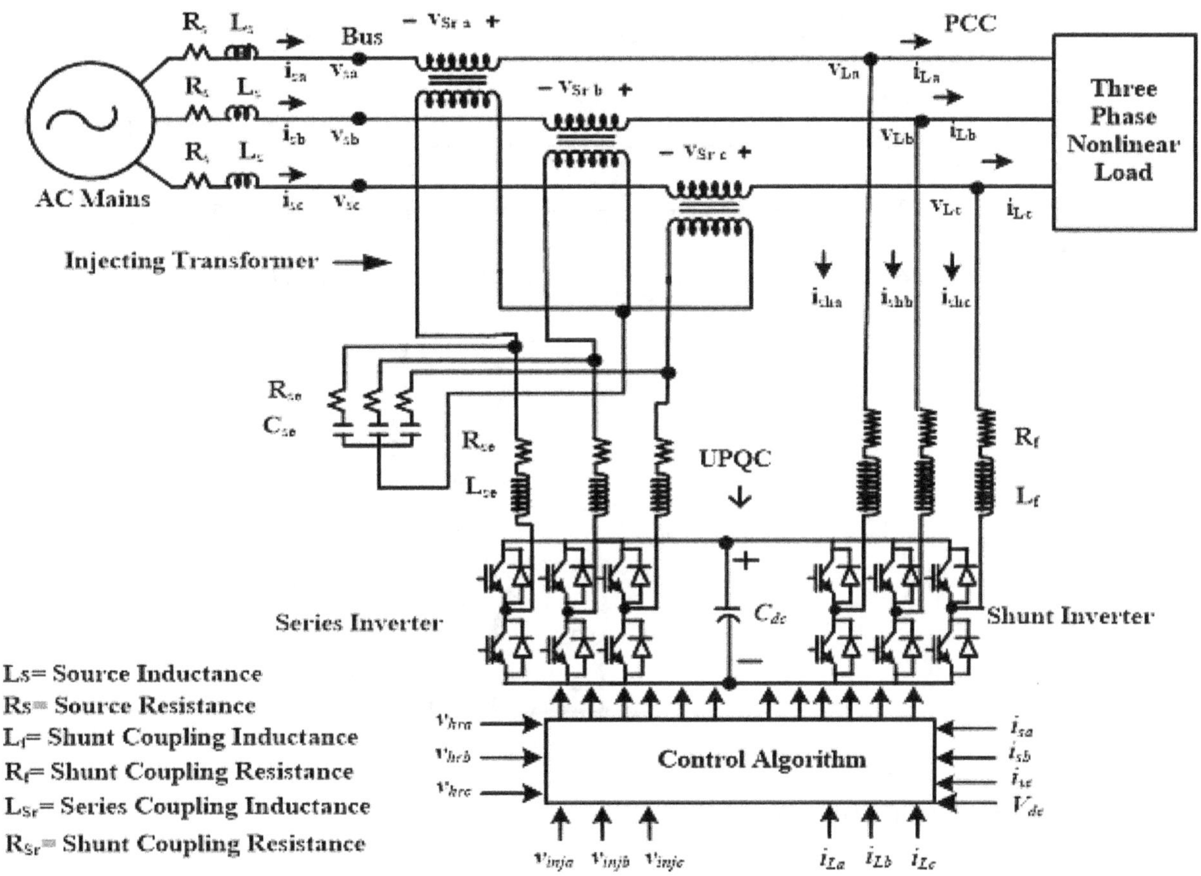

A large size of coupling inductor provides good isolation and protection from the AC power system. On the other hand, a large size of coupling inductor limits the higher order harmonics elimination, hence, shunt converter loses its ability.

Figure 4 shows the connection of the shunt converter it also illustrates the waveform of load current and generated compensating current. The shunt converter is the bilateral converter, it generates the compensating currents which is difference value of reference current, generated by proper control algorithm and actual load current. The mathematical expression of current injection through shunt converter is given as follows,

$$i_{injabc} = i^{*}_{sabc} - i_{Labc} \tag{1}$$

where, i^{*}_{sabc} is three phase reference current generated by the shunt control algorithm and i_{Labc} is the three-phase load current. Therefore, the basic function of the shunt converter can be stated as: it generates a current equal and opposites in polarity to the harmonic current drawn by the load and injects it to the point of coupling thereby forcing the AC mains current to be pure sinusoidal.

Figure 3. Three phase three wire system configuration of shunt converter

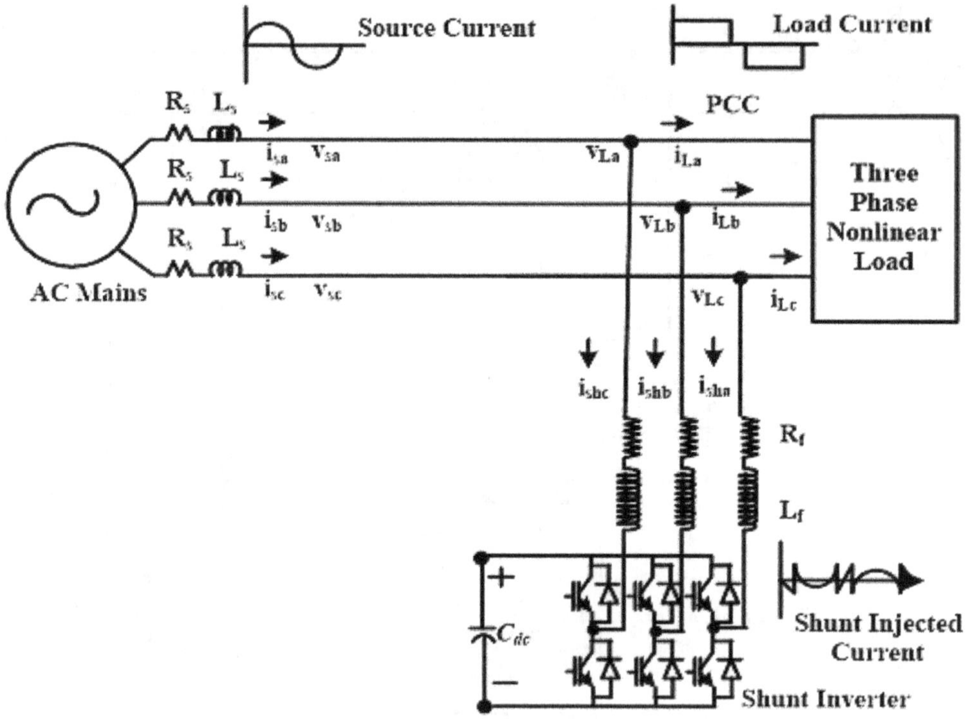

Operation of Series Converter

The series converter when realized as a part of UPQC is shown in Figure 2. The series converter can mitigate voltage sag, voltage swell, voltage flicker, voltage unbalanced and eliminates voltage harmonics. The series converter of UPQC is connected in series between AC mains and connected load.

Figure 4 illustrates at the source end side desired voltage which is dropping as the impedance of the line which has resistance and inductance. The fundamental operating principle of series converter is to maintain the desire voltage at load side independent to the variation in the AC mains voltage. The switches of series converter are operated as a voltage controller. It injects voltage which is equal to desired value of set reference voltage as generated by control algorithm. The actual load voltage is reduced voltage, therefore, series converter injects a voltage to desire load terminal voltage, it can be estimated as follows,

$$V_{inj.abc} = V^{*}_{Labc} - V_{sabc} \tag{2}$$

where V^{*}_{Labc} is the set reference voltage at load terminal. If the AC mains voltage contains the harmonics, the series converter injects harmonic of voltage in order to maintain sinusoidal voltage at PCC. There is certain voltage control law which is made to follow, when series converter injects a voltage in series with the AC mains. The sum of these voltage must be sinusoidal at PCC. This means that if AC mains voltages are distorted or non-sinusoidal, proper selection of injected voltage magnitude makes sure the voltages at PCC to be undistorted and sinusoidal. Moreover, the load voltage can also be isolated from

Figure 4. Three phase three wire system configuration of series converter

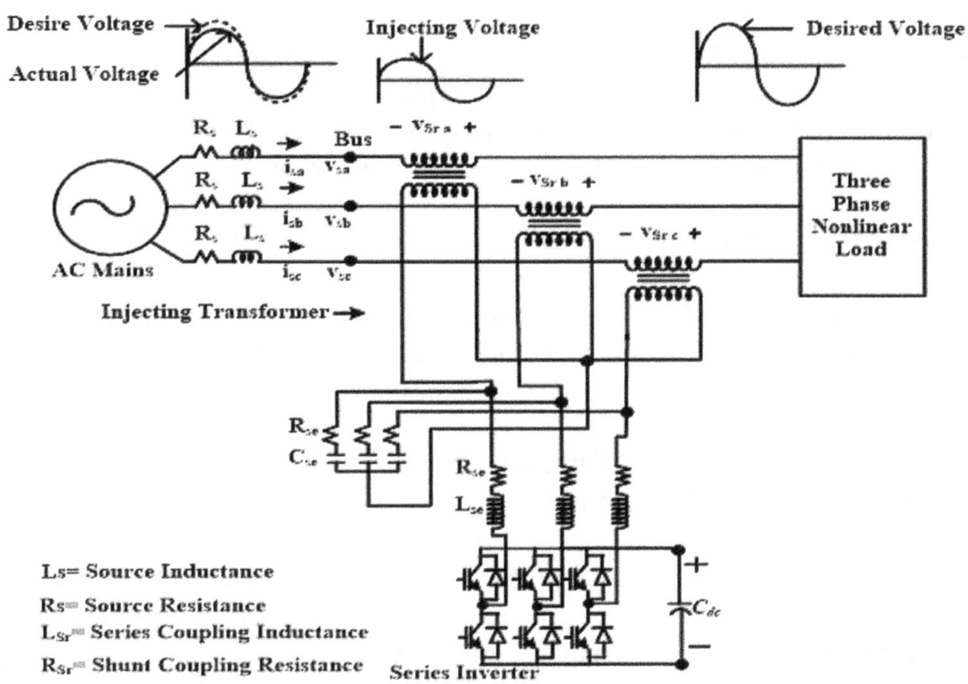

voltage sags, swells and unbalanced. Therefore, the series converter can assist to compel discipline at PCC and thus, the sensitive loads connected to the PCC always observe a stiff sinusoidal AC mains voltage. Thus, main objective of series converter can be stated as follows. The series converter injects a voltage in series with the lines of AC mains such that at the PCC voltage is always sinusoidal of desired voltage magnitude. While performing the aforesaid tasks the injected phase can vary from 0° to 360°. This means that the series converter could have to inject or absorb real power besides the reactive volt-amperes. It means that the series converter could not have a self-supporting DC link. Thus, series converter has to be connected with shunt converter sharing a common DC link to have real power balance. This combination of the UPQC gives the satisfactory performance of UPQC. The shunt converter not only injects or absorb the real power as demanded by series converter but it also compensates the power quality of load current.

MATLAB Modelling of Three Phase Three Wire UPQC

The MATLAB modelling of three phase three wire is required parameters that could be considered while designing a UPQC are DC link voltage, capacitor value, shunt inverter interfacing inductance, series inverter interfacing inductance and series resistance and capacitance value. As the UPQC incorporates inverter thus, it is also required to select an appropriate switching frequency and a ripple filter for both the inverters whereas an auxiliary device and series injection transformer are required to design for series inverter. The summary of design of three-phase three-wire (3P3W) UPQC is given in Table-1 for the linear load of 20kVA at 0.9 lagging power factor. The voltage and the current ratings of the loads are

239 V and 27.3 A per phase respectively. The voltage sag is also considered upto 20% for the design of 3P3W UPQC.

In Figure 5, a complete MATLAB model is developed for three phase three wire (3P3W) configuration of the UPQC. The model consists of model of three phase AC source, series injection transformers, 3P3W linear and nonlinear loads, series and shunt inverters of the UPQCs and the model of UPQC, 3 leg VSC of the series part and 3-leg VSC of the shunt part with back to back connection of DC link capacitor.

MATLAB Modeling of Controller for Series and Shunt Inverter

A MATLAB model of synchronous reference frame (SRF) theory based control algorithm to generate gating pulses of the series inverter which is shown in Figure 6 whereas a MATLAB model of SRF control algorithm to generate gating pulses of shunt VSC is shown in Figure 7.

Performance of UPQC for Critical Balanced Load Under Normal and Voltage Sag Condition

The maximum load demand is 20kVA at 0.9 lagging power factor under normal and voltage sag condition of the AC mains. The source voltages (v_{sabc}), source currents (i_{sabc}), load voltages (v_{Labc}) load currents (i_{Labc}), series injection voltages (v_{srabc}), shunt injecting currents (i_{shabc}), DC link voltage (v_{dc}), and source voltage for phase 'a' (v_s) verses source current for phase *'a'*(i_s) are shown in Figure 8 The simulation results are shown for the period of 0.2s to 0.35s (t_1 to t_4) for normal and voltage sag condition of the AC mains under linear load.

From t_1=0.2s to t_2=0.25s and t_3=0.3s to t_4=0.35s, UPQC is working under normal operating condition. Both the series VSC and shunt VSC of the UPQC share reactive power of the loads and reduce the burden of reactive power from AC mains. The series VSC injects the voltages (v_{srabc}) in quadrature to the

Figure 5. MATLAB based model of UPQC for 3P3W distribution system

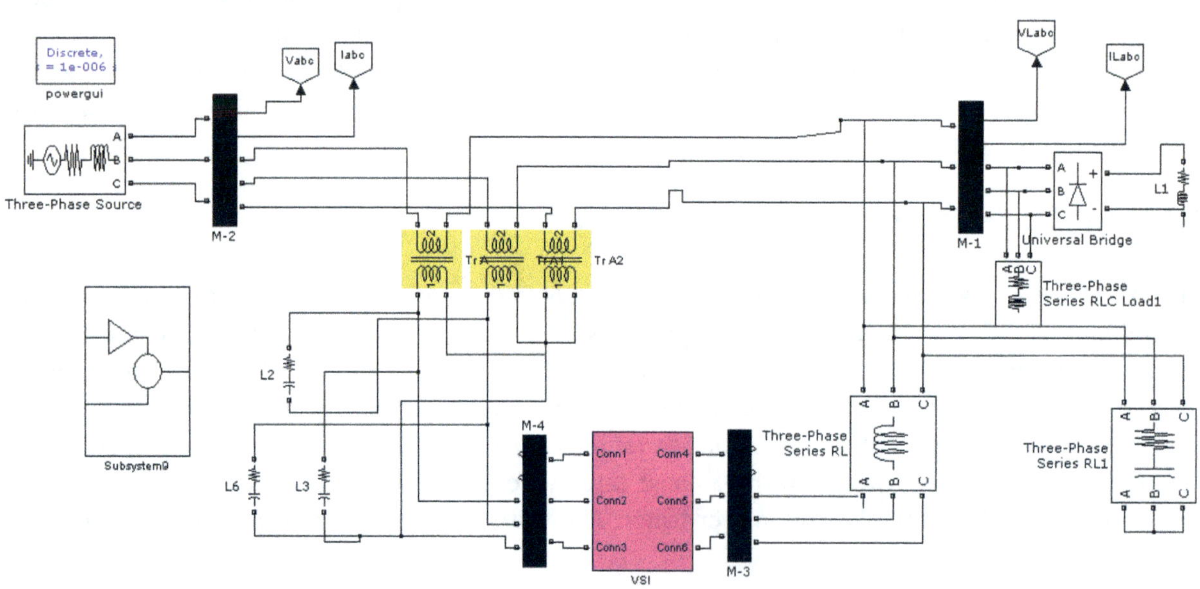

Figure 6. MATLAB model of SRF based control algorithm for pulse generation of series inverter

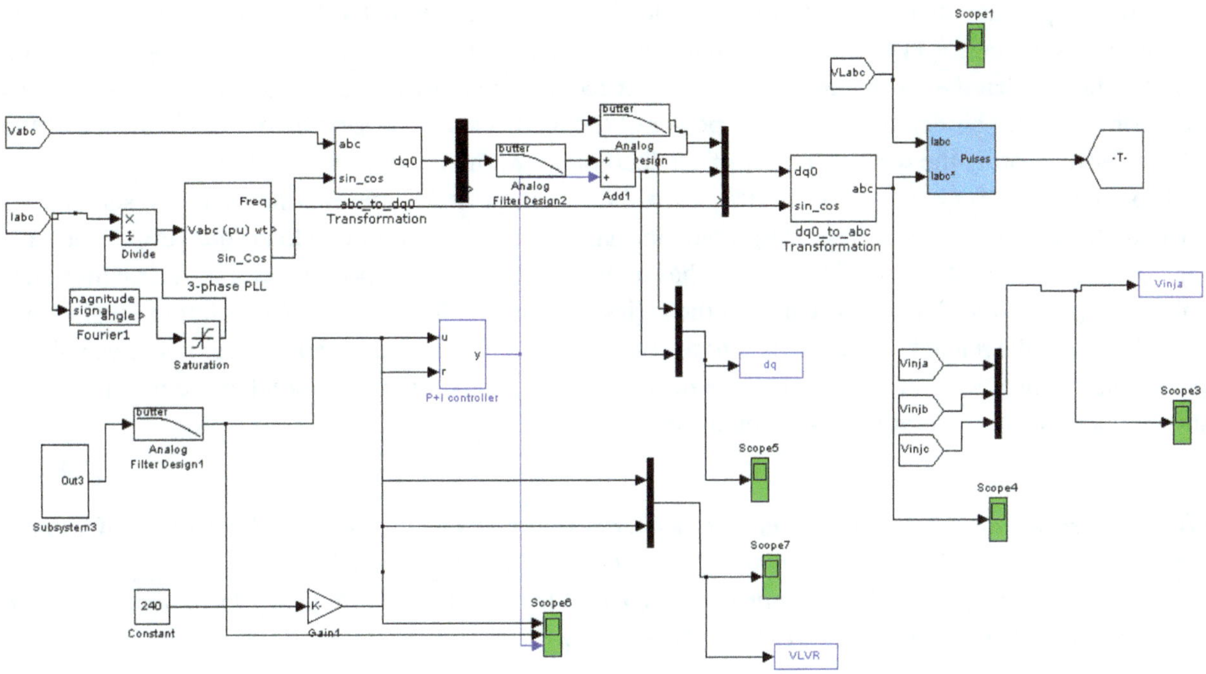

Figure 7. MATLAB model of SRF based control algorithm for pulse generation of shunt inverter

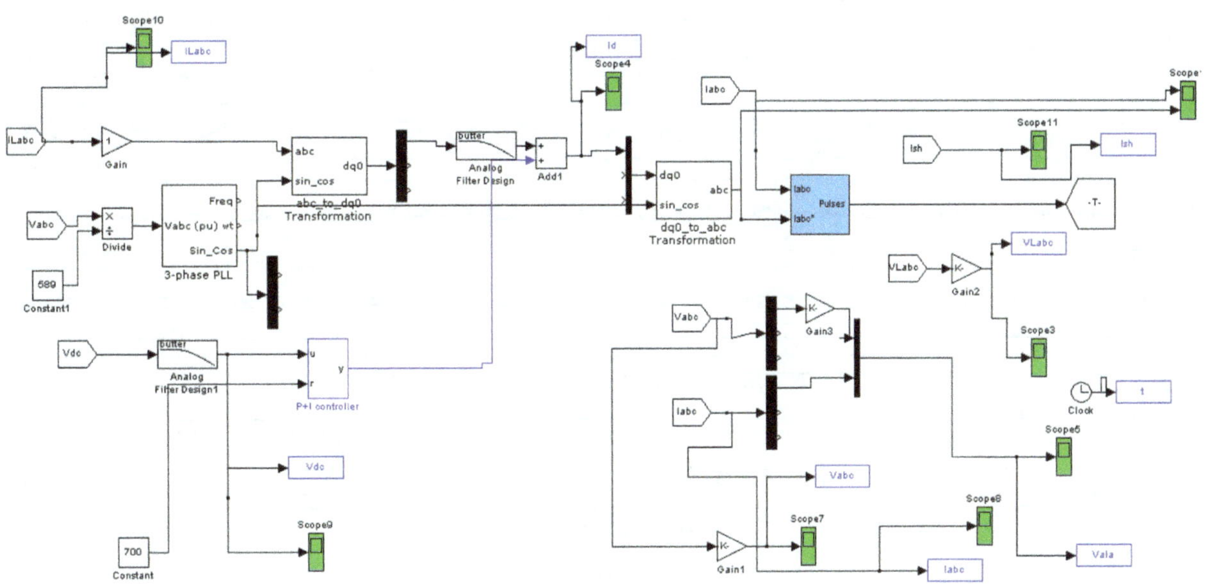

source currents (i_{sabc}) to compensate the reactive power of the loads whereas shunt VSC injects currents (i_{shabc}) to compensate the reactive power of the loads while maintaining the DC link voltage (v_{dc}) at 700 V. The source voltage v_s *(V)* of phase 'a' source current i_s *(A)* of phase 'a' are shown as a v_s *(V)* verses i_s *(A)*, it indicates that the AC mains has unity power factor. The unity power factor means the AC mains free from the extra burden of the reactive power of the load. This is the satisfactory performance of the UPQC for mitigating the reactive power of the load under normal condition.

From t_2=0.25s to t_3=0.3s, there is 20% sag of PCC voltage at AC mains condition. The series VSC injects extra demands of the voltage of the loads and has maintained the constant load voltage under voltage sag condition of the AC mains. Thus UPQC has maintained the at unity power factor at AC mains under voltage sag condition. The performance of the series and shunt VSCs of the UPQC is found satisfactory.

In Figure 9, the reference direct axis source voltage (v_d*) and reference quadrature axis source voltage (v_q*), peak voltage V_L at PCC and reference peak voltage V_L*, and reference direct axis current (i_d*) are shown under normal and voltage sag condition.

Figure 8. Performance of the UPQC under normal operating and voltage sag condition of the AC mains. Source voltage (v_{sabc}(V)), Source current (i_{sabc}(A)), Load voltage (v_{Labc}(V)), Load current (i_{Labc}(A), Series injected voltage (v_{Srabc}(V)), Shunt compensating current (i_{shabc}(A)), DC voltage (V_{dc} (V)), Source voltage for phase 'a' (v_s) verses source current for phase 'a'(i_s).

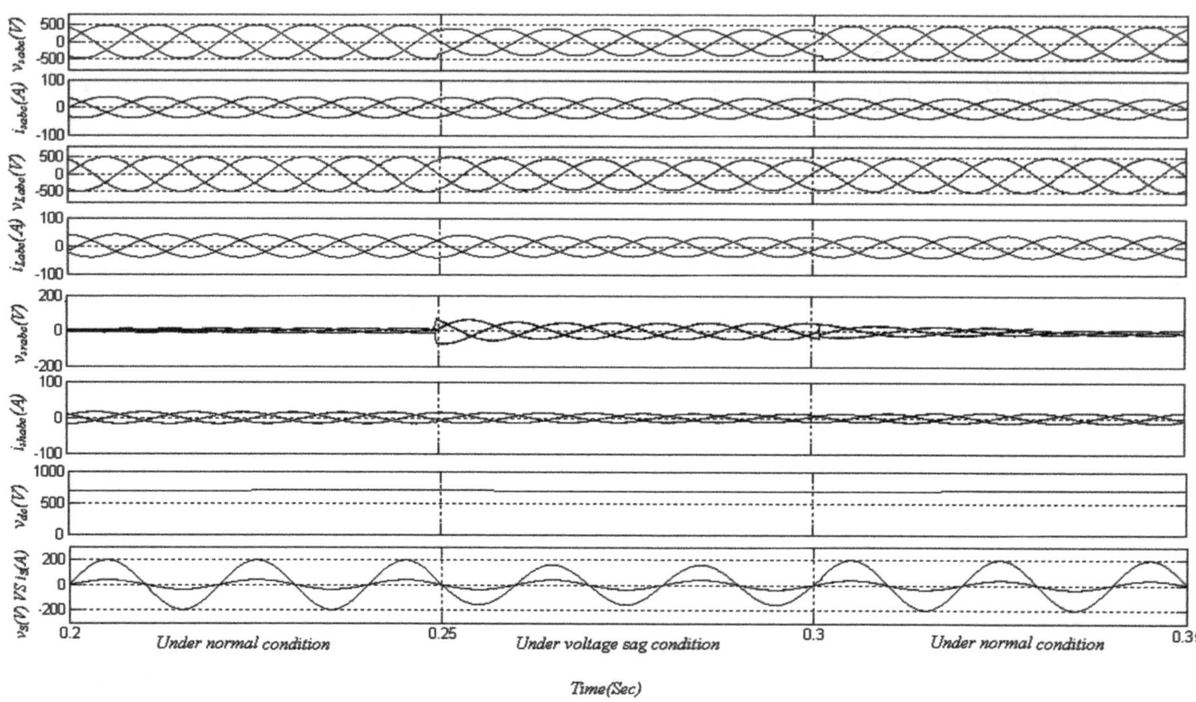

Figure 9. Performance of UPQC under normal operating and voltage sag condition at the AC mains under linear load. Reference direct axis voltage (v_d) and reference quadrature axis voltage (v_q*), Peak voltage at PCC (V_L) and reference peak voltage at PCC (V_L*), Reference direct axis current (v_d*).*

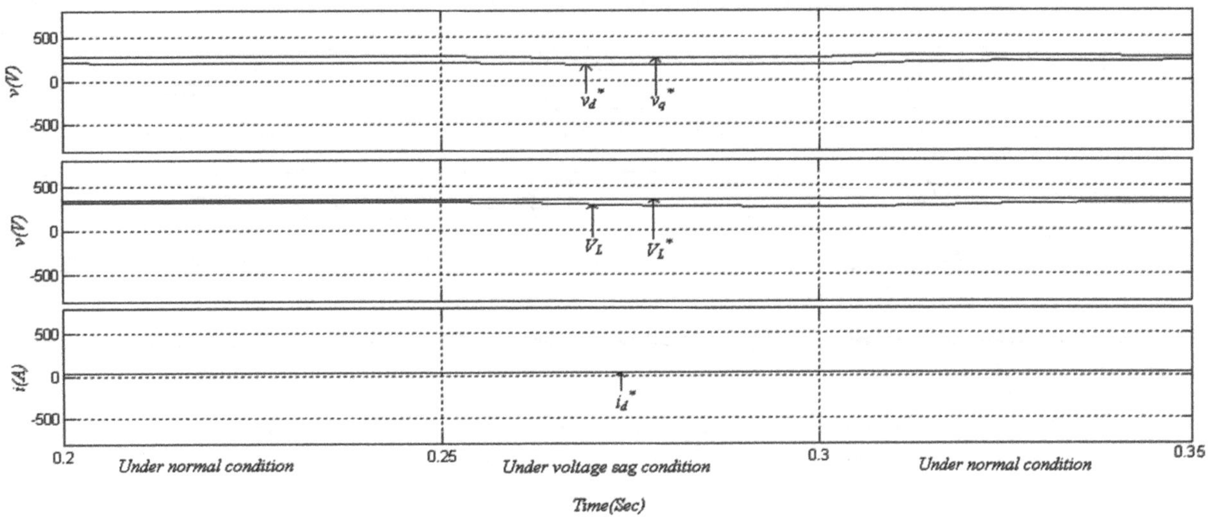

Performance of the UPQC for Critical Balanced Nonlinear Load Under Normal and Voltage Sag Condition

In Figure 10, the source voltages (v_{sabc}), source currents (i_{sabc}), load voltages (v_{Labc}) load currents (i_{Labc}), series injection voltages (v_{srabc}), shunt injecting currents (i_{shabc}), DC link voltage (v_{dc}), and source voltage for phase 'a' (v_s) verses source current for phase 'a' (i_s) are given for a 3P3W nonlinear load. The 3P3W nonlinear load based on a diode rectifier with a resistance (R) of 40Ω and an inductance (L) of 200mH under normal and 20% voltage sag condition in the AC mains., the simulation results are shown for the period of 0.35s to 0.45s (t_5 to t_8) for normal and voltage sag condition of the AC mains under nonlinear load. From t_5=0.35s to t_6=0.4s and t_7=0.45s to t_8=0.5s, UPQC is working under normal operating condition. Both the series VSC and shunt VSC of the UPQC share the reactive power of the loads and are reducing the burden of reactive power from AC mains under nonlinear load. Moreover, series VSC and shunt VSC eliminate the voltage and current harmonics of the loads. Thus, AC mains has a unity power factor while the current of the AC source and load voltage are free from harmonics or its maintaining IEEE-519 standard. At t_6=0.4s to t_7=0.45s, AC mains has 20% voltage sag condition. The series VSC converter injects extra voltage in quadrature to the source current to suppress the voltage sag of AC mains. Under the voltage sag conditions, both the series and shunt VSCs of the UPQC share reactive power of the load. The waveform of phase "a" ac source current (i_{sa}), ac source voltage (v_{sa}), load current (i_{La}) and voltage at PCC (v_{La}) are shown in Figure 11 (a)-(d), respectively. The total harmonic distortion (THD) of the source current, source voltage, load current and voltage at PCC are found to be 4.13%, 0.03%, 30.72% and .06%, respectively. It has been seen that UPQC is able to maintain THD of source current and load voltage within an IEEE-519 standard.

Figure 10. Performance of the UPQC under normal operating and voltage sag condition of the AC mains under nonlinear load. Source voltages (v_{sabc}(V)), Source currents (i_{sabc}(A)), Load voltages (v_{Labc}(V)), Load currents (i_{Labc}(A), Series injected voltages (v_{Srabc}(V)), Shunt compensating currents (i_{shabc}(A)), DC voltage (V_{dc} (V)), Source voltage for phase 'a' (v_s) verses source current for phase 'a'(i_s).

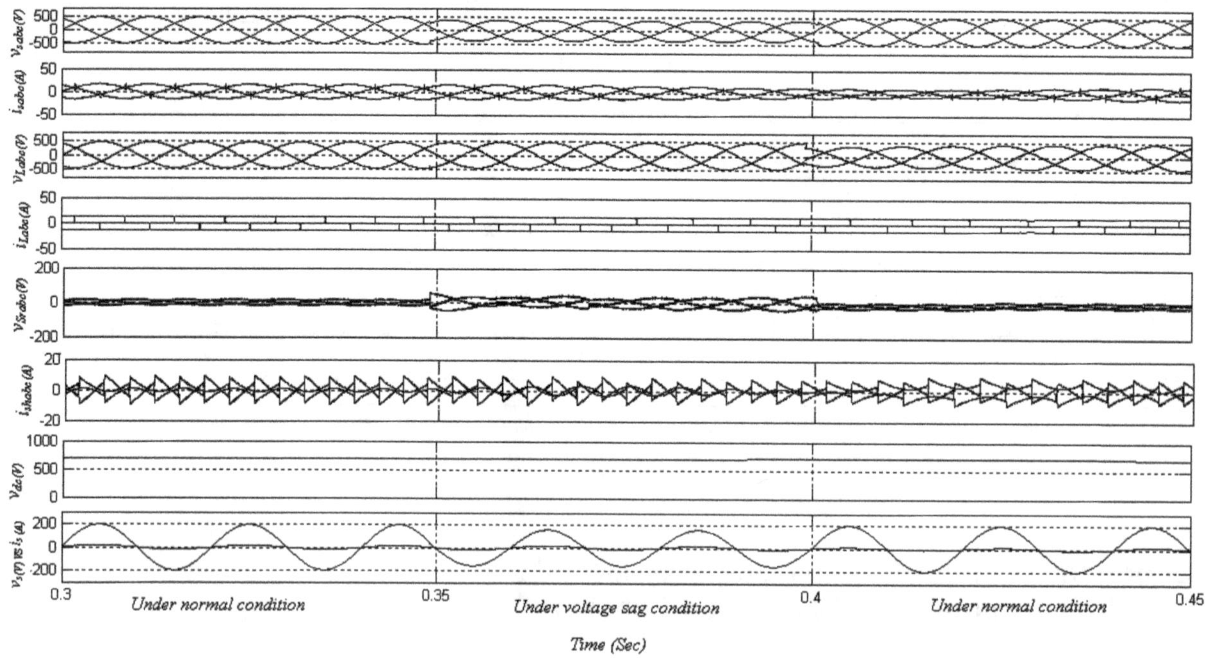

Figure 11. Waveform and harmonic spectra of (a) source current of phase "a", (b) source voltage "a", (c) load current of phase "a", (d) voltage at PCC of phase "a"

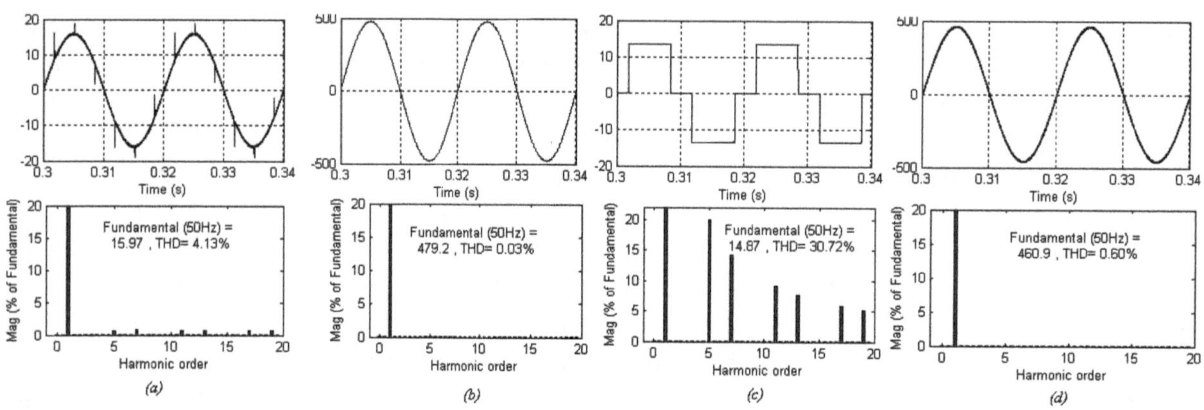

Table 1. Summary of design of 3P3W UPQC based on given specification of the load

S. No.	Parameters	Design Expression	Calculated Value	Selected Value
1	DC link voltage	$V_{dc} = 2\sqrt{2V_{LL}} /(\sqrt{3m})$	677.69 V	700 V
2	DC link capacitance	$C_{dc} = 6 \times \frac{V_{rms}(\mathrm{aI}t_{rec})}{V_{dc.ref}^2 - V_{dc.act}^2}$	2600μF	3000 μF
3	Shunt AC inductance	$L_f = \frac{mV_{dc}}{2*\grave{A}*f_{sw} * a * (\%I_{rip})}$	1.85mH	2mH
4	Voltage rating of switch	$V_{Sw} = V_{dc}$ +10% of V_{dc}	770 V	1200 V
5	Current rating of switch	$I_{Sw} = 1.25*(I_{rip} + I_{s\,peak})$	79.9 A	100 A
6	Series AC inductance	$L_{se} = \frac{V_{DC}}{\sqrt{3}*4*f_{sw} * (\%I_{rip})}$	2.02mH	2.5mH
7	Injection transformer	10kVA, 120V/240V		

SCOPE OF UPQC

The power quality improvement of power distribution system is achieved using the unified power quality conditioner (UPQC). The scope of work covers classification of the UPQC. The classification of the UPQC is based on various parameters which is shown in Figure 12. It addresses the control and implementation features of UPQC for practical applications.

Classification of UPQC

The tree of the UPQC is classified, which is shown in Figure 12. It is a pictorial view of UPQC classification as a tree. The tree of the UPQC has main two branches, structural based and series voltage injection angle based control. The structural based UPQC is further classified as AC supply system based, configuration based, and converter topology based. Each branch has a different topology and configuration.

Classification Based on Structure

The UPQC has been organized on the basis of the physical structure. It is useful to improve the power quality in AC distribution system. The important parameters that contribute to these classifications are based on numbers of phases, back to back connected energy storing devices, and physical location of the series and shunt converter.

Figure 12. Tree of the UPQC

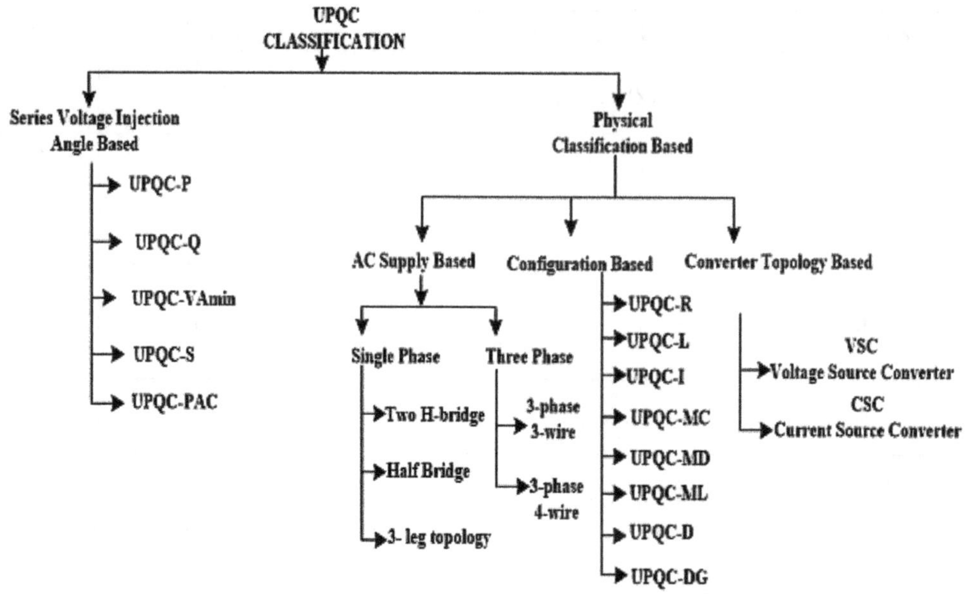

Classification Based on Supply System

The AC loads are mainly classified based on single phase and three phase or three phase four wire AC mains supply. In Figure 13, a single phase two wire (1P2W) H–bridge converter based UPQC configuration with eight semiconductor switches is connected to AC distribution system. Correa et al (2003, 2005) have worked on single phase based UPQC. It represents the VSC-based 1P2W UPQC topology.

Figure 13. Two H-bridge configurations of single phase 2-wire UPQC

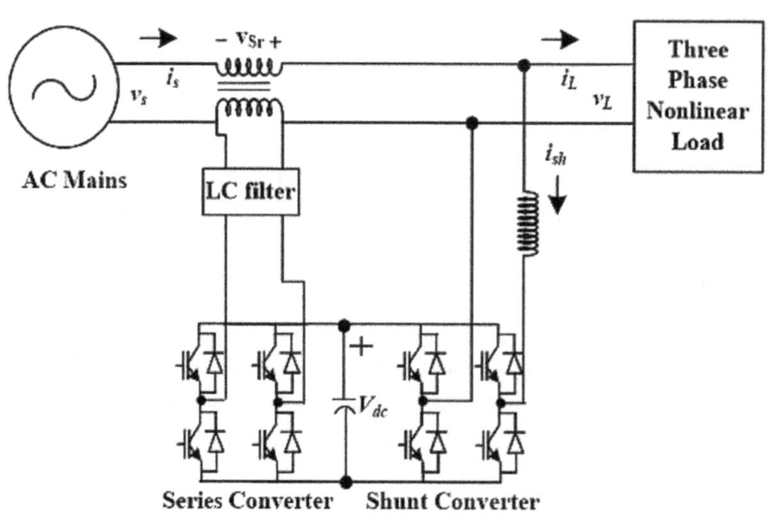

The same topology is also available for current source converter (CSC) based UPQC in (Simoes, 1989). Nasiri and Emadi (2003) presented a modified H–bridge converter based UPQC configuration with six semiconductor switches which is shown in Figure 14.

In addition, Prieto et al. (2005) have given the half-bridge topology based UPQC which is shown in Figure 15. The similar topology can be also found in (Kwan et al., 2009; Khoor & Machmoum, 2007) Zhang et al (2008) have considered a bidirectional two H-bridge DC/DC isolated converter topology to isolate UPQC shunt and series converters from each other. The reduced switching devices may affect the compensation performance of UPQC. These topologies are used for low power and low-cost applications.

A 3P3W voltage source converter based UPQC is depicted in Figure 16. Muthu and Kim (1997) have proposed three phase UPQC to enhance the power quality. This configuration is highly recognized. Despite the 3P3W loads, the demand of industrial plants is the combination of single phase and three phases three wire loads supplied by 3P4W AC mains which is shown in Figure 17. The fourth wire is the neutral conductor of the 3P4W AC mains system. Under unbalanced condition, an excessive neutral current flows in the fourth wire, thus an additional compensator is required. Zhili et al (2006) have introduced 3P4W UPQC to mitigate the neutral current in three phase four-wire AC system. Many shunt converter configuration have been attempted, namely, two split capacitor. Pal et al. (2010) have proposed 3 H-bridge, 4 H-bridge topologies. The 2C topology comprises of two split capacitors on the DC side. The midpoint of the split capacitors is treated as neutral point where potential difference expected to be zero. The neutral point is used as a neutral conductor for 3P4W load. In 2C topology, it is essential to maintain equal voltages at both the capacitors to avoid the flow of circulating current. Thus, it needs an extra control algorithm for DC link capacitor to regulate the voltage in 2C topology.

In Figure 18, a 3P4W topology or 4L topology of UPQC is illustrated. It requires a supplementary leg (two semiconductor switches) to compensate the neutral current.

In Figure 19, the 3H-bridge (3HB) topology consists of single phase H-bridge converters connected back to back DC link where shunt converter have 3H-bridge. Jindal et al (2007) have introduced the shunt converter and series converter as 3HB topology (twenty-four semiconductor switches). However,

Figure 14. Three leg configuration of single phase 2-wire UPQC

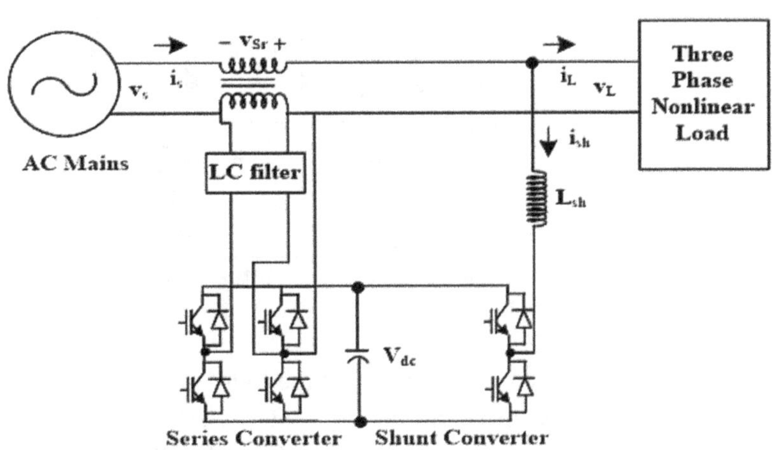

Figure 15. Half H bridge configuration of single phase 2-wire UPQC

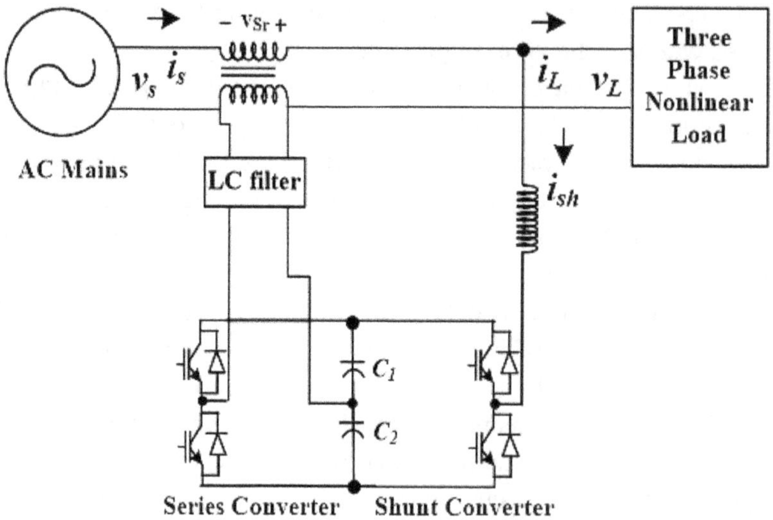

Figure 16. Three phase three wire UPQC

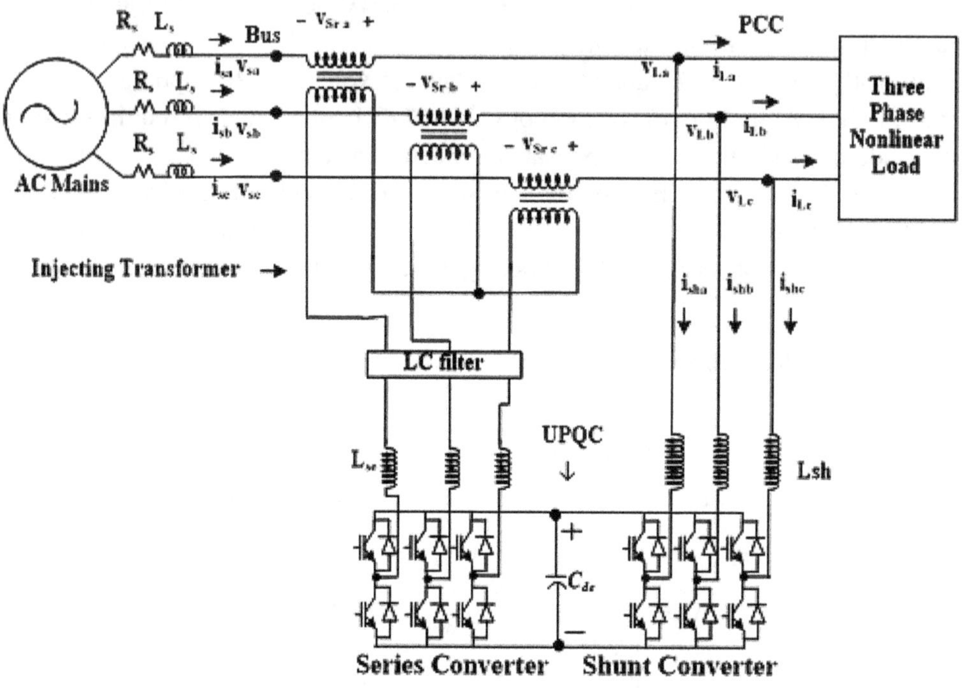

such type of configuration is complicated and it increases the total number of semiconductor switches. Moreover, this topology of UPQC system would increase the losses, overall size, and the cost of the system In Pal et al. (2010), the neutral current compensation of 3P3W UPQC comprises star–hexagon/T-connected transformer to circulate the neutral current.

Figure 17. Three phase four wire based on 2C shunt converter topology

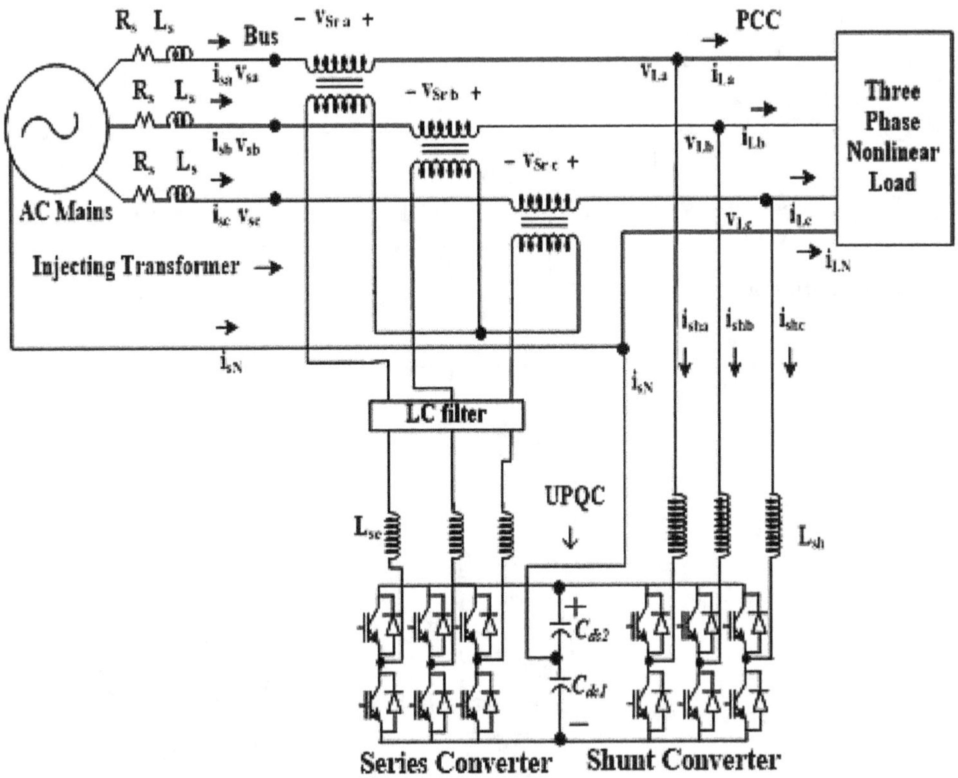

Figure 18. Three phase four wire four line shunt converter topology

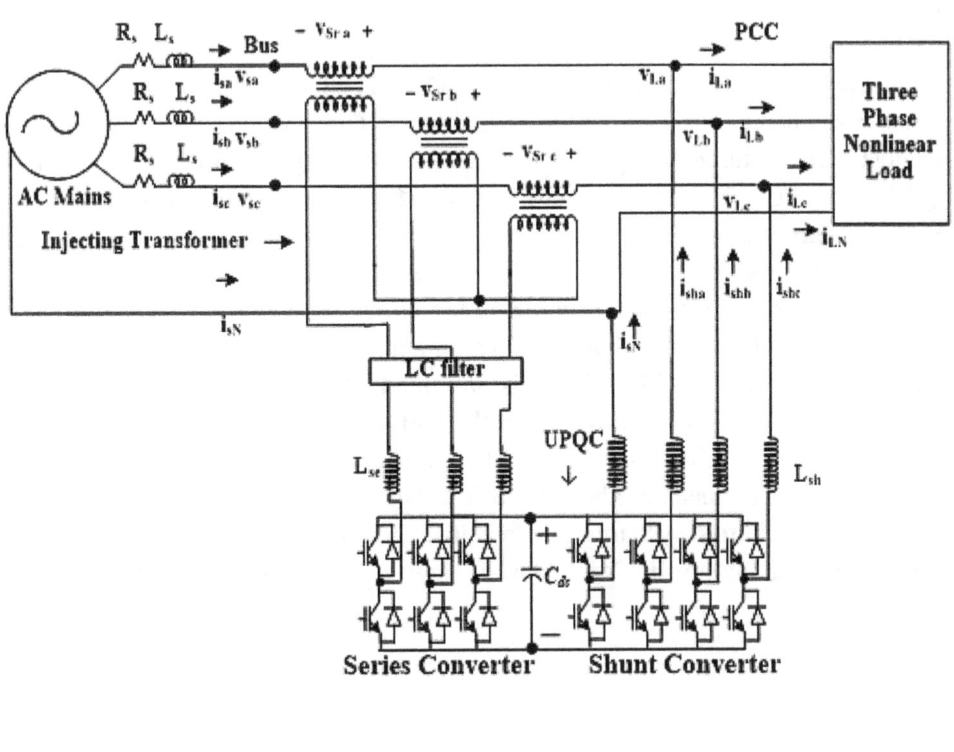

Figure 19. Three phase four wire based on 3HB shunt converter topology

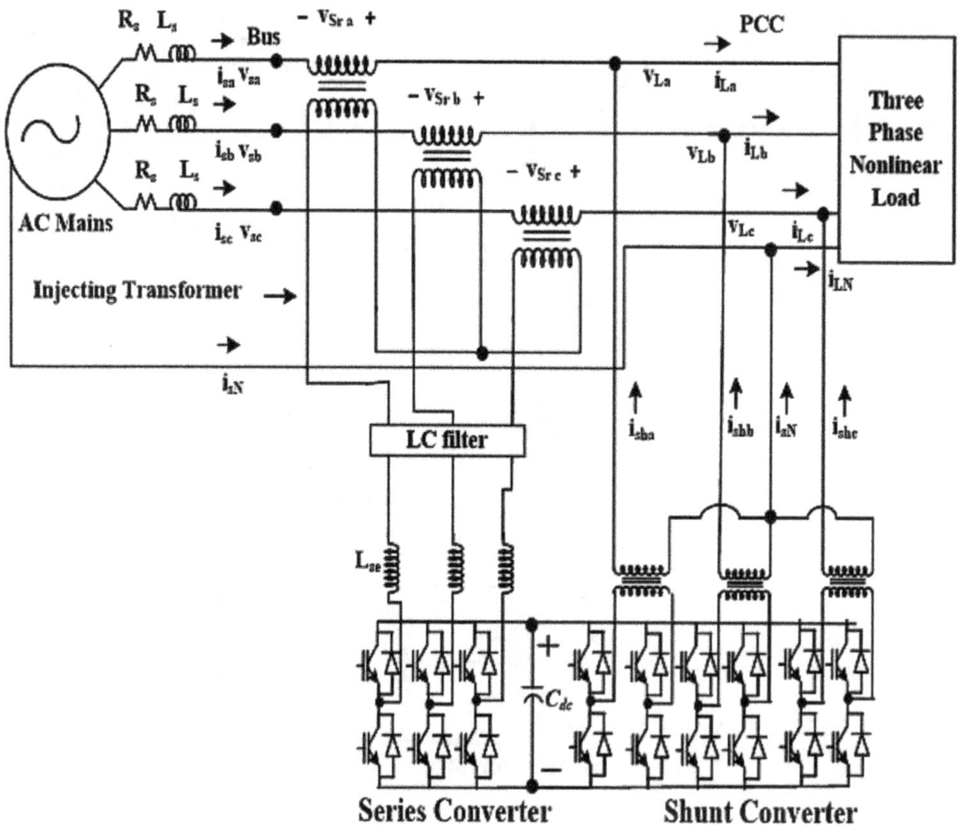

Classification Based on Configuration

This section classifies on different configuration of UPQCs. The UPQC has two parts of converters, one is called as series part whereas second is called as shunt part. The placement of the series and shunt part can be interchanged. This interchanging of shunt converter of UPQC is knows as Right and Left Shunt UPQC (UPQC-R and UPQC-L). Figures 22-26 represent UPQC-R system configuration, while Figure 20 shows UPQC-L configuration. Presently UPQC-R is well known among two configuration, since in UPQC-R currents flow through series transformers which are nearly sinusoidal under any load. Thus, UPQC-R shows the better performance as compared to UPQC-L. The UPQC-L configuration is used in to avoid the interference between the shunt converter and passive filters.

In Figure 21, an interline UPQC (UPQC-I) is presented which is proposed by Jindal et al (2007), where the series converter and shunt converter of the UPQC are connected between two AC distribution feeders named as interline UPQC (UPQC-I).The series converter is connected in one of the feeders whereas shunt converter is connected other of the AC distribution feeders. In this configuration, both the feeders of the AC distribution system regulate voltage simultaneously. Moreover, real power of two feeders can be managed and controlled by UPQC-I. However, UPQC-I has certain limitations such as

Figure 20. Configuration of UPQC-L

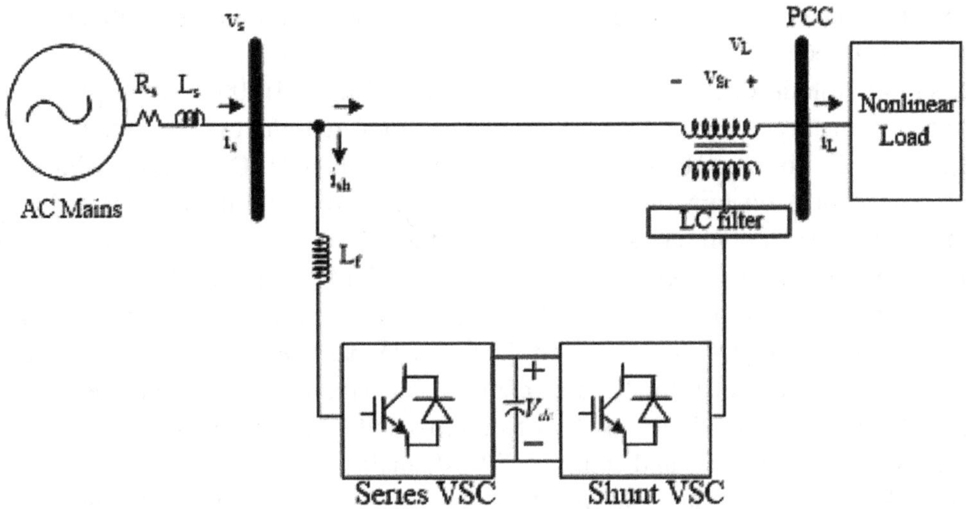

Figure 21. An interline configuration of the UPQC (UPQC-I)

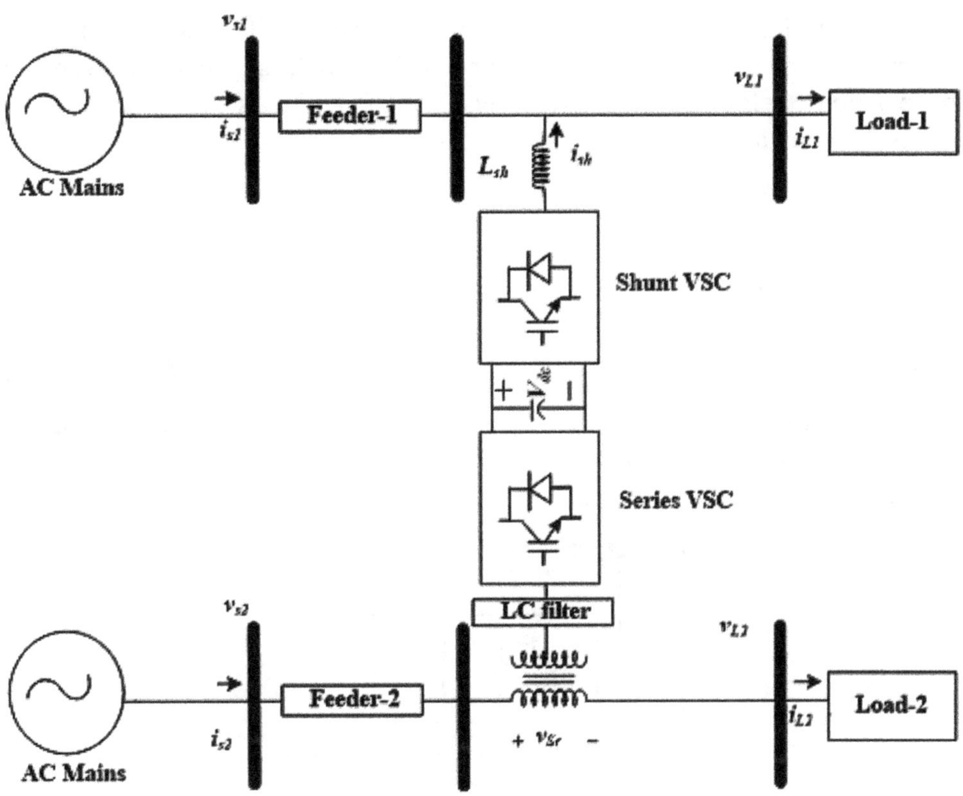

harmonics and unbalanced load which may be effectively compensated only on the feeder in which the shunt VSC is connected. Alternatively, the voltage harmonics can only be mitigated in the series converter connected feeder.

Figure 22 shows the multi-converter UPQC (UPQC-MC). Wong et al (2000) have explored the potential to improve the performance of UPQC by introducing third converter unit to support to DC link. In Wong et al (2000), the authors discussed enhancing the performance of the UPQC-MC by using a storage battery or a super capacitor. Mohammadi et al. (2009) have proposed the third converter which is placed in parallel with the same feeder or in series or parallel with the adjacent feeder. This configuration is called as MC-UPQC (Multi-converter UPQC). In MC-UPQC, the third converter is connected in series with the adjacent feeder. Similar to UPQC-I, the MC-UPQC can be connected between two different feeders.

In Figure 23, a modular UPQC configuration (UPQC-MD) has been introduced by Han et al. (2006). In this configuration, several UPQC modules are connected in parallel to each other. The shunt converters of the UPQC-MD are connected in series through multi-winding transformer whereas series converters are connected directly in series and connect with AC mains without a series transformer. It is also presented in the literature the series converter H-bridges are connected in parallel and inserted in the AC mains through series transformers. In UPQC-MD as the numbers of modules are more, the voltage regulated by each individual H-bridge may reduce, and therefore, it can be useful in the medium voltage application to achieve higher power levels.

In Figure 24, a UPQC-ML system configuration is presented by Rubilar et al. (2007) based on a three-level neutral point clamped (NPC) topology. Similar to UPQC-MD, the UPQC-ML can be configured as an alternative option to achieve higher power levels. The UPQC-ML is also realized for several levels like 3 level, 5 level, 7 level and so on.

Figure 22. UPQC-MC system configuration

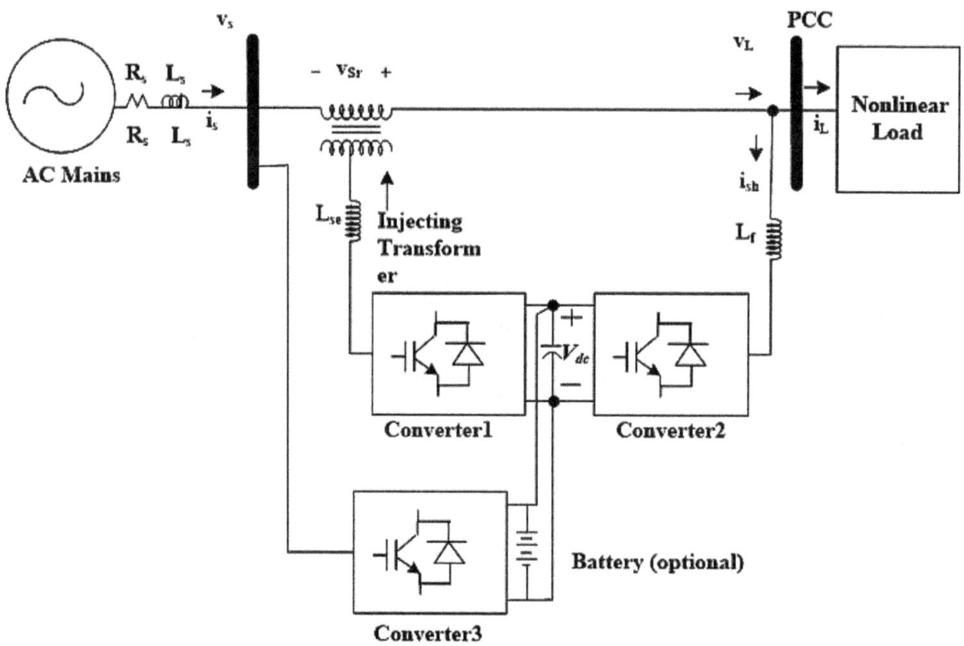

Figure 23. UPQC-MD system configuration

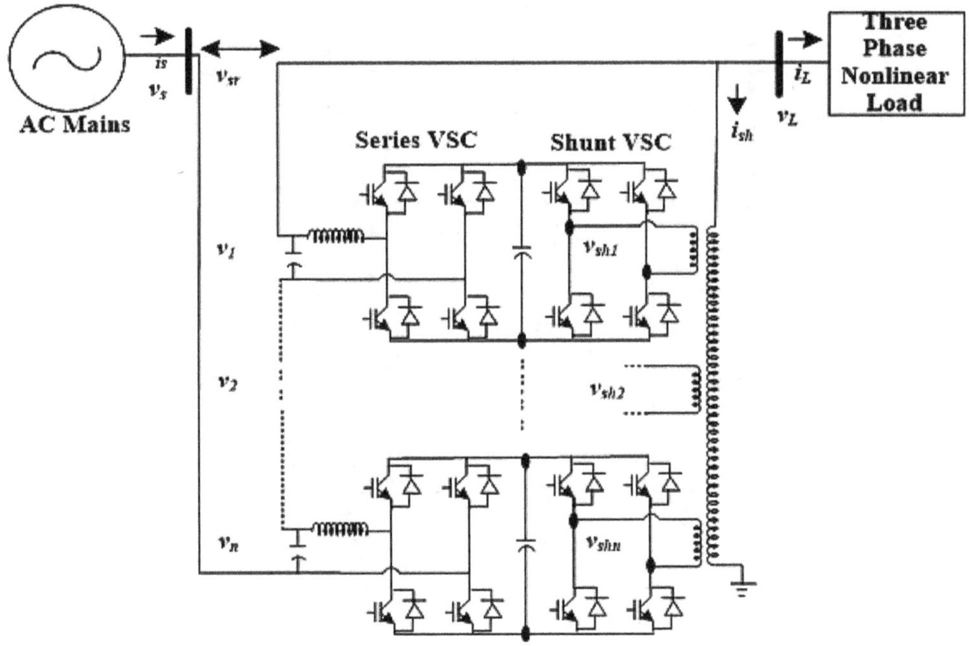

Figure 24. UPQC-ML system configuration

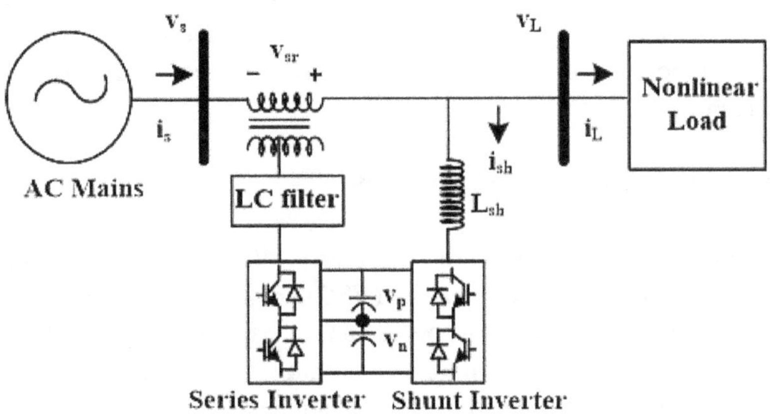

In Figure 25, distributed three phase four-wire UPQC (UPQC-D) is presented by Khadkikar and Chandra (2009, 2011). A 3P4W AC distribution system is mostly running a neutral conductor along with power lines from substation or by utilizing a delta–star transformer at the AC distribution level. While a new topology for 3P4W UPQC-based AC distribution system is possible to extend the UPQC-based 3P43 AC system to a 3P4W AC system. The neutral of series transformer, used in the series part of UPQC, is considered as a neutral for 3P4W AC system.

In Figure 26, the UPQC is integrated with distributed generation (DG) systems (Melı'in et al., 2010). This configuration is represented as UPQC-DG system. It has many additional advantages by injecting the active power to the grid and the loads. It is also used as islanding mode. On the other hand, this

Figure 25. UPQC-D system configuration

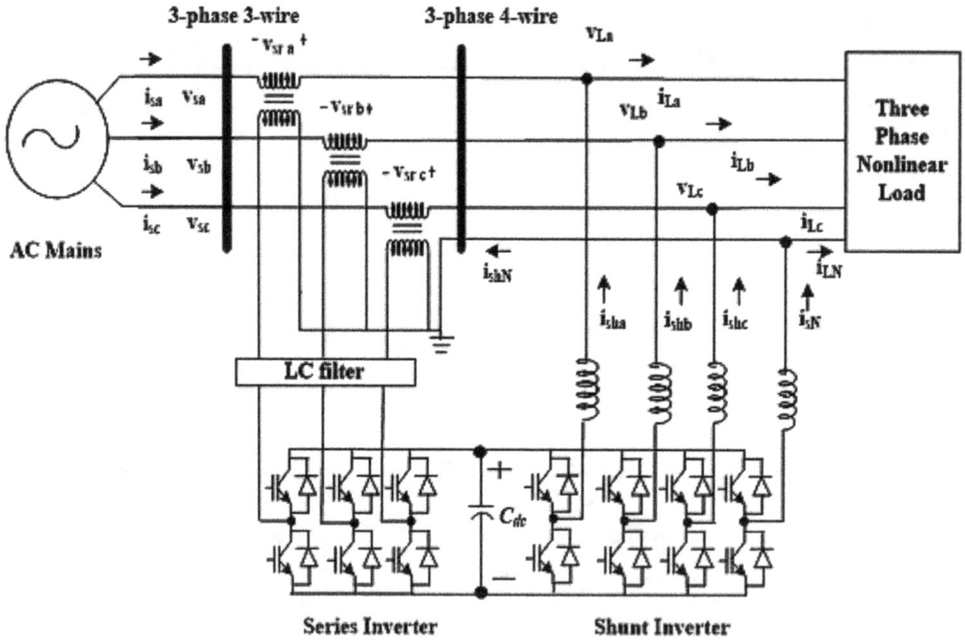

configuration has some complexity to control the active power injection under normal operating condition of the AC mains. Moreover, complexity of the DG output control circuit is increased the cost of UPQC-DG system. These features must to be addressed effectively for practical feasibility of UPQC-DG system. However, these topologies give alternative options to appreciate the UPQC-DG based system configuration in different ways.

Classification Based on Converter Topologies

In this section, the classification of UPQC based on converter topologies is two types. The most of the UPQC topologies based on voltage source converter (VSC) are used in the literature. On the other hand, some of the researchers have proposed current source based UPQC topology in (Akagi, 1994; Basu et al 2008; Basu et al., 2004). The connection of the DC link that shares a common energy storage inductor L_{dc} is in series with the back to back connection of CSC which is shown in Figure 27. The DC current in the inductor is regulated such that the average input power is equal to the average output power plus the power losses in the UPQC. The CSC-based UPQC topology has increased losses, high cost, and the fact that it cannot be used in multilevel configurations. In Figure 20, a single-line diagram of a VSC-based UPQC system is presented. This topology is offered many advantage over CSC based topology such as lighter in weight, no blocking diode, low cost, ability of multilevel operation, and flexible overall control.

Classification Based on Angle of Series Voltage Injection

The power quality in AC distribution system is so important, which depends on the voltage and current quality. The voltage of AC mains is frequently changing as load is changing, therefore, the series converter

Figure 26. UPQC-DG system configuration

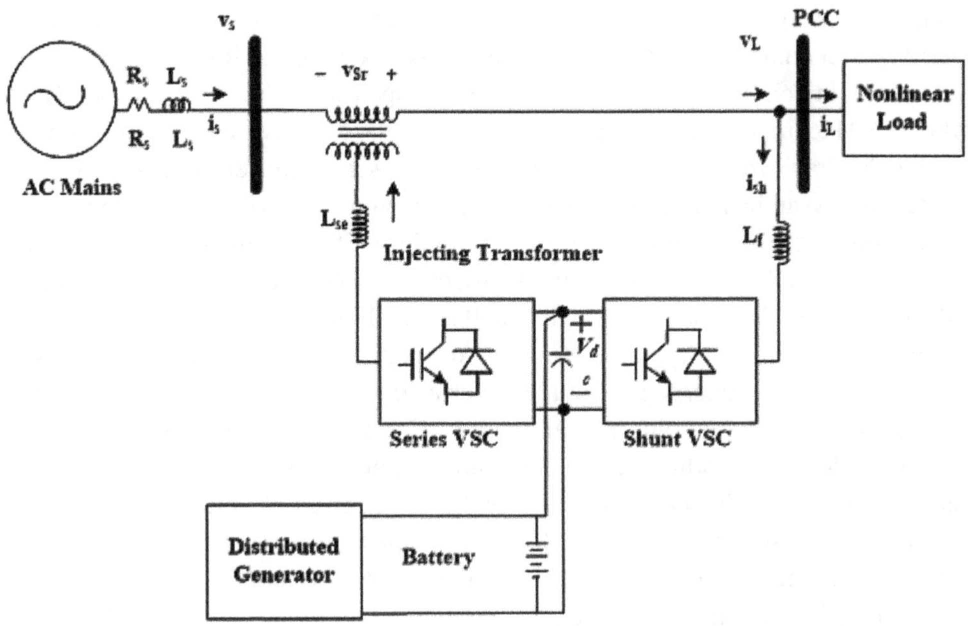

Figure 27. Current source converter (CSC) based UPQC

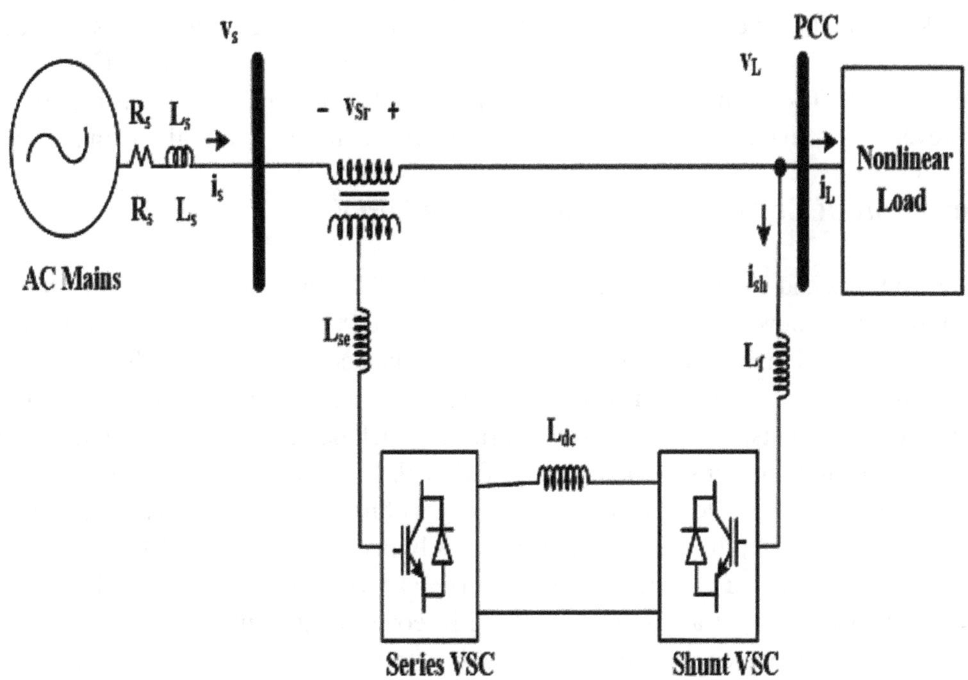

always busy to maintain the load voltage. It has been seen in the literature voltages sag is major power quality problems. There are four modes of voltage injection to the AC mains to mitigate the voltage sag. First one is active power control mode (UPQC-P) under this mode voltage injects in phase with AC mains current, this modes of injection requires active power to compensate voltage sag, thus UPQC is named as UPQC-P (Khoor & Machmoum, 2005). In order to achieve the effective voltage sag compensation, the shunt VSC of UPQC draws the required active power by the series converter plus the losses associated with UPQC. Due to this, an increased AC mains current magnitude during sag compensation is seen in UPQC-P. Second one is the reactive power injection mode (UPQC-Q) under this mode, the voltage is injected in quadrature with the AC mains current. It also compensates the reactive power of the load (Basu et al, 2008; Basu et al., 2004). In such a case, it is called as UPQC-Q (Q for reactive power). The shunt converter of UPQC maintains a unity power factor at AC mains side. To compensate same percentage of voltage sag, the UPQC-Q needs large magnitude of series injection voltage than the UPQC-P. Besides, the UPQC-Q cannot mitigate the voltage swell of the AC mains. The third mode of voltage injection is to minimize the UPQC VA loading during sag compensation (Kolhatkar & Das, 2011; Kisck et al., 2007; Khadkikar & Chandra, 2011). Under this condition injection of voltage neither in phase nor in quadrature of AC mains current but it is injected at a certain optimal angle with respect to AC mains current. This method to compensate the voltage sag using UPQC is abbreviated as UPQC-VAmin. Fourth mode of voltage injection is simultaneous active and reactive power control (UPQC-S). This mode is same as to UPQC-VAmin, where the series converter injects both active and reactive powers. Unlike the UPQC-VAmin, in this method, the efforts are made to use the available series converter VA loading to its maximum value. The series converter of UPQC is controlled to perform simultaneous voltage sag/ swell compensation and load reactive power sharing with the shunt VSC. Since, the series converter of UPQC in this case, injects both active and reactive powers, it is given the name UPQC-S (S for complex power) (Tlusty, 2004). The control UPQC-S has many control loops and, thus, looks relatively complex to employ. Nevertheless, it can simply be implemented when controlled digitally using a DSP.

Implementation of Control Algorithms of the UPQC

The reference voltages and currents are extracted using the different control algorithms for controlling the switching of series converter and shunt converter of the UPQC. The control algorithms can shape the load voltages and AC mains currents sinusoidal and balanced at PCC. Thus the control algorithms are the heart of power electronics based custom power devices which plays the significant role in power electronics based equipments. It is the control algorithm which decides the performance of the custom power devices in AC distribution system. Since UPQC is also a power electronics based system, therefore, the effectiveness of UPQC depends on control algorithm. There are so many control algorithms presented in the available literature which have effectively been used with UPQC systems. Frequency domain methods, based on the Fast Fourier Transform (FFT) which is rarely used as it takes large computational time and delay in operation of controller. Moreover, control methods for UPQC in the time domain are based on instantaneous derivation of compensating commands in the form of voltage and current signals. Thus, time domain based control algorithm is fast in operation than the FFT. There are a large number of control methods in the time domain. The various control algorithms are categorized and explained to get the basic knowledge of their applications.

Frequency Domain Based Control Algorithms

These methods are based on extraction of harmonics using techniques such as the Fast Fourier Transformation (FFT). The harmonic content of voltage or current is obtained by extracting AC fundamental component using a low pass filter. Then the FFT of harmonics is carried out and coefficients for each harmonic are computed. The disadvantage of this method is large computation time hence, delay for the calculation of FFT. In many cases, a minimum one whole period of waveform is needed to perform the FFT. Therefore, the controllers based on frequency domain methods are not popular for these application, because the response time plays a crucial role. Tlusty and Valouch (2004) have used FFT based controller to flicker mitigation of distribution system and Kwan et al. (2007) have also realized control algorithm using frequency domain method.

Time Domain Based Control Algorithms

A large numbers of time domain based control algorithms are further classified in transform based control algorithms, simple control algorithms and model predictive control algorithms for UPQC which are being implemented.

Transform Based Control Algorithms

The instantaneous active and reactive power method or pq theory and synchronous reference method or dq theory are most popular time-domain based control algorithms which are most widely used for UPQC. The fundamental active and reactive components of load voltages and load currents can be extracted using transformation in time based control algorithms. The quantities of these signals can be extracted using Park's transform and Clark's transformation or any other transformation methods. Therefore, these types of algorithms are known as transform based control algorithms. In dq theory voltage and current signals in three phase ABC rotating reference frame are transformed to stationary reference frame, it is used to extract fundamental and harmonic quantities. The basic difference in pq and dq theory is that pq theory based on instantaneous active and reactive component, which depends on AC mains current whereas, the dq theory is independent of AC mains currents. These two theories associate active and reactive powers of the fundamental components. The fundamental components are DC quantities which can be extracted using low pass filter (LPF) or high pass filter (HPF). Since DC signal extraction is the filtering of signals in the α-β reference frame is in sensitive to any phase shift errors established by LPF. On the other hand, the cutoff frequency of these LPF or HPF can affect the dynamic performance of the controller. Akagi et al (1984) have proposed pq theory based controller to generate pulse of series and shunt converterss of the UPQC and then numbers of researchers have used which are reported in literatures (Watanabe & Aredes, 2002; Khoor & Machoum 2005; Devassy & Singh, 2017; Khadkikar et al., 2011; Bhattacharya & Divan 1995). Bhattacharya and Divan (1995) have proposed dq based control algorithm, it can be also found in (Watanabe & Aredes, 2002; Chen & Joos, 2001; Mohammadi et al., 2009; Brenna et al., 2009). The pq theory based control algorithm under distortion of AC mains voltage and currents exhibits limitations. These limitations can overcome by introducing the modified theory referred as pqr theory. This modified theory can be found in (Han et al., 2006).

Classical Control Algorithms

In classical control algorithms, estimation of fundamental component of the load currents and voltages are totally based on a concept of a simple mathematics like divisions/subtractions and triangle rules. The various control algorithms in this category, are unit vector template generation (UVTG), phase locked loop (PLL), enhanced phase locked loop (EPLL), one cycle control (OCC) and the power angle controller (PAC). G. Chen et al. (2004) have given one cycle control (OCC) of switching converters concept based control algorithm to generate the pulses of UPQC. Ghosh et al. (2003) have proposed a pole shift control algorithm to extract the reference voltage and current for UPQC. It is a discrete-time control algorithm where the closed-loop poles are taken by radially shifting the open-loop poles toward the origin. In Javadi et al. (2016) unit vector template generation (UVTG) has proposed to extract the sinusoidal voltage and current for generation of the gate pulses of UPQC. Moreover, Khoor and Machmoum (2005) have introduced an analogical technique for perturbation detection of voltage and current signals. It does not require PLL to synchronize AC mains signals. Ansari et al. (2015) have proposed a modified power angle controller. This controller is capable to compensate the power quality in the AC distribution system. The PAC algorithm helps to reduce the overall rating of the shunt VSC of the UPQC.

Model Predictive Control (MPC) Algorithm

The model predictive control (MPC) methods are under the time domain based control algorithms that consider system dynamics. Zhang et al. (2010) have proposed MPC based control model for UPQC. The MPCs can handle multivariable control problems and have relatively simple online calculations. Li et al. (2005) have proposed a H∞-based model matching control to track the converter output waveforms for robust control of UPQC. Besides, Kwan et al. (2009) have introduced a model-based solution via H∞ loop shaping for effective control of UPQC. Kwan et al. (2005) have proposed a filter that can be combined to extract the harmonics in AC mains voltage and load current, it is also reported in Zhang et al. (2010) and Bai et al. (2005).

Additional Significant Control Algorithms

The significant control algorithms are based on the soft computing techniques and it also contains DC link voltage controllers. Estimation of fundamental active and reactive power components of load voltages and the load currents are based on training of input signals, used functions, constant parameters and other factors in soft computing techniques. Constant parameters and other factors depend on selected control algorithms. Shen et al. (2009) have given particle swarm optimization technique to develop the control algorithm of UPQC. Furthermore, an ANN technique can also handle the multi-input multioutput control system effectively. Tey et al. (2002) have suggested artificial neural network (ANN) technique used to develop the control algorithm of the UPQC. Banaei and Hosseini (2006) have given feed forward ANN scheme to extract the harmonics components under the nonlinear load. Zhou et al. (2006) have reported a Levenberg Marquardt back propagation ANN technique is used for UPQC. There are certain drawbacks in the time-domain and frequency-domain methods. Thus, a wavelet technique is introduced by Elnady et al. (2001) to generate pulse of UPQC.

The best performance of the UPQC also depends on DC link voltage which plays key role to get desired output. Many control algorithms have been introduced to maintain the DC link voltage under the dynamic condition of the AC mains. Vilathgamuwa et al. (1998) have given a PI controller to restore the DC link voltage at set reference value with a minimum delay as well as lower overshoot. Singh et al. (1998) have suggested a fuzzy logic based controller which fast in response time. Zhang et al. (2007) have suggested the optimized controller of DC link controller.

FUTURE RESEARCH DIRECTIONS

The renewable energy is the emerging in the area source of electricity production. Most popular renewable energy is the solar and the wind. The most of the renewable source required a means to make the system in desire operation such as AC to DC and DC to AC operation. The voltage source converter (VSC) plays a key role to convert the generated DC power to AC power and extracted power transfer directly to the load or grid. The distributed generation (DG) can be used as DC source connected back to back with UPQC. Thus DG system can be fed to DC link to the UPQC. The UPQC can be used to regulate the DC power of the DG system directly and supply AC loads at PCC with improved power quality. The active power of the DG system can provide to compensate the load reactive power, load harmonics, voltage sag and swell under normal operating condition. Moreover, excess DG source can be supplied to DC storing device as a backup. Such kind of provision gives the additional support by providing uninterruptible power supply to the load under voltage interruption of AC mains. The DG system can be used as an interconnected mode or islanding mode. The low energy generation of DG system can also be integrate with UPQC while the generation of DC voltage is required high step up voltage under such condition. Thus integration of DG system with UPQC depends on various factors which is to be addressed adequately. However, the integration of UPQC and renewable energy may play a important role in modern AC distribution system to simultaneously control the consumer demand and improve the quality of power.

CONCLUSION

The application of the power electronics in power system has been explored in this chapter. The evolution of power electronics has presented as a hierarchy development for power system. The FACTs controlling devices and custom power devices (CPDs) are mostly used to improve the power quality. One of the custom devices a UPQC has been introduced. The two-converter connected back to back with DC link is known as the UPQC. The Matlab model of UPQC and its control approaches based on synchronous reference theory with results are discussed. This chapter has also presented main concepts of operating principle of the series and shunt converters of UPQC. It is also presented various types of configurations based on interesting features of the UPQC by organizing an acronymic list. These acronyms can be utilized to clearly recognize particular function, utilization, configuration, and characteristic of the UPQC system under normal condition. The up to date literature on UPQC is also highlighted. Moreover, the various controlling approaches of the UPQC have been discussed in deliberately to understand the how to generate the pulses of the gate of the power electronics based equipments. Overall discussion concludes that desired performance of UPQC be achieved with the maintaining the DC link voltage at constant level which is achieved by shunt converter.

REFERENCES

Akagi, H. (1994). Trends in active power line conditioners. *IEEE Transactions on Power Electronics*, *9*(9), 263–268. doi:10.1109/63.311258

Akagi, H. (1996). New trends in active filters for power conditioning. *IEEE Transactions on Industry Applications*, *l*(32), 1312–1322. doi:10.1109/28.556633

Akagi, H., Kanazawad, Y., & Nabae, A. (1984). Instantaneous reactive power compensators comprising switching devices without energy storage components. *IEEE Transactions on Industry Applications*, *IA-20*(20), 625–630. doi:10.1109/TIA.1984.4504460

Ansari, A. Q., Singh, B., & Hasan, M. (2015). Algorithm for power angle control to improve power quality in distribution system using unified power quality conditioner. *IET Generation, Transmission & Distribution*, *9*(9), 1439–1439. doi:10.1049/iet-gtd.2014.0734

Aredes, M., Heumann, K., & Watanabe, E. H. (1998). An universal active power line conditioner. *IEEE Transactions on Power Delivery*, *1*(13), 545–551. doi:10.1109/61.660927

Bai, P., Li, Q., Zhao, B., & Yang, Y. (2005). Power quality control center and its control method. *Proc. Transmiss. Distrib. Conf. Exhib.*, 1–6.

Banaei & Hosseini. (2006). Mitigation of current harmonic using adaptive neural network with active power line conditioner. *Proc. IEEE 5th International Conference on Power Electronics and Motion Control*, 1–5.

Basu, M., Das, S. P., & Dubey, G. K. (2008). Investigation on the performance of UPQC- for voltage sag mitigation and power quality improvement at a critical load point. *IET Generation, Transmission & Distribution*, *l*(2), 414–423. doi:10.1049/iet-gtd:20060317

Basu, M., Farrell, M., Conlon, F., Gaughan, K., & Coyle, E. (2004). Optimal control strategy of UPQC for minimum operational losses. In *Proc. 39th Int* (pp. 246–250). Univ. Power Eng.

Bhattacharya, S., & Divan, D. (1995). Synchronous frame based controller implementation for a hybrid series active filter system. *Proc. 30th Ind. Appl. Soc. Annu. Meet. Ind. Appl. Conf.*, 2531– 2540. 10.1109/IAS.1995.530625

Brenna, M., Faranda, R., & Tironi, E. (2009). A new proposal for power quality and custom power improvement: OPEN UPQC. *IEEE Transactions on Power Delivery*, *l*(24), 2107–2116. doi:10.1109/TPWRD.2009.2028791

Chen, G., Chen, Y., & Smedley, K. M. (2004). Three-phase four-leg active power quality conditioner without references calculation. *Proc. Appl. Power Electron. Conf.*, 587–593.

Chen, S., & Joos, G. (2001). A unified series–parallel deadbeat control technique for an active power quality conditioner with full digital implementation. *Proc. IEEE 36th Ind. Appl. Soc. Annu. Meet. Ind. Appl. Conf.*, 172–178.

Correa, J., Farret, F., & Simoes, M. (2005). Application of a modified single phase P-Q Theory in the control of shunt and series active filters in a 400Hz microgrid. *Proc. Power Electron. Spec. Conf.*, 2585–2591.

Correa, M., Chakraborty, S., Simoes, G., & Farret, A. (2003). A single phase high frequency AC microgrid with an unified power quality conditioner. *Proc. IEEE Int. Symp. Ind. Electron.*, 956–962.

Devassy, S., & Singh, B. (2017). Modified p-q Theory Based Control of Solar PV Integrated UPQC-S. IEEE Transactions on Industry Applications, 1(1) 1-11.

Elnady, A., Goauda, A., & Salama, M. M. A. (2001). Unified power quality conditioner with a novel control algorithm based on wavelet transfor. *Annual IEEE Canadian Conference on Electrical and Computer Engineering*, 1041-1045.

Elnady, A., Goauda, A., & Salama, M. M. A. (2001). Unified power quality conditioner with a novel control algorithm based on wavelet transform. *Proc. Can. Conf. Electr. Comput. Eng.*, 1041–1045. 10.1109/CCECE.2001.933586

Fujita, H., & Akagi, H. (1998). The Unified Power Quality Conditioner: The Integration of Series- and Shunt-Active Filters. *IEEE Transactions on Power Electronics*, *13*(12), 315–322. doi:10.1109/63.662847

Ghosh, A., Jindal, A. K., & Joshi, A. (2003). Inverter control using output feedback for power compensating devices. *Proc. Convergent Technol. Conf.*, 48–52. 10.1109/TENCON.2003.1273212

Ghosh, A., Jindal, K., & Joshi, A. (2004). A unified power quality conditioner for voltage regulation of critical load bus. IEEE Power Engineering Society Meeting, 471-476.

Han, B., Bae, B., Baek, S., & Jang, G. (2006). New configuration of UPQC for medium-voltage application. *IEEE Transactions on Power Delivery*, *l*(21), 1438–1444. doi:10.1109/TPWRD.2005.860235

Hasan, Dinesh, & Zafer. (2013). Multimodules of diode clamped multilevel converter: A novel option for high power facts controller. *International Journal of Advance Research in Electrical Electronics and Instrumentation Engineering*, *l*(2), 109–112.

Hasan, M., Ansari, A. Q., & Singh, B. (2015). Parameters estimation of a series VSC and shunt VSC to design a unified power quality conditioner (UPQC). *39th National System Conference*, 1-6. 10.1109/NATSYS.2015.7489111

Hingorani, N. G. (1995). Introducing Custom Power. *IEEE Spectrum*, *32*(6), 41–48. doi:10.1109/6.387140

Javadi, A., Hamadi, A., Woodward, L., & Al-Haddad, K. (2016). Experimental Investigation on a Hybrid Series Active Power Compensator to Improve Power Quality of Typical Households. *IEEE Transactions on Industrial Electronics*, 63.

Jindal, K. A., Ghosh, A., & Joshi, A. (2007). Interline unified power quality conditioner. *IEEE Transactions on Power Delivery*, *1*(22), 364–372. doi:10.1109/TPWRD.2006.881581

Kazemi, A., Sarlak, M., & Barkhordary, M. (2006). An Adaptive Noise Canceling Method for Single-Phase Unified Power Quality Conditioner. *I[st] IEEE Conference on Industrial Electronics and Applications,* 6.

Khadkikar, V. (2012). Enhancing Electric Power Quality Using UPQC: A comprehensive overview. *IEEE Transactions on Power Electronics, 27*(5), 2284–2297. doi:10.1109/TPEL.2011.2172001

Khadkikar, V., & Chandra, A. (2009). A novel structure for three-phase four wire distribution system utilizing unified power quality conditioner (UPQC). *IEEE Transactions on Industry Applications, 45*(5), 1897–1902. doi:10.1109/TIA.2009.2027147

Khadkikar, V., & Chandra, A. (2011). UPQC-S: A Novel Concept of Simultaneous Voltage Sag/Swell and Load Reactive Power Compensations Utilizing Series Inverter of UPQC. IEEE Transactions on Power Electronics, 1(26), 2414-2425.

Khadkikar, V., Chandra, A., & Singh, B. (2011). Digital signal processor implementation and performance evaluation of split capacitor, four-leg and three H-bridge-based three-phase four-wire shunt active filters. *IET Power Electronics, 4*(4), 463–470. doi:10.1049/iet-pel.2010.0198

Khoor, M., & Machmoum, M. (2005). Simplified analogical control of a unified power quality conditioner. *Proc. Power Electron. Spec. Conf.*, 2565–2570. 10.1109/PESC.2005.1581994

Khoor, M., Machmoum, M., & Abdelli, Y. (2005). Using a simple analogical identification method and PID controllers for unified power quality conditioning. *European Conference on Power Electronics and Applications,* 10.

Khoor, S. M., & Machmoum, M. (2007). A novel single-phase reduced parts on-line UPS with power quality conditioning capability. *Proc. Power Electron. Spec. Conf.*, 1170–1175. 10.1109/PESC.2007.4342158

Kisck, D., Navrapescu, V., & Kisck, M. (2007). Single-phase unified power quality conditioner with optimum voltage angle injection for minimum VA requirement. *Proc. IEEE Int. Symp. Ind. Electron.,* 2443–2448.

Kolhatkar, Y., & Das, S. (2007). Experimental investigation of a single-phase UPQC with minimum VA loading. *IEEE Transactions on Power Delivery,* (22): 371–380.

Kolhatkar, Y., Errabelli, R., & Das, S. (2005). A sliding mode controller based optimum UPQC with minimum VA loading. *Proc. Power Eng. Soc. Gen. Meet.,* 871–875.

Kwan, H. K., Chu, C. Y., & So, L. P. (2009). Model-based H-∞ control of a unified power quality conditioner. *IEEE Transactions on Industrial Electronics, 56*(7), 2493–2504. doi:10.1109/TIE.2009.2020705

Kwan, K., So, P., & Chu, Y. (2005). Unified power quality conditioner for improving power quality using MVR with Kalman filters. *7th Int. Power Eng. Conf.*, 980–985. 10.1109/IPEC.2005.207050

Kwan, K. H., Chu, Y. C., & So, P. L. (2009). Model-based H-∞ control of a unified power quality conditioner. *IEEE Transactions on Industrial Electronics, 56*(7), 2493–2504. doi:10.1109/TIE.2009.2020705

Kwan, K. H., So, P. L., & Chu, Y. C. (2007). A harmonic selective unified power quality conditioner using MVR with Kalman filters. *Power Engineering Conference, International, 1* (3), 332-337.

Manimekalai, P., Harikumar, R., & Raghavan, S. (2016). SOGI Algorithm-Based Shunt Active Power Filter for Grid Integration of Photovoltaic Systems. *Journal of Circuits, Systems, and Computers*, (25): 20.

Melı'in, P. E., Espinoza, J. R., Mu˜noz, J. A., Baier, C. R., & Espinosa, E. E. (2010). Decoupled control of a unified power quality conditioner based on a current source topology for fast AC mains disturbance compensation. *Proc. IEEE Int. Conf. Ind. Technol.*, 730–736. 10.1109/ICIT.2010.5472711

Mohammadi, R. H., Varjani, Y. R., & Mokhtari, H. (2009). Multi-converter unified power-quality conditioning system: MC-UPQC. *IEEE Transactions on Power Delivery*, *l*(24), 1679–1686. doi:10.1109/TPWRD.2009.2016822

Muthu, S., & Kim, S. M. J. (1997). Steady-state operating characteristics of unified active power filters. *Proc. IEEE Appl. Power Electron. Conf.*, 199–205. 10.1109/APEC.1997.581454

Nasiri, & Emadi, A. (2003). Different topologies for single-phase unified power quality conditioners. *Proc. 38th Int. Appl. Soc. Annu. Meet. Ind. Appl. Conf.*, 976–981.

Pal, Y., Swarup, A., & Singh, B. (2010). A comparative analysis of three-phase four-wire UPQC topologies. *Proc. Int. Conf. Power Electron. Drives Energy System*, 1–6. 10.1109/PEDES.2010.5712513

Prieto, J., Salmeron, P., & Herrera, S. R. (2005). A unified power quality conditioner for wide load range: Practical design and experimental results. *Proc. IEEE Russia Power Technol.*, 1–7.

Rubilar, I., Espinoza, J., Munoz, J., & Moran, L. (2007). DC link voltage unbalance control in three-phase UPQCs based on NPC topologies. *Proc. 42nd Ind. Appl. Soc. Annu. Meet. Ind. Appl. Conf.*, 597–602.

Shen, H., Wan, J., Yuan, C., Liu, Y., & Li, G. (2009). Harmonic signal detection algorithm in parallel of UPQC studies based on PSO-FUZZY. *Asia-Pacific Power and Energy Engineering Conference*, 1–4. 10.1109/APPEEC.2009.4918399

Simoes, M. (1989). A line voltage regulator/conditioner for harmonic-sensitive load isolation. *Proc. IEEE Ind. Appl. Soc. Annu. Meet. Conf.*, 947–951.

Singh, B., Al-Haddad, K., & Chandra, A. (1999). A review of active filters for power quality improvement. *IEEE Transactions on Industrial Electronics*, *46*(5), 960–971. doi:10.1109/41.793345

Singh, B. N., Chandra, A., Al-Haddad, K., & Singh, B. (1998). Fuzzy control algorithm for universal active filter. *Proc. Power Quality Conf.*, 73–80. 10.1109/PQ.1998.710347

Tey, L. H., So, P. L., & Chu, Y. C. (2002). Neural network-controlled unified power quality conditioner for system harmonics compensation. Proc. IEEE/PES Transmiss. Distrib. Conf. Exhib., 1038–1043.

Tlusty, J., & Valouch, V. (2004). Effectiveness of unified power quality conditioner for flicker mitigation. *Proc. 4th Int. Power Electron. Motion Control Conf.*, 902–907.

Vilathgamuwa, M., Zhang, H. Y., & Choi, S. S. (1998). Modelling, analysis and control of unified power quality conditioner. In *Proc Harmonics Quality Power*. doi:10.1109/ICHQP.1998.760182

Watanabe, E. H., & Aredes, M. (2002). Power quality considerations on shunt/series current and voltage conditioners. *Proc.10th Int. Conf. Harmonics Quality Power*, 595–600. 10.1109/ICHQP.2002.1221503

Wong, C. M., Zhan, J. C., Han, D. Y., & Zhao, B. L. (2000). A unified approach for distribution system conditioning: Distribution system unified conditioner. *Proc. Power Eng. Soc. Winter Meet*, 2757–2762.

Zhang, H., Liu, J., Huang, X., & Wang, Z. (2007). Design of a new DC link voltage controller for universal power quality controllers. *Proc. Appl. Power Electron. Conf.*, 473–477.

Zhang, X., Zhang, W., Liu, Y., Lv, W., & Wang, Q. (2010). Unified power quality conditioner with model predictive control. *Proc. 5th Int. Conf. Comput. Sci. Educ.*, 1239–1244.

Zhang, Y., Zhang, S., & Chen, J. (2008). The applications of bidirectional full bridge DC–DC isolated converter in UPQC. *Proc. Int. Conf Electr. Mach. Syst.*, 1916–1921.

Zhili, T., Xun, L., Jian, C., Yong, K., & Shanxu, D. (2006). A direct control strategy for UPQC in three-phase four-wire system. *Proc. Power Electron. Motion Control Conf.*, 1–5.

Zhou, M., Wan, J. R., Wei, Z. O., & Cui, J. (2006). Control method for power quality compensation based on Levenberg–Marquard optimized BP neural networks. *Proc. IEEE 5th Int. Conf. Power Electron. Motion Control Conf.*, 1–4.

KEY TERMS AND DEFINITIONS

Custom Power Devices: Are the D-FACTS that are used in distribution systems like DVR, D-STATCOM, UPQC, and interline power quality conditioner (IPQC).

D-FACTS: Is the FACTS used in distribution system to enhance the power quality of the customers and utility.

FACTS: Is the flexible AC transmission system that is used to improve the reliability and stability of power system.

SVCs: Is static VAR compensators (SVCs) that are used to improve the quality and control of voltage by supplying or absorbing reactive power using inductors and capacitors.

TSC: Is thyristor switched capacitors that can control real and reactive in transmission system.

TSR: Is thyristor switched reactor that is self-commuted switching converter that controls the active and reactive power of power system.

UPQC: Is the custom power device used to enhance the multiple power quality in distribution systems.

Chapter 11
Performance of the Multicarrier Sinusoidal PWM-Based Multilevel Inverter With Reduced Power Loss and Fault Diagnosis Using Wavelets Transformation:
Hybrid Multilevel Inverter

Durga Prasad Garapati
Shri Vishnu Engineering College for Women, India

Jegathesan V.
Karunya Institute of Technology and Sciences, India

Moorthy Veerasamy
Vishnu Institute of Technology, India

ABSTRACT

The performances of multilevel inverters (MLIs) are of high competence when compared to the conventional two-level inverters due to reduced harmonic distortions, lower electromagnetic interference, and higher dc link voltages. However, the increased number of components, complex PWM control, and voltage-balancing problem, component failure in the circuit are some of the disadvantages. The topology preferred in this chapter provides a dc voltage in the shape of a staircase which approximates the rectified shape of a commanded sinusoidal wave to the bridge inverter, which in turn alternates the polarity to produce an AC voltage with lesser total harmonic distortion. This topology requires fewer components and hence it leads to a reduction of overall cost and complexity particularly for higher output voltage levels. The component fault diagnostic procedure is developed using wavelets transform tool. Finally, the experimental prototype is developed and validated with the simulation results for different loading conditions.

DOI: 10.4018/978-1-5225-3935-3.ch011

INTRODUCTION

The world appears to move through a period of resurgence in its efforts to attract a greater sense of automation in the use of industrial and domestic utilities. The direction foresees a path of modulating the available source of power to suit the exhilarating needs of the society. The immediate benefits stretch to conserve the conventional energy resources and thus mitigate the environmental impact. In this scenario the inevitable role of power converters and the indomitable actions involved in the operation of the semiconductor switches challenges the quality of power supply. The philosophy centers on the astute choice of power modulators together with the appropriate etiquettes to ensure a higher level of power quality.

The industries exhibit a trend to nurture an undistorted sinusoidal supply with variable voltage and frequency for supporting adjustable speed drives, variable frequency induction furnaces and related applications. The subservient accolades facilitate to foster their demand in terms of appropriateness, easier operating mechanisms and acquiring a higher efficiency (Jih sheng and Peng, 1996). The perspectives migrate for developing high power semiconductor device modules for the classical topologies in the process of matching the end user specifications. The second plane of thought owes to articulate newer power converter structures to explore a reduction in the switch count and profess to lay down innovative techniques for enjoying the full use of the available power. The third plane is to identify the faulty device in the preferred topology through wavelet transformation tool.

RESEARCH BACKGROUND

The emergence of new semiconductor devices translates to realize high power requirements in preference to the use of multiple connected low voltage devices. Though the series connection favors to increase the voltage withstanding capacity, still the current handling phenomena infuse rhetoric issues. Besides the problem of voltage sharing during switching in tune with the variation in switching speed explodes with a larger number of devices connected in series. The theory of special gate drive circuits along with snubber networks carry relevance and add to the burden in the smooth functioning of the circuit.

The inverters fall in the category of power conversion systems that owe to synthesize the variable alternating voltage from the fixed dc input by appropriately turning on the power switches in the modular architecture. The classical two level inverters tailor to satisfy higher power applications through a single and or two devices at any point in time. The resultant shape of the voltage and current waveforms cause very large sized steps and introduce lower order harmonic content to create deleterious impact on the performance of the products.

Though a variety of filters prorogue to suppress the lower order harmonics at the output side, still the filter size keeps worrying the designers and venture to contemplate on alternatives. The enhancements in the filter configurations and the advent of active filters bring to light a fresh space for their presence in power converter circuits. However, their bulky nature and the increasing losses leave much to be desired in the role of filters especially in these areas. These are addressed with suitable power converter technology through enduring research activity.

Elevation of Multilevel Inverter Technology

With the imminent drawbacks of classical two-level inverters overpowering its merits, an alternative path to meet the high-power demand endeavors the upsurge of multilevel inverters (MLIs). The MLIs refer to the state-of-the-art power conversion systems through the use of medium power semiconductor devices and capacitor voltage sources to generate a stepped output voltage waveform as shown in Figure 1.

Figure 2 illustrates a host of ways exists to produce multiple levels of output voltages depending on the combination of power semiconductor devices and the number of independent capacitive direct current sources in the structure. In which the number of levels increases to produce a stair case wave that approaches a desired waveform. As more steps are added to the waveform, the harmonic distortion of the output wave decreases, approaching zero as the number of levels increases.

In the genus of MLI a diode clamped inverter (DCI) is a primary topology that the m-level DCI typically consists of (m-1) capacitors on the DC bus and produces m-levels from a common DC source. Though it synthesizes a near sinusoidal voltage, the harmonic content is low enough to avoid the need of filters; the requirements of excessive clamping diodes have complicated the real power flow control of the individual converter in multi converter system.

Another member is a flying capacitor inverter (FCI) wherein a large amount of storage capacitors are used unlike diode clamped inverter. High level inverters are more difficult to package with bulky power capacitors and are more expensive too. The switching frequency and switching loss are a great concern for real power transmission.

A cascaded multilevel inverter consists of a series of H-bridge inverter units known as a cascaded H - bridge (CHB). The general function of this multilevel inverter is to synthesize a desired voltage from several DC sources and is well suited for various renewable energy sources. Unlike the DCI and FCI, the cascaded inverter does not require any voltage-clamping diodes or voltage-balancing capacitors and also the required level is achieved with minimal components in contrast to DCI and FCI. Further, the CHB is categorized based on the rating of DC supply and the relation established among each supply while arrangement. Namely symmetrical and asymmetrical CHB's, both can be generated high level response with fewer components.

Figure 1. Block diagram representation of MLI technology

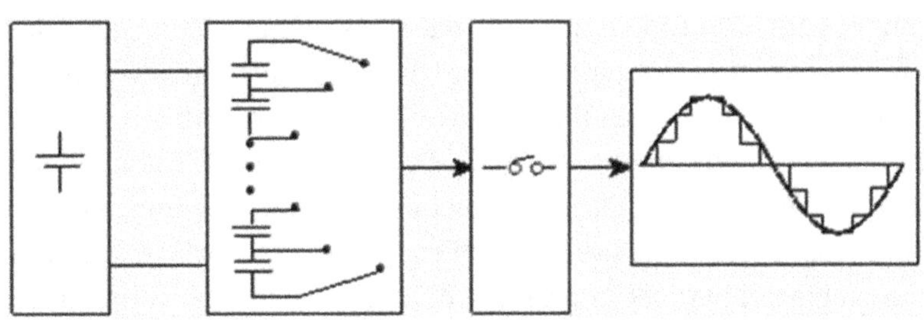

Figure 2. Stemma of multilevel inverters

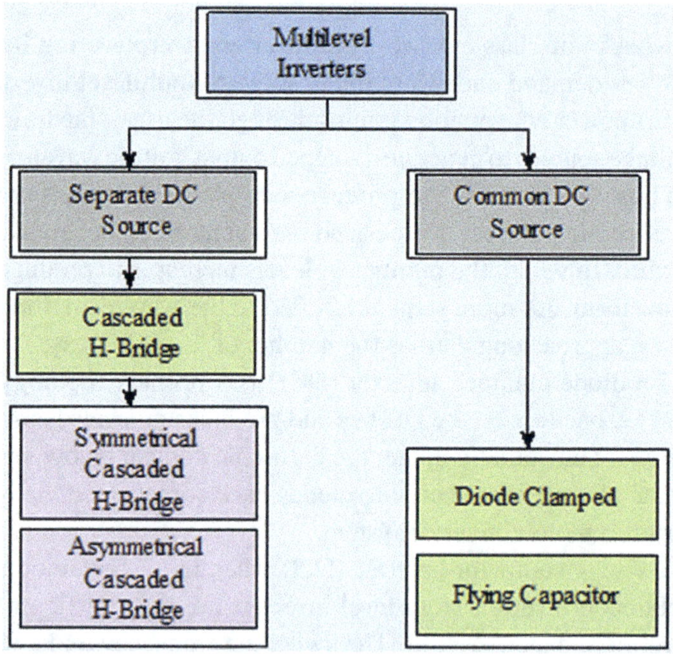

Evolution of Pulse Width Modulation Techniques

The trend in the evolution of pulse width modulation (PWM) concept bestows to suit different converter topologies with the main objective of reducing the harmonic distortion and increasing the fundamental output magnitudes for a given switching frequency. The philosophy in a modulation scheme intends to generate a train of pulses that is guided by the reference signal at any instant.

The focus spreads to calculate the converter switch on times that produce the desired output. The second feature enters to ascertain switching states effectively to reduce total harmonic distortion (THD) and switching losses. The procedure orients to minimize a suitable objective function for allowing the switching pulses in every cycle to change in a discrete fashion as the frequency decreases to maintain a good quality waveform. A large number of different modulation algorithms, see the light of the day to characterize the gamut of converter topologies with each bestowing unique advantages and drawbacks.

The classification of modulation algorithms depends on the average switching frequency at which it operates is shown in Figure 3. It evinces that the growth of innumerable topologies has demanded to broaden the classical modulation techniques to the operation of MLIs, which inherently complicate the control mechanism of switching devices, but simultaneously exploits the degrees of freedom of switching states generated by these topologies. The present ambience revolves on the extradition of PWM methods to pre-calculate the switching functions in order to up-heave harmonic mitigation and fulfill specific grid codes in the formulation of new control strategies.

Table 1 figure out the important MLI applications from low power to high power, and it also shows that they are potentially enough to use in a wide range of applications like; low power devices, home appliances, electric vehicles, photovoltaic, transport (train traction, ship propulsion, and automotive applications), and energy conversion, manufacturing, mining, and petrochemical applications.

Figure 3. Classification of PWM techniques

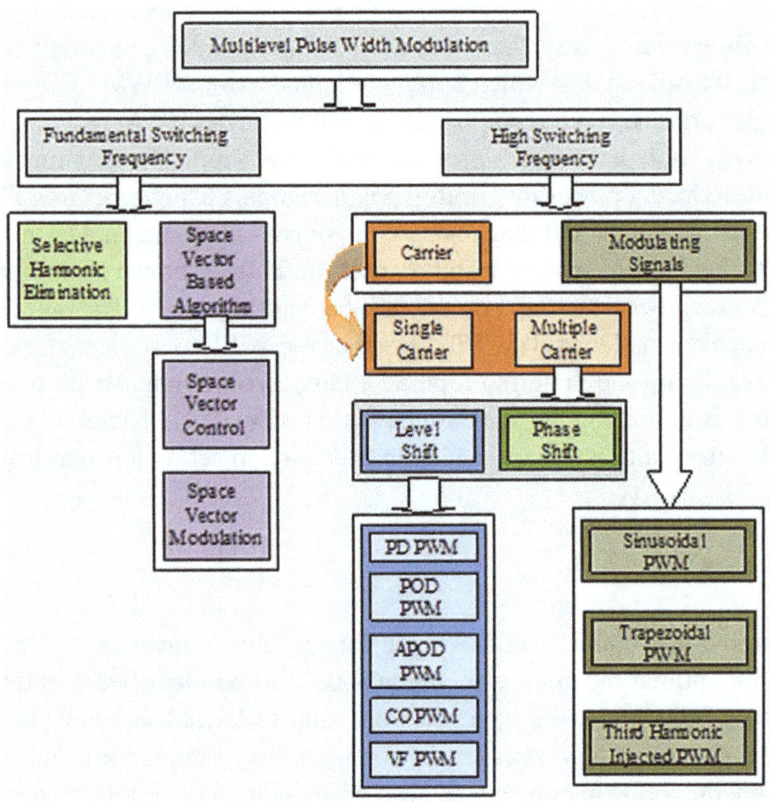

Table 1. Summary of power inverters

	Low Power	Medium Power	High Power
Range	Up to 2 KW	2-500 KW	More than 500 KW
converter Topologies	AC/DC, DC/DC	AC/DC, DC/DC, DC/AC	AC/DC, DC/AC
Power Semiconductor device	MOSFET	MOSFET, IGBT	IGBT, IGCT, Thyristors
Applications of MLI	 Low Power Devices  Home appliances	 Electric Vehicle & solar energy conversion	  Renewable Energy conversion

Motivation

From the extensive discussion, it is realized that the MLI has synthesized small voltage levels into a nearer sinusoidal output response. For which pulse width modulation (PWM) using space vector PWM, sine PWM with single carrier and reference signals and multicarrier PWM technique with single reference has been employed. Though the MLI's provides qualitative supply, formulating new topology using fewer components for reducing power loss, mitigating harmonics using emerging PWM techniques of controlling controllable parameter and diagnosing component failure are endless research task.

The semiconductor device failure significantly causes the fault in a power electronic circuit. Therefore a suitable fault diagnostic process for the topology has become necessary. The output of the topology is subjected to wavelet transformation tool (WTT). Based on the WTT coefficients for the fault-free circuit as well as simulated faults for the preferred topology, fault dictionary has been framed. Fault dictionary is being generated by extracting the standard deviation of WTT coefficients. Extracted parameters are utilized to develop the fault dictionary, which is later used to set up a generalized process for fault classification.

Aim and Objectives

Efficient energy conversion is a function of switching states of the converter. Based on the application, power converters with optimal modulation techniques need to be identified and developed to deliver maximum output power. Therefore, this research work aims to formulate a newfangled cascaded MLI with reduced number of active components for the compatibility of the scope of this chapter. In order to achieve the stated aim, the following objectives have been deliberated in this research:

- To investigate the preferred topology using level-shift PWM technique for various loading conditions.
- To examine the feasibility of hardware implementation of preferred MLI using FPGA.

To develop an algorithm for the fault diagnosis of the power semiconductor device in the topology.

BACKGROUND

An extensive literature survey on research area concern provides a solid foundation to a chapter. The survey focused on the development of multilevel inverter topologies, emerging pulse width modulation techniques and finally, wavelet transformation based fault analysis also included. These are comprehensively discussed in this section.

Multilevel Inverter Topologies

The survey on MLI topology is initiated from the statement of (Jih Sheng and Peng, 1996) that it is a new breed of power converter which is based on the process of amalgamation of several levels of DC capacitor voltages to obtain the staircase voltage wave. In this work, a diode-clamp (DC), flying-capacitors (FC), and cascaded H-Bridge (CHB) inverters with separate DC sources are deliberately used to derive desired voltage in multi-level. In which voltage unbalancing occurs due to different voltage level capacitors that could be balanced using voltage clamping or capacitor charge control.

The cutting edge multilevel innovation, considering settling and developing topologies and their control methods are broadly reviewed. Extraordinary consideration is committed to the most recent and most important modern applications of these converters. At long last, the conceivable outcomes for future improvement are tended (Rodriguez, Jih Sheng, and Peng, 2002). A state-of-the art performance of conventional MLI topologies like DC MLI, FC MLI and cascaded cell multilevel inverters have been compared to different PWM techniques (Panagis, Stergiopoulos, Marabeas and Manias, 2008). A medium voltage MLI and their control scheme with few applications on targeting minimum THD and high efficiency at low carrier frequencies (CFs) has investigated (Abu-Rub, Holtz, Rodriguez and GeBaoming, 2010) whereas (Colak, Kabalci and Bayindir, 2011) has explored the usage of MLI on medium and high power applications and also the role of switching technique for eliminating the harmonics.

A cascaded H-bridge MLI is preferred in which isolated DC sources are replaced by a single DC source and capacitors. To eliminate harmonics in the multilevel output the seven and fifteen level output voltage is analyzed by considering a particular switch angle shift (Zhong, Tolbert, Chiasson and Ozpineci, 2006). Later, a DC–AC CHB multilevel inverter for electric vehicle application is proposed without inductors utilize a standard three-leg inverter and a HB in arrangement with every inverter leg. A line frequency PWM scheme is utilized for modulation control, to yield five-level output voltages. (Zhong, Ozpineci, Tolbert and Chiasson, 2009). An isolated CHB MLI is proposed by utilizing low-frequency three-phase transformers and a single dc input control source. This setup can diminish various transformers contrasted and customary three-phase MLI's utilizing single-phase transformers. It manages changing phase angles to get an ideal exchanging design related to the line frequency of the desired output. Inferable from this run procedure, harmonic components of the waveform and switching losses can be decreased impressively (Song, Kang and Sung-Jun Park, 2009).

A new Static Frequency Converter (SFC) with CHB MLI is implemented for the starting of reversible pumped generating units and explained its superior performance by comparing it with conventional SFC type (Wang and Jiang, 2011). It is well known CHB topology has been popularly used in the power converter that has facilitated the selection of variety of cascaded cell numbers and dc source ratios. Nevertheless, if all the cells are not fed by equal voltages, the inverter becomes an asymmetrical one and the inverter arm cells have different effect on the output voltage. Hence, (Khoucha, Lagoun, Kheloui and Benbouzid, 2011) has investigated the symmetrical and asymmetrical arrangement of five and seven level HBI to find an optimum arrangement with low switching losses and optimized output voltage quality for direct torque control of induction motor drive.

Another MLI with the capacity of the creation of all levels at the output is determined by (Nilkar Mohammad, Ebrahim Babaei and Mehran Sabahi, 2012). This inverter comprises of a few essential three-level cells with bypass capacity and a HB for delivering positive and negative levels at the load side. In the analysis of MLI advances, another MLI topology with decreased number of devices has been offered by (Farhadi Kangarlu, Babaei and Laali, 2012). Decrease of the quantity of devices prompts diminishes the control many-sided quality. Furthermore, the topology is better than other non-ordinary MLIs regarding number of utilized IGBTs. A new topology based on associated a several independent units is implemented. Contrasted with customary CHBI, the recommended topology lessens the quantity of switches while utilizes the same number of DC sources with similar values. The recommended topology linked with an appropriate method to identify the ratio of DC voltage value for asymmetrical MLI (Ali Ajami, Ataollah and Mohammad Reza, 2013).

A multilevel VSI can be framed by associating one of the Multilevel DC link inverter (MLDCL) with a single phase HBI. The MLDCL furnishes a dc voltage with the state of a staircase approximating the rectified shape of a sinusoidal wave, with or without PWM, to the bridge inverter, which thus substitutes the extremity to deliver an AC voltage (Su Gui-Jia, 2003).

Another topology, which is made out of floating input DC sources. These sources are on the other hand associated in inverse polarities through the power switches. With controlled switching, the input sources supply the load either independently or in arrangement blends of at least two sources. The proposed approach prompts decreased number of switches and along these lines gate driver units. It can be utilized in medium voltage drives applications where a phase moving transformer with numerous optional windings is by and large utilized for the most part for the diminishment of line-current distortion, therefore giving separated DC sources (Akbar Ahmad and Rajesh Gupta, 2013).

Toupchi Khosroshahi, (2014) inspected a novel CMLI topology with established CHB converter on different execution parameters and the benefit of the proposed CMLI is finalized. The recommended topology by (Mokhberdoran Ataollah and Ajami Ali, 2014) is summed up utilizing arrangement association of basic blocks. Each block additionally can be considered as a MLI. Initially, the essential block and summed up topology are introduced; from there on a point by point investigation of power loss and estimation of PIV are given. Additionally, in proceed with an examination between proposed topology, the customary one and an as of recently created topology is completed. The CMLI design is a proper decision for pay of current harmonic in distribution networks. For all intents and purposes, the particular format and bundling empowers any number of HB modules to be stacked with a specific end goal to get the required output and makes it appropriate for medium-to-high power, high voltage applications. In addition, a noteworthy reason the MLI serves is decrease of stress over the semiconductor devices. The reference compensation filters current for three-stages are extricated using i_d–i_q conspire and MCPWM is done to produce switching signals for MLI. Optimal load compensation is guaranteed by the proposed usage of BFO-based PI controller for fruitful direction of inverter dc-link voltage, in this manner limiting the undesirable power loss in charge of degradation of APF performance (Anup and Sushree, 2015).

In view of reducing active component a inventive symmetrical MLI topology was proposed by (Edwin Jose and Titus, 2016), which expends lesser device check. The proposed level dependent sources concoction MLI (LDSCMLI) is fundamentally a multilevel dc interface MLI (MLDCMLI), which initially combines a stepped dc link voltage utilizing a sources creation module and after that grasps the AC waveform through an H-Bridge.

PWM Techniques

The speculations of the PWM sub harmonic technique to control single-phase or three-phase MLI has been portrayed in (Carrara, Gardella, Marchesoni, Salutarin and Sciutto, 1990). The expository articulation of the phantom segments of the waveforms covering all the working conditions has been determined in light of an augmentation of the Bennet's technique. The enhancements in the harmonic spectrum have been indicated out to demonstrate the legitimacy of the multilevel modulation.

A novel centroid-based PWM switching strategy for HBI has been proposed by (Ali Yazdian Varjani, Sarath Perera and Joe Chicharo, 1998) with a dual objective to echo reduction in THD and lower order harmonics. A hybrid switching sequences have also been developed to extract further reduction in the switching losses. The performance has been compared with earlier PWM switching strategies. At the same time a family of MCPWM techniques suitable for high power converter structures has been discussed by (Vassilios Agelidis and Martina Calais, 1998). The distinctive characteristics of the MCPWM techniques have been revealed. Moreover, a hybrid MCPWM techniques combining the disposition and phase shifted concept has been evolving and its performance reported in terms of the THD of the output line voltage in the liner over modulation region.

Siriroj Sirisukprasert, Jih-Sheng Lai and Tian-Hua Liu, (2002) has suggested a novel modulation technique to generate an output stepped waveforms with a wide range of modulation indices and minimized total voltage harmonic distortion. The method has been designed to involve the switch with switching only once per cycle and require only the polarities and the number of levels of different modulation levels. The result of the cascaded converter based hardware has been shown to the effective for reduction of harmonics for a MLI.

Hyoung-Chang Kim et al. (2003) has determined the conducting angle using voltage second areas of the divided reference voltage method for calculating conducting angles of switches for cascaded multilevel inverter. This method overcomes the disadvantages (like large amount of calculation and time, off line operation) of conventional methods. Followed, a slew of carrier based pulse width modulation (CBPWM) techniques based on the description of controllable redundant parameters in the modulating signals have been proposed by (Nguyen Van, Gun Woo Moon and Myung Joong Youn, 2004). A generalized mathematical formulation has been presented for CBPWM methods that offers took to provide outputs similar to that of SVPWM. The method has been echoed complete control of the fundamental voltage, vector redundancies and phase redundancies possess in the CBPWM.

Azli and Choong (2006) has analyzed the performance of three-phase five-level cascaded H-bridge MLI based on the MCSPWM control technique with a single reference, Behera, Dixit and Shyama, (2006) has explained the analytical solutions of different multilevel PWM such as in phase sinusoidal PWM, bipolar modulation and phase opposite sinusoidal PWM for 3-level neutral point clamped inverter fed induction motor drive. An adjusted PD PWM is one that consolidates the advantages of PD and PS PWM strategies, accomplishing great output voltage and input current quality (Angulo, Lezana, Kouro, Rodrıguez and Bin Wu, 2007).

Dahidah et al. (2008) has executed another variety of selective harmonic elimination (SHE) PWM procedure appropriately for a five-level converter utilized as a part of the consistent frequency utility application. The administering arrangement of conditions related to the disposal of particular harmonic is characterized in view of an equivalent number of switching transitions when looked at against the single carrier (SC) SPWM procedure. The proposed SHE PWM gave a superior harmonic performance contrasted with its SCSPWM equivalent.

A new algorithm has been projected to achieve different goals like extending the modulation index and reducing the number of switching. It has been tested using a five level five phase inverter to bring out the low computational complex (Oscar Lopez, Jacobo Alvarez, Jesus Doval-Gandoy and Francisco Freijedo 2009).

Recently, a single phase five level inverter with a dual reference modulation technique i.e., two identical rectified sine waves with a level shift for PV application (Rahim and Selvaraj, 2010), with appropriate adaptations, the fractional part of the general space vector multilevel multiphase PWM as a stand-alone PWM method in conjunction with two-level voltage-source converters, sub harmonic pulse width modulation strategy (SHPWM) (Ramkumar et al. 2012), multilevel modulation phase disposition (PD) SPWM where, carriers do not have any coincidence and have definite offset from each other (Najafi E, Abdul Halim and Mohamed Yatim, 2012), phase disposition (PD) multi-carrier pulse width modulation (MC-PWM) strategy (Thamizharasan et al. 2013) and once again SHE-PWM (Khounjahan,

Banaei. M. and Amir Farakhor, 2014) techniques have been practiced to synthesis high quality output voltage. Finally, an audit was presented by Balamurugan et al. (2016) has referred that creates diverse switching techniques for MLI. The switching techniques proposed in this paper are to contrast different strategies and with anticipate the correct switching strategy for various applications in view of its nature of the yields.

Fault Diagnosis of Multilevel

Debaprasad Kastha and Bimal K. Bose (1994) portrayed a precise examination concerning the different fault methods of a voltage-fed PWM inverter framework for acceptance motor drives. In the wake of recognizing all the fault modes, a preparatory numerical examination has been made for the key fault sorts, to be specific, input supply single line to ground fault, transistor base drive open, and short-circuit conditions. The review has been utilized to decide stresses in power circuit segments and to assess satisfactory post-blame enduring state working locales. The outcomes are similarly helpful for better protection framework design and simple blame conclusion.

In this connection an easy fault analyze technique and the neutral-point voltage control strategy that can protect a structure from the unbalancing voltage of the DC-connect capacitors when the switching component faults in the 3-level inverter happen that is exceptionally productive in high voltage and high power applications. Numerous MLI faults have been recognized utilizing the over voltage and the over current in their structures and the prevalence of the proposed work has illustrated by Ho-In Son et al. (2004).

Alian Chen et al. (2005) projected another topology with fault tolerant capacity that enhances the dependability of MLI's. This topology is created through examination of various power devices failure modes. With the proposed plot, the capacity of the power stage can be kept up even some part of it comes up fails. Its fault tolerant capacity comes about because of the repetitive way of the multi switching state topology and from control signal alteration.

Followed, a testing of a wavelet packet transform (WPT)- based system for protecting and controlling threephase VSI PWM under extreme load conditions has been introduced by Saleh et al. (2007). At the same time, (Frederic Richardeau and Pham TTL, 2013) examined that MLI's have many power devices and drivers in the circuit. An immediate unwavering quality count construct just with respect to the principal failure event on one of the parts plainly drives them to be debased contrasted with two-level converters.

Finally, a study of fault tolerant methods for three-phase VSI was presented by Behrooz Mirafzal et al. (2014). Accessibility and unwavering quality of inverters have turned out to be progressively critical. Taking after early stage fault detections in inverters, therapeutic activities can expand ordinary operation of inverters and, at times, derate the system to counteract sudden shutdowns. The fundamental reason for this paper was to give an informative overview of existing adaptation to internal failure and remedial methods for MLI's.

This section presents a review of different MLI topologies with various PWM control strategies and fault diagnosis procedures. This literature review covered a variety of topics, techniques, methods and approaches of constructing and analyzing the MLI topologies.

CASCADED H-BRIDGE MULTILEVEL INVERTER

The energy efficient operation is the endless process and emerged power electronic converters and modulation techniques provide a platform to utilize available renewable resources efficiently. This chapter deals the multi-carrier sinusoidal PWM (MCSPWM) with different control strategies to obtain switching pulses for the preferred converter and these are elongated to multiple voltage levels. Further, it has examined using field programmable gate arrays (FPGA) controller. The suitability of the preferred reduced switch cascaded MLI with an appropriate control strategy for different loading conditions is investigated for different switching frequencies and modulation index (MI).

Preferred Topology

At conventional MLI, high frequency power semiconductor devices are adapted to create an alternating waveform with the fundamental frequency. Nevertheless, there is no compelling reason to use the greater part of the switches for creating all levels. Therefore, a new hybrid MLI is preferred in this study and the same is represented as block diagram in Figure 4. It comprises fundamentally two parts, one is utilized for creating the voltage levels of positive extremity using HF power semiconductor devices to produce the required number of levels, by interfacing n-number of segregated DC voltage sources. Whereas, second is known as the polarity generation part (PGP) which creates an alternating waveform and requires low frequency switches which work as a single phase full bridge (SPFB) inverter at fundamental frequency.

These two parts jointly generate the required voltage levels at the output side. In order to generate a complete multilevel output, the zero and positive levels are generated by the level generation part (LGP) and these are in turn fed to a PGP i.e. SPFB inverter for generation of the alternating output voltage waveform. In the suggested topology, the number of IGBTs used is lesser when compared to the cascaded H-bridge (CHB) for any number of levels; the standing voltage on the switches of the suggested topology is lesser than the topology presented (Ramkumar et al. 2012; Thamizharasan et al. 2013). Nevertheless, in CHB a variety of switches in terms of voltage rating are required for topology considered in this work. Figure 5 Illustrates a fifteen level suggested symmetrical topology, its left half is a level generator comprises seven switches, six diodes, and seven isolated DC sources which produce required output voltage levels, whereas the right half which is the SPFB inverter provides the alternating waveform with required levels. Therefore, there is no need for all switches to work on high-frequency, thereby leading to simpler and more reliable control of the inverter and this leads to reduction of power loss in a circuit.

In order to achieve the next level, the whole module is duplicated as in Figure 5. Further, it can be adapted to the required number of voltage levels by adding the required number of modules. It also provides further scope to 3-ϕ applications. Najafi et al., (2012) revealed that the voltage source requirement would be the ratio of 1:3 with traditional CHB in the perspective of 3-ϕ application. The maximum output voltage is obtained by adding the amplitude of all the DC sources connected in the circuit. Therefore, the output voltage levels are derived from the selection of individual DC source and from the proper switching action of the power semiconductor switches.

$$V_{Maximum\ Output} = \sum_{k=1}^{n} V_k \tag{1}$$

Figure 4. Block diagram of the preferred 15-level cascaded MLI topology

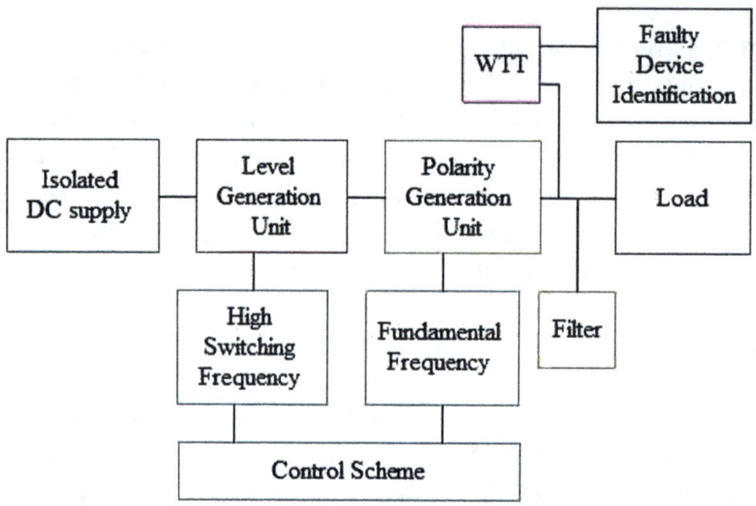

Figure 5. Schematic diagram of fifteen-level preferred MLI in single phase

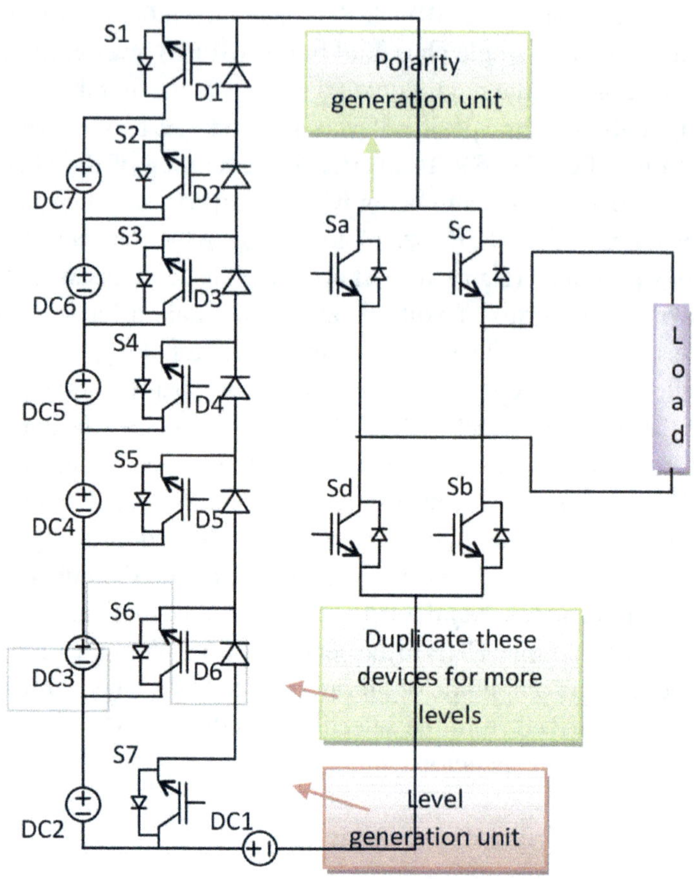

The LGP gives a DC bus output, with the state of a staircase that estimates the amended waveform of the commanded sinusoidal voltage, to the PGP, which exchanges the voltage polarity to deliver an AC voltage of a staircase form with necessary levels. In particular, the LGP formed by the 'n' series connected sub-cells to give staircase formed DC bus output of n steps, to the PGP which thus substitutes the voltage polarity to create an AC voltage of a staircase form with (2*n+1) levels.

$$\text{The active number of IGBTS in preferred topology} = \left[\left(\frac{L-1}{2}\right) + 4\right] \quad (2)$$

$$\text{Number of dc sources} = \frac{(L-1)}{2} \quad (3)$$

$$\text{Number of Diodes} = \frac{(L-3)}{2} \quad (4)$$

where, L is the number of levels.

As said, one of the real favorable facts of this topology is that it requires LF switches in PGP, where a couple of switches just supplies stack at fundamental frequency (50Hz). Thus, the reliability of preferred topology is better than CHB as per MIL-HDBK-217F (Society of Reliability Engineers).

Modes of Operation

The switching states of the preferred symmetric 15-level inverter are shown in Figure 6.

Modes of operation: The preferred topology has developed to provide nearly sinusoidal waveform in 15 levels with eight modes of operation, using possible combinations of the switches S_1-S_7 in LGP and S_a, S_b, S_c and S_d in PGP and described in the Figure 6.

Mode 0-0': Primarily, none of the switches allowed turning ON, but levels 1st and 15th are mandatory for zero crossing.

Mode 1-1': This mode performs to obtain 2nd and 14th stairs of positive and negative half cycle of the output waveform. S_7, S_a and S_b are closed, thus $+1V_{dc}$ has come across the load. Whereas, S_7, S_c and S_d are closed to complete $-1V_{dc}$ across the load.

Mode 2-2': In this incident S_6, S_a and S_b have provided a path to raise the output voltage into $+2V_{dc}$. Similarly, S_6, S_c and S_d have accepted the $-2V_{dc}$ to complete 3rd and 13th steps of both cycles.

Mode 3-3': In this event S_5, S_a and S_b, S_5, S_c and S_d have turned to active state, thus the output lead to $+3V_{dc}$ and $-3V_{dc}$ respectively. This seems to be 4th and 12th staircase of the load voltage.

Mode 4-4': Progressively, S_a and S_b, S_c and S_d along with S_4 have come to an active state in order to fulfil the next stage of output, i.e., 5th and 11th levels with $+4V_{dc}$ and $-4V_{dc}$.

Mode 5-5': Consequently, S_3, S_a and S_b, S_c and S_d have turned into ON, thus source voltage $+5V_{dc}$ and $-5V_{dc}$ have added to the respective cycle intended to attain 6th and 10th levels.

Figure 6. Switching states of preferred MLI topology

Output Voltage	Switch Status of the LG Unit							Switch Status of the LG Unit							Switch Status of PG Unit				Fundamental Cycle
	S1	S2	S3	S4	S5	S6	S7	S7	S6	S5	S4	S3	S2	S1	Sa	Sb	Sc	Sd	
0								0	0	0	0	0	0	0	1	1	0	0	
1*V_{DC}								1	0	0	0	0	0	0	1	1	0	0	
2*V_{DC}								0	1	0	0	0	0	0	1	1	0	0	
3*V_{DC}								0	0	1	0	0	0	0	1	1	0	0	
4*V_{DC}								0	0	0	1	0	0	0	1	1	0	0	
5*V_{DC}								0	0	0	0	1	0	0	1	1	0	0	
6*V_{DC}								0	0	0	0	0	1	0	1	1	0	0	
7*V_{DC}								0	0	0	0	0	0	1	1	1	0	0	0-π
6*V_{DC}								0	0	0	0	0	1	0	1	1	0	0	
5*V_{DC}								0	0	0	0	1	0	0	1	1	0	0	
4*V_{DC}								0	0	0	1	0	0	0	1	1	0	0	
3*V_{DC}								0	0	1	0	0	0	0	1	1	0	0	
2*V_{DC}								0	1	0	0	0	0	0	1	1	0	0	
1*V_{DC}								1	0	0	0	0	0	0	1	1	0	0	
0								0	0	0	0	0	0	0	1	1	0	0	
0								0	0	0	0	0	0	0	0	0	1	1	
-1*V_{DC}	0	0	0	0	0	0	1								0	0	1	1	
-2*V_{DC}	0	0	0	0	0	1	0								0	0	1	1	
-3*V_{DC}	0	0	0	0	1	0	0								0	0	1	1	
-4*V_{DC}	0	0	0	1	0	0	0								0	0	1	1	
-5*V_{DC}	0	0	1	0	0	0	0								0	0	1	1	
-6*V_{DC}	0	1	0	0	0	0	0								0	0	1	1	
-7*V_{DC}	1	0	0	0	0	0	0								0	0	1	1	π-2*π
-6*V_{DC}	0	1	0	0	0	0	0								0	0	1	1	
-5*V_{DC}	0	0	1	0	0	0	0								0	0	1	1	
-4*V_{DC}	0	0	0	1	0	0	0								0	0	1	1	
-3*V_{DC}	0	0	0	0	1	0	0								0	0	1	1	
-2*V_{DC}	0	0	0	0	0	1	0								0	0	1	1	
-1*V_{DC}	0	0	0	0	0	0	1								0	0	1	1	
-0	0	0	0	0	0	0	0								0	0	1	1	

Mode 6-6': Followed, S_2, S_a and S_b have formed a positive group to withstand $+6V_{dc}$, alternatively S_2, S_c and S_d have grouped for enduring $-6V_{dc}$. Thus, the output voltage is attained 7th and 9th levels.

Mode 7-7': In order to curve the output voltage at $+7V_{dc}$ and $-7V_{dc}$, S_a and S_b, S_c and S_d respectively with S_1 have completed the final stage.

In Figure 7 preferred and existing topologies are compared; (a) addresses the number of active switch requirement with respect to level attainment whereas (b) indicates necessity of the driver circuit. It can be clearly inferred that the preferred topology requires less number of components. So, an added advantage of this topology is that it will occupy less space and develop the required end product with lesser distortions. The maximum output voltage is obtained by adding the amplitude of all the DC sources connected in the circuit. Thus, the output voltage levels are derived from a selection of individual DC source and from the proper switching action of the power semiconductor switches.

PWM Techniques

It is well-known that the successful operation of any topology depends on the selection of control strategy. In general, space vector modulation (SVM) and MCSPWM will be used as modulation techniques. Among these, SVM technique is more constructive from the point of view of switching timings and it requires additional effort for implementation. Hence, MCSPWM is identified as commonly used control strategy for CHB inverters because of its robustness where a sinusoidal reference voltage waveform is standardized with a carrier waveform to produce a gate signals for the switches in the topology.

Figure 7. Comparison of topologies requirement

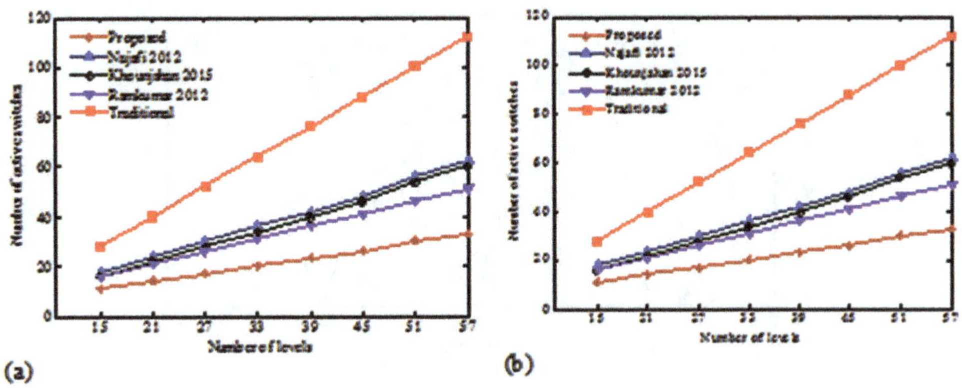

Particularly in this chapter, an absolute sinusoidal reference wave is compared with triangular carrier signal with phase disposition (PD), carrier-overlapping (CO) and variable frequency (VF) arrangement for performing desired operation. Moreover, it is explored that the requirement of these carrier signals to the preferred topology is half of the requirement to the conventional. Here, reduction of the carrier signal is a commendable achievement for inverter control as it only generates positive polarity multilevel output voltage. So as to keep away from undesirable voltage levels amid switching cycles, the switching patterns ought to be chosen so that the switching transitions wind up plainly insignificant at every mode. Likewise, this will diminish switching power dissipation.

Phase Disposition Level-Shift SPWM (PDLSPWM)

Figure 8 illustrates a carrier arrangement for an m-level inverter using PDLSPWM technique, (m-1) carriers with the same frequency f_c and same peak to peak amplitude A_c are disposed. If the magnitude of the reference signal is more than a carrier signal, then the switching devices corresponding to that carrier are switched ON. Otherwise, the devices are switched OFF. The amplitude modulation index (MI_a) and frequency modulation index (MI_f) are given as:

$$MI_a = 2A_m / ((m-1)A_c) \tag{5}$$

$$MI_f = f_c / f_m \tag{6}$$

Carrier Overlapping Level-Shift SPWM (COLSPWM)

Let, m is the level of inverter, f_c and A_c are the frequency and peak to peak amplitude of the (m-1) carrier signal and are disposed vertically each other with a distance of $A_c/2$. The modulating waveform has A_m and f_m and the signal arrangement is shown in Figure 9.

Figure 8. Carrier arrangement of PDLSPWM for MI=1

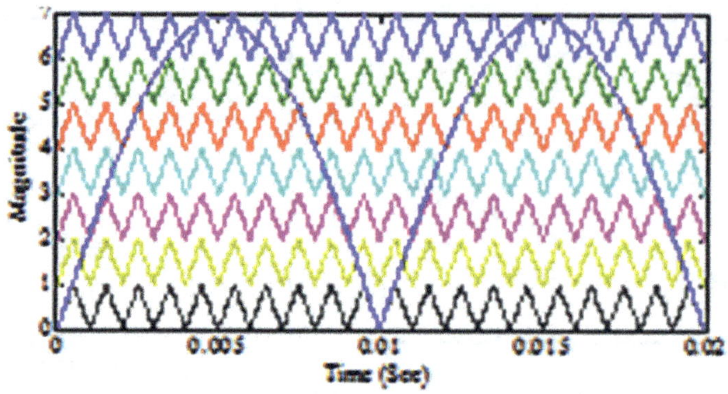

Figure 9. Carrier arrangement of COLSPWM for MI=1

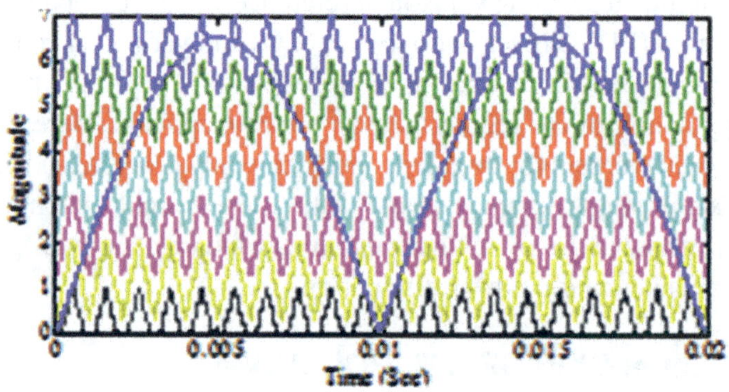

$$MI_a = A_m / 2A_c \tag{7}$$

$$MI_f = f_c / f_m \tag{8}$$

Variable Frequency Level-Shift SPWM (VFLSPWM)

In this type of carrier arrangement, the required number of signals with different carrier frequencies (f_c's) are level shifted in PD manner and the same is shown in Figure 10. The amplitude modulation index (MI_a) and frequency modulation index (MI_f) are same as in PDLSPWM.

SIMULATION RESULTS AND DISCUSSION

The topology preferred in above section has been extended to 15 levels and simulation is done using MATLAB/Simulink software 7.9.0 (R2009b) version. The Sim-power systems block in the MATLAB/

Figure 10. Carrier arrangement of VFLSPWM for MI=1

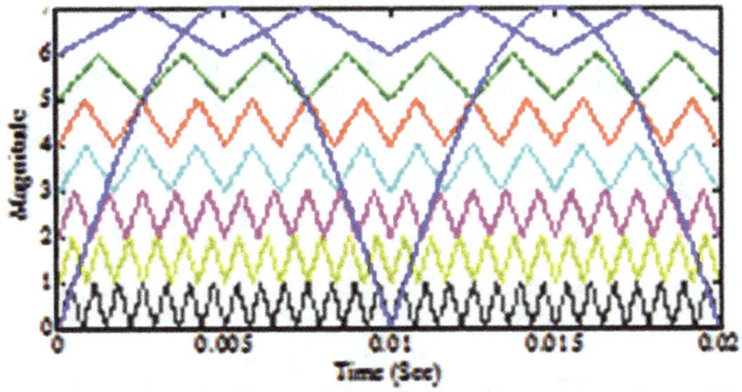

Simulink is used to test the prototype of MLI and the results are studied. The preferred topology is simulated in the presence of different MCSPWM for R and RL loads in which the MI and switching frequency (f_{sw}) are varied, and the observed performance indices are discussed in this section.

Triangular Carrier Signal

To verify the applicability of preferred topology, a strategy based on carrier arrangement is included in the modulation technique and the performance is deliberately analyzed in this section.

PDLSPWM

The simulation is performed with two different loads separately. i.e., R and RL of valued 10 Ω and 1mH. The responses of the preferred topology for R and RL loads are presented in Figures 11 and 12 respectively. It is clearly identified from the trace (a) that the LGP has generated only positive polarity in seven levels and the same is fed to the PGP thus, a complete fifteen level alternative supply has arrived. In order to enhance the synthesized output voltage, a filter arrangement has included and the differences are traced in both the cases for R and RL loads. Since the inductance value is small, there exists slight variation in the current phase, but considerably smoothen the load current.

The concernment of the preferred topology for different f_{sw} is verified by varying MI and the test results are listed in Tables 2 and 3 successively for R and RL loads. It is inferred from the tables that the content of percentage THD in case of 'with filter' reduces nearly half of the value than 'without filter'. Moreover, the preferred topology performs well for MI equal to 1 and also the performance has been turned well at the switching frequency of 5 kHz.

COLSPWM

With COLSPWM strategy, the simulation is executed for the same loading condition as discussed in the earlier case. Figure 13 shows the effect of CO in topology output response for R load that has included level generation, voltage and current waveform without and with filter.

Figure 11. Output response of preferred topology with PD for R load

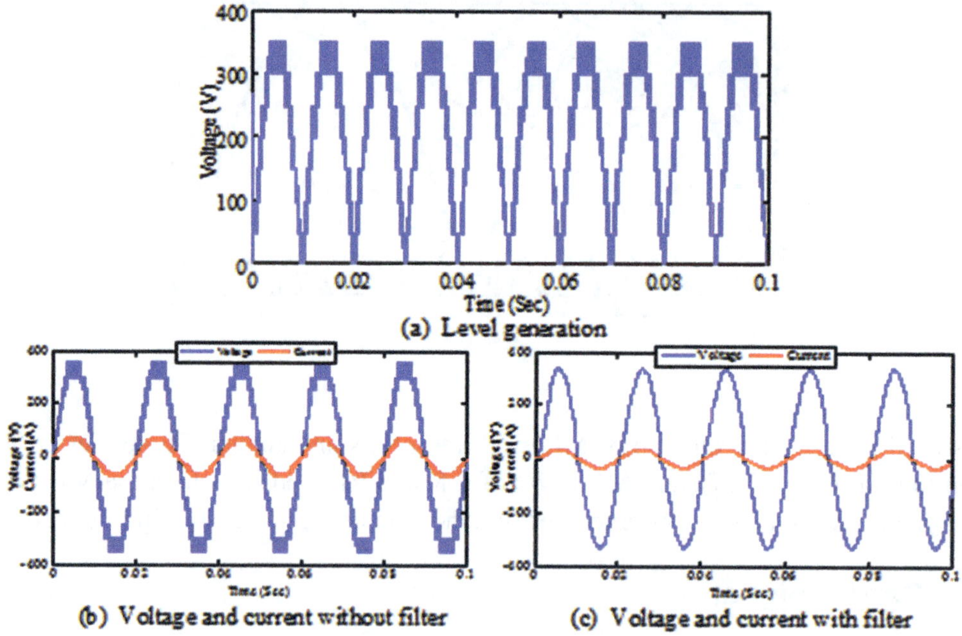

Figure 12. Output response of preferred topology with PD for RL load

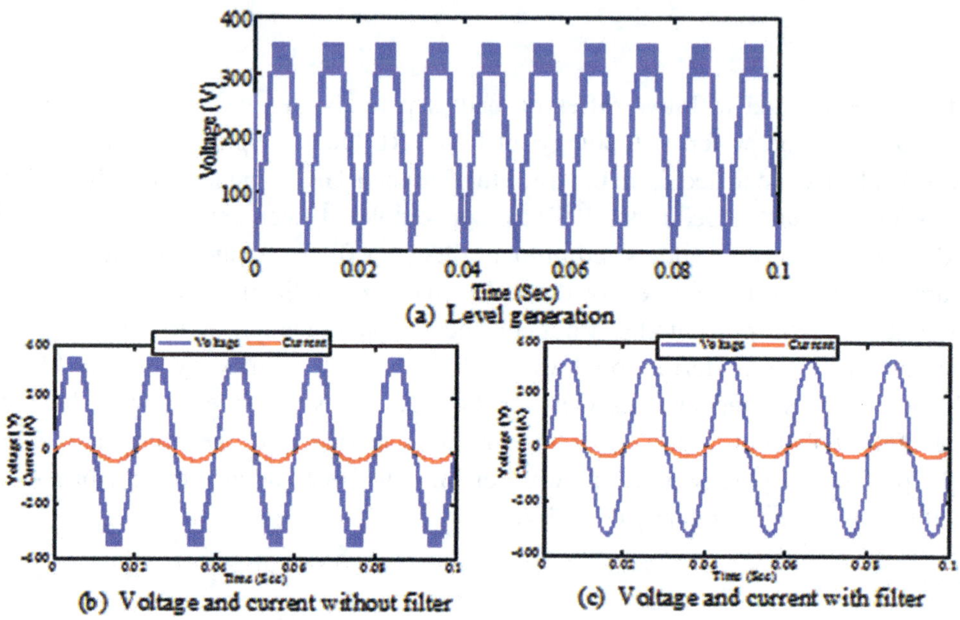

Though, the voltage and current response of PD and CO are looking like similar, there will be a change in performance indices. Therefore, a THD analysis is conducted and the test results were presented in Table 4. Where, the MI is varied from 0.6 to 1 for the f_{sw} 2 and 5 kHz. It is inferred that the %THD was decreased as the MI has increased.

Table 2. Performance indices of preferred topology with PD for R load

MI	Switching Frequency=2kHz				Switching Frequency =5kHz			
	Voltage % THD		Current % THD		Voltage % THD		Current % THD	
	WOF	WF	WOF	WF	WOF	WF	WOF	WF
1	4.16	1.92	4.16	1.92	2.95	1.05	2.95	1.05
0.8	4.87	2.02	4.87	2.02	3.02	1.98	3.02	1.98
0.6	5.10	2.33	5.10	2.33	3.19	2.04	3.19	2.04

Table 3. Performance indices of preferred topology with PD for RL load

M.I	Switching Frequency=2kHz				Switching Frequency =5kHz			
	Voltage % THD		Current % THD		Voltage % THD		Current % THD	
	WOF	WF	WOF	WF	WOF	WF	WOF	WF
1	4.4	1.16	3.06	1.67	3.78	1.09	3.14	1.11
0.8	5.23	1.22	5.06	1.97	4.89	1.17	4.27	1.52
0.6	6.38	1.46	5.91	2.09	5.53	1.69	5.91	2.04

Figure 13. Role of CO in output response of preferred topology for R load

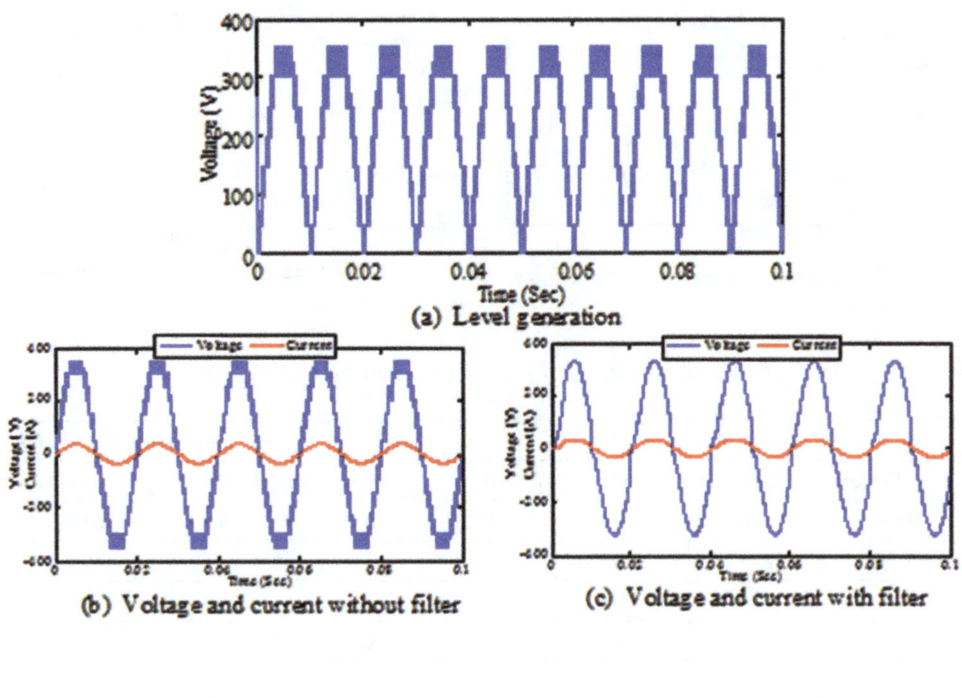

For the diversity, R load is replaced by RL and the output response, and performance indices are presented in Figure 14 and Table 5 respectively. It is confirmed that the current waveform is better than R load and the preferred topology is more suitable for practical case since most of the loads are an inductive nature.

Table 4. Role of CO on performance indices of preferred topology for R load

MI	Switching Frequency=2kHz				Switching Frequency =5kHz			
	Voltage % THD		Current % THD		Voltage % THD		Current % THD	
	WOF	WF	WOF	WF	WOF	WF	WOF	WF
1	4.96	2.21	4.96	2.21	3.37	1.64	3.37	1.64
0.8	5.23	2.97	5.23	2.97	4.82	2.39	4.82	2.39
0.6	6.10	3.03	6.10	3.03	5.13	2.96	5.13	2.96

Figure 14. Role of CO in output response of preferred topology for RL load

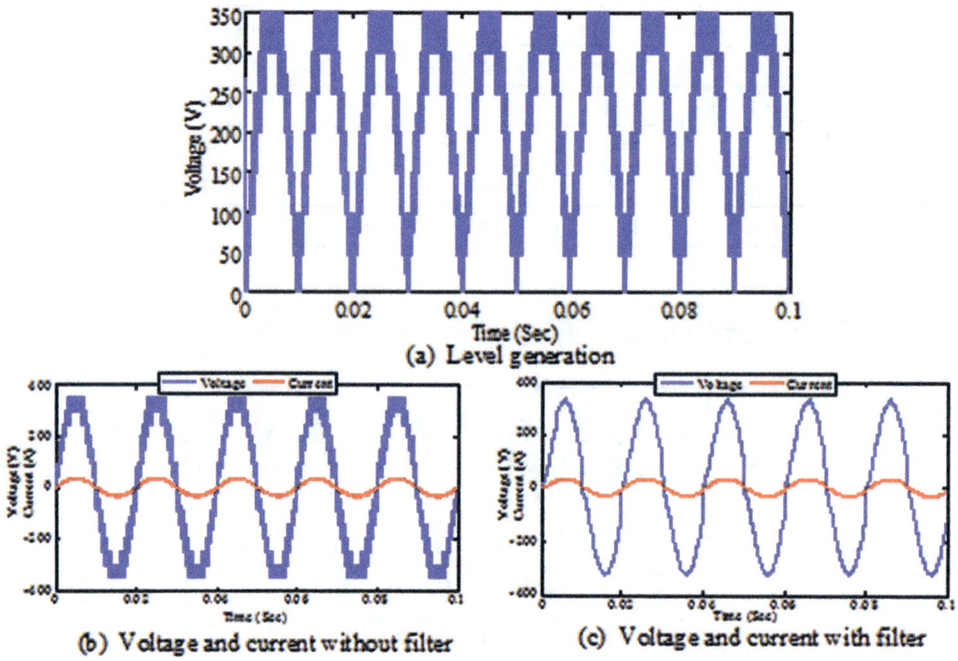

Table 5. Role of CO on performance indices of preferred topology for RL load

MI	Switching Frequency=2kHz				Switching Frequency =5kHz			
	Voltage % THD		Current % THD		Voltage % THD		Current % THD	
	WOF	WF	WOF	WF	WOF	WF	WOF	WF
1	5.26	2.28	4.26	2.76	4.89	2.34	4.19	2.21
0.8	6.14	2.79	5.93	2.98	5.42	2.89	5.17	2.22
0.6	7.37	3.19	6.98	3.32	6.34	3.46	6.19	2.94

VFLSPWM

Similarly, the applicability of preferred topology is examined with VFLSPWM for the same loading condition as discussed in earlier two cases. Figures 15 and 16 have illustrated the output response for R and RL load without and with filter respectively. Once again, it is confirmed that the topology has performed well in both the cases. Additionally, it provides very smooth current waveform in case of RL load without a filter.

As usual, the output response is analyzed to obtain the % THD for both load conditions including without and with filter. The investigations are recorded in Tables 6 and 7 in which the MI is varied from 0.6 to 1 for the f_{sw} 2 and 5 kHz. It is understood that the %THD has varied in contrast to MI.

EXPERIMENTAL FRAMEWORK

With the aim of results that have been discussed in the previous section, the preferred topology was constructed using physical components and an experimental setup was arranged as shown in Figure 17. The entire experimental setup was divided into four modules and was briefly discussed here.

- **Pulse Generation Module:** SPARTAN-3A DSP from the Field Programmable Gate Arrays (FPGA) family is considered to be a pulse generation module which developes driving pulse in required patterns using VHDL. FPGAs are semiconductor devices that are based on a matrix of configurable logic blocks (CLBs) connected via programmable interconnects. The FPGA can be programmed by different stages. Design entry is the stage of implementing a required design of

Figure 15. Effect of VF in output response of preferred topology for R load

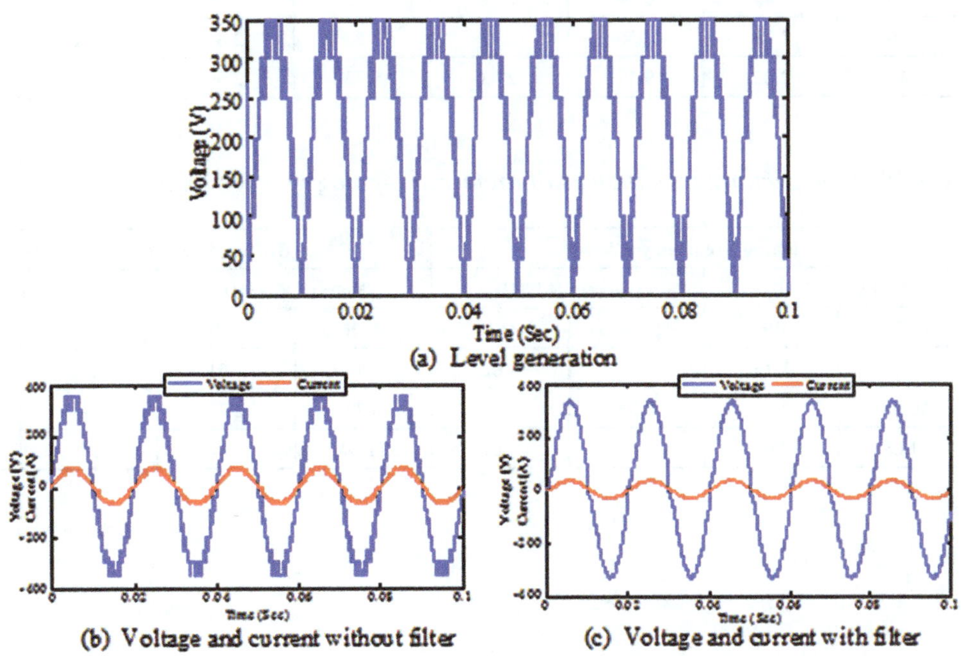

Figure 16. Effect of VF in output response of preferred topology for RL load

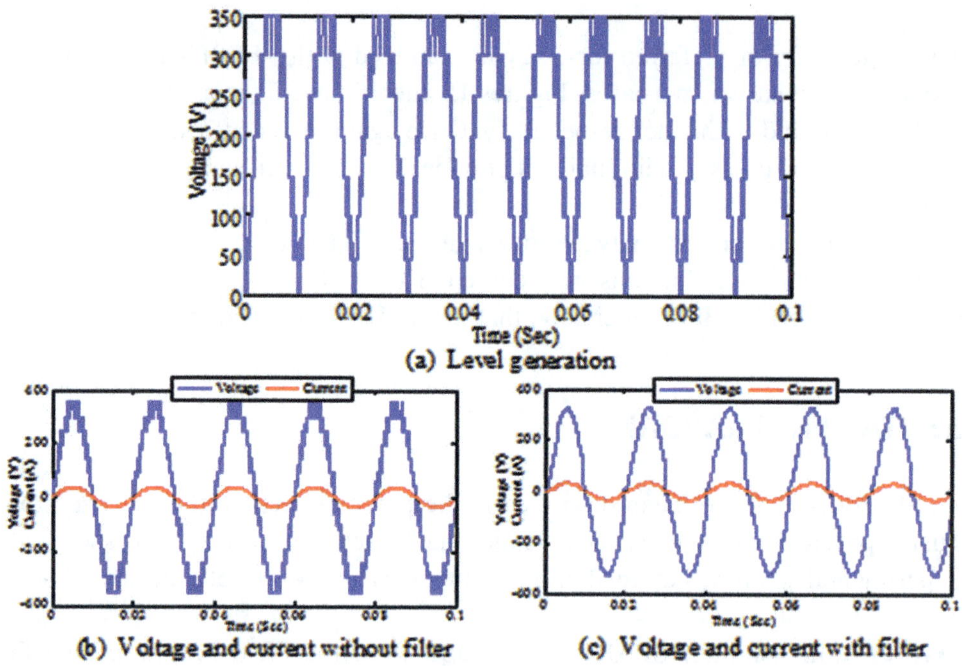

Table 6. Effect of VF on performance indices of preferred topology for R load

MI	Switching Frequency=2kHz				Switching Frequency =5kHz			
	Voltage % THD		Current % THD		Voltage % THD		Current % THD	
	WOF	WF	WOF	WF	WOF	WF	WOF	WF
1	6.29	2.49	6.29	2.49	5.82	1.96	5.73	1.96
0.8	7.56	2.38	7.56	2.38	7.52	2.84	7.46	2.84
0.6	9.19	2.83	9.19	2.83	9.73	2.69	9.31	2.69

Table 7. Effect of VF on performance indices of preferred topology for RL load

MI	Switching Frequency=2kHz				Switching Frequency =5kHz			
	Voltage % THD		Current % THD		Voltage % THD		Current % THD	
	WOF	WF	WOF	WF	WOF	WF	WOF	WF
1	8.32	1.75	7.19	1.02	6.48	1.61	7.32	1.54
0.8	10.68	2.38	8.53	2.48	8.72	2.73	8.47	2.12
0.6	11.08	2.93	10.76	2.87	11.04	2.92	11.23	2.75

Figure 17. Experimental setup of the preferred MLI

the programmer in the soft format on FPGA. Xilinx ISE 12.1 software is used to write a program to generate PWM pulses using VHDL (Very High-Speed Integrated Chip Hardware Description Language). Once, the code is written the entire code is verified for syntax. After the code is synthesized, it will convert into equivalent logic gates and produces a net-list. In the implementation stage, the design is used for Placing and Routing which places the logic blocks of the design into FPGA and route them together, so that they occupy the minimum area and meet timing requirements. Then the code is converted into digital bits as FPGA can understand only digital language. This process is used to transfer the code from the system to kit, using joint test action group cable.

- **Gate Driver Module:** TLP250 acts not only as a gate driving circuit of IGBT but also optically isolates input from output.
- **Power Module:** The power module is constructed using IGBT switches, forward path diodes, bridge rectifier and capacitor whose particulars are furnished in *Appendix 1*.
- **Load and Filter Module:** This module consists of R and RL loads, and LC filter whose values given in *Appendix 1,* removes high-frequency switching ripple present in the output.

The experimental procedure to generate the switching pulses begins from the execution of FPGA unit. This includes step-by-step initialization process, absolute sinusoidal wave and carrier signal generator block (e.g. triangular wave), followed by comparator unit. SPARTAN-3A DSP from the FPGA family is considered to be a pulse generation module with which required patterns have been developed using VHDL. The derived switching pulses are presented in Figure 18.

Experimental Results

The experimental procedures are executed and the measurements have been carried out with a scale division of 5A/div and 15 Volt/div for R and RL Loads. For illustration, the output waveform of LGP of preferred topology for triangular carrier signal with a PDLSPWM strategy in the presence of R load is shown in Figure 19. This is obtained experimentally with MI and the f_{sw} are equal to 1 and 5 kHz respectively.

Figure 18. Switching pulse derived from the FPGA

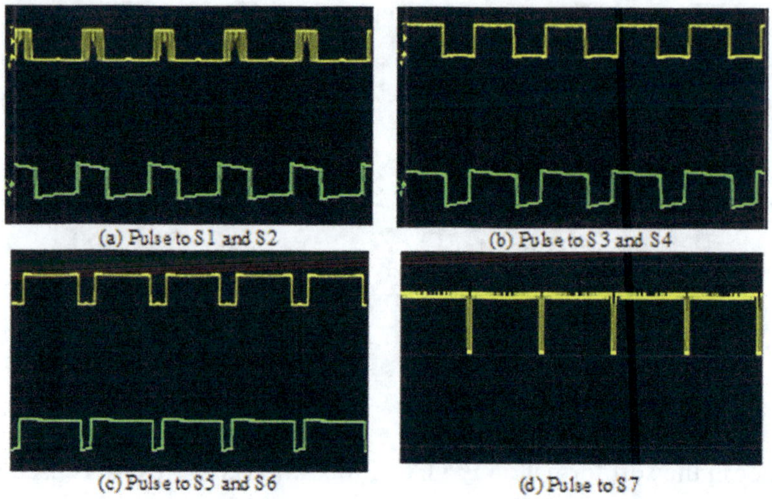

(a) Pulse to S1 and S2 (b) Pulse to S3 and S4 (c) Pulse to S5 and S6 (d) Pulse to S7

Figure 19. Multilevel DC link output voltage

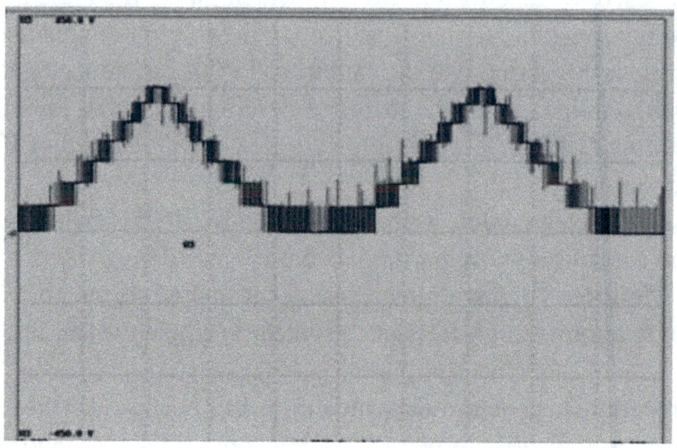

R Load

With the knowledge of developed topology in Simulink environment and simulation results, the hardware implementation is carried out and the applicability of the newfangled topology with R load is verified practically by considering triangular carrier signal with PDLSPWM, COLSPWM and VFLSPWM techniques. The test results of the topology for considering load and f_{sw} = 5 KHz are presented sequentially from the Figure 20 to the Figure 22.

Figure 20 represents the voltage across load along with its current waveform by using PDLSPWM strategy for MI=1. Figure 21 shows the effect of variation for MI=0.8 that has included the voltage and current response for COLSPWM, whereas in Figure 22, the applicability of the preferred topology for same load is tested for MI=0.6 in case of VFLSPWM. It is confirmed that topology has performed well for all the MI values.

Figure 20. Test results of voltage and current for PDLSPWM with R load

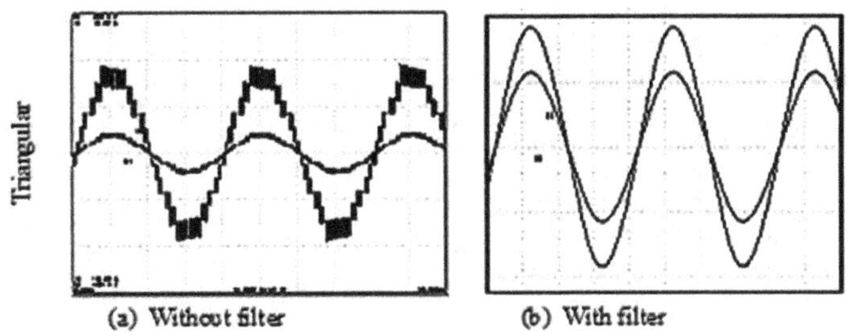

Figure 21. Test results of voltage and current for COLSPWM with R load

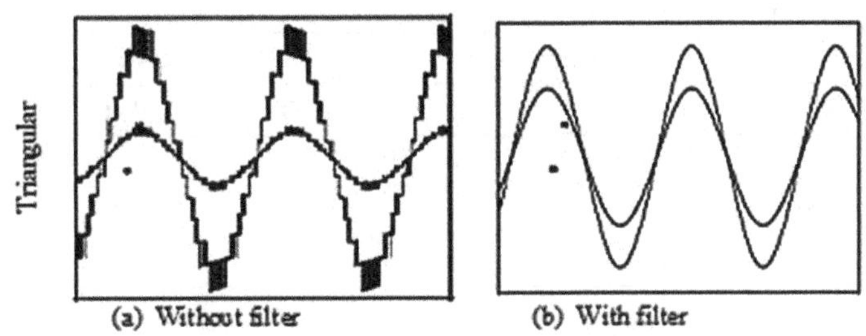

Figure 22. Test results of voltage and current for VFLSPWM with R load

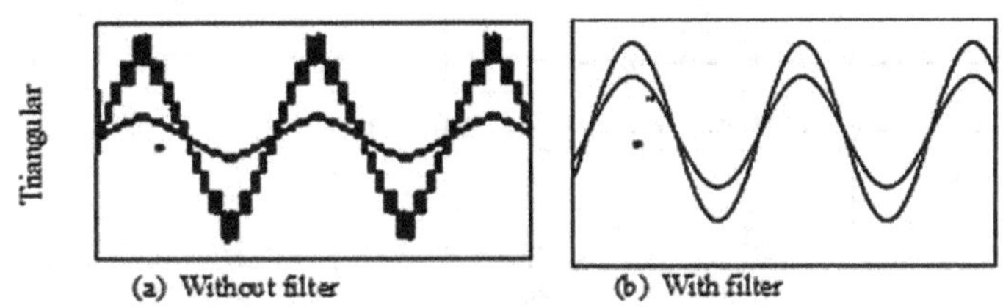

The significance of the preferred topology for different f_{sw} is tested by varying MI and the test results are listed in Table 8 for triangular carrier signals. It is implicated from the tables that the content of percentage THD in case of 'with filter' reduces nearly half of the value than 'without filter'. Moreover, the preferred topology performs well for MI which is equal to 1 and also the performance has been turned well at the switching frequency of 5 kHz.

Predictably, the output response is verified to obtain the % THD for different PWM strategies including without and with filter. The observations are recorded in Table 9 in which the MI is vary from 0.6 to 1 for the f_{sw} =2 and 5 kHz. It is understood that the %THD has varied in contrast to MI.

The voltage and current THD for the different MI values (1, 0.8 and 0.6) and the switching frequencies of 2 kHz and 5 kHz are tabulated in Table 10. The comparison reveals that the preferred topology works well for all the LSPWM techniques and the percentage THD is reduced to maximum extent in the case of 'with filter' than 'without filter'. Similarly, it exhibits better performance at MI =1 for all selected switching frequencies.

Table 8. THD % of MLI with PDLSPWM using triangular carrier

MI	Switching Frequency=2kHz				Switching Frequency =5kHz			
	Voltage % THD		Current % THD		Voltage % THD		Current % THD	
	WOF	WF	WOF	WF	WOF	WF	WOF	WF
1	4.58	2.02	4.67	2.06	3.05	1.15	3.01	1.63
0.8	5.02	2.25	5.04	2.29	3.13	2.03	3.07	2.07
0.6	5.80	2.65	5.90	2.67	3.79	2.22	3.79	2.27

Table 9. THD % of the MLI with COLSPWM using triangular carrier

MI	Switching Frequency=2kHz				Switching Frequency =5kHz			
	Voltage % THD		Current % THD		Voltage % THD		Current % THD	
	WOF	WF	WOF	WF	WOF	WF	WOF	WF
1	5.68	2.72	5.67	2.61	3.76	1.72	3.81	2.31
0.8	6.42	3.15	6.34	3.26	4.16	2.16	4.08	3.39
0.6	7.01	3.89	6.96	3.41	5.32	2.95	5.12	3.57

Table 10. THD % of the MLI with VFLSPWM using triangular carrier

MI	Switching Frequency=2kHz				Switching Frequency =5kHz			
	Voltage % THD		Current % THD		Voltage % THD		Current % THD	
	WOF	WF	WOF	WF	WOF	WF	WOF	WF
1	7.29	2.06	7.29	2.06	6.76	2.08	6.76	1.19
0.8	8.56	3.24	8.56	3.24	8.16	3.09	8.16	2.32
0.6	11.14	3.89	11.14	3.89	10.03	3.66	10.03	2.97

RL Load

To know the appropriateness of the preferred topology, it has been tested practically in the presence of RL load. For diversity, this study was conducted with different scenario, i.e., three types such as PDLSPWM, COLSPWM and VFLSPWM modulation strategy based on carrier arrangement were under trialed for vertical variation whereas, each of the technique has involved triangular, carrier signal for horizontal variation.

The output response of each scenario at f_{sw} is equal to 5 kHz and has been presented in this section and moreover, effect of filter module is also included. Figure 23 shows the load voltage and current waveform for PDLSPWM at MI is equal to 1, Figure 24 illustrates the output voltage and current response for COLSPWM at MI is equal to 0.8 and Figure 25 presents the voltage across the load and its current flow for MI is equal to 0.6 for VFLSPWM.

Figure 23. Test results of voltage and current for PDLSPWM with RL load

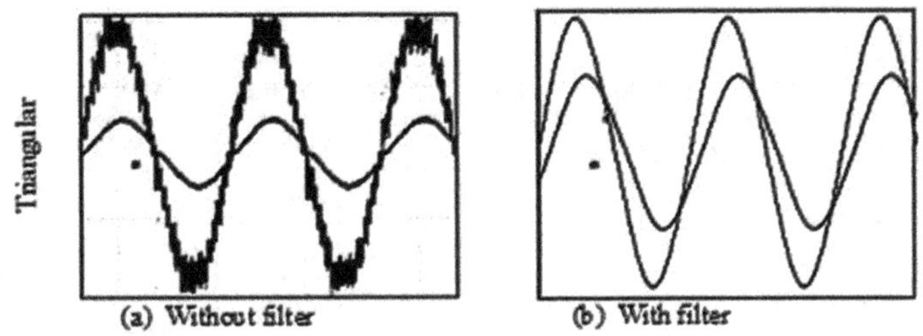

Figure 24. Test results of voltage and current for COLSPWM with RL load

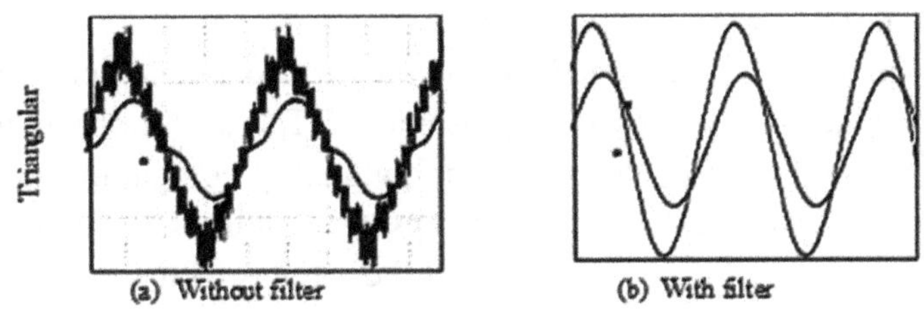

Figure 25. Test results of voltage and current for VFLSPWM with RL load

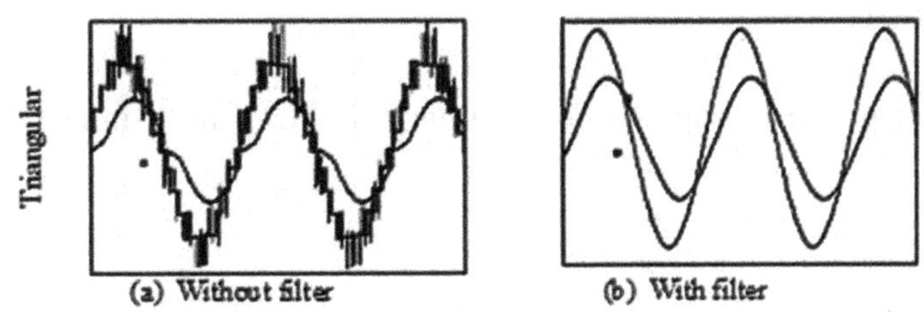

A THD analysis was performed to reveal the quality of power supply that was derived using PDL-SPWM for triangular carrier. In which the MI is varied each of f_{sw} 2 and 5 kHz and the test results are presented from Table 11. Additionally, the effect of filter module is also compared nicely.

Similarly, the same THD analysis was extended to the output response of new MLI while controlling by COLSPWM. The performance indices are demonstrated in Table 12 for triangular carrier. The test

Table 11. THD % of the MLI of RL-Load with PD using triangular carrier

MI	Switching Frequency=2kHz				Switching Frequency =5kHz			
	Voltage % THD		Current % THD		Voltage % THD		Current % THD	
	WOF	WF	WOF	WF	WOF	WF	WOF	WF
1	13.86	1.46	10.36	2.29	12.48	1.39	9.17	2.54
0.8	13.93	1.62	11.06	1.74	14.99	1.70	10.61	2.87
0.6	16.48	2.85	11.31	2.19	17.43	1.57	11.68	3.14

Table 12. THD % of MLI for RL Load with CO using triangular carrier

MI	Switching Frequency=2kHz				Switching Frequency =5kHz			
	Voltage % THD		Current % THD		Voltage % THD		Current % THD	
	WOF	WF	WOF	WF	WOF	WF	WOF	WF
1	15.71	1.78	11.54	1.14	13.76	2.89	11.81	1.89
0.8	16.43	2.23	12.15	2.21	15.62	3.10	12.59	2.06
0.6	17.98	2.96	12.98	2.83	17.93	3.48	14.04	2.31

results that have been provided in this table were explored the supply quality comparatively including LC filter for the variation of MI and f_{sw} 2 and 5 kHz.

Finally, the percentage THD was obtained for the output response of preferred topology with the help of VFLSPWM which has been successively presented in Table 13 for triangular carrier signal. The test results provided in the table was explored the supply quality comparatively by using LC filter for the variation of MI and f_{sw} 2 and 5 kHz. To bring a consensus, it has been compared exclusive as well as including LC filter modules and it is observed that the indices of the later outperforms the indices of predecessor.

LOSS CALCULATION

Total loss incurred in the preferred topology constitutes switching and conduction losses.

Conduction Loss

Let on-state saturation voltage is $v_{on}(t)$ and current flow is $i(t)$ then, power dissipation during conduction is given by Equation (9) and the same have to be extended to overall preferred inverter. V_T and V_D are the threshold voltages of power devices, R_T and R_D are the equivalent resistances of the voltage drop across the power devices, and β is a constant related to the specification of the device. The instantaneous conduction loss of IGBT ($p_{CT}(t)$) and diode ($p_{CD}(t)$) are given in Equations (10) and (11) respectively (Mohammad, et al. 2013).

Table 13. THD % of MLI for RL Load with VF using triangular carrier

MI	Switching Frequency=2kHz				Switching Frequency =5kHz			
	Voltage % THD		Current % THD		Voltage %THD		Current %THD	
	WOF	WF	WOF	WF	WOF	WF	WOF	WF
1	16.28	2.62	12.14	2.14	14.32	2.32	12.19	2.38
0.8	17.74	3.42	13.65	2.96	15.94	3.47	13.74	3.21
0.6	18.23	3.86	13.98	3.13	16.36	4.71	14.87	4.01

$$p(t) = v_{on}(t) * i(t) \quad (9)$$

$$p_{CT}(t) = \left[V_T + R_T i^{\beta}(t)\right] i(t) \quad (10)$$

$$p_{CD}(t) = \left[V_D + R_D i(t)\right] i(t) \quad (11)$$

The number of on state IGBTs and diodes are $N_T(t)$ and $N_D(t)$ respectively. The output current of MLI can assume to be sinusoidal. Therefore, average conduction loss of LG unit is represented by Equation (12).

$$P_{C,LG} = \frac{1}{\pi}\int_0^{\pi}\left[N_T(t) * P_{CT}(t) + N_D(t) * P_{CD}(t)\right] d(\omega t) \quad (12)$$

The output current of PG unit is considered to be sinusoidal as $i(t)=I_m \sin(\omega t)$, the conduction loss of the PG unit is given by (Mohammad, et al. 2013):

$$P_{C,PG} = \left\{ \begin{array}{l} \frac{2}{\pi}\left[V_{D,H} I_m \left(1 - \cos\phi\right) + \frac{R_{D,H} I_m^2}{4}\left(2\phi - \sin\left(2\phi\right)\right) \right. \\ \left. + V_{T,H} I_m \left(1 + \cos\phi\right) + R_{T,H} I_m^{\beta+1} \int_{\phi}^{\pi} \left(\omega t\right) d\left(\omega t\right) \right] \end{array} \right\} \quad (13)$$

where, ϕ is the power factor angle and H indicates the parameter related to PG unit.

Switching Loss

The switching loss of a typical power semiconductor device can be evaluated with the linear approximation of current and voltage during switching period (t) that includes turn-on period (t_{on}) and turn-off period (t_{off}) (Mohammad, et al. 2013). Then turn-on loss ($E_{on,J}$) of switch J is given by:

$$E_{on,J} = \int_0^{t_{on}} v(t) i(t) dt = \int_0^{t_{on}} \left[\frac{V_{sw,J}}{t_{on}} t\right]\left[-\frac{I}{t_{on}}\left(t - t_{on}\right)\right] dt$$

$$E_{on,J} = \frac{1}{6} V_{sw,J} I t_{on} \quad (14)$$

where, V_{sw} is on state voltage on the switch and I is the current through the switch after turn-on.

Similarly turn-off loss ($E_{off,J}$) is given below

$$E_{off,J} = \frac{1}{6} V'_{sw,J} I' t_{off} \tag{15}$$

where, V'_{sw} is off state voltage on the switch and I' is the current through the switch before turn-off.

The switching losses are relevant with the number of switching transitions and also to the control scheme. Commonly the average switching loss can be given for i^{th} time as:

$$P_{sw} = 2f_{sw} \left[\sum_{J=1}^{N_{switch}} \left(\sum_{i=1}^{N_{on,J}} E_{on,Ji} + \sum_{i=1}^{N_{off,J}} E_{off,Ji} \right) \right] \tag{16}$$

where, f_{sw} is the switching frequency, N_{switch} is switch number, $N_{on,J}$ and $N_{off,J}$ are the number of turning on and off of the J_{th} switch during a half fundamental cycle.

Therefore,

$$\text{Total power loss Ploss} = P_{C,LG} + P_{C,PG} + P_{sw} \tag{17}$$

The Figure 26 compares with the total power loss in the preferred and classical CHB Inverter for the different modulation index (MI). The total power loss is reduced in the case of preferred topology as compared with classical inverter for the same MI, as it requires less number of components and very few switches will be in on-state at a time, so the total power loss are also minimized compared to CHB.

Figure 26. Comparing total loss

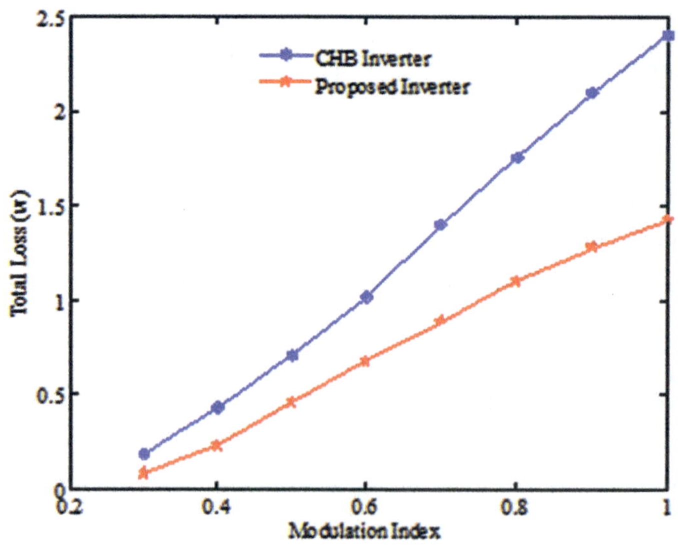

FAULT DIAGNOSIS OF THE PREFERRED TOPOLOGY

In recent years, multilevel inverters (MLI) provide greater opportunity for industrial applications, but the occurrence of a fault perhaps prompts the system's collapse. Open or short circuit is the most common fault occurs in power electronic switches of the MLI. But the open circuit fault doesn't really bring about the circuit shutdown and can stay undetected for longer time. This may prompt auxiliary fault in the inverter or in the remaining parts of the system, bringing about the entire system into failure. On the other hand, short circuit fault is very destructive and requires special measures to shut down the entire drive immediately.

To improve the reliability of the MLI, a new topology with fault tolerant ability is identified by Alian Chen et al. (2005) with various power device failures. At the same time, Xiaomin Kou et al. (2004) have designed flying capacitor type multilevel inverters with fault-tolerant features. Further, Mingyao Ma et al. (2007) discussed the fault diagnosis procedure for phase disposition PWM strategy (PDPWM) and phase shifted PWM strategy (PSPWM). Frederic Richardeau et al. (2013) extended to examine the faulty device and the reliability of converter through the theoretical procedure, while Vinothkumar et al. (2013) applied a new fault identification procedure based on discrete wavelet transforms for MLI. Followed, Ui-Min Choi et al. (2014) extended diagnosis method of an open-switch fault control strategy for T-type MLI configurations.

It is the tendency of MLI that is used to operate under faulty conditions; however, fault should be detected as soon as possible. Otherwise, it causes an unbalanced output on the load side. In case of an induction motor (IM) load, the unbalanced supply causes severe loss in the machine, which leads to reduce the motor insulation life due to overheating. Present work deals, open circuit fault and the knowledge of fault behavior, prediction, and diagnosis are nicely exhibited for better understanding of various transformation tools.

Transformation Tools

There are different transformation tools to analyze the signal in which it is transformed into another form without affecting the original information that has been carried by it, is shown in Figure 27.

Generally a Fourier analysis (FA) is applied to transform the time domain signal into the frequency domain. During this process the time domain information is lost and the same is treated as a serious drawback of FA. Therefore, Dennis Gabor (1946) introduced a short-time Fourier Transform (STFT) to overcome that shortcoming in which a technique called windowing is adapted to analyze a small portion of the signal at a time and gives both time and frequency information. The STFT gives the complete information about the signal, but the window size has limited precision. Hence, a wavelet transformation tool (WTT) is emphasized analyzing the signal. In wavelet analysis (WA) the use of time - scale region seems to be an inherent advantage of local window analysis. If any small discontinuity happens in the signal as shown in the Figure 28 can be easily identified and it is recognized as a fault in the signal.

Figure 27. Functionality of the transformation tools

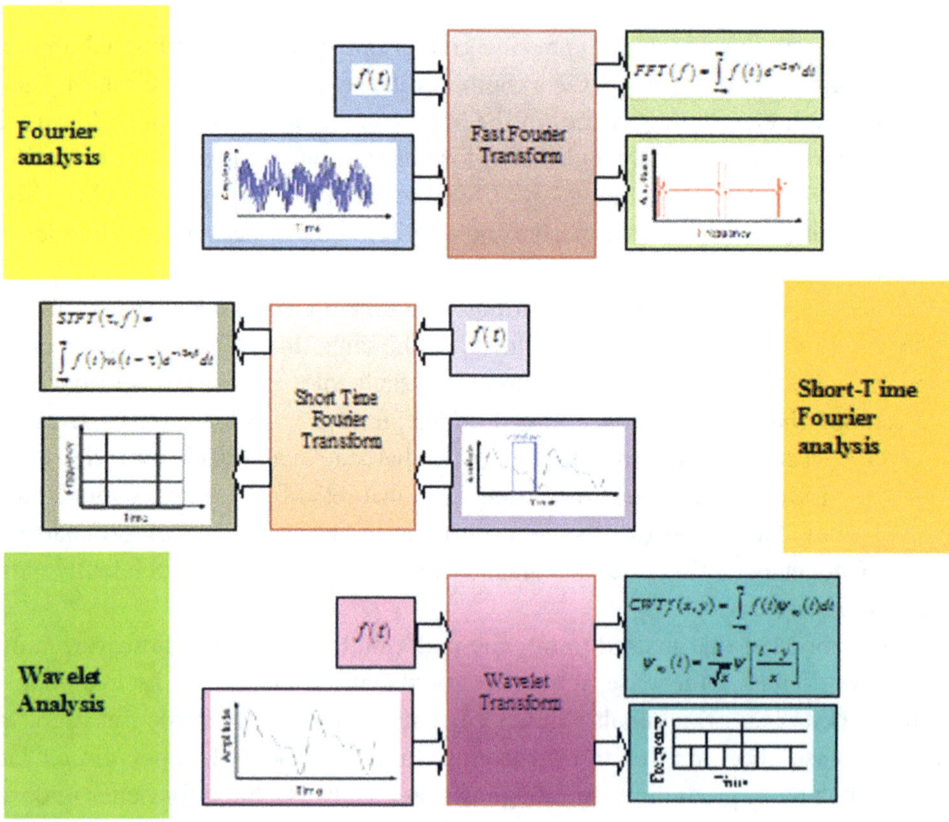

Figure 28. Discontinuity sinusoid signal

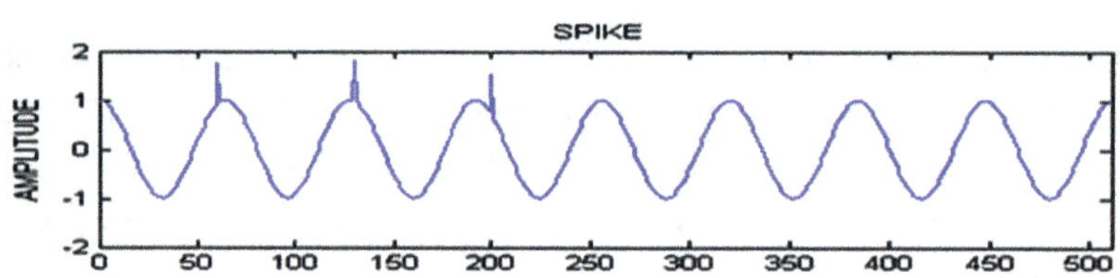

Further, WA accomplishes the enlightening aspects of information such as trends, breakdown points and discontinuities in higher derivatives. It gives a more diverse view of data than conventional techniques and it can frequently compress or de-noise a signal without substantial degradation. Moreover, data analysis using WT is categorized based on dimensions such as one and two dimensional. Particularly, the two-dimensional data analysis is sub-categorized as continuous and discrete WT.

Continuous Wavelet Transform

The continuous wavelet transform (*CWT*) is defined as the sum over all time of the signal (*f(t)*) multiplied by scaled, shifted versions of the wavelet function (ψ). The *CWT* as the function of translation (*x*) and dilation (y) is given in equation (18). Further, it can be multiplied by the appropriately scaled and shifted wavelet, giving the constituent wavelets of the original signal and shown in Figure 29.

$$CWT_f^{\psi}(x,y) = \int_{-\infty}^{\infty} f(t)\psi_{xy}(t)dt \tag{18}$$

$$\psi_{xy}\left(t\right) = \frac{1}{\sqrt{x}}\psi\left[\frac{t-y}{x}\right];x,y \in R;x \neq 0 \tag{19}$$

where, R is the real continuous number system.

Discrete Wavelet Transform

In discrete wavelet transform (*DWT*) the wavelet coefficients are scaled and translated in discrete steps at dyadic intervals.

Let,

f(t)- is the function of discretized signal
$\psi\left(t\right)$-is the mother wavelet

Then the *DWT* can be defined as:

$$DWT_f^{\psi}(x,y) = \frac{1}{\sqrt{a^x}}\int_{-\infty}^{\infty} f(t)\psi_{xy}^{*}\left[\frac{t-yba^x}{a^x}\right]dt;a>1,b>0,x,y \in Z \tag{20}$$

where, Z is a set of positive integers.

Mallat, (1988) developed an efficient algorithm to calculate the wavelet coefficients using filters. It yields a fast wavelet transform, a box into which a signal passes, and out of which wavelet coefficients quickly emerge. The approximations are high scale and low-frequency components of selected signal, whereas, the details are the low scale and high-frequency components. The filtering process closely looks as shown in Figure 30. It shows that detailed coefficients (D) mainly consist of high-frequency noise and approximation coefficients (A), mainly consist of low-frequency noise than the original signal.

Figure 29. Generation of wavelet coefficients

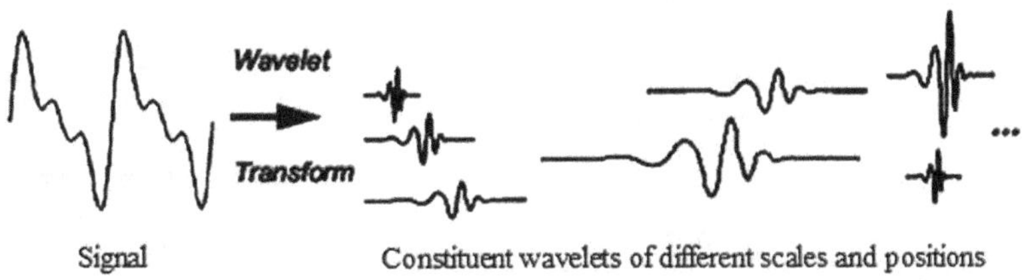

The decomposition or filtering process can be iterated to many levels based on the other information in the original signal which is known as wavelet decomposition tree process and is shown in Figure 31. At stage one A1 and D1 constitute the signal, whereas A1, A2 and D2 comprises the signal at stage two, while A2, A3 and D3 involves in reconstruct of the signal at stage three.

In practice, a suitable number of iterations or decomposition levels can be selected based on the nature of signal or a suitable criteria can be made such as energy and entropy.

Implementation of WTT for Fault Analysis

This section presents the fault analysis using WTT on the preferred inverter with all types of loads. The analysis was carried out using WTT in MATLAB/Simulink. The comparison of detailed coefficients of an output voltage and current waveforms under faulty and healthy conditions is done and an algorithm is developed to identify the faulty device with respect to a healthy condition. The procedure used to develop an algorithm is depicted below.

Figure 30. Overview of signal decomposition using DWT

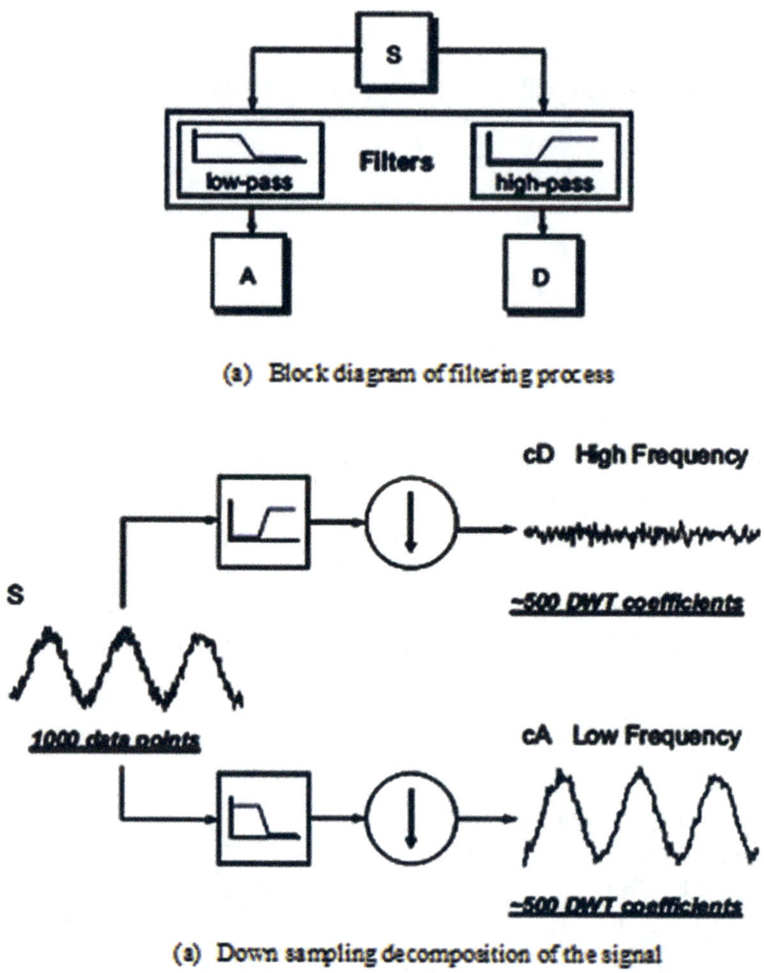

Figure 31. Wavelet decomposition tree of a signal

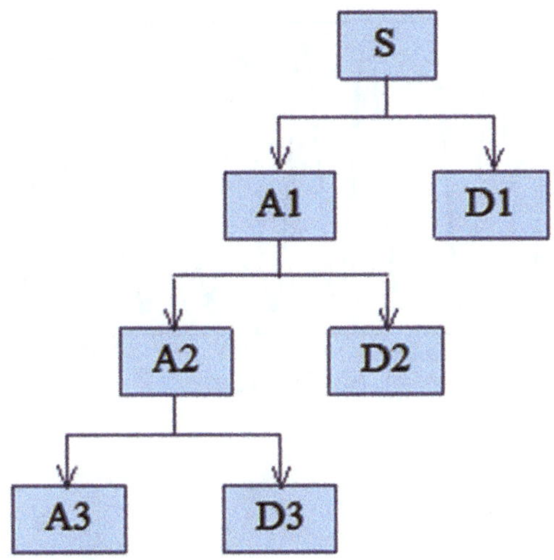

Step 1: Initialize the fault dictionary that contains information about fault free (FF) and different possible fault conditions of level generation (LG) and polarity generation (PG) unit components. i.e., healthy condition is designated as 1, otherwise 0.

Step 2: Developed 15-level inverter is simulated using MATLAB/Simulink as per the fault dictionary initialized in step 1.

Step 3: Simulated output is imported to WTT (one-dimensional discrete wavelet analysis tool).

Step 4: The mother wavelet and decomposition levels are chosen to analyze the signal without losing its original information.

Step 5: The analysis is executed for above stated operating conditions; the corresponding standard deviation (SD) value is imported to the same fault dictionary and fault identification (ID) has been given for classification.

Further, the test result is very much useful to do the post fault analysis to identify the type of fault and faulty device.

Let's consider two case studies based on fault free and open circuit fault.

Case I: Fault Free Diagnosis

The preferred topology is simulated without fault in the circuit, i.e., S1~S7, D1~D6 and Sa ~ Sb are conducted properly to generate the 15-level output.

The signal shown in Figure 32 is imported to WTT for analysis. The selected signal has been decomposed upto 5 levels without missing any information of the original signal. Figure 33 shows the approximate and detail coefficients separately. Figure 34 represents the reconstructed signal after decomposition levels.

After analyzing the selected signal 'S', the WTT provides SD values for selected decomposition levels such as SD1=0.182 SD2=0.3073 and SD3=0.553. Based on the developed algorithm the following test results are identified:

Figure 32. Fault free output voltage of MLI

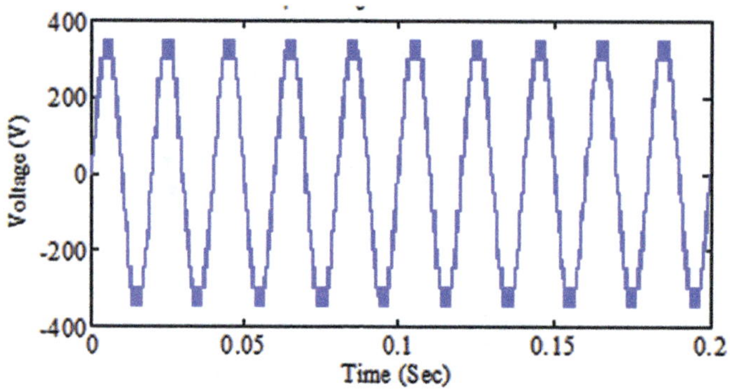

Figure 33. Signal decomposition using Wavelets

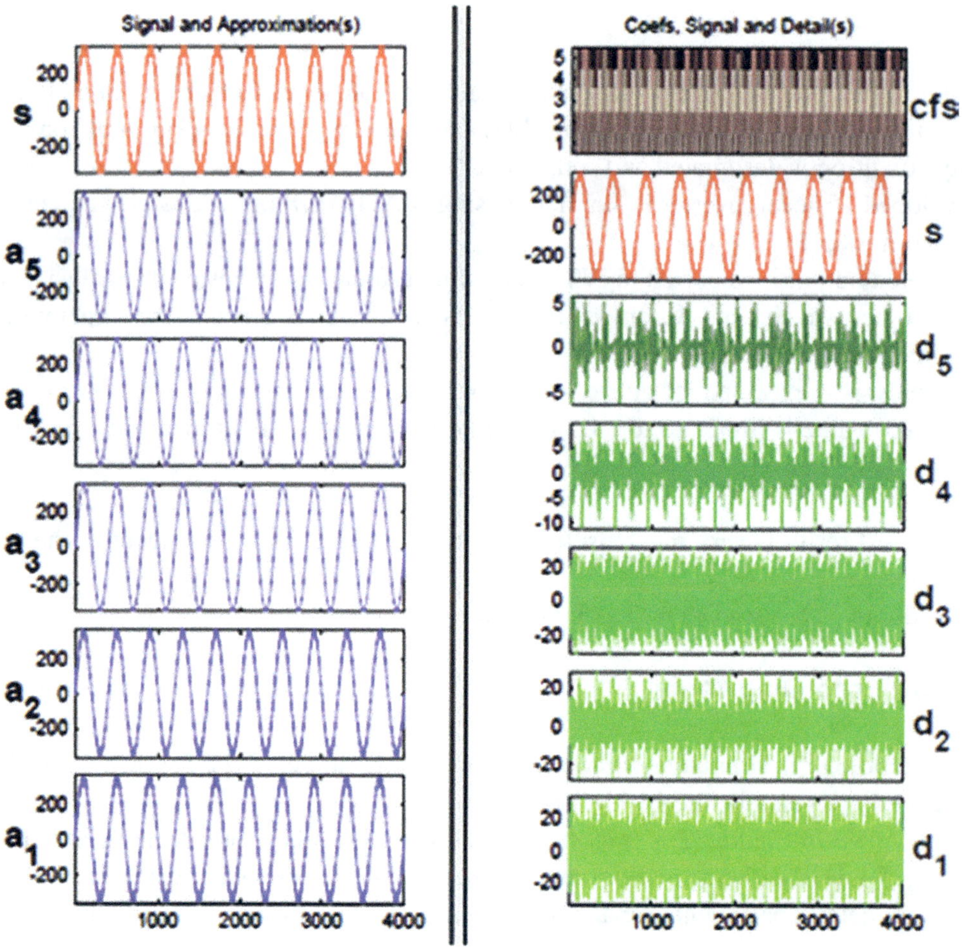

Figure 34. Healthy signal in DWT tree mode

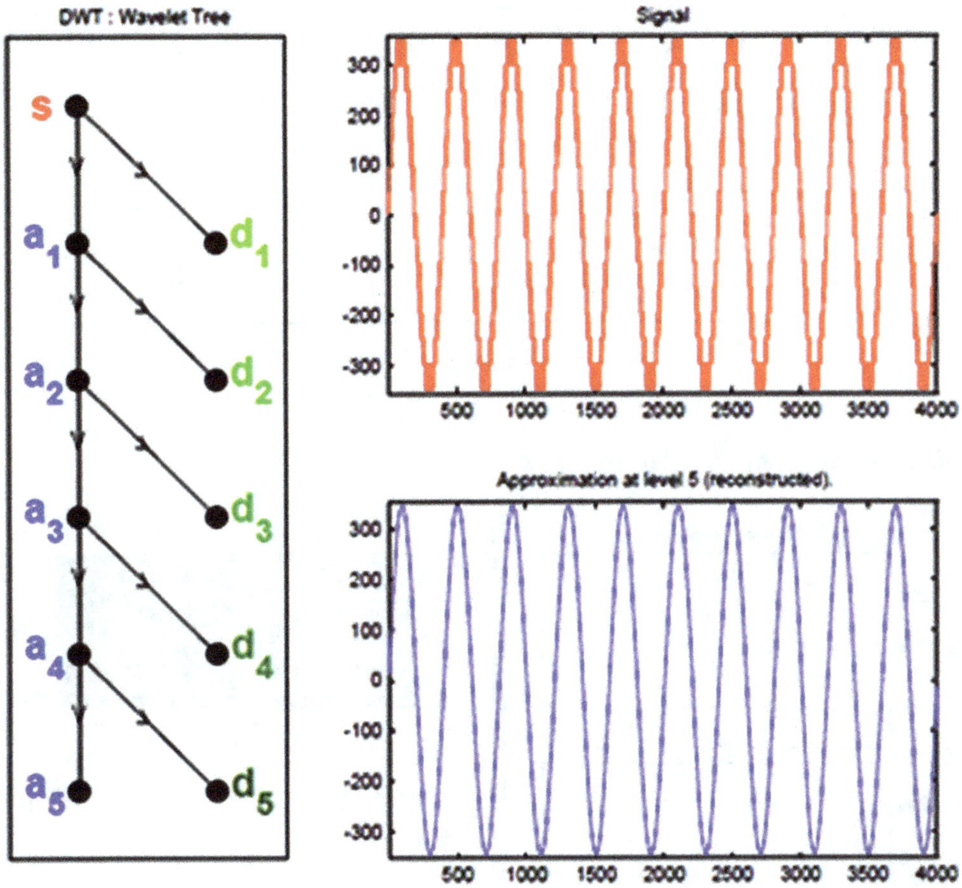

S1~S7 [1 1 1 1 1 1 1], D1~D6 [1 1 1 1 1 1] and Sa~Sd [1 1 1 1]

FAULT ID: FF (see *Appendix 2*)

Further, the test result confirms the fault free operation.

Case II: Fault Diagnosis

Let's assume S3, D1 in LG unit where as Sa in PG unit is open circuited. At the end of analysis, the SD values are SD1=29.81, SD2=38.64 and SD3=103.6. The Figure 35 and Figure 36 illustrate the corresponding output voltage under a fault condition and WTT analyzer respectively. The fault device identification algorithm reflects the following test results:

S1~S7 [1 1 0 1 1 1 1], D1~D6 [0 1 1 1 1 1] and Sa~Sd [0 1 1 1]

Figure 35. Faulty output voltage of MLI

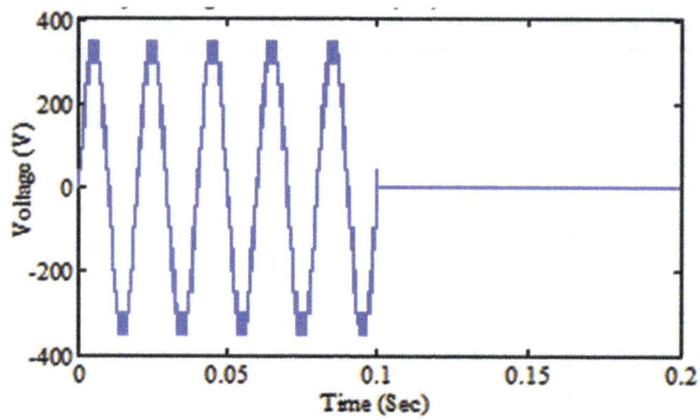

Figure 36. Signal decomposition using Wavelets

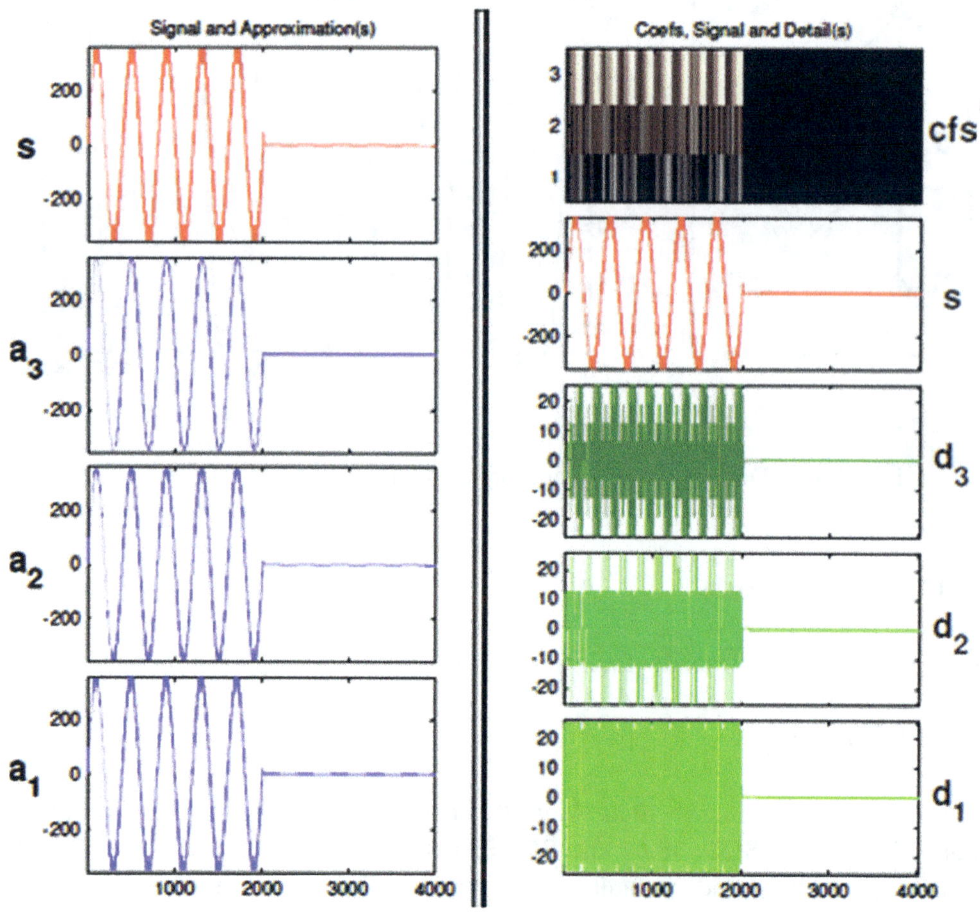

FAULT ID: OF36 (see *Appendix* 2)

Moreover, the test result confirms the faulty devices.

Additionally, the % THD can be computed using an FFT analyzer for all possible open circuit faults and the same has been compared with the value that was obtained from WTT, the THD of the signal using WTT is calculated using the following equation.

$$\%THD = \frac{\frac{1}{D_i}\sqrt{d_1^2 + d_2^2 + d_3^2 + \ldots.. + d_i^2}}{\frac{1}{D_i}\sqrt{a_1^2 + a_2^2 + a_3^2 + \ldots.. + a_i^2}} \qquad (21)$$

where,

D_i - Decomposition at i[th] level
d_i - Detailed coefficient at i[th] level
a_i - Approximate coefficients at i[th] level

The result shows that WTT computes the %THD as accurate as it is shown in Figure 37.

Figure 37. Comparison of % THD for open-circuit fault concern FFT and WTT

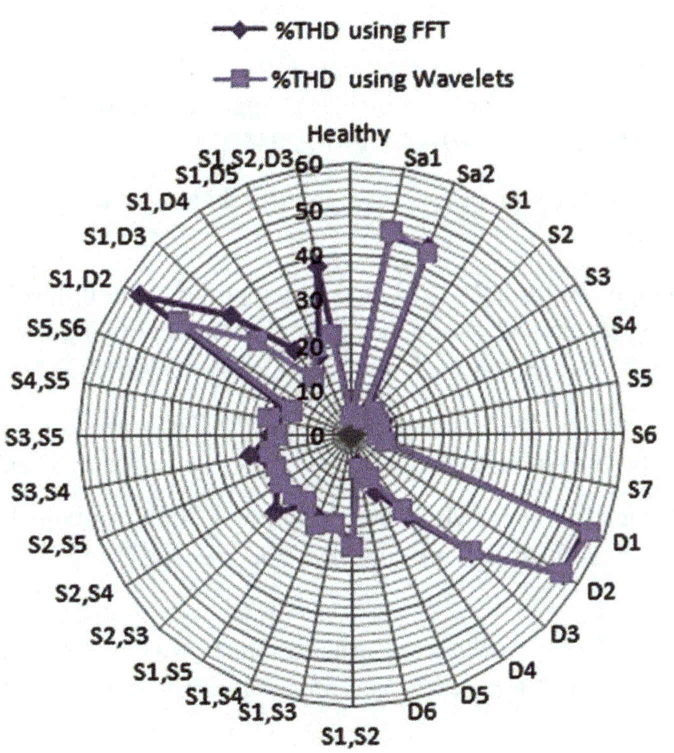

FUTURE RESEARCH DIRECTIONS

The topology discussed in this chapter finds application in converting available renewable energy into required alternating form. The topology can use in speed control based on V/f technique and find application in irrigation pumps.

CONCLUSION

A symmetric Multi Level Inverter with reduced number of switches has been preferred in this chapter, where required higher voltage levels is achieved easily by duplicating the level generator module and less number of gating pulses is required to generate the required output levels as the comparisons to traditional MLI. The topology is simulated to observe the performance for R and RL loads with and without filters and the results shows that the percentage THD are reduced to an acceptable standard. The same was investigated experimentally for the same loading condition to validate the simulation results, and the test results prove that the experimental setup is realized the simulated one. The performance by designing topology was analyzed for different MI at selected switching frequencies, and suitable operation condition was derived. Additionally, the possible fault conditions and faulty components can be identified using a selective algorithm with the help of WTT. Therefore, it is concluded that a MLI topology has designed and performance has been analyzed for different test conditions using different MCPWM techniques.

REFERENCES

Abu-Rub, H., Holtz, J., Rodriguez, J., & Ge Baoming. (2010). Medium- voltage multilevel converters-state of the art, challenges and requirements in industrial applications. *IEEE Transactions on Industrial Electronics*, *57*(8), 2581–2596. doi:10.1109/TIE.2010.2043039

Ahmad & Gupta. (2013). Digital pwm of cascaded multilevel voltage source inverter using fpga. *Engineering and Systems*, 12-14.

Ajami, A., Mokhberdoran, A., & Reza, M. (2013). A new topology of multilevel voltage source inverter to minimize the number of circuit devices and maximize the number of output voltage levels. *Journal of Electrical Engineering & Technology*, *8*(6), 1321–1329. doi:10.5370/JEET.2013.8.6.1328

Angulo, M., Lezana, P., Kouro, S., Rodrıguez, J., & Wu, B. (2007). Level shifted pwm for cascaded multilevel inverters with even power distribution. *IEEE Power Electronics Specialists Conference*, 2373-2378.

Ataollah, M. (2014). Symmetric and asymmetric design and implementation of new cascaded multilevel inverter topology. *IEEE Transactions on Power Electronics*, *29*(12), 6712–6724. doi:10.1109/TPEL.2014.2302873

Azli, N. A., & Choong, Y. C. (2006). Analysis on the performance of a three-phase cascaded h-bridge multilevel inverter. *First International Power and Energy Conference*, 405-410. 10.1109/PECON.2006.346685

Balamurugan, C. R., Natarajan, S. P., Bensraj, R., & Shanthi, B. (2016). A review on modulation strategies of multi level inverter. *Indonesian Journal of Electrical Engineering and Computer Science*, *3*(3), 681–705. doi:10.11591/ijeecs.v3.i3.pp681-705

Behera, R. K., Dixit, T. V., & Das, S. P. (2006). Analysis of experimental investigation of various carrier-based modulation schemes for three level neutral point clamped inverter-fed induction motor drive. *International Power Electronics, Drives and Energy Systems Conference*, 1-6.

Carrara, G., Gardella, S., Marchesoni, M., Salutarin, R., & Sciutto, G. (1990). A new multilevel PWM method: a theoretical analysis. *Power Electronics Specialists Conference*, 363-371. 10.1109/PESC.1990.131211

Chen, A., Hu, L., Chen, L., Deng, Y., & He, X. (2005). A multilevel converter topology with fault-tolerant ability. *IEEE Transactions on Power Electronics*, *20*(2), 405–415. doi:10.1109/TPEL.2004.842983

Colak, I., Kabalci, E., & Bayindir, R. (2011). Review of multilevel voltage source inverter topologies and control schemes. *Energy Conversion and Management*, *52*(2), 1114–1128. doi:10.1016/j.enconman.2010.09.006

Dahidah, M. S. A., & Agelidis, V. G. (2008). Single-carrier sinusoidal PWM equivalent selective harmonic elimination for a five-level voltage source converter. *Electric Power Systems Research*, *78*(11), 1826–1836. doi:10.1016/j.epsr.2008.01.021

Du, Z., Tolbert, L. M., Chiasson, J. N., & Ozpineci, B. (2006). A Cascade Multilevel Inverter using a Single DC Source. *Twenty-First Annual IEEE Applied Power Electronics Conference and Exposition*, 426-430. 10.1109/APEC.2006.1620573

Edwin Jose, S., & Titus, S. (2016). A level dependent source concoction multilevel inverter topology with a reduced number of power switches. *Journal of Power Electronics*, *16*(4), 1316–1323. doi:10.6113/JPE.2016.16.4.1316

Farhadi Kangarlu, M., Babaei, E., & Laali, S. (2012). Symmetric multilevel inverter with reduced components based on non-insulated dc voltage sources. *IET Power Electronics*, *5*(5), 571–581. doi:10.1049/iet-pel.2011.0263

Gabor, D. (1946). Theory of communication. *J IEEE*, *93*(3), 429–457.

Kastha, D., & Bose, B. K. (1994). Investigation of Fault Modes of Voltage-Fed Inverter System for Induction Motor Drive. *IEEE Transactions on Industry Applications*, *30*(4), 1028–1038. doi:10.1109/28.297920

Khoucha, F., Lagoun, M. S., Kheloui, A., & Benbouzid, M. E. H. (2011). A comparison of symmetrical and asymmetrical three-phase h-bridge ltilevel inverter for dtc induction motor drives. *IEEE Transactions on Energy Conversion*, *26*(1), 64–72. doi:10.1109/TEC.2010.2077296

Khounjahan, Banaei, & Farakhor. (2014). A new low cost cascaded transformer multilevel inverter topology using minimum number of components with modified selective harmonic elimination modulation. *Ain Shams Engineering Journal*, *6*, 67–73.

Kim, Kim, Kang, & Hyun. (2003). A simple method for conducting angle calculation in cascaded inverters using step pulse waves. *29th IEEE Annual Conference, Industrial Electronics Society*, *3*, 2735-2740.

Kou, X., Corzine, K. A., & Familiant, Y. L. (2004). A unique fault-tolerant design for flying capacitor multilevel inverter. *IEEE Transactions on Power Electronics*, *19*(4), 979–987. doi:10.1109/TPEL.2004.830037

Lai. (1996). Multilevel converters - a new breed of power converter. *IEEE Transactions on Industry Applications*, *32*(3), 509–517. doi:10.1109/28.502161

Lopez, O., Alvarez, J., Doval-Gandoy, J., & Freijedo, F. D. (2009). Multilevel multiphase space vector pwm algorithm with switching state redundancy. *IEEE Transactions on Industrial Electronics*, *56*(3), 792–804. doi:10.1109/TIE.2008.2004390

Ma, M., Hu, L., Chen, A., & He, X. (2007). Reconfiguration of carrier-based modulation strategy for fault tolerant multilevel inverters. *IEEE Transactions on Power Electronics*, *22*(5), 2050–2060. doi:10.1109/TPEL.2007.904249

Mirafzal, B. (2014). Survey of fault-tolerance techniques for three-phase voltage source inverters. *IEEE Transactions on Power Electronics*, *61*(10), 5192–5202.

Mohammad, Babaei, & Sabahi. (2012). A new single-phase cascade multilevel inverter topology using four-level cells. *International conference in Electrical Engineering*, 348–53.

Mohammad, F. K., & Ebrahim, B. (2013). A generalized cascaded multilevel inverter using series connection of sub-multilevel inverter. *IEEE Transactions on Power Electronics*, *28*(2), 625-636.

Najafi, E., & Yatim, A. H. M. (2012). Design and implementation of a new multi level inverter topology. *IEEE Transactions on Industrial Electronics*, *59*(11), 4148–4154. doi:10.1109/TIE.2011.2176691

Panagis, P., Stergiopoulos, F., Marabeas, P., & Manias, S. (2008). Comparison of state of the art multilevel inverters. *Electronics Specialists Conference*, 4296-4301. 10.1109/PESC.2008.4592633

Panda, A. K., & Patnaik, S. S. (2015). Analysis of cascaded multilevel inverters for active harmonic filtering in distribution networks. *Electrical Power and Energy Systems*, *66*, 216–226. doi:10.1016/j.ijepes.2014.10.034

Rahim, N. A., & Selvaraj, J. (2010). Multistring five-level inverter with novel pwm control scheme for PV application. *IEEE Transactions on Industrial Electronics*, *57*(6), 2111–2123. doi:10.1109/TIE.2009.2034683

Ramkumar, S., Kamaraj, V., Thamizharasan, S., & Jeevananthan, S. (2012). A new series parallel switched multilevel dc-link inverter topology. *International Journal of Electrical Power & Energy Systems*, *36*(1), 93–99. doi:10.1016/j.ijepes.2011.10.028

Richardeau, F., & Pham, T. T. L. (2013). Reliability calculation of multilevel converters: Theory and applications. *IEEE Transactions on Industrial Electronics*, *60*(10), 4224–4233. doi:10.1109/TIE.2012.2211315

Rodriguez, Lai, & Peng. (2002). Multilevel inverters: a survey of topologies, controls and applications. *IEEE Transactions on Industrial Electronics*, *49*(4), 724-738.

Saleh, S. A., Radwan, T. S., & Rahman, M. A. (2007). Real-Time Testing of WPT-Based Protection of Three-Phase VS PWM Inverter-Fed Motor. *IEEE Transactions On Power Delivery, VOL.*, *22*(4), 2108–2115. doi:10.1109/TPWRD.2007.905559

Sirisukprasert, S., Jih-Sheng Lai, & Tian-Hua Liu. (2002). Optimum harmonic reduction with a wide range of modulation indexes for multilevel converters. *IEEE Transactions on Industrial Electronics*, *49*(4), 875–881. doi:10.1109/TIE.2002.801226

Son, H.-I., Kim, T.-J., Kang, D.-W., & Hyun, D.-S. (2004). Fault Diagnosis. 4nd Neutral Point Voltage Control When The 3-Level Inverter Faults Occur. *35th Annual IEEE Power Electronics Specialists Conference*, 4558-4563.

Song, Kang, & Park. (2009). Cascaded multilevel inverter employing three-phase transformers and single DC input. *IEEE Transactions on Industrial Electronics*, *56*(6), 2005-2014.

Su, G.-J. (2003). *Multilevel DC link inverter.* US Patent 6577087.

Thamizharasan, S., Baskaran, J., Ramkumar, S., & Jeevananthan, S. (2013). New dual bridge multilevel dc-link inverter topology. *Electrical Power and Energy Systems*, *45*(1), 376–383. doi:10.1016/j.ijepes.2012.09.024

Toupchi Khosroshahi, M. (2014). Crisscross cascade multilevel inverter with reduction in number of components. *IET Power Electronics*, *7*(12), 2914–2924. doi:10.1049/iet-pel.2013.0541

Ui-Min Choi, Kyo-Beum Lee, & Blaabjerg, F. (2014). Diagnosis and tolerant strategy of an open-switch fault for T-type three-level inverter systems. *IEEE Transactions on Industry Applications*, *50*(1), 495–508. doi:10.1109/TIA.2013.2269531

Van, Moon, & Youn. (2004). Analysis of carrier based PWM methods in relation to common mode voltage for multilevel inverter. *30th Annual Conference of IEEE Industrial Electronics Society*, *2*, 1811 – 1816.

Vassilios & Calais. (1998). Application specific harmonic performance evaluation of multicarrier PWM techniques. *Power Electronics Specialists Conference*, 172-178.

Vinothkumar, V. (2013). Fault diagnosis in h-bridge multilevel inverter drive using wavelet transforms. *International Journal of Advanced Research in Electrical Electronics and Instrumentation Engineering*, *2*(4), 1255–1262.

Wang, F., & Jiang, J. (2011). A novel static frequency converter based on multilevel cascaded h-bridge used for the start-up of synchronous motor in pumped-storage power station. *Energy Conversion and Management*, *52*(5), 2085–2091. doi:10.1016/j.enconman.2010.12.018

Yazdian-Varjani, A., Perera, S., & Chicharo, J. F. (1998). A centroid-based pwm switching technique for full-bridge inverter applications. *IEEE Transactions on Power Electronics*, *13*(1), 115–124. doi:10.1109/63.654965

Zhong, Ozpineci, Tolbert, & Chiasson. (2009). DC–AC cascaded h-bridge multilevel boost inverter with no inductors for electric hybrid electric vehicle applications. *IEEE Transactions on Industry Applications*, *45*(3), 963-970.

APPENDIX 1

Description of Hardware Components

Table 14. List of hardware components

Sl. No.	Description of Components
1	IGBT switches: CT60
2	Forward path diodes: MUR3060
3	Bridge rectifier: KBPC608
4	Capacitor: 2350F/100V
5	LC filters: L=15mH, C=4.7 μF
6	R Load: 100 Ω
7	RL Load: 100 Ω & 120mH

APPENDIX 2

Description of Post Fault Dictionary

Table 15. Sample fault dictionary for post fault analysis

S1	S2	S3	S4	S5	S6	S7	D1	D2	D3	D4	D5	D6	Sa	Sc	SD1	SD2	SD3	FAULT ID
1	1	1	1	1	1	1	1	1	1	1	1	1	1	1	0.182	0.3073	0.553	FF
0	1	1	1	1	1	1	1	1	1	1	1	1	1	1	21.03	38	68.32	OF1
1	0	1	1	1	1	1	1	1	1	1	1	1	1	1	22.39	21.34	57.42	OF2
1	1	0	1	1	1	1	1	1	1	1	1	1	1	1	16.80	18.94	52.80	OF3
1	1	1	0	1	1	1	1	1	1	1	1	1	1	1	18.71	15.62	43.60	OF4
1	1	1	1	0	1	1	1	1	1	1	1	1	1	1	17.24	19.93	48.24	OF5
1	1	1	1	1	0	1	1	1	1	1	1	1	1	1	18.81	16.8	48.04	OF6
.						.	.					.			.			.
.						.	.					.			.			.
.						.	.					.			.			.
1	0	1	1	1	1	1	0	1	1	0	1	1	1	1	34.02	40.03	111.5	OF33
1	1	0	1	0	1	1	1	1	1	1	0	1	1	1	64.49	126	243.5	OF34
1	1	0	1	1	0	1	1	1	0	1	1	1	1	0	63.58	106.5	186.9	OF35
1	1	0	1	1	1	1	0	1	1	1	1	1	0	1	29.81	38.64	103.6	OF36
.						.	.					.			.			.
.						.	.					.			.			.
.						.	.					.			.			.

Section 2

Energy Management and Conversion

Chapter 12
Physics, Modelling, and Optimization Studies of Photon-Enhanced Thermionic Emission-Based Hybrid Energy Conversion System:
A Review

S. C. Kaushik
Indian Institute of Technology Delhi, India

Ravita Lamba
Indian Institute of Technology Delhi, India

S. K. Tyagi
Indian Institute of Technology Delhi, India

ABSTRACT

The sustainable development of clean and efficient electricity generation techniques accelerated the research for invention of alternative electricity generation methods. In this chapter, the conceptual analysis of newly invented photon-enhanced thermionic emission (PETE) energy conversion process has been presented. It is a promising option for harvesting solar energy in terms of capturing photon as well as thermal energy simultaneously and converting solar energy into electrical energy based on photovoltaic and thermionic emission processes of energy conversion. Thus, the PETE process utilizes photons for PV conversion and heat of radiation for thermionic emission process. The main objective of this chapter is to review and analyze the performance of PETE converters including thermal modeling, choice of materials, and parametric optimization. The appropriate choice of material requirements for cathode and anode of PETE converters is necessary for practical design of PETE converters. The PETE converter may be an efficient future option for electricity generation using solar energy.

DOI: 10.4018/978-1-5225-3935-3.ch012

NOMENCLATURE

A_0: Area (m^2)
N_C: Effective density of states in conduction band (# per cm^3)
c: Speed of light (m/s)
n_{id}: Diode ideality factor
E_g: Energy band gap of semiconductor (eV)
N_V: Effective density of states in valence band (# per cm^3)
$E_{g,PV}$: Energy band gap of semiconductor material used in solar cell (eV)
N_A: Electron concentration for ionized acceptor (# per cm^3)
$E_{F,p}$: Quasi-Fermi levels for valence band (eV)
N_A: Concentration of p-type acceptor (# per cm^3)
$E_{F,n}$: Quasi-Fermi levels for conduction band (eV)
n_{eq}: Equilibrium electron concentration without photo-excitation (# per cm^3)
E_F: Equilibrium Fermi level (eV)
n: Electron concentration with photo-excitation (# per cm^3)
E_A: Ionization energy of p-type acceptor (eV)
δn: Excess carrier concentration from equilibrium condition (# per cm^3)
E_C: Conduction band minimum energy level (eV)
P: Holes concentration with photo-excitation (# per cm^3)
E_V: Valence band maximum energy level (eV)
p_{eq}: Equilibrium hole concentration without photo-excitation (# per cm^3)
Ex: Exergy (W)
P: Power (W)
En: Energy (W)
Q: Energy (W)
G: Solar irradiation (W/m^2)
R_0: Rate of photon emission at equilibrium conditions (# per m^2 per sec)
H: Cathode height (cm)
R: Rate of photon emission at non-equilibrium conditions (# per m^2 per sec)
H: Planck constant (J-s)
R_r: Rate of photon-enhanced recombination at non-equilibrium conditions (# per m^2 per sec)
I_0: Reverse saturation current (A)
R_C: Rate of thermionic emission in cathode (# per m^2 per sec)
Irr: Irreversibilities (W)
R_A: Rate of reverse thermionic emission from anode (# per m^2 per sec)
I_{ph}: Photocurrent (A)
R_S: Rate of photon induced electrons (# per m^2 per sec)
I: Electrical current (A)
R: Resistance (Ω)
J: Current density (A/cm^2)
S: Entropy (W/K)
K_{RD}: Richardson Dushmann constant (A/cm^2K^2)
s_r: Aspect ratio

k_B: Boltzmann constant (J/K)
T: Temperature (K)
m_0: Electron mass (9.1×10^{-31} kg)
U: Overall heat transfer coefficient (W/m^2K)
m_n: Effective electron mass (kg)
V: Voltage (V)
m_p: Effective hole mass (kg)
W: Width (m/cm)

Greek Letters

Φ: Work function (eV)
Λ: Wavelength of photon flux (nm)
φ: Photon flux (# per m^2 per sec.)
υ: Frequency of photon flux (per sec)
ε: Emmisivity
χ: Electron affinity (eV)
σ: Stefan Boltzamm constant (W/m^2K^4)
η: Energy efficiency
r: Reflection coefficient
ψ: Exergy efficiency

Subscript

A: Anode
Net: Net
C: Cathode
Out: Output
Des: Destruction
OC: Open circuit
Gen: Generation
PV: Photovoltaic
In: Input
PETE: Photon enhanced thermionic emission
Lost: Lost
R: Radiative
SC: Short circuit

INTRODUCTION

The existence of human life is affected directly by the consumption of energy. In recent decade, excessive growth in consumption of fossil fuels due to advancements in the quality of life and escalation of the population results in rapid scarcity of fossil fuel resources and adverse effect on the climate. So, it is desirable to find out the solutions of energy related problems. The energy supply security and various environmental challenges such as greenhouse gas emissions, global warming and climate changes, causing carbon emissions into the atmosphere, have emerged concern among the scientific community. Thus, the escalating energy demand and growing concern of global environmental issues result in development of more and more new technologies to harness energy from renewable energy resources. The energy resources have been classified mainly into three categories which are fossil fuels, nuclear resources and renewable resources. The electricity generation from renewable energy sources such as solar, wind, biomass, geothermal and hydro can play a significant role to combat these environmental issues and to improve the living standards. Therefore, identifying the potential of renewable energy resources has generated a resurgence of interest in promoting clean renewable sources to meet the world's increasing energy demand. Among all renewable energy resources, solar energy has gained significant attention due to high potential, environmentally benign, abundantly available, clean and green source of energy for generating electricity as well as thermal energy (Smestad, 2008; Baetens, Jelle, & Gustavsen, 2010). However, it is of dilute intensity and intermittent in nature and hence, its availability and durability varies from place to place and time to time. Solar energy can be harvested basically by two methods: one is direct route by using solar photovoltaic (PV) cells in which the incident solar radiation is converted directly into electricity (Bundgaard, & Krebs, 2007; Shah et al., 2003; Currie et al., 2008) and another is indirect route in which solar energy is first converted into work and then subsequent conversion of work into electricity by the solar thermal systems (Qu, Yin, & Archer, 2010; Kraemer et al., 2011; Chen, 1992).

In the photovoltaic energy conversion process, electrons are excited from valance band to conduction band upon the illumination of PV cell with sunlight. The photons having energy greater than or equal to the band gap of the semiconductor are absorbed in semiconductor surface and then give their energy to valence band electrons to knockout them from valance band into conduction band and thus, the electron will be excited to higher energy states depending on photon's energy. Although, the only energy, equal to the band gap is utilized and the additional photon energy is dissipated as heat in PV cell which increases the temperature of PV system, resulting in decrease in efficiency of PV system.

To produce electricity and heat simultaneously, photovoltaic thermal (PVT) hybrid systems are used which are combination of PV and solar thermal systems. However, these systems have low efficiency (Meillaud, Shah, Droz, Vallat-Sauvain, & Miazza, 2006; Melosh, 2013), because the efficiency of PV systems decreases with increase in temperature of PV cell and that of thermal system increases with increase in PV cell temperature.

Another process to generate electricity is by means of standard thermionic (TI) emission process using thermionic converters in which the electrons are emitted from a metal surface which is heated to a sufficiently high temperature of order of 2500 K. The sources of thermal energy, supplied to the cathode may be solar energy, chemical energy or nuclear energy. However, to get such high temperatures of order of 2500 K using solar thermal systems is not an easy task and it is not cost effective. The efficiency of standard thermionic (TI) devices increases with increase in temperature similar to solar thermal systems. Therefore, to take the advantage of both PV and TI processes simultaneously, both devices are combined into a single device which is called photon enhanced thermionic emission (PETE) converter.

The motivation of this chapter is to understand the physics and thermal modelling of photovoltaic, thermionic and photon enhanced thermionic emission energy conversion processes followed by parametric optimization and possible material used for cathode and anode of thermionic energy converter and photon enhanced thermionic emission converter.

BACKGROUND

Photon Enhanced Thermionic Emission converter is a newly invented promising option for harvesting solar energy. It is based on the combined photovoltaic and thermal charge carrier excitation along with thermionic emission in a semiconductor. It harvests both photon as well as thermal energy of incident solar radiation, photon energy for photovoltaic process and thermal energy for thermionic process, overcoming the fundamental limits on the efficiency of photovoltaic process. Unlike PV, the efficiency of PETE converter increases with increase in temperature. The decrease in efficiency at higher temperatures in photovoltaic devices can be overcome by using PETE converters which results in PETE converter efficiency higher than the Shockley-Queisser (SQ) limits for single junction photovoltaic device. Since PV systems use semiconductor materials to convert solar energy into electricity, but only a small portion of solar spectrum can be converted into electricity and most part of this spectrum is generating heat into PV systems. This heat and other inherent losses in the PV system result in losses of more than 60% of input solar energy to the PV systems. The part of solar spectrum, which is wasted as heat in PV systems can be harnessed in PETE converters. Apart from this, the PV systems never generate high temperature heat to get hot enough for this waste heat to be harnessed in solar and other thermal energy conversion systems. However, the operating temperature of PETE converter is high enough for generating high temperature heat which can be harnessed in thermal energy conversion systems. Melosh (2013) has suggested that the PETE systems alone have nearly 50% efficiency with concentrated solar radiation flux and if these systems are cascaded with thermal energy conversion systems such as heat engine or thermoelectric generator, then the efficiency of combined system could achieve more than 55% which is almost three times the efficiency of present existing PV systems. Thus, the efficiency of PETE conversion process is more than that of individual single junction photovoltaic (Singh, & Ravindra, 2012; Radziemska, 2003) and thermionic (Mahan, 1994; Goldsmid, 2010) conversion process as predicted by theoretical analyses. Therefore, based on theoretical performance limits, the potential of very high conversion efficiencies of this promising technology has motivated the research in this field. Further, to understand the physics of PETE converters, the physics of photovoltaic and thermionic systems has been explained in this section followed by physics of PETE converter.

Physics of Photovoltaic Power Generation

A PV cell is a light illuminated pn-junction diode which directly converts solar energy into electricity (Daher, Schmid, & Antunes, 2008; Casaro, & Martins, 2008). The different type of materials used for solar cells are Silicon (crystalline and amorphous), Copper Indium Diselenide ($CuInSe_2$), Indium Phosphide, Gallium Arsenide, Cadmium Telluride etc. Based on atomic structure, PV cells can be monocrystalline, multicrystalline and amorphous. Since the manufacturing cost of monocrystalline, multicrystalline technology based solar cells is very high, so now a days, thin film technology based solar cells such as CdTe are manufactured which are cost effective due to less material requirement and less manufacturing technology cost. There are three basic steps for generation of electricity using PV cells which are following (Kjaer, Pedersen, & Blaabjerg, 2005; Myrzik, & Calais, 2003): First is absorption of solar radiation, second is generation of free charge carriers and third is transport and then collection of charge carriers at PV cell terminals. Solar cells use energy of solar spectrum as their input energy source. Solar spectrum is the electromagnetic radiation from the sun. It provides heat, light and energy required for all living inhabitants. Out of entire solar spectrum from sun, only a small portion of solar spectrum reaches on earth's surface which has wavelength from 100 nm to 1 mm. This wavelength range contains three bands which are ultraviolet (UV), visible and infrared radiation (IR). According to Gibson (2003), the UV portion is 8%, visible is 42.3% and IR is 49.4% of total solar spectrum reaching on earth's surface. A typical PV cell generally absorbs only visible portion of solar spectrum reaching on earth's surface. The solar spectrum, incident on solar cell has photons of different energy. The photons with energy lesser than the band gap energy of cell material are not absorbed by cell and thus they are unusable for solar cell. The incident photons with energy greater than or equal to the band gap of PV cell material are absorbed by cell. The absorption of incident solar radiation depends on intrinsic carrier concentration of semiconductor material, temperature of cell, recombination rate of free charge carriers, diffusion length and mobility of charge carriers. When these photons strike the cell surface and consequently, covalent electrons of semiconductor materials, the photon energy will be given to electrons and these electrons are excited to higher energy states. Thus, free charge carriers are generated and rate of generation depends on wavelength of incident solar spectrum and semiconductor material properties such as absorption coefficient etc. For photons with energy greater than the band gap of material, only energy equal to the band gap is used and the excess energy is converted into heat through thermalization process. After absorbing the photons with energy greater than band gap, the electrons will be excited to higher energy states in the conduction band. However, these electrons are not stable in higher energy states and return quickly to either conduction band minimum or valence band and release the excess energy in the form of vibrating phonons which will increase the temperature of cell or in the form of photons. The generated charge carriers will exist only for minority carrier life time before recombination. Therefore, it is necessary to collect these free charge carriers before recombination. The electric field present at p-n junction is responsible for separation of charge carriers. If these minority charge carriers that are electrons in p-type and holes in n-type reach the p-n junction, they are swept across the p-n junction by the electric field present at the junction where these carriers become a majority charge carriers. Now if both the terminals of solar cell are connected together through external load, then, these free charge carriers flow across the cell through external load. The practical model of a solar cell consists of combination of a photo current source coupled to a diode in anti-parallel, a series resistor for internal resistance and a parallel resistor for leakage current as shown in Figure 1 (Tsai, Tu, & Su, 2008; Wasynczuk, 1989; Messenger, & Ventre, 2000).

Figure 1. Equivalent circuit of solar cell

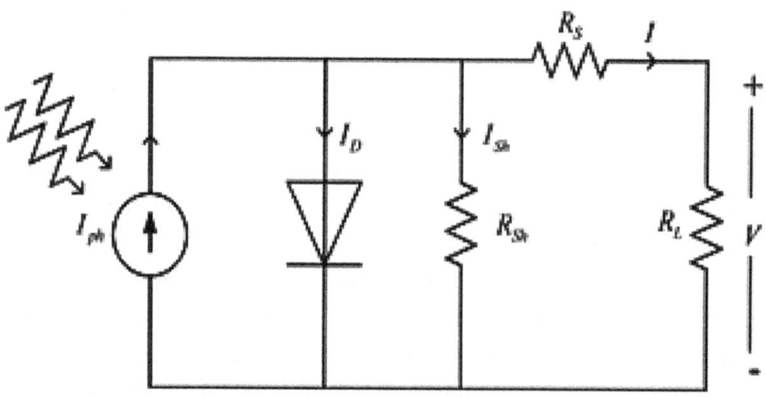

Physics of Thermionic Emission Power Generation

Electron emission is the process of discharge of electrons from a metal surface. At room temperature, the electrons of an atom are free to move randomly in the same atom or from one atom to another, but they cannot escape metal surface to make electron emission. Since, the free electrons at the metal surface find a strong attractive force towards nuclei which prevents them to escape from the metal. Thus, the surface of the metal acts as a barrier which is called surface barrier. However, if the free electrons at metal surface are provided with enough energy then, due to increase in kinetic energy of free electrons, they will cross the surface barrier to escape from the metal. This electron emission process has been studied as an electricity generation method for last two centuries. According to Jensen (2007), depending upon the energy source, the electron emission can be classified into following categories:

1. **Thermionic Emission:** If the metal is heated to a sufficiently high temperature of order of 2500 K, electrons will emit out from the metal surface. The higher temperature results in more number of electrons emitted out from metal surface and thus, electron emission depends on temperature. Thermionic emission occurs in vacuum tubes.
2. **Field Emission:** In this type of emission, a strong electric field, applied to metal surface results in free electron emitted out from metal surface. The electron emission is proportional to applied electric field.
3. **Photoelectric Emission:** In this type of emission, the energy of photons, incident on metal surface is given to electrons to make free electrons emitted out from metal surface. Higher light intensity results in more photoelectric emission.
4. **Secondary Emission:** The emission of electrons from a solid surface, bombarded by the charged particles from a primary source.

Figure 2 shows a basic energy diagram for different types of electron emission from metals.

The thermionic emission phenomenon is widely used as the source of electrons in electronics and communication. Becquerel (1853) first noticed electric current flowing between two, one hot and other cold, platinum wires which were placed in different gases. The research on thermionic emission started back to 1853. Many researchers paid attention for understanding thermionic emission process in 1873; the relation between heat and static electricity has been studied by Guthire (1873) who measured the positive charged, red hot sphere discharge into air that supported understanding of thermionic emission process further more. The thermionic phenomenon was first witnessed by Thomas A. Edision as the passage of electricity between a metal plate and filament inside an incandescent lamp in 1883. Thus, Edision discovered the discharge of electrons from a hot body. After that, many researchers worked in this field and Flaming created thermionic diode rectifier in 1904. The idea of converting heat into electricity by thermionic emission was first proposed by Schlicter in 1915. In 1950s decade, research on thermionic emission devices started rigorously. Elster and Geitel examined sealed devices in which electric current generated between one hot and one cold electrode which were measured through electrometer (Angrist, 1965). Crookes (1896) first developed and studied cathode rays and they presented the understanding of thermionic emission phenomenon and plasma, known as fourth state of matter. Murphy, & Good (1956) presented a study of thermionic and field emission rigorously. Hatsopoulos, & Kaye (1958) performed theoretical and experimental work on thermo electron engine and presented a new thermionic converter. He introduced two types of thermionic energy converters (TECs), one is vacuum thermionic energy converter (VTEC) and another is vapor TEC. The first proposed thermionic converter has experimental conversion efficiency of 10-15% (Hatsopoulos, & Kaye, 1958; Hatsopoulos, & Gyftopoulos, 1973). Thermionic energy conversion has been analysed for years long back by many researchers (Houston, 1959; Wilson, 1959). Although thermionic energy conversion has been promised good results back in 1950s, even though the development in thermionic energy conversion devices has been limited due to lack of efficient and advanced manufacturing technology. The main challenges for thermionic converters are very high temperature and current density which are essential for efficient converter operation (National Research Council, 2001; Adams, 2006).

Being a direct energy conversion solid state device, converting heat directly into electricity, they have the advantage of no mechanically moving parts, noiseless operation, light weight, compactness, requires less maintenance, high reliability and long life. The high power density of TIGs has attracted the researchers to explore the thermionic devices and many prototype systems have been developed and tested (Wilson, 1959; Hatsopoulos, & Gyftopoulos, 1979). The vacuum gap between cathode and anode in a vacuum TIG is of order of few micron to reduce the effect of space charge accumulation (Lee, Bargatin, Melosh, & Howe, 2012). Usually, the practical vacuum TIG operate at cathode temperatures higher than 1000 K to achieve high efficiencies (Hatsopoulos, & Gyftopoulos, 1973).The net current density depends on the net emission of electrons from the cathode and anode. The high energy electrons emitted from the cathode flow from cathode to anode against the electrical potential differences. The concentrated solar radiation can be used as the input heat supply for TIG. Therefore, TIGs can operate like a PV cell to overcome the upper efficiency limits of single junction crystalline PV cells (Schwede et al., 2010). Moreover, TIG can be used in hybrid systems also such as combined TIG-TEG (thermoelectric generator) system, TIG-heat engine, fuel cell- TIG to enhance the overall conversion efficiency of hybrid systems (Segev, G., Kribus, A., & Rosenwaks, 2013; Schwede et al., 2013; Yaghoobi, Vahdani-Moghaddam, & Nojeh, 2012; Wang, Dai, & Wang, 2013; Xuan, & Li, 2003; Meng, Zhang, & Wang, 2014; Cheng, & Liang, 2012; Hsiao, Chang, & Chen, 2010; Wang, Su, Liu, Su, & Chen, 2015). In past few years, several models of TIGs have been proposed. Furthermore, the performance of TIGs has been analyzed based on various parameters. The detailed analysis of tunneling-thermionic emission based power generating device has been done by Zeng (2006). The effect of electron energy spectrum shape on the performance of thermionic and thermoelectric devices has been analyzed by O'Dwyer, Lewis, Zhang, & Humphrey (2005). Wang et al. (2013) proposed a model for an irreversible vacuum thermionic generator in which both internal as well as external irreversible heat losses have been considered (Wang, Su, Lin, & Chen, 2013). Mahan, Sofo, & Bartkowiak (1998) developed a new model of multilayer thermionic generator and refrigerator and determined the efficiencies for both devices. To achieve maximum power, Hemphrey (2005) derived an ideal energy spectrum for both thermionic generator and refrigerator (Humphrey, O'Dwyer, & Linke, 2005). Luo et al. (2013) described the effect of position and width of energy spectrum of energy selective electron mechanism on thermionic heat engine. To model the effects of space charge accumulation in vacuum thermionic devices, Smith (2009) presented a theory in which negative electron affinity diamond has been considered (Smith, Bilbro, & Nemanich, 2009). Ito & Cappelli (2012) suggested a solution to overcome the space charge effects in photon enhanced thermionic emission (PETE) converters, in which cesium filling is added into vacuum gap by continuous laser to produce space charge neutralizing plasma. To suppress the reflection of electrons from anode, Islam et al. (2006) adopted thin film deposition technique which improves the performance of thermionic energy converter (Islam, Inal, & Luke, 2006). For room temperature applications of thermionic energy converters, Tavkhelidze (2010) developed an electrode, consisting of a metal ridged quantum wells layer and a base substrate, which has low work function.

The performance of thermionic energy converters has been investigated based on finite time thermodynamics theory by some researchers. The finite time thermodynamics analysis of solar powered thermionic engine has been studied first by Wu (1992). He analysed the characteristics of the engine at maximum power conditions. Bhattacharyya (1999) has obtained the optimum temperature of cathode and anode of a thermionic generator at maximum power output using finite time thermodynamics theory for a given heat source and electrode material. Erbay, & Yavuz (1997) have described the design parameters of an endoreversible vacuum thermionic generator with external heat transfer at maximum power output conditions. They concluded that the performance of the device is affected significantly at room temperature applications by considering the external heat transfer. They have also analysed the performance of a combined thermionic-thermoelectric refrigerator by considering external heat transfer. They have replaced the metal conductor by thermoelectric refrigerator. Thus, the Joule heating and heat conduction losses in the metal conductor have been reduced and the performance of the combined device has been improved.

Physics of Photon Enhanced Thermionic Emission Power Generation

PETE is a newly invented promising technology for generation of electricity using solar energy. It is a novel method of harvesting solar energy combing photovoltaic and thermionic processes. The solar energy can be converted into electricity by two methods either by quantum or thermal approach. The quantum method of harvesting the solar energy is used in PV cells in which photons of solar radiation with energy equal to or greater than band gap of PV cell material are absorbed by cell material to excite electrons from valence band into conduction band for electricity generation. It is a direct energy conversion approach. This quantum method possesses higher theoretical efficiency but in practical, due to collection of only a small portion of spectral energy density, quantum method has low conversion efficiency. In thermal method of solar energy conversion, the concentrated solar radiation flux is used as the source of thermal energy for heat engine to generate electricity indirectly. The thermal process has the advantage of using entire solar energy spectrum. In PV systems, the efficiency decreases rapidly at elevated temperatures due to increase in thermally generated reverse saturation current at higher temperatures. Since, in PV systems, the energy of photons which is in excess of band gap of semiconductor material is wasted as heat which results in increase of the temperature PV system. These losses are called thermalization losses Also, the sub band gap photons are not absorbed in PV cell which is called absorption losses. In single junction Silicon solar cells, these thermalization and absorption losses are nearly 50% of incident solar radiation flux as given by Wurfel (2009). These losses can be recovered by utilizing the wasted heat in a heat engine which comprises thermal cycle. According to Luque, & Matri (1999), the efficiency of combined PV and thermal cycle is predicted to be more than 60% but these systems are not feasible practically because of decrease in PV system's efficiency at elevated temperatures as suggested by Green (2003). So the combination of both PV and heat engine simply results in failure because of decrease in PV system's efficiency at elevated temperature and decrease in efficiency of heat engine at lower temperatures given by Kribus, & Mittleman (2008). However, in thermionic converters, the efficiency increases with increase in temperature. Therefore, the benefit of both quantum and thermal approach of solar energy harvesting methods can be taken simultaneously in PETE. Unlike standard thermionic converters, which consists of discharge of free charge carriers over energy barrier from a heated metallic cathode, PETE combines PV and thermionic to take advantage of both techniques simultaneously. Thus, PETE converters operate at elevated temperature attuned with solar heat engines while simultaneously using quantum method of harvesting solar energy. The first PETE converter was proposed in 2010 by Schwede et al. (2010), they first presented the thermionic emission of photoexcited electrons from PETE cathode at higher temperature. They showed that for ideal photon enhanced thermionic emission converter using Gallium Nitride (GaN) as cathode has theoretical efficiency limit of more than the Shockley-Queisser (SQ) limits for single junction Silicon photovoltaic device. They proposed the PETE converter which operates at more than 200 ^{0}C and the waste heat from converter's anode is used for secondary heat engine connected to back side of anode. The theoretical efficiency of combined system was reported to be more than 50%. Segev, Rosenwaks, & Kribus (2012) have calculated the cathode temperature by considering the energy balance for PETE system. They have calculated the cathode temperature and efficiency at maximum power point for different concentrations. At the cathode temperature of 700 ^{0}C and flux concentration of 163, the PETE efficiency is 35%. To get the efficiency compatible with reported efficiency for state of art multi junction cells at several hundred-flux concentration, the cathode temperature of 921 ^{0}C and flux concentration of 433 is required which results in 40% efficiency as given by Kinsey et al. (2008) and Wojtczuk et al. (2010). Further, at temperature of 1136 ^{0}C and flux concentration of 1000, the efficiency is 44.7%. However, such a high temperature is not appropriate for semiconductor, as Silicon and GaAs have melting point of 1410 and 1240 ^{0}C respectively. So, we should ensure that to get efficiency of more than 40%, the semiconductor cathode material should be thermally stable at very high temperature. They also have calculated the conduction band electron concentration of cathode for different operating voltages. They have showed that efficiency of PETE converter increases with increase in temperature and decreases with increase in electron affinity. Varpula, & Prunnila (2012) have developed theoretical models for PETE converter. They have considered effect of electron diffusion, surface recombination and non-uniform photo generation. At cathode temperature of 800 K and flux concentration of 1000, the surface recombination reduces the converter efficiency to less than 10%. Their numerical model always gives lower efficiency than that of rate equation model of Schewede et al. (2010). Segev, Rosenwaks, & Kribus (2013) have presented one dimensional model for PETE converter by considering various recombination processes, carrier concentration, potential distribution and surface effects. They calculated the converter efficiency for Silicon and GaAs cathode under different operating conditions which was around 31% with homo junction back surface field layers. They have analysed various loss mechanisms and calculated that recombination of electrons at the cathode contact results in significant losses. Therefore, to get higher efficiency, an electron blocking junction at back contact of cathode is required. Their analysis can be adopted for cathode material selection and electron affinity of back surface junction. Segev, Kribus, & Rosenwaks (2013) have presented a new model for PETE converter. In previous models, the cathode and anode are maintained at different temperature, but in this model, both cathode and anode are operating at same temperature with side illuminated cathode instead of front illuminated cathode of previous models. This configuration has the advantage of higher electron emission area, series connection of multiple units to get higher voltage. The wasted heat in this configuration is recovered from both cathode and anode. The wasted heat is at higher temperature compared to previous models. For moderate temperature and flux concentration, the conversion efficiency of 30-40% can be achieved. If the waste heat is also recovered, then the conversion efficiency of this PETE model will be 40-50%. At temperature of 810 K and flux concentration of 1000, this isothermal model results in overall efficiency of 52%. Su, Zhang, Chen, Kang, & Chen (2013) have discussed the various losses such as radiation, conduction from cathode and anode and radiation between cathode and anode. They have derived the expressions for power output and conversion efficiency analytically. They have calculated the optimum values of cathode band gap and electron affinity at maximum efficiency. For these optimum values of band gap and electron affinity, they have calculated the cathode and anode operating temperatures. They also have calculated the maximum efficiency of PETE converter numerically. Based on the optimum values of cathode electron affinity and band gap, the appropriate cathode material can be selected. They have considered the inner surface properties for both electrodes. The found that when the cathode band gap is between 1.36-1.39 eV, then the maximum efficiency is greater than or equal to 37.8% which is greater than that of standard thermionic devices (Smestad, 2004; Lee, Jeong, No, Hannebauer, & Yoo, 2009) or photovoltaic devices (Chow, 2010; Green, Emery, Hishikawa, Warta, & Dunlop, 2012; Schultz, Glunz, & Willeke, 2004). Su, Wang, Wang, Xu, & Chen (2014) have proposed a new solar hybrid system which consists of concentrated solar system, PETE converter and thermoelectric generator (TEG) to harvest solar energy more efficiently by considering various irreversible losses in PETE and TEG. The waste heat from PETE anode is used as source for TEG. They have derived the power output and efficiency expressions for hybrid system analytically. They have analysed the effect of various parameters on the performance of hybrid system. They have calculated the maximum efficiency of hybrid system and concluded that hybrid systems have increased efficiency. They have determined the optimum values for band gap and electron affinity of PETE cathode, TEG reduced current, structural parameters and solar concentration of hybrid system at optimum efficiency which are useful for providing guidance to optimally design hybrid systems. They determined that efficiency of hybrid system do not increases monotonically with solar flux concentration. Su, Wang, Liu, Su, & Chen (2014) have discussed the effects of space charge on performance of PETE converters using Langmuir theory. These effects become effective when the gap between cathode and anode is of order of microns. According to Wien's displacement law, when the thermal radiation characteristic wavelength is smaller than the gap between electrodes which is of order of micron or more, then the effects of near field heat transfer become negligible. The efficiency of PETE converter decreases due to space charge effects. For voltage lower than saturation voltage or greater than critical voltage, the performance of PETE converter is similar to the model described by Segev, Rosenwaks, & Kribus (2012). For voltage greater than saturation voltage or lower than critical voltage, the space charge effects will be significant, and they will affect the performance of PETE. They concluded that the gap between cathode and anode should be similar to that of thermionic converters. However, the efficiency and power output of PETE converters is greater than that of thermionic converters.

Varpula, Tappura, & Prunnila (2015) have described the use of Si, GaAs and InP for cathode materials in PETE converters using numerical model. They showed that if electron affinity, surface recombination and thermal stability of these materials are optimized then these materials are most promising cathode materials. GaAs and InP have higher absorption coefficient compared to Silicon so they give better efficiency regardless of having higher radiative recombination. They showed that for GaAs and InP, the increment in efficiency saturates at nearly 100 suns and for Silicon the efficiency has a peak value at 100 suns. They also showed that GaAs and InP decomposes above 870 K (Tersoff, Jesson, & Tang, 2009; Haynes, Chu, Aselage, & Picraux, 1986) and 750 K (Riesz, Dobos, & Karanyi, 1998) respectively so they are not thermally stable at higher temperatures and can be used as cathode materials below these temperatures. Also InP is the most promising material due to its higher absorption properties. Si is better cathode material at higher temperatures due to having better thermal stability. Segev, Rosenwaks, & Kribus (2015) have analysed the upper limit for efficiency of PETE converters for different configurations and compared that limit with the upper limit of PV and solar thermal converters. They showed that isothermal PETE converter configuration has efficiency less than S-Q limit and non-isothermal PETE converter has efficiency exceeding the S-Q limit. The efficiency limit increases with increase in flux concentration and it is 52% at 1000 suns. The conversion efficiency of 56% is obtained at 1000 suns with spectral splitting. For isothermal and dual band gap PETE configuration, the overall efficiency including the thermal cycle efficiency is 69.8 and 70.4% respectively. The upper limit for PV cells with single junction is limited to S-Q limit given by Shockley, & Queisser (1961). This limit can be surpassed by reducing the thermalization losses by using two or more band gap materials. Matri, &Araujo (1996) have suggested that in multi junction PV cells, several cells are connected optically in series with decreasing order of band gap so that the photons transmitted from the material of higher band gap can be absorbed in the next material with lower band gap. For dual junction cells, the upper limit of efficiency is 54.4% for 1000 suns. For intermediate band gap PV cells, the band gap is divided into two sub band gaps (Green, 2006). The upper limit of efficiency for solar thermal conversion systems is given by Fletcher, & Moen (1977). Reck, & Hansen (2014) have presented a thermodynamic model for evaluating the PETE converter efficiency analytically using entropy production due to thermionic emission. They have calculated the efficiency of PETE converter numerically and grouped into four regions based on efficiency range. For maximum concentration of 49198 suns, the efficiency has maximum value of 67.3%. The maximum efficiency of PETE converter including the thermal cycle with Carnot efficiency is nearly 61% at 1000 suns. The overall efficiency of PETE converter in tandem with Carnot heat engine can reach Landsberg limit given by Landsberg, & Tonge (1980) if internal entropy flux is considered to be zero.

THERMAL MODELING OF PHOTOVOLTAIC ENERGY CONVERSION SYSTEMS

A generalised mathematical model description of current-voltage characteristics of a typical solar cell has been developed over last four decades (Angrist, 1982; Wasynezuk, 1983; Phang, Chan, & Phillips, 1984). According to Tsai et al., (2008), the I-V characteristic equation for general model of the cell can be given as:

$$I = I_{ph} - I_0 \left[\exp\left\{ \frac{q\left(V_{PV} + IR_S\right)}{kT_{PV} n_{id}} \right\} - 1 \right] - \frac{V_{PV} + IR_S}{R_{Sh}} \tag{1}$$

The value of diode ideality factor n_{id} depends on type of PV technology, adopted for manufacturing the solar cell as suggested by Hua, & Shen (1998). The light generated current basically depends on the solar radiation flux and operating temperature of the PV cell which is given as:

$$I_{ph} = \left[I_{SC} + K_I \left(T_{PV} - T_{ref} \right) \right] \frac{G}{G_n} \tag{2}$$

The performance of PV cell depends on many parameters. Fill factor is one of the important parameter among them. Graphically, it depicts the closeness of the I-V characteristic curve of solar cell to a square wave. It can be defined as ratio of the maximum power output from the cell to the product of short circuit current and open circuit voltage. The area of the largest rectangle which can fit in the I-V characteristics curve is a measure of fill factor. Mathematically, it is given as:

$$FF = \frac{P_{\max}}{V_{OC} I_{SC}} = \frac{V_{mp} I_{mp}}{V_{OC} I_{SC}} \tag{3}$$

The current-voltage and power-voltage curve of a typical PV cell is shown in Figure 2. Since the output voltage of a single solar cell is very low, so, many cells are connected together in series or parallel to form a module to get higher voltage and current. A string is defined as the series combination of PV modules. The strings are connected in parallel to form the PV array. The relationship between output current and voltage of a PV array consisting of N_s cells in series and N_p strings in parallel is given as (Luque, & Hegedus, 2003; Hongmei et al., 2012):

$$I = N_P I_{ph} - N_P I_0 \left[\exp\left(\frac{q\left(\frac{V_{PV}}{N_S} + \frac{IR_S}{N_P} \right)}{kT_{PV} n_{id}} \right) - 1 \right] - \frac{\frac{N_P V_{PV}}{N_S} + IR_S}{R_{Sh}} \tag{4}$$

Generally, efficiency is used to compare the performance of solar cells. It is defined as the ratio of power output from the solar cell to input power to cell. The efficiency of a cell depends on temperature of cell, solar spectrum and intensity of incident solar radiation. The efficiency (energy efficiency or first law efficiency) of a solar cell using some maximum power point tracking (MPPT) technique for getting maximum power output, is given as:

$$\eta_{PV} = \frac{P_{\max}}{P_{in}} = \frac{V_{mp} I_{mp}}{A_{PV} G} = \frac{V_{OC} I_{SC} FF}{A_{PV} G} \tag{5}$$

Figure 2. Energy band diagram for electron emission in metals

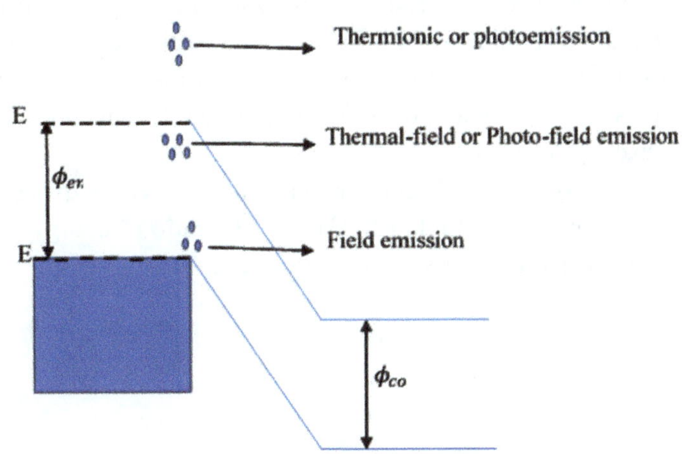

Figure 3. (a) Current-voltage and (b) power-voltage curve of a PV cell

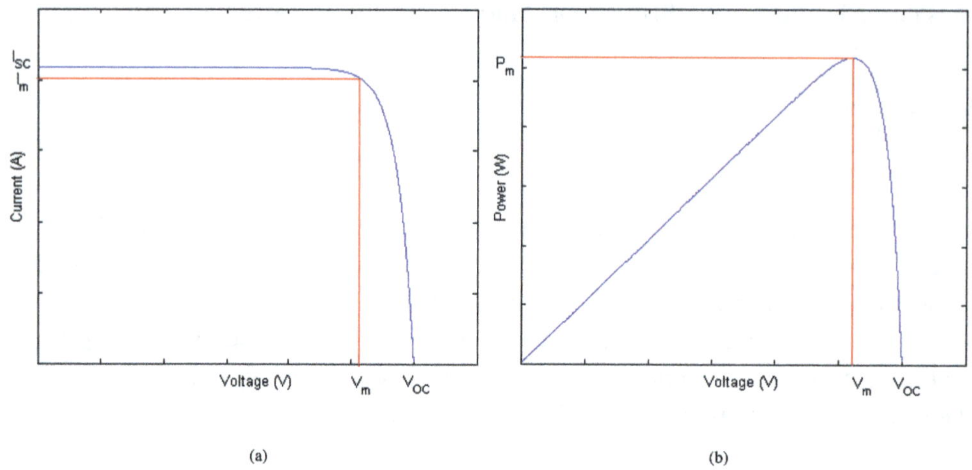

Although, energy analysis is based on first law of thermodynamics and it is the quantitative analysis of energy which pays no attention towards reduction in energy potential. Therefore, to calculate the maximum potential of PV systems, exergy analysis is required which is based on second law of thermodynamics and it takes into consideration the quality of energy. Bejan (1988) and Cengel, & Boles (2006) have defined exergy as the maximum work potential that can be extracted from energy. It is a powerful tool for evaluating the performance and thermodynamic analysis of PV and many other systems (Rosen, Hooper, & Barbaris, 1988; Rosen, & Hooper, 1996; Rosen, 1999; Rosen, & Dincer, 1999). Larson, & Cortez (1995) have provided that exergy analysis takes both usable and unusable energy which are called availability and irreversibility respectively.

According to Sahin, Dincer, & Rosen (2006) and Joshi, Dincer, & Reddy (2009), the efficiency of solar cell given in equation (5) is called energy efficiency of PV system. For a steady flow process, the exergy balance equation for PV system can be given by Wong (2000):

$$Exergy\ Input = Exergy\ Output + Exergy\ Loss + Irreversibility \tag{6}$$

Exergy or availability loss is defined as degradation in quality of energy. According to Hepbasli (2006), this exergy loss is also called irreversibility. The PV system absorbs solar radiation flux and converts it into electrical and thermal energy. If this thermal energy is utilized by some means then it is called useful energy and if it is dissipated as heat to outside environment through conduction, convection and radiation then it is called exergy destruction.

The exergy efficiency of PV systems is defined by many researchers in different ways (Bejan, 1988; Kotas, 195; Wong, 2000). According to Petela (2003), it is the ratio of exergy output from PV system to exergy input to PV system which is exergy of solar radiation.

$$\psi_{PV} = \frac{\dot{Ex}_{out}}{\dot{Ex}_{in}} = \frac{\dot{Ex}_{out}}{\dot{Ex}_{sun}} \tag{7}$$

The exergy of solar radiation has been calculated by many researchers using different methods (Petela, 2003; Bejan, 2006). Among these methods, the most frequently used methods are those proposed by Petala (1964), Spanner (1964) and Jeter (1981) respectively:

$$\dot{Ex}_{sun} = A_{PV}G\left[1 - \frac{4}{3}\left(\frac{T_{amb}}{T_{sun}}\right) + \frac{1}{3}\left(\frac{T_{amb}}{T_{sun}}\right)^4\right] \tag{8}$$

$$\dot{Ex}_{sun} = A_{PV}G\left[1 - \frac{4}{3}\left(\frac{T_{amb}}{T_{sun}}\right)\right] \tag{9}$$

$$\dot{Ex}_{sun} = A_{PV}G\left[1 - \frac{T_{amb}}{T_{sun}}\right] \tag{10}$$

Now, the exergy output of PV system includes only exergy associated with electrical energy from PV system. If the thermal energy output part from PV system is used as gain then this hybrid system is called as photovoltaic thermal (PVT) system and its exergy output include exergies associated with both electrical and thermal energies as given by Sarhaddi et al. (2010):

$$\dot{Ex}_{out} = \dot{Ex}_{ele} + \dot{Ex}_{Th} \tag{11}$$

The thermal exergy is given as:

$$\dot{Ex}_{Th} = \dot{Q}_{Th}\left(1 - \frac{T_{amb}}{T_{PV}}\right) \tag{12}$$

where, Q is given as:

$$\dot{Q}_{Th} = UA_{PV}\left(T_{PV} - T_{amb}\right) \tag{13}$$

The overall heat transfer coefficient of PV system, U includes convection as well as radiation losses from the PV system which is given as:

$$U = h_{con} + h_{rad} \tag{14}$$

According to Boyle (2004), the convective heat transfer coefficient is expressed as:

$$h_{con} = 2.8 + 3 * v_{air} \tag{15}$$

The radiative heat transfer coefficient between the PV system and outside environment is given by Watmuff, Charters, & Proctor (1977):

$$h_{rad} = \varepsilon\sigma\left(T_{sky} + T_{PV}\right)\left(T_{sky}^2 + T_{PV}^2\right) \tag{16}$$

The sky temperature is given by Watmuff, Charters, & Proctor (1977):

$$T_{sky} = T_{amb} - 6 \tag{17}$$

The electrical exergy is same as the electrical power output PV system which is given by Joshi, Dincer, & Reddy (2009):

$$\dot{E}x_{ele} = \dot{E}_{ele} = V_{OC} I_{SC} FF = V_{mp} I_{mp} \tag{18}$$

Thus, the exergy efficiency of PV and PVT systems is mathematically given as:

$$\psi_{PV} = \frac{\dot{E}x_{out}}{\dot{E}x_{sun}} = \frac{V_{mp} I_{mp}}{\dot{E}x_{sun}} \tag{19}$$

$$\psi_{PVT} = \frac{\dot{E}x_{out}}{\dot{E}x_{sun}} = \frac{V_{mp} I_{mp} + Q_{Th}\left(1 - \frac{T_{amb}}{T_{PV}}\right)}{\dot{E}x_{sun}} \tag{20}$$

THERMAL MODELING OF THERMIONIC ENERGY CONVERSION SYSTEMS

A thermionic energy converter is a solid state static device that directly converts heat into electricity through thermionic emission process. From electronics point of view, it can be considered as an electronic diode that converts heat into electricity using thermionic emission process. From thermoelectric point of view, it can be considered as a thermocouple in that an evacuated space has been replaced by one of conductor. From the thermodynamic point of view, it can be considered as a heat engine, operating between two heat reservoirs that takes an electron rich gas as its working medium. The possibility of converting the highest fraction of heat input into electricity for thermionic energy converters can be given by Carnot efficiency based on second law of thermodynamics.

Since, the electrons of the outermost shell a metal have sufficient kinetic energy, so these electrons try to be emitted out from the metal surface, but due to attractive forces of nuclei, these electrons cannot escape from the metal surface. However, if the temperature of the metal is high enough of order of 1500 to 2500 K, some electrons gain sufficient thermal energy to escape out from the metal surface. Thus, electrons are emitted from metal surface which is called thermionic emission. The emission of electrons from a hot metal surface is similar to discharge of steam from heated water.

A thermionic converter consists of two metallic electrodes, separated by either a small, highly evacuated gap or plasma. One of the electrodes is maintained at sufficiently high temperature to act as a thermionic electron emitter or hot plate which is called cathode. Another electrode, called anode is maintained at a significantly lower temperature to collect the electrons emitted by the cathode. The higher the temperature of metal, more number of electrons will be emitted out from metal surface and thus, electron emission depends on temperature. The spacing between the cathode and anode may by either evacuated or filled with a gas at low pressure to limit the number of electrons travelling through the gap. The thermal input energy to the cathode is supplied from a heat reservoir which is connected to the cathode and the excess heat from the collector is removed using heat sink which is connected to the anode. The electrons are escaped from the cathode surface by taking the thermal energy from the input heat source which is required for overcoming potential energy barrier (work function). The kinetic energy of electrons increases by supplying thermal energy to the cathode and the electrons are escaped out form the surface by gaining the sufficient energy. The electrons emitted by cathode flow towards anode via vacuum or plasma gap and they are collected at anode and thus, creating negative charge at the anode which causes voltage difference between both electrode. If the cathode and anode are connected by external load, then these electrons flow from anode to cathode due to potential difference via external load circuit where some portion of thermal energy supplied to cathode to emit electrons is directly converted into electricity in the form of dissipation of electrons' energy into external load and some portion of it is used to heat up the anode. After dissipation, these electrons again gain energy at the cathode from the input heat and emits again from the cathode surface. The current density emitted from cathode is given by the Richardson Dushman equation which is given as:

$$J = K_{RD} T^2 \exp\left(-\frac{\phi}{kT}\right) \tag{21}$$

From this equation, it is clear that emission current density increases with cathode temperature and decreases with cathode work function exponentially. So, it is essential for cathode material to have small work function and that can operate well at very high temperatures.

Since the electron escaping from the cathode surface have total energy equal to work function of cathode plus some surplus kinetic energy. When these electrons strike the anode, some part of this energy is used for flow of current through external load circuit. Thus, converts the thermal energy into electricity. The part of energy converted into heat in the anode must be taken away to maintain the anode at a sufficient lower temperature to minimize back emission of electrons from anode to cathode. The work function of anode material should also be small.

Thermionic converters can be of vacuum, plasma or gas filled and auxiliary discharge type converter. In vacuum thermionic energy converters (VTECs), the gap between cathode and anode is in very small range of 0.025 to 0.038 mm to suppress the space charge effects. For these types of converters at a cathode temperature of 1100 K, the electrical power output varies from 0.1 to 1 W/cm^2 of cathode surface. Since manufacturing of such converters with such a small spacing is difficult. Furthermore, due to unavailability of materials with stabilized low work function and problems of achieving a thermal isolation between cathode and anode, the practical implementation of VTECs is very difficult. So, these converters have only limited applications. In vapour/gas filled or plasma thermionic converters, the charged positive ions are generated and added with negative charged electrons in spacing between cathode and anode to give plasma with neutral space charge. Due to this, the electrons escaped from cathode surface encounter some electrostatic resistive force during passage from cathode to anode. Caesium due to its low ionization potential is used mostly. Potassium, Rhodium and alkali metals are also used to generate an ionization vapour. When ionization potential is less than cathode work function, contact ionization occurs. The vapour pressure inside the gap is of order of 100 Pascal's. However, vapour TECs cannot achieve optimal performance due to insufficient positive ions required to neutralize the space charge effects.

In an auxiliary discharge thermionic converter, inert gases such as Neon, Argon or Xenon are used in gap between electrodes. By applying voltage to a third electrode, positive ions are generated. Due to spark plug type configuration, the operating temperature of these converters may be relatively low of order of 1500 K which results in conventional fossil fuels to be used as heat source for cathode.

Energy Analysis of Thermionic Generator

The schematic diagram of thermionic converter is shown in Figure 4. Thermodynamic modelling of TEC is required for understanding the energy distribution in the converter. The metallic cathode is maintained at temperature T_C by transferring heat flux q_{input} from external source and anode is maintained at lower temperature T_A by transferring heat away from anode. The cathode and anode are separated by vacuum gap having width w. Due to different temperatures of cathode and anode; both emit electrons at different rates so a potential difference is established between them. If cathode and anode are connected through external load, then current will flow from anode to cathode through external circuit. The current density emitted from cathode and anode at their respective temperatures is given by the Richardson Dushman equation. The emitted current density will be a fraction of saturation current density. The only electrons having enough energy to overcome the potential barrier will be emitted. The potential barrier for electrons emitted by cathode is givens as (Angrist, 1982):

$$\varphi_C \geq \varphi_A + \phi_A + V - \phi_C \tag{22}$$

The current density emitted from cathode is given by the Richardson Dushman equation as:

$$J_C = K_{RD} T_C^2 \exp\left(-\frac{\phi_C}{kT_C}\right) \exp\left(-\frac{\varphi_A + \phi_A + qV - \phi_C}{kT_C}\right) \tag{23}$$

or

$$J_C = K_{RD} T_C^2 \exp\left(-\frac{qV_C}{kT_C}\right) \tag{24}$$

where,

$$qV_C = \varphi_C + \phi_C = \varphi_A + \phi_A + qV \tag{25}$$

Similarly, the current density emitted from anode is given by the Richardson Dushman equation as:

$$J_A = K_{RD} T_A^2 \exp\left(-\frac{\phi_A}{kT_A}\right) \exp\left(-\frac{\varphi_A}{kT_A}\right) \tag{26}$$

or

$$J_A = K_{RD} T_A^2 \exp\left(-\frac{qV_A}{kT_A}\right) \tag{27}$$

where,

$$qV_A = \phi_A + \varphi_A \tag{28}$$

Now, the net current density from cathode is given as:

$$J_{net} = J_C - J_A \tag{29}$$

or

$$J_{net} = K_{RD} T_C^2 \exp\left(-\frac{qV_C}{kT_C}\right) - K_{RD} T_A^2 \exp\left(-\frac{qV_A}{kT_A}\right) \tag{30}$$

It is clear from above equation that the net current density from cathode to anode depends on temperature of cathode and anode, space charge potential barrier and work function of cathode and anode such that $\varphi_A + \phi_A + qV > \phi_C$ or $qV_C > \phi_C$.

The net voltage output is given as (Angrist, 1982):

$$qV = q\left(V_C - V_A\right) = \left[\left(\phi_C - \phi_A\right) + \left(\varphi_C - \varphi_A\right)\right] \tag{31}$$

The net voltage across the external load is given as:

Figure 4. Schematic diagram of thermionic converter

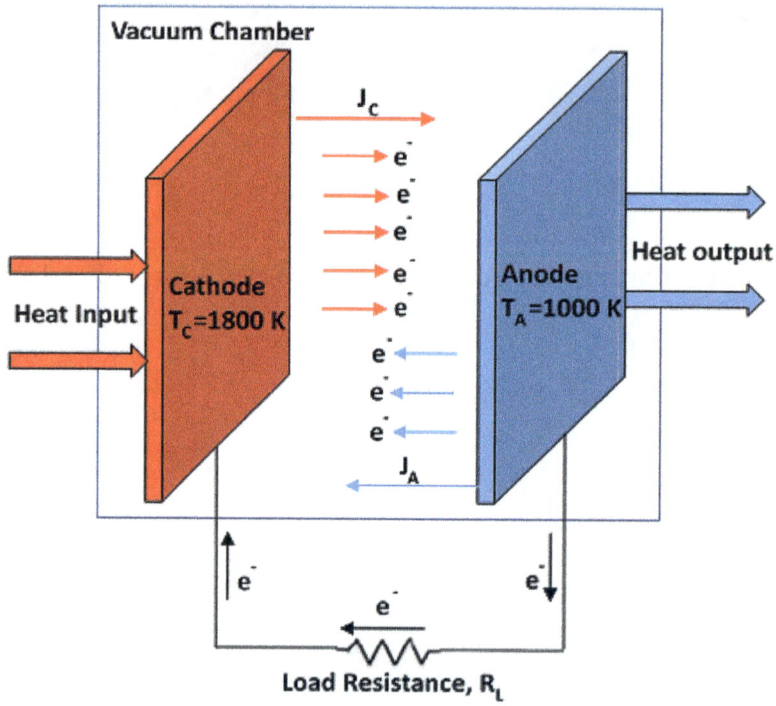

$$V_{load} = V_C - V_A - IR_l \tag{32}$$

where, IR_l is the voltage drop across the lead wire. The open circuit voltage can be found using equation (30) by making net current density equal to zero and solving for V which is given as (Angrist, 1982):

$$qV_0 = 2k_BT_C \ln\left(\frac{T_C}{T_A}\right) + \left(\phi_C + \varphi_A\right)\left(\frac{T_C}{T_A} - 1\right) \tag{33}$$

The energy band diagram for thermionic converter is shown in Figure 5. To calculate the efficiency of thermionic converter, the heat input flux, supplied to the cathode has to be known. This heat input flux can be found by writing the energy balance equation based on first law of thermodynamics for the TEC. According to the first law of thermodynamics, the energy balance for the thermionic converter system is given as (Angrist, 1982):

$$En_{output} = En_{input} \tag{34}$$

$$q_{cooling} + q_{evap} + q_{cond} + q_{rad} = q_A + q_{Joule} + q_{input} \tag{35}$$

where, $q_{cooling}$ is energy flux associated with cooling of cathode due to energy carried away from by electrons emitted from cathode which is given as (Angrist, 1982):

$$q_{cooling} == J_C\left(\varphi_A + \phi_A + qV - \phi_C + 2kT_C\right) \tag{36}$$

This cathode cooling is similar to Peltier cooling of thermoelectric converter. The electrons emitted from cathode have average kinetic energy of $2kT_C$. However, only a fraction of electrons have enough kinetic energy to cross the potential barrier of magnitude $\varphi_A + \phi_A + qV - \phi_C$.

q_{evap} is the energy flux, associated with electrons leaving the cathode due to vaporization which is called as latent heat of vaporization. This energy flux is required to overcome the potential barrier of work function which is given as (Angrist, 1982):

$$q_{evap} = J_C\phi_C \tag{37}$$

The energy flux carried away from cathode due to conduction through lead wires is given as (Angrist, 1982):

$$q_{cond} = \left(T_C - T_A\right)\left[\frac{\lambda_l A_l}{A_C l} + \frac{\lambda_w A_w}{A_C w}\right] \tag{38}$$

Figure 5. Energy band diagram for thermionic converter

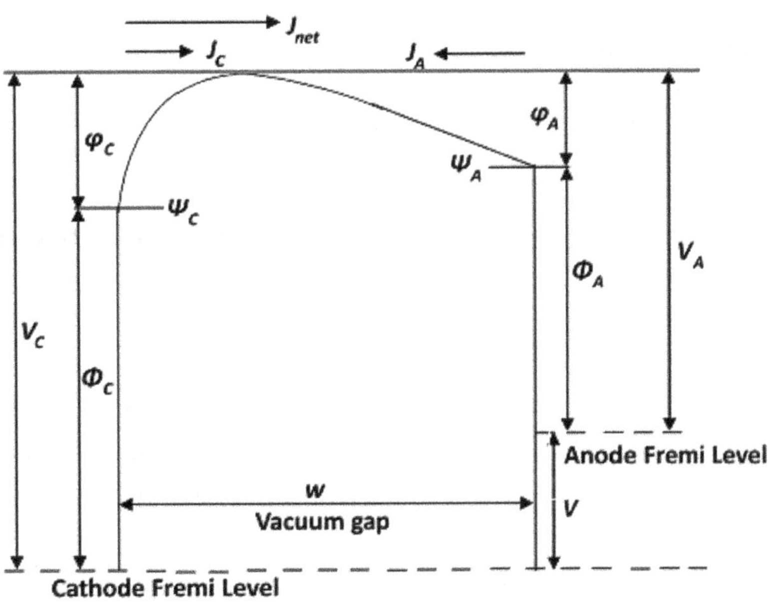

Since there is vacuum gap between cathode and anode so, there will be heat transfer between cathode and anode due to radiation. The net energy flux carried away from cathode due to radiative heat transfer is given as (Angrist, 1982):

$$q_{rad} = \sigma\left(\varepsilon_C^{-1} + \varepsilon_A^{-1} - 1\right)^{-1}\left(T_C^4 - T_A^4\right) \tag{39}$$

The energy flux associated with electrons emitted from anode and travelling towards the cathode is given by (Angrist, 1982):

$$q_A = J_A\left(\varphi_A + \phi_A + qV + 2kT_A\right) \tag{40}$$

These electrons emitted from anode have to overcome the potential barrier of magnitude $\left(\varphi_A + \phi_A + qV + 2kT_A\right)$. The Joule heating flux, taking place in lead wires is given by following expression (Angrist, 1982):

$$q_{Joule} = \frac{1}{2}\left(J_C - J_A\right)^2\left(R_l + R_w\right) \tag{41}$$

It is assumed that half of this heat will go towards cathode and half will go towards anode so half factor is coming in the above equation. Now, the input heat flux, supplied to cathode can be found by substituting equations (36) to (41) in equation (35) and then solving equation (35) for q_{input} which is given as:

$$\begin{aligned} q_{input} &= J_C\left(qV_C + 2kT_C\right) - J_C\left(qV_C + 2kT_C\right) + \sigma\left(\varepsilon_C^{-1} + \varepsilon_A^{-1} - 1\right)^{-1}\left(T_C^4 - T_A^4\right) \\ &+ \left(T_C - T_A\right)\left[\frac{\lambda_l A_l}{A_C l} + \frac{\lambda_w A_w}{A_C w}\right] - \frac{1}{2}\left(J_C - J_A\right)^2\left(R_l + R_w\right) \end{aligned} \tag{42}$$

The first law efficiency or thermal efficiency of thermionic converter can be defined as:

$$\eta = \frac{V\left(J_C - J_A\right)}{q_{input}} \tag{43}$$

To calculate the second law or exergy efficiency of converter, the exergy associated with heat input supplied to cathode has to be known. So the exergy efficiency of converter is defined as:

$$\psi = \frac{V\left(J_C - J_A\right)}{Ex_{input}} \tag{44}$$

where,

$$Ex_{input} = q_{input}\left(1 - \frac{T_{amb}}{T_C}\right) \tag{45}$$

Thermodynamic Model of a Practical Irreversible Vacuum Thermionic Generator

The schematic diagram of a vacuum TIG operating between two heat reservoirs has been shown in Figure 6. It consists of two parallel metal plates having vacuum gap between them. The cathode of the TIG absorbs heat from the higher temperature heat reservoir maintained at temperature, T_H. The anode releases heat to the lower temperature heat reservoir maintained at temperature T_L. The temperatures of cathode and anode can be determined by writing energy balance equations for cathode and anode using first law of thermodynamics. To generalize this analysis for a more practical irreversible TIG, the heat exchange between the electrodes and heat reservoirs is considered to obey both Newton's law and Stefan-Boltzmann's law of radiation. Therefore, the heat transferred from hot reservoir to cathode and heat released from anode to cold reservoir respectively can be expresses as:

$$Q_H = U_H A_H \left(T_H - T_C\right) + A_H \varepsilon\sigma \left(T_H^4 - T_C^4\right) \tag{46}$$

$$Q_L = U_L A_L \left(T_A - T_L\right) + A_L \varepsilon\sigma \left(T_A^4 - T_L^4\right) \tag{47}$$

where, U_H and U_L are heat transfer coefficients for heat flowing from hot reservoir to cathode and from anode to cold reservoir respectively, A_H and A_L are the effective heat transfer areas for heat flowing from hot reservoir to cathode and from anode to cold reservoir respectively.

The losses due to heat leakage, Q_{loss} between hot and cold reservoir is also considered to obey both Newton's law and Stefan-Boltzmann's law of radiation which can be expressed as:

$$Q_{loss} = U_{loss} A_{loss} \left(T_H - T_L\right) + A_{loss} \varepsilon\sigma \left(T_H^4 - T_L^4\right) \tag{48}$$

where, U_{loss} is heat transfer coefficient between hot and cold reservoir and A_{loss} is the effective heat exchange area between hot and cold reservoir.

Under normal operating mode, the heat is absorbed by the cathode of TIG from hot reservoir. After absorbing heat, the electrons of cathode are thermalized rapidly and spread out over the cathode surface based on the temperature of cathode. The electrons having (sum of both kinetic and potential) energy greater than the work function barrier are emitted out of cathode surface and flow towards anode through vacuum gap where they are collected by anode. These electrons then flow back from anode to cathode through external electrical load and thus thermal energy is converted into electrical energy directly by TIG.

Figure 6. Schematic diagram of an irreversible vacuum TIG

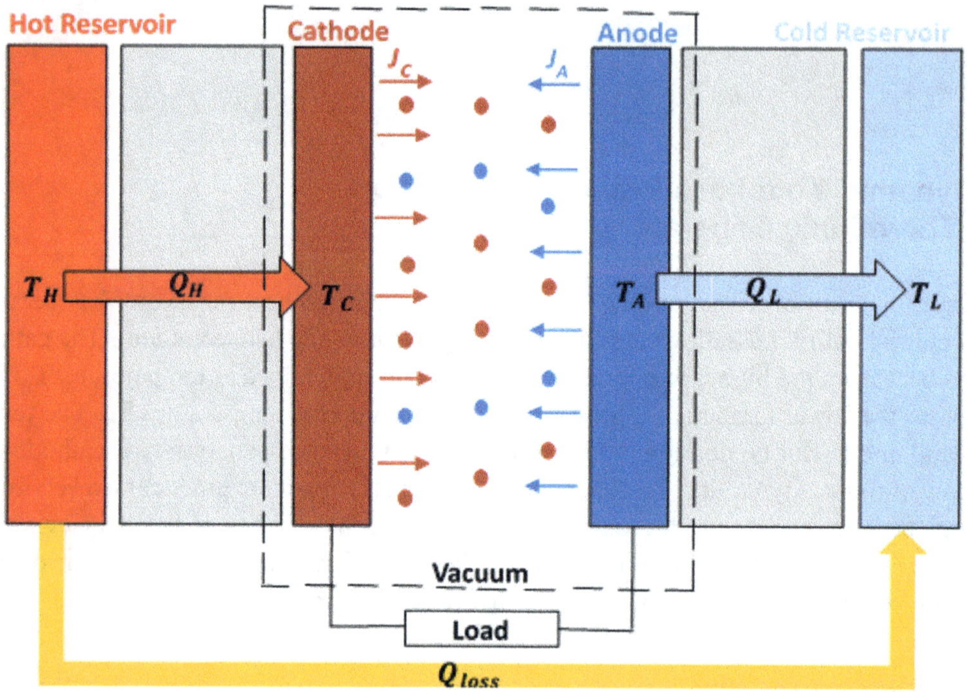

The electrical current densities from cathode and anode can be expressed by well-known Richardson equation as (Hatsopoulos, & Gyftopoulos, 1979):

$$J_C = K_{RD} T_C^2 \exp\left(-\frac{\phi_C}{k_B T_C}\right) = K_{RD} T_C^2 \exp\left(-\frac{\phi_A + qV}{k_B T_C}\right) \tag{49}$$

$$J_A = K_{RD} T_A^2 \exp\left(-\frac{\phi_A}{k_B T_A}\right) \tag{50}$$

where, K_{RD} is Richardson-Dushmann constant which is given as $K_{RD} = (4\pi m k_B^2 q/h^3)$, and ϕ_C and ϕ_A are the work functions of cathode and anode respectively. The output voltage of TIG depends on the work functions of cathode and anode which is given as:

$$V = \frac{\phi_C - \phi_A}{q} \tag{51}$$

The net electrical current density can be calculated from Equations. (49) and (50), which is given as:

$$J_{net} = J_C - J_A \tag{52}$$

$$J_{net} = K_{RD} T_C^2 \exp\left(-\frac{\phi_A + qV}{k_B T_C}\right) - K_{RD} T_A^2 \exp\left(-\frac{\phi_A}{k_B T_A}\right) \tag{53}$$

Now, based on first law of thermodynamics, the energy balance equations of heat flow rates for the cathode and anode of TIG can be written respectively as:

$$Q_H = A_0 \left[\left(V + \frac{\phi_A + 2k_B T_C}{q}\right) J_C - \left(V + \frac{\phi_A + 2k_B T_A}{q}\right) J_A + \varepsilon_{in} \sigma \left(T_C^4 - T_A^4\right)\right] \tag{54}$$

$$Q_L = A_0 \left[\left(\frac{\phi_A + 2k_B T_C}{q}\right) J_C - \left(\frac{\phi_A + 2k_B T_A}{q}\right) J_A + \varepsilon_{in} \sigma \left(T_C^4 - T_A^4\right)\right] \tag{55}$$

Substituting the expression of J_C, and J_A from Equations (49) and (50) respectively in Equation (54) results in following equations:

$$Q_H = A_0 \begin{bmatrix} K_{RD} T_C^2 \left(V + \frac{\phi_A + 2k_B T_C}{q}\right) \exp\left(-\frac{\phi_A + qV}{k_B T_C}\right) \\ -K_{RD} T_A^2 \left(V + \frac{\phi_A + 2k_B T_A}{q}\right) \exp\left(-\frac{\phi_A}{k_B T_A}\right) + \varepsilon_{in} \sigma \left(T_C^4 - T_A^4\right) \end{bmatrix} \tag{56}$$

Now, equating Equations (46) and (56) and further simplifying results in following expression:

$$\begin{aligned} & k_1 T_C^4 + k_2 T_C^3 \exp\left(-\frac{\phi_A + qV}{k_B T_C}\right) + k_3 \left(\phi_A + qV\right) T_C^2 \exp\left(-\frac{\phi_A + qV}{k_B T_C}\right) \\ & + T_C - k_4 T_A^4 - k_2 T_A^3 \exp\left(-\frac{\phi_A}{k_B T_A}\right) - k_3 \left(\phi_A + qV\right) T_A^2 \exp\left(-\frac{\phi_A}{k_B T_A}\right) - k_5 = 0 \end{aligned} \tag{57}$$

Similarly, substituting the expression of J_C, and J_A from Equations (49) and (50) respectively in Equation (55) results in following equations:

$$Q_L = A_0 \begin{bmatrix} K_{RD} T_C^2 \left(\frac{\phi_A + 2k_B T_C}{q} \right) \exp\left(-\frac{\phi_A + qV}{k_B T_C} \right) \\ -K_{RD} T_A^2 \left(\frac{\phi_A + 2k_B T_A}{q} \right) \exp\left(-\frac{\phi_A}{k_B T_A} \right) + \varepsilon_{in} \sigma \left(T_C^4 - T_A^4 \right) \end{bmatrix} \tag{58}$$

Now, equating Equations (47) and (58) and, further simplifying results in following expression:

$$\begin{aligned} & k_6 T_A^4 + k_7 T_A^3 \exp\left(-\frac{\phi_A}{k_B T_A} \right) + k_8 \phi_A T_A^2 \exp\left(-\frac{\phi_A}{k_B T_A} \right) + T_A - k_9 T_C^4 \\ & -k_7 T_C^3 \exp\left(-\frac{\phi_A + qV}{k_B T_C} \right) - k_8 \phi_A T_C^2 \exp\left(-\frac{\phi_A + qV}{k_B T_C} \right) - k_{10} = 0 \end{aligned} \tag{59}$$

where, k_1-k_{10} are constants which are defined in the Appendix. The Equations (57) and (59) cannot be solved analytically, therefore, some numerical method is used to find the temperature of cathode and anode. After calculating the cathode and anode temperature and then net current densities, the power output and energy efficiency of the vacuum TIG can be obtained. The expressions for power output and efficiency are given as:

$$P_{out} = A_0 J_{net} V \tag{60}$$

$$P_{out} = K_{RD} A_0 V \left(T_C^2 \exp\left(-\frac{\phi_A + qV}{k_B T_C} \right) - T_A^2 \exp\left(-\frac{\phi_A}{k_B T_A} \right) \right) \tag{61}$$

and,

$$\eta = \frac{Q_H - Q_L}{Q_H + Q_{loss}} = \frac{P_{out}}{Q_H + Q_{loss}} \tag{62}$$

$$\eta = \frac{K_{RD} A_0 V \left(T_C^2 \exp\left(-\frac{\phi_A + qV}{k_B T_C} \right) - T_A^2 \exp\left(-\frac{\phi_A}{k_B T_A} \right) \right)}{\begin{aligned} & \left(V + \frac{\phi_A + 2k_B T_C}{q} \right) J_C - \left(V + \frac{\phi_A + 2k_B T_C}{q} \right) J_A + \varepsilon_{in} \sigma \left(T_C^4 - T_A^4 \right) \\ & - \frac{U_{loss} A_{loss}}{A_{TI}} \left(T_H - T_L \right) - \frac{A_{loss} \varepsilon \sigma}{A_{TI}} \left(T_H^4 - T_L^4 \right) \end{aligned}} \tag{63}$$

The exergy efficiency of the system can be determined by calculating the exergy input, entropy generation and irreversibility of the thermionic generator. The exergy associated to input heat can be calculated as:

$$Ex_{Qh} = Q_H \left(1 - \frac{T_0}{T_H}\right) \tag{64}$$

where T_0 is the environment temperature.

The entropy generation, S_{gen} can be calculated by writing the entropy balance equation based on second law of thermodynamics.

$$S_{gen} = \Delta S_{sysetm} + \Delta S_{surrounding} \tag{65}$$

However, the entropy associated to thermal radiation is different from that of other heat transfer processes which is given as (Chen, Wu, & Sun, 1999):

$$S_{gen,rad} = \varepsilon_{in} \sigma T_C^4 \left(\frac{1}{T_A} - \frac{1}{T_C}\right) \tag{66}$$

Therefore, the total entropy generated in the system can be given as:

$$S_{gen} = A_0 \left[\begin{array}{l} \left[\left(\frac{\phi_A + 2k_B T_C}{q}\right)\frac{J_C}{T_L} - \left(\frac{\phi_A + 2k_B T_A}{q}\right)\frac{J_A}{T_L} + \varepsilon_{in}\sigma T_C^4 \left(\frac{1}{T_A} - \frac{1}{T_C}\right)\right] \\ - \left[\left(\frac{\phi_A + 2k_B T_C}{q}\right)\frac{J_C}{T_H} - \left(\frac{\phi_A + 2k_B T_A}{q}\right)\frac{J_A}{T_H} + \varepsilon_{in}\sigma T_C^4 \left(\frac{1}{T_A} - \frac{1}{T_C}\right)\right] \end{array} \right] \tag{67}$$

After simplifying the above expression,

$$S_{gen} = K_{RD} A_0 \left[\begin{array}{l} \left(\frac{\phi_A + 2k_B T_C}{q}\right)\left(\frac{1}{T_L} - \frac{1}{T_H}\right) T_C^2 \exp\left(-\frac{\phi_A + qV}{k_B T_C}\right) \\ - \left(\frac{\phi_A + 2k_B T_A}{q}\right)\left(\frac{1}{T_L} - \frac{1}{T_H}\right) T_A^2 \exp\left(-\frac{\phi_A}{k_B T_A}\right) \end{array} \right] \tag{68}$$

Based on first and second law of thermodynamics, the exergy balance equation of a thermodynamic system at steady state conditions can be written as:

$$Ex_{input} = Ex_{output} + Ex_{lost} + Ex_{des} \tag{69}$$

where, Ex_{input} is the exergy input to the system, Ex_{output} is the exergy output of the system, Ex_{lost} is the exergy lost in output stream if it is present and Ex_{des} is the exergy destroyed in the process. The exergy destroyed is the irreversibilities present in the system. In the thermionic generator, the exergy input is the thermal exergy and exergy output is the electrical power output which is purely exergy. Therefore, the irreversibilities of the thermionic generator can be expressed as:

$$Irr = T_0 S_{gen} \tag{70}$$

Substituting the expression of S_{gen} from Equation (68) in Equation (70) results in following expression for irreversibilities in the generator:

$$Irr = K_{RD} A_0 T_0 \left[\begin{array}{l} \left(\frac{\phi_A + 2k_B T_C}{q}\right)\left(\frac{1}{T_L} - \frac{1}{T_H}\right) T_C^2 \exp\left(-\frac{\phi_A + qV}{k_B T_C}\right) \\ -\left(\frac{\phi_A + 2k_B T_A}{q}\right)\left(\frac{1}{T_L} - \frac{1}{T_H}\right) T_A^2 \exp\left(-\frac{\phi_A}{k_B T_A}\right) \end{array} \right] \tag{71}$$

The exergy efficiency of the thermionic generator can be given as:

$$\psi = \frac{Ex_{out}}{Ex_{in}} = \frac{P_{out}}{Ex_{Qh}} = \frac{E_{Qh} - Irr}{Ex_{Qh}} = 1 - \frac{T_0 S_{gen}}{Ex_{Qh}} \tag{72}$$

$$\psi = 1 - \frac{K_{RD} T_0 \left[\begin{array}{l} \left(\frac{\phi_A + 2k_B T_C}{q}\right)\left(\frac{1}{T_L} - \frac{1}{T_H}\right) T_C^2 \exp\left(-\frac{\phi_A + qV}{k_B T_C}\right) \\ -\left(\frac{\phi_A + 2k_B T_A}{q}\right)\left(\frac{1}{T_L} - \frac{1}{T_H}\right) T_A^2 \exp\left(-\frac{\phi_A}{k_B T_A}\right) \end{array} \right]}{\left[\begin{array}{l} K_{RD} T_C^2 \left(V + \frac{\phi_A + 2k_B T_C}{q}\right) \exp\left(-\frac{\phi_A + qV}{k_B T_C}\right) \\ -K_{RD} T_A^2 \left(V + \frac{\phi_A + 2k_B T_A}{q}\right) \exp\left(-\frac{\phi_A}{k_B T_A}\right) + \varepsilon_{in} \sigma \left(T_C^4 - T_A^4\right) \end{array} \right] \left(1 - \frac{T_0}{T_H}\right)} \tag{73}$$

Parametric Optimization Studies of Irreversible Thermionic Generator

The parametric optimization of TIG has been studied based on numerical and graphical methods. It is clear from Equations (57) and (59) that the temperatures of cathode and anode depend on cathode and anode work functions. Therefore, the temperatures of cathode and anode at different values of work function of anode and voltage can be obtained by solving Equations (57) and (59) with all other parameters known which are given in Table 1.

Table 1. Parameters used in vacuum TIG model

Parameter	Value	Parameter	Value
A_H	0.0025 m^2	T_H	1500 K
A_L	0.0250 m^2	T_L	300 K
A_{Loss}	2.5 x 10^{-4} m^2	T_0	300 K
A_{TI}	2.5 x 10^{-4} m^2	U_H	1.1022 x10^4 W/m^2K
h	6.626 x 10^{-34} Js	U_L	2.2045 x10^4 W/m^2K
k_B	1.38 x 10^{-23} J K^{-1}	U_{Loss}	1.1022 x10^3 W/m^2K
K_{RD}	1.2 x 10^6 A/ m^2K^4	σ	5.67 x 10^{-8} W/m^2K^4
q	1.6 x 10^{-19}	ε_{in}	0.18
ε	0.9		

However, exponential terms are present in Equations (57) and (59), so the solution of these two equations cannot be obtained analytically. Therefore, numerical methods are used to obtain the solution of Equations (57) and (59). The effects of output voltage and anode work function on cathode and anode temperature have been shown in three-dimensional graphs of Figure 7(a)-(b) respectively. It is seen from this figure that cathode temperature, T_C increases monotonically with increase of output voltage and anode temperature, T_A decreases monotonically with increase of output voltage for a given work function value. After calculating the cathode and anode temperature, the current densities of cathode and anode and then net current density can be determined using Equations. (49), (50) and (53) respectively.

The three-dimensional variation of normalized net current densities with the output voltage and anode work function have been plotted in Figure 8 where the normalized current density is defined as J/J_{max}. It is clear graphically from the Figure 8 and mathematically from Equation (53) that the net current density decreases exponentially with increase in output voltage at a fixed value of work function and

Figure 7. Three-dimensional graphs of (a) cathode and (b) anode temperature varying with voltage and anode work function

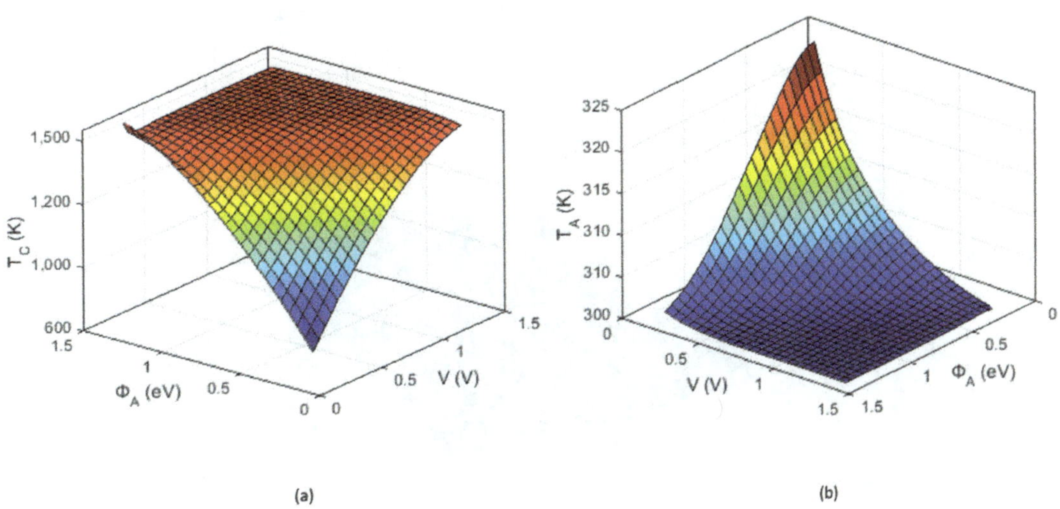

similarly, decreases exponentially with increase in work function at a fixed value of output voltage. At lower value of voltage or work function, more electrons will be having sufficient thermal energy to be emitted out of cathode surface and thus current density increases.

After calculating net current density, the power output of the TIG can be calculated from Equation. (61). The power output is zero when either the output voltage is zero or the net current density is zero which implies that the power output is maximum at some fixed value of output voltage and anode work function. The three dimensional graph of normalized power (P/P_{max}) varying with voltage and work function has been plotted in Figure 9. It is seen from Figure 9 that the normalized power is unity which means the power is maximum when the anode work function is 0.2 eV and the output voltage is 0.3 V.

Similar to power output plot of Figure 9, the energy and exergy efficiencies are zero at either zero output voltage or zero net current density implying maximum energy and exergy efficiencies at some fixed value of anode work function and output voltage. The energy and exergy efficiencies varying with anode work function and output voltage are plotted in Figure 10(a)-(b) respectively. The values of work function and output voltage, corresponding to maximum energy efficiency value of 0.606, are 0.25 eV and 0.95 V respectively. The maximum value of exergy efficiency is 0.666 at which the anode work function and output voltage values are 0.25 eV and 1.1 V respectively. It is clear from the Figure 10(a)-(b) that the energy and exergy efficiencies are not maximum at the same values for anode work function and output voltage. Also, the power output is not maximum at maximum energy and exergy efficiency values for anode work function and output voltage.

The variation of normalized irreversibilities of the TIG with the anode work function and output voltage has been plotted in Figure 11. As it is known that the irreversibility is proportional to entropy generated in the system and the entropy generation in the TIG is more at lower values of anode work function and output voltage. Therefore, the irreversibilities will also be more at the lower anode work function and output voltage.

Figure 8. Three-dimensional graph of net normalized current density varying with voltage and anode work function

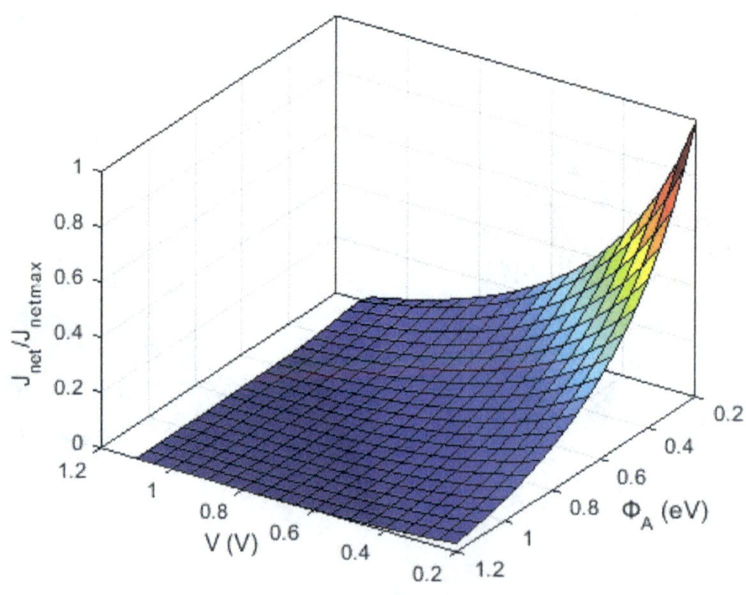

Figure 9. Three-dimensional graph of normalized power varying with voltage and anode work function

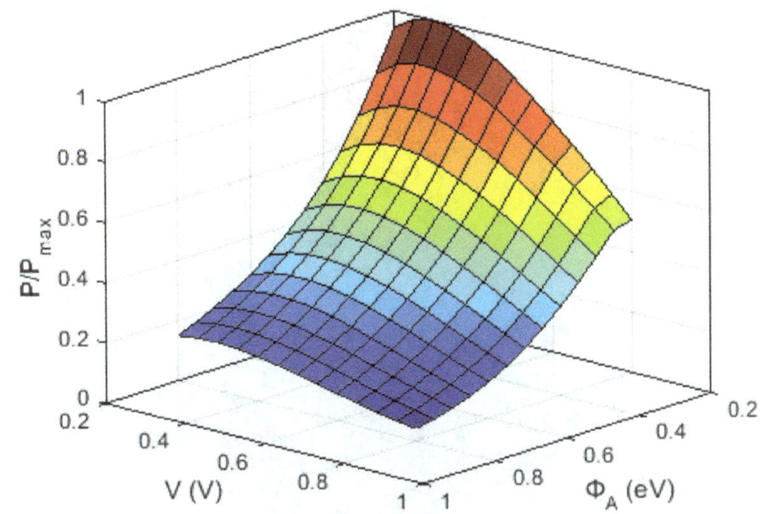

Figure 10. Three-dimensional graphs of (a) energy and (b) exergy efficiency varying with voltage and anode work function

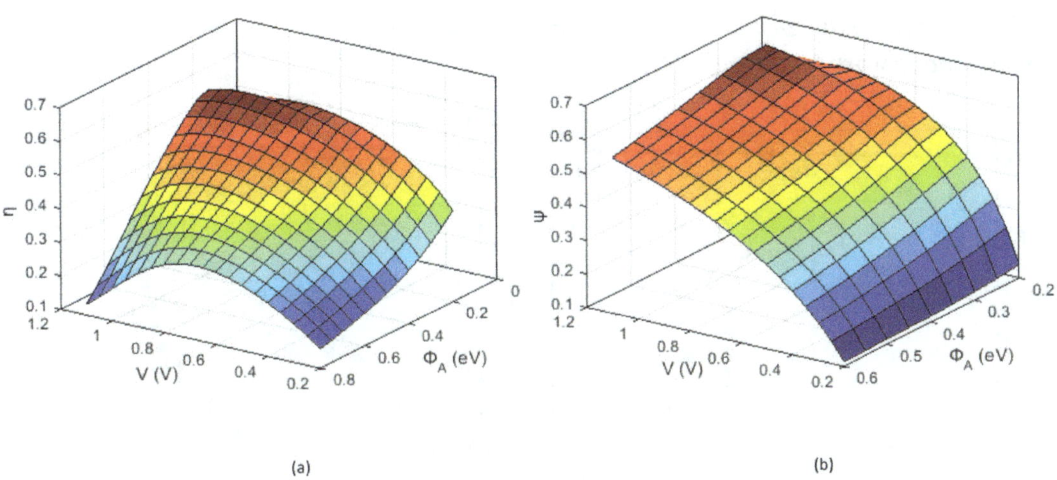

The plots of dimensionless power which is defined for some given anode work function and given as $(P/P_{max})_A$, varying with energy and exergy efficiency have been shown in Figure 12 (a) and (b) respectively. Due to the loop shape of the curve, it is clear from figure 15 (figure 16) that there is an optimum value of energy (exergy) efficiency at which the power output is maximum and similarly, there is an optimum value of power output at which the energy (exergy) efficiency is maximum. Thus, the power output and energy (exergy) efficiency are not globally maximum simultaneously. The curves are shown for anode work function values of 0.2, 0.25 and 0.3 eV. It can be seen from Figure 12 that the power output is maximum for anode work function value of 0.2 eV and the energy (exergy) efficiency is maximum for anode work function value of 0.25 (0.20) eV. Therefore, the work function should be chosen between this range only to get optimum power output and energy (exergy) efficiency. Table 2 shows the values of different parameters obtained at different optimum conditions.

Figure 11. Three-dimensional graph of normalized irreversibility varying voltage and anode work function

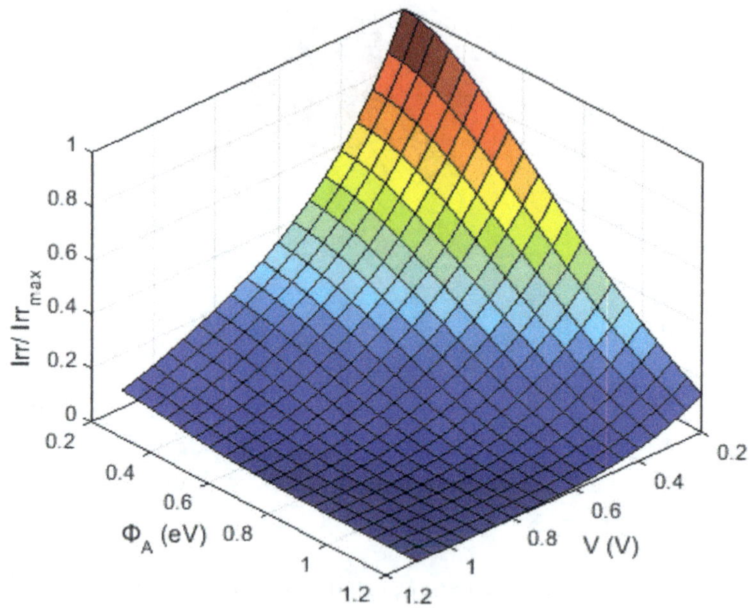

Figure 12. The curves of dimensionless power varying with (a) energy and (b) exergy efficiency at some fixed value of anode work function

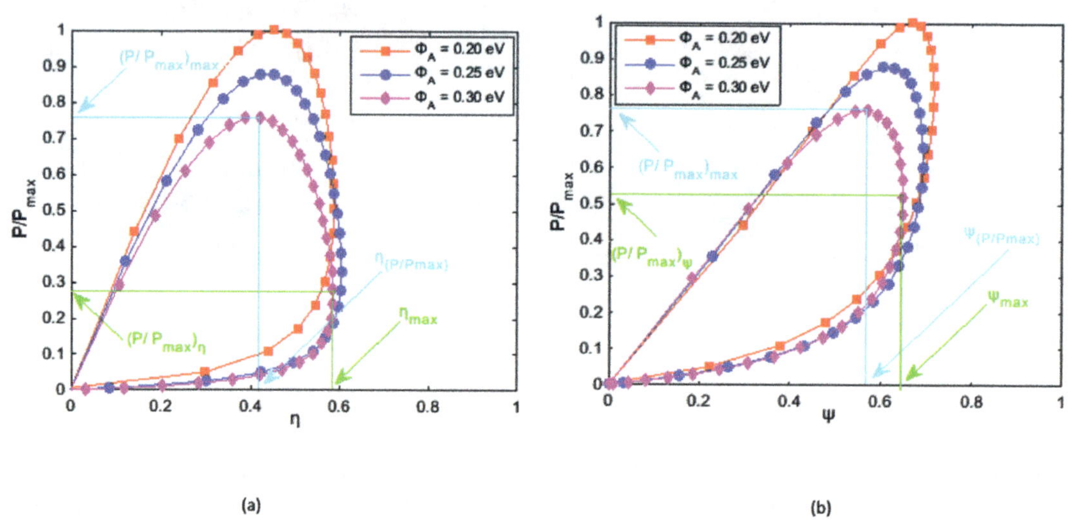

Major Challenges and Their Solutions

For developing practical thermionic converters, the major problem are the space charge effects and the availability of low work function materials for cathode and anode. The space charge effects are created in the gap between the electrodes due to electron cloud formation in the gap which reduces the performance of TEC. This space charge effect imposes a limit on maximum current density that can be achieved from

Table 2. Values of different parameters of vacuum thermionic generator at different optimum conditions

Parameter	$(J_{net}/J_{net,max})_{max}$	$(P/P_{max})_{max}$	$\eta_{en,max}$	$\eta_{ex,max}$
Φ_A (eV)	0.2	0.2	0.25	0.25
Φ_C (eV)	0.4	0.5	1.20	1.35
V (V)	0.2	0.3	0.95	1.1
T_C (K)	713.94	817.31	1342.6	1434.5
T_A (K)	324.66	318.59	302.76	301.45
$(J_{net}/J_{net,max})$	1.00	0.707	0.074	0.037
(Irr/Irr_{max})	1.00	0.754	0.11	0.059
(P/P_{max})	0.943	1.00	0.330	0.187
η_{en}	0.370	0.451	0.606	0.587
η_{ex}	0.140	0.317	0.643	0.666

the thermionic converter and thus, reducing the efficiency. The negative charges of electrons emitted from both electrodes repel each other and reduce the current. Thus, space charge effects forms additional potential barrier for further emission of electrons from cathode. This space charge problem can be resolved either by reducing the spacing between the electrodes to micrometer or by insertion of positive ions into negatively charged electron's cloud near the cathode to neutralize the space charge effects. These ions are produced from caesium as caesium has low ionization potential. However, caesium being a highly reactive material makes the system complex and the ionization process takes some energy during thermionic conversion. Another way to supress space charge effects is by modifying the cathode material properties such as use of negative electron affinity materials for cathode which reduces the vacuum level below the conduction level.

Another major problem in developing the TECs is the availability of low work function electrode. Since, the performance of thermionic converters depends on many parameters. Among them, the work functions of both the electrodes, cathode and anode are the main surface parameters. It is defined as the minimum amount of energy required for removal of an electron from the material into the vacuum just outside its surface. The work function is the characteristic of the electrodes and may vary between 1 to 5 eV for different materials. The electrons in the cathode having energy greater than the work function of cathode can escape from electrodes depending upon their temperature. Thus, the work function acts as a barrier for evaporation of electrons from electrode surface. The work function of both electrodes should be minimum so that less energy is required for emission of electrons from electrode surface and the difference between the work functions of cathode and anode should be maximum, cathode work function at least 1 eV higher than that of anode, for better performance of TECs. However, the anode work function should be as low as possible to avoid the operation of TEC at high temperature which causes significant heat losses to the environment. Therefore, the cathode and anode materials should be chosen by taking these points into account.

THERMAL MODELING OF PHOTON ENHANCED THERMIONIC EMISSION ENERGY CONVERSION SYSTEMS

The schematic diagram and energy band diagram of PETE converter is shown in Figure 13 and 14 respectively. The PETE converter consists of two electrodes, cathode and anode separated by vacuum gap similar to thermionic energy converter. However, in PETE converters, the cathode is made of semiconductor material instead of metallic cathode in thermionic converter. The temperature of cathode is maintained very high to emit electrons from the cathode surface. The temperature of anode should be maintained low to prevent back emission from the anode. Thus, the physical separation of both electrodes gives a large temperature difference between cathode and anode. Apart from this, the heat of anode can also be harnessed by attaching a thermoelectric generator or heat engine to the back side of the PETE converter because the operating temperature of thermoelectric generator is in the range of that of anode of PETE converter. Thus, at high concentrated solar applications, the PETE converter is a perfect device due to the operation of PETE efficiently at elevated temperature. PETE is also used as a topping cycle for solar heat engines which uses anode as its heat source in which, PETE converter harvests some fraction of incident solar radiation flux and rest of incident radiation flux is used as heat source for thermal engine. A PETE converter system with efficiency of 20% in conjunction with Stirling solar heat engine of 30% efficiency would give combined system efficiency of 44%.

Figure 13. Schematic model of the PETE converter

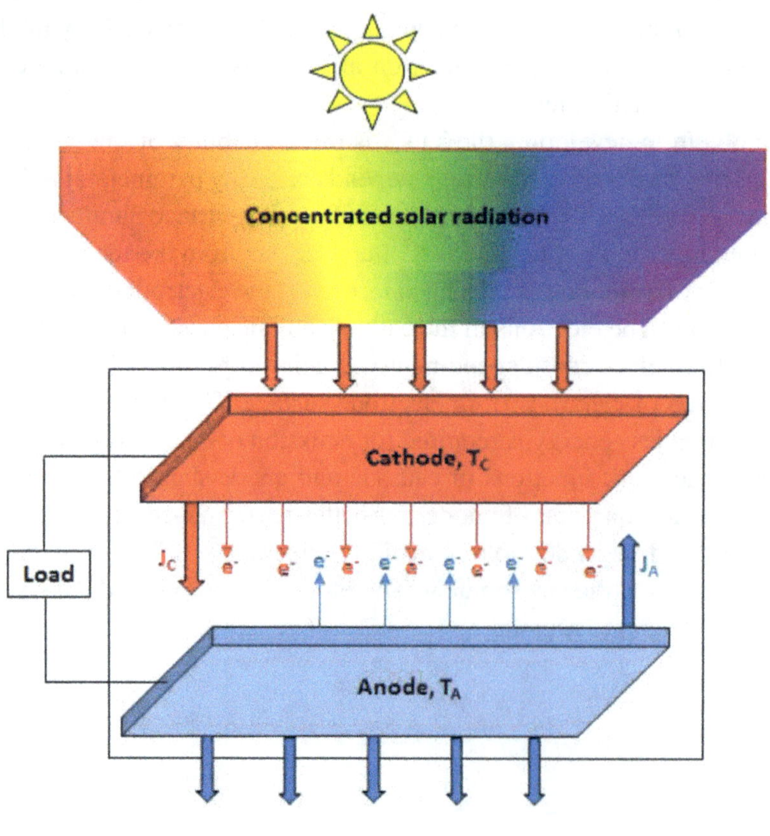

For generation of electricity, PETE basically involves three steps. First of all, the electrons are emitted into conduction band after absorption of photons with energy equal to or greater than band gap of material. If absorbed photon has energy greater than band gap, then the electrons, absorbing such a photon is excited to higher energy state. However, electrons are not stable in higher energy states so, they thermalize rapidly within the conduction band according to the equilibrium thermal distribution depending on temperature. After that, electrons utilize the energy lost during thermalization and absorption losses for driving thermionic emission. Thus, electrons having energy greater than electron affinity escape from the cathode surface. The emitted electrons then flow towards anode via vacuum gap and are collected at the anode. So each electron emitted from semiconductor cathode harvest photon as well as thermal to overcome band gap and electron affinity respectively. Therefore, due to thermal boosting, the output voltage produced by PETE converter is higher than that of PV system of same band gap. Since the efficiency of photovoltaic devices decreases and that of thermionic devices increases with increase in temperature of the device. Therefore, there will be an optimum temperature at which the PETE converter efficiency will be maximum.

The main issues for development of practical PETE converter is the availability of materials with high melting temperature and desired bandgap. InAs having bandgap of 0.35 eV and melting temperature of 1220 K and PbS with band gap of 0.41 eV and melting temperature of 1390 K are the best suitable materials reported so far in the literature for PETE converters. Schwede et al. (2010) have provided caesiated GaN which is an ultraviolet photocathode with caesium coating for high thermal stability based proof-of-concept devices as the first example of photon enhanced thermionic emission device.

Figure 14. Energy band diagram of PETE converter

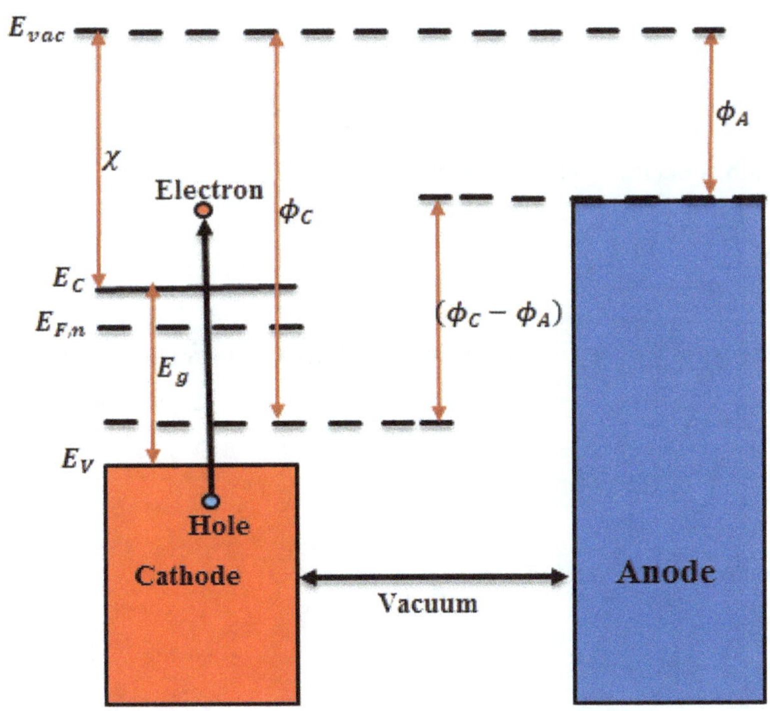

The key factors which should be considered for improving the efficiency of PETE converters are carrier recombination and absorption, and chemical and temperature stability.

Energy Analysis of Photon Enhanced Thermionic Emission Converter

The output current density in PETE converter can be calculated from photo excited electron flux having sufficient energy to be emitted from cathode surface, analogous to standard thermionic converter but the concentration of photo-excited electrons in PETE cathode conduction band is distributed based on quasi-Fermi level and it is more than that of equilibrium state. The distribution of electrons according to quasi Fermi level which is given by Wurfel (2009) for non-degenerate semiconductors as:

$$E_{F,n} = E_F + kT_c \ln\left(\frac{n}{n_{eq}}\right) \tag{74}$$

Thus, the energy barrier for cathode electrons is reduced due to solar illumination in PETE converter. The equilibrium Fermi level, E_F can be calculated by considering the basic charge neutrality equation given in Sze (1981):

$$n_{eq} + N_A^- = p_{eq} + N_D^+ \tag{75}$$

However, for p-type semiconductor cathode, the donor concentration is zero so the above equation becomes:

$$n_{eq} + N_A^- = p_{eq} \tag{76}$$

where, n_{eq}, the electron concentration is expressed as:

$$n_{eq} = N_C \exp\left\{-\left(E_g - E_F\right)/ kT_C\right\} \tag{77}$$

The ionization energy of p-type acceptor using the Boltzmann statistics is given as:

$$N_A^- = N_A \frac{1}{1 + 4\exp\left\{-\left(E_A - E_F\right)/ kT_C\right\}} \tag{78}$$

The holes' concentration is given by:

$$p_{eq} = N_V \exp\left\{-E_F / kT_C\right\} \tag{79}$$

E_A is the ionization energy of p-type acceptor in intrinsic semiconductor at a typical doping level as given in Schwede et al. (2010) and Sze (1981).

The emission current density is derived according to Boltzmann statistics theory and can be given by Schwede et al. (2010):

$$J_C = qn\sqrt{\frac{kT_C}{2\Pi m_n}}\exp\left\{-\left(\frac{\chi}{kT_C}\right)\right\};V \leq \phi_C - \phi_A \tag{80}$$

$$J_C = qn\sqrt{\frac{kT_C}{2\Pi m_n}}\exp\left\{-\left(\frac{\chi + \left(\phi_A + qV - \phi_C\right)}{kT_C}\right)\right\};V > \phi_C - \phi_A \tag{81}$$

where, $\langle v_x \rangle = \sqrt{\frac{kT_C}{2\Pi m_n}}$ is the average electron velocity perpendicular to the cathode surface.

The operating voltage, V is the difference between cathode's and anode's equilibrium Fermi level. Equations (80) and (81) shows that the cathode illumination by sunlight reduces the electron emission barrier by the energy difference between quasi-Fermi level with photon excitation and Fermi level without excitation. Thus, the conduction band electron concentration increases over equilibrium concentration.

The reverse emission current density from the anode is given by famous Richardson's equation for standard thermionic emission as given in Mahan (1994); Hatsopoulos, & Gyftopoulos (1973) and Lee, Bargatin, Melosh, & Howe (2012):

$$J_A = K_{RD}T_A^2\exp\left\{-\left(\frac{\phi_C - qV}{kT_A}\right)\right\};V \leq \phi_C - \phi_A \tag{82}$$

$$J_A = K_{RD}T_A^2\exp\left\{-\left(\frac{\phi_A}{kT_A}\right)\right\};V > \phi_C - \phi_A \tag{83}$$

At steady state condition, the conservation of charge equation is given as by Schwede et al. (2010):

$$\frac{R_S}{H} - \frac{R_C}{w} - \frac{\left(2 - \Gamma\right)R_r}{H} + \frac{R_A}{w} = 0 \tag{84}$$

The rate of photon induced electrons R_S can be calculated using photon flux density $\varphi(\lambda)$ given by Schwede et al. (2010) and Su et al. (2013) as:

$$R_s = \int_0^{\lambda_g} \varphi(\lambda)\, d\lambda \tag{85}$$

The rate of thermionic emission, R_C in cathode is given Schwede et al. (2010):

$$R_C = n\sqrt{\frac{kT_C}{2\Pi m_n}} \exp\left\{-\left(\frac{\chi}{kT_C}\right)\right\}; V \le \phi_C - \phi_A \tag{86}$$

$$R_C = n\sqrt{\frac{kT_C}{2\Pi m_n}} \exp\left\{-\left(\frac{\chi + \left(\phi_A + qV - \phi_C\right)}{kT_C}\right)\right\}; V > \phi_C - \phi_A \tag{87}$$

This emission rate is a linear varying linearly with electron concentration so it can be written as:

$$R_C = nK_{PETE} \tag{88}$$

The rate of back emission from PETE anode, R_A can be given as:

$$R_A = \frac{K_{RD} T_A^2}{q} \exp\left\{-\left(\frac{\phi_C - qV}{kT_A}\right)\right\}; V \le \phi_C - \phi_A \tag{89}$$

$$R_A = \frac{K_{RD} T_A^2}{q} \exp\left\{-\left(\frac{\phi_A}{kT_A}\right)\right\}; V > \phi_C - \phi_A \tag{90}$$

The photon-enhanced recombination rate at non-equilibrium conditions, R_r can be calculated by rate of photon emission equilibrium conditions, R_0 given by Wurfel (2009) and Schwede et al. (2010) as:

$$R_0 = \frac{2\Pi}{h^3 c^2} \int_{E_g}^{\infty} \frac{(hv)^2 d(hv)}{\exp(hv / kT) - 1} \tag{91}$$

Under non-equilibrium conditions, R_0 increases exponentially given by Wurfel (2009) which is given as:

$$R = R_0 \exp\left\{\frac{E_{F,n} - E_{F,p}}{kT}\right\} \tag{92}$$

$$R = R_0 \left(\frac{np}{n_{eq} p_{eq}} \right) \tag{93}$$

Here, $E_{F,n}$ and $E_{F,p}$ are quasi-Fermi levels for valence band and conduction band respectively and, n_{eq} and p_{eq} are the electrons and holes concentrations at equilibrium conditions. Therefore, rate of recombination at non-equilibrium conditions is given by Wurfel (2009) and Schwede et al. (2010) as:

$$R_r = R_0 \left(\frac{np}{n_{eq} p_{eq}} - 1 \right) \tag{94}$$

$$R_r = K_R \left(np - n_{eq} p_{eq} \right) \tag{95}$$

Thus, combining Equations from (85) to (95) and substituting in Equation (84) results in following equation:

$$\frac{R_S \left(E > E_g \right)}{H} - \frac{nK_{PETE}}{w} - \frac{\left(2 - \Gamma \right) K_R \left(np - n_{eq} p_{eq} \right)}{H} + \frac{R_A}{w} = 0 \tag{96}$$

The electrons and holes' excess carrier concentration is assumed to be equal and is given as follows:

$$\delta n = n - n_{eq} = p - p_{eq} \tag{97}$$

Combining Equations (96) and (97) results in quadratic equation in δn:

$$\left(2 - \Gamma \right) K_R \left(\delta n \right)^2 + \left[\left(2 - \Gamma \right) \left(n_{eq} + p_{eq} \right) K_R + K_{PETE} s_r \right] \delta n - \left(R_S + R_A s_r - K_{PETE} n_{eq} s_r \right) = 0 \tag{98}$$

The concentration of electrons in PETE cathode can be calculated by solving Equation (98). For ease of calculation, the cathode and anode surfaces have been assumed to be perfectly reflective with unity reflection coefficient. The non-equilibrium electron concentration can be found from Equation (97). After calculating concentration of electrons, the current densities from cathode and anode can be calculated using Equations (80), (81) and (82), (83) respectively. The net power output and efficiency of the PETE converter can be expressed as follows:

$$P_{PETE} = A_0 \left(J_C - J_A \right) \left(\phi_C - \phi_A \right) \tag{99}$$

$$\eta = \frac{\left(J_C - J_A\right)\left(\phi_C - \phi_A\right)}{P_{in}} \tag{100}$$

The input power, P_{in} can be obtained from the concentrated AM 1.5 direct circumsolar radiation fluxes. Since, the direct spectrum can be concentrated, so the incident radiation flux is summation of both diffuse solar radiation flux and concentrated direct solar radiation flux.

Parametric Optimization Studies of Photon Enhanced Thermionic Emission Converter

Su et al. (2013) have carried out the parametric optimization of PETE converter with some fixed parameters of $\Phi_A = 0.9$ eV, $T_{amb} = 300$ and $\Gamma = 0$. The operating temperatures of cathode and anode have been calculated by solving energy balance equations for cathode and anode for fixed bandgap and electron affinity. After knowing the cathode and anode temperature, the current density and consequently the power output can be calculated. They have maximized the efficiency for different bandgap and electron affinity values. The optimum values of different parameters calculated by Su et al. (2013) have been given in Table 3.

Su et al. (2013) have plotted the efficiency varying with bandgap as shown in Figure 15. They have considered the various heat losses between cathode and anode, and between anode and ambient. The optimum values of the band gap ($E_{g,opt}$) and electron affinity (χ_{opt}) corresponding to maximum efficiency

Table 3. Optimum values of different parameters of PETE converter at maximum energy efficiency conditions for given bandgap value

E_g (eV)	η_{max}	χ_{opt} (eV)	$T_{C,opt}$ (K)	$T_{A,opt}$ (K)	V_{opt} (V)	J_{opt} (A/cm^2)
0.600	0.256	1.010	997	600	0.519	44.500
0.800	0.303	0.950	996	578	0.649	42.100
1.000	0.348	0.900	1007	556	0.794	39.500
1.200	0.374	0.960	1098	532	1.030	32.800
1.300	0.375	1.000	1156	522	1.150	29.400
1.360	0.379	1.010	1170	518	1.210	28.200
1.380	0.378	1.010	1179	517	1.230	27.800
1.390	0.378	1.010	1183	516	1.240	27.500
1.400	0.377	1.020	1189	515	1.260	27.000
1.410	0.376	1.020	1196	515	1.270	26.700
1.420	0.375	1.030	1202	514	1.290	26.300
1.440	0.373	1.040	1214	513	1.310	25.700
1.460	0.370	1.050	1226	511	1.340	24.900
1.500	0.363	1.060	1251	509	1.380	23.700
1.600	0.347	1.100	1300	504	1.510	20.800

Su et al., 2013.

Figure 15. Variation of PETE efficiency with bandgap
Su et al., 2013.

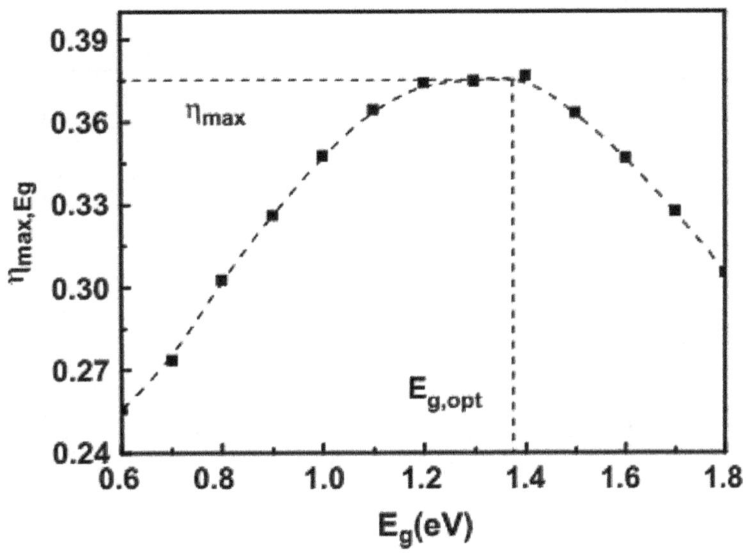

($\eta_{max,Eg}$) are 1.36 eV and 1.01 eV respectively as given in Table 3. The cathode and anode temperatures corresponding to maximum efficiency are found to be 1170 K and 518 K, respectively.

Varpula, Tappura, & Prunnila, (2015) have plotted the current–voltage curves for Si, GaAs and InP as cathode material for PETE converter as shown in Figure 16 with fixed electron affinity of 0.4 eV and cathode temperature of 700 K at 1000 suns. The flat band voltages for these cathode material have been shown by dashed lines and the maximum power points are shown by plus sign in the figure. They found that the power output of Si is lesser than that of GaAs and InP due to larger band gap of GaAs and InP which causes higher flat band (V_{fb}) and consequently higher output voltage for same electron affinity of all materials. Although the flat band voltage of InP is lower, however, the operating temperature of InP is higher than that of GaAs due to higher photon absorption coefficient of InP. The photon absorption efficiency and distance travelled by photoelectron before emission depend on the thickness of the cathode.

They also have maximized the efficiency of PETE converter with Si, GaAs and InP as cathode material, with respect to cathode thickness for different acceptor density values as shown in Figure 17 with cathode temperature of 700 K and electron affinity of 0.4 eV. The efficiencies of GaAs and InP are higher than that of Si due to higher absorption coefficient and band gap values. The efficiencies for three cathode material are smaller both at lower and higher values of cathode thickness due to low absorption coefficient and higher recombination of photo generated electrons respectively.

Based on this analysis, it can be stated that, although the radiative recombination is very high in GasAs and InP, they have higher efficiencies than Si due to strong their absorption coefficient. The efficiency of InP is even higher than that of GaAs due to better photon absorption coefficient. Thus, InP seems a better candidate as PETE cathodes. At higher operating temperatures, Si seems to be a better candidate due to its higher thermal stability even though the performance of Si is lower.

Figure 16. Variation of current density with voltage for PETE devices with Si, GaAs, and InP as cathode material
Varpula, Tappura, & Prunnila, 2015.

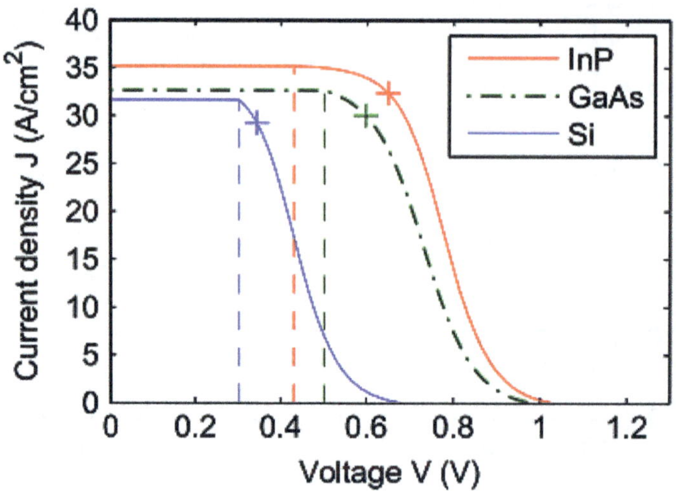

FUTURE RESEARCH DIRECTIONS

The PETE system can be an advanced future competitive power generation option with research in advanced materials having low work function. The progress in the direction of better cathode and anode materials having lower work function and thermally stable at higher temperatures can improve the power output and efficiency of thermionic and PETE systems. With improvement in solar concentrating techniques, higher concentration ratio can be obtained and the heat losses from the top side can be reduced by using wavelength selective solar absorbers. The problem of space charge effect can also be overcome with the application of gate voltage or by inserting positive ions with improved properties in the vacuum gap. Apart from this, the overall performance of thermionic and PETE converters can be improved by adding a topping cycle at the anode of thermionic and PETE converter to harness the rejected heat from the anode. The exergy analysis of thermionic and PETE converters can be carried out which will be help in quantifying the true irreversibilities happening in different part of the system and corrective action can be taken to reduce the irreversibilities. Further, various single and multiobjective optimization techniques can be applied to these systems to obtain the maximum power output and efficiency.

CONCLUSION

In this review article, the basic thermodynamic analysis of photovoltaic and thermionic energy conversion systems have been presented for understanding of PETE system. The physics and thermal modelling of photovoltaic, thermionic and PETE energy conversion systems have been presented. Apart from this, the materials, major challenges in TEC and PETE converter and their possible solutions have been discussed. Apart from this, the optimization of TEC and PETE converter have also been studied. The power output, energy and exergy efficiencies have been taken as objective function for optimization using graphical and numerical methods.

Figure 17. Variation of PETE efficiency of (a)Si, (b) GaAs and (c) InP with cathode thickess for different accepter density values
Varpula, Tappura, & Prunnila, 2015.

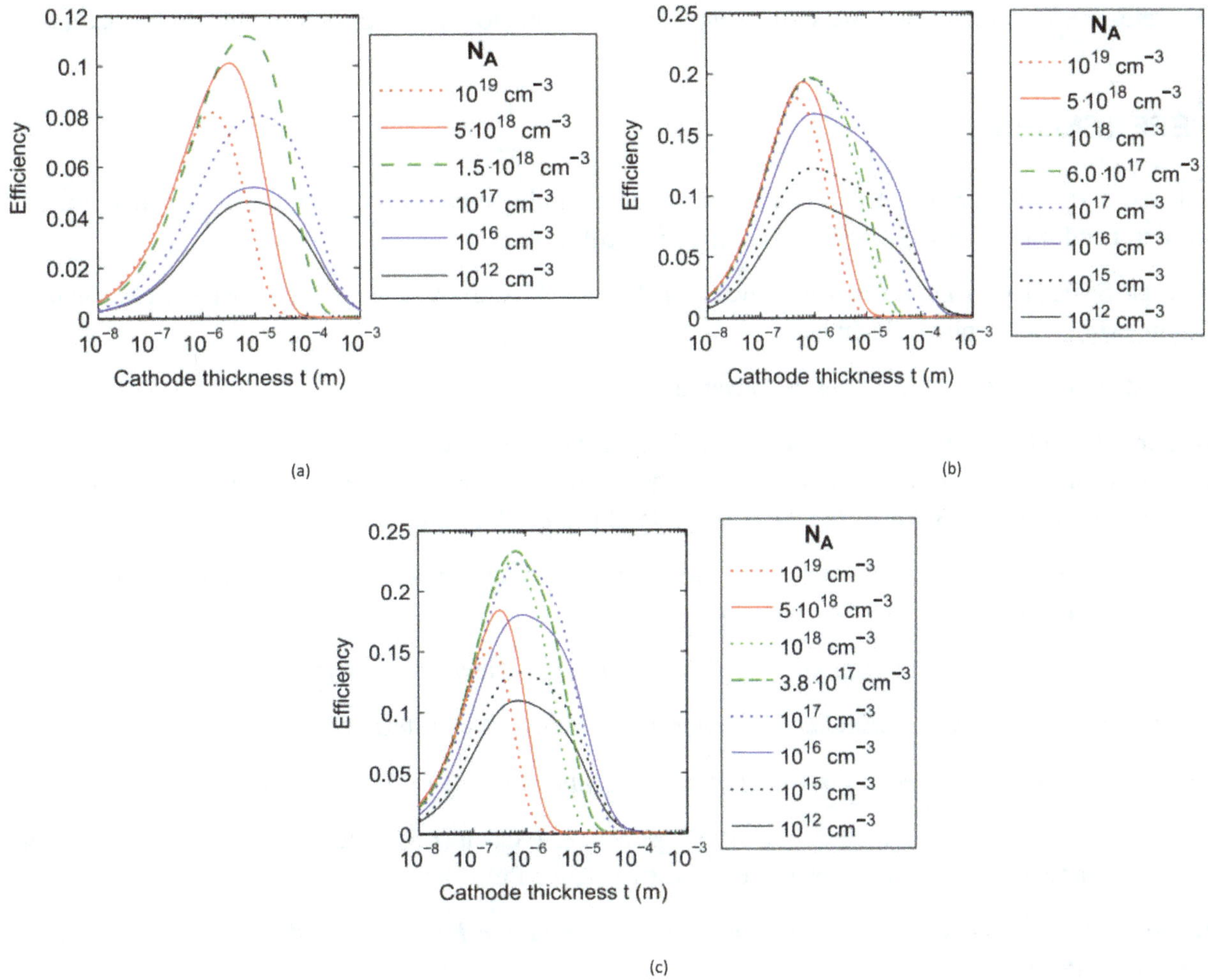

PETE has higher efficiency compared to that of individual PV or thermionic systems. So it can be a better future option to utilize solar energy more efficiently by combining PV and thermionic technologies to simultaneously harness photon as well as thermal energy of solar radiation flux. Further better thermophysical characteristics of cathode and anode materials of PETE systems have to be searched for practical designing of PETE systems. In spite of several challenges for implementing TEC and PETE converters, these are clean mechanism for power generation with noiseless operation, no moving parts, high reliability and long life. Further to increase the light absorption and carrier concentration for improving the efficiency, research in development of new material, use of nanostructures is required.

Thus, this study is helpful in design, performance optimization and selection of materials for cathode and anode of a practical irreversible vacuum TEC and PETE converter.

ACKNOWLEDGMENT

The authors gratefully acknowledgement the researchers in this field whose work has been consulted and referred. Thanks are also due to the editors and reviewers for their suggestions and kind cooperation.

REFERENCES

Adams, S. F. (2006). Solar thermionic space power technology testing: A historical perspective. In. AIP Conference Proceedings: Vol. 813. *No. 1* (pp. 590–597). AIP.

Angrist, S. W. (1965). *Direct energy conversion* (4th ed.). Boston: Allyn and Bacon Series in Mechanical Engineering and Applied Mechanics.

Angrist, S. W. (1982). Direct energy conversion (4th ed.). Allyn and Bacon.

Baetens, R., Jelle, B. P., & Gustavsen, A. (2010). Properties, requirements and possibilities of smart windows for dynamic daylight and solar energy control in buildings: A state-of-the-art review. *Solar Energy Materials and Solar Cells*, *94*(2), 87–105. doi:10.1016/j.solmat.2009.08.021

Becquerel, M. E. (1853). Researches on the electrical conductibility of gases at high temperatures. *Philosophical Magazine Series 4*, *6*(41), 456–457.

Bejan, A. (1988). *Advanced Engineering Thermodynamics*. New York: Wiley Interscience.

Bejan, A. (1998). *Advanced Engineering Thermodynamics*. Chichester: Wiley.

Bejan, A. (2006). *Advanced engineering thermodynamics* (3rd ed.). New York: Willey.

Bhattacharyya, S. (1999). Chapter. In C. Wu, L. Chen, & J. Chen (Eds.), Recent advances in finite-time thermodynamics (pp. 105–113). New York: Nova Science Publishers.

Boyle, G. (2004). *Renewable Energy Power for a Sustainable Future* (2nd ed.). Oxford, UK: Oxford University Press.

Bundgaard, E., & Krebs, F. C. (2007). Low band gap polymers for organic photovoltaics. *Solar Energy Materials and Solar Cells*, *91*(11), 954–985. doi:10.1016/j.solmat.2007.01.015

Casaro, M. M., & Martins, D. C. (2008, June). Grid-connected PV system: Introduction to behavior matching. In *Power Electronics Specialists Conference, 2008. PESC 2008. IEEE* (pp. 951-956). IEEE. 10.1109/PESC.2008.4592052

Cengel, Y. A., & Boles, M. A. (2006). *Thermodynamics: An Engineering Approach* (5th ed.). New York: Tata McGraw Hill.

Chen, J. (1992). Optimization of a solar-driven heat engine. *Journal of Applied Physics*, *72*(8), 3778–3780. doi:10.1063/1.352273

Chen, L., Wu, C., & Sun, F. (1999). Finite time thermodynamic optimization or entropy generation minimization of energy systems. *Journal of Non-Equilibrium Thermodynamics*, *24*(4), 327–359.

Cheng, X., & Liang, X. (2012). Entransy loss in thermodynamic processes and its application. *Energy*, *44*(1), 964–972. doi:10.1016/j.energy.2012.04.054

Chow, T. T. (2010). A review on photovoltaic/thermal hybrid solar technology. *Applied Energy*, *87*(2), 365–379. doi:10.1016/j.apenergy.2009.06.037

Crookes, W. (1896). The True Cathode Rays. *The New York Times*.

Currie, M. J., Mapel, J. K., Heidel, T. D., Goffri, S., & Baldo, M. A. (2008). High-efficiency organic solar concentrators for photovoltaics. *Science*, *321*(5886), 226–228. doi:10.1126cience.1158342 PMID:18621664

Daher, S., Schmid, J., & Antunes, F. L. (2008). Multilevel inverter topologies for stand-alone PV systems. *IEEE Transactions on Industrial Electronics*, *55*(7), 2703–2712. doi:10.1109/TIE.2008.922601

Erbay, L. B., & Yavuz, H. (1997). Design parameters of a thermionic energy converter. In *Proceedings of the International Conference on Thermodynamics Analysis and Improvement of Energy Systems (TAIES'97)* (pp. 20-24). Academic Press.

Fletcher, E. A., & Moen, R. L. (1977). Hydrogen and oxygen from water. *Science*, *197*(4308), 1050–1056. doi:10.1126cience.197.4308.1050 PMID:17836063

Gibson, J. H. (2003). *UVB radiation: Definition and characteristics*. Retrieved from http://uvb. nrel. colostate. edu

Goldsmid, H. J. (2010). *Introduction to thermoelectricity* (Vol. 121). Heidelberg, Germany: Springer. doi:10.1007/978-3-642-00716-3

Green, M. A. (2003). General temperature dependence of solar cell performance and implications for device modelling. *Progress in Photovoltaics: Research and Applications*, *11*(5), 333–340. doi:10.1002/pip.496

Green, M. A. (2006). *Third generation photovoltaics: Advanced Solar Energy Conversion*. Berlin: Springer.

Green, M. A., Emery, K., Hishikawa, Y., Warta, W., & Dunlop, E. D. (2012). Solar cell efficiency tables (Version 39). *Progress in Photovoltaics: Research and Applications*, *20*(1), 12–20. doi:10.1002/pip.2163

Guthrie, F. (1873). On a relation between heat and static electricity. *The London, Edinburgh and Dublin Philosophical Magazine and Journal of Science*, *46*(306), 257–266.

Hatsopoulos, G. N., & Gyftopoulos, E. P. (1973). Thermionic energy conversion: Vol. 1. Processes and devices. MIT Press.

Hatsopoulos, G. N., & Gyftopoulos, E. P. (1973). *Thermionic energy conversion* (Vol. 1). Cambridge, MA: MIT Press.

Hatsopoulos, G. N., & Gyftopoulos, E. P. (1979). Thermionic energy conversion: Vol. 2. Theory, technology, and application (No. DOE/TIC-11035). MIT Press.

Hatsopoulos, G. N., & Kaye, J. (1958). Analysis and Experimental esults of a Diode Configuration of a Novel Thermoelectron Engine. *Proceedings of the IRE*, *46*(9), 1574–1579. doi:10.1109/JRPROC.1958.286932

Hatsopoulos, G. N., & Kaye, J. (1958). Measured thermal efficiencies of a diode configuration of a thermo electron engine. *Journal of Applied Physics*, *29*(7), 1124–1125. doi:10.1063/1.1723373

Haynes, T. E., Chu, W. K., Aselage, T. L., & Picraux, S. T. (1986). Initial decomposition of GaAs during rapid thermal annealing. *Applied Physics Letters*, *49*(11), 666–668. doi:10.1063/1.97562

Hepbasli, A. (2008). A key review on exergetic analysis and assessment of renewable energy resources for a sustainable future. *Renewable & Sustainable Energy Reviews*, *12*(3), 593–661. doi:10.1016/j.rser.2006.10.001

HongmeiT.FernandoM. D.KevinE.PeterJ.EduardM. (2012). Retrieved from http://www.nrel.gov/docs/fy12osti/54601.pdf

Houston, J. M. (1959). Theoretical efficiency of the thermionic energy converter. *Journal of Applied Physics*, *30*(4), 481–487. doi:10.1063/1.1702392

Hsiao, Y. Y., Chang, W. C., & Chen, S. L. (2010). A mathematic model of thermoelectric module with applications on waste heat recovery from automobile engine. *Energy*, *35*(3), 1447–1454. doi:10.1016/j.energy.2009.11.030

Hua, C., & Shen, C. (1998, May). Study of maximum power tracking techniques and control of DC/DC converters for photovoltaic power system. In *Power Electronics Specialists Conference, 1998. PESC 98 Record. 29th Annual IEEE* (Vol. 1, pp. 86-93). IEEE.

Humphrey, T. E., O'Dwyer, M. F., & Linke, H. (2005). Power optimization in thermionic devices. *Journal of Physics. D, Applied Physics*, *38*(12), 2051–2054. doi:10.1088/0022-3727/38/12/029

Islam, M., Inal, O. T., & Luke, J. R. (2006). Development of electron reflection suppression materials for improved thermionic energy converter performance using thin film deposition techniques. *Journal of Applied Physics*, *100*(8), 084903. doi:10.1063/1.2356785

Ito, T., & Cappelli, M. A. (2012). Optically pumped cesium plasma neutralization of space charge in photon-enhanced thermionic energy converters. *Applied Physics Letters*, *101*(21), 213901. doi:10.1063/1.4767349

Jensen, K. L. (2007). General formulation of thermal, field, and photoinduced electron emission. *Journal of Applied Physics*, *102*(2), 024911. doi:10.1063/1.2752122

Jeter, S. M. (1981). Maximum conversion efficiency for the utilization of direct solar radiation. *Solar Energy*, *26*(3), 231–236. doi:10.1016/0038-092X(81)90207-3

Joshi, A. S., Dincer, I., & Reddy, B. V. (2009). Thermodynamic assessment of photovoltaic systems. *Solar Energy*, *83*(8), 1139–1149. doi:10.1016/j.solener.2009.01.011

Kinsey, G. S., Hebert, P., Barbour, K. E., Krut, D. D., Cotal, H. L., & Sherif, R. A. (2008). Concentrator multijunction solar cell characteristics under variable intensity and temperature. *Progress in Photovoltaics: Research and Applications*, *16*(6), 503–508. doi:10.1002/pip.834

Kjaer, S. B., Pedersen, J. K., & Blaabjerg, F. (2005). A review of single-phase grid-connected inverters for photovoltaic modules. *IEEE Transactions on Industry Applications*, *41*(5), 1292–1306. doi:10.1109/TIA.2005.853371

Kotas, T. J. (1995). *The Exergy Method of Thermal Plant Analysis*. Malabar, FL: Krieger Publish Company.

Kraemer, D., Poudel, B., Feng, H. P., Caylor, J. C., Yu, B., Yan, X., ... McEnaney, K. (2011). High-performance flat-panel solar thermoelectric generators with high thermal concentration. *Nature Materials*, *10*(7), 532–538. doi:10.1038/nmat3013 PMID:21532584

Kreith, F. (1974). *Principles of heat transfer* (3rd ed.). Scranton: International Textbook Company.

Kribus, A., & Mittelman, G. (2008). Potential of polygeneration with solar thermal and photovoltaic systems. *Journal of Solar Energy Engineering*, *130*(1), 011001. doi:10.1115/1.2804618

Landsberg, P. T., & Tonge, G. (1980). Thermodynamic energy conversion efficiencies. *Journal of Applied Physics*, *51*(7), R1–R20. doi:10.1063/1.328187

Larson, D. L. (1995). Exergy analysis: Essential to effective energy management. *Transactions of the ASAE. American Society of Agricultural Engineers*, *38*(4), 1173–1178. doi:10.13031/2013.27937

Lee, J. H., Bargatin, I., Melosh, N. A., & Howe, R. T. (2012). Optimal emitter-collector gap for thermionic energy converters. *Applied Physics Letters*, *100*(17), 173904. doi:10.1063/1.4707379

Lee, J. I., Jeong, Y. H., No, H. C., Hannebauer, R., & Yoo, S. K. (2009). Size effect of nanometer vacuum gap thermionic power conversion device with CsI coated graphite electrodes. *Applied Physics Letters*, *95*(22), 223107. doi:10.1063/1.3266921

Luo, X., Li, C., Liu, N., Li, R., He, J., & Qiu, T. (2013). The impact of energy spectrum width in the energy selective electron low-temperature thermionic heat engine at maximum power. *Physics Letters. [Part A]*, *377*(25), 1566–1570. doi:10.1016/j.physleta.2013.04.045

Luque, A., & Hegedus, S. (2003). *Handbook of photovoltaic science and engineering*. John Wiley & Sons. doi:10.1002/0470014008

Luque, A., & Martí, A. (1999). Limiting efficiency of coupled thermal and photovoltaic converters. *Solar Energy Materials and Solar Cells*, *58*(2), 147–165. doi:10.1016/S0927-0248(98)00199-8

Mahan, G. D. (1994). Thermionic refrigeration. *Journal of Applied Physics*, *76*(7), 4362–4366. doi:10.1063/1.357324

Mahan, G. D., Sofo, J. O., & Bartkowiak, M. (1998). Multilayer thermionic refrigerator and generator. *Journal of Applied Physics*, *83*(9), 4683–4689. doi:10.1063/1.367255

Marti, A., & Araújo, G. L. (1996). Limiting efficiencies for photovoltaic energy conversion in multigap systems. *Solar Energy Materials and Solar Cells*, *43*(2), 203–222. doi:10.1016/0927-0248(96)00015-3

Meillaud, F., Shah, A., Droz, C., Vallat-Sauvain, E., & Miazza, C. (2006). Efficiency limits for single-junction and tandem solar cells. *Solar Energy Materials and Solar Cells*, *90*(18), 2952–2959. doi:10.1016/j.solmat.2006.06.002

Melosh, N. (2013). *Photon Enhanced Thermionic Emission for Solar Energy Harvesting Final Report to the Global Climate and Energy Project* (No. SLAC-PUB-15455). SLAC National Accelerator Laboratory (SLAC).

Meng, J. H., Zhang, X. X., & Wang, X. D. (2014). Multi-objective and multi-parameter optimization of a thermoelectric generator module. *Energy*, *71*, 367–376. doi:10.1016/j.energy.2014.04.082

Messenger, R. A., & Ventre, J. (2000). Photovoltaic systems engineering. CRC Press.

Murphy, E. L., & Good, R. H. Jr. (1956). Thermionic emission, field emission, and the transition region. *Physical Review*, *102*(6), 1464–1473. doi:10.1103/PhysRev.102.1464

Myrzik, J. M., & Calais, M. (2003, June). String and module integrated inverters for single-phase grid connected photovoltaic systems-a review. In *Power Tech Conference Proceedings, 2003 IEEE Bologna* (Vol. 2). IEEE. 10.1109/PTC.2003.1304589

National Research Council. (2001). *Thermionics Quo Vadis? An Assessment of the DTRA's Advanced Thermionics Research and Development Program*. National Academies Press.

O'Dwyer, M. F., Lewis, R. A., Zhang, C., & Humphrey, T. E. (2005). Electronic efficiency in nanostructured thermionic and thermoelectric devices. *Physical Review B: Condensed Matter and Materials Physics*, *72*(20), 205330. doi:10.1103/PhysRevB.72.205330

Petela, R. (1964). Exergy of heat radiation. *ASME J. Heat Transfer*, *86*(2), 187–192. doi:10.1115/1.3687092

Petela, R. (2003). Exergy of undiluted thermal radiation. *Solar Energy*, *74*(6), 469–488. doi:10.1016/S0038-092X(03)00226-3

Phang, J. C. H., Chan, D. S. H., & Phillips, J. R. (1984). Accurate analytical method for the extraction of solar cell model parameters. *Electronics Letters*, *20*(10), 406–408. doi:10.1049/el:19840281

Qu, M., Yin, H., & Archer, D. H. (2010). A solar thermal cooling and heating system for a building: Experimental and model based performance analysis and design. *Solar Energy*, *84*(2), 166–182. doi:10.1016/j.solener.2009.10.010

Radziemska, E. (2003). Thermal performance of Si and GaAs based solar cells and modules: A review. *Progress in Energy and Combustion Science*, *29*(5), 407–424. doi:10.1016/S0360-1285(03)00032-7

Reck, K., & Hansen, O. (2014). Thermodynamics of photon-enhanced thermionic emission solar cells. *Applied Physics Letters*, *104*(2), 023902. doi:10.1063/1.4862535

Riesz, F., Dobos, L., & Karanyi, J. (1998). Thermal decomposition of bulk and heteroepitaxial (100) InP surfaces: A combined in situ scanning electron microscopy and mass spectrometric study. *Journal of Vacuum Science & Technology. B, Microelectronics and Nanometer Structures: Processing, Measurement, and Phenomena: An Official Journal of the American Vacuum Society*, *16*(5), 2672–2674.

Rosen, M. A. (1999). Second-law analysis: Approaches and implications. *International Journal of Energy Research*, *23*(5), 415–429. doi:10.1002/(SICI)1099-114X(199904)23:5<415::AID-ER4893.0.CO;2-7

Rosen, M. A., & Dincer, I. (1999). Thermal storage and exergy analysis: the impact of stratification. *Transactions of the Canadian Society for Mechanical Engineering, 23*(1B), 173-186.

Rosen, M. A., & Hooper, F. C. (July, 1996). Second Law Analysis of Thermal Energy Storage Systems. In *Proceedings of the First Trabzon International Energy and Environment Symposium* (pp. 361–371). Trabzon, Turkey: Academic Press.

Rosen, M. A., Hooper, F. C., & Barbaris, L. N. (1988). Exergy analysis for the evaluation of the performance of closed thermal energy storage systems. *Journal of Solar Energy Engineering, 110*(4), 255–261. doi:10.1115/1.3268265

Sahin, A. D., Dincer, I., & Rosen, M. A. (2007). Thermodynamic analysis of solar photovoltaic cell systems. *Solar Energy Materials and Solar Cells, 91*(2), 153–159. doi:10.1016/j.solmat.2006.07.015

Sarhaddi, F., Farahat, S., Ajam, H., & Behzadmehr, A. (2010). Exergetic performance evaluation of a solar photovoltaic (PV) array. *Australian Journal of Basic and Applied Sciences, 4*(3), 502–519.

Schultz, O., Glunz, S. W., & Willeke, G. P. (2004). Multicrystalline silicon solar cells exceeding 20% efficiency. *Progress in Photovoltaics: Research and Applications, 12*(7), 553–558. doi:10.1002/pip.583

Schwede, J. W., Bargatin, I., Riley, D. C., Hardin, B. E., Rosenthal, S. J., Sun, Y., ... Melosh, N. A. (2010). Photon-enhanced thermionic emission for solar concentrator systems. *Nature Materials, 9*(9), 762–767. doi:10.1038/nmat2814 PMID:20676086

Schwede, J. W., Sarmiento, T., Narasimhan, V. K., Rosenthal, S. J., Riley, D. C., Schmitt, F., & Melosh, N. A. (2013). Photon-enhanced thermionic emission from heterostructures with low interface recombination. *Nature Communications, 4*, 1576. doi:10.1038/ncomms2577 PMID:23481384

Segev, G., Kribus, A., & Rosenwaks, Y. (2013). High performance isothermal photo-thermionic solar converters. *Solar Energy Materials and Solar Cells, 113*, 114–123. doi:10.1016/j.solmat.2013.02.002

Segev, G., Rosenwaks, Y., & Kribus, A. (2012). Efficiency of photon enhanced thermionic emission solar converters. *Solar Energy Materials and Solar Cells, 107*, 125–130. doi:10.1016/j.solmat.2012.08.006

Segev, G., Rosenwaks, Y., & Kribus, A. (2013). Loss mechanisms and back surface field effect in photon enhanced thermionic emission converters. *Journal of Applied Physics, 114*(4), 044505. doi:10.1063/1.4816256

Segev, G., Rosenwaks, Y., & Kribus, A. (2015). Limit of efficiency for photon-enhanced thermionic emission vs. photovoltaic and thermal conversion. *Solar Energy Materials and Solar Cells, 140*, 464–476. doi:10.1016/j.solmat.2015.05.001

Shah, A. V., Meier, J., Vallat-Sauvain, E., Wyrsch, N., Kroll, U., Droz, C., & Graf, U. (2003). Material and solar cell research in microcrystalline silicon. *Solar Energy Materials and Solar Cells, 78*(1), 469–491. doi:10.1016/S0927-0248(02)00448-8

Shockley, W., & Queisser, H. J. (1961). Detailed balance limit of efficiency of p-n junction solar cells. *Journal of Applied Physics, 32*(3), 510–519. doi:10.1063/1.1736034

Singh, P., & Ravindra, N. M. (2012). Temperature dependence of solar cell performance—an analysis. *Solar Energy Materials and Solar Cells, 101*, 36–45. doi:10.1016/j.solmat.2012.02.019

Smestad, G. P. (2004). Conversion of heat and light simultaneously using a vacuum photodiode and the thermionic and photoelectric effects. *Solar Energy Materials and Solar Cells, 82*(1), 227–240. doi:10.1016/j.solmat.2004.01.020

Smestad, G. P., Krebs, F. C., Lampert, C. M., Granqvist, C. G., Chopra, K. L., Mathew, X., & Takakura, H. (2008). Reporting solar cell efficiencies in solar energy materials and solar cells. *Solar Energy Materials and Solar Cells*, *92*(4), 371–373. doi:10.1016/j.solmat.2008.01.003

Smith, J. R., Bilbro, G. L., & Nemanich, R. J. (2009). Theory of space charge limited regime of thermionic energy converter with negative electron affinity emitter. *Journal of Vacuum Science & Technology. B, Microelectronics and Nanometer Structures: Processing, Measurement, and Phenomena: An Official Journal of the American Vacuum Society*, *27*(3), 1132–1141.

Spanner, D. C. (1964). *Introduction to thermodynamics*. London: Academic Press.

Su, S., Wang, Y., Liu, T., Su, G., & Chen, J. (2014). Space charge effects on the maximum efficiency and parametric design of a photon-enhanced thermionic solar cell. *Solar Energy Materials and Solar Cells*, *121*, 137–143. doi:10.1016/j.solmat.2013.11.008

Su, S., Wang, Y., Wang, J., Xu, Z., & Chen, J. (2014). Material optimum choices and parametric design strategies of a photon-enhanced solar cell hybrid system. *Solar Energy Materials and Solar Cells*, *128*, 112–118. doi:10.1016/j.solmat.2014.04.038

Su, S., Zhang, H., Chen, X., Kang, J., & Chen, J. (2013). Parametric optimum design of a photon-enhanced thermionic solar cell. *Solar Energy Materials and Solar Cells*, *117*, 219–224. doi:10.1016/j.solmat.2013.05.001

Sze, S. M. (1981). *Physics of Semiconductor Devices* (2nd ed.). New York: John Wiley.

Tavkhelidze, A. N. (2010). Nanostructured electrodes for thermionic and thermotunnel devices. *Journal of Applied Physics*, *108*(4), 044313. doi:10.1063/1.3464256

Tersoff, J., Jesson, D. E., & Tang, W. X. (2009). Running droplets of gallium from evaporation of gallium arsenide. *Science*, *324*(5924), 236–238. doi:10.1126cience.1169546 PMID:19359583

Tsai, H. L., Tu, C. S., & Su, Y. J. (2008, October). Development of generalized photovoltaic model using MATLAB/SIMULINK. In *Proceedings of the world congress on Engineering and computer science* (*Vol. 2008*, pp. 1-6). Academic Press.

Varpula, A., & Prunnila, M. (2012). Diffusion-emission theory of photon enhanced thermionic emission solar energy harvesters. *Journal of Applied Physics*, *112*(4), 044506. doi:10.1063/1.4747905

Varpula, A., Tappura, K., & Prunnila, M. (2015). Si, GaAs, and InP as cathode materials for photon-enhanced thermionic emission solar cells. *Solar Energy Materials and Solar Cells*, *134*, 351–358. doi:10.1016/j.solmat.2014.12.021

Wang, Y., Dai, C., & Wang, S. (2013). Theoretical analysis of a thermoelectric generator using exhaust gas of vehicles as heat source. *Applied Energy*, *112*, 1171–1180. doi:10.1016/j.apenergy.2013.01.018

Wang, Y., Su, S., Lin, B., & Chen, J. (2013). Parametric design criteria of an irreversible vacuum thermionic generator. *Journal of Applied Physics*, *114*(5), 053502. doi:10.1063/1.4817084

Wang, Y., Su, S., Liu, T., Su, G., & Chen, J. (2015). Performance evaluation and parametric optimum design of an updated thermionic-thermoelectric generator hybrid system. *Energy*, *90*, 1575–1583. doi:10.1016/j.energy.2015.06.120

Wasynczuk, O. (1989). Modeling and dynamic performance of a line-commutated photovoltaic inverter system. *IEEE Transactions on Energy Conversion*, *4*(3), 337–343. doi:10.1109/60.43233

Wasynezuk, O. (1983). Dynamic behavior of a class of photovoltaic power systems. *IEEE Transactions on Power Apparatus and Systems*, *PAS-102*(9), 3031–3037. doi:10.1109/TPAS.1983.318109

Watmuff, J. H., Charters, W. W. S., & Proctor, D. (1977). *Solar and wind induced external coefficients for solar collectors. Comples, 2*. Marseille: Rev. Inter. Heliotech.

Werner, J. H., Kolodinski, S., & Queisser, H. J. (1994). Novel optimization principles and efficiency limits for semiconductor solar cells. *Physical Review Letters*, *72*(24), 3851–3854. doi:10.1103/PhysRevLett.72.3851 PMID:10056313

Wilson, V. C. (1959). Conversion of heat to electricity by thermionic emission. *Journal of Applied Physics*, *30*(4), 475–481. doi:10.1063/1.1702391

Wojtczuk, S., Chiu, P., Zhang, X., Derkacs, D., Harris, C., Pulver, D., & Timmons, M. (2010, June). InGaP/GaAs/InGaAs 41% concentrator cells using bi-facial epigrowth. In *Photovoltaic Specialists Conference (PVSC), 2010 35th IEEE* (pp. 001259-001264). IEEE. 10.1109/PVSC.2010.5614196

Wong, K. F. V. (2000). *Thermodynamics for Engineers*. Boca Raton, FL: CRC Press LLC.

Wu, C. (1992). Optimal power from a radiating solar-powered thermionic engine. *Energy Conversion and Management*, *33*(4), 279–282. doi:10.1016/0196-8904(92)90119-H

Würfel, P. (2009). *Physics of solar cells: from basic principles to advanced concepts* (2nd ed.). Wiley-VCH.

Xuan, X. C., & Li, D. (2003). Optimization of a combined thermionic–thermoelectric generator. *Journal of Power Sources*, *115*(1), 167–170. doi:10.1016/S0378-7753(02)00620-1

Yaghoobi, P., Vahdani Moghaddam, M., & Nojeh, A. (2012). Solar electron source and thermionic solar cell. *AIP Advances*, *2*(4), 042139. doi:10.1063/1.4766942

Zeng, T. (2006). Thermionic-tunneling multilayer nanostructures for power generation. *Applied Physics Letters*, *88*(15), 153104. doi:10.1063/1.2192973

ADDITIONAL READING

Bejan, A., & Mamut, E. (Eds.). (2012). *Thermodynamic optimization of complex energy systems* (Vol. 69). Springer Science & Business Media.

Borel, L., & Favrat, D. (2010). *Thermodynamics and energy systems analysis: From Energy to Exergy*. EPFL press.

Dincer, I., & Rosen, M. A. (2012). *Exergy: energy, environment and sustainable development*. Newnes.

Häberlin, H. (2012). *Photovoltaics system design and practice*. John Wiley & Sons. doi:10.1002/9781119976998

Kribus, A., & Segev, G. (2016). Solar energy conversion with photon-enhanced thermionic emission. *Journal of Optics*, *18*(7), 073001. doi:10.1088/2040-8978/18/7/073001

Szargut, J., Morris, D. R., & Steward, F. R. (1988). Energy analysis of thermal, chemical, and metallurgical processes.

KEY TERMS AND DEFINITIONS

Absorption Coefficient: It describes how far the light of a particular wavelength can penetrate into a material of a given thickness before it is absorbed by the material.

Carnot Efficiency: It describes the maximum thermal efficiency that a heat engine can achieve as permitted by the second law of thermodynamics.

Current Density: The amount of electric current travelling per unit cross-section area is called as current density.

Diffusion Length: It is the average distance that a carrier covers between generation and recombination.

Electron Affinity: Electron affinity of a semiconductor is the minimum energy required to move an electron from the conduction band bottom to the vacuum level.

Energy Band Gap: It is defined as the minimum energy required to excite an electron up to a state in the conduction band where it can participate in conduction.

Fermi Level: the energy level (if it exists) which has a 50% probability of getting occupied by an electron for the given temperature of the solid and at absolute zero temperature occupancy is 100%.

Ionization Energy: It is the minimum energy required to remove an electron from the ground state of an atom.

Mobility: It is the characteristic of an electron to move through a metal or semiconductor upon application of electric field.

Photon: A photon is the quantum of electromagnetic radiation. A photon is a particle of light defined as a discrete bundle or quantum of electromagnetic radiation or light.

Radiative Recombination: When an electron from the conduction band directly combines with a hole in the valence band and releases a photon, it is called radiative recombination.

Vacuum Level: It is the energy of a free electron (an electron outside the semiconductor) which is at rest with respect to the semiconductor.

Work Function: It is defined as the minimum amount of thermodynamic work or energy required to remove an electron from a solid to a point in the vacuum immediately outside the solid surface.

APPENDIX

The constants required to calculate the cathode and anode temperatures of the irreversible vacuum thermionic generator, derived by the authors are given as.

$$k_1 = \left[\frac{A_0 \sigma \varepsilon_{in}}{U_H A_H} + \frac{\varepsilon \sigma}{U_H}\right] \tag{101}$$

$$k_2 = \left[\frac{2 k_B K_{RD} A_0}{U_H A_H}\right] \tag{102}$$

$$k_3 = \left[\frac{K_{RD} A_0}{q U_H A_H}\right] \tag{103}$$

$$k_4 = \left[\frac{A_0 \sigma \varepsilon_{in}}{U_H A_H}\right] \tag{104}$$

$$k_5 = \left[T_H + \frac{\varepsilon \sigma T_H^4}{U_H}\right] \tag{105}$$

$$k_6 = \left[\frac{A_0 \sigma \varepsilon_{in}}{U_L A_L} + \frac{\varepsilon \sigma}{U_L}\right] \tag{106}$$

$$k_7 = \left[\frac{2 k_B K_{RD} A_0}{U_L A_L}\right] \tag{107}$$

$$k_8 = \left[\frac{K_{RD} A_0}{q U_L A_L}\right] \tag{108}$$

$$k_9 = \left[\frac{\varepsilon_{in} A_0}{U_L A_L}\right] \tag{109}$$

$$k_{10} = \left[T_C + \frac{\varepsilon \sigma T_C^4}{U_L}\right] \tag{110}$$

Chapter 13
Demand-Side Management:
Energy Efficiency and Demand Response

Pawan Kumar
Thapar University, India

Ikbal Ali
Jamia Millia Islamia, India

Dip V. Thanki
Lovely Professional University, India

ABSTRACT

Growing demands are causing increased pressure on the electrical infrastructure and perpetually escalated energy prices. Utilities around the world have been considering demand-side management in their strategic planning. The costs of constructing and operating a new capacity generation unit are increasing every day as well as transmission, distribution, and land issues for new generation plants, which force the utilities to search for other alternatives. Here, demand-side management has been implemented as it is less expensive to intelligently influence a load than to build a new power plant or install electrical based storage device. In this chapter, the author has discussed energy efficiency and demand response fulfilling the criteria of energy management which usually tries to take influence onto the energy consumption of a number of energy consumers. The explained demand-side management technical objectives are peak clipping, valley filling, load shifting, load building, energy conservation, and flexible load shape.

INTRODUCTION

The chapter gives an overview of demand side management (DSM) which reduces electricity utilization through programs that promote conservation or more efficient load management. It is basically a portfolio of selection, planning, and implementation of measures intended to have an influence on the demand or customer side of the electric meter, either caused directly or stimulated indirectly by the utility. DSM programs are classified into EE and DR programs. Energy audit and load research is an important part of DSM for precise understanding. Load variations in electrical consumption patterns

DOI: 10.4018/978-1-5225-3935-3.ch013

have been described with an example of a commercial body. Different technical load-shape objectives are defined which can be used to alter the load duration. The main objective of this chapter is to understand DSM in broad perspectives in order to modify the energy consumption of different customers by various programs based on load management.

BACKGROUND

DSM involves planning, implementation, evaluation and activity monitoring which can be selected from a variety of technical alternatives. The wide range of different technical alternatives mandates DSM as there are a large number of technical alternatives and there is a wide variety of choice for appropriate alternative selection. The selection of a suitable alternative is little difficult especially when the alternatives are influenced in a strong manner due to the specific local and regional factors. These factors include mix generation, expected growth in load demand, plans for expanding capacity, reserve margins, climate regulation, load factor, and shape of the load for normal and extreme conditions. As a result, transfer of the above factors from one region to another is inappropriate without proper adjustments.

DSM is the planning and implementation of those activities that are designed to modify the energy consumption that would further reflect desired changes in energy magnitude and time pattern. A program of DSM encompasses load management, new uses, strategic conservation, electrification, distributed generation and market share adjustments of energy consuming appliances. DSM extends beyond load management and conservation of load to design programs particularly to change energy use in both on-peak and off-peak time periods. Cost control or customer options include a fair partnership between energy service supplier and the customers. But ultimately, it is the customer who decides whether to participate in a program or not. Load managing programs totally focus on the reduction on consumer's premises at the time of the rise in utility system loads. The main aspiration is to avoid buildup of generation, production and the delivery facilities that would be required for a small time operation in hours per year.

On the other hand, EE focuses on the annual reduction of loads but load management techniques target specific moments and it has fewer impacts on the total energy consumption. Load management programs are proven to be more stable and reliable over time periods than any other customer programs. The main components of load management activities are direct load control, pricing based choices, and thermal energy storage. There were several important objectives that were to identify the technologies that were efficient in peak clipping, valley filling and load shifting and shift them in market places. Moreover, the conservation efforts were also met that could focus on annual load reduction.

The decrease in the energy demand and consumption in power distribution systems, can free up generation, transmission and distribution capacity. Sinha *et al.* (2011) observed that with DSM measures the cost of energy saved has been estimated to be as low as 10% of the cost of the added capacity. For a more efficient future planning process and better service, there should be support for EE at customer's installations which also bring the utility in closer contact with the consumers. It does encourage the consumers to make decisions prior to their energy consumption contributing to energy providers to reduce the maximum load demand. Hence, the reshaping of load profile can be done in an effective manner (Esther, Priya, and Kumar, 2016). Also, Kumar et al. (2015) have developed a simulation model of the synchronized operation of the grid power, solar power and battery for energy management at home level.

To lessen the system instabilities caused due to the increasing electricity demand, an appropriate objective of DSM activities can modify the shape of load curve (Logenthiran, 2011). This can be achieved by total load demand reduction of power distribution system during maximum load periods. Further, the load demand curve can be modified by shifting the loads to be served during appropriate time periods in order to minimize overall planning and operational cost of the network (Logenthiran, 2012). Implementation of DSM leads to greater flexibility in facing rapid change in today's business environment. DSM is not a complete solution for all the problems related to the energy industry and its stakeholders but it does provide additional alternatives for meeting future challenges. Rather than using hardware/technology as the driving force behind programs, DSM is an important strategy in the market that particularly promotes focus of the consumers.

Kumar et al. (2011, 2013, 2016 and 2017) presented a detailed analysis of the energy efficiency under different load models which are voltage dependent. In these studies author have observed that the system voltage profile not only affect the power loss rather it majorly changes the system loadability and load profile and hence it has direct effect on the DSM schemes. The fundamental explanation behind executing DSM in the power segment is that it is more financially cost-effective and socially useful to oversee power demand. It is done through interest in efficiency and other demand-side measures than to expand power supply or transmission limit (Abogaleela, Elsobki, and El Marsafawy, 2012). Different projects for DSM are drilled to dispose of or moderate the desire for extra peak or base-load generating limit as well as distribution facilities. It additionally grants existing generation to satisfy the requirements of a bigger variety of customers to reduce the requirement for a brand new capacity. On the other hand, in the ever growing demand scenario a comprehensive stability analysis, reactive power injection, loss minimization under different loading pattern for distribution system are appears in the work presented by Kumar P. and Singh S., (2014a, 2014b, 2014c)

Utilities, after all, can profit by these mitigations or moves in consumer electrical consumption. For a few utilities, DSM programs will facilitate them cut back their peak power purchases on the discount advertise, in this manner bringing down their general price of the operations. Within the short term, DSM programs will cut back energy prices for utilities, and within the long run, DSM programs will facilitate limit the necessity for utilities to make new power plants, distribution, and transmission lines. To put it plainly, a DSM program is less costly to actualize than building a brand new generating plant. (Abogaleela, Elsobki and ElMarsafawy, 2012).

DSM is a technique executed to control the load-profile by implication keeping in mind the end goals to accomplish. These goals are to have the load-factor near to unity and to include the maximum load inside the framework limit. By accomplishing these precedent objectives, the utility would get the greatest conceivable energy from the introduced units, which further maximizes the aggregate benefit and minimizes the normal cost per kWh. On occasion visible as a separate category, electricity conservation will be seen as a part of DSM. Energy conservation focuses on users and behavioral modifications to reap more efficient energy usage (Palensky and Dietrich, 2011). Ali *et al.* (2015, 2016) and Kumar *et al.* (2014a, 2014b, 2014c and 2017) shows that loads are voltage dependent and they vary with time of the day, season and state of economy. Therefore, combining DSM and Time of Use price lists drastically increases safety, lowers the price and the emissions of power system. However, dynamic DSM doesn't always reduce energy consumption rather affects the consumption patterns. In this scenario, if a procedure is interrupted for some motive it would have to seize up once it is permitted. An instance is a water pumping machine that is easily shed for thirty minutes due to its storage tanks.

Evaluating the various flavors of DSM, EE is the most required one. It saves energy and emissions, while most of the other alternative type simply shifts the load in different time-periods. So, the primary objective is to improve efficiency and then the dynamics can be optimized. Depending at the boundary situations of the machine, one dynamic DSM method is probably chosen. Evidently, the ruin-even point of automation investments and financial incentives determines how some distance you'll be able to move.

Loads that can react to cost or reliability signs, are alluded to as demand response (DR) which brings down the energy demand amid times of high costs or the requirement for generation capacity amid times of high load. Therefore, network administrators are investigating the utilization of DR for subordinate services like spinning reserve. Demand response (DR) for spinning reserve might be suitable for clients whose operational limitations block cooperation in energy and limit DR programs. The inherent attributes of DR are a characteristic match to the prerequisites of spinning reserve assets. Load assets can give higher incline rates at lower costs than traditional generation. Moreover, numerous loads that are individually less reliable than a generator may give total reliability in excess of that provided by a couple of huge generators. The timescale on which spinning reserve works is very much served by DR (Palensky and Dietrich, 2011).

However, modern DSM tools can switch off non-essential loads quickly for short time and with the reduction in costs of storage batteries having storage improves the amount of renewable that can accommodated without much spinning reserve. Designing the system properly with Energy Conservation is a must at the design state. Arora and Kumar, (2012), have presented a sample infrastructural development of energy planning and energy management for smart grid implementation for future requirements and energy management.

SCOPE OF DSM

Figure 1 represents the scope of DSM as a mutual benefit for the customer and the utility. DSM involves systematic utility and government activities which are designed to alter the amount and timing of the customer demand for the collective benefits to the society. Modifications in the ultimate load profile will rely on the planning intentions and utility company operations (Logenthiran, Srinivasan and Shun (2012)). In addition, it also changes the electricity consumption patterns of the customers to produce the desired variation in load shape curve of the end user premises for enhancing the load -profile.

Depending on the timing and the effect of the applied measures on the consumer process, DSM program can be classified into:

1. **Energy Efficiency (EE) Programs:** Which are intended to lessen power utilization during the time by concentrating on diminishing energy utilization and overall demand.

Figure 1. Scope of DSM

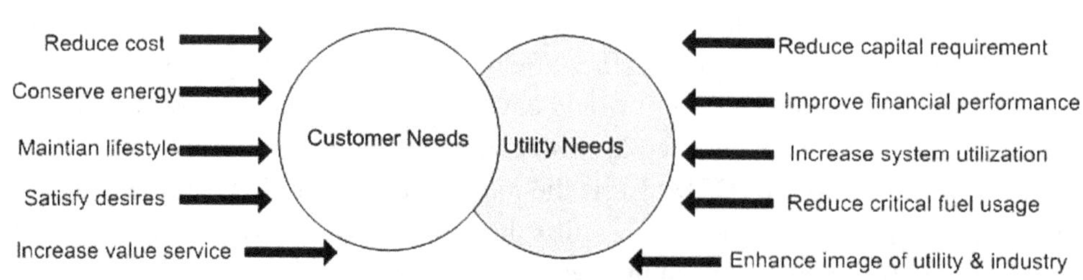

2. **Demand Response (DR) Programs:** Which are programmed with a procedure unit having the privilege to limit or close down bound apparatuses (e.g. ventilation systems, pool pumps, clothes washers, and so forth) for a concise measure of time at consumer's premises.
3. **Time of Use (ToU):** Is the strategy in which utilities have different charges for electrical consumption during different time of operation.
4. **Spinning Reserve (SR):** Which is the reserve capacity that is on-line and can be made available within a 10 minute period in order to avoid cut-off of the electricity supplied by the system.

From the figure, it is observed that the bottom edge of the DSM spectrum represents the EE measures. They include all the permanent changes on system like replacing inefficient ventilation system with a better one or investment inside the building by adding extra insulation. Such measures bring about instant and everlasting electricity and emissions savings and are therefore the maximum welcome method.

ENERGY EFFICIENCY MEASURES

Energy efficiency (EE) in electrical appliances is comprehended to mean the use of energy in the cost effective manner such that energy waste is limited and the overall utilization of prime energy assets is decreased without affecting the required output level. In other words, energy efficient practices or frameworks will try to utilize less energy while directing any energy-dependent activity and at the same time, the negative ecological effects of energy consumption are limited. The desired objectives for EE are reaching past simply specialized solutions and go significantly further in promoting cost-viability, financing, acknowledgment, development and ecological effect evaluation.

In practice, EE is a wide term and it can be described in different manner where the particular definition relies upon the specific circumstance and in the way it is utilized it speaks to a proportion of energy output to energy input.

The various institutional, monetary and specialized barriers to EE programs, either genuine or perceived, include:

- Policy and administrative obstructions;
- Lack of data and consciousness of the potential for EE;

Figure 2. Impact of improved EE vs. DR

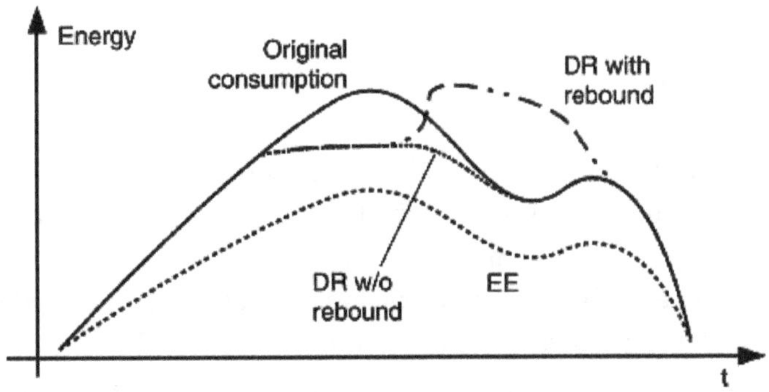

- Lack of industry activities to stress energy management as a basic piece of aggregate management frameworks;
- Lack of specialized ability to distinguish, evaluate, create and execute EE ventures;
- Financial and speculation boundaries;
- Technological boundaries.

Efficient technical solutions widely available under different applications are described as:

Lighting

Lighting speaks to more than 35% of energy utilization in structures relying upon the business. It is considered as one of the easiest demanding approaches to save the energy costs for low speculation and is a standout amongst the most well-known energy saving measures.

Good illumination is essential for everyday tasks. For several applications, natural lighting is insufficient and needs to be increased with artificial lighting systems. Consequently, lighting systems are widely utilized in buildings, and they account for a considerable portion of total building energy consumption. It is more important than ever that lighting systems are designed to provide for optimum energy use, furthermore to give human comfort and productiveness. Proper planning and design allows the efficient operation of new and the existing lighting system. Regularly expending and consuming at least 10% of modern plant power, lighting ordinarily offers great opportunities for savings. A portion of the things recorded underneath will likewise apply to the household segment:

- Use energy proficient light sources;
- Utilizing energy effective electronic balances;
- Clean illuminating presences routinely;
- Use proper lighting levels for various parts of the work territory;
- Install lights at working level where conceivable, not really on a high roof;
- Use translucent rooftop boards or sky facing windows;
- Paint dividers and roofs white for better light reflection.

Expanded utilization of natural and task lighting has been appeared to build efficiency in institutions and workplaces. Recently, light emitting diode based lighting systems utilize 75% less energy and may last 10 to 20 times longer than incandescent light bulbs. They create a characteristic light and in many applications they are cost effective despite their higher initial cost with payback periods as low as a couple of months.

Space Cooling and Space Heating

Space cooling and space heating comprise the majority of energy consumption in buildings particularly in industrialized and cooled countries. However, the breakdown of energy use for heating, ventilation, and cooling vary significantly with the climate, building characteristics, use and economic conditions. For instance, northern areas in Republic of India have higher heating demands while southern areas need a lot of cooling in different environmental conditions. Additionally, the well-insulated building might have nominal heating and cooling and they may require a bigger portion of forced ventilation.

On the other hand, for residential buildings, the house space conditioning needs generally arise from the exterior environment. The ways in which this can be done in an appropriate manner is by keeping good measure and control relating to improve maintenance and operation of existing equipment. This option is inexpensive and easy to implement or it can be replaced with high- efficiency alternatives on modifying existing equipment to optimize efficiency. However, this is expensive and complex to implement but can result in a higher energy saving scheme. There are additionally area warming and cooling systems which can utilize city wastes. For instance, replacing an older 50% efficient natural gas furnace with a new 95% productive one will significantly decrease energy utilization, carbon outflows, and winter gas bills.

Refrigeration

This is an appliance which is always on and does contribute to base and peak load. In India, five-star units result in noteworthy decrease of the peak and also general energy utilization. There are two sorts of units on advertise:

- Direct cool type
- Frost free type

Direct cool type is proven more energy efficient as compared to frost free type whereas, the choice of the type depends on customers operation. The customers using frost-free type clients more energy for same limits due to programmed defrost instrument.

The issue related to refrigeration is that utilities execute pilot programs for supplanting of old units with five-star units however clients are slanted to move for the twofold door units, frost-free of higher limit which some leads to higher energy utilization. In this situation, the replacement of the refrigerators with five-star refrigerators is proven more energy efficient. However, commercial refrigeration refers to the use of the refrigeration process in course of profit-seeking businesses like multiplexes and food-points operating at a large scale. Non-commercial refrigeration is the use of refrigerators in either a small office or in a house where food-service is not the industry.

Water Heating

Hot water used in the domestic households, industrial sector and commercial facilities for various purposes. Therefore, choice of hot water technology differs for various load categories in terms of its ability to fulfill the capacity requirements as well as cost effectiveness. Electric water heaters have a major share in the water heater market for residential and commercial sector. With availability of energy efficient alternatives for water heating in market, there exists a potential for energy savings in this end use category that can be tapped to replace the energy intensive electric water heaters.

The monthly water heating bills can be reduced by selecting the appropriate water heater for home or pool and by using some energy-efficient water heating strategies. Some simple projects

are insulating hot water pipes and lowering water heating temperature which also saves money and energy on water heating.

An energy efficient hot water heater is correctly sized for household, well positioned, and appropriate according to the climate and it has the potential to reduce household energy bills significantly.

The energy efficient hot water heaters can be chosen based upon the following,

- Fuel source
- Household size
- Budget
- Preference for instantaneous or storage systems.

The selection of best hot water system according to the needs and using it wisely can reduce the operating cost and improve EE significantly.

Motor and Pumping

Worldwide, motors are the major component for power consumption. In industry, motors and drive systems consume over 70 percent of the total electricity used. Therefore, poor motor performance is typically a major source of energy losses.

There exists an opportunity in many situations to reduce electrical loads with several measure. Particularly for locations where motors run continuously for long periods of time it would be cost effective to replace conventional motors by high efficiency motors. This may be carried out at any time provided but, in practice, companies typically find that the shortest paybacks period and replace a conventional motor when it fails in service. The following activities ought to be considered to enhance motor performance:

- Use legitimately measured motors and just run when required.
- Prefer electronic variable speed controls under variable loading conditions.
- Utilize cogged belts or enhanced gears, as smooth belts slip and are not proficient.
- Install enhanced orientation and lubricate it consistently.
- Check power factor frequently and use capacitor banks near the load point.

A motor driven framework includes an electric motor, motor drive or controls, mechanical transmission framework and end-utilize stack. Whereas, the fundamental power expending parts of the framework are an electric motor and their drivers. Therefore, mere adopting a systematic and energy efficient approach including energy efficient motors, efficient coupling system, optimum pump selection and sizing, efficient flow control mechanism will reduce the losses to significant levels.

Smart Meters

Smart meters are gradually being adopted by the business part to concentrate staff on energy issues and for interior checking purposes the building's energy utilization in a dynamic respectable configuration. The utilization of Power Quality Analyzers can be acquainted into a current working with survey use, harmonic distortion, peaks, swells and interruptions amongst others EE is an easy solution for achieving immediate energy savings.

Smart meter infrastructure, sensors, and communication-enabled devices and controls give electricity consumers and utilities new abilities to monitor electricity consumption and potentially lower usage in response to time-of-use, local distribution, or price constraints. Smart meters also provide a number of other benefits, including enhanced outage management and restoration, improved distribution system monitoring, and utility operational savings.

Smart-Grid and Micro-Grid

Advances in communications, metering, sensors, controls, and storage technologies are enabling consumers, utilities, and other service providers to manage electrical loads in response to price and other constraints. Small-scale distributed electricity storage is becoming more widely available and can reduce peak load, improve electrical stability, reduce power quality disturbances, and facilitate increased penetration of variable wind and solar resources.

Technological developments offer a future scope of specialized answers for enhancing EE with hindrances such as authoritative, money related and behavioral. Therefore, in recent years, micro grids are also becoming more prevalent as distributed generation, storage, and demand management technologies have decreased in price and the public begins to place greater emphasis on ensuring system reliability during grid outages and natural disasters. However, the capacity of micro grids is fairly small but it is growing rapidly.

DEMAND RESPONSE

Demand response (DR) is the primary instrument of DSM. The consumers are motivated in order to take part to decrease the maximum demand of the framework by modifying normal power consumption. In return, consumers will be benefitted by cost or administration motivating incentives. DR help to reduce the peak load as the consumers shift their load operation time from crest to off-peak hours.

The majority of the DR programs are developed to expand the framework reliability and decrease the energy generation which is based on fossil fuel power plants. Further, the DR concept is used to mitigate the electricity prices with service while simultaneously maintaining reliability. Therefore, DR is a better alternative solution instead of expanding the system.

CLASSIFICATION OF DEMAND RESPONSE

Demand Response can be classified as:

Incentive Based DR

These types of progress are developed by the utility in which contracts are made with the consumers for participating during the maximum demand periods. The contracts contain the detailed description of the sort and greatness of the incentive provided to the clients. There are penalties on customers if the contracts are violated by them. The various types of incentive based DR are:

1. **Direct Load Control (DLC):** It is the most usually drilled program by coordinate utility control of the customer's load. It is generally convenient for private clients.

Figure 3. DR classification

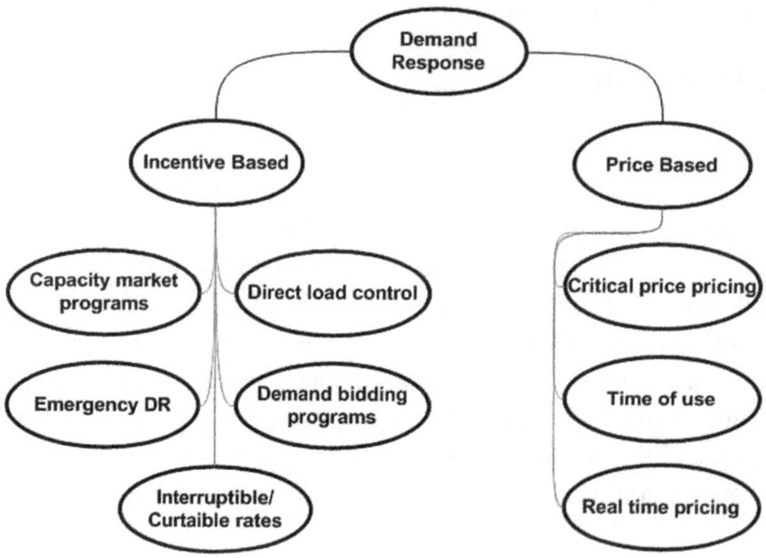

2. **Interruptible/Curtail (I/C) Service:** In this type, the clients get special contracts with limited sheds. The customers are offered discount rates to participate during the system outages in order to adjust the system demand. In case of violation penalties will be imposed over the customer. However, these programs are appropriate for large industrial consumers.
3. **Demand Bidding Programs:** Customers can bid for curtailing at attractive prices. The clients put their enthusiasm to decrease their electrical utilization in the discount power showcase. These programs are suitable for bulk load consumption in megawatts (MW).
4. **Emergency DR:** It involves a voluntary response to emergency signals. The consumers are positively motivated and encouraged to reduce their electricity consumption under short reserves to prevent shut-down.
5. **Capacity Market Programs:** In this process customer assurance is required. The end users offer load curtailment as system capacity. This is cost-effective and less expensive than to expand the existing infrastructure which requires higher financial investment.

Price Based DR

This is a circuitous technique for customer motivation to participate in DR programs. Electricity programs go about as an inspiration for the clients to participate in the programs of system and the cost relies upon the energy available in the system. The higher electrical costs speak to top hours and lower power speaks to off-crest hours. In view of the price variations consumers modify the operation of their equipment. Thus, the peak demand reduces on the system and a flexible load curve is obtained.

1. **Time of Use (ToU):** In ToU, a static price schedule is applied. The cost of electricity depends on the time of load operation. Price scheme is proposed where the different price for different time

duration depending upon usage of appliances is developed. Customers shift their loads to off-peak in order to reduce the cost of energy consumption.

2. **Real-Time Pricing (RTP):** The discount showcase market costs are sent to end clients. Based on the supply demand in the wholesale electricity market the energy prices vary for each interval. The prices are communicated with the customers on an hourly basis. These prices act as a standard for the participation of the customers.
3. **Critical Peak Pricing:** It is a combined plan of ToU and RTP in which cost has three pieces and this is a less predetermined variant of time of use scheme. However, prices during peak hours are subjected to market dynamics.

LOAD SHAPE OBJECTIVES IN CONTEXT TO DSM

It is generally more convenient for the system utilities to inquire DSM in terms of broad load shaping objectives. These shapes indicate the daily and/or seasonal electricity demands of the residential, commercial and industrial consumers by the time of the day, day of the week, and season of the year. The electrical consumption between peak and off-peak times can be modified by six broad categories namely:

1. Peak clipping,
2. Valley filling,
3. Load shifting,
4. Strategic conservation,
5. Strategic load growth, and
6. Flexible load shape.

Load shifting, strategic conservation and strategic load-growth are the advanced objectives that are derived from the rest of the basic objectives. Basically, these are the appropriate DSM techniques that can be used to amend electricity consumption patterns. Figure 4-9 represents the time *vs* load demand curves under the above six categories of load-shape.

Peak Clipping

It is the first classical load management technique in which maximum load on a system can be decreased by direct load control method. Direct load control method is generally implemented by the power utilities. This technique primarily focuses on mitigating the burden of maximum demand as well as increase the system security. A good example for peak clipping technique is the utilization of curtailable rates for commercial as well as the industrial customers.

Valley Filling

After peak clipping, the second classical form of load management technique is valley filling which does construct the off-peak load demand by application of direct load control method. Valley filling technique gives importance to build off-peak loads for desirable times. This technique is appropriate for time periods when the average price of electrical consumption is more than the long-term incremental cost.

Figure 4. Peak clipping objective

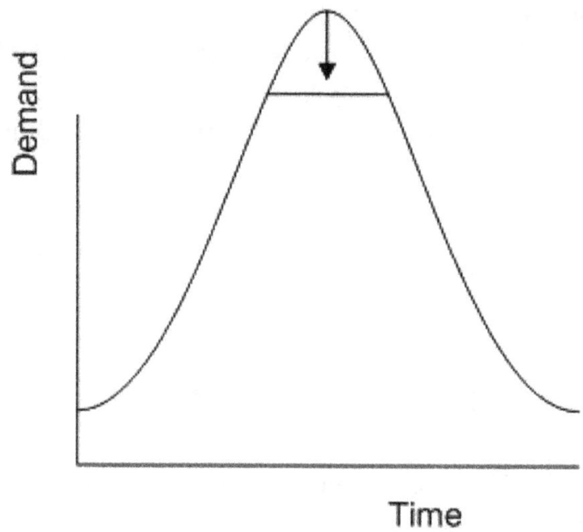

Figure 5. Valley filling objective

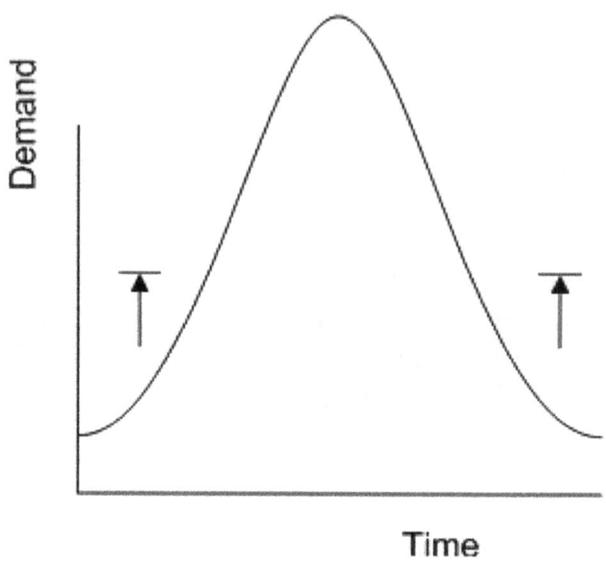

Load Shifting

After valley filling, the third and the last classical form of load management technique is load shifting which does take advantage of time independence of loads. This technique is the most widely used for effective load management in current power distribution networks. It is a blend of both the classical form i.e. peak-clipping and valley filling. It involves utilization of storage devices like water storage heating, space storage heating to displace it with conventional appliances like electrical water heaters and change the load pattern of the customer's premises.

Figure 6. load shifting objective

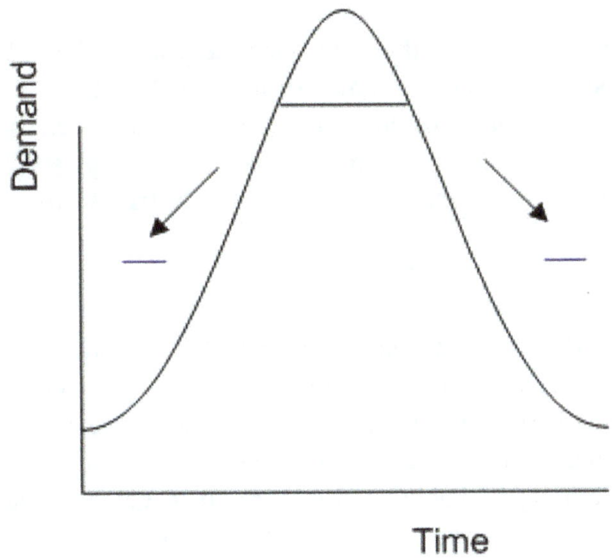

Strategic Conservation

It uses utility-simulated programs that are directed at consumer's premises which do easily modify the load shape reflecting in deals reduction and additionally changing the consumption design. This technique gives importance for employment of energy conservation to change the load shape. The difference between the naturally occurring and induced changes is taken into consideration for load shape and energy consumption. A common example for strategic conservation is improving the efficiency of any appliance and weatherproofing.

Figure 7. Strategic conservation

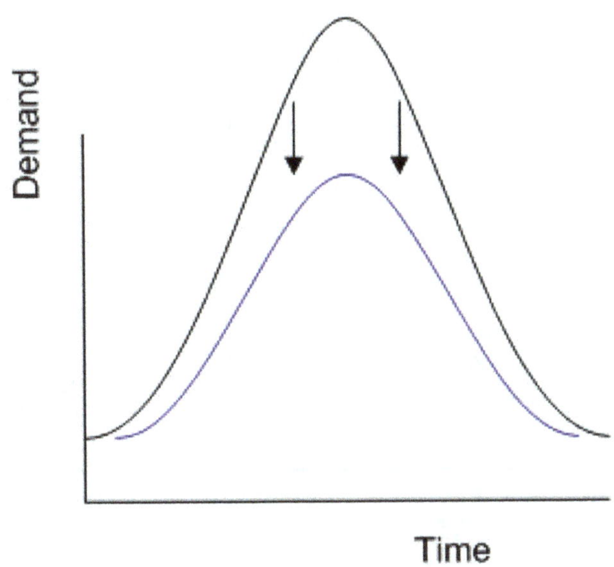

Strategic Load Growth

It modifies the load-shape in such a way that sales increase prior to the utility stimulation exceeding valley-filling. This technique involves desired increment in market share of the loads served by competing fuels along with the economic development in the area of service. These activities increase the employment of electrification. Common examples include heat pumps and industrial process heating which have potential to increase the electricity intensity of industrial sector.

Flexible Load Shape

This technique directly related to reliability and planning. In this technique initially, an overview forecast is done which does include the DSM activities; further, the utility planner does an analysis of the ultimate supply-side options that is flexible and optimum. The load shape can be flexible if the customers are ready to co-operate with the options available as to the variations in service quality in exchange for incentives. The different programs available involve energy management systems or direct load control of individual consumer premises offering different service options.

LOAD RESEARCH

According to DSM perspective, load research enables the utility to analyze the patterns in which their clients consume electricity. These analyses can be done on a total basis or by individual electrical consumption of the end user's premises. Load research is a process which includes data collection of load consumption and the demand requirement over a period of a day, a month or a year. This is essential because it is used as a tool in DSM planning and can be sometimes used an input to program design.

Figure 8. Strategic load growth

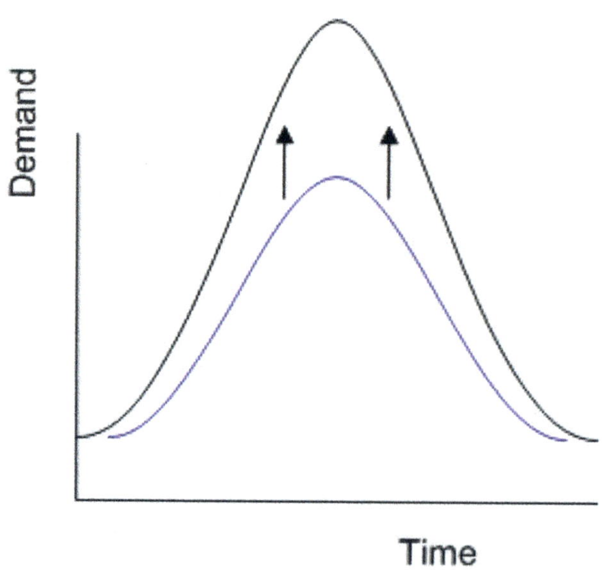

Figure 9. Flexible load shape

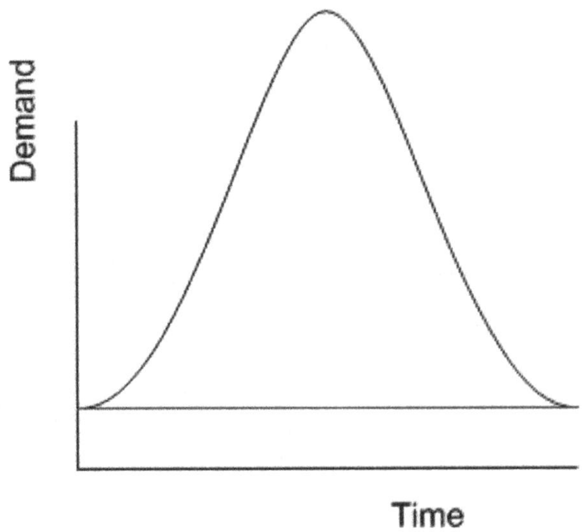

The main purpose of load research is to give a comprehensive breakdown and analysis of system level load profile. To successfully implement DSM, load research is necessary to:

- Understand end-use consumption pattern
- Identify potential DSM programs and associated targets
- Evolve a multiyear DSM implementation plan
- Estimate expected energy savings and demand shifted

DIFFERENT ASPECTS OF LOAD RESEARCH

1. Load research is beneficial for utility for:
    a. Financial planning
    b. Improving customer service
    c. Transmission and distribution upgrade

Metering of loading patterns along with the survey of sample customers is manipulated to estimate the end-use load patterns and its variation over days in a week, months or seasons. The survey of customers assists in collecting information about the number of installed customer's equipment together with capacity, make, technology and time of operation. This is a helping hand in estimating end-use load profiles i.e. lighting, space cooling, water heating, etc. Handling the information from billing and sales records of utilities and the data from meters and survey of sample customers, segment load profiles are developed. Examining the substation generation data along with sector load profiles and annual sales record, system load profiles are estimated.

Table 1. Sample for a load research strategy

Customer	LR Method	LR Objective	Target Consumers	Sample Size	Implementing Body
HT Industrial	Metering	Understand convergence with utility peak	>500KVA	30	Utility/service provider
HT Commercial	Energy audits	Field-monitored data on peak-contributing end- uses, investment analysis	>100KVA	Open-ended	Energy auditors
LT Residential	Load survey	Equipment census, WTP, load profile	Stratified random sampling	300	Market research agency
LT Commercial	Energy audits	Field-monitored data on peak-contributing end-uses, investment analysis	>100KVA	Open-ended	Energy auditors
LT Industrial	Load survey	Equipment census, WTP, load profile	Stratified random sampling	150	Market research agency

Note: LR - Load research
HT - High tension / High voltage LT - Low tension/Low voltage WTP - Willingness to pay

2. Characteristics of load research:
    a. It should have statistically valid sample size
    b. Sample selection includes different consumer groups
    c. It should include seasonal variations like weathers, festivals, elections or other events that impact the load
    d. It should be able to combine different approaches depending on utility load-profile
    e. It should be repeated at regular intervals
3. Approach for load-research are:
    a. **Top Down:** Developing utility load profile and conducting metering, suitable for utilities with a smaller consumer base, mainly heavy industries.
    b. **Bottom Up:** Surveying a sample of customers; suitable for utilities with a larger portion of small commercial and residential customers.
4. The load research strategy includes the various points:
    a. Development of sampling methodology and sampling plan
    b. Preparation of questionnaires for administering to consumers during the consumer survey
    c. Undertaking the consumer survey
    d. Analysis of inputs received in the consumer survey
    e. Undertaking energy audits for large customers
    f. Analysis of energy audits results
    g. Preparation of appliance-level and category- level load curves using consumer survey and energy audit data
    h. Preparation of consumer-level load curves using distribution transformer hourly load-data
    i. Collecting information on EE and vendors.

LOAD RESEARCH TECHNIQUES

- Energy audits and metering
- Automated meter reading (AMR) data
- Load variations and market survey

ENERGY AUDIT AND METERING

In the energy management energy system, energy audit is the necessary investigation, overview and examination of energy streams to a systematic approach for decision making, energy preservation in a building, process or framework to lessen the measure of energy input into the system. The energy auditing of distribution practice includes the energy accounting followed by the investigation of energy flow data for determination of losses and variations at different strides in sub-transmission & distribution. The profit of power distribution depends on the minimization of losses. The distributor having minimal losses has a facet over the competition because of better profitability & reasonably-priced fees.

WHY ENERGY AUDITING?

Distribution losses accommodate technical losses & commercial losses. The technical losses intrinsic capabilities of the distribution enterprise and are losses in diverse additives of power distribution system up to the supply point i.e. no-load &full-load losses in transformers, losses in conductors of overhead lines or underground cables, switchgears, fuses, meters, dielectric losses in capacitors, cables etc. Those losses cannot be absolutely removed yet might be lessened by means of devising proper technical specification whilst purchasing transformers, switchgears, cables, overhead conductors, capacitors, and meters.

The commercial losses accommodates wrong charging, mistakes in meter studying, faulty meters, pilferage because of unmetered unapproved supply (coordinate), altered meters. These losses may be removed by means of right &consistent monitoring. The practice of energy audit aims intends to adjust the aggregate energy inputs with its outflows in different energy streams. To accomplish this, one needs to have energy meters at all incoming & outgoing energy flow path i.e. 66 KV /11 KV feeders and 11KV/415 Volts Transformers, & all consumers. In commercial organizations like institutes and colleges, the top operating expense is mostly found to be electrical energy. In many evaluations of the manageability of price or cost savings, energy audit would constantly develop as a first-hand priority. Energy audits constitute a key zone for decrease in cost.

A successful and well-done energy audit can invariably facilitate homeowners to know additional regarding the ways in which energy is employed in their organizations, and facilitate to spot areas wherever waste will occur and wherever scope for improvement exists. The primary aspiration of the energy review is to see different approaches to decrease vitality utilization per unit of product output or to bring down operational prices alternatively. The energy audit gives a benchmark, or reference frame, for overseeing and getting to energy use over the association and supply the idea for making compelling utilization of energy.

CASE STUDY: ENERGY AUDITING OF INSTITUTIONAL BUILDING

For a detailed explanation and illustration of the power distribution system a large institutional building has been cultivated. It gets power from two feeders of 66/11KV and 33/11KV as shown in Fig. 5. The distribution system incorporates a main station of 66KV with 13 sub-stations of 11KV each, in order to feed supply to 58 blocks.

The electrical meters are installed at 66/11KV (i.e. main substation), distribution substations and blocks.

In addition, there are 32 step-down transformers of 630 KVA at the voltage level 11KV/415V and 27 distributed generators of 250KVA and 500KVA for local power generation, installed at different sub-stations. However, the load is distributed among blocks at the level of 415/240V.

For the above system, the energy audit has been carried out from the primary sub-station till 11 KV feeders in first period of energy review. However, in the second stage, from the feeder to the distribution transformers is considered. The collected data is analyzed for variations in different parameters in the distribution systems.

In order to segregate the losses and observe the load variations in technical & commercial category, it is important to have energy meters at entry areas in distribution system.

The electrical utilization recorded by the meters introduced in different blocks of the distribution system up to every 11 KV feeders in substations for the determination of load variations. The manual readings of meters are recorded in log books of particular substation which is utilized as a source for this activity. The manual reading system has inherent shortcoming of slight error because of time delay in taking readings from first meter to last meters. However, the error is compensated to reasonable extent as the audit period (12 months) is sufficiently large as compared to time delay. The following table rep-

Figure 10. Interconnection of sub-stations

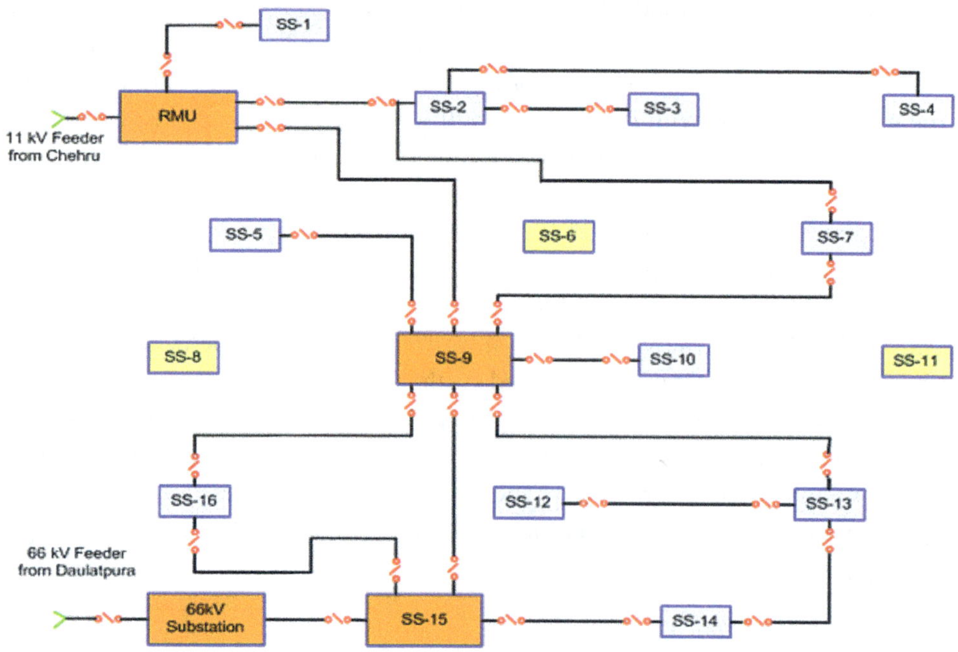

Figure 11. Single line diagram of power distribution system in institutional building

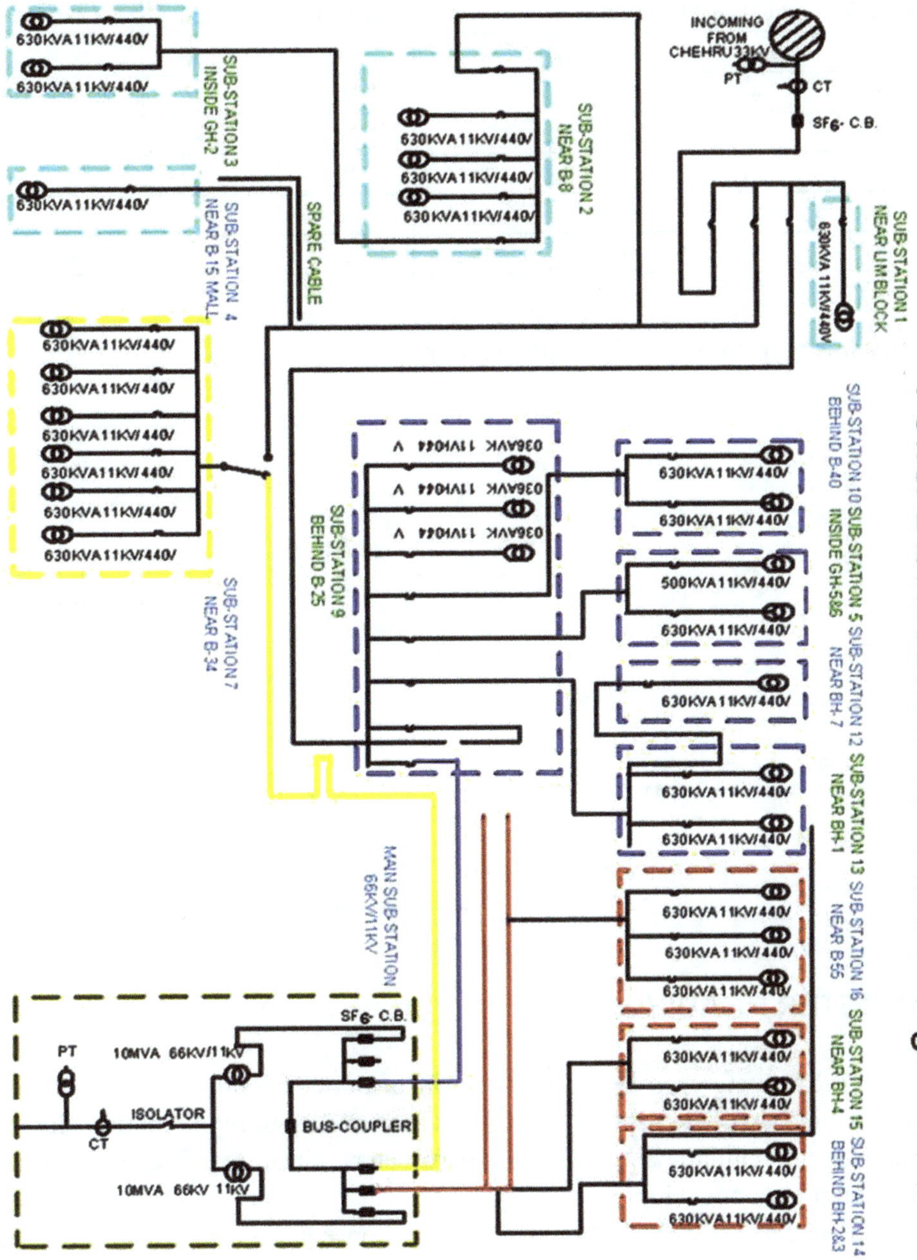

resents the energy consumption in KWh of sample block, of institutional building under consideration, from 9AM to 5PM.

Figure 12 represents the graph which is evolved to examine the load-variation on a weekly basis. It has been spotted that the maximum load occurs on Tuesday and Wednesday which has to be mitigated. The load on Wednesday from 12PM to 2PM, is relatively less and a dig in the graph is observed. These things are to be intensified in order to have a reliable and a better load profile.

Table 2. Hourly energy consumption (in KWh) in sample block

Time Days	9 -10am	10-11am	11-12pm	12-1pm	1-2pm	2-3pm	3-4pm	4-5pm
Mon	29	39	42	38	31	36	35	31
Tue	33	48	46	37	30	38	34	38
Wed	35	43	47	31	23	38	32	33
Thu	37	41	45	34	28	33	35	37
Fri	32	38	43	36	32	41	33	35
Sat	13	15	11	7	9	5	4	3

Figure 12. Energy saving potential for Monday to Saturday

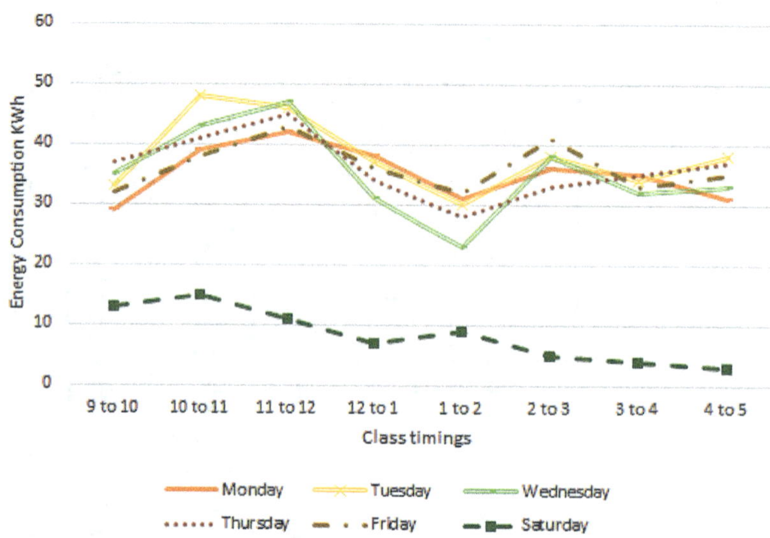

AUTOMATIC METER READING

Advanced Metering Infrastructure, or AMI is the new term instituted to represent networking technology of settled network meter frameworks that go past AMR into remote utility administration. The meters in an AMI framework are regularly alluded to as smart-meters, since they frequently utilize gathered information based on programmed logic.

Automatic meter reading (AMR), is the innovation of consequently gathering utilization, indicative, and status information from energy metering devices and exchanging that information to a central database for charging, investigating, and breaking down. AMR enables billing based on near real-time consumption rather than on estimates based on past or predicted consumption. This auspicious data combined with investigation can help both the utility suppliers and customers to have better control on the utilization and production of electric energy.

AMR innovations incorporate handheld, mobile and system advances based on telephony platforms (wired and remote), radio frequency (RF), or electrical cable transmission. It includes events alarms such as tamper, leak detection, low battery, or reverse flow. AMR devices can also capture interval data, and log meter events. The logged data can be used to collect or control time of use or rate of use data that can be used for energy usage profiling, time of use billing, demand forecasting, DR, rate of flow recording, leak detection, flow monitoring, water and energy conservation enforcement, remote shutoff, etc.

BENEFITS OF ADVANCED METERING

Advanced metering systems frameworks can give advantages to utilities, retail suppliers and clients. Advantages will be perceived by the utilities with expanded efficiencies, blackout recognition, alter notice and lessened work cost because of automating reads, connections and disconnections. Retail suppliers will have the capacity to offer new imaginative items in addition to modifying packs for their clients. Also, with the meter information being promptly accessible, more adaptable charging cycles would be accessible to their clients instead following the standard utility read cycles. With convenient utilization of data which is accessible to the clients, advantages will be seen through chances to deal with their energy utilization and change starting with one REP then onto the next with real meter information. As a result of these advantages, numerous utilities are moving towards actualizing a few sorts of AMR arrangements.

The benefits of smart metering for the utility:

- Accurate meter-reading, leading to no more estimates
- Improvement in billing system
- Definite profile classes and measurement classes, genuine costs applicable
- Enhanced security and tamper detection for equipment
- Energy-management by profile information charts
- Minimized financial burden by redirection
- Low expenditure
- Transparency of "cost to read" metering .

The advantages of smart metering for the client are the Enhanced charging and tracking of use.

Automated Meter Reading (AMR) is an empowering innovation, which is fit for producing exact multi- segment metering information and totals them on local network administrator side for expansive zone and in-house investigation of current energy utilizations and network load-conditions. Moreover, current endeavors in setting of the web of things plan to associate more gadgets in the family to make a more astute home area network (HAN), including parts of client energy management systems (CEMS) like distributed energy resources (DER) and storages, DSM, personal electrical vehicle-charging and interaction of client. In setting of AMR, this includes an extra method for home building mechanization

by consolidating the energy utilization of agreeing parts with the present status of the energy network to enhance its strength and to be stable by load-shifting adjusted with the local area network.

In conjunction with accessible price data, inspiration for lessening general electrical power utilization can be accomplished. Additionally to the basic functionality of the AMR deployment, a more balanced utilization of volatile renewable energy sources (RES) and shift-able and controllable load system (CLS) in the distribution grids is achievable by dynamic integration of parts on the consumer side. Moreover, wind farms, photovoltaic systems, and energy-aware households, employ AMR to get a detailed forecast of the energy demand and extra transparency in energy utilization on the customer's premises.

LOAD VARIATIONS AND MARKET SURVEY

Load Survey

Commercial and institutional buildings spend a lot of money on energy bills. In this chapter energy auditing of a large institutional building is conducted to study the energy consumption pattern and identify the energy saving measures thereof. The energy audit for one day is the index for electrical consumption that normalizes things of the energy crisis by providing various conservation schemes. The organization proposes to adopt the appropriate technology or theme of energy conservation to attenuate, in order to minimize the unwanted power shutdown either incidentally or by load shedding. The similar exercise may be obsessed by any other commercial body also, to reap the benefits of energy conservation.

Figure 13 represents the average variation in the loading pattern of the main sub-station (66KV) of a sample building under consideration. By inspecting the data of the main sub-station, a load profile has been elaborated. It has been observed that the peak occurs daily on an average at 9AM and the load is less at 4PM and 5PM as compared to the morning time therefore, by implementation of DSM measures a good load-profile can be achieved which is evenly distributed.

On the other hand, Figure 14 represents the average load variation at sub- station 7 (i.e. 11KV). From this substation power is distributed among different blocks. For this sub- station, the maximum load occurs at 1PM and 2PM and the load is less at 9AM and 5PM. Therefore, there is good scope of DSM using different techniques including load shifting, peak-clipping and valley- filling in order to ensure a better load-profile.

Figure 15 shows the variation of load of a particular block (sample building) connected to the substation 7 and here it can be observed that the load is maximum at 12PM. Therefore, in this case, peak-clipping technique is adequate to reduce the peak and to ensure an overall good load- profile.

Figure 13. Average variation in load profile of 66KV sub-station

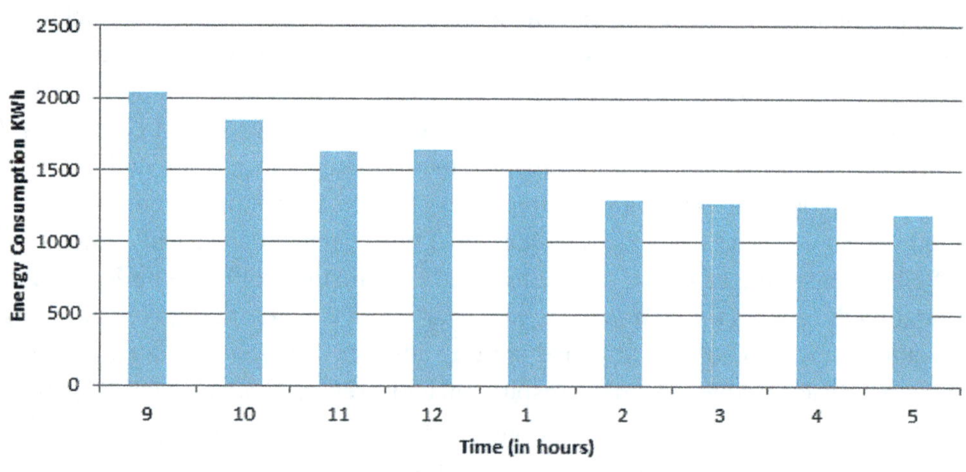

Figure 14. Average variation in load profile of sub-station 7

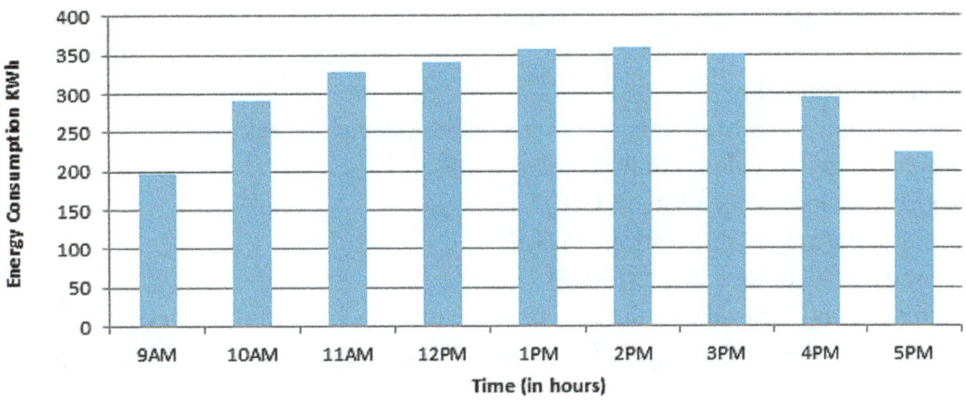

Figure 15. Average variation in load profile of sample building connected to substation 7

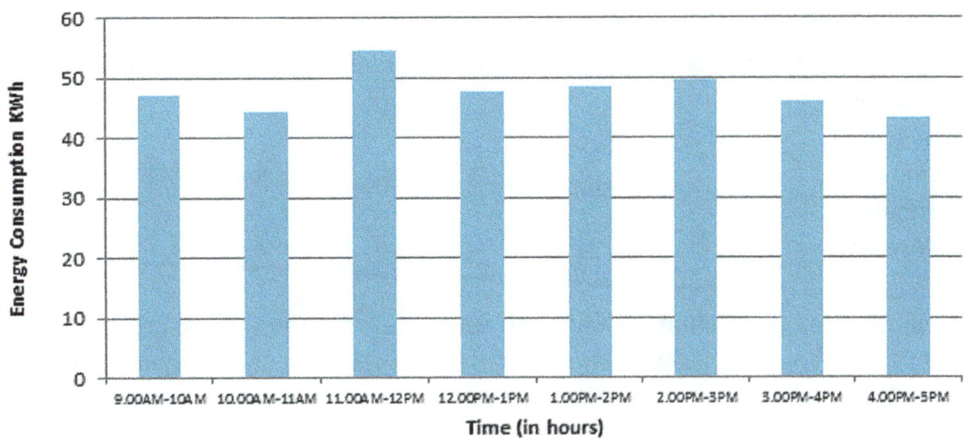

Load Forecasting

Load forecast is an essential part of the power industry. It is mandatory to determine the future-estimate of the long-term interest for electrical power as far ahead of time as could be expected under the circumstances. As indicated by the normal load, the corporate does investments and choices, on purchasing energy from the generating firms, and coming up plans for upkeep and enlargement. It is necessary to claim a few information of future power utilization. Electrical power distributors require an apparatus that allows the load in order to help its administration.

Precise forecasting of an electric load is challenging. An expansive scope of the established expectation models is unseemly for this displaying attributable to their demand. The forecast of the electric load at a future time is a testing issue in light of the various attributes of the electric load and the uncertain factors related with them. The attributes of the electric load rely on the character of the clients and the hardware used, for example, pumps, aeration and cooling systems, lighting, and so on.

1. Importance of load forecasting in the deregulated market is considered as follows:
    a. Purchasing, generation, sales, contracts
    b. DSM
    c. Local-area planning
    d. Infrastructure improvement/capital expenditure decision making
2. Load Forecasting is divided into three main Types:
    a. Short-term load forecasting (STLF)
    b. Medium-term load forecasting (MTLF)
    c. Long-term load forecasting (LTLF)
3. **Load Forecasting Methods:** In the course of the most recent couple of decades, various forecasting strategies have been developed. The diverse forecasting strategies could be expressed as takes after:
    a. Regression Models.
    b. Time serried methodologies.
    c. Comparative day approach.
    d. Artificial Neural Networks (ANN).
4. **Load Classes:** From various points of view, the different electric load can be classified into following categories:
    a. Residential
    b. Commercial
    c. Agricultural
    d. Industrial

To put it plainly, load determining is imperative nowadays for the service organizations especially those utilizing the DSM options, as DSM choices will be connected on the estimated load demand rather than the current load.

FUTURE RESEARCH DIRECTIONS

Essentially, in creating nations, there is a low level of awareness of EE and demand-side management programs, and in this manner showcasing is a need to these advances. In the service area of a service organization, the segments and end-customers that can profit by DSM should be recognized, customized programs created alongside their cost adequacy assessed and afterward a scheme to market and execute the projects should be studied.

Numerous industrial and business organizations still have not done energy auditing to accumulate solid data on their present energy streams. While this might be because of disappointment by the organization to welcome the potential favorable circumstances EE; a few organizations will require talented staff prepared to perform audits importance is given to utilizing outer faculty, as the cost will typically be all around advocated.

Load administration projects to build energy productivity needs to consider the accompanying elements:

- The cost to the client
- Variations in the costs of power and different energies

- The estimation of stayed away from misfortunes coming about because of enhanced power framework dependability
- Any potential misfortunes underway while executing DSM programs.

Therefore, in future managing ratepayer-supported energy efficiency DSM projects should give careful consideration to the particular market destinations of the projects and to the harmony amongst open and private premiums.

CONCLUSION

This chapter presented the overview of the demand side management (DSM), energy efficiency and demand response. DSM in its different structures is a fundamental apparatus for enabling a more effective use of the energy assets accessible to a zone or a locale. For instance, DSM applied to power systems can reduce electrical system emergencies, limit power outages and decrease reliance on costly imports, lessen vitality costs, give help to the power network and generation units, transmission and distribution systems. Comparative advantages can be accomplished from DSM when connected to the utilization of different sorts of energy.

Energy auditing of a sample building is presented and it has been observed that the load variation can be studied at three different levels namely; institutional level, substation level and building (block) level. This is due to the fact that the load variation in these cases of study is varying different at different time of the day. Therefore, the usage of EE measures at all phases of the supply/demand chain could lessen altogether the negative effects of utilization of energy on the environment and human prosperity, and increment the accessibility of essential energy reserves while accomplishing most extreme advantages in terms of output from the available energy source. In consequence, DSM can offer significant economical and environmental benefits as well.

REFERENCES

Abogaleela, M., Elsobki, M., & ElMarsafawy, M. (2012). *A two level optimal DSM load shifting formulation using genetics algorithm, case study: residential load, Power Africa*. IEEE PES. doi:10.1109/PowerAfrica.2012.6498651

Ali, I., Thomas, M. S., & Kumar, P. (2011). Optimal capacitor placement in smart distribution systems to improve its loadability and energy efficiency. *International Journal of Engineering Science and Technology*, *3*(8), 271–284.

Ali, I., Thomas, M. S., & Kumar, P. (December 2012), Effect of loading pattern on the performance of reconfigured medium size distribution system. *Proc. of IEEE 5th Power India Conference*, 1-6.

Ali, I., Thomas, M. S., & Kumar, P. (2013). Distributed resource planning for improved voltage stability of radial distribution system. *International Journal of Advanced Research in Electrical, Electronics and Instrumentation Engineering*, *2*(1), 95-101.

Ali, I., Thomas, M. S., & Kumar, P. (2015, August). Energy efficient reconfiguration for practical load combination in distribution system. *IET Generation, Transmission & Distribution*, *9*(11), 1051–1060. doi:10.1049/iet-gtd.2014.0915

Arora, S., & Kumar, P. (2012). Infrastructural development of energy planning and energy management for smart grid implementation. *Journal of Electrical and Electronics Engineering, 2*(1), 26-31.

Esther, P. B., & Kumar, K. S. (2016). A survey on residential Demand Side Management architecture, approaches, optimization models and methods. *Renewable & Sustainable Energy Reviews*, *59*, 342–351. doi:10.1016/j.rser.2015.12.282

Kumar, P., Ali, I., & Thomas, M. S. (2015). *Synchronizing solar cell, battery and grid supply for development of smart power system for home*. Presented in IEEE International Conference INDICON-2015 held at Jamia Millia Islamia, New Delhi, India.

Kumar, P., Ali, I., & Thomas, M. S. (2016). Energy efficiency analysis of reconfigured distribution system for practical loads. *Elsevier Journal of Perspectives in Science*, *8*, 498–501. doi:10.1016/j.pisc.2016.05.007

Kumar, P., Ali, I., Thomas, M. S., & Singh, S. (2017, March). Imposing voltage security and network radiality for reconfiguration of distribution system using efficient heuristic and meta-heuristic approaches. *IET Generation, Transmission & Distribution*, *11*(10), 2457–2467. doi:10.1049/iet-gtd.2016.0935

Kumar, P., & Singh, S. (2014a). Comprehensive stability analysis of radial distribution system with load growth. *IEEE International Conference IICPE-2014*.

Kumar, P., & Singh, S. (2014b). Energy efficiency performance of radial distribution system with reactive power injection. *IEEE International Conference IICPE-2014*.

Kumar, P., & Singh, S. (2014c). Reconfiguration of radial distribution system with static load models for loss minimization. *IEEE International Conference PEDES-2014*.

Logenthiran, T. (2011). Multi-Agent system for Demand Side Management in smart grid. *2011 IEEE Ninth International Conference on Power Electronics and Drive Systems*. 10.1109/PEDS.2011.6147283

Logenthiran, T., Srinivasan, D., & Shun, T. Z. (2012). Demand side management in smart grid using heuristic optimization. *IEEE Transactions on Smart Grid*, *3*(3), 1244–1252. doi:10.1109/TSG.2012.2195686

Logenthiran, T., Srinivasan, D., & Vanessa, K. W. M. (2014). Demand side management of smart grid: Load shifting and incentives. *Journal of Renewable and Sustainable Energy*, *6*(3), 033136. doi:10.1063/1.4885106

Palensky, P., & Dietrich, D. (2011, August). Demand Side Management: Demand Response, Intelligent Energy Systems, and Smart Loads. *IEEE Transactions on Industrial Informatics*, *7*(3), 381–388. doi:10.1109/TII.2011.2158841

Sinha, A., Neogi, S., Lahiri, R. N., Chowdhury, S., Chowdhury, S. P., & Chakraborty, N. (2011). Role of Demand Side Management for power distribution utility in India. *2011 IEEE Power and Energy Society General Meeting*. 10.1109/PES.2011.6039352

KEY TERMS AND DEFINITIONS

Demand Response: It is an amendment or modification in the electrical power utilization of an associate electrical utility customer to appropriately coordinate the demand for power with the supply provision.

Demand-Side Management: It is the modification of consumer demand for energy through various methods. It is typically used in the specific sense of actions to modify electricity loads to achieve higher system efficiency. Also used to improve energy efficiency and reduce energy consumption for a given output.

Direct Load Control: The measure of the magnitude of the consumer demand that can be hindered at the day and age of the occasional or seasonal peak load by coordinate control of the framework administrator by intruding the power supply to individual equipment or appliances at intervals within the end-user's premises.

Energy Audit: Evaluation of the energy-related performance of a manufacturing plant or building, used to identify energy efficiency opportunities.

Energy Efficiency: It is the utilization of energy in a cost effective way to carry out a manufacturing process or provide a service.

Energy Management: The judicious and effective use of energy to maximize profits and enhance competitive positions.

Load Forecast: It is a technique used by power or energy providing companies to predict the power/energy needed to meet the demand and supply equilibrium. The accuracy of forecasting is of great significance for the operational and managerial loading of a utility company.

Load Management: It is the process of balancing the supply of electricity on the network with the electrical load by adjusting or controlling the load rather than the power station output.

Load Research: An activity embracing the measurement and study of the characteristics of electric loads to provide a thorough and reliable knowledge of trends, and general behavior of the load characteristics of the customers serviced by the electrical industry.

Load Shape: An approach of describing the maximum load demand on the system and the relationship of power supplied at the time of occurrence.

Load Shifting: It is a load management technique that aims to move demand from peak hours to off-peak hours of the day. Loads are shifted from peak to valley times (achieving clipping and filling).

Time of Use: Utilities have different charges for power use during different periods.

Chapter 14
Market Power in Deregulated Power System

Mohammad Quadeer Fahad
Jamia Millia Islamia, India

Mohd Tauseef Khan
Jamia Millia Islamia, India

Anwar Shahzad Siddiqui
Jamia Millia Islamia, India

ABSTRACT

In today's competitive market, deregulation of power industry is inevitable. The aim of deregulating the power markets is to bring competition into them and thereby make them more economically efficient. In an economically efficient market, no consumer or producer has the ability to impact on prices by itself or by collaborating with any other participant. However, the electricity wholesale market is not a perfect market and the potential for market power exploitation is an issue. Sometimes private companies collaborate with each other to get more profit, driving the prices to a higher level and thus acquiring a market power which is an anti-competitive practice. Thus, market power is the capability of a seller or a group of sellers to profitably maintain the prices above a competitive level and control the total output for a noteworthy period of time.

INTRODUCTION

In recent years, large evolution of electrical power system has been observed around the world. Traditional electricity market was a Vertically Integrated Utility (VIU) i.e., a single company runs the whole chain of production on its own and is being indicted in Figure 1. In the vertically integrated market, the company serves as a producer, a transmission owner, a distributor and as the system operator in every geographical area. They serve as the only electricity provider in a region and consumers have no choice but to purchase the electrical power from its local electric company. However, in order to prevent companies from abusing their market positions, this market structure is regulated thereby making certain rules

DOI: 10.4018/978-1-5225-3935-3.ch014

and obligations which companies have to follow. Since, every company is responsible of all elements in their own system, investments in production, transmission and distribution are more easily coordinated. Also, the technical solution, in order to keep a secure transmission of electricity in the system, is handled with fewer complications.

The vertically integrated utility only accounts for the information flow between the generators and the transmission system, it does not account for the information flow between the consumers and distribution system nor it does consider the pattern of consumption of the consumers. The basic objective of an operator in a VIU is to minimize total cost considering the system constraints. VIU had a centralized system of planning for long-term generation, transmission expansion, maintenance, production and fuel scheduling.

The drawbacks of this type of market are:

1. The company exerts its monopoly in the market.
2. Cost of energy is high.
3. Consumers are not given any choices to choose from.
4. It has no incentives to operate efficiently.
5. Money flow is unidirectional, i.e. from the consumer to the electric company.

Figure 1. Vertically integrated utility

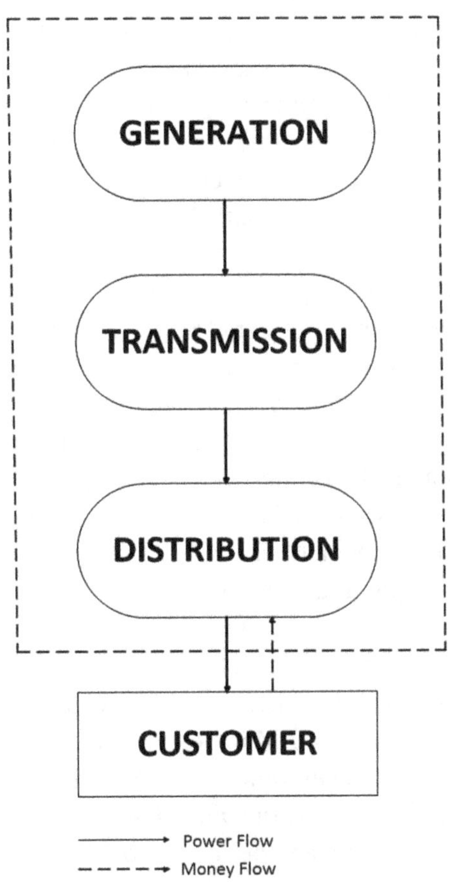

6. Information exchange exists only between the generation and the transmission systems.
7. Customers are often charged on a regular tariff rate depending on the accumulated cost during a period.
8. Customers are not given any incentives to use the electrical power in the off-peak period.

Because of these drawbacks, vertically integrated utility model of the electrical industry has to be changed. With the success in privatization of the airline, telecommunication industry, power industry has also been motivated to restructure and adopt deregulation. Although, privatization on electricity markets was introduced already in the 1970s, it took until the 1990s before a larger trend in deregulation was established. Chile was first to bring competition into the electricity market, but the model they emphasized had its flaws and the market suffered from its poor structure. One key event that really started deregulation worldwide came in 1989 when the UK Government decided to privatize its vertically integrated electricity industry (Lai, 2001). Norway and California followed in 1990 and 1996 respectively.

LITERATURE BACKGROUND

Introduction of competition in power market was one of the many salient features of deregulation. However, market power may prevent competition in electricity industry (Shahidehpour & Alomoush, 2001). Market power is an anti-competitive practice that exists in deregulated power systems. With market power, Generation Company exerts its influence on market pricing or on the availability of electrical energy. It results in driving the price of electricity over a competitive level or exclusion of certain competitors from a relevant market for a significant period of time. Generators with global market power can manipulate the spot price or locational marginal price due to congestion of transmission networks.

Market power can be created or strengthened by elimination of potential competitors through mergers or collusion. Successful collusion requires mutually profitable terms and ability of detecting and punishing deviants, undermining the coordinated action (Shahidehpour & Alomoush, 2001). However, several aspects of electricity markets discourage collusive behavior viz. variation of considerable demand conditions throughout the day, and secrecy of the bid prices of each participant of electricity market.

Wolak measured the unilateral market power in California wholesale electricity market (Wolak, 2003). He used the actual bids submitted to California Independent System Operator's (CAISO) real time energy market for computing the hourly price elasticity during four month period from 1 June to 30 September of 1998, 1999 and 2000. Knittel and Metaxoglou extended the work done by Wolak by calculating a lower bound for seller's price-cost margins using inverse elasticities of their residual demand curves (Knittel & Metaxoglou, 2008). Their calculations establish markup lower bounds in the neighborhood of 20 percent for regulation up and spinning reserves.

Shafie-khah et al has proposed new indices for ex-ante assessment of market power (Miadreza Shafie-khah, Moghaddam, & Sheikh-El-Eslami, 2016). The authors have also taken renewable energy resources into consideration. A short term market power assessment in a long term dynamic modeling of capacity investment is proposed in (Hasani-Marzooni & Hosseini, 2013). Earlier studies of market power considers a region with multiple generation companies and a simplified structure in which each operator is associated with a unique node in a constrained network and the market power is considered as a static concept. However in real system, an operator is concerned with more than one node. Diaz et al has demonstrated that market power is a dynamic quantity and can vary with the weather pattern and fuel prices (Díaz, Webster, Villar, & Campos, 2016).

Horizontal market power present in restructured power system is highlighted in (Borenstein, Bushnell, & Knittel, 1999). Authors discussed the weaknesses of concentration measures used for estimation and prediction of market power and proposed a market simulations based method taking advantage of existing plant level data. They also analyzed the impact of divestitures of California's largest producers on market power. Borenstein and Bushnell used demand and plant-level cost data for simulating the competition in California Electricity Market (Borenstein & Bushnell, 1998). They recognized that firms might have an incentive to restrict output for raising the electricity price and analyzed each firm's ability to do so. They found the potential of significant market power in high demand hours. For determining the severity of market power, levels of hydroelectric production and elasticity of demand are most important factors. These factors are having greater impact on equilibrium prices than the proposed divestitures in California electricity market. They concluded that policies promoting responsiveness to price fluctuations for both producers and consumers will result in significant reduction of the market power problem.

Potential of market power can be checked by market monitoring and adopting different indices. Khajeh et al proposed new ex-ante indices of market power that consider competitor offer price as an incentive to generator to exercise market power (Khajeh, Foroud, & Firoozi, 2016). Market power with transmission constraints presents more complex situation. This case is discussed in (Khan, Siddiqui, Sarwar, & Iqbal, 2015), where HHI index is used for the analysis of market concentration with and without transmission constraints. Nash Equilibrium is used for the mitigation of market power. Lee et al has introduced four new indices of market power which are based on economic principles and representation of transmission network as per Kirchhoff's law (Lee, Hur, Baldick, & Pineda, 2011). Reactive power is one of the most important parameter for the efficient operation of a network. Its pricing mechanism is discussed for a long time even before the deregulation of power system. Feng et al has investigated the issues of reactive power using Must-Run indices (Feng, Zhong, & Gan, 2008). The authors have proposed an index called Reactive Must-Run index (RMRI) which takes into account voltage profile of the system, reactive power outputs and network topology.

Cho and Kim investigated the market power and welfare performance of the Californian wholesale electricity market between 1998 and 2000 with transmission constraints (Cho & Kim, 2007). They decomposed the difference between market price and marginal production cost of generators into market power and inefficiency arising from network constraints. Joskow and Tirole analyzed whether the behavior of electricity generators and consumers with market power is affected by allocation of transmission rights associated with the use of electric power networks (Joskow & Tirole, 2000).

Lesieutre et al proposed a sensitivity approach to detect the potential of local market power (Lesieutre, Rogers, Overbye, & Borden, 2011). They used the sensitivity information from linear programming optimal power flow (LP OPF) for identification of generators having potential for market power either individually or within a small group. They group the generators having potential of local market power by analyzing the impact of network constraints on admissible price perturbations. They found certain specific price perturbation vectors that highlight a constraint-induced locational advantage for these suppliers. They informed that local market power is commonly observed in load pockets which are mitigated by ISO policies.

Shukla and Thampy analyses competition and market power in the wholesale electricity market in India (Shukla & Thampy, 2011). They discussed Electricity Act 2003 and other policy initiatives taken by government of India in order to deregulate the electrical power sector of India. They mentioned the felicitation of competition by Central Electricity Regulatory Commission (CERC) through regulatory framework of inter-state trading, power exchanges, availability based tariff, open access in inter-state transmission and the Indian Electricity Grid Code. Electricity prices increased, despite these initiatives, in the Wholesale Electricity Market in India (WEMI). So, authors analyzed the structure of power market and competitiveness in WEMI to investigate whether market power was one of the reasons for increase in market prices. Based on the analysis, they concluded that market power of firms may be one of the reasons for increase in electricity prices in WEMI. Further, they suggested certain measures for increasing the competition in WEMI.

DEREGULATION

The term deregulation does not mean that the system is not regulated. Electrical power system has always been regulated since its existence. Deregulation means that the system is re-regulated and in doing so, the system is unbundled and competition is introduced. One of the principal forms of a competitive structure is the identification and parting of the various tasks which are normally carried out by a traditional organization so that these tasks can be open to competition in an applied and cost-effective manner. The aim of deregulating markets is to bring competition into them and thereby make them more economically efficient.

The first step in deregulation of power industry is the separation of transmission activity from the generation activity. Transmission sector has always been a natural monopoly and is always controlled by the public authorities so as to prevent it from overpricing for its facilities. The second step was to introduce competition in generation and distribution activities. Competition in generation sector is introduced by inviting private industries to set up their own generating station and to sell the energy in open market under the guidelines of Independent System Operator (ISO).

Deregulation in generation sector is also facilitated by the introduction of Distributed Generation (DG). Distributed generation, also known as on-site generation, employs small-scale technologies like photovoltaic, wind, diesel, tidal, small hydro plants to produce electrical power close to the load centre. DG technologies often consist of modular generators, and they offer a number of prospective benefits. In most of the cases, DG provide lower-cost of electricity, higher power reliability and security with fewer environmental consequences than traditional power generators.

Deregulation in distribution sector is introduced by privatizing the distribution and the metering department. Different DIStribution COmpanies (DISCOs) operate in a given geographical area, all coordinating with a central dispatch centre. Small companies are given the contract to collect the billing of use. One of the important points in deregulation is that restructuring process is not uniform in all countries. In most of the countries, restructuring started with the disintegration of large vertically integrated utility, while in some countries it started with setting up small municipal monopolies to create competition.

Figure 2. Deregulated electricity system showing information and money flow between various players
Abhyakar & Khaparde.

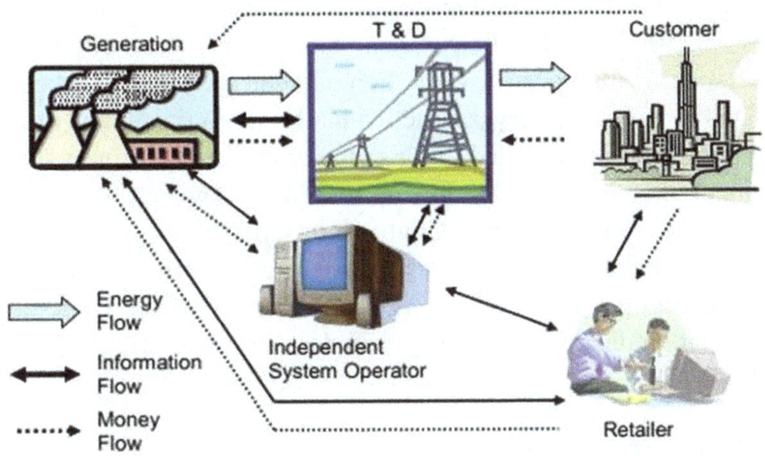

Deregulated Market Structure

There are mainly two kinds of deregulated market structures present; one is the centralized market and the other is bilateral market. An overview of these two structures is shown in Fig 3 and Fig 4.

Centralized Market

The main feature of a centralized market is that producers and consumers do not deal directly with each other. Instead, supply and demand bids are submitted to a centralized auction which is held by the market operator. With respect to the restrictions on the transmission network, the bids are cleared and the market price is determined by the most expensive unit sold. Then everything produced is bought and sold to that price.

Figure 3. Centralized market; all participants are obligated to buy or sell energy on the centralized exchange market

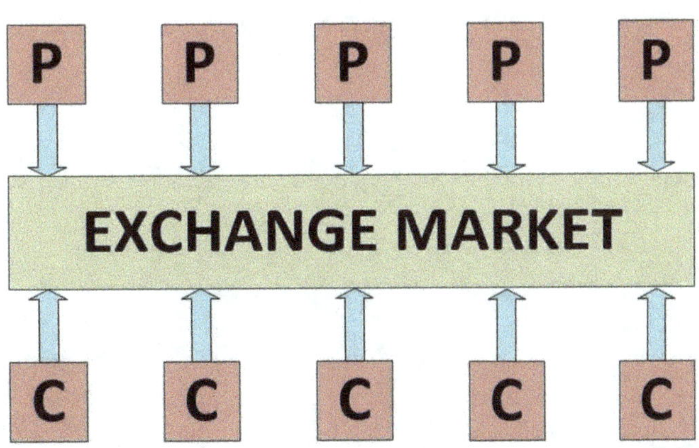

Bilateral Market

Just like the centralized market, the bilateral market also has an exchange market. What separates these two market structures is that the market participants on a bilateral market are allowed to buy and sell freely to one another and no specific market price exist. In a bilateral electricity market, the market operator plays a much more supervising role. Every transaction has to be reported to the market operator, so they can check that all commitments are fulfilled. In this type of market, a business for retailers arises. The business concept is simply to buy electricity directly from the producer or on the exchange market and then sell it to the consumers. This middleman might seem superfluous, but the retailer's work leads to gain in freedom of choice for the consumer which in turn improves the market competition. And when the competition become stronger, the service gets better. Some retailers can even offer to undertake some of the price risk, that producers and consumers undergoes, by offering stable electricity prices for a longer time period.

Components of Deregulated Market

In this section, the main entities participating in the deregulated electricity markets, their various roles, responsibilities and objectives will be briefly introduced. This is based on the market structure of the bilateral wholesale market in the U.S. These understandings come from reading (Bose, Wu, Xu, Wierman, & Mohsenian-Rad, 2015; M Shafie-khah et al., 2015).

Independent System Operator

The Independent System Operator (ISO) is the supreme entity in the control of the transmission system. The basic requirement of an ISO is the disassociation from all market participants and absence from all financial interest in the generation and distribution business. The ISO has a crucial role in the market and operates under the authority of Electricity Regulatory Commission (ERC). The ISO has a supervising, coordinating and synchronizing role towards the companies competing in the electricity market. The

Figure 4. Bilateral market, all participants are free to act with one another

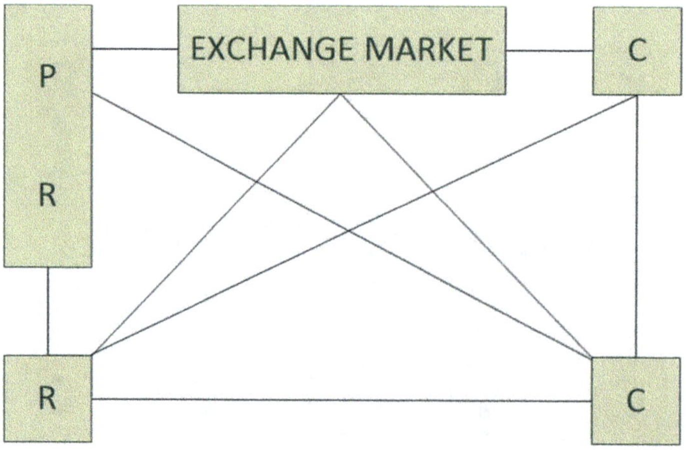

ISO's objectives are similar to the operational ones of an air-traffic controller, coordinating the aviation market. They both ensure that market operates efficiently and competitively while being independent and non-favorable to any of the market's participants. The ISO schedules generation with the objective to maximize social welfare. To accomplish this, the ISO needs powerful network management tools and scheduling software. The ISO also forwards payments in the market and monitors entered contractual agreements to ensure that commitments, such as energy deliverance are kept. Examples of ISOs are the CAISO and NYISO, operating in California and New York City respectively.

Power Exchange

The Power Exchange (PX) is an independent, non-government and non-profit entity that handles the electric power pool, which provides a forum to match electric energy supply and demand based on bid prices. The PX calculates the market-clearing price (MCP) based on the highest price bid in the market. The time horizon of the pool market may range from half an hour to a week or a month or a year. The most usual time horizon is day-ahead market to facilitate energy trading one day before each operating day. An hour-ahead market is also useful since it provides additional opportunities for energy trading to redress short-term imbalance. In the England & Whales (E & W) system, the ISO and PX functions both exist within the NGC. In the evolving Asian systems the arrangements for the future are still under intense discussion, e.g. in China, India, Thailand, Indonesia and also in most parts of Africa.

Power Markets

Based on trading activities, the power market types include the energy market, ancillary services market, and transmission market. Also markets can be classified as forward market (day-ahead or hour-ahead) and real-time market. Energy market is a place where the competitive trading of electricity occurs. The energy market is a centralized mechanism that facilitates energy trading between buyers and sellers. Ancillary services are needed for the power system to operate reliably. In the regulated industry, ancillary services are bundled with energy. In the restructured industry, ancillary services are mandated to be unbundled from energy. Ancillary services are procured through the market competitively. In the transmission market, the commodity traded is the transmission rights. This may be the right to transfer power, the right to inject power into the network, or the right to extract power from the network. The holder of a transmission right can either physically exercise the right by transferring power or be compensated financially for transferring the right for using the transmission network to others.

Forward market includes day-ahead market which is used to schedule resources at each hour of the following day. An hour-ahead forward market is a market for deviations from the day-ahead schedule. To ensure the reliability of power systems, the production and consumption of electric power must be balanced in real-time. However, real-time values of load, generation, and transmission system can differ from forward market schedules. Therefore, the real-time market is established to meet the balancing requirement.

FERC

Since energy is such a critical resource in modern society, governmental involvement is unavoidable. Market rules and regulations are necessary in order to accomplish competition on the deregulated market.

The Federal Energy Regulatory Commission (FERC) is the governmental body that has the ultimate authority regarding these matters. FERC determines and approves new market rules. The main objective for FERC is that electricity is delivered in an economically efficient and reliable way. Central Electricity Regulatory Commission (CERC) is synonymous to FERC in Indian power sector.

GENCO

The GENeration COmpanies (GENCOs) are the companies that produce and sell energy. A GENCO owns a plant or a portfolio of plants that may have different technologies, such as fuel types (coal, gas, nuclear, wind, hydro etc.). The main objective for a GENCO is to maximize profit from selling energy. GENCOs have the opportunity to sell electricity to entities with which they have negotiated sales contracts. GENCOs may also opt to sell electricity to the PX from which large customers such as DISCOs and aggregators may purchase electricity to meet their needs. In addition to real power, GENCOs may trade reactive power and operating reserves.

Retailer

A retailer buys energy on the wholesale market and resells it to the customer. This middleman exists when the bilateral market structure is applicable. Retailers provide the customers with a freedom of choice which is good for competition. Still, the objective remains in maximizing profit from differences in wholesale energy prices and retail prices. A retailer may deal indirectly with end-use customers through aggregators.

Transmission Owners

In the regulated electricity market transmission network was state owned and this integrity will be retained after the restructuring of electricity market. The basic requirement for open transmission access is that transmission operators treat all users on a non-discriminatory basis in respect of access and use of services. This requirement cannot be ensured if transmission owners (TOs) have financial interests in energy generation or supply.

Ancillary Services

Ancillary services are generally referred to as those services other than energy that are essential for ensuring the reliable operation of the electrical grid. The reliable operation of a power system requires generation reserves to be available in order to cover generation and transmission contingencies. As restructuring evolves, determining the cost of supplying ancillary services and finding out how these costs would change with respect to operating decisions is becoming a major issue. According to FERC, ancillary services are necessary to support the transmission of power from sellers to buyers given the obligation of control areas and transmission utilities to maintain a reliable operation of the interconnected transmission system.

MARKET POWER

There are various factors in the electric power industry that adversely affects the industry but non-competitive practices especially in the generation industry gives rise to very undesirable phenomenon known as market power. When a generation facility is able to influence the price of electricity, market power is established. Competition and the choices available to a customer in a restructured power system are hampered by market power practice.

In other words, market power can be defined as the ability of a seller, or a group of sellers, to control the total availability of power, to raise the prices above a competitive level and thus excluding the competition from the market for a significant period of time (Shahidehpour & Alomoush, 2001). Market power can reduce the competition in power generation, service quality, and can even halts the technological innovation.

Market power may be exercised deliberately or accidentally. For example, market power in the generation sector may be exercised when a company is committing its costly generating units instead of the other units which could have been less expensive for offering an excessive amount of generation to a market (deliberately). On the other hand, there are some transmission constraints that could limit the transfer capability in a certain area (accidental). Because of these transmission constraints, some of the generating units are unable to supply power and thus results in local dominant providers which drive the market prices by offering more costly units to the market.

Another reason for the exercise of market power is the unavailability of demand response. Demand Response is a phenomenon of change in pattern of the use of electricity by a customer from his normal consumption pattern in response to the incentives provided to him to use the electrical power in the off peak period. Demand response is only possible if hourly pricing is available to the customers. This could encourage customers to manage their loads in response to the fluctuating prices. However, unavailability of hourly information could motivate the generating companies to drive the market prices up to their own benefit in the peak load period and providing little information to the customers. In the past, transmission companies could exercise market power by providing relevant transmission information to their associated generating companies and withholding it from other competitors.

MARKET POWER IN DEREGULATED ELECTRICITY MARKET

In general, market power is referred as the ability of a market participant to profitably maintain prices above a competitive level for a significant period of time. When defining market power in this way we need to be careful in the attempt of measuring its deployment. In electricity markets, just because a supplier gets paid a price that is well over the competitive level it does not mean that this market participant is exercising any market power (Elfstadius & Gecer, 2008). When the demand exceeds the supply in an area, prices can sometimes rise above any marginal cost of production even though no producer actually exploits market power. Also, sometimes it is impossible to import more power from other regions due to transmission constraints; customers who are willing to pay prices that surpass the highest competitive price will offer to do so. This leads to a continuous rise in price that will keep on going until the supply meets the demand. Therefore, by simply looking at market prices no assumption can be made whether market power is actually being abused or not.

Economists talk about two types of market power, horizontal and vertical ("Horizontal Market Power in Restructured Electricity Markets," 2000). A company having control over a single activity, such as power generation, is said to have horizontal market power if it succeeds in driving the prices above a competitive level. A company involved in two related activities, such as power generation and transmission, exploit vertical market power when it uses its dominance to drive the prices and increase profits for the overall portfolio. In the deregulated electricity market, problems related to vertical market power are mitigated by operating the transmission system independently. However, horizontal market power is still a big concern. Therefore, when discussing market power in deregulated electricity markets, it is horizontal market power that is concerned.

Market power can be exercised or reinforced by the elimination of participants and potential competitors through horizontal mergers or by agreement (Dalton, 1997). For agreement of sellers to be successful, sellers must be able to reach terms that are mutually profitable. Factors that tend to facilitate agreement are:

- Repeated auction of a similar product under similar demand and supply conditions.
- Intimate information of competitor's operational costs.
- Instantaneous knowledge of competitor's actions

Causes of Market Power

Basically, there are two main reasons which brought the potential of market power in the electricity market (David & Wen, 2001). These are:

- Market Dominance
- Transmission Constraints

Market Dominance

Market dominance is a scenario which is present in every imperfect market including the electricity market. A supplier which has large generation capacity with respect to others in the electricity market can exploit market power by either continuously bidding above the market price of the production which drives the price up or by withholding some of the generation capacity which reduces the effective supply and in turn drives up the price for rest of the seller's portfolio. If market power is exploited this way, profit gained due to the higher priced portfolio will exceed the loss obtained from the withheld capacity and this withheld capacity will be made available to the market in the peak demand period. An example is the England and Wales (E&W) pool where a highly concentrated market may have allowed the two dominant suppliers, National Power and PowerGen, to selectively withdraw capacity during peak periods and increase profit.

Transmission Constraints

The presence of transmission constraints like congestion results into local or locational market power maker. Before starting any plan for the detection of market power one must have full and clear understanding of the topology of the transmission system. This characteristic of the electricity market increases the opportunities to exploit market power compared with other industries. A scenario of this kind is easiest explained by the load pocket. A load pocket is an area where transmission constraints make it impossible to import less expensive power from elsewhere (Zhong & Bhattacharya, 2002). A distributor placed within such an area will possess locational market power and will find itself in a state of monopoly. Since the supplier is the only producer of the electricity in that area, he can gain profit either by increasing the production or by creating congestion and increasing the price above the marginal price. Thus, even a small supplier can exploit market power due to the transmission constraints.

Impacts of Market Power

Market power severely affects the restructured electricity market. Some of the impacts of market power are listed below:

1. Reduces competitiveness
2. Gives rise to monopoly
3. Reduces quality
4. Adversely affect technological innovation

Factors Affecting Market Power

1. **Market Share:** The amount of generation owned by a single company with respect to the total demand over a specified period of time in the market is known as market share. This metric is used to give a general idea of the size of a generation company in relation to its market and the competitors.
2. **Market Concentration:** Dispersion of market shares among the companies on the market is known as market concentration. It is a function of the number of firms and their respective shares of the total production in a market.
3. **Elasticity of Demand:** It measure how well demand responds to a sudden change in price. Typically, electricity markets have very inelastic demand due to the importance of electricity in the modern society.
4. **Typical Contractual Arrangements:** Commitments regarding deliverance of energy are bound through contractual agreements, valid over a period of time. These contracts can be either bilateral or open market contract. Bilateral contracts are made between two parties with mutual understanding and accordingly prices are fixed. In open market, every participant's spot price is displayed on a notice board and participant can approach one another and then they can form bilateral contracts under the supervision of Independent System Operator (ISO).

MARKET POWER INDICES

Market power is calculated using numerous indexes. Some of these indexes are as follows:

Herfindahl Hirschman Index (HHI)

Market concentration is represented by Herfindahl Hirschman Index (HHI). It shows the number of contributors in the market and the variation (inequality) in their market shares. Mathematically, it is expressed as the summation of the squares of their individual market share percentages. This is shown in equation (1).

$$HHI = \sum_{j=1}^{N} S_j^2 \tag{1}$$

where,

N = Number of contributors
Sj = jth contributor's market share in per unit or in percentage

This index is simple and easy to understand but it fails to provide the complete information when transmission constraints and load in the system is varied. The following spectrum has been introduced by the US department of Justice ("US Department of Justice," 2015).

$HHI < 1000$: Market is not concentrated

$1000 < HHI < 1800$: Market is moderately concentrated

$HHI > 1800$: Market is highly concentrated

Consider a market with five participants having market share of 43%, 23%, 16%, 14% and 4%. The resultant HHI would be $43^2 + 23^2 + 16^2 + 14^2 + 4^2 = 2846$. This represents highly concentrated energy market. Suppose a market having 50 participants with equal market share, without considering transmission constraints, HHI would be 200. Thus, no market power exists in this highly simplified conceptual case, ignoring completely the transmission constraints and transportation charges, according to HHI measure.

Lerner Index (LI)

The Lerner index is named after the economist Abba Lerner (S Prabhakar Karthikeyan, Raglend, & Kothari, 2013). It is used to measure the relative deviation of price at the firms' profit-maximizing output from the firms marginal cost at that output. The mathematical formula is:

$$LI = \frac{Price - Marginal\ Cost}{Price}$$

$$LI_j = \frac{\left(P_j - MC_j\right)}{P_j} = \frac{1}{E_j^d} \tag{2}$$

where,

LI_j = Lerner Index for firm j,
P_j = Competitive Price at the firm,
MCj = Marginal cost at the firms profit-maximizing output and
E_j^d = Elasticity of demand seen by the firm.

The value of LI exists between 0 and 1. Value 0 indicates that there is no market power while values closer to 1 reflect the presence of large market power. This index includes the effect of fringe firm's elasticity of supply in the form of the market clearing price 'P' and the elasticity of demand based on market power. For a generation company, if value of LI is larger than 0, then it possesses some market power. Ineffectiveness in reflecting the price response on demand is the basic limitation of this index. Both the indices, HHI and LI fail to address the impacts on the market price due to the constraints of transmission line.

Must Run Ratio (MRR)

The main objective of Must Run Ratio is to consider the transmission line constraints (Estebsari, Sadeghzadeh, & Fotouhi, 2009). The MRR provides necessary information about the capacity that is to be generated by a generating company to supply a given load through a congestion zone. It can also provide useful market power signals in a congestion zone, which refers to a simple configuration in which one transmission line (or a set of lines in a ''corridor'') can be filled to its limit by exporting generation from a low-cost region to a high-cost region.

The MRR for supplier 'n' in the transmission zone 'm' is defined as follows:

$$MRR = \frac{P_l - P_i - \left(\sum_{j=1}^{N_g} Po_{j,\max} - \sum_{j=1}^{N_{g,n}} Po_{j,\max}\right)}{\sum_{j=1}^{N_{g,n}} Po_{j,\max}} \tag{3}$$

where,

Pi = Import limit of the zone
Pl= Total load demand of the zone
$Po_{j,\max}$ = output limit of generator j in the zone
Ng is the number of generators in the zone
Ng,n is the number of generators owned by supplier K in the zone

Must Run Generation (MRG)

It is the least generation which must be provided by a generator to the total load taking into account generation and transmission constraints. It can be found by solving the optimization problem as given below:

$$\mathrm{Min}(P_{gn}) \tag{4}$$

Such that

$$a^T = \left(P_{gen} - P_{load}\right) = 0 \tag{5}$$

$$0 \leq P_{gen} \leq P_{g\max} \tag{6}$$

$$-Pl_{\max} \leq F\left(P_{gen} - P_{load}\right) \leq Pl_{\max} \tag{7}$$

where,

a = vector with all ones,
P_{gen} = power generation vector,
P_{load} = demand vector,
$Pl_{\max}$ = line limit vector,
F = distribution factors matrix.

Equation (5) denotes power balance equation and equations (6) and (7) denotes the generator output limits and transmission line limits respectively.

For a GENCO 'G', the MRG of each unit in the company can be determined by solving the following equations:

$$\min \sum_{n \in G} Pg_n \tag{8}$$

Subject to the constraints mentioned in equations (5) to (7).

Solving equations (5) to (8), the total must run generation of GENCO 'G' then can be calculated using

$$Pg_G^{must} = \sum_{n \in G} Pg_n^{must} \tag{9}$$

Alternatively, the MRG of a generator can also be found by running the load flow and finding out the respective power generation and line flows of the system.

Must Run Share (MRS)

The variation in market power with load level is represented by an index known as MRS (S Prabhakar Karthikeyan et al., 2013). It represents the minimum market share required for generator in a power market to supply a given market load. Mathematically, it is defined as:

$$MRS_n = \frac{Pg_n^{must}}{P_{load}} \tag{10}$$

where,

P_{load} = Total demand in the power market.

In an ideal market, the MRS of each generator is expected to be zero which ensures fair competition. Generally, a generator is considered to have market power if its MRS is greater than zero.

Nodal Must Run Share (NMRS)

In order to consider the geographical difference of market power, the concept of the MRS is applied to each load bus to define NMRS (S Prabhakar Karthikeyan et al., 2013). The NMRSn,i which represents the minimum capacity that must be provided by the must run generator n to supply a given load at node i is defined as:

$$NMRS_{n,i} = \frac{Pg_{n,i}^{must}}{P_d} \text{ i=1,2,...,N} \tag{11}$$

where,

N is the number of buses in a power system,
Pdi is the load at bus i,
$Pg_{n,i}^{must}$ is the contribution of the must run generator n to Pdi.

The Pg_n^{must} and the generator outputs and line flows under the Pg_n^{must} can be obtained after solving Eqns4–7. The $NMRS_{n,i}$ for node can be calculated based on these results using the topological analysis of power flow. Considering a general node i under the Pg_n^{must} as shown in Fig. 5.

The total power inflow PI_i into the node is

$$PI_i = \sum_{j \in N_s} Pl_{ji} + Pg_i \tag{12}$$

Figure 5. General node i in a power system

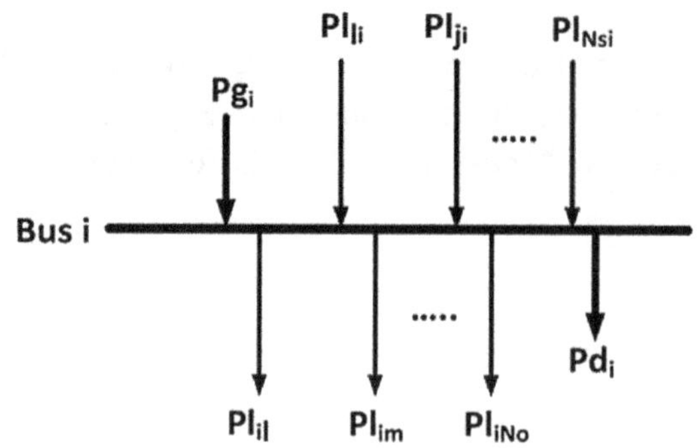

where N_s is the number of inflows at node *i*, Pg_i and Pd_i are the generation and load at node *i*, respectively, Pl_{ji} is the inflow from bus *j* to *i* and Pl_{im} is the outflow from bus *i* to m .

Considering a system with nodes, equation can be expressed as

$$Pl_i - \sum_{j\in N_s}\left(\frac{Pl_{ji}}{PI_j}PI_j = Pg_i, i = 1,2,\ldots,N\right) \tag{13}$$

or in matrix form

$$M * PI = P_g \tag{14}$$

where PI is the vector of nodal supplying power, Pg is the vector of nodal generations, and M is the distribution matrix with element M_{ij} .

$$M_{ij} = \begin{cases} 1 & j = i \\ -\frac{Pl_{ji}}{PI_j} & j \in N_S \\ 0 & otherwise \end{cases} \tag{15}$$

If [M^{-1}] exists, then

$$PI = \left[M^{-1}\right]P_g \tag{16}$$

The *i*th element in PI is

$$PI_i = \sum_{j\in N}\left[M^{-1}\right]_{ij}P_{gj} \tag{17}$$

This equation shows how the supplying power at node i is contributed from all the generators in a system.

The load at node i can be expressed as

$$Pd_i = \frac{Pd_i}{PI_i} PI_i = \frac{Pd_i}{PI_i} \sum_{j \in N} [M^{-1}]_{ij} Pg_j$$

$$Pd_i = \sum_{j \in N} \frac{Pd_i}{PI_i} [M^{-1}]_{ij} Pg_j$$

$$Pd_i = \frac{Pd_i}{PI_i} ([M^{-1}]_{i1} Pg_1 + ... + [M^{-1}]_{ik} Pg_k^{must} + ...) \quad (18)$$

The kth element in above equation is the minimum capacity which must be provided by generator k to supply the load at node i.

$$Pg_{k,i}^{must} = \frac{Pd_i}{PI_i} [M^{-1}]_{ik} Pg_k^{must} \quad (19)$$

Substituting equation 18 and 19 into equation 11, the NMRS of generator n at bus i can be obtained using

$$NMRS_{n,i} = \frac{Pg_{n,i}^{must}}{Pd_i} = \frac{[M^{-1}]_{ik} Pg_n^{must}}{\sum_{j \in N} [M^{-1}]_{ij} Pg_j} \quad (20)$$

Expected Nodal Must Run Share (ENMRS)

Due to random failures, under contingency state, a generation company has a chance to exercise its nodal market power. Therefore, system contingency states have to be included in the NMRS. Considering a power system with n independent components, for system state's with exactly m failed components, the state probability is given as:

$$p_s = \prod_{i=1}^{m} U_i \times \prod_{i=m+1}^{n} A_i \quad (21)$$

where,

A_i = Availability of component i

U_i = Unavailability of component i

After the NMRS and probability for each state are determined, the ENMRS of GENCO A at bus i can be calculated using

$$ENMRS_{A,i} = \sum_{s=1}^{Nc} p_s * NMRS_{A,i}^{s} \tag{22}$$

where

Nc is the number of failure contingencies considered.

IMPACT OF FACTS DEVICES ON MARKET POWER

In electrical power system, Flexible Alternating Current Transmission Systems (FACTS) are the power electronic based devices that provide control of one or more AC transmission system parameters to enhance power quality and increase power transfer capability. Though FACTS devices are used to relieve the transmission congestion and increase the power flow through the transmission lines but it indirectly assists the generation company to implement its market power. The enhancement of market power due to FACTS device can be explained in the following manner: When a FACTS device is employed in a region where a generating station is located, then congestion and power transfer capability of that region is improved. This encourages the generating station to transfer more amount of power which it was unable to transmit due to the network constraints.

Thus, while finding the optimal location of FACTS devices through any optimization technique, market power should be considered as one of the important factors.

OPTIMAL LOCATION OF TCSC USING REACTIVE POWER LOSS SENSITIVITY INDEX

This method is based on the sensitivity of the total system reactive power loss, $\sum Q_L$ with respect to the control variable of TCSC, X_{mn} (Gad, Shinde, & Kulkarni, 2012). Where X_{mn} is the net line series reactance with TCSC, placed between the buses m and n, and a_{mn} is the loss sensitivity with respect to control parameter X_{mn} (Rajderkar & Chandrakar, 2009; Reddy, Kumari, & Sydulu, 2010). The loss sensitivity with respect to X_{mn} can be expressed as:

$$a_{mn} = \frac{\partial Q_L}{\partial X_{mn}} = [V_m^2 + V_n^2 - 2V_m V_n \cos\delta_{mn}] \frac{r_{mn}^2 - x_{mn}^2}{(r_{mn}^2 + x_{mn}^2)^2}$$

The loss sensitivity for each line can be calculated using above equation by substituting the value of $X_{TCSC} = 0$. More positive is the value of a_{mn}, more is the sensitivity of that line. Hence a TCSC is placed in the line having the most positive sensitivity index.

In this analysis, IEEE- 14 bus system with five generators at bus number 1, 2, 3, 6 and 8 having maximum generation capacity 32, 88, 63, 72 & 31MW respectively is taken. These generators will exert their market power on the loads located at bus number 2, 3, 4, 5, 6, 9, 10, 11, 12, 13 and 14. To determine the optimal location of TCSC, MATLAB software has been used for finding the load flows and sensitivity index. The sensitivity index, a_{mn} for the IEEE-14 bus system is shown in Table 1.

Based on sensitivity index, the lines in decreasing order of sensitivities are: lines 12-13, 4-9, 9-14. Hence, line 12-13 is the optimal location for the placement of the TCSC. Based on the sensitivity ranking TCSC with 70% compensation is placed in the sensitive line and market power is calculated using the NMRS index.

Two cases have been formulated for the comparative study of market power with the TCSC location. These are given as follows:

1. When TCSC is placed in line 6-13 as given in (S P Karthikeyan et al., 2012).

Table 1. Sensitivity indices

From Bus	To Bus	a_{mn}
1	2	-0.0572
1	5	-0.0187
2	3	-0.0646
2	4	-0.0379
2	5	-0.0183
3	4	-0.0074
4	5	-0.0695
4	7	-0.0181
4	9	-0.0009
5	6	-0.0344
6	11	-0.0173
6	12	-0.0049
6	13	-0.0301
7	8	-0.1268
7	9	-0.0937
9	10	-0.0022
9	14	-0.0013
10	11	-0.0116
12	13	0.0001
13	14	-0.0081

2. When TCSC is placed in the line 12-13 having most positive sensitivity index.

Case 1: TCSC Is Placed In-Line 6-13 With Xc = 0.0132 p.u.

Based on NMRS values for each generator from Table 2 and Table 3, MRG share is plotted for different loads due to individual generators.

Following are the MRG share plots for generator 1,2,3,6 and 8.

Figure 6 shows the MW share of generator at bus 1 to the loads at different buses. Since TCSC is placed in line 6-13 which is far from the generator 1, therefore the market power is slightly increased when compared to the base case. It is observed that the share of this generator to load at bus 5 is increased from 12.7% to 13.5% (0.9672 MW to 1.026MW). It is also observed that the MW share of this generator to

Table 2. NMRS values of each generator to each load without any TCSC in the system

	Pg1	Pg2	Pg3	Pg6	Pg8
Pd2	0	1	0	0	0
Pd3	0.006698	0.57956	0.413742	0	0
Pd4	0.061008	0.938992	0	0	0
Pd5	0.12727	0.87273	0	0	0
Pd6	0.019728	0.135279	0	0.844994	0
Pd9	0.034107	0.522398	0	0.012848	0.430646
Pd10	0.019728	0.135279	0	0.844994	0
Pd11	0.019728	0.135279	0	0.844994	0
Pd12	0.019728	0.135279	0	0.844994	0
Pd13	0.019728	0.135279	0	0.844994	0
Pd14	0.025013	0.277577	0	0.539111	0.158298

Table 3. NMRS values of each generator to each load with TCSC in the 6-13 line

	Pg1	Pg2	Pg3	Pg6	Pg8
Pd2	0	1	0	0	0
Pd3	0.007029	0.575297	0.417673	0	0
Pd4	0.064684	0.935316	0	0	0
Pd5	0.135007	0.864993	0	0	0
Pd6	0.021056	0.134904	0	0.84404	0
Pd9	0.035722	0.51464	0	0.009384	0.440255
Pd10	0.021056	0.134904	0	0.84404	0
Pd11	0.021056	0.134904	0	0.84404	0
Pd12	0.021056	0.134904	0	0.84404	0
Pd13	0.021056	0.134904	0	0.84404	0
Pd14	0.026244	0.269254	0	0.54874	0.155762

Figure 6. MRG share of the generator at bus-1 on different load

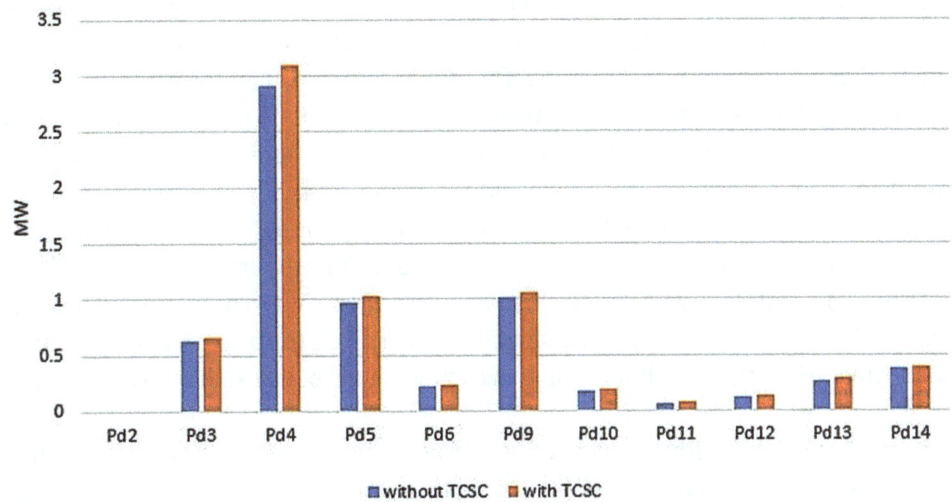

load at bus 4 is increased from 2.916 MW to 3.0918 MW. There is also slight increment of load at bus 3 and 9 because these buses are far from line 6-13, so contribution of generator 1 increased to these buses. However, buses near to line 6-13 will experience market power of generator 6 as it is nearby the TCSC.

Figure 7 indicates the market power of the generator at bus 2 is not much altered with respect to TCSC. They are almost similar to the base case because change in MRG is almost negligible and it is almost equal to the maximum capacity of that generator. The MW share has slightly decreased with respect to the base case for load at bus 3 (54.594 MW to 54.192 MW), bus 4 (44.88 MW to 44.70 MW) and bus 9 (15.41 MW to 15.18 MW).

Figure 7. MRG share of generator at bus-2 on different loads

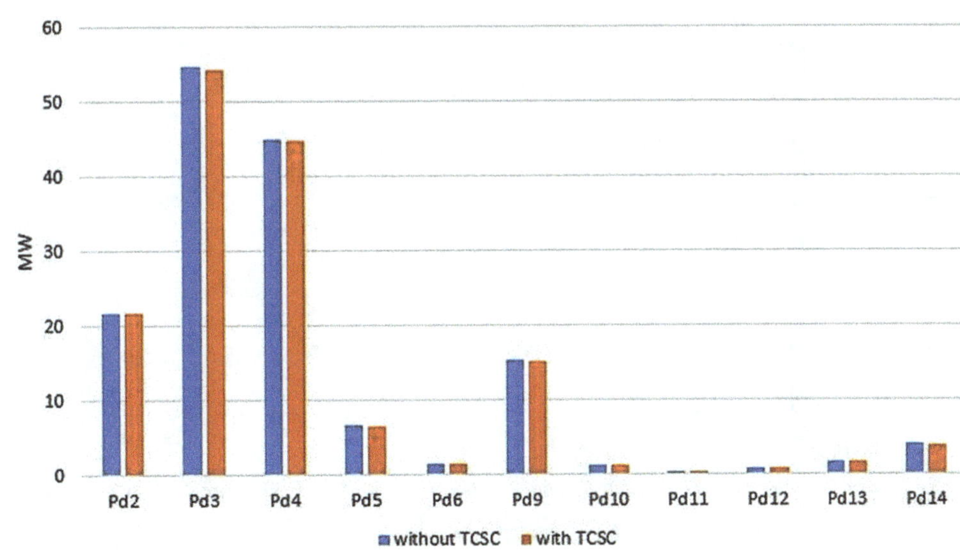

Figure 8 shows that the MW share of the generator at bus 3 is totally met by the load at the same bus and the share of MRG to the other loads is zero. This is because the load at bus 3 (i.e. 94.2 MW) is more than the generation capacity of the generator at bus 3 (i.e. 63 MW). In case of TCSC, the generator share to the load has increased from 41.3% to 41.7% (38.97 MW to 39.344MW).

The generator at bus 6 is very near to the line 6-13 where TCSC is placed, so the shares from this generator to the loads, which are nearby to this line, are more as indicated in Figure 9. This shows the dominance of this generator over these loads and it can exercise its market power on these loads. Generator 6 is located in a position such that majority of the loads are connected with this bus directly. Therefore we can see that the shares has increased for load at 6 bus from 8.47 MW to 10.86 MW, for load at 11 bus from 2.33 MW to 2.95 MW, for load at 12 bus from 5.036 MW to 5.637 MW, for load at 13 bus from 9.37 MW to 11.39 MW. Also the shares has increased for bus 10 and 14 which though are not directly connected to bus 6, but still there shares has increased because the MW losses have reduced when TCSC is connected.

Figure 10 shows that contribution of load at bus 9 from generator at bus 8 is increased in the present case. This is due to the reason that bus 9 is very near to generator at bus 8. However, contribution to load at bus 14 is reduced as generator at bus 6 has increased its contribution to load at bus 14. This is due to the reason that TCSC is placed very near to bus 6, so its market power has increased significantly.

Case 2: When TCSC Is Placed in the Line 12-13 as Per the Sensitivity Index

TCSC is now placed in line 12-13 as per the sensitivity index obtained in Table 1.

Based on NMRS values for each generator from Table 2 and Table 4, MRG share is plotted for different loads due to individual generators. Following are the MRG share plots for generator 1, 2, 3, 6 and 8.

Figure 8. MRG share of generator at bus-3 on different loads with TCSC in line 6-13

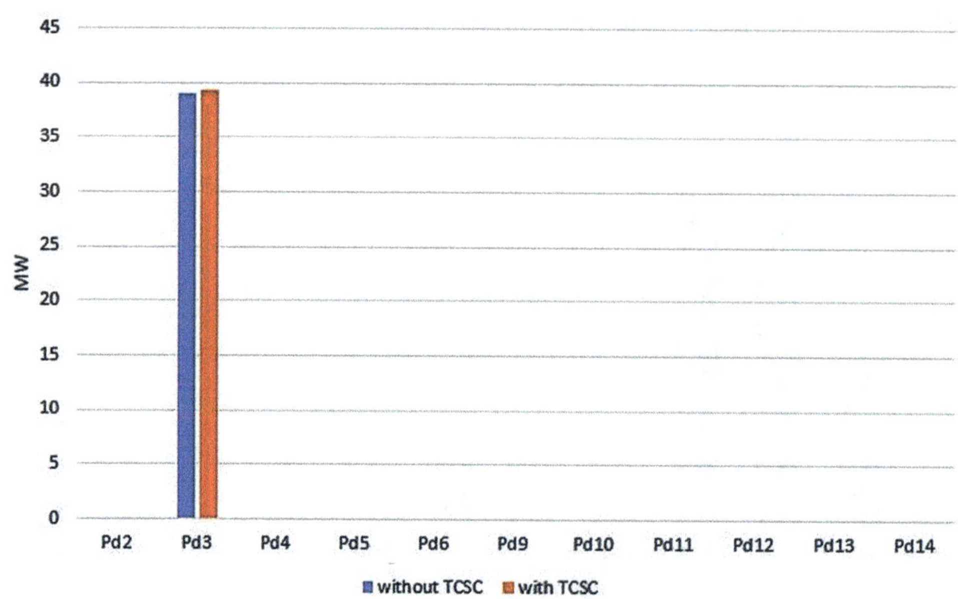

Figure 9. MRG share of the generator at bus-6 on different loads

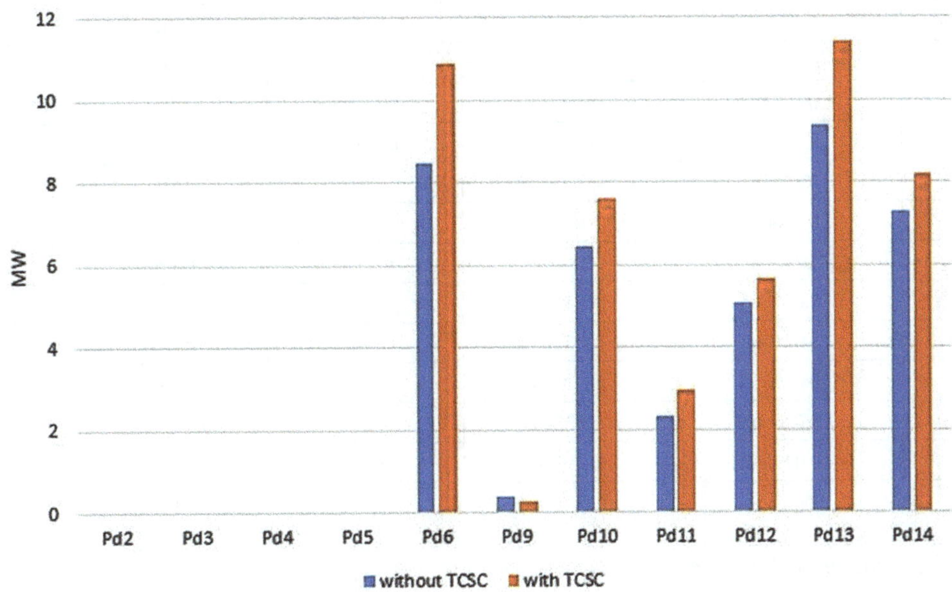

Figure 10. MRG share of generator at bus-8 on different loads with TCSC in line 6-13

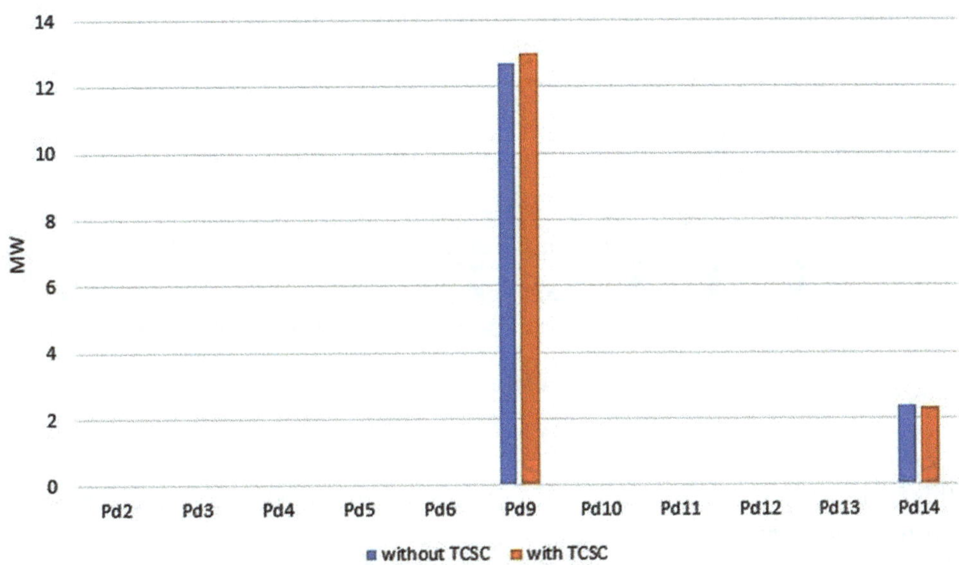

Figure 11 depicts that when TCSC is placed at its optimal location then overall generation of the generator at bus 1 is increased. Earlier there was no MW share to the load at bus 2, but in this case generator 1 is providing 1.68 MW capacity to the load at bus 2. This increase in capacity is due to the reduction in losses between lines 1-2. Similarly the MW share for bus 4, 5 has increased from 2.916 and 0.967 to 4.99 and 3.39 MW respectively. The MW share has also increased for load located at bus 9, 12,

Table 4. NMRS values of each generator to each load with TCSC in the 12-13 line

	Pg1	Pg2	Pg3	Pg6	Pg8
Pd2	0.123652	0.876348	0	0	0
Pd3	0.006472	0.045867	0.947662	0	0
Pd4	0.167196	0.52323	0.309574	0	0
Pd5	0.446632	0.553368	0	0	0
Pd6	0.026065	0.032294	0	0.941642	0
Pd9	0.06341	0.19841	0.11738	0.000548	0.620251
Pd10	0.026065	0.032294	0	0.941642	0
Pd11	0.026065	0.032294	0	0.941642	0
Pd12	0.026065	0.032294	0	0.941642	0
Pd13	0.026065	0.032294	0	0.941642	0
Pd14	0.039547	0.092263	0.042375	0.6019	0.223915

Figure 11. MRG share of the generator at bus-1 on different loads

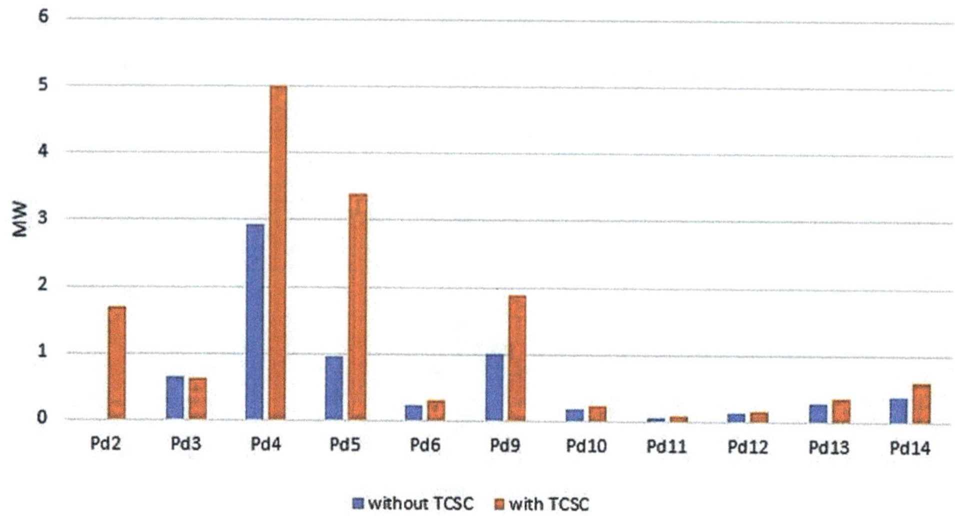

13, 14, but this increase in share is quite less significant as these loads are located at the buses which are very far from the generator 1.

Figure 12 shows that market power of the generator at bus 2 has reduced significantly because of the reduction in MRG. The MW share for load at bus 3 has reduced from 54.59 MW to 28.32 MW, because now the load at bus 3 is taking its majority share from generator 3 which is 62.29 MW. Previously, load at bus 3 from generator at bus 3 was 39.344 MW. The MW share for load 4 has decreased from 44.88 MW to 25.01 MW, as generator 1 is now supplying the capacity which was earlier supplied by the generator 2. That means generator 1 is exerting its market power on loads 2, 3 and 4 and has taken over the competition from generator 2.

Figure 12. MRG share of generator at bus-2 on different load with TCSC in line 12-13

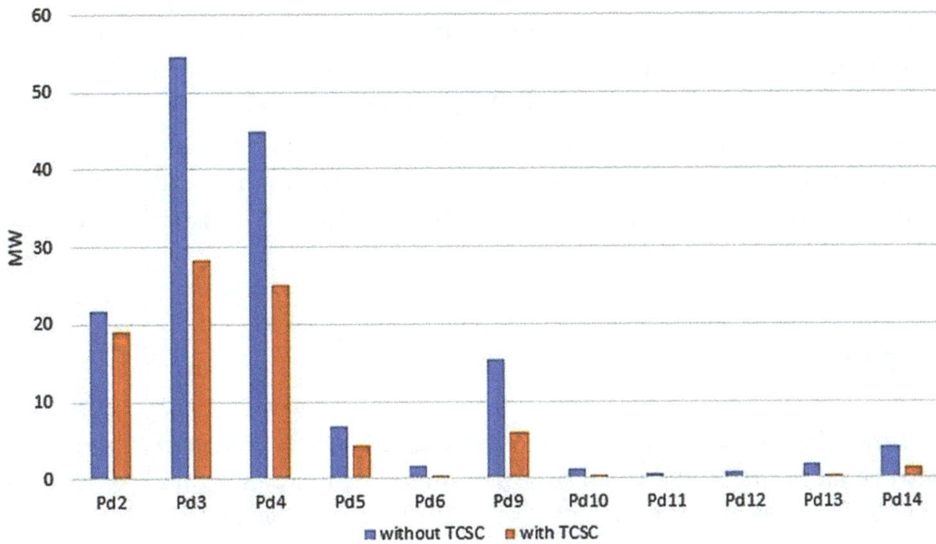

Figure 13 shows that the MW share of the generator at bus 3 is increased from 38.97 MW to 62.264 MW. In the previous case, load at bus 3 was fulfilling its requirement from generator 2 (54.19 MW) and generator 3 (39.44 MW). But now its requirement is mainly fulfilled by the generator 3 which is supplying 62.264 MW and generator 2 which is supplying 28.32 MW of power. Hence now generator 3 is exercising its market power on this load.

Figure 13. MRG share of generator at bus-3 on different load with TCSC in line 12-13

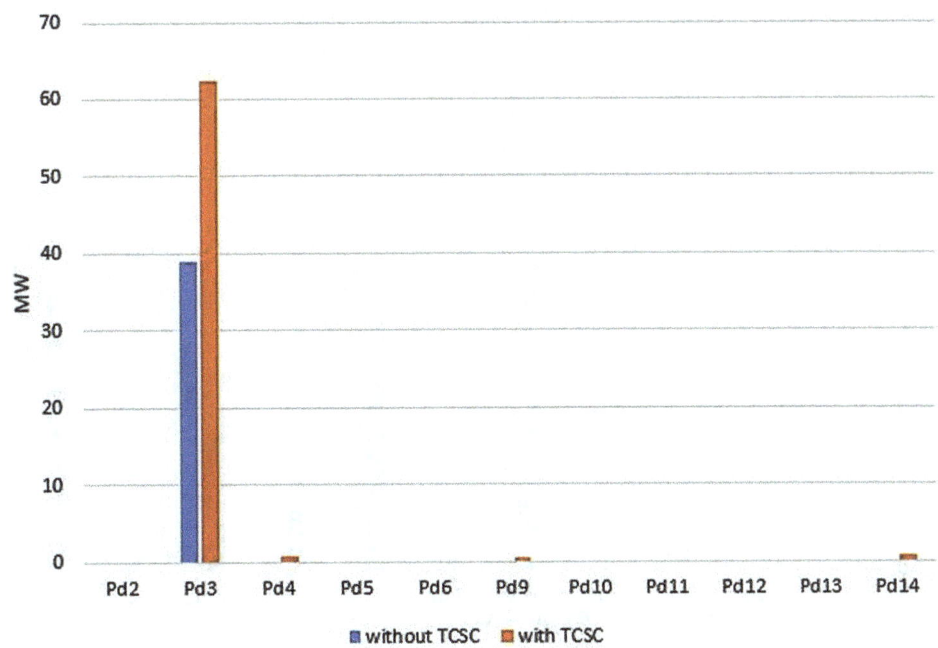

Figure 14 shows that the MW share from generator to the loads which are nearby to the line are more in case of optimal location of TCSC. This shows the dominance of this generator over these loads and it can exercise its market power on these loads. However these changes are similar to the changes observed in case-I, because for case-I TCSC was placed in line 6-13 and for case-II TCSC is placed in line 12-13. So the values of MW share are quite close in both the cases.

Figure 15 shows the share of the generator at bus 8 to the load at bus 9 has increased from 43% to 62.05% (12.7 MW to 18.29 MW). Since only load at bus 9 is near to this generator therefore this generator can exercise its market power on this load only. Share of load 14 is also increased from 2.35 MW to 3.33 MW.

FUTURE SCOPE

In this chapter only TCSC is taken as the FACTS device to see the effect on market power. Other FACTS devices can also be taken to calculate market power. Moreover more than one FACTS device can also be placed and its effect on market power can be seen. Cost analysis can also be done for each generator to observe the real profit which a generator is making.

CONCLUSION

This chapter outlines various factors that should be considered when assessing the competitiveness of a generation market. Market power is an anti-competitive practice which should be regularly checked

Figure 14. MRG share of the generator at bus-6 on different loads

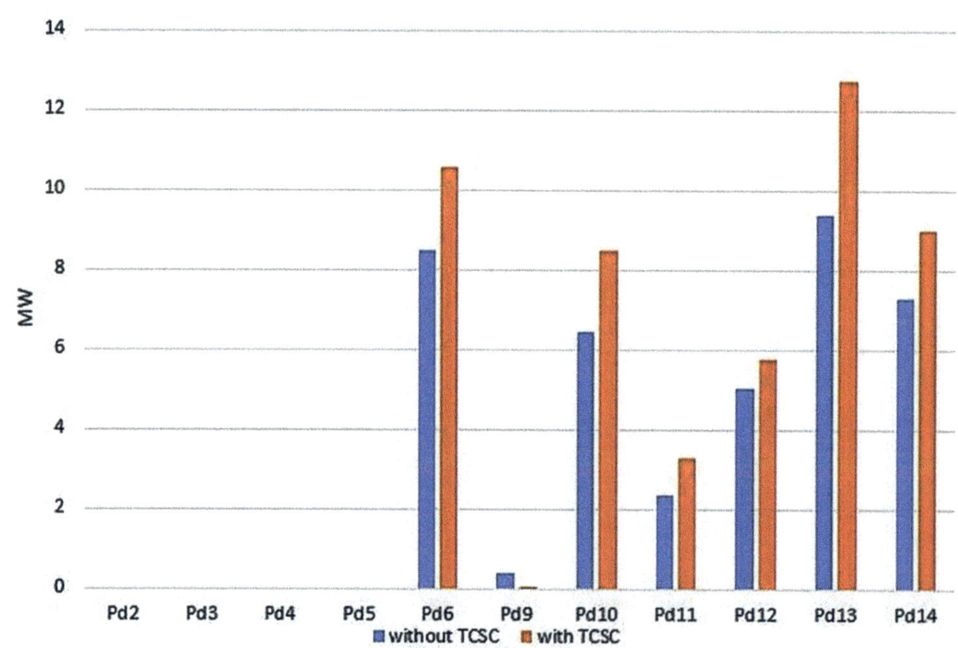

Figure 15. MRG share of the generator at bus-8 on different loads

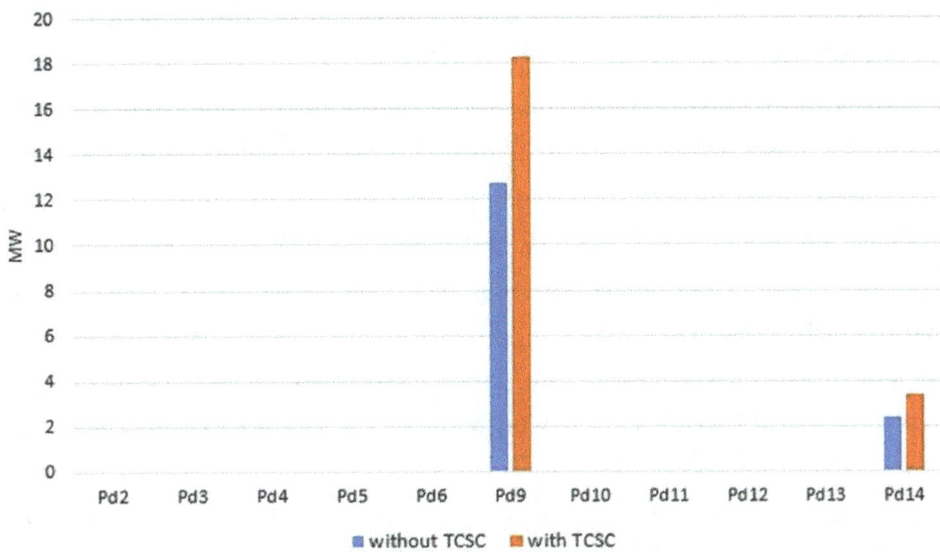

in a deregulated market. Presence of market power results in reduced competition, lower power quality and monopoly. Different indices have been discussed in this chapter, each having its own importance and application. MRS takes into account the load variation on market power. The NMRS accounts for the geographical difference of market power due to transmission constraints. Impact of availability and unavailability of a component due to random failures is expressed by ENMRS.

Mitigation of market power is an important issue to ensure the competition and fair practice in a generation market. One of the mitigation strategy is the establishment of bidding rules and the bidding should be carried out by an independent market participant. Bilateral contracts in which the prices are discussed and agreed prior to the dispatch of power should also help in mitigating market power.

Two cases have been discussed in this chapter to see the effect of FACTS devices on the market power. While placing FACTS devices one must decide which factor to consider, i.e. whether to opt for minimum losses (optimal location of TCSC) or minimum market power.

REFERENCES

Abhyakar, A. R., & Khaparde, S. (n.d.). *Introduction to Deregulation in Power Industry*. Retrieved from http://www.nptel.ac.in/courses/108101040/Module 7/L01-Introduction to Deregulation-1.pdf

Borenstein, S., & Bushnell, J. (1998). *An empirical analysis of the potential for market power in California's electricity industry.* Working Paper 6463. NBER Working Paper Series, National Bureau of Economic Research; March. Retrieved from http://www.nber.org/papers/w6463

Borenstein, S., Bushnell, J., & Knittel, C. R. (1999). Market Power in Electricity Markets: Beyond Concentration Measures. *Energy Journal*, *20*(4), 65–88. doi:10.5547/ISSN0195-6574-EJ-Vol20-No4-3

Bose, S., Wu, C., Xu, Y., Wierman, A., & Mohsenian-Rad, H. (2015). A Unifying Market Power Measure for Deregulated Transmission-Constrained Electricity Markets. *IEEE Transactions on Power Systems, 30*(5), 2338–2348. doi:10.1109/TPWRS.2014.2360216

Cho, I.-K., & Kim, H. (2007). Market power and network constraint in a deregulated electricity market. *Energy Journal, 28*(2), 1–34. doi:10.5547/ISSN0195-6574-EJ-Vol28-No2-1

Dalton, J. C., & markets. (1997). Assessing the competitiveness of restructured generation service markets. *The Electricity Journal, 10*(3), 30–39. doi:10.1016/S1040-6190(97)80375-2

David, A. K., & Wen, F. (2001). Market power in electricity supply. *IEEE Transactions on Energy Conversion*. doi:10.1109/60.969475

Díaz, C. A., Webster, M. D., Villar, J., & Campos, F. A. (2016). Dynamics of Market Power in ERCOT System: A Fundamental CSFE With Network Constraints. *IEEE Transactions on Power Systems, 31*(2), 861–871. doi:10.1109/TPWRS.2015.2411051

Elfstadius, M., & Gecer, D. (2008). *Method to Detect and Measure Potential Market Power Caused by Transmission Network Congestions on Electricity Markets*. Linköping, Sweden: Department of Electrical Engineering, Linköpings University.

Estebsari, A., Sadeghzadeh, S. M., & Fotouhi, M. A. (2009). Transmission constraints impacts on monopolistic production capacity under strategic bidding. *2009 International Conference on Electrical Machines and Systems*. 10.1109/ICEMS.2009.5382644

Feng, D., Zhong, J., & Gan, D. (2008). Reactive Market Power Analysis Using Must-Run Indices. *IEEE Transactions on Power Systems, 23*(2), 755–765. doi:10.1109/TPWRS.2008.919416

Gad, M., Shinde, P., & Kulkarni, S. U. (2012). Optimal Location of TCSC by Sensitivity Methods. *International Journal of Computational Engineering Research, 2*(6), 162–168.

Hasani-Marzooni, M., & Hosseini, S. H. (2013). Short-term market power assessment in a long-term dynamic modeling of capacity investment. *IEEE Transactions on Power Systems, 28*(2), 626–638. doi:10.1109/TPWRS.2012.2217990

Horizontal Market Power in Restructured Electricity Markets. (2000). Washington, DC: Office of Economic,Electricity and Natural Gas Analysis, U.S. Department of Energy.

Joskow, P. L., & Tirole, J. (2000). Transmission rights and market power on electric power networks. *The Rand Journal of Economics, 31*(3), 450–487. doi:10.2307/2600996

Karthikeyan, S. P., Raglend, I. J., & Kothari, D. P. (2013). A review on market power in deregulated electricity market. *International Journal of Electrical Power & Energy Systems, 48*, 139–147. doi:10.1016/j.ijepes.2012.11.024

Karthikeyan, S. P., Sahoo, S. K., Roopesh, R., Mathur, S., Raglend, I. J., & Kothari, D. P. (2012). Utilizing FACTS devices in enhancement of market power under congestion-a threat to the deregulated electricity market. *2012 Annual IEEE India Conference (INDICON)*. 10.1109/INDCON.2012.6420632

Khajeh, H., Foroud, A. A., & Firoozi, H. (2016). New ex-ante indices of market power. In *24th Iranian Conference on Electrical Engineering (ICEE)*. 10.1109/IranianCEE.2016.7585391

Khan, M. T., Siddiqui, A. S., Sarwar, M., & Iqbal, F. (2015). Mitigation of Market Power in deregulated power system using Nash Equilibrium. *2015 Annual IEEE India Conference (INDICON)*, 1–5. 10.1109/INDICON.2015.7443645

Knittel, C. R., & Metaxoglou, K. (2008). Diagnosing unilateral market power in electricity reserves market. *Journal of Energy Markets*, *1*(1), 65–95. doi:10.21314/JEM.2008.001

Lai, L. L. (2001). *Power System Restructuring and Deregulation: Trading, Performance and Information Technology*. John Wiley & Sons, Ltd. doi:10.1002/0470846119

Lee, Y. Y., Hur, J., Baldick, R., & Pineda, S. (2011). New Indices of Market Power in Transmission-Constrained Electricity Markets. *IEEE Transactions on Power Systems*, *26*(2), 681–689. doi:10.1109/TPWRS.2010.2060738

Lesieutre, B. C., Rogers, K. M., Overbye, T. J., & Borden, A. R. (2011). A Sensitivity Approach to Detection of Local Market Power Potential. *IEEE Transactions on Power Systems*, *26*(4), 1980–1988. doi:10.1109/TPWRS.2011.2105893

Rajderkar, V. P., & Chandrakar, V. K. (2009). Comparison of Series FACTS Devices Via Optimal Location in a Power System for Congestion Management. *2009 Asia-Pacific Power and Energy Engineering Conference*. 10.1109/APPEEC.2009.4918227

Reddy, S. S., Kumari, M. S., & Sydulu, M. (2010). Congestion management in deregulated power system by optimal choice and allocation of FACTS controllers using multi-objective genetic algorithm. *IEEE PES T&D*, *2010*, 1–7. doi:10.1109/TDC.2010.5484520

Shafie-khah, M., Heydarian-Forushani, E., Golshan, M. E. H., Moghaddam, M. P., Sheikh-El-Eslami, M. K., & Catalão, J. P. S. (2015). Strategic Offering for a Price-Maker Wind Power Producer in Oligopoly Markets Considering Demand Response Exchange. *IEEE Transactions on Industrial Informatics*, *11*(6), 1542–1553. doi:10.1109/TII.2015.2472339

Shafie-khah, M., Moghaddam, M. P., & Sheikh-El-Eslami, M. K. (2016). Ex-ante evaluation and optimal mitigation of market power in electricity markets including renewable energy resources. *IET Generation, Transmission & Distribution*, *10*(8), 1842–1852. doi:10.1049/iet-gtd.2015.0981

Shahidehpour, M., & Alomoush, M. (2001). *Restructured Electrical Power Systems: Operation, Trading, and Volatility*. Taylor & Francis. Retrieved from https://books.google.co.in/books?id=U7WonDR_b00C

Shukla, U. K., & Thampy, A. (2011). Analysis of competition and market power in the whole sale electricity market in India. *Energy Policy*, *2*(39), 2699–2710. doi:10.1016/j.enpol.2011.02.039

US Department of Justice. (2015). Retrieved from https://www.justice.gov/atr/15-concentration-and-market-shares

Wolak, F. A. (2003). Measuring unilateral market power in wholesale electricity markets: The California market 1998–2000. *The American Economic Review*, *93*(2), 425–430. doi:10.1257/000282803321947461

Zhong, J., & Bhattacharya, K. (2002). Toward a competitive market for reactive power. *IEEE Transactions on Power Systems*, *17*(4), 1206–1215. doi:10.1109/TPWRS.2002.805025

ADDITIONAL READING

Lai, L. L. (2001). *Power System Restructuring and Deregulation*. London: John Wiley & Sons, Ltd. doi:10.1002/0470846119

Shahidehpour, M., Yamin, H., & Li, Z. (2002). *Market Operation in Electric Power System*. New York: Wiley Press. doi:10.1002/047122412X

KEY TERMS AND DEFINITION

Deregulation: Disintegrating or unbundling the existing power system into the basic components of generation, transmission, and distribution and offering each component separately for sale with separate rates.

DG: Distributed generation (DG) also referred by the name of on-site generation is a small capacity (1 KW to 10,000 KW) generating plant which provides an alternative route for the enhancement of traditional power system.

FACTS: A flexible alternating current transmission system (FACTS) is a system composed of power electronics components used for the AC transmission of electrical energy to enhance controllability and increase power transfer capability of the network.

Market Power: Market power is the ability to raise the prices above the competitive level without affecting the generator's dispatch.

Regulation: The rules or set of instructions that are followed by an operator or organization for the efficient and reliable operation of a power system.

STATCOM: Static synchronous compensator (STATCOM) is a shunt device of the flexible AC transmission systems (FACTS) family using power electronics to control power flow and improve transient stability on power grids.

TCSC: Thyristor controlled series compensation is a well-established technology which is used to reduce the transfer reactance in the transmission line in order to provide sufficient load compensation.

Chapter 15
Seamless Control and Unified Dynamic Energy Management in a Renewable/Clean Energy Integrated Self-Reliant DC Microgrid:
Integration of Renewable Sources With High-Gain Power Processing Stages

Vulisi Narendra Kumar
National Institute of Technology Meghalaya, India

Gayadhar Panda
National Institute of Technology Meghalaya, India

Bonu Ramesh Naidu
Indian Institute of Technology Kharagpur, India

ABSTRACT

The growing demand for electrical energy calls for the assimilation of renewable energy sources to the main utility grid. Multiple renewable energy sources (RESs) like solar PV array, wind turbine, micro-hydro plant, etc. can be combined and controlled to form a microgrid. In spite of the availability of different microgrid topologies, DC microgrid largely facilitates the injection of DC power from various renewable energy sources into the stabilised DC power pool. The requirement for a minimal number of conversion stages, simple structure, economic operation, and numerous localised applications are driving factors for the DC microgrid technology. The mettle of the DC microgrid technology lies in choosing the appropriate microgrid participants for energy interchange and the suitable supervisory control to tap power from the microgrid partakers even after respecting their operating constraints. The use of high gain DC-DC converters is inevitable in DC microgrid due to the low terminal voltage levels of different RESs.

DOI: 10.4018/978-1-5225-3935-3.ch015

NOMENCLATURE

L_1 : Primary inductance of the coupled inductor (Henry).
L_2 : The secondary inductance of the coupled inductor (Henry).
L_3 : The inductance of the bi-directional high gain converter (Henry).
L_k : Leakage inductance of the coupled inductor (Henry).
L_m : Magnetizing inductance of the coupled inductor (Henry).
C_1 : The capacitance of the passive clamp circuit (Farad).
D_1 : The diode of the passive clamp circuit.
D_2 : Feedback diode of the high gain converter.
D_3 : Output diode of the bi-directional high gain converter.
D_4 : The diode used for the step-down operation of the bi-directional high gain converter.
C_o : Output capacitor of the high gain converter.
n : Turns ratio of the coupled inductor.
d : Duty ratio supplied to MOSFET switch of the Converter.
P_{pv} : The instantaneous power generated by PV source (Watt).
P_{wn} : The instantaneous power generated by Wind source (Watt).
P_{sc} : The instantaneous power supplied/absorbed by supercapacitor (Watt).
P_{bat} : The instantaneous power supplied/absorbed by a battery (Watt).
I_{D1} : Current flowing through the diode (D_1) (Ampere).
I_{Co} : Current flowing through the output capacitor (C_o) (Ampere).
V_i : The voltage supplied to the high-gain converter (Volt).
V_{DC} : The voltage supplied to the bi-directional high-gain converter (Volt).
V_{D1} : Voltage Across the diode (D_1) (Volt).
V_{C1} : Voltage Across the capacitor (C_1) (Volt).
V_{C2} : Voltage Across the capacitor (C_2) (Volt).
V_{Co} : Voltage Across the capacitor (C_o) (Volt).
V_{L1} : Voltage Across the inductor (L_1) (Volt).
V_{L2} : Voltage Across the inductor (L_2) (Volt).
I_D : Dark Current of the PV cell (Ampere).
I_o : Saturation Current of the diode (Ampere).
V_c : The voltage of the PV cell (Volt).
I : Photovoltaic Current of the PV cell (Ampere).
R_s : The series resistance of the PV cell (Ohm).
R_{sh} : Shunt resistance of the PV cell (Ohm).
T_w : Mechanical input torque of the wind turbine (N-m).

V_w : Wind speed striking the wind turbine
(Rpm).
β : Pitch Angle of the rotor blades.
C_p : Power coefficient of the wind turbine.
E_j : Energy requirement per day (Ah).
N : Number of sunshine hours.
$P_{\max}$: Maximum power produced from the PV array (Watt).
E_{worst} : Energy produced during the cloudy weather from PV array $\left(Wh / m2 / day\right)$.
C_l : Factor to include losses due to pitch and alignment in PV cell
N_{Pv} : A number of PV cell required.
N_{wt} : A number of wind turbine required.
V_{error} : DC link voltage deviation (volt).
$I_{ref}\left(s\right)$: Reference Current generated from the controller.
$I_{sc}\left(s\right)$: Reference Current supplied to a supercapacitor.
$I_{bat}\left(s\right)$: Reference Current supplied to the battery.

INTRODUCTION

DC microgrid with centralized solar/wind generation and decentralized energy storage is the simple, reliable, cost-effective, scalable and highly efficient solution to provide access to electricity in remote places. The primary advantages possessed by the DC microgrid are

- Electrification of loads is more economical compared to AC Microgrid.
- Higher efficiency compared to AC Microgrid.
- Can supply power for household appliances (i.e. Light, TV, Fan, Mobile charger, etc.), community need (i.e. water pumping) and needs of small business establishments.
- Less investment for entrepreneurs compared to AC microgrid.

The DC microgrid with different energy sources and energy storage systems (ESS) is shown in Figure 1. To maintain the high voltage at the common DC link bus bar with the available low voltage renewable technologies, a high-gain DC-DC power processing stages are used in this chapter. The higher voltage DC microgrid has the following benefits

- Improves efficiency because of the higher voltage levels.
- Easy integration of the renewable sources with the main utility grid.
- Cost of the storage devices like battery and supercapacitor bank reduces drastically.
- High survivability during transient conditions.
- Provides improved power quality and reliability.
- Electric vehicles can be used for backup/emergency power demand.

The high voltage DC microgrid is becoming more significant in the field of transportation, commercial buildings, telecom and data centres due to its economic benefits to both the operator and the consumers. The increasing demand for high voltage DC microgrid demands for innovative power processing units.

BACKGROUND

A microgrid contains distributed energy resources (DERs) and local loads that act as a single controllable entity and is connected to the utility grid at the distribution level. It can be operated in an islanded mode at the time of grid failure (Mahmoud, Saif Ur Rahman& Sunni, 2015). DC microgrid is one of the most focused area as it facilitates the inclusion of conventional or non-conventional energy resources of different operating characteristics with a common DC power pool (Ghareeb & Mohammed, 2013; Xu & Cheng, 2011). The renewable sources are connected to common bus bar through the power electronic boost converter for stepping up the voltage level. Power electronic converters along with their control perform a significant part in the efficient addition of renewable energy sources into the microgrid. Improved efficiency, negligible power quality problems and sophisticated control are the driving factors for the increased usage of DC microgrids (Sannino, Postiglione & Bollen, 2003; Hammerstrom, 2007).

Figure 1. DC microgrid with renewable energy resources

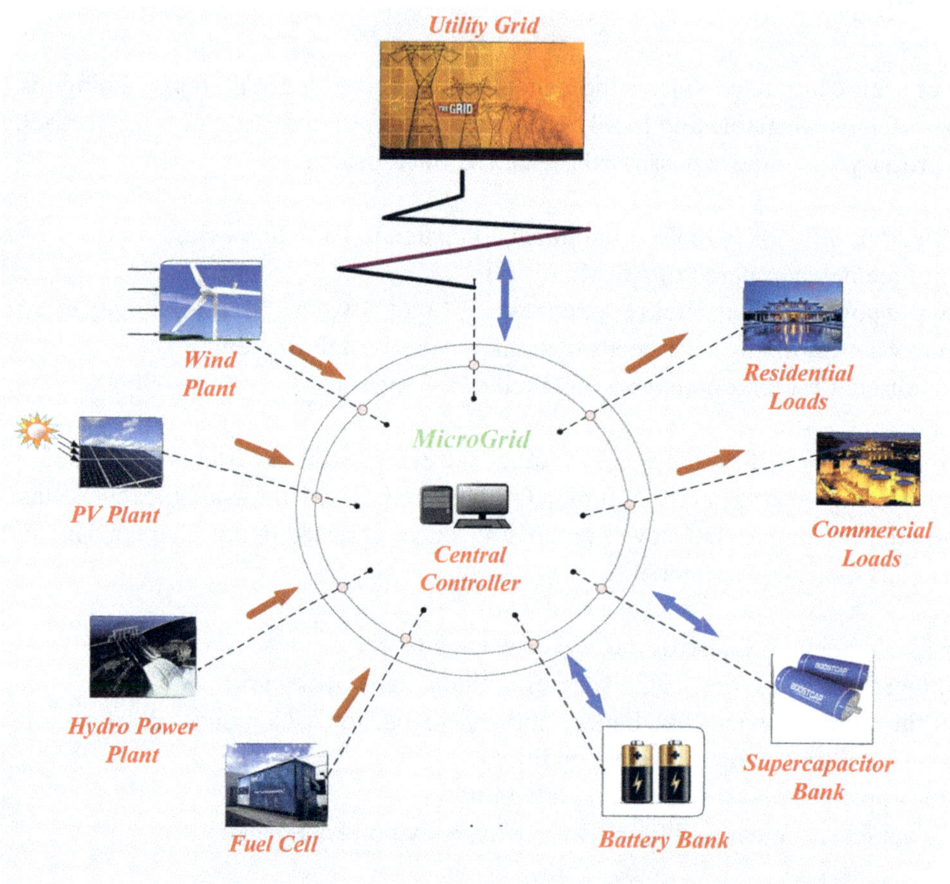

Renewable energy sources are clean and environment-friendly in nature. Hence these sources are finding the maximum application in power industry irrespective of their sporadic nature and low voltage generation limitations. Solar PV generation falls into the group of most established and technologically advanced renewable energy generation. It can be installed near the loads and thus transmission losses, cost per kWh are minimized (Jewell, 1989). Though minimal running cost is incurred in PV generation, its dependence on weather condition makes it unavailable throughout the day ensuing reduction in the reliability of the system. PV power output is intermittent in nature due to the sporadic nature of the solar irradiance. The overall PV power output depends on the climate and site location of the PV plant (Raygani, Sharma & Saha, 2015). When a large amount of PV power is transmitted into the weak utility grid, it may pose stability and voltage swing problems for short time duration. For example, a typical weak utility grid with 40% PV penetration can cause a voltage swing of 4V. Whereas, it creates only 0.14V deviation for a strong network with similar penetration levels (Sayeef et al., 2012). The cost of energy storage devices can be reduced by extracting maximum power from the PV array. Hence, it is always implicit to operate the PV array at maximum power point (MPP) to facilitate the abstraction of maximum power. This requires a power processing stage controlled by a maximum power point tracking (MPPT) algorithm. The perturb and observe algorithm and incremental conductance algorithm are the basic MPPT techniques used for PV array (Dorofte, Borup & Blaabjerg, 2005).

Wind power generation is also one of the clean, reliable & lowest-cost renewable energy technologies available today. It is widely used in microgrid technology around the globe (DeMeo, Grant, Milligan & Schuerger, 2005). The wind power produced depends on the wind velocities. As the wind speed is unpredicted and fluctuating in nature, the wind generation is unpredictable. The maximum power from the wind turbine can be extracted by using Maximum Power Point Tracking Technique. The multitude of MPPT algorithms available for WECS include tip speed ratio (TSR) algorithm, hill-climb search (HCS) algorithm, optimal torque control algorithm, power signal feedback control algorithm, perturb and observe algorithm. TSR algorithm depends on the exact wind speed to maintain proper tip speed ratio. The HCS based control algorithm is simple and can be implemented easily but it takes a longer duration to settle for new value (Kadri, Marzougui & Bacha, 2016).

ESS's come with many advantages like overall capacity boosting, intermittency handling, transmission cramming relief, improved power quality and many other ancillary services when being applied to microgrids (Kreczanik, Venet, Hijazi& Clerc, 2014; Graditi, Ippolito, Telaretti & Zizzo, 2015). ESS can be categorised into two groups i.e., transitory response and stress-free response devices. Among the multitude of devices available under transitory response, the supercapacitor is a well-established technology finding many applications. Supercapacitor modelling, sizing and application in a hybrid system are described in (Parvini, Siegel, Stefanopoulou & Vahidi, 2016; Phatiphat Thounthong, StephaneRael & Bernard Davat, 2009). Under stress-free response devices, the battery is known for its maximum storage capacity. Modelling and application of battery storage system are dealt in (Kawakami *et al.*, 2014). Due to multiple causes like the requirement of voltage balancing circuits for high voltage operation, space constraints, etc., ESS also come in low voltage ratings. The combination of battery and supercapacitor offers a better power density as well as improved energy density. Therefore, to establish an effective and reliable PV power transferring capability, energy storage systems (ESSs) such as battery and supercapacitor bank are used. The energy storage system absorbs/delivers the power depending on the system requirement. However, this methodology improves the power transferring capability but the cost of energy storage system increases drastically. Therefore, to reduce the cost of storage devices, the wind turbines are used with the PV array. The combination of PV and wind turbine are capable of adjusting output power quickly. Consequently, transmission capability and reliability also improves (Long, Eghlimi & Zhang, 2017).

The Solar produces power at a very low terminal voltage, which is not adequate to drive some specified loads. The wind turbine connected through conventional converters suffers from more loss when operated at high voltage ratings. The DC-DC boost converter connects the PV array with the common DC bus. The conventional DC-DC converter is subjected to high stress and more losses to achieve high voltage at their output terminals. The increasing priority for inclusion of low voltage renewable generation and ESS demands employment of high gain DC-DC power processing units. The topology with an interleaved coupled inductor, an intermediate buffer capacitor and a passive clamp circuit results in loss minimization, high efficiency and low-cost. Owing to its relative advantages high gain DC-DC boost converter presented in (Moumita Das & Vivek Agarwal, 2016) is used as a power processing stage for the solar and wind source in this chapter. Similarly, to achieve the benefits of high-gain, high-efficiency and reduced complexity in control, a bi-directional DC-DC converter operated with two switches is introduced in (Shreelakshmi, Das & Agarwal, 2013). This topology has been adopted to interconnect the ESS to the common DC bus in this chapter.

MAIN FOCUS OF THE CHAPTER

The significant increase in the usage of electronic devices demands restructuring DC microgrid and a new supply network. A microgrid is a system which supplies power to customers irrespective of the availability/non-availability of the utility grid. It comprises of different distributed energy resources (DERs) and various energy storage systems which are connected to local loads as well as a centralized grid (Lu & Wang 2017). A microgrid can operate in standalone and also in a grid-connected mode based on the operating scenarios. The concept of microgrids expedites the growth of renewable integration in the traditional power grid. The researchers foresee that future power systems embrace a number of microgrids (Rocabert, Azevedo, Luna, Guerrero, Candela & Rodríguez, 2011). More important is the exchange of electricity among microgrids with/without centralized grid that can provide a major revenue income for the power sector. This feature attracts the private investors into the microgrid technology and leads to innovative technologies. Besides this, the factors driving for choosing a microgrid are

- **Improving Local Energy Delivery:** The microgrid enables the energy delivery to the local loads so that the transmission and distribution losses can be reduced greatly.
- **Increasing Reliability:** Microgrids help to maintain the stable power supply to the consumer without interruption.
- **Saving Money:** As the generation by microgrid is possible within close proximity of loads, hence the cost intricacies in transmission and distribution lines are reduced.
- **Making the Grid More Resilient:** Microgrids can be isolated during abnormal fault conditions occurred in the local centralized grid.

Issues associated with existing microgrids

- Implementation of MPPT technique for renewable energy sources like wind/solar becomes cumbersome due to the low voltage levels of these devices.

- Design and control of proper high gain interfacing converters are yet another challenging task.
- The co-existence of different energy sources and storage devices makes it difficult to address the issues of DC-link voltage control and power sharing among the microgrid participants.

The main focus of this chapter is listed as follows.

- To implement incremental conductance (INC) MPPT algorithm for PV array and perturb & observe (P&O) MPPT algorithm for the wind turbine.
- To employ high-gain high-efficient DC-DC converters for PV array and a wind turbine in the DC microgrid.
- To use power processing stages with the high-gain high-efficient bi-directional DC-DC converters for battery and supercapacitor bank system in the DC microgrid.
- To maintain a constant voltage at common DC bus of the microgrid by executing the proper power-sharing among the microgrid participants. This task is done by the help of dynamic energy management algorithm.
- To check the real-time feasibility of the dynamic energy management algorithm by the help of FPGA kit.

SOLUTIONS AND RECOMMENDATIONS

The difficulties associated with the conventional DC-DC converters like low terminal voltage, more stress and high losses can be overcome by designing the high gain DC-DC converters. There are different types of high gain DC-DC converter topologies are available in the literature to achieve the high voltage levels at the terminals of renewable sources. Direct voltage levels boosting by the help of high-frequency transformer operated DC-DC converter and also called as the isolated current fed DC-DC converter is proposed in (Lee, Liang & Chen, 2014). This method of voltage boosting is convenient, safe, ground leakage currents can be eliminated and also the converter can be controlled with less complexity. A coupled inductor based DC-DC converter can be designed to achieve high voltage gains and can be designed with less cost, low rating components and without much complexity (Witulski, 1995). The interleaved coupled inductor employed boost converter with the application to high powers is proposed in (Garcia, Pomilio & Spiazzi, 2013). To reduce the losses associated with the leakage inductance, an active clamp network consisting of switch and capacitor can be used to provide a different path for the leakage energy to recover it (Kwon & Kwon, 2009). The passive clamp network containing passive components like diode, capacitor and resistor is proposed in (Lee, Kim & Choi, 2013). The passive clamp network is able to reduce voltage stress at the switch terminals by recovering leakage energy. An intermediate capacitor is used in combination with the coupled inductor can be used to achieve the high voltage gains without much increase in duty ratio dealt in (Hsieh, Chen, Liang & Yang, 2013). Besides that, a coupled inductor based boost converter with a passive clamp network is used to get high voltages presented in (Zhao & Lee, 2003). A coupled inductor in combination with the voltage multiplier circuit dealt in (Young, Chen, Chang, Ko & Jen, 2013). Also coupled inductor in combination with the voltage multiplier circuit and the intermediate capacitor is proposed in (Liang, Chen, Yang, L. S., Chen & Ioinovici, 2012). The demerits of the above-proposed topologies are high voltage spikes at the switch terminals with an increased turns ratio of the high-frequency transformer and also produces a large ripple in the transformer primary current. Also, this technique incurs with more cost, more losses and occupies more space (Prudente, Pfitscher, Emmendoerfer, Romaneli, & Gules, 2008). The drawbacks associated with coupled inductor are large leakage inductance, switching losses and reduced overall efficiency. Due to the presence of switch in active clamp network, it suffers from conduction losses across the switch. To overcome the demerits associated with the existing high gain DC-DC converter topologies, a new topology of high-gain converters employing interleaving coupled inductor, an intermediate storage capacitor and a passive clamp network with a reduced number of switches is proposed in (Moumita Das & Vivek Agarwal, 2016) is used in this chapter.

In the same manner, Conventional bi-directional DC-DC converters fail to enhance the low terminal voltage of the ESS to a much higher level. In literature, different high gain bi-directional DC-DC converter topologies are proposed to overcome the above-mentioned problem. Among them, a high-gain bi-directional DC-DC converter employing four switches is proposed in (Chen, Yim-Shu Lee, Hui, Dehong Xu & Wang, 2000). Also, a high gain bi-directional DC-DC converter employing three switches is proposed in (Wai, Duan & Jheng, 2012). To eliminate the losses associated with the existing bi-directional high gain DC-DC converter topologies, a new topology of high-gain bi-directional converter topology with a reduced number of switches is presented in Shreelakshmi, Das & Agarwal, 2013) is implemented in this chapter. The conventional DC-DC converter is replaced by high-gain, high-efficient DC-DC converter due to having the advantage of soft switching capability. The solution for the problem of matching the low voltage renewable sources with the high voltage DC bus is overcome by designing high-gain high-efficient converters. The proposed high-gain high-efficient DC-DC boost converter topology contains a coupled inductor, an intermediate storage capacitor and a passive clamp network. The interleaved coupled inductor is used to charge the intermediate storage capacitor and this stored energy is recovered through the passive clamp network. The detailed modelling and sizing of the proposed high-gain high-efficient converter are presented in the next section.

The DC microgrid with solar and wind sources supported by battery and supercapacitor banks is shown in Figure 2. All devices are interfaced to the high voltage DC link through the high gain power processing stages. Owing to the intermittent nature of solar/wind, a hybrid ESS is employed to assist the PV array and wind turbine. The PV array and WECS are interfaced to the DC-link via high-gain high-efficiency DC-DC converters. The energy storage devices are allied through high-gain bi-directional converters. The power balance equation in the DC microgrid can be expressed as follows

$$P_{pv} + P_{wn} = P_{load} + P_{SC} + P_{bat} \tag{1}$$

where P_{pv} is the instantaneous power generated by PV source: P_{wn} is the instantaneous power generated by the wind turbine; P_{load} is the power demanded by the load; P_{sc} and P_{bat} denote the amount of power flowing from/to the supercapacitor and battery bank respectively (positive power represents discharging operation and vice-versa). All powers are in watts.

Modelling and Sizing of Major DC Microgrid Participants

The mathematical modelling of the DC microgrid components is necessary to analyze the steady-state and dynamic behaviour of the proposed system. The primary sources of the DC microgrid are PV array and wind turbine. The DC microgrid is assisted by ESS (i.e., battery and supercapacitor) to maintain the

Figure 2. DC microgrid under study

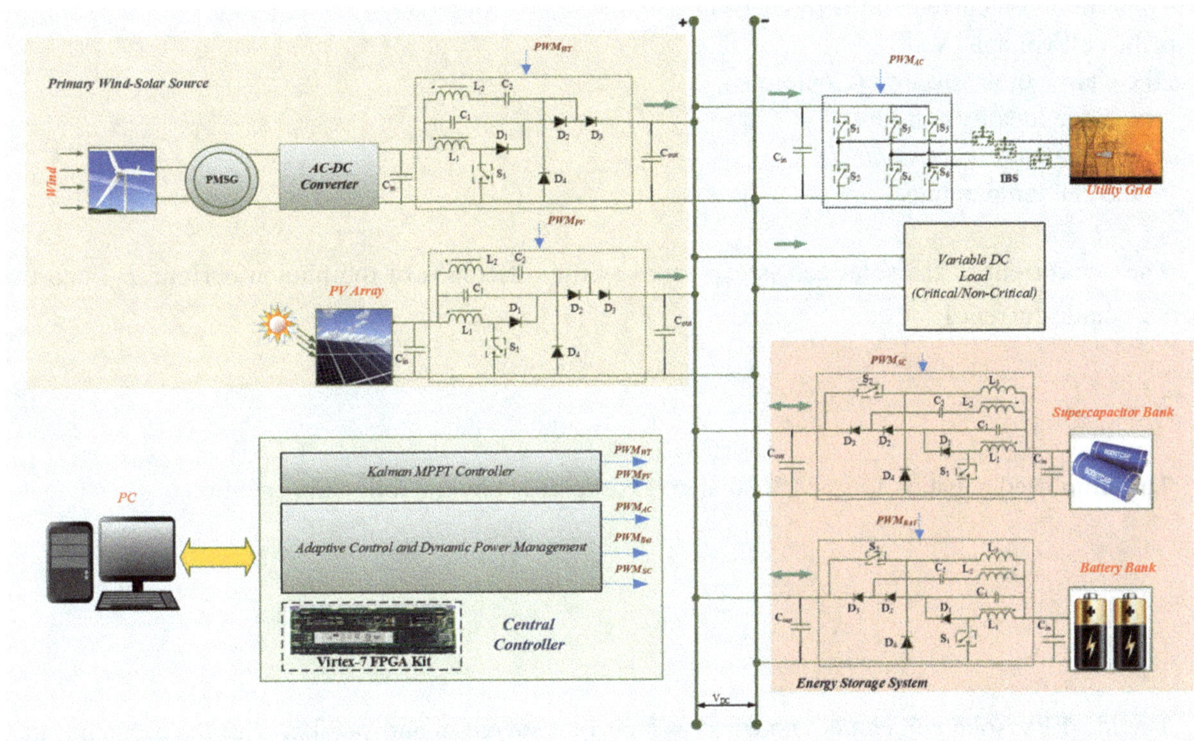

grid stability. The coupling between high voltage common bus and low voltage microgrid sources done through high gain DC-DC converters. Hence, the modelling of the PV array, wind turbine, ESS and the high gain DC-DC converter is an essential task. The prerequisite for designing the DC microgrid is to size the major components of the system accurately. Therefore the modelling and sizing of the major DC microgrid participants are presented here

PV Array

The PV array is the combination of a number of series/parallel solar cells. The modelling of PV array is presented in (Xiao, Dunford & Capel, 2004; Marcelo Gradella Villalva, Jonas Rafael Gazoli & Ernesto Ruppert Filho 2009). The solar cell is a p-n junction device that produces electricity by means of the photovoltaic effect.

The I-V characteristics of the p-n junction diode is nonlinear and it can be represented by the equation as follows

$$I_D = I_o\left(e^{\frac{V_c q}{akT_{cK}}} - 1\right) \qquad (2)$$

where

I_D is the diode current (A),
I_0 is the saturation current of the diode (A),
v_C is the cell voltage (V),
q is the charge of an electron (Coulombs),
a is the diode ideality constant,
k is the Boltzmann's constant (j/K),
T_{cK} is the cell temperature.

The net current of the solar cell is expressed as the difference of the photon current I_{ph} and the normal diode current I_D

$$I = I_{ph} - I_D \tag{3}$$

The simplified equation for the PV/solar cell is expressed by the following equation

$$I = I_{ph} - I_o \cdot \left(e^{\frac{(V_0 + IR_s) \times q}{akT_{cK}}} - 1 \right) \tag{4}$$

The ideal PV/solar cell can be modelled as a current source in anti-parallel with the diode but in a practical case, the losses are also to be included. The PV/ solar cell is modelled with series resistance R_s and parallel resistance R_{sh} to consider the losses. The complete solar PV module with several resistances can be modelled by the equation shown below

$$I = I_{ph} - I_o \cdot \left(e^{\frac{(V_0 + IR_s) \times q}{akT_{cK}}} - 1 \right) - \frac{\left(V_0 + IR_s\right)}{R_{sh}} \tag{5}$$

For sizing microgrid participants, the first step is to forecast the energy requirement per day, which is expressed in ampere-hours (Ah) (Grouz & Sbita, 2014). The Ampere-hours (Ah) can be calculated by the knowledge of the appliances power rating (W) and their mean operating time (h) per day and the solar-wind source operating voltage (V). The required energy in Ampere-hours (Ah) per day can be expressed as follows

$$E_j = \frac{Energy\ requirements\ per\ Day\ in\ Wh}{Operating\ voltage\ of\ microgrid\left(V\right)} Ah / Day \tag{6}$$

In order to find the solar panels to be fixed, the maximum energy that can be produced from each PV module has to be calculated. The energy produced per day under ideal conditions is given by

$$E_{elec} = N \times P_{\max} Wh / Day \tag{7}$$

where N is number of hours of shining sunshine per day and $P_{\max}$ is the maximum power captured working under Standard Test Conditions (STC). But in the real case, the energy produced per day depends on many parameters like seasonal variations, local estimated patterns and non-conventional source installation. The number of sun shining hours N_e can be calculated by the knowledge of lowest solar energy E_{worst} (Wh/m²/day) available per month all over the year. At a particular place, solar insolation is not same and it changes depending upon the climatic conditions. A change in insolation on the PV panel will affect the energy produced from it. To consider the insolation effect on the energy produced, the number of sun shining hours is calculated from the lowest solar energy E_{worst} (Wh/m²/day) using the following equation

$$N_e = \frac{E_{worst}}{1000} \tag{8}$$

where 1000 is the standard solar flux i.e., 1000 W/m².The real energy produced per day from a solar panel can be calculated as

$$E_{elec} = N_e \times P_{\max} \times C_p Wh / Day \tag{9}$$

where N_e is the number of sun shining hours and C_p is the factor to include losses due to pitch and alignment of the solar system and cable losses. The value of C_p varies from 0.65 to 0.9. The number of PV panel N_{PV} required is calculated by the equation as follows

$$\therefore N_{PV} = \frac{Total\ energy\ requirements\left(E_{jt}\right)}{Produced\ energy / Day\left(Eelec\right)} \tag{10}$$

Wind Turbine

The wind turbine produces electricity by utilizing the kinetic energy available with the wind. The permanent magnet synchronous generator (PMSG) is mostly used in wind generation due to its high power concentration. The rotor of the permanent magnet synchronous generator (PMSG) is coupled with the wind turbine through the gear mechanism to match the speed of the PMSG with that of the wind turbine. The turbine that is often used in wind plants is horizontal-axis wind turbine (HAWT) owing to its capability to extract maximum power from the winds. The wind turbine modelling is presented in (Alexandridis, A. T. & Marmidis, G. E. (2010). In a horizontal-axis wind turbine, the axis of rotation of rotor blades is parallel to the wind speed. The mechanical input torque T_w accessible for the wind speed (V_w) is calculated from the aerodynamics as

$$T_w = \frac{\rho \pi C_p(\lambda, \beta) V_w}{\lambda} R^3 \tag{11}$$

$$\lambda = \frac{R\omega_r}{V_w} \tag{12}$$

$$C_p(\lambda, \beta) = c_1 \left(c_2 / \lambda_i - c_3\beta - c_4 \right) e^{\frac{-c_5}{\lambda_i}} + c_6\beta \tag{13}$$

where c_1 =0.5176; c_2 =116; c_3=0.4; c_4 =5; c_5=21; c_6=0.0068

$$\frac{1}{\lambda_i} = \frac{1}{\lambda + 0.08\beta} + \frac{0.035}{\beta^3 + 1} \tag{14}$$

To extract maximum power from the wind turbine, the pitch angle β should be zero. In PMSG, the excitation is provided by the permanent magnets in place of separate field coil and the stator consists of the armature winding to produce electricity. The modelling of PMSG is given in (Schmidlin Junior & Araujo Lima, 2011). The wind turbine driven PMSG can be represented by the rotor reference frame equations as

$$V_{dg} = -\left(R_s + \frac{d}{dt} L_q \right) i_q - \omega_r L_d i_d + \omega_r \lambda_m \tag{15}$$

$$V_{qg} = -\left(R_s + \frac{d}{dt} L_d \right) i_d + \omega_r L_q i_q \tag{16}$$

$$T_e = \frac{3}{2}\frac{P}{2}\left[i_d i_q \left(L_q - L_d \right) + \lambda_m i_q \right] \tag{17}$$

$$\frac{d\omega_r}{dt} = \frac{P}{2J}\left(T_m - T_e \right) \tag{18}$$

The relationship between rotor angular velocity of the generator ω_r and the mechanical angular velocity ω_t of the turbine rotor is given by

$$\omega_t = \frac{\omega_r}{\frac{PG}{2}} \tag{19}$$

Similarly, the turbine rotor torque T_w and the input torque of the generator T_m is given by

$$T_m = \frac{T_w}{G} \tag{20}$$

The d-q components of generator currents of the PWM rectifier can be calculated as follows

$$\frac{di_q}{dt} = \frac{1}{L_{sr} + L_q}\left[-R_s i_q - \omega_r i_d \left(L_{sr} + L_d\right)\right] + \omega_r \lambda_m \tag{21}$$

$$\frac{di_d}{dt} = \frac{1}{L_{sr} + L_d}\left[-R_s i_d + \omega_r i_q \left(L_{sr} + L_q\right)\right] + V_{dr} \tag{22}$$

The three-phase rectifier is used to convert the pulsating AC supply produced from PMSG into constant DC. The modelling of the three-phase rectifier is done in the following manner. The average value of the output voltage can be calculated as

$$V_{dc} = \frac{6}{2\pi}\int_{\frac{\pi}{3}}^{\frac{2\pi}{3}} \sqrt{3}V_m \sin \omega t d\omega t \tag{23}$$

$$V_{dc} = \frac{3\sqrt{3}}{\pi} V_m \tag{24}$$

where ωt is the phase angle of the generated voltage and V_{dc} is the DC-link output voltage and V_m is the maximum value for the generator output phase voltage. The wind energy produced by wind turbine per day is given by

$$E_d = P_w \times 24Wh / Day \tag{25}$$

where P_w is power produced by the wind turbine corresponding to the wind speed. The required number of wind turbines N_{wt} can be easily computed by the following equation

$$N_{wt} = \frac{Total\ energy\ requirements\left(E_{jt}\right)}{Produced\ wind\ energy\ /\ Day\left(Ed\right)} \tag{26}$$

Energy Storage System (ESS)

As the Solar/wind energy is intermittent in nature it should be assisted by the ESS to make the grid reliable and stable. The battery and supercapacitor combination act as an effective ESS for the DC microgrid. The battery is an electrochemical device, it converts chemical energy into electrical energy through an electrochemical process. The modelling of the battery bank can be done according to (Tremblay, Dessaint & Dekkiche, 2007). The battery is represented as a controlled voltage source in series with a fixed resistance. The equation representing the battery voltage is given by

$$V_{bat} = E_o - K_{bat}\frac{Q_{\max}}{Q_{\max} - \int i_{bat}dt} + A_{bat.}e^{-B_{bat}\int(i_{bat}dt)} - i_{bat.}R_i \tag{27}$$

where V, E_o are the terminal voltage of the battery and constant voltage respectively. K_{bat} is the polarization constant. $Q_{\max}$ is the maximum capacity of the battery. A_{bat} is the exponential zone voltage in volts. B_{bat} is the exponential zone capacity $(Ah)^{-1}$. i_{bat} is the battery current in amperes. dt is the time in seconds. R_i is the internal resistance of the battery.

A supercapacitor is an electrostatic energy storage device with the negligible amount of material degradation. It has high specific power capability.

The minimum value of capacitance required can be calculated as

$$C_{sc} = \frac{\left(2 \times P_{sc} \times t_d\right)}{\left(V^2_{\max} - V^2_{\min}\right)} \tag{28}$$

where P_{sc} is supercapacitor power in watts supplied in time t_d. $V_{\max}$ and $V_{\min}$ are the maximum and minimum voltages. The state of charge of the supercapacitor is determined by

$$SOC_{sc} = \frac{Q_{ini} - \int_0^T i_{sc}\left(t\right)dt}{Q_T} \times 100\% \tag{29}$$

Energy storage systems are considered to be one of the most cost-effective and reliable devices that provide reserve energy capacity in a standalone microgrid system and also helps in maintaining safe grid operation during shadowing and low wind speeds. To achieve long-running period for ESS, it is recommended to operate them below 70% of its rated capacity (Lee, Kim, Shin & Kim, 2014). The number of energy storage devices required can be found by the following relation

$$Number\ of\ energy\ storage\ systems\left(ESS\right) = \frac{Energy\ requirements\ expressed\ in\ Ah}{Capacity\ of\ each\ ESS \times 0.7} \quad (30)$$

High-Gain Boost Converter

The high-gain high-efficient DC-DC boost converter topology chosen in this chapter for the low voltage renewable sources is shown in Figure 3. It comprises of one passive clamp network, an interleaved coupled inductors (L_1 and L_2), and an intermediate capacitor C_2. The solar/wind output DC voltage is applied at the input side of the converter for boosting the voltage level. The switching ON/OFF of MOSFET is done to control the output voltage (V_o). L_1 and L_2 are the primary and secondary inductances of the coupled inductor respectively. Whereas, L_k and L_m are leakage and magnetizing inductances of the coupled inductor respectively. The combination of C_1 and D_1 forms a passive clamp circuit across L_1 to recover energy from leakage inductor. The output capacitor of the converter is C_{out} and the output diode of the converter is D_3. In addition, C_2 and D_2 represent the intermediate energy storage capacitor and feedback diode of the proposed converter respectively. The turn ratio 'n' of a coupled inductor can be found from the relation as follows.

$$n = \frac{V_{L2}}{V_{L1}} \quad (31)$$

where V_{L1} and V_{L2} denote the voltages across the primary L_1 and secondary L_2 respectively. Similarly, V_{C1} and V_{C2} denote the voltages across passive clamp capacitor C_1 and intermediate storage capacitor C_2 correspondingly. The operation of a high-gain high-efficient topology can be explained in different modes as follows.

Figure 3. Circuit diagram for high gain DC-DC boost converter for low voltage application

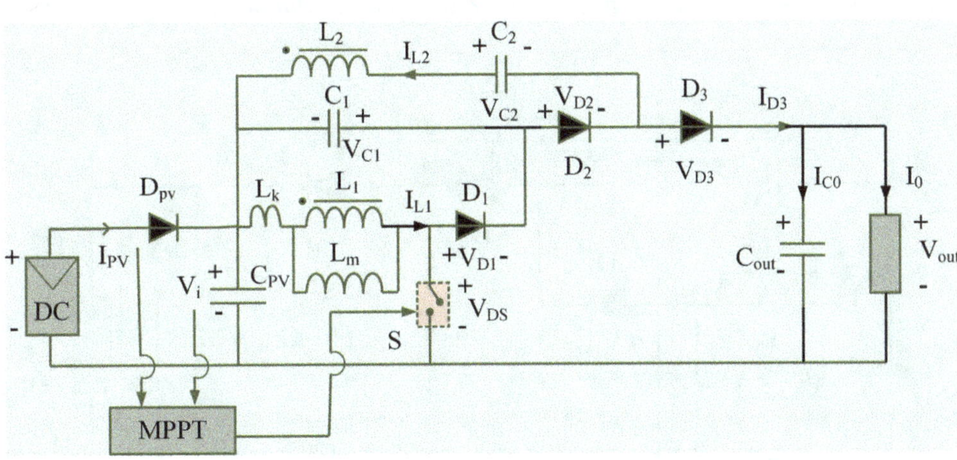

Mode 1 [t_0- t_1]

Initially, the MOSFET is turned ON by supplying the switching pulse. Therefore, the source current starts flowing through the path as shown in Figure 4. It energizes the magnetizing inductance (L_m) of the coupled inductor. In this mode of operation, the voltage appears across passive clamp circuit comprising of D_1 and D_3, which makes them reverse biased. The feedback diode D_2 works in forwarding conduction mode. Therefore the intermediate capacitor C_2 and passive clamp circuit capacitor C_1 is charged by the energy stored in the secondary side of the coupled inductor L_2 through the diode D_2 as shown in Figure 4. The passive clamp circuit diode D_2 will be turned off when the sum of the voltage across secondary of the coupled inductor L_2 and the passive clamp circuit capacitor C_1 is equal to the voltage across the intermediate storage capacitor (C_2). The current flowing through the magnetizing inductance L_m (i_{Lm}) can be found by using the following relation:

$$i_{Lm}\left(t\right)=\frac{V}{L_k+L_m}\left(t-t_0\right)+i_{Lm}\left(t_0\right) \tag{32}$$

where t_0-t_1 is the time duration of operation in mode 1.

Mode 2 [t_1- t_2]

At the starting of this mode of operation, the MOSFET switch (S) is made to turn-OFF. The current flowing through the primary of the coupled inductor L_1 charges the parasitic capacitance of the MOSFET switch S. The current continues to flow through the feedback diode D_2 as it continues to work in forwarding conduction mode. The current path through the feedback diode D_2 is shown in Figure 5. The magnetizing current flowing to charge the parasitic capacitance of the MOSFET switch (S) is expressed by the following equation:

Figure 4. Operation of the high gain converter for continuous conduction during mode 1

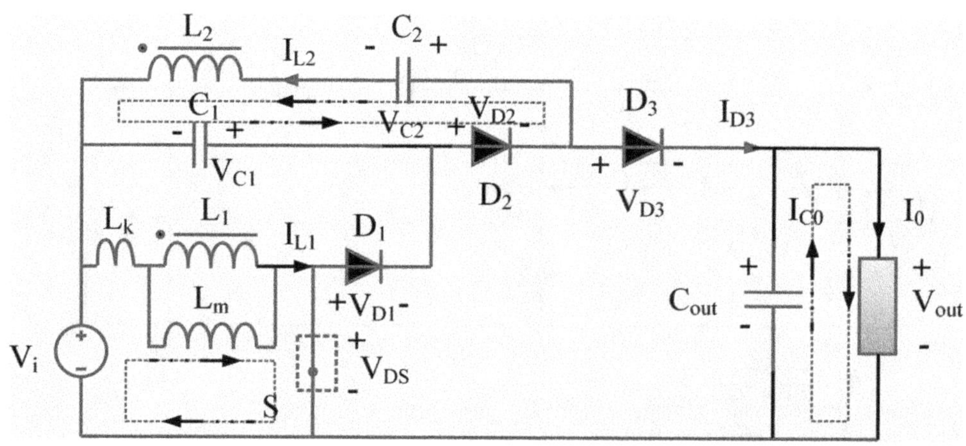

Figure 5. Operation of high gain converter for continuous conduction in mode 2

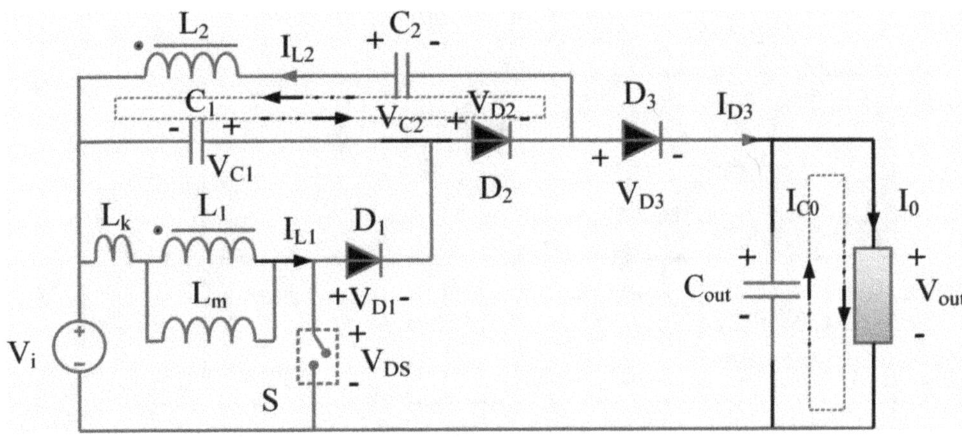

$$i_{Lm}\left(t\right)=\frac{V}{L_k+L_m}\left(t-t_1\right)+i_{Lm}\left(t_1\right) \tag{33}$$

where t_1-t_2 is the time duration in mode 2 operation.

Mode 3 [t_2- t_3]

In this mode of operation the diodes D_1 and D_3 goes to forward conduction mode when the D_2 becomes reverse biased. The diode D_2 turns off. The clamp capacitor (C_1) charges through the diode D_1 to recover the leakage energy stored in the primary of the coupled inductor (L_1). Also, the diode D_3 transfers energy from the input side of the converter to the output of the load side in the path shown in Figure 6. The leakage current (i_{Lk}) charging the passive clamp circuit is given by the relation as:

$$i_{Lk}\left(t\right)=\frac{Vc1}{L_k+L_m}\left(t-t_2\right)+i_{Lm}\left(t_2\right) \tag{34}$$

where $t_2 - t_3$ is the time duration in mode 3 operation.

Mode 4 [t_3- t_4]

This mode begins when the leakage energy stored in the coupled inductor L_1 has been recovered. The diodeD$_1$ becomes reverse biased and the diode D_3 remains in forwarding conduction during this mode. The current path from the source side to the load side will be as shown in Figure 7. The magnetizing current i_{Lm} flowing through the secondary of the inductor (L_2) is given by the following relation:

$$i_{Lm}\left(t\right)=\frac{\left(V_O-V_{C2}-V_i\right)}{L_k+L_m}\left(t-t_3\right)+i_{Lm}\left(t_3\right) \tag{35}$$

Figure 6. Operation of high gain converter for continuous conduction in mode 3

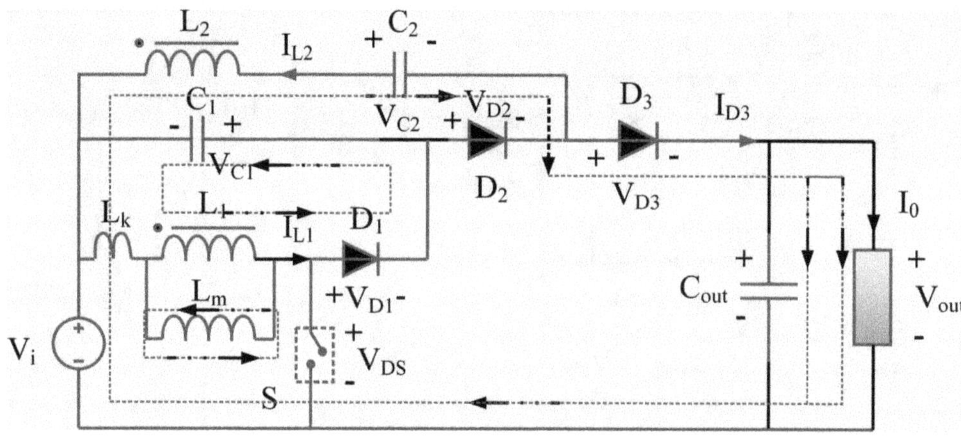

Figure 7. Operation of high gain converter for continuous conduction in mode 4

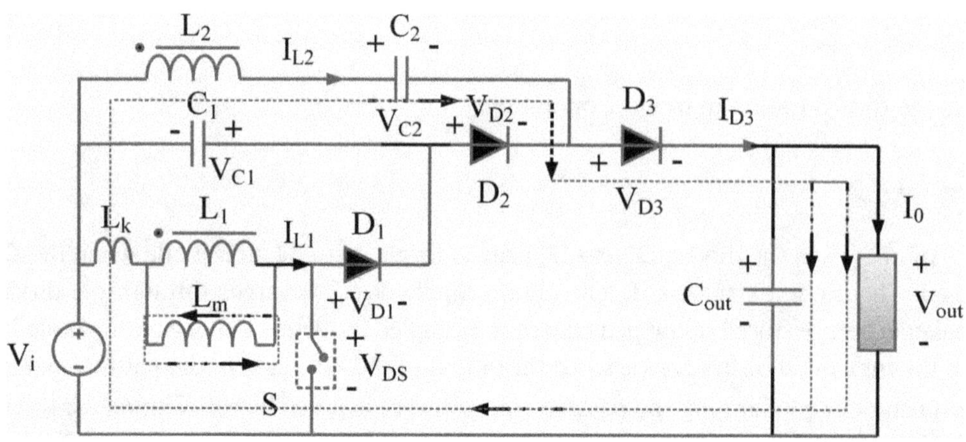

where $t_3 - t_4$ is the time duration in mode 4 operation.

Mode 5 [$t_4 - t_0$]

In this mode, the switch S is turned on at the starting. The energy stored in the parasitic capacitances of the switch is released and spent in magnetizing the leakage inductor in a short period. The two diodes D_1 and D_2 become reverse biased throughout this mode. The current flow path in this mode is shown in Figure 8. When the diode D_3 becomes reverse biased, the current flowing through the inductor L_2 changes its path. The inherent property of an inductor that does not allow an instantaneous change in current, the voltage across the switch also cannot change abruptly. Therefore, there exists a small overlap during the rise and fall of the inductor current which causes a negligible switching loss at the time of turning-ON. The secondary inductor current (i_{Lm}) continues to flow in this mode and current is given by

$$i_{Lm}(t) = \frac{(V_O - V_{C2} - V_i)}{L_k + L_m}(t - t_4) + i_{Lm}(t_4) \quad (36)$$

where $t_4 - t_0$ is the time duration in mode 5 operation.

The waveforms for the high gain converter for continuous conduction in all the modes of operation are given below,

The above Figure 9(a)-(h) depicts the voltage (V_{GS}) applied to the gate of the switch (S), the voltage (V_{DS}) across the switch (S), the current (I_{Lm}) flowing through the magnetizing inductance (L_m) the current (I_{D3}), flowing through the diode (D_3), the voltage (V_{D3}) across the diode (D_3), the current (I_{CO}) flowing through the output capacitor (C_O), the current (I_{D1}) flowing through the diode (D_1), the voltage (V_{D1}) across the diode (D_1), the voltage (V_{C1}) across the diode (C_1), respectively.

Assuming, all the elements used in the converter to be ideal, for designing and analyzing the high-gain DC-DC converter. (Kazimierczuk, 2008).

When switch S is turned-on: The Voltage across L_1 is given by

$$V_{L1(ON)} = V_i \quad (37)$$

The voltage across L_2 is given by

$$V_{L2} = V_{C2} - V_{C1} \quad (38)$$

$$V_{L2} = nV_i \quad (39)$$

where n is the ratio of the inductor voltages. It is specified as

Figure 8. Operation of the high gain converter for continuous conduction in mode 5

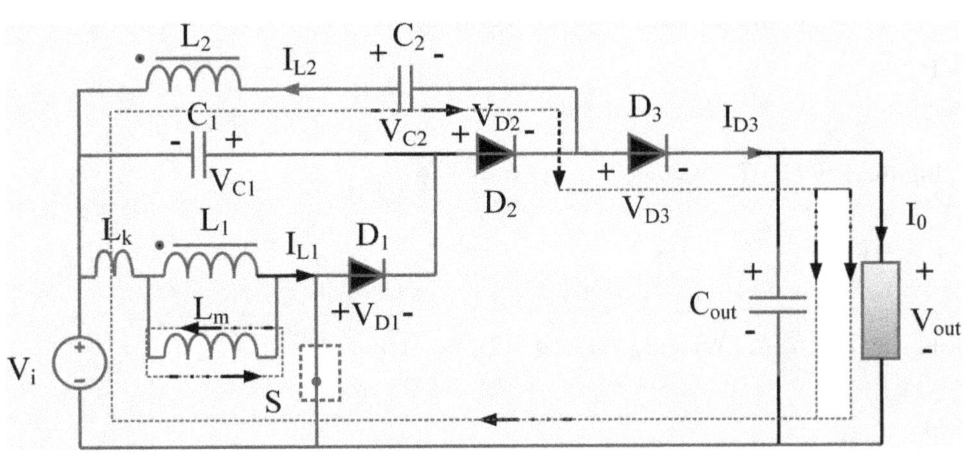

Figure 9. Waveforms of high gain converter for continuous conduction in all modes

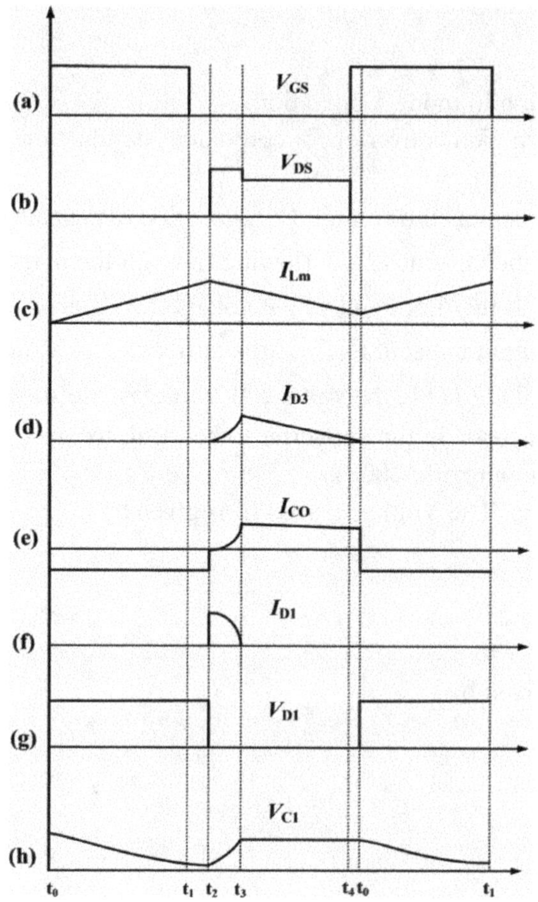

$$n = \frac{V_{L2}}{V_{L1}} \tag{40}$$

When switch S is turned – OFF: The Voltage across L_1 is given by

$$V_{L1(OFF)} = -V_{C1} \tag{41}$$

By applying the Kirchhoff's Voltage law in Mode 3 gives

$$V_{L2} = V_i + V_{C2} - V_o \tag{42}$$

By substituting V_{C2} from (39) and (41) into (42), we arrive at

$$V_{L2} = V_i - V_{L1(OFF)} + nV_i - V_o \tag{43}$$

By substituting (43) into (40), voltage across L_1 during switch OFF condition becomes

$$V_{L1(OFF)} = \frac{V_i - V_{L1(OFF)} + nV_i - V_o}{n} \tag{44}$$

(Or)

$$V_{L1(OFF)} = \frac{V_i - nV_i - V_o}{(n+1)} \tag{45}$$

Voltage Gain

By applying the voltage-sec balance across L_1 in order to obtain the voltage gain

$$V_{L1(ON)}d + V_{L1(OFF)}(1-d) = 0 \tag{46}$$

Substituting the values of $V_{L1(ON)}$ and $V_{L1(OFF)}$ from equations (37) and (45), respectively into (46) results in

$$V_i d + \frac{V_i - nV_i - V_o}{(n+1)}(1-d) = 0 \tag{47}$$

By simplifying the above expression, the voltage gain is given by

$$\frac{V_o}{V_i} = \frac{(n+1)}{(1-d)} \tag{48}$$

where d- is the duty ratio applied to switch (S)
Substituting (48) into (45) results in

$$V_{L1(OFF)} = -\frac{d}{(1-d)}V_i \tag{49}$$

Switch Voltage

When the switch is turned off, the voltage across it is

$$V_{DS} = -V_{L1(OFF)} + V_i \tag{50}$$

Substituting (49) into (50) yields

$$\therefore V_{DS} = \frac{V_i}{\left(1-d\right)} \tag{51}$$

The voltage across the clamp capacitor C_1 is constant for the entire switching period can be obtained from (41) and (48)

$$\therefore V_{C1} = \frac{d}{\left(1-d\right)} V_i \tag{52}$$

Intermediate Capacitor Voltage (V_{C2})

The voltage across the intermediate storage capacitor C_2 is almost constant over the entire period and can be found from (39) as follows

$$V_{C2} = V_{L2} + V_{C1} \tag{53}$$

By substituting (42) and (52) into (53) results in

$$\therefore V_{C2} = \frac{d\left(1-n\right)+n}{\left(1-d\right)} V_i \tag{54}$$

Reverse Bias Voltage Across Diodes (V_{D1}, V_{D2}, V_{D3})

Voltage across diode (D_1)

$$V_{D1} = V_{L1(ON)} + V_{C1} \tag{55}$$

By substituting (39) and (52) into (55), the voltage across diode (D_1) becomes

$$\therefore V_{D1} = \frac{V_i}{(1-d)} \tag{56}$$

Voltage across diode (D_2)

$$V_{D2} = V_{L2} + V_{C2} - V_{C1} \tag{57}$$

By substituting (42), (52) and (54) into (57), the voltage across diode (D_2) is

$$\therefore V_{D2} = nV_i \tag{58}$$

Voltage across diode (D_3) is

$$V_{D3} = V_{L2} - V_{C2} - V_i + V_o \tag{59}$$

By substituting V_{L2} from (42), V_o from (48) and V_{C2} from (54) into (59), the voltage across diode (D_3) becomes

$$\therefore V_{D3} = \frac{n}{(1-d)} V_i \tag{60}$$

Forward Bias Diode Currents (I_{D1}, I_{D2}, I_{D3})

The current flowing through diode (D_1) can be found as follows

$$L_1 \frac{dI_{D1}}{dt} = V_{C1}$$

$$\therefore I_{D1} = \frac{V_{C1} \times d_{lk}}{L_1 \times f_s} \tag{61}$$

where $\frac{d_{lk}}{f_s}$ is the period during which the leakage energy is recovered. The current flowing through diode (D_2) can be found as given below (Switch ON)

$$L_2 \frac{dI_{Lm}}{dt} = V_{C2} - V_{C1}$$

where $I_{Lm} = nI_{L2}; I_{L2} = I_{D2}$

$$\therefore I_{D2} = \frac{(V_{C2} - V_{C1}) \times d}{n \times L_2 \times f_s} \tag{62}$$

Similarly, the current flowing through diode (D_3) can be found as follows (Switch OFF)

$$L_2 \frac{dI_{Lm}}{dt} = V_o - V_{C2} - V_i$$

where $I_{Lm} = nI_{L2}; I_{L2} = I_{D3}$

$$\therefore I_{D3} = \frac{\left(V_o - V_{C2} - V_i\right) \times \left(1 - d - d_{lk}\right)}{n \times L_2 \times f_s} \tag{63}$$

Coupled Inductor Parameters (L_1, L_2, L_3)

The value of primary side coupled inductor $\left(L_1\right)$ from (61) is given as

$$\therefore L_1 = \frac{V_{C1} \times d_{lk}}{I_{D1} \times f_s} \tag{64}$$

The value of secondary side coupled inductor $\left(L_2\right)$ from (33) is given as

$$\therefore L_2 = \frac{(V_{C2} - V_{C1}) \times d}{n \times I_{D2} \times f_s} \tag{65}$$

The voltage across the primary winding and the magnetizing inductance L_m is

$$L_m \frac{dI_{Lm}}{dt} = V_i \tag{66}$$

Hence, the current through the magnetizing inductance L_m is

$$I_{Lm} = \frac{1}{L_m} \int_0^t V_i dt + I_m\left(0\right)$$

$$I_{Lm} = \frac{V_i}{L_m} t + I_m(0) \tag{67}$$

where $I_m(0)$ is the initial current flowing through the magnetizing inductance L_m at t = 0. The peak-to-peak ripple current of the magnetizing inductance is

$$\Delta I_{Lm} = I_{Lm}(dT) + I_{Lm}(0) = \frac{V_i \times d \times T}{L_m} = \frac{V_i \times d}{L_m \times f_s} \tag{68}$$

The current flowing through the magnetizing inductance at t = dT is

$$I_{Lm}(dT) = \frac{\Delta I_{Lm}}{2} = \frac{V_i \times d \times T}{2 \times L_m} = \frac{V_i \times d}{2 \times L_{m \times} f_s} \tag{69}$$

The magnetizing inductance (L_m) can be made by the relation

$$\therefore L_m = \frac{V_i d}{2 \Delta I_{Lm} f_s} \tag{70}$$

The value of clamp capacitor (C_1) can be found by the relation

$$C_1 \frac{dV_{C1}}{dt} = I_{C1}$$

where $I_{C1} = I_{Lm}$ for the period of $\frac{d_{lk}}{f_s}$, then

$$\therefore C_1 = \frac{I_{Lm} d_{lk}}{\Delta V_{c1} f_s} \tag{71}$$

Similarly, the intermediate capacitor (C_2) value can be found by the relation

$$C_2 \frac{dV_{C2}}{dt} = I_{C2}$$

where $I_{C2} = nI_{Lm}$ for the period of $\frac{d}{f_s}$, then

$$\therefore C_2 = \frac{I_{Lm} d}{n\Delta V_{c2} f_s} \tag{72}$$

Finally, the output capacitor (C_o) value can be found by the relation

$$C_o \frac{dV_o}{dt} = I_o$$

$$\therefore C_o = \frac{I_o d}{\Delta V_o f_s} \tag{73}$$

To store the power produced from the renewable sources during off-load periods the storage systems are necessary for the DC microgrid system. The conventional bi-directional converters used for ESS incurred with many losses for accomplishing the high voltage levels. Different types of high-gain bi-directional converters that are proposed with four and three switches suffers from more loss as discussed in the above literature. The high-gain high- efficient bi-directional DC-DC boost converter the converter topology with two switches proposed in (Shreelakshmi, Das & Agarwal, 2013) is shown in below figure 10. The working of the proposed bi-directional converter in the step-up mode is similar to the working of high-gain high-efficient DC-DC boost converter presented in the previous section. The step-down mode of the proposed bi-directional converter is given as follows.

BI-DIRECTIONAL HIGH GAIN CONVERTER

The circuit diagram of the proposed bi-directional DC-DC high gain converter is shown in Figure 10. DC stands for the ESS and V_{out} is the DC-bus voltage. L_1 And L_2 are the primary and secondary of the coupled inductor and S_1 is the MOSFET switch used in step-up operation. The passive clamp circuit formed by the diode D_1 and capacitor C_1 are used to recover the leakage energy of the coupled inductor. The intermediate storage capacitor C_2 helps to boost the output voltage with a less duty ratio. The MOSFET switch S_1 is used for boost operation whereas switch S_2, inductor L_3 and diode D_4 forms the circuit for the step-down operation.

Mode 1

This mode starts with turning on the switch S_2 and therefore the current starts flowing from the DC link to the ESS. The path of the current is shown in Figure 10. The current flowing through the inductor L_3. stores certain energy in the form of magnetic field. From the Kirchhoff's voltage law the output of the DC bus voltage V_{out} can be given as follows.

$$V_{out} = V_i + V_{L3} \tag{74}$$

Figure 10. Operation of bi-directional high gain converter during mode 1 of step down operation

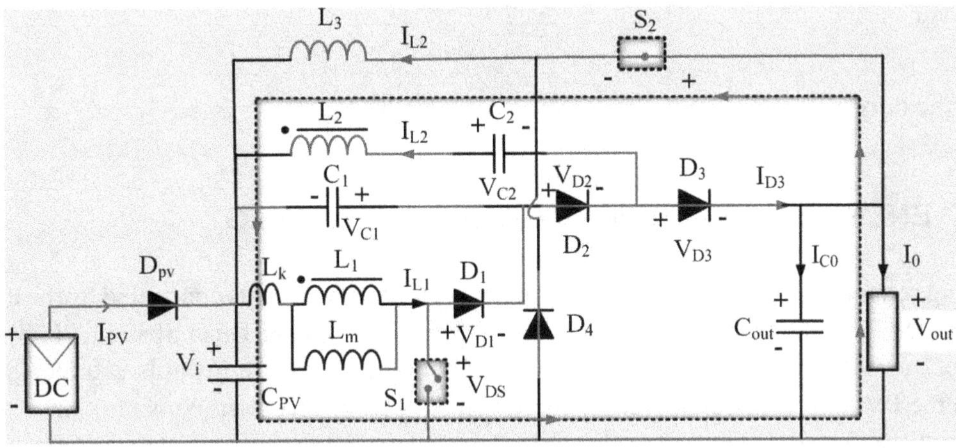

Mode 2

In this mode, the MOSFET power switch S_2 is turned OFF. The stored energy in the inductor L_3 in the mode 1 is recovered to DC input side through the path of the diode D_4 as shown in Figure 11. The DC bus voltage V_{out} is given as

$$V_{out} = V_{L3} \tag{75}$$

The value of inductor $\left(L_3\right)$ can be calculated from (46) as follows

Figure 11. Operation of Bi-Directional high gain converter during mode 2 of step down operation

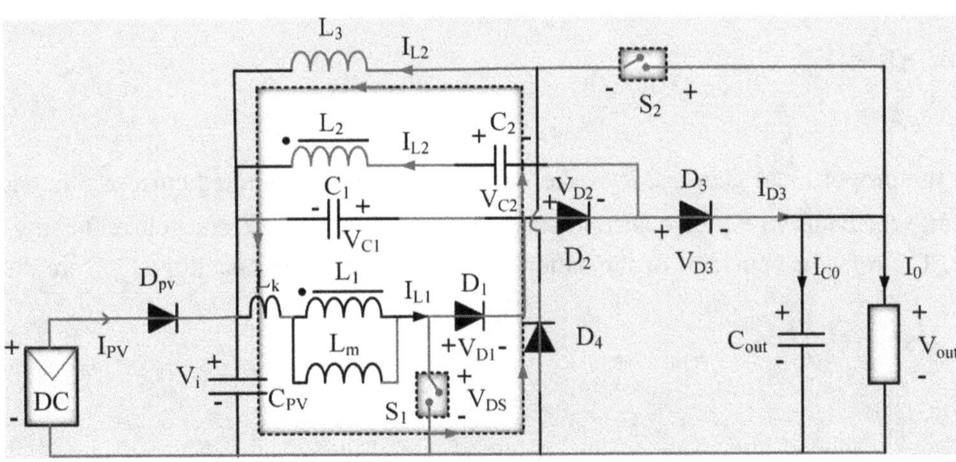

$$\therefore L_3 = \frac{V_{out} \times d}{I_O \times f_s} \tag{76}$$

where d- is the duty ratio supplied to the converter of switch (S_2).

DYNAMIC ENERGY MANAGEMENT CONTROL STRATEGY

The switching sequence of the boost converter connected to the PV array is controlled by the incremental conductance (INC) MPPT algorithm (Zhou Yan, Liu Fei, Yin Jinjun & Duan Shanxu, 2008). There are different types of MPPT techniques available in the literature that work towards extracting maximum power from the PV array. Among them, the incremental conductance method is very accurate and has quick settling time. The INC algorithm always tries to make the dynamic resistance of the converter is equal to the static resistance such that maximum power can be transferred to the load. Generally, perturb & observe (P&O) MPPT algorithm is used to operate the DC-DC converter of the wind turbine to extract the maximum power from it (Lahfaoui, Zouggar, Elhafyani & Seddik, 2015). The P&O MPPT algorithm is mainly used because of its easy implementation and low cost.

The control algorithm shown in Figure 12, is the primary control employed to the ESS in order to maintain a constant voltage at the DC link. To generate the control signals for the ESS converters, the measurements that are needed are the DC link voltage, battery state of charge (SOC) and supercapacitor SOC.

Hence, the main function of the controller is to maintain a constant rated voltage at the DC-link bus bar during all stages of microgrid operation. The underlying task that guarantees the above characteristic is a proper power balance among the microgrid participants. The concept of dual loop control is applied skillfully to estimate the amount of current that is to be fed to or taken away from the common DC link in order to maintain the rated voltage across it. A PI control based voltage controller is employed to generate reference current that is to be supplied/absorbed by the participating equipment. Further, an individual current control is chosen to tap controlled power flow from battery and supercapacitor bank. The deviation in DC link Voltage (V_{error}) is passed on to the voltage controller and the corresponding reference current is given by

$$I_{ref}\left(s\right) = \left(K_p + \frac{K_i}{s}\right) V_{error} \tag{77}$$

where k_p is the proportional gain and k_i is the integral gain. The generated current reference (I_{ref}) is segregated into the battery reference current and the supercapacitor reference current using a low-pass filter (LPF). The reference currents of the battery (I_{bat}) and the supercapacitor (I_{SC}) are thus given by

$$I_{bat}\left(s\right) = I_{ref}\left(s\right) \times H\left(s\right) \tag{78}$$

Figure 12. The control algorithm flowchart of microgrid for ESS

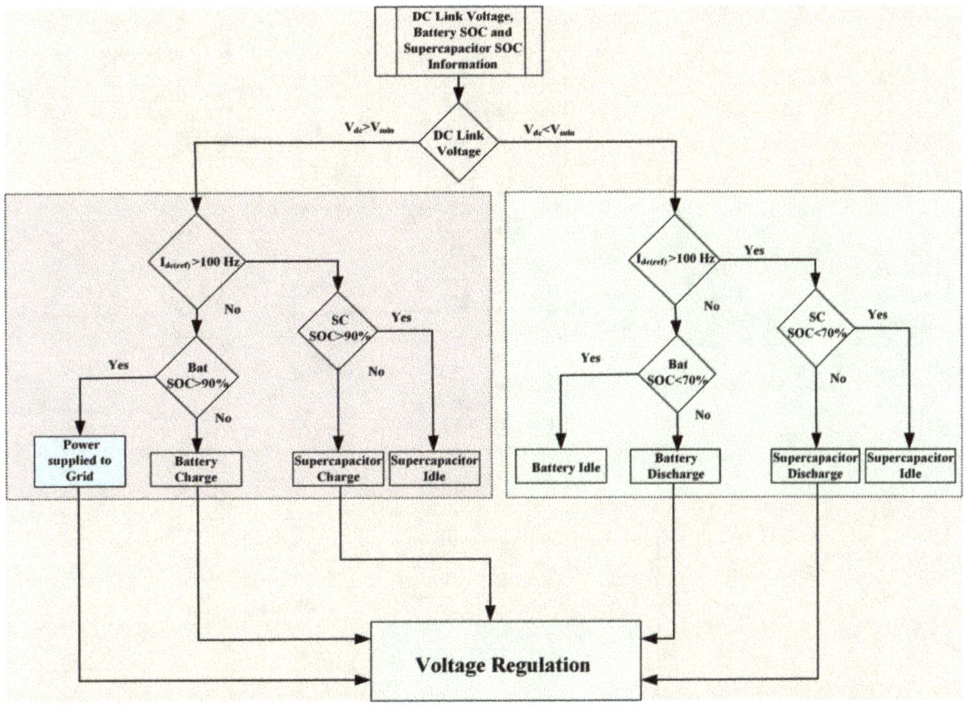

$$H(s)=\left(\frac{\omega}{s+\omega}\right) \tag{79}$$

$$I_{SC}(s)=I_{ref}(s)-I_{bat}(s) \tag{80}$$

where H(s) is the transfer function of a low-pass filter with a cut-off frequency of ω rad/s.

Upon generation and separation of the individual reference currents, it is supplied to respective devices such that the required power can be tapped/supplied from/to them. Many current controller techniques are available in the literature which can be directly applied to DC-DC converters. Generally, current controller is cascaded with a PWM generator to generate the switching pulse for the converters. Doing away with this extra stage, a set-reset flip-flop is employed for the purpose of the current control in this work. Upon receiving the current command from the outer loop, the actual current of the device is compared with the reference current to generate the reset signal for the set-reset flip-flop. The entire controller design is shown in Figure 13. A reset pulse at regular intervals is given to the set port of the flip-flop to put a constraint on the switching frequency of the switches.

Figure 13. The controller employed for ESS

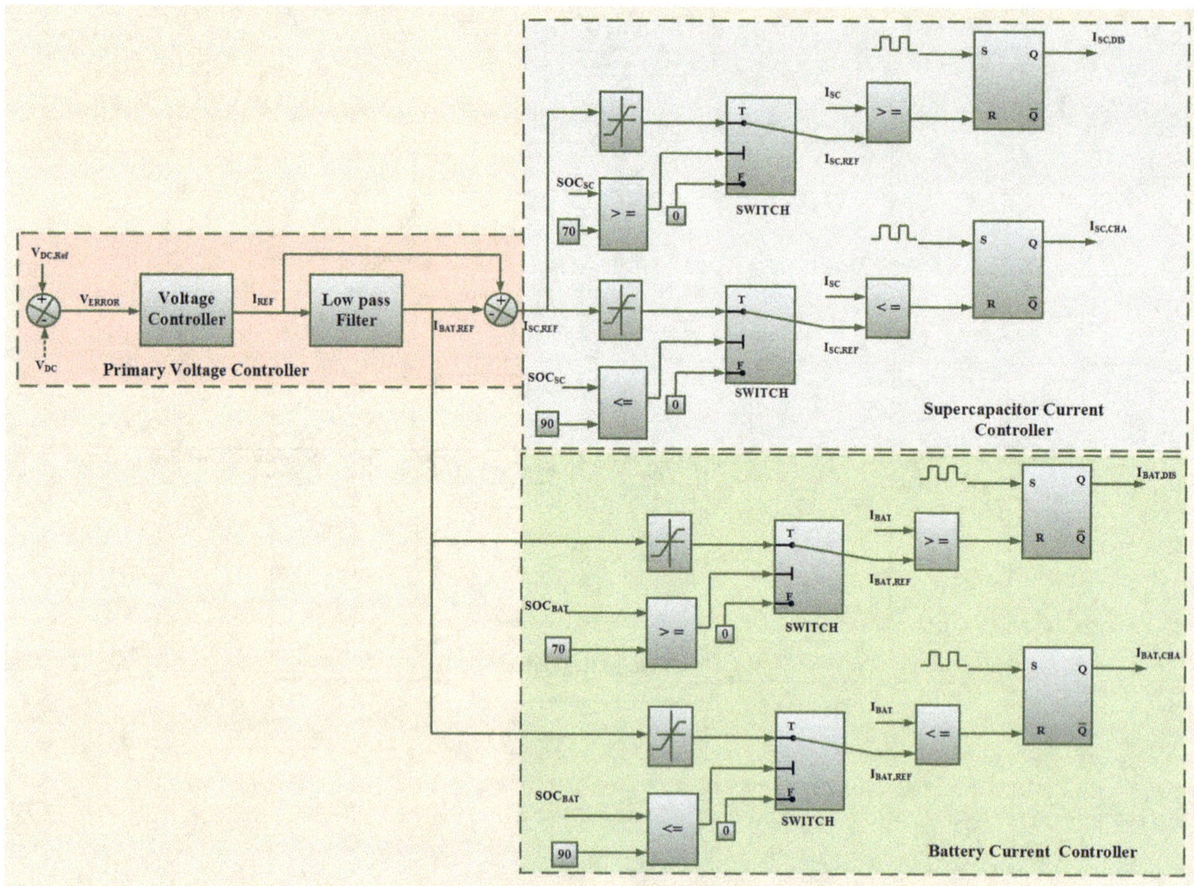

RESULTS AND DISCUSSIONS

The system under consideration is modelled and analysed using MATLAB/Simulink. The system specifications and important control parameters considered for the time domain analysis via MATLAB/Simulink are shown in Table 1 & Table 2 respectively. To analyze the dynamic performance of the controller, the following generation and load profiles were tested.

- The wind speed is maintained at 11 m/s up to 6 s and is changed gradually from 11 m/s to 12 m/s in the duration of 6 s to 8 s. Finally, it is maintained at 12 m/s from 8 s to 10 s.
- The solar irradiance on PV array is maintained at 1000 W/ m^2 and it is reduced to 900 W/ m^2 in the duration of 3 s to 5 s. lastly, the irradiance of 900 W/ m^2 maintained during 5 s to 10 s.
- Initially, a DC load of 400 watts is applied to the system at 0.3 s and the load is increased to 720 watts at 1.5 s. At, 2.5 s the load is further increased to 1520 watt. Finally, the load is reduced to 720 watts at 9 s.

From Figure 14, it is observed that the power sharing among the microgrid partakers is maintained suitably throughout the variable operating scenarios. As the PV and wind sources are generating the power to meet the load demand, the excess power available to the common DC bus is stored in the ESS. During nights and cloudy weather conditions, the load demand is supplemented by the ESS as shown in Figure 15. The load current and common DC bus voltage are displayed in Figure 16, it is observed that the load voltage is maintained constant irrespective of the load change. The battery and supercapacitor parameters like current, voltage and SOC during the presence the wind generation alone are displayed in Figure 17 & Figure 18. Lastly, we can conclude that the high gain converters chosen in this chapter are able to boost the voltage level to 400 V and the dynamic power management controller employed for the DC microgrid is able to maintain the grid variables within the limits.

The proposed high gain DC-DC boost converter displays a better performance compared with the conventional converter. Different performance indices are compared and tabulated in Table 3.

HARDWARE-IN-LOOP ANALYSIS

Nowadays the real-time analysis of the controller is a vital task. Such an analysis requires hardware components of the equipment being controlled which incurs the huge cost and poses a threat to the devices being controlled (if the control algorithm fails to function properly). Doing away with this hardship, FPGA devices are now being employed to perform hardware-in-loop analysis of the controller alone. Xilinx system generator (XSG) performs the task of modelling, converting and programming the FPGA

Table 1. System parameters

Parameter	Specification
PV array voltage (MPP)	70 V
PV array current (MPP)	13 A
Wind turbine voltage (MPP)	150 V
Wind turbine current (MPP)	5 A
Supercapacitor module capacitance	58 F
Supercapacitor module voltage (nominal)	32 V
Supercapacitor module current(nominal)	100 A
Battery voltage (nominal)	40 V
Battery current (nominal)	20 A

Table 2. Control parameters

Parameter	Specification
$V_{DC,Ref}$	400 V
Switching frequency	20KHz
ω	2×3.14×100 rad/s
Sampling period	5μs

Figure 14. Power sharing among the microgrid partakers in the presence of solar-wind sources

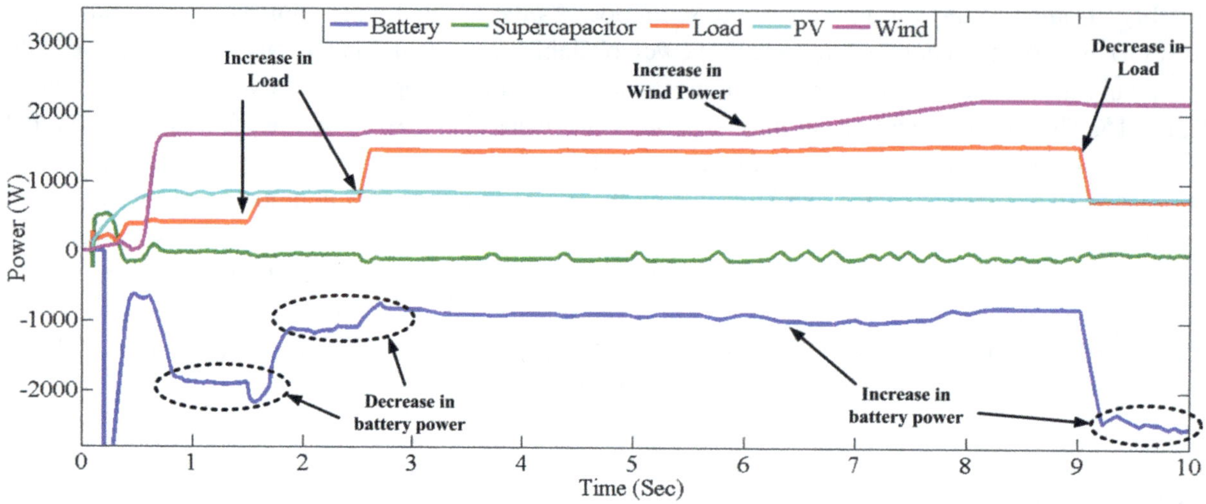

Figure 15. Power sharing among the microgrid partakers in the presence of wind source alone

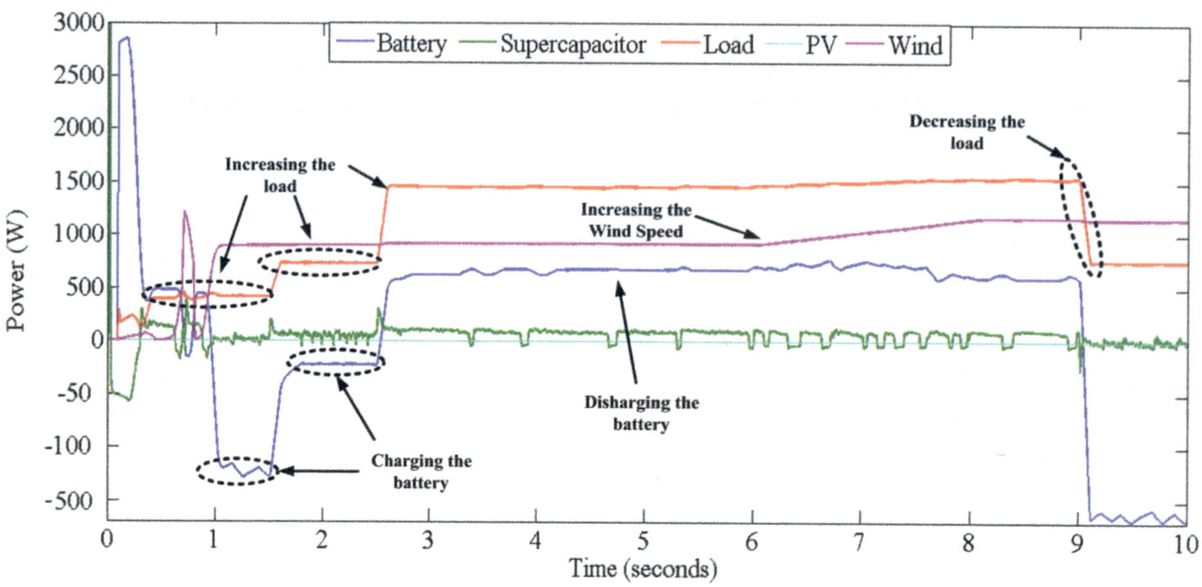

kit with the required control algorithm (Pinto, Panda & Peesapati, 2017). XSG is interfaced with the MATLAB/Simulink software. Then, Simulink inputs are then communicated to the controller through the USB JTAG interface and are processed to generate the controller output.

Xilinx modelling of the controller employed for the proposed DC system is shown in Figure 19. Xilinx blocks like constant, addition, gateway-in, gateway-out, m-code, delay, counter, assert, cast, cmult and relational blocks are utilized in the Xilinx modelling of the presented dynamic control. Xilinx modelling of the DC voltage controller employing PI algorithm is documented in (Naidu, Panda & Ganesh, 2016). Important sections in the Xilinx modelling of the controller are marked in Figure 19. Xilinx modelling

Figure 16. DC Common bus bar parameters (a) Load current (b) Bus voltage

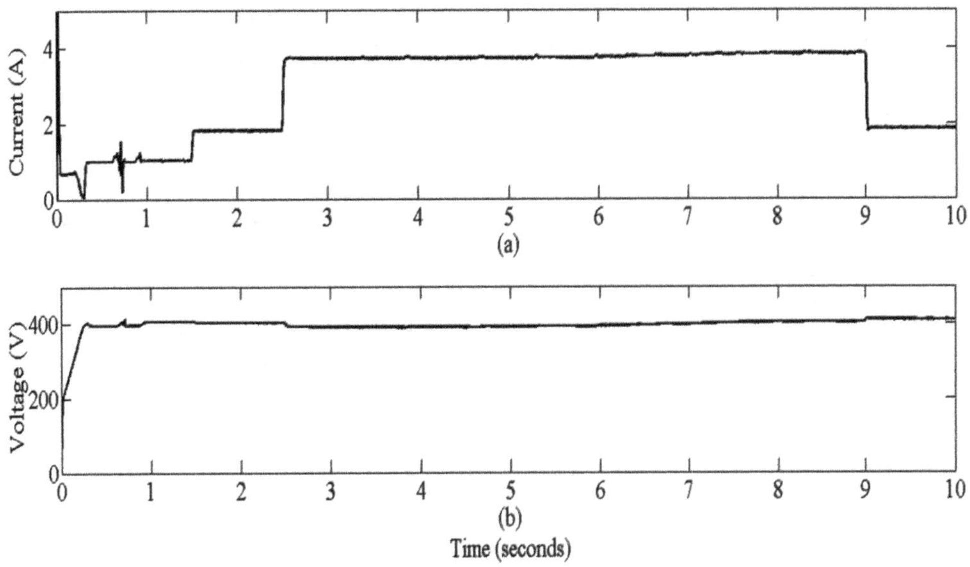

Figure 17. Battery parameters with wind source alone (a) Current (b) Voltage (c) SOC

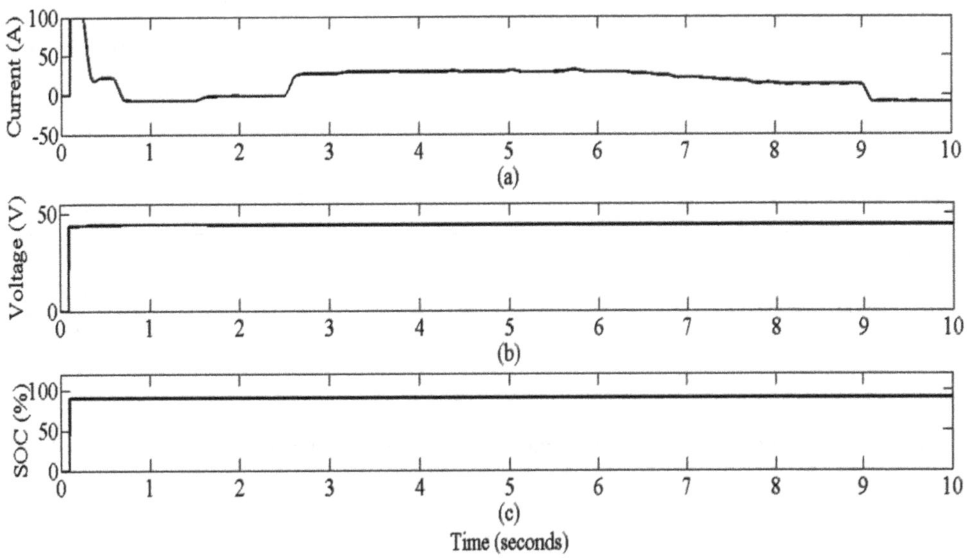

of the supercapacitor discharging current controller and battery charging current controller are not shown in Figure 19, as it replicates the "Battery discharge current controller" and "Supercapacitor charging current controller" presented in Figure 19. The bilinear transformation has been utilized to convert the low-pass filter from Laplace domain into discrete domain. The discrete low pass filter has been realised using direct form-I as shown in Figure 19. The bilinear transformation has been utilized to convert the low pass filter from analogue domain to discrete domain. The discrete low pass filter has been realized

Figure 18. Supercapacitor parameters with wind source alone (a) Current (b) Voltage (c) SOC

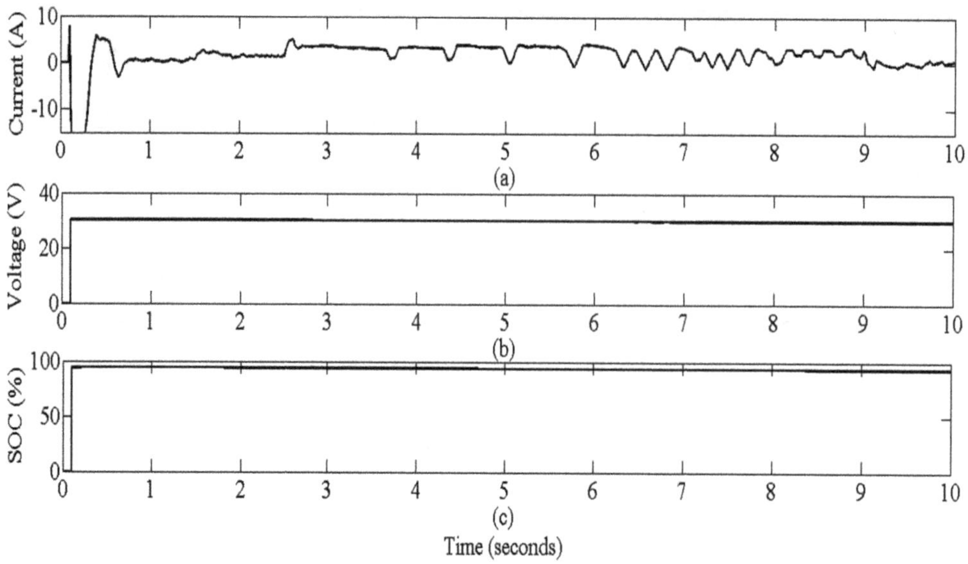

Table 3. Comparison between conventional and proposed boost converter

	Boost Converter (Duty cycle = 0.9, Input voltage=45 V, Supply frequency=50Hz)	**Proposed Converter (Duty cycle =0.9, Input voltage=45 V, Supply frequency=50Hz)**
Voltage Gain	10	10
Switch conduction loss	6.12 W	0.18 W
Switch Voltage drop (V_{DS})	450 V	90 V
Current through inductor	18 A	9 A
Leakage energy	0.38 W	0.004 W
Switch current	18 A	9 A

using direct form-I as shown in Figure 19. According to bilinear transformation, the relation between Laplace domain (analogue) and z-domain (digital) is given by

$$S = \frac{2}{T}\frac{\left(Z-1\right)}{\left(Z+1\right)} \tag{81}$$

The equation for Direct form- I realization of digital filter is given by

$$H\left(z\right) = \frac{b_0 + b_1 z^{-1}}{1 + a_1 z^{-1}} \tag{82}$$

where $b_0 = \dfrac{\omega T}{2 + \omega T}$, $b_1 = \dfrac{\omega T}{2 + \omega T}$ and $a_1 = \dfrac{\omega T - 2}{\omega T + 2}$

The important FPGA resources utilized during the implementation of the given controller is tabulated in Table 4. Figure 20 displays the Zynq ZC702 FPGA-based hardware-in-loop experimental setup used for analysis. The typical total path delay occurred in the implementation of the programmed logic is around 3.09 ns resulting in a possibility to operate the controller at 250 MHz against the 100MHz operation performed in this analysis.

FUTURE RESEARCH DIRECTIONS

My future work deals with the

- Implementation of dual Kalman filter based P&O MPPT controller for solar and/or wind generation in order to extract maximum power from it.
- To design new adaptive nonlinear control algorithms in order to get better performance and to perform the Hardware-in-loop analysis of the controller to validate the real-time performance of the controllers.
- To use the proposed wind-solar system in the field of transportation system.

Figure 19. Xilinx modelling of the important sections of the controller

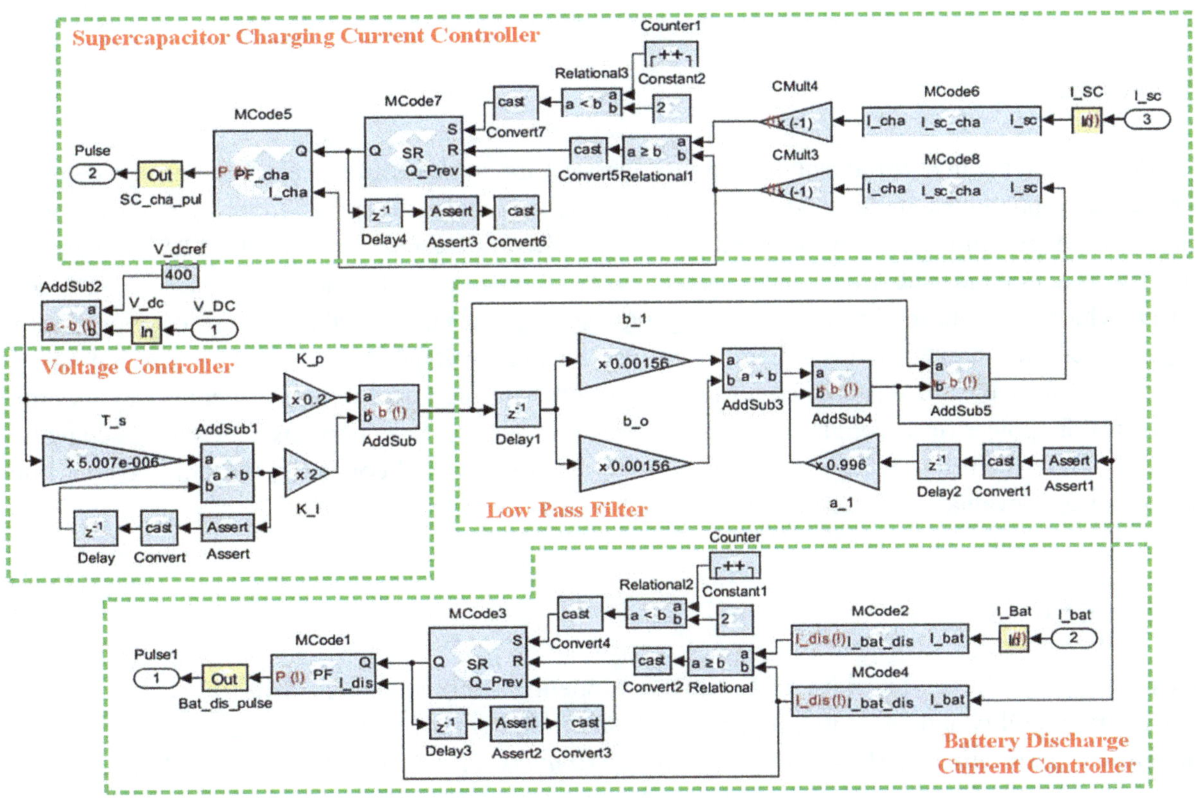

Figure 20. Experimental test bench for hardware-in-loop analysis

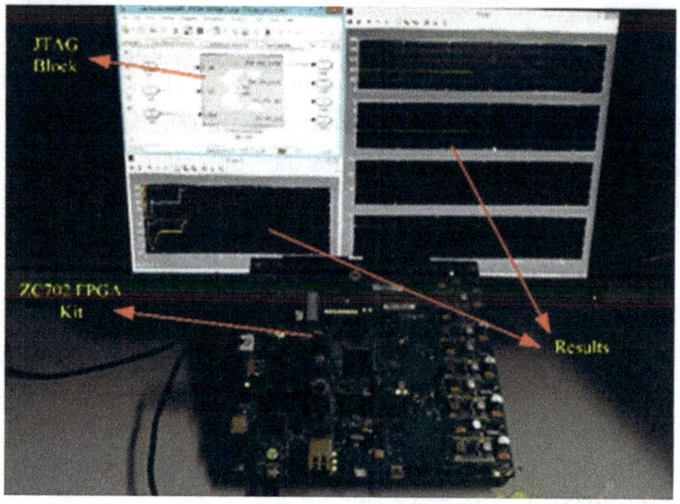

Table 4. FPGA resource utilization

Resource	Available	Utilized	% of Utilisation
Slice LUT's	53200	3037	5.71
Slice Registers	106400	1286	1.21
Block RAM tiles	140	2	1.43

CONCLUSION

The drastic increase in environmental pollutants liberated during generation of electrical energy demands for the inclusion of clean/renewable energy into the conventional utility grids. In this perspective, a DC microgrid incorporating solar/wind based renewable generation escorted with hybrid ESS comprising of supercapacitor and battery bank has been analysed in this work. Unlike the conventional DC-DC converters which are prone to high stresses during high gain operation, high gain DC-DC power processing stages are employed in this paper to boost up the low terminal voltage of the energy sources and storage devices. Dual loop control is applied for the control of these high gain interfacing converters in order to control the power flow from/to the devices under varying system conditions. The chosen converter topology gives the better efficiency compared with the conventional converter. Further, the controller performance is evaluated in real-time using FPGA based control HIL analysis.

REFERENCES

Alexandridis, A. T., & Marmidis, G. E. (2010). Modeling wind generators with full-scale frequency converters: Stability and passivity properties. *7th Mediterranean Conference and Exhibition on Power Generation, Transmission, Distribution and Energy Conversion*, 1-5. 10.1049/cp.2010.0870

Chen, G. (2000, August). Actively clamped bi-directional fly back converter. *IEEE Transactions on Industrial Electronics, 47*(4), 770–779. doi:10.1109/41.857957

Das, M., & Agarwal, V. (2016, May). Design and Analysis of a High-Efficiency DC–DC Converter with Soft Switching Capability for Renewable Energy Applications Requiring High Voltage Gain. *IEEE Transactions on Industrial Electronics, 63*(5), 2936–2944. doi:10.1109/TIE.2016.2515565

DeMeo, E. A., Grant, W., Milligan, M. R., & Schuerger, M. (2005). Wind plant integration. *IEEE Power & Energy Magazine, 3*(6), 38–46. doi:10.1109/MPAE.2005.1524619

Dorofte, C., Borup, U. & Blaabjerg, F. (2005). A combined two-method MPPT control scheme for grid-connected photovoltaic systems. *Power Electronics and Applications conference*, 10.

Garcia, F. S., Pomilio, J. A., & Spiazzi, G. (2013). Modeling and control design of the interleaved double dual boost converter. *IEEE Transactions on Industrial Electronics, 60*(8), 3283–3290. doi:10.1109/TIE.2012.2203770

Ghareeb, T., & Mohammed, O. A. (2013). DC Microgrids and Distribution Systems: An Overview. *IEEE Power and Energy Society General Meeting*, 1 - 5.

Graditi, G., Ippolito, M., Telaretti, E., & Zizzo, G. (2015, April). An innovative conversion device to the grid interface of combined RES-based generators and electric storage systems. *IEEE Transactions on Industrial Electronics, 62*(4), 2540–2550. doi:10.1109/TIE.2014.2336620

Grouz, F., & Sbita, L. A. (2014). Safe and easy methodology for design and sizing of standalone hybrid PV-wind system. *International Conference on Electrical Sciences and Technologies in Maghreb (CISTEM)*, 1-8.

Hammerstrom, J. (2007). AC Versus DC Distribution Systems: Did we get it Right? *IEEE Power Engineering Society General Meeting*, 1 - 5.

Hsieh, Y. P., Chen, J. F., Liang, T. J., & Yang, L. S. (2013, April). Novel high step-up DC-DC converter for distributed generation system. *IEEE Transactions on Industrial Electronics, 60*(4), 1473–1482. doi:10.1109/TIE.2011.2107721

Jewell, W. T. (1989, June). Electric utility experience with solar photovoltaic generation. *IEEE Transactions on Energy Conversion, 4*(2), 166–171. doi:10.1109/60.17907

Kadri, A., Marzougui, H., & Bacha, F. (2016). MPPT control methods in wind energy conversion system using DFIG. *4th International Conference on Control Engineering & Information Technology (CEIT)*, 1-6. 10.1109/CEIT.2016.7929115

Kawakami, N., Ota, S., Kon, H., Konno, S., Akagi, H., Kobayashi, H., & Okada, N. (2014, November). Development of a 500-kW Modular Multilevel Cascade Converter for Battery Energy Storage Systems. *IEEE Transactions on Industry Applications, 50*(6), 3902–3910. doi:10.1109/TIA.2014.2313657

Kazimierczuk, M. K. (2008). *Pulse-Width Modulated DC-DC Power Converters* (1st ed.). Hoboken, NJ, USA: Wiley. doi:10.1002/9780470694640

Kreczanik, P., Venet, P., Hijazi, A., & Clerc, G. (2014, September). Study of supercapacitor aging and lifetime estimation according to voltage, temperature and rms current. *IEEE Transactions on Industrial Electronics*, *61*(9), 4895–4902. doi:10.1109/TIE.2013.2293695

Kumar, D., Zare, F., & Ghosh, A. (2017). DC Microgrid Technology: System Architectures, AC Grid Interfaces, Grounding Schemes, Power Quality, Communication Networks, Applications, and Standardizations Aspects. *IEEE Access: Practical Innovations, Open Solutions*, *5*, 12230–12256. doi:10.1109/ACCESS.2017.2705914

Kwon, M., & Kwon, B. H. (2009). High step-up active-clamp converter with input-current doubler and output-voltage doubler for fuel cell power systems. *IEEE Transactions on Power Electronics*, *24*(1), 108–115. doi:10.1109/TPEL.2008.2006268

Lahfaoui, B., Zouggar, S., Elhafyani, M. L., & Seddik, M. (2015). Experimental study of P&O MPPT control for wind PMSG turbine. *3rd International Renewable and Sustainable Energy Conference (IRSEC)*, 1-6. 10.1109/IRSEC.2015.7455020

Lee, J. H., Liang, T. J., & Chen, J. F. (2014, July). Isolated coupled-inductor- integrated ADC–DC converter with nondissipative snubber for solar energy applications. *IEEE Transactions on Industrial Electronics*, *61*(7), 3337–3348. doi:10.1109/TIE.2013.2278517

Lee, N. E., Kim, W. w., Shin, J. h., & Kim, J. O. (2014). Optimal SOC limit estimation of ESS considering renewable energy sources. *International Conference on Information Science, Electronics and Electrical Engineering*, 1927-1931.

Lee, S., Kim, P., & Choi, S. (2013, July). High step-up soft-switched converters using voltage multiplier cells. *IEEE Transactions on Power Electronics*, *28*(7), 3379–3387. doi:10.1109/TPEL.2012.2227508

Li, W., & He, X. (2011, April). Review of non-isolated high-step-up DC/DC converters in photovoltaic grid-connected applications. *IEEE Transactions on Industrial Electronics*, *58*(4), 1239–1250. doi:10.1109/TIE.2010.2049715

Liang, T. J., Chen, S. M., Yang, L. S., Chen, J. F., & Ioinovici, A. (2012, April). Ultra- large gain step-up switched-capacitor DC-DC converter with a coupled inductor for alternative sources of energy. *IEEE Transactions on Circuits and Systems*, *59*(4), 864–874. doi:10.1109/TCSI.2011.2169886

Long, H., Eghlimi, M., & Zhang, Z. (2017, January). Configuration Optimization and Analysis of a Large-Scale PV/Wind System. *IEEE Transactions on Sustainable Energy*, *8*(1), 84–93. doi:10.1109/TSTE.2016.2583469

Lu, X., & Wang, J. (2017, June). A Game Changer: Electrifying Remote Communities by Using Isolated Microgrids. *IEEE Electrification Magazine*, *5*(2), 56–63. doi:10.1109/MELE.2017.2685958

Ma, K. W., & Lee, Y. S. (1996, March). An integrated flyback converter for DC uninterruptible power supply. *IEEE Transactions on Power Electronics*, *11*(2), 318–327. doi:10.1109/63.486182

Mahmoud, M. S., Saif Ur Rahman, M., & Sunni, F. M. A. L. (2015). Review of microgrid architectures – a system of systems perspective. *IET Renewable Power Generation*, *9*(8), 1064–1078. doi:10.1049/iet-rpg.2014.0171

Mendis, N., Muttaqi, K. M., Perera, S., & Uddin, M. N. (2015, March). Remote Area Power Supply System: An Integrated Control Approach Based on Active Power Balance. *IEEE Industry Applications Magazine*, *21*(2), 63–76. doi:10.1109/MIAS.2014.2345817

Naidu, B. R., Panda, G., & Ganesh, R. (2016). Modelling and Hardware-in-Loop validation of a modified controller for intermittent operation of a standalone low voltage DC fuel cell system. *IEEE Region 10 Conference (TENCON)*, 1429-1434.

Pan, C. T., Chuang, C. F., & Chu, C. C. (2014, September). A novel transformer-less adaptable voltage quadruple DC converter with low switch voltage stress. *IEEE Transactions on Power Electronics*, *29*(9), 4787–4796. doi:10.1109/TPEL.2013.2287020

Parvini, S., Siegel, J. B., Stefanopoulou, A. G., & Vahidi, A. (2016, March). Supercapacitor Electrical and Thermal Modeling, Identification, and Validation for a Wide Range of Temperature and Power Applications. *IEEE Transactions on Industrial Electronics*, *63*(3), 1574–1585. doi:10.1109/TIE.2015.2494868

Pinto, S. J., Panda, G., & Peesapati, R. (2017). An Implementation of Hybrid Control Strategy for Distributed Generation System Interface Using Xilinx System Generator. *IEEE Transactions on Industrial Informatics*.

Prudente, M., Pfitscher, L. L., Emmendoerfer, G., Romaneli, E. F., & Gules, R. (2008, March). Voltage multiplier cells applied to non-isolated DC-DC converters. *IEEE Transactions on Power Electronics*, *23*(2), 871–887. doi:10.1109/TPEL.2007.915762

Raygani, S. V., Sharma, R. & Saha, T. K. (2015). PV power output uncertainty in Australia. *IEEE Power & Energy Society General Meeting*, 1-5.

Rocabert, J., Azevedo, G. M. S., Luna, A., Guerrero, J. M., Candela, J. I., & Rodríguez, P. (2011, October). Intelligent Connection Agent for Three-Phase Grid-Connected Microgrids. *IEEE Transactions on Power Electronics*, *26*(10), 2993–3005. doi:10.1109/TPEL.2011.2116126

Sannino, A., Postiglione, G., & Bollen, M. H. J. (2003, October). Feasibility of a DC network for commercial facilities. *IEEE Transactions on Industry Applications*, *39*(5), 1499–1507. doi:10.1109/TIA.2003.816517

Sayeef, S. (2012). *Solar Intermittency: Australia's Clean Energy Challenge: Characterising the Effect of High Penetration Solar Intermittency on Australian Electricity Networks*. Sydney, Australia: CSIRO.

Schmidlin, C. R. Junior, & Araujo Lima, F. K. (2011, October). Wind Turbine and PMSG Dynamic Modelling in PSIM. *IEEE Latin America Transactions*, *14*(9), 4115–4120. doi:10.1109/TLA.2016.7785941

Shreelakshmi, M. P., Das, M., & Agarwal, V. (2013). High gain, a high efficiency bi-directional DC-DC converter for battery charging applications in stand-alone photovoltaic systems. *IEEE 39th Photovoltaic Specialists Conference (PVSC)*, 2857 – 2861.

Silveira, G. C., Tofoli, F. L., Bezerra, L. D. S., & Torrico-Bascope, R. P. (2014, December). A non-isolated DC-DC boost converter with high voltage gain and balanced output voltage. *IEEE Transactions on Industrial Electronics*, *61*(12), 6739–6746. doi:10.1109/TIE.2014.2317136

Thounthong, P., Rael, S., & Davat, B. (2009, March). Analysis of Supercapacitor as Second Source Based on Fuel Cell Power Generation. *IEEE Transactions on Energy Conversion*, *24*(1), 247–255. doi:10.1109/TEC.2008.2003216

Tremblay, O., Dessaint, L. A., & Dekkiche, A. I. (2007). A Generic Battery Model for the Dynamic Simulation of Hybrid Electric Vehicles. *IEEE Vehicle Power and Propulsion Conference*, 284-289. 10.1109/VPPC.2007.4544139

Villalva, M. G., Gazoli, J. R., & Filho, E. R. (2009, May). Comprehensive Approach to Modeling and Simulation of Photovoltaic Arrays. *IEEE Transactions on Power Electronics*, *24*(5), 1198–1208. doi:10.1109/TPEL.2009.2013862

Wai, R. J., Duan, R. Y., & Jheng, K. H. (2012, February). High-efficiency bidirectional dc-DC converter with high-voltage gain. *IEEE Transactions on Power Electronics*, *5*(2), 173–184. doi:10.1049/iet-pel.2011.0194

Witulski, A. F. (1995, May). Introduction to modeling of transformers and coupled inductors. *IEEE Transactions on Power Electronics*, *10*(3), 349–357. doi:10.1109/63.388001

Xiao, W., Dunford, W. G., & Capel, A. (2004). A novel modelling method for photovoltaic cells. *IEEE 35th Annual Power Electronics Specialist Conference (PESC), 3*, 1950–1956.

Xu, D. &Cheng, K. W. E. (2011). A survey of distributed power system-- AC versus DC distributed power system. *IEEE Power Electronics Systems and Applications (PESA),* 1 - 12.

Yan, Z., Fei, L., Yin, J., & Duan, S. (2008). Study on realizing MPPT by improved Incremental Conductance method with variable step-size. *3rd IEEE Conference on Industrial Electronics and Applications*, 547-550. 10.1109/ICIEA.2008.4582575

Young, C. M., Chen, M. H., Chang, T. A., Ko, C. C., & Jen, K. K. (2013, February). Cascade Cockcroft–Walton voltage multiplier applied to transformerless high step-up DC-DC converter. *IEEE Transactions on Industrial Electronics*, *60*(2), 523–537. doi:10.1109/TIE.2012.2188255

Zhao, Q., & Lee, F. C. (2003, January). High-efficiency, high step-up DC-DC converters. *IEEE Transactions on Power Electronics*, *18*(1), 65–73. doi:10.1109/TPEL.2002.807188

Zhao, Q., & Lee, F. C. (2003, January). High-efficiency, high step-up DC-DC converters. *IEEE Transactions on Power Electronics*, *18*(1), 65–73. doi:10.1109/TPEL.2002.807188

KEY TERMS AND DEFINITIONS

Energy Storage Systems: The combination of storage systems (like battery and supercapacitor) used to store the excess energy and supplies when it is desired.

FPGA: Field programmable gate arrays, used to implement the designed algorithm.

Maximum Power Point Tracking: The particular voltage and current at which the maximum power can be drawn from the source.

Microgrid: The combination of renewable sources and local loads that acts as a single controllable entity.

SOC: State of the charge of the storage devices such as battery and supercapacitor.

Wind Energy Conversion System: The system used to produce electricity by using kinetic energy from the winds.

XSG: Xilinx system generator, used to interface MATLAB/Simulink with the FPGA kit.

Chapter 16
Design and Analysis of a Cost-Effective Standalone Solar: Diesel Hybrid Power System

Fahad Iqbal
Jamia Millia Islamia, India

Ankur Singh Rana
Jamia Millia Islamia, India

Shufali Ashraf Wani
Jamia Millia Islamia, India

ABSTRACT

The foremost issues of the twenty-first century are the ever-increasing challenging demand of electrical energy and controlling the emission of greenhouse gases (GHG). Along with these issues and with limited energy resources, it is imperative to look for non-conventional methods of power generation like from renewable energy resources. Microgrid has emerged as a new field that can meet the energy demand with a special emphasis on good power quality, reliability, and security. A major concern with the use of renewable energy resources is their intermittent nature which makes their integration and operation a challengeable task. Energy storage devices like batteries can be used to overcome the problem of intermittent nature of renewable energy resources. This chapter focusses on different aspects of renewable energy resources in detail. It analyzes the effectiveness of the proposed topology of the microgrid for health clinic load profile with the help of PVSYST software.

INTRODUCTION

The exponential rise in demand for electrical energy and hazardous gas emissions due to conventional methods of power generation has provided an impetus to switch from conventional to non-conventional methods of power generation. Rapid depletion of fossil fuels, insecurity and uncertainty in power supply, inefficient and uneconomical methods of power generation are some other undesirable factors of

DOI: 10.4018/978-1-5225-3935-3.ch016

conventional power plants that aid and motivate power engineers to shift their focus from conventional to non-conventional methods of power generation. Also, sometimes power supply from the traditional grid is not possible in some remote and rural areas. This may be due to lack of transmission infrastructure, unfavorable terrestrial condition, uneconomical condition etc. All these problems can be sorted out by the use of renewable energy resources. Renewable energy resources are available in-plenty in nature, free of cost and environmental friendly and can be used to generate electrical power either in a standalone or in hybrid form. These sources can be connected near or far from the load depending upon the availability and production from these renewable energy resources. Microgrids, nowadays an emerging technology can be used to connect different types of renewable energy resources in hybrid fashion and cater all the required load demand. Microgrids have the capability to operate either in a grid connected and stand-alone mode. This feature of microgrid helps in improving the performance, peak shaving, reliability and security of the entire system along with huge environmental benefits. The same type of renewable energy resource can be connected on large scale far from the load center for large scale production like in solar farms, wind farms etc.

Sensitive loads like schools, colleges, central jails, health clinics, hospitals, etc. where the electrical power supply is very essential are available here and there. Such types of loads can also be present in remote and rural areas where power supply from the conventional grid is not possible due to various above-mentioned factors. But these areas may be rich in certain type of renewable energy resources. So, in such locations available renewable energy resource can be used to produce electricity for essential and sensitive loads without depending on conventional grid power supply. Energy storage devices play a very important role in such cases to cater intermittent nature of renewable energy resources and to maintain voltage and frequency profile of the network.

In this chapter, the importance of renewable energy resource and the sensitive load is reflected and how these renewable energy resources can be optimally placed together to cater the load demand of sensitive load is demonstrated with the help of PVSYST software. PVSYST is a software package for study, sizing, simulation and data analysis for a complete PV system. It mainly consists of four basic parts, preliminary design, project design, databases and tools. Preliminary design provides a quick evaluation of the potential of a project designed in a situation under predefined constraints. Project design is the main part where all designing of a project is done. Various types of project are associated with PV system such as Standalone PV system, grid connected PV system, etc. Databases consist of two basic parts one is associated with metrological data of monthly along with weekly information for a specific location and another part of the database is the information of the components available in the software. Also, some tools are available which helps in quickly estimating the behavior of the solar based system. The results obtained from this software should reflect how optimal and economical is the size of the considered configuration, how much this configuration is environmental friendly and how much this configuration is reliable. First, the size of the test system is analyzed, then its location, the availability of renewable energy resource at this particular location, then the optimal size of the source to cater overall load demand, then the best optimal configuration and at last the findings and results of this configuration are provided.

This chapter is based on the use of microgrid for monitoring the load of a health clinic. It is sub-divided into following subsections: the background is followed by the overview of PVSYST software in which simulation is carried out. Next section is dedicated to the understanding of microgrid. This section is followed by implementation of microgrid based topology to meet the load demand of health clinic in

Jodhpur city. Results section provides the simulation results and costs analysis of proposed topology. Conclusion section summarizes the proposed system and findings.

BACKGROUND

Electricity is an essential commodity for any developing and developed nation. In today's modern world almost each and everything is dependent on electricity. Most of the industries are completely based on the electric power supply for their proper functioning. The demand for electricity is increasing globally every passing day (Xoubi, 2015) and conventional methods of power generation are not able to meet this rapidly increasing demand for electricity due to depleting nature of fossil fuels (Buragohain, Mahanta, & Moholkar, 2010). Also, some disadvantages of power generation from conventional methods of power generation such as power generation from thermal power plants contribute to global warming because of excessive GHG emissions like CO_2, SO_2 and NO_2 (Pali & Vadhera, 2016). The problem with power generation from the nuclear power plant is that the handling of nuclear fuel is extremely dangerous and disposal of radioactive waste is very expensive. Also, the problem of skin disease and cancer is associated with radioactive rays. Other factors that shows the generation from conventional methods is not advisable are as follows: i) Fossil fuels are limited in nature and their available quantity is depleting day by day. So, it is hard to rely on them. ii) Emissions like Green House Gasses (GHG), CO_2 footprints, oxides of Sulphur and Nitrogen are environmentally hazardous in nature. The problem of ozone depletion and acid rain can be attributed to these hazardous emissions. This ultimately forces towards methods of power generation which are free from such type of emissions. iii) Also in some remote and rural areas, the power supply from conventional methods of power generation is found to be expensive either due to lack of transmission infrastructure or due to unfavorable terrestrial conditions in such locations since, they are located far away from the utility grid (Bhattacharyya, 2012).

All these issues reflect the need for the use of alternative resources of power generation that are highly efficient and environmentally friendly. Renewable energy resources can be considered as one of best alternative of power generation as they possess the above-mentioned qualities. Renewable energy resources are those energy sources which are continuously and naturally replenished. Renewable energy resources are available in plenty, free of cost, sustainable, emissions free and environmental friendly (Singal, Varun, & Singh, 2007). Due to all these features, renewable energy resources also considered as green and clean energy. Many benefits are associated with the use of renewable energy resource like: i) Pollution reduction: Conventional energy sources emit poisonous gases that are hazardous to human beings as well as to flora and fauna. On the other hand, use of renewable energy resources is free from such hazardous and polluting gaseous emissions. So, usage of renewable energy resources will considerably diminish overall pollutants present in the atmosphere. ii) Distributed generation: Renewable energy sources are distributed in nature i.e. usually connected near-by to load (Ackermann, Andersson, & Soder, 2001). Unavailability of renewable resources will affect only the dedicated load but all other loads remain unaffected. In this way the overall reliability of the entire system will enhance and also helps in peak shaving (Rana, Thomas, & Senroy, 2017). iii) Modular system: Usually more than one wind turbine or solar panel is connected to the particular area. Any damage to the equipment won't affect the supply. So, the back-up equipment can supply to the partial load (Kumar & Singh, 2014; Kumar, Ali, & Thomas, 2015) . iv) Less water requirement: In conventional energy resources, water is required in large quantities for cooling purposes whereas for power generation with renewable energy resources water is not required (Lopes, Hatziargyriou, Mutale, Djapic, & Jenkins, 2010). Hence the quality and quantity of water will be maintained and scarcity of water does not affect power generation. v) Because of zero emissions, it can boost public health. vi) the power generation from renewable energy resources is a new technology which ultimately creates several job opportunities (Pimentel et al., 2002). However, there are also some disadvantages associated with the use of renewable energy resources like i) High initial investment: Electricity generation via renewable energy sources is still in developing stage so the equipment required is costly (Pepermans, Driesen, Haeseldonckx, Belmans, & D'haeseleer, 2005). Also, it requires skilled professional expertise for the installation of setup which inherently increases the investment cost. ii) Space required: Both solar panels/ arrays, as well as wind turbines for large or medium power generation, require huge space for installation. iii) Unreliability in power generation: Renewable energy sources are sporadic and hence it's hard to get uninterrupted power supply from them (Ali, Thomas, & Kumar, 2015; Kumar, Ali, Thomas, & Singh, 2017). iv) Power quality: For maintaining stable voltage and frequency supply, power converters play an important role which again adds the cost of generation, as well as malfunctioning of these converters, may generate problems in grid integration.

With the advancement in technology and power electronics, all the problems associated with the use of renewable energy resources will be eliminated very soon in future. With the rapid use and enormous advantages associated with renewable energy resources, two other technologies Distributed Generation (DG) and Microgrid (MG) are also emerging technologies nowadays. Distributed generations are small energy sources that are used to meet load demand. Generally, these sources are local and renewable and can integrate with the utility grid (Al, Hassan, Pelov, & Nuaymi, 2015). Individually these distributed generators are small in capacity but all together are quite effective and can reduce a large amount of load demand from the conventional grid. Similarly, Microgrids are the small power subsystem comprising sensitive and non-sensitive loads, one or more distributed generators and energy storage devices working together in a controlled and coordinated manner. A microgrid can operate either in a grid connected and isolated mode. The benefits of these two technologies are manifolds like these technologies can increase system reliability and power quality, they can provide emergency power supply, they can help in peak power shaving, Microgrids can be used to provide ancillary services, MGs can provide reduction in vulnerability to terrorism and improvements in infrastructure resilience (Sahoo, Abhitharan, Kalaivani, & Karthik, 2015).

Different locations of different countries are enriched in different types of renewable energy resources like some locations may be rich in solar energy and wind energy, some locations may be rich only in wind energy, some may be rich in geothermal energy and so on. For example, Jodhpur location which comes in Rajasthan state is good in solar energy. In such locations, microgrids prove to be a very important and effective technology as it allows for effective integration of renewable energy resources with the existing grid and meets existing load demand in an optimum way(Nazir, Laksono, Waldi, Ekaputra, & Coveria, 2014). Nowadays, the trend is to meet the load demand of an area with available local distributed energy sources like as in solar parks and microgrids which ultimately helps in congestion management and transmission reinforcement. Many countries have already minimized the use of conventional methods which is proved by observing in the rapid growth of electric vehicles, microgrid, solar plants, solar parks etc. However, the problem with the use of renewable energy resources is their intermittent nature i.e. they are dependent on weather conditions, time of use and many more things (Liang, 2017). For example, solar energy is available during day time but one cannot produce any electricity from solar PV system during night times. Also, there is a possibility of partial and full shading of solar panels due to clouds which ultimately reduces or stops the power generation. Similarly, wind is available only at some time but at other times it may not be available. Also, the power generation from wind turbines is not advisable if the speed of wind is below or above cut-in and cut-off region respectively. This intermittent nature problem of renewable energy resources can also sorted out by the use of high density energy storage devices (N. Li & Hedman, 2015). These energy storage devices aid in storing surplus energy and supports deficit energy during high demand periods. Different sections of this chapter give detailed description of all the equipment and technology used in the proposed configuration for the considered test system.

MICROGRID ARCHITECTURE

A microgrid can be considered as a small grid consisting of electrical loads, distributed generations, renewable energy resources and energy storage devices like battery banks, supercapacitor, flywheels etc. Microgrids are different from the traditional power systems because they can store surplus generated energy using energy storage devices and help in shaving peak load demands (Hafez & Bhattacharya, 2012). Smart microgrids offer bidirectional communication flow and control devices which ultimately helps in sending and receiving information, reduces excess energy requirements and enhance reliability and security of the system (Caldognetto, Tenti, & Brandao, 2017). Microgrids offers two modes of operations: Grid connected mode and islanded mode. In an Islanded mode, microgrids continue to serve local loads and maintain its reliability. In the grid connected mode, microgrids operate while connected to the conventional grid and they can purchase deficit power from the conventional grid, similarly, they can sell the surplus to the conventional grid (Y. Li & Nejabatkhah, 2014). By interconnecting number of microgrids together, microgrids can participate in power exchanges. In this way, microgrids can sell/purchase their excess/deficit power requirements respectively from power exchanges. The resulting network helps in efficient and economic operation (Khodaei & Shahidehpour, 2013). If microgrid serves a large number of loads using many Distributed Energy Resources (DER's) then such type of individual microgrids can also participate in power exchanges. Figure 1 represents a typical microgrid architecture where a number of microgrids are interconnected together. Point of Common Connection (PCC) is required to connect microgrid network with the main grid network (L. Che, Zhang, Shahidehpour, Alabdulwahab, & Abusorrah, 2017). Depending on the requirement of MG network, it can be multi-dimensional, double or single type of connection. Energy Management Unit (EMU) is an interfacing device used to connect various units available in microgrids with a global central controller. In the islanded mode of operation, EMU has the ultimate responsibility of ensuring optimized operation of the microgrid (Sun, Zhou, Guerrero, & Zhang, 2015). It has the capability to receive or send control signals with various available units such as ESDs, communicable loads etc. It is also responsible for calculating and forecasting the change in load demand, change in power generation from renewable energy resources and variation in the energy prices on hourly, minute and second basis (Liang Che & Shahidehpour, 2014). EMU sends all such types of information to the Global Central Controller (GCC), which ultimately helps in reducing the effect of power variations from these sources on the microgrid distribution system.

SYSTEM DESCRIPTION

A location of Jodhpur city and a load profile of health clinic is taken here to check the effectiveness of the proposed microgrid topology. Jodhpur is a city in the Thar Desert of the northwest Indian state of

Figure 1. Typical microgrid architecture

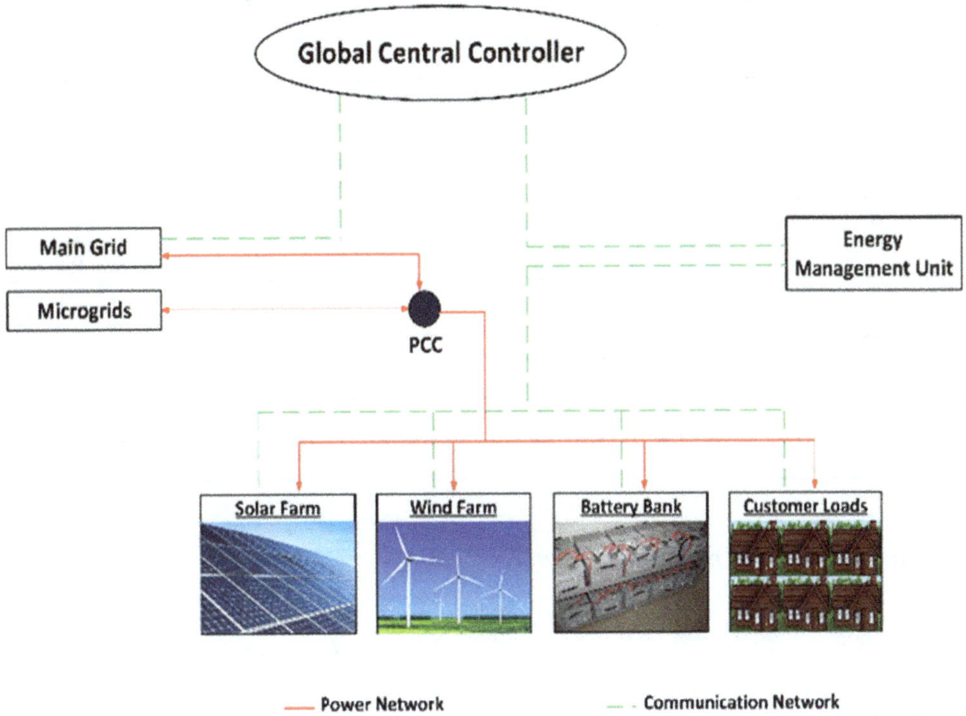

Rajasthan. The latitude and longitude of this location are 26.3 and 73.02 respectively. The architecture of this health clinic includes general wards, reception room, wards, dispensary room, pathology labs, wash rooms, cafeteria, store room, doctors room and some other miscellaneous facilities. Table 1 shows some commonly used electrical equipment's along with their ratings in kW. The timings of the health clinic are assumed to be open from 09:00 am to 05:00 pm. So, most of the load will be present during this time while in the remaining times there will be only emergency load present.

Figure 2 shows the general layout of the proposed microgrid topology for the assumed health clinic. Solar Photovoltaic (PV) system is the main source of power supply to this health clinic. However, because of its intermittent nature, the battery bank is considered here to supply deficit power to the health clinic during deficiency in power generation from solar PV system and similarly to store excess power from the solar PV system. Since health clinic is open during the day and during day time solar PV system is also affective because of the continuous availability of solar radiation. A diesel generator is considered here as a backup power supply. So, during unavailability of solar radiation, fault conditions and during the maintenance period, diesel generator can be used as the main source of power supply. Converters are necessary to convert DC into AC and vice-versa respectively. These converters are necessary because solar PV system generates DC power whereas many connected loads require AC power. Table 2 shows the availability of the daily solar horizontal radiation for this particular location.

Figure 3 shows the load profile of the health clinic. As it can be seen from the graph that the load demand is higher during 09:00 hour to 17:00 hour. Since during these timings, health clinic is assumed to be open. Also, load demand is higher during June and July because these are the months of summer and during these months some additional electric loads are added like fans, air coolers, refrigerators etc. as compared to the winter months.

Table 1. Electrical equipment's along with their ratings in kW

Items	Rating
LED Bulb	0.014
Tube light	0.04
Ceiling Fan	0.08
Refrigerator	0.2
Computer	0.12
Water Cooler	0.6
Air Cooler	0.2
Halogen Emergency Lights	0.2
Exhaust Fan	0.16
Room Heater/Blower	1
Miscellaneous	-

Figure 2. General layout of the proposed microgrid topology

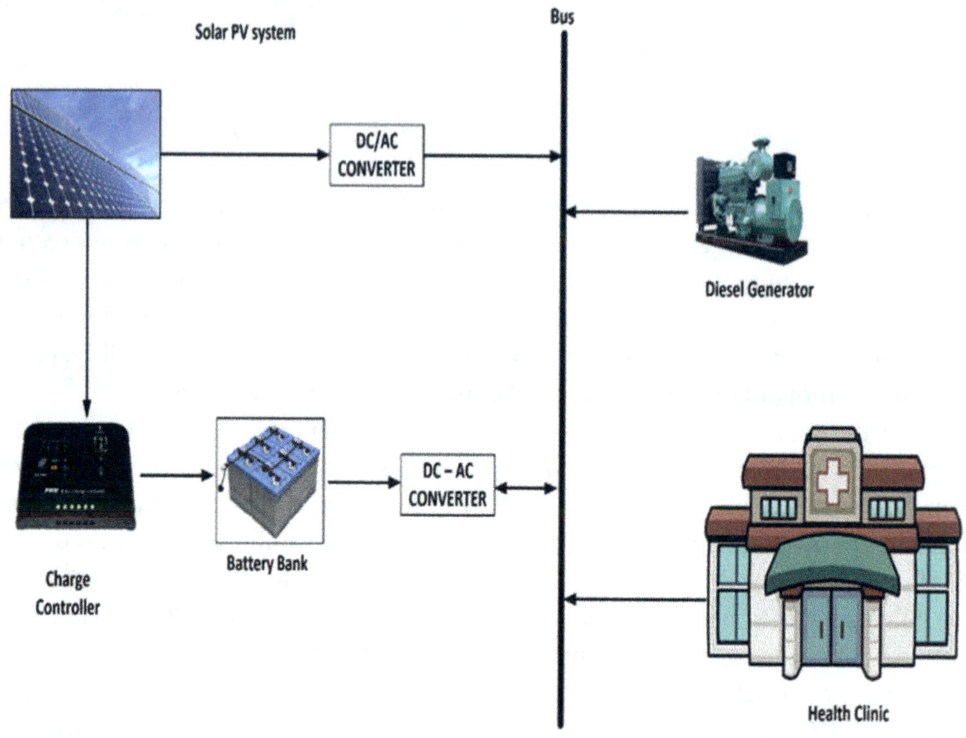

Table 2. Availability of the daily solar horizontal radiation for this particular location

Month	Radiation (kWh/m2/d)
January	3.95
February	4.58
March	5.54
April	6.40
May	6.82
June	6.63
July	5.60
August	5.30
September	5.31
October	4.68
November	4.08
December	3.60

Figure 3. Hourly load profile of health clinic of a whole year

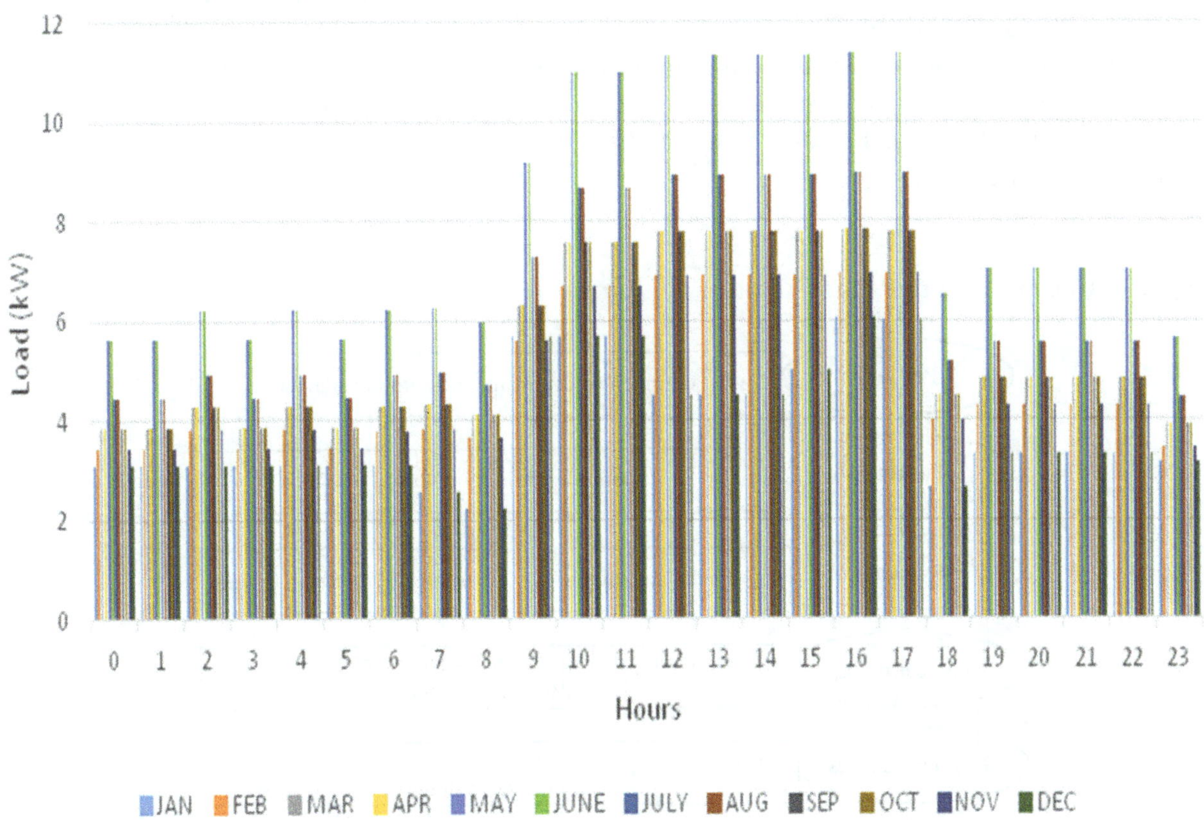

OVERVIEW OF PVSYST SOFTWARE

PVSYST was initially developed at the University of Geneva as PV-centric simulation tool but now works as a standalone company. It mainly focuses on designing and deployment in engineering aspect (Mermoud & Wittmer, 2014). PVSYST is licensed software although it gives a free trial for 30 days. It supports a number of PV system types such as Grid-tie, off-grid, and DC grid. Meteonorm, TMY2, TMY3, EPW, PVGIS, WRDC, Retscreen, Helioclim, SolarGIS are different types of weather files supported by PVSYST. Also, support PV Modules, Charge Controller, Inverter, Battery, and Generator with full flexibility. But sometimes customers face difficulty while doing economic modeling with PVSYST.

PVSYST is an efficient tool which can be used for predicting the potential of solar energy at particular sites. This helps in optimally designing a PV technology i.e. different type of configurations with a best possible solution in harvesting solar energy. PV technology needs highly qualified professionals for its installation, but with the help of PVSYST software technology, it becomes easy to study the behavior of PV system. The process of designing a system in PVSYST includes the following basic steps shown in Figure 4.

While designing a project in PVSYST, locations are usually associated with meteorological files and if not given then locations can be defined in the interactive maps. It helps in finding climate conditions for that latitude and longitude. After location defining, orientation is defined with array's azimuth and tilt. Then the main system parameters, including module and inverter model, system capacity, string lengths etc., are defined in the Array Design screen. A string is considered homogeneous if it contains only one orientation, tilt and module type. Currently, only homogeneous strings may be designed with PVSYST. In practice, PV system can be designed with non-homogenous strings, meaning that different orientations, module tilts and module types can be accommodated within a string, increasing system flexibility and ease of design.

Figure 4. General steps in PVSYST software

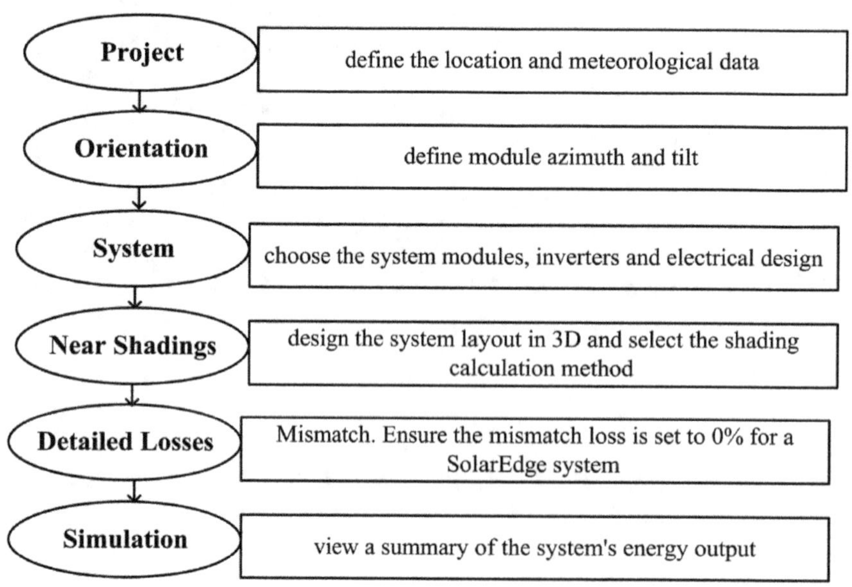

METHODOLOGY

This section represents the methodology used to design proposed microgrid configuration for the assumed test system (health clinic). Initially the geographical coordinates of the test system are identified which includes latitude and longitude of the health clinic. Then the availability of various renewable energy sources for the particular location are checked. Now collect all the required data related to the available energy resource like solar radiation, wind-speed etc. Renewable energy resources are sporadic in nature and different locations may be rich either in one renewable source or more than one renewable source. In order to make system economical those energy sources should be chosen which are available throughout the year and their quantity should be good enough for electrical energy production. The location selected for the proposed microgrid topology in this work is the city of Jodhpur situated in Rajasthan, India. In Jodhpur city solar radiation is profusely available so solar PV system is considered here for electrical energy production so that an economical system can be designed.

Next step is to design and analyse the load profile of the considered test system. Load profile has to be designed for entire year showing load demand on monthly and hourly basis. From this load profile, operator can take an idea of overall energy demand as well as peak load demand. From this analysis, operator estimate the required quantity of available renewable energy resource, which should be greater than the peak load including the losses. Since renewable energy resources are intermittent in nature so back power supply and energy storage device is required to maintain uninterrupted power supply and to maintain voltage and frequency profile.

Now next step includes designing of proposed system configuration in PVSYST software. Here operator should define all the required components like solar panels, battery, inverter, back-up power source etc. The rating of each component and load demand will decide the required number of components as well as their series and parallel combinations e.g. In case of solar panels and battery. The manufacturer of each component has to selected smartly by the operator because cost of the components and their efficiency vary with manufacturing company.

After defining all the pre-requisites, the simulation is run. As per load demand if any of the constraints like voltage, current, power etc. are not satisfied based on the defined pre-requisites then a warning message is flashed. This indicates that the operator needs to recheck and redefine all the entries until the desirable result is achieved.

Once the energy production is able to meet the load demand while satisfying all the required constraints, the configuration thus achieved is the desirable one. After this, operator can perform economical analysis in order to check the effectiveness of the proposed configuration either mathematically or through some software like RETSCREEN. Figure 5 shows the step by step procedure for designing an effective microgrid configuration using PVSYST software.

RESULTS AND DISCUSSIONS

The results of the described test set up are discussed in this section. The instantaneous peak load of the health clinic load profile is found to be 12 kW. The proposed microgrid configuration intends to meet the entire load demand using PV system along with battery bank system to overcome the intermittent nature of solar energy. Further diesel generator is also taken as a backup source for supplying energy during maintenance period and emergency conditions. Figure 6 provides the simulated system circuit

Figure 5. Flow chart for designing microgrid using PVSYST software

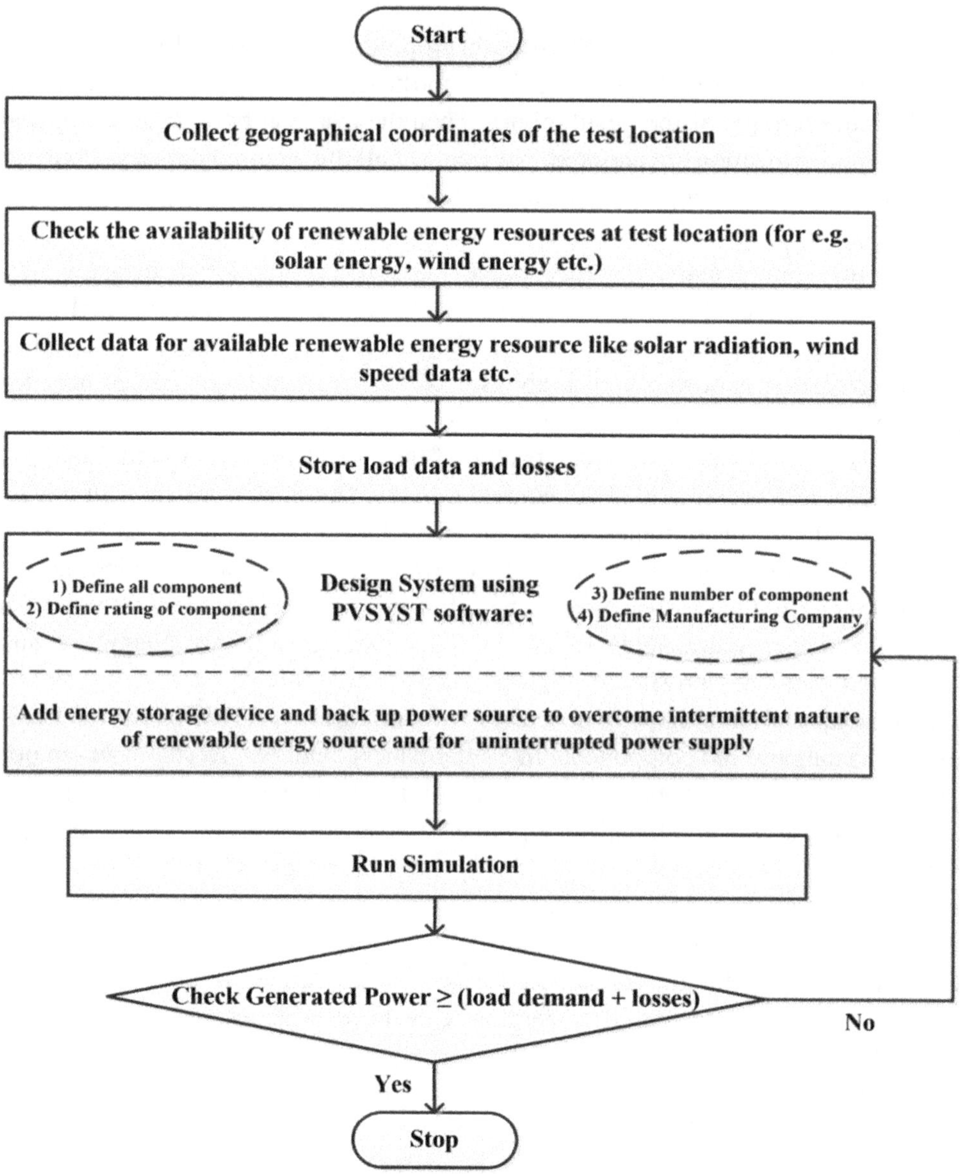

layout. Simulated system is broadly divided into three sub sections, such as PV array, System, and user-load. Under normal system conditions, the regulator switch will allow power to flow from the PV array towards load and towards battery bank. However, when proper sunlight is not available for the PV array working, regulator switch will allow the supply power to load from the batteries.

There may arise some chances when neither the battery nor the PV array is available to supply power to load, in those cases diesel generators will be used as the backup. The direction of power transfer from the PV array and the diesel generator is unidirectional as they both supply power whenever required, whereas batteries will be connected to bidirectional switch as they need to be charged when generation from the PV array is more and need to be discharged when PV array is not working properly.

Figure 6. Simulated system circuit for health clinic

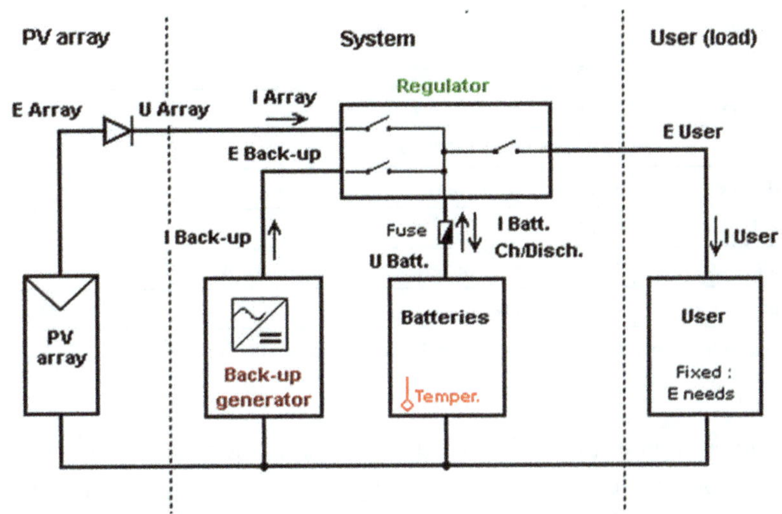

The overall system wiring for the proposed microgrid configuration comprising various solar PV modules, battery bank, diesel generator, protection devices like a fuse, MCB, converters, etc. are shown in Figure 7 in a detailed way. This system wiring is found to be very effective in designing and implementing the proposed microgrid configuration in a real world.

Table 3 represents some important parameters which decided the overall cost of the proposed system. For the given total estimated load and energy demand, the other important parameters like type of solar inverter, type and number of solar modules required, battery bank capacity required etc. are selected in such a way that results in overall system more economical and more efficient.

Figure 7. System wiring for power supply to health clinic

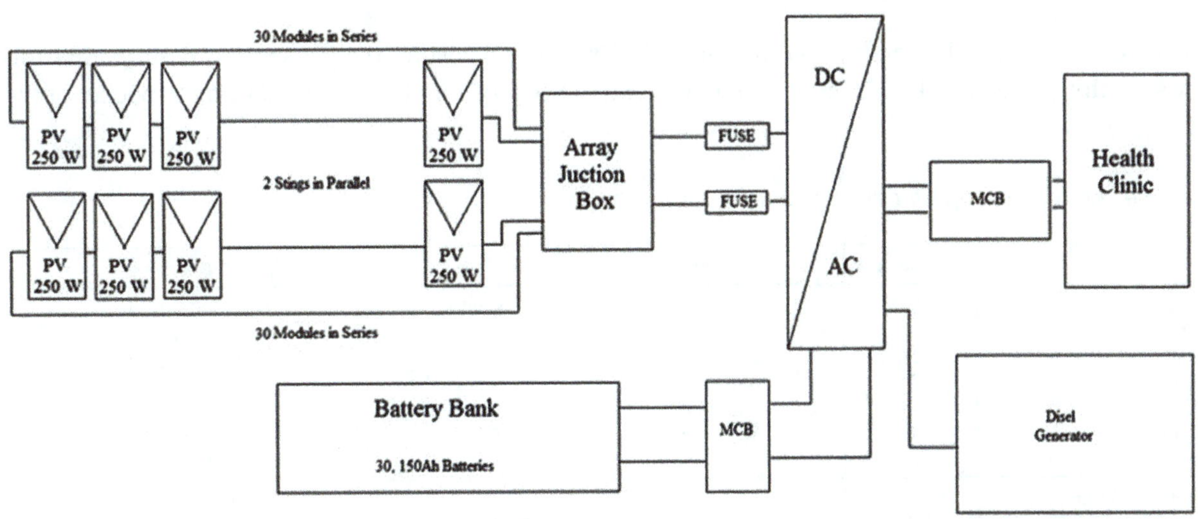

Table 3. Parameters of test system

Total Estimated Load	12kW
Total Daily Energy	286966Wh/day
Total Monthly Energy	8609kWh/month
Selected Solar Inverter	Delta RPI-M15 (15000 Watt)
Selected Solar Module	Tata Power (250 Watt)
Total No of Modules Required	60 Modules
Battery Bank	12V, 150 Ah, 30 Batteries (4.5 Hours Backup)
Diesel Generator	12kV (Standby)

*According to MPPT range of Solar inverter, two stings each of 30 modules in parallel are needed.

Table 4 shows different types of electrical equipment used in a health clinic along with their rating, quantity, usage hours in order to compute overall energy requirements per hour per day.

Since, solar PV system provides DC output and most of the load equipment's are AC based, the inverter is required to convert the DC power of solar energy into AC power suitable for various installed AC loads. Table 5 represents the detailed specifications of the inverter used in the proposed test system.

Solar energy being intermittent in nature, the efficiency of solar PV system depends on various factors like time of use, ambient temperature, weather conditions, solar radiation, performance ratio etc. In PVSYST there is a list of a number of vendors who deal in the manufacturing and selling of solar panels, batteries, inverters, controllers etc. The designer should select the required item and vendor based on the system requirements. The decision of the designer will depend on various factors like the efficiency of the item, its power output, its market cost, its operating constraints and so on. In this project, a solar module of Tata Power is selected for designing purpose. Table 6 represents different parameters of the selected solar module whereas Table 7 lists out other important factors on which the efficiency of solar panel depends.

COST ANALYSIS

This section deals with overall economic analysis. The system cost for the PV setup is described as follows: In this project, 15 kW inverter is required and the cost of this 15kW inverter on the pocket of the

Table 4. Load breakup and energy consumption

	Number	Power	Use	Energy
Lamps	56		8 h/day	17920 Wh/day
PC	2	40W/lamp	8 h/day	1920 Wh/day
Fan	28	120W/app	8 h/day	13440 Wh/day
Fridge	2	60 W/app	24 h/day	240000 Wh/day
Cooler	16	60 W/Halogen	6 h/day	10560 Wh/day
Halogen Light	2	1W/Exhaust	10 h/day	2400 Wh/day
Exhaust	11		6 h/day	726 Wh/day
Total Daily Energy				286966 Wh/day

Table 5. Specifications of inverter

Model		RPI-M8	RPI-M10	RPI-M15
DC Input	Max. DC Voltage	1000V	1000V	1000V
	Operating voltage Range	200-1000V	200-1000V	200-1000V
	MPP Voltage Range	280-850V	350-850V	416-850V
	Start Voltage	>250V	>250V	>250V
	Rated Voltage	635V	635V	635V
	Max. Input Current	Each MPPT:20A Total: 30A	Each MPPT: 20A Total: 30A	Each MPPT: 20A Total: 30A
	MPP Tracker	2	2	2
	Connection Type	4 Pairs MC4	4 Pairs MC4	4 Pairs MC4
AC Output	Max. Output Power	8400VA	10500VA	15000VA
	Max. Output Current	12.8A	16A	19.2A
	Rated Voltage	3Ph, 230/400V	3Ph, 230/400V	3Ph, 230/400V
	Operating Voltage Range	±20%	±20%	±20%
	Operating Frequency Range	50/60Hz, ±5Hz	50/60Hz, ±5Hz	50/60Hz, ±5Hz
	Power Factor (adjustable)	0.8 ind- 0.8 Cap	0.8 ind- 0.8 Cap	0.8 ind- 0.8 Cap
	THD	<3%	<3%	<3%
	Night Time Consumption	<2W	<2W	<2W

Table 6. Specifications of modules

National Power Output (W)	235	240	245	250
Power Tolerance (%)	±2.5	±2.5	±2.5	±2.5
Module Efficiency (η%)	14.10	14.40	14.70	15.00
Voltage at P_{MAX} V_{MPP}(V)	29.6	30.0	30.3	30.7
Current at P_{MAX} A_{MPP}(A)	7.95	8.03	8.11	8.16
Open-Circuit Voltage V_{OC} (V)	37.0	37.3	37.7	38.1
Short-Circuit Current I_{SC} (A)	8.48	8.52	8.55	8.58

user is found to be approx. 1.5 lacs. 40 INR/Watt are taken as the cost of the solar module, so the total module cost on the pocket of the user is assumed to be approx. 6 lacs. Similarly, the final cost of the required battery bank comes out to be 3 lacs. The net installation cost is assumed to be approx. 2.5 lacs. If all the above-mentioned costs are added up then the total cost sums up to be approx. 13 lacs. Now, the per watt generation cost for solar PV setup is 13,00,000/15,000 i.e. approx. 87 INR. The monthly bill for a 15kW installation for January month comes out to be17600 INR. After installation of solar PV, the monthly bill for January month reduces to 10611 INR leading to a saving of approx. 40%. The overall savings in cost as discussed above is represented by Figure 8.

Table 7. Simulation results

Month	GlobHor kWh/m²	DiffHor kWh/m²	T Amb °C	GlobInc kWh/m²	GlobEff kWh/m²	E-Need kWh/m²	EArray kWh/m²	PR
January	128.9	35.2	17.16	178.9	174.9	1760	1722	0.837
February	144.2	37.2	20.49	184.2	180.1	1771	1733	0.818
March	187.0	56.8	26.71	213.0	207.8	1978	1936	0.790
April	201.9	69.3	31.13	206.2	200.6	1874	1833	0.773
May	218.3	84.6	34.69	204.4	198.1	1839	1798	0.765
June	190.2	96.1	33.49	173.0	167.2	1589	1554	0.781
July	169.9	102.2	31.09	157.0	151.6	1462	1429	0.791
August	161.0	93.6	29.62	156.7	151.5	1472	1439	0.798
September	169.6	73.9	30.08	181.4	176.3	1674	1638	0.785
October	163.3	48.8	28.67	199.6	194.7	1840	1800	0.784
November	126.9	43.3	22.98	168.8	164.8	1611	1576	0.812
December	123.5	32.2	19.10	177.3	173.3	1728	1691	0.829
Year	1984.7	773.2	27.13	2200.6	2141.1	20600	20149	0.796

Legends:
GlobHor: Horizontal global irradiation
DiffHor: Horizontal diffuse irradiation
T Amb: Ambient Temperature
GlobInc: Global incident in coll. plane
GlobEff: Efficient Global, corr. For IAM and Shadings
EArray: Effective energy at the output of the array
E-Need: Energy injected into grid
PR: Performance Ratio

Figure 8. Cost reduction

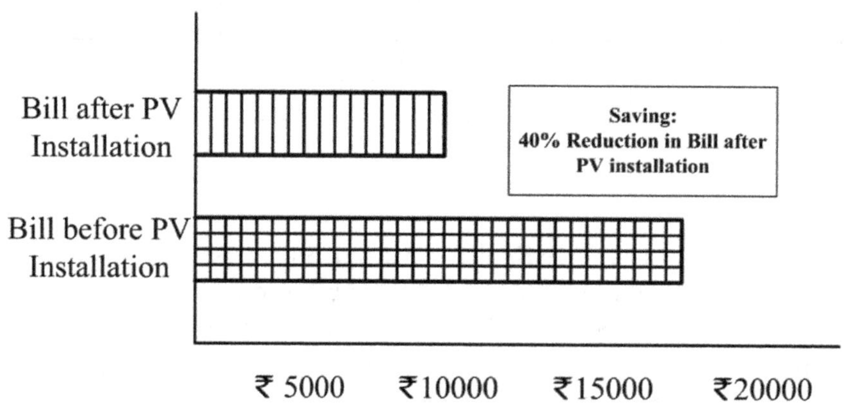

FUTURE RESEARCH DIRECTIONS

As microgrid is one of the emerging technology nowadays, there is lot of scope for further research in this emerging field. It is very difficult to consider one perfect algorithm and design which can cover all the different possible situations. Some areas are suggested here which can be explored further to improve microgrid operation. i) There are number of software available in the market like PVSYST, HOMER, MATLAB, RET Screen etc. for renewable energy resources and microgrid, it is always advisable to cross check the obtained results from more than one software in order to get best and accurate techno-economical and feasibility results, parameter values, system configuration. ii) Microgrid technology is generally deployed for sensitive loads like hospitals, data centers, military base etc. where the outage of power supply can cause severe damages and high cost. In order to avoid such possibilities, it is always advisable to perform reliability analysis for different possible configurations of microgrid. There are various techniques available in literature for reliability analysis like Monte-Carlo simulation, Markov model, Series parallel combination, fault tree analysis etc. iii) As renewable energy sources are intermittent in nature, their power output depends on the availability of their energy resource, weather conditions etc. DG's and energy storage devices are also very expensive and require huge amount of initial investment so it is always advisable to design a system after long term forecasting. This way, system owner can get required amount of power output and face minimal amount of power outages for which he/she will already ready. iv) when power generation from DG's are high due to all possible favorable conditions and when load demand is low then there is excess power generation which can either be sold to grid or to power exchanges. In India, recently there are two power exchanges: Indian Energy Exchange and Power Exchange India Limited. The owner of microgrid or cluster of microgrids can get better selling price if he/she takes wise decision that from where he/she wants to sell or purchase the excess or deficit power respectively. v) The conventional methods of power system protection are no longer valid for microgrids, mainly due to two reasons. DG's can be connected to different possible location in microgrids which results in bidirectional power flow. Secondly, the magnitude of fault current from DG's is low as compared to fault current produced by utility generators. So, when microgrid switches its operation from grid-connected to standalone mode, it is very hard for over current relays to properly sense the fault present in the system. Huge amount of research is also required in this field.

CONCLUSION

Electricity plays a most critical role in revolutionizing the modern era. The sources involved in the generation of electricity can be categorized as conventional and non-conventional sources of energy. Conventional energy sources are limited and not environment-friendly. So, turning the attention towards renewable energy sources is the need of the hour. Microgrids are emerging as one of the best technology that can take care of the optimum use of renewable energy resources for some particular load demand and location subjected to various specified constraints. Some load demands are time dependent like school's load, shop's load, health clinic's load etc. These load demands occur only for some hours of the day and not for all 24 hours. In such cases, some particular type of renewable energy resources like solar energy, wind energy etc. along with some type of energy storage devices like battery banks alone is capable to meet that particular load demand. In this chapter, the effectiveness of one microgrid topology for health clinic load profile is analyzed using solar PVSYST software. The solar photovoltaic system is considered here as a DG source along with battery bank in order to compensate the intermittent nature of DG source as well as for maintaining the voltage and frequency profile of the system. Results show that the proposed configuration is good enough to meet the load demand of health clinic. Also, proposed configuration is also found to be cost effective and environmental friendly.

REFERENCES

Ackermann, T., Andersson, G., & Soder, L. (2001). Distributed Generation : A Definition. *Electric Power Systems Research*, *57*, 195–204.

Al, H., Hassan, H. A. J., Pelov, A., & Nuaymi, L. (2015). Integrating Cellular Networks, Smart Grid, and Renewable Energy: Analysis, Architecture, and Challenges. *IEEE Access: Practical Innovations, Open Solutions*, *3*, 2755–2770.

Ali, I., Thomas, M. S., & Kumar, P. (2015). Energy efficient reconfiguration for practical load combination in distribution system. *IET Generation, Transmission & Distribution*, *9*(11), 1051–1060.

Bhattacharyya, S. C. (2012). Energy for Sustainable Development Energy access Programmes and Sustainable Development : A Critical Review and Analysis. *Energy for Sustainable Development*, *16*(3), 260–271. doi:10.1016/j.esd.2012.05.002

Buragohain, B., Mahanta, P., & Moholkar, V. S. (2010). Biomass gasification for decentralized power generation : The Indian perspective. *Renewable & Sustainable Energy Reviews*, *14*(1), 73–92. doi:10.1016/j.rser.2009.07.034

Caldognetto, T., Tenti, P., & Brandao, D. I. (2017). Fully-Dispatchable Microgrid: Architecture, Implementation and Experimental Validation. In *IEEE 26th International Symposium on Industrial Electronics (ISIE)* (pp. 93–98). IEEE.

Che, L., & Shahidehpour, M. (2014). DC Microgrids : Economic Operation and Enhancement of Resilience by Hierarchical Control. *IEEE Transactions on Smart Grid*, *5*(5), 2517–2526.

Che, L., Zhang, X., Shahidehpour, M., Alabdulwahab, A., & Abusorrah, A. (2017). Optimal Interconnection Planning of Community Microgrids With Renewable Energy Sources. *IEEE Transactions on Smart Grid*, *8*(3), 1054–1063.

Hafez, O., & Bhattacharya, K. (2012). Optimal Planning and Design of a Renewable Energy Based Supply System for Microgrids. *Renewable Energy*, *45*, 7–15. doi:10.1016/j.renene.2012.01.087

Khodaei, A., & Shahidehpour, M. (2013). Microgrid-based Co-Optimization of Generation and Transmission Planning in Power Systems. *IEEE Transactions on Power Systems*, *28*(2), 1582–1590. doi:10.1109/TPWRS.2012.2224676

Kumar, P., Ali, I., & Thomas, M. S. (2015). Synchronizing Solar Cell, Battery and Grid Supply for Development of Smart Power System for Home. In *IEEE International Conference INDICON-2015* (pp. 1–6). IEEE.

Kumar, P., Ali, I., Thomas, M. S., & Singh, S. (2017). Imposing Voltage Security and Network Radiality for Reconfiguration of Distribution System Using Efficient Heuristic and Meta-heuristic approaches. *IET Generation, Transmission & Distribution*, *11*(10), 2457–2467.

Kumar, P., & Singh, S. (2014). Reconfiguration of Radial Distribution System with Static Load Models for Loss Minimization. In *IEEE International Conference PEDES-2014* (pp. 1–5). IEEE.

Li, N., & Hedman, K. W. (2015). Economic Assessment of Energy Storage in Systems With High Levels of Renewable Resources. *IEEE Transactions on Sustainable Energy*, *6*(3), 1103–1111.

Li, Y., & Nejabatkhah, F. J. (2014). Overview of Control, Integration and Energy Management of Microgrids. *Journal of Modern Power Systems and Clean Energy*, *2*(3), 212–222. doi:10.100740565-014-0063-1

Liang, X. (2017). Emerging Power Quality Challenges Due to Integration of Renewable Energy Sources. *IEEE Transactions on Industry Applications*, *53*(2), 855–866.

Lopes, J. A. P., Hatziargyriou, N., Mutale, J., Djapic, P., & Jenkins, N. (2010). Integrating distributed generation into electric power systems : A review of drivers, challenges and opportunities. *Electric Power Systems Research*, *77*(9), 1189–1203. doi:10.1016/j.epsr.2006.08.016

MermoudA. A.WittmerB. (2014). *PVSYST user's manual.* Retrieved from www.pvsyst.com/images/pdf/PVsyst_Tutorials.pdf

Nazir, R., Laksono, H. D., Waldi, E. P., Ekaputra, E., & Coveria, P. (2014). Renewable Energy Sources Optimization : A Micro-Grid Model Design. *Energy Procedia*, *52*, 316–327. doi:10.1016/j.egypro.2014.07.083

Pali, B. S., & Vadhera, S. (2016). Renewable Energy Systems for Generating Electric Power : A Review. In *IEEE 1st International Conference on Power Electronics, Intelligent Control and Energy Systems (ICPEICES)* (pp. 1–6). IEEE.

Pepermans, G., Driesen, J., Haeseldonckx, D., Belmans, R., & D'haeseleer, W. (2005). Distributed Generation : Definition, Benefits and Issues. *Energy Policy*, *33*(6), 787–798. doi:10.1016/j.enpol.2003.10.004

Pimentel, D., Herz, M., Glickstein, M., Zimmerman, M., Becker, A. R. K., Evans, J., ... Seidel, T. (2002). Renewable Energy : Current and Potential Issues. *American Institute of Biological Sciences*, *52*(12), 1111–1120.

Rana, A. S., Thomas, M. S., & Senroy, N. (2017). Reliability Evaluation of WAMS using Markov-based Graph Theory Approach. *IET Generation, Transmission & Distribution*, *11*(11), 2930–2937.

Sahoo, A. K., Abhitharan, K. P., Kalaivani, A., & Karthik, T. J. (2015). Feasibility Study of Microgrid Installation in an Educational Institution with Grid Uncertainty. *Procedia Computer Science*, *70*, 550–557. doi:10.1016/j.procs.2015.10.099

Singal, S. K., Varun, & Singh, R. P. (2007). Rural Electrification of a Remote Island by Renewable Energy Sources. *Renewable Energy*, *32*(15), 2491–2501. doi:10.1016/j.renene.2006.12.013

Sun, Q., Zhou, J., Guerrero, J. M., & Zhang, H. (2015). Hybrid Three-Phase/Single-Phase Microgrid Architecture With Power Management Capabilities. *IEEE Transactions on Power Electronics*, *30*(10), 5964–5977.

Xoubi, N. (2015). Viability of a Utility-Scale Grid-Connected Photovoltaic Power Plant in the Middle East. *Journal of Applied Sciences (Faisalabad)*, *15*(11), 1278–1287. doi:10.3923/jas.2015.1278.1287

KEY TERMS AND DEFINITIONS

Distributed Generation: Distributed generation also known as distributed energy or on-site generation can be defined as power generation near to the point of consumer load.

Energy Storage Devices: Energy storage devices are the devices that have the ability of storing and releasing of electrical energy when needed.

Microgrid: A microgrid can be defined as a small grid that interconnects local loads, distributed generation sources, and energy storage devices.

PVSYST: PVSYST is an energy modelling tool that helps in analyzing how much solar energy can be harvested into an electrical energy from a particular site or location.

Renewable Energy Sources: These are the energy sources which are naturally and continuously replenished such as wind, sunlight, water, etc.

Solar PV System: A solar photo voltaic (PV) system is a power system that converts sunlight or solar energy into an electrical energy by means of solar cells.

Chapter 17
Optimization of Dynamic Power for System on Programmable Chip SOPC:
Power Optimization

Mehdi Jemai
University of Monastir, Tunisia

Siwar Ben Haj Hassine
National School of Engineers of Sousse, Tunisia

Bouraoui Ouni
National School of Engineers of Sousse, Tunisia

Abdellatif Mtibaa
University of Monastir, Tunisia

ABSTRACT

In this chapter, the authors present a new scheduling algorithm that brings a reduction in dynamic power consumption by achieving components scheduling while holding the global latency of the application. The main idea of that algorithm is to augment the latency of some components without impacting the dependency constraint and degrading the global latency of the system. There exist many solutions that manage to increase component's latency; one of them is through decreasing the frequency of their corresponding clocks. Generally, such a method leads to an augmentation in global latency of a system. However, this algorithm manages to reduce the consumed power and hold the same global latency of the system. The presented algorithm has been tested and it provides a significant gain in power at both simulation and physical levels.

DOI: 10.4018/978-1-5225-3935-3.ch017

INTRODUCTION

Currently, the implementation of electronic systems leads to two main directions.The first one is summed up by the increasing complexity and the second one is reflected in significant functional variability. According to the famous law of Moore, a migration towards geometries of technological smaller processes is noticed. Therefore, it is expected today to have more than 12 billion transistors per chip when using a standard 20 nm technology. The number of transistors reached a level of 7.1 billion and 2.6 billion in Graphics Processing Units (GPUs) and general purpose microprocessors, respectively. The second direction, attempts to integrate various complex functions with higher performance. Applications Specific Integrated Circuits (ASICs) with a number of transistors greater than 200 million per chip and standard parts (ASSP) applications with transistors greater than 100 million per chipwill be replaced by System in Package(SiP) andSystem on Chip (SoC). Practically speaking, 70% to 80% of today's integrated circuits are System on Chips, ranging from targeted applications to new microprocessors and integrated telecommunications systems.

The quality / performance of these devices is not only measured by what types of features these devices are capable of, but also how long these gadgets survive without being plugged into the supply voltage to be recharged. There is an increased demand for a reduced form factor or a smaller size of such devices, which is generally determined by the power consumption of such devices.

Since power dissipation is one of the most important parameters that improve the reliability and performance of these systems, power minimization has become a vital and critical aspect while creating a system on chip. Although embedded systems seem to be able to meet the demand for processing power and flexibility of complex applications, such systems are very complicated to design and optimize. Thus, the design methodology plays a major role to ensure the product success.

Without a well-established methodology for design and verification, the complexity of these circuits results in an increasing number of functional errors, which can lead to failures and multiple re-designs. Previously, the conventional design of a heterogeneous embedded system consists on carrying out the algorithm design, the hardware design and the software design separately and sequentially. Given that the expected growth rate of conventional design productivity is much lower than the complexity of the system, a new approach has emerged as a new design methodology in recent years. It is called hardware / software (HW / SW) co-design.

This methodology is divided into two levels: high level and low level. Traditionally, the hardware design flow begins with the register transfer level (RTL) arriving to the logic level and ending with the realization of the circuit. The methods of power estimation and optimization at the low level have been widely studied. However, the increase in current demand and the reduction of time-to-market of electronic equipment has forced designers to adopt high-level models in their design flow. Thus, high-level techniques have become the vital today's requirement.

Several techniques have been proposed in order to have a less consumable and more reliable systems. These techniques are situated on several levels of abstraction while impacting with a different way the final consumption of the circuit

High level synthesis is a process of translating a behavioral description into a set of connected storage and functional units. The basic operations executed in high-level synthesis are partitioning, scheduling and allocation. The partitioning algorithms divide a behavioral or structural description into subscriptions in order to reduce the size of the problem or to satisfy certain external constraints. Scheduling algorithms schedule variable assignments and time slot operations, and distribute partitions into storage and functional units.

Scheduling is an important task in high-level synthesis because it has an impact on the trade-off between design cost and performance. To do this, the scheduling algorithms must be adapted to the different target architectures used for the implementation. Certain priority functions expected in the control step are considered. There are a variety of achievements of this approach (Thomas et al., 1989). With regard to force-based scheduling, the objective is to create a balanced distribution of operations between the control stages. Using the mobility of each operation to define the possible intervals of execution, the potential demand for each control step is determined. The assignment of the operation to the control step which contributes to the most homogeneous distribution is accepted at each step. This approach has been integrated into high-level synthesis systems (Paulin. & Knight, 1987; Camposano, 1991). Another type of scheduling method is called path scheduling (Fisher, 1981). This method has been proposed for ordering control flow graphs. The execution paths in the control / data flow are managed individually. Each of these possible paths is programmed independently. Then, the final schedule is built by imposing resource constraints and overlapping the path plans accordingly. Other popular techniques of scheduling with the control flow are the scheduling of traces from compacting the microcode (Potasman, Lis, Nicolau, & Gajski, 1991) and scheduling percolation (Memik, Kastner, Bozorgzadeh, & Sarrafzadeh, 2005). Although most of the aforementioned scheduling heuristics produce a scheduling decision for one operation at a time, the algorithm proposed by Luo & Jha (2000) generates an assignment among several heuristics.In the aforementioned path-based scheduling algorithms, any sequence of operations that can optionally be executed in accordance with the control flow is a path. These operations do not need to create a string with direct data dependencies. Therefore, path-based scheduling algorithms consider an exponential number of possible execution paths and combine their schedules eventually.

The evolution of integrated circuit technology has made it possible to realize more complex systems and even their integration on a single chip. This development requires increasingly high levels of integration, driven by the needs of more efficient, lightweight, compact and low power systems. In such circumstances, the reasoning on a transistor or gate level circuit becomes difficult and a costly task when considering the marketing constraints of a product. Moreover, by disengaging from the most basic levels of specification or description, the functionality is more easily understood and can therefore be more easily corrected or modified. Therefore, in order to reduce the cost and design time of circuits, automation of the design process and abstraction at the design level are preferred (Belhadj, 1994).

Automatic synthesis is one of the most promising approaches. It consists on the translation of an abstract description of a circuit into a more concrete description. It allows evaluating several possibilities of implementation. It also contributes to improving the quality of the circuits (Belhadj, 1994; Tchoumatchenko, 1998).

Hence, the optimization of dynamic power has become an important research area for electronic engineering and researchers. As a consequence, researchers have always been interested in discovering new methods or designs to achieve the goal of optimizing power. A digital system is efficient if it consumes less power without influencing other constraints such as time. A digital system contains several interconnected components. In this book chapter, a new algorithm has been proposed based on scheduling between the components of a target system in order to reduce power consumption. The major idea of the proposed algorithm is to increase the latency of some components whenever it is possible without increasing the latency of the global system. There are many possible ways to increase the latency of components; one of the most used options is the decrease of frequency value of their corresponding clock.

LEVELS OF ABSTRACTION

The abstraction of a componentis a succinct description that removes unnecessary layout details to understand its functionality. Thus, in each abstraction, two levels are identified: the highest level, or specification, and the lowest detailed level, or realization (Laurent, 1999). There are three areas of representation of a circuit (Tchoumatchenko, 1998):

- **Behavioral Domain:** The behavior of each output is described as a function of inputs and time. Circuit operation can be specified by Boolean equations, input and output value tables, or algorithms written in standard high-level languages or special hardware description languages (HDLs) such as VHDL, Verilog.
- **Structural Domain:** It is the link between the behavioral domain and the physical domain. The circuit is represented as an interconnection of basic components (gates, flip-flops, etc.).
- **Physical Domain:** Specifies how to build a structure. Generally, the circuit presents itself in this domain in the form of a mask drawing. The functionality is ignored, as far as possible, the actual objects (for example the transistors of a circuit) are described.

The Y-diagram of Gajski and Kuhn (Belhadj, 1994; Tchoumatchenko, 1998; Ellervee, 2000) illustrates the different types of synthesis and the different levels of abstraction. The concentric circles present the different levels of abstraction and their representations at each of the three domains. A synthesis task can be seen as a transformation from one axis to another and / or a transformation from one level of abstraction to another (Ellervee, 2000). A level of abstraction is characterized by the objects that it manipulates. An object can be a rectangle (Layout level), a transistor, a complex operator, etc. (Aichouchi, 1994).Different levels of abstraction have been identified (Belhadj, 1994; Ellervee, 2000; Jervan, 1998).

- **System Level:** The system level is used to specify integer systems, including software parts and hardware parts. Behavioral descriptions of the system level are generally given in terms of communicating processes. In the structural domain, the descriptions are in terms of interconnections of elements such as processors, memories, peripheral units, etc. The basic step is the communication transaction to control the flow of processes.
- **Algorithmic Level:** The description at this level attempts to describe the behavior (behavior) of a model without worrying about the structure. The objects manipulated at this level are variables and operators. The basic step is the calculation step. A calculation step consists of a set of operations executed between two successive points of input / output and / or synchronization. It can last several clock cycles.
- **Register Transfer Level:** The behavior at this level is described by transfers and operations between registers. The structure is represented by the interconnection of registers, arithmetic and logical operators, etc. The time is divided into intervals called control states.
- **Gate Level:** At this level, the behavioral description is done with Boolean equations, which are represented in the structural domain by logic blocks (flip-flops, gates).
- **Circuit Level:** This is the lowest level of abstraction. The behavior at this level is described by transfer functions. The structure is represented by the interconnection of electronic components such as transistors, diodes, etc.

Decomposition into levels of abstraction may vary according to authors. Some authors (Tchoumatchenko, 1998; Aichouchi, 1994 ; Berrebi, 1997)use only four levels of abstraction . These levels of abstraction correspond to the stages of the design of a circuit on silicon. The transition from one level of representation to a lower level is achieved by so-called "silicon compilers".

DESIGN FLOW OF A CIRCUIT

The current circuits are too complex that it has been necessary to develop a set of abstraction levels in order to divide the synthesis stream into simplified sub-problems. Depending on the level of abstraction of the input description and the output, several types of synthesis can be considered: synthesis at the system level, behavioral synthesis, synthesis at the register transfer level, logical synthesis,and physical synthesis.

Synthesis at System Level

The first stage of the synthesis flow is the partitioning step. The partitioning consists on cutting the system into communicating subsystems (processes). A process can be partitioned again into a set of communicating processes. The system level is used to specify integer systems, including software parts and hardware parts. Partitioning can be guided by many parameters such as functionality, synchronization and task competition (Berrebi, 1997)and hardware / software partitioning (Belhadj, 1994), etc.

High Level Synthesis

The high-level synthesis (also known as architectural or behavioral synthesis) takes as input an algorithmic specification of the function performed by the system to be designed. This description will be translated, through an architectural synthesis tool, into a structural description at the register transfer level. Behavioral synthesis can be decomposed into three major stages: scheduling, allocation, and resource assignment.

Synthesis at the Register Transfer Level

The RTL synthesis translates a register transfer level description, generally composed of a control part and a data path part, into a description at the gate level. It transforms the specification of a circuit into a set of logical equations by also performing a resource allocation (Laurent, 1999).

Logical Synthesis

The logical synthesis tools takes as input a description in the form of interconnected logical blocks. The aim is to optimize the logical blocks in terms of area of the generated circuit (Belhadj, 1994 ; Berrebi, 1997).

Physical Synthesis

Physical synthesis consists in generating the geometrical patterns of circuit fabrication masks. This step of placing and routing the descriptions of the resulting architectures on the silicon terminates the design flow.

High Level Synthesis

The complexity of current circuits has increased exponentially (Aichouchi, 1994), hence the need to automate the design process of integrated circuits as well as the use of higher levels of abstraction have become necessary. Today, synthesis tools starting with a behavioral description offer the automation and acceleration of the design process (Friedman, & Yang, 1969; Zimmermann, 1979 ; Parker, Pizarro, & Mlinar, 1986; Note, Geurts, Catthoor, & De Man, 1991). High-level synthesis has many advantages (Ellervee, 2000; Aichouchi, 1994; Sugar, 2000; Ding, 1996; Lin, 1997). In fact, the overall design time will be greatly reduced: Estimation of performance at a higher level of abstraction, and more thorough architectural exploration before choosing the solution with the best compromise (surface, speed, consumption, etc.). High-level synthesis tools link systems and circuit design domains andautomatic design tools can surpass human designers by satisfying the design constraints.

Research in behavioral synthesis has been intensive for more than twenty years. The ALERT Friedman, & Yang, 1969) and MIMOLA (Zimmermann, 1979) tools are the precursors in the field of architectural synthesis tools. Today, several academic and commercial behavioral synthesis tools are available.

General Principle of High Level Synthesis

High-level synthesiscorresponds to a set of tasks of refinements or transformations that start from a behavioral specification generate a target architecture described at the transfer level of registers, while satisfying a certain Number of constraints (Kuchcinski, 1998; Gajski, Dutt, Wu, & Lin, 2012; Govindarajan, 2000; Silc, 1994). McFarland and Al were the first to give a definition to high level synthesis: "The task of high level synthesis is to take a specification of the desired behavior of a system and a set of constraints and objectives to satisfy and to find a structure that implements this behavior while satisfying the objectives and constraints ".

In addition, architectural synthesis requires a library of components (Kission, 1996). These elements are used to carry out the operations of the description . Possible objectives include: minimum area, minimum energy consumption, or maximum flow. Constraints are often defined in terms of execution order, completion time, and surface. Constraints and objectives depend on the type of application to be synthesized (Heijligers, 1996). With this combination of behavioral description, the problem of high level synthesis is to find the best architectural solution.

The restriction of a behavioral synthesis tool to a specific field of application makes it possible to achieve effective results by reducing the complexity of the synthesis process (Berrebi, 1997 ; Friedman, & Yang, 1969). In the literature there exist two domains of application that are data-oriented applications and control flow-oriented applications.

Data flow-dominated applications handle regular data flows. The behavior of such application is generally specified as a periodic set of operations to be performed on each new data set. Typical applications are digital filters, encoders / decoders, circuits making encryption algorithms.

Applications dominated by the control flow are controlled by a set of commands to be interpreted. Each command may involve reading or writing complex data structures and executing a specific algorithm for that command. The instructions executed may depend on data, communication with the environment (query / acknowledgment protocols), interrupts, etc. Typical applications are controllers, telecommunications applications, protocol converters, etc.

The design of complex systems often requires the combination of stream-oriented data components and flow-oriented components. In this case, the system must be decomposed into data flow oriented modules and control flow oriented modules. This is usually done by the designers to specifya system. Thus, to allow the use of synthetic tools oriented to specific application domains (Sugar, 2000)..

High Level Synthesis Flow

High level synthesis transforms a behavioral description at the algorithmic level into a structural description at the register transfer level. The behavioral description describes the functionality that must be performed by the synthesized system. In such a description, the calculations are described using atomic elements called operations. Operations manipulate data storage elements called variables. The algorithmic specification includes relational, arithmetic and logical operations applied to the variables; and control structures such as loops, branch instructions, and procedure calls. This description is written in ordinary programming language or in special hardware description language (Silc, 1994) such as VHDL, Verilog (Thomas, & Moorby, 2008), HardwareC (Ku, & De Micheli, 1988) and Silage (De Man, Rabaey, Six, & Claesen, 1986). A behavioral description can be written by the designer himself, or generated by synthesis tools of higher levels of abstraction. The resulting architecture consists of a controller and a data path (Kission, 1996; Heijligers, 1996) The data path is described by functional unit instances (adders, multipliers, UALs and logical units), storage units (RAMs, ROMs, registers, etc.), and interconnected communication units (bus And multiplexers). Thus, the operative part of a circuit is represented as a set of interconnected components "netlist". All of these components are called resources. The controller is defined as a finite state machine which determines at each clock cycle which operations of the operative part are to be performed and by which functional units. The high level synthesis process maps the variables and operations of the behavioral description to the functional units and registers in the RTL description respectively.

A set of algorithms are applied to transform this initial specification gradually into target architecture (Sugar, 2000). The high-level synthesis is often composedof 5 steps (Berrebi, 1997; Sugar, 2000; Govindarajan, 2000): compilation of the behavioral description into an internal representation, scheduling, allocation, assignment of Resources, and the generation of architecture. This decomposition is not adopted by all authors, some of them add a pre-processing step called "partitioning" (ben haj hassine et al.,2017; jemai et al.,2017), although others consider it a task of synthesis at system level ; Other authors neglect the assignment step and consider it as a subtask of the allocation (Silc, 1994; Kission, 1996). It is important to note that only the scheduling step is specific to behavioral synthesis, the other steps can be accomplished using synthesis tools at the register transfer level (Berrebi, 1997; Sugar, 2000). Most approaches use Graph-based Representationspresented by the CFG, DFG and the CDFG. The three following stages form the core of the high-level synthesis: the scheduling of operations, the allocation of resources, and the allocation of resources. Given their importance, these tasks have been the subject of several studies and different algorithms have been published (Canis et al.,2014; Canis et al., 2013; Chang et al., 2016; Li et al., 2016; Jadhav, 2016; Sugar, 2000; Lin, 1997; Kuchcinski, 1998; Govindarajan,

2000; Silc, 1994; Heijligers, 1996; Haynal, 2000; Timmer, & Jess, 1995; Narasimhan, & Ramanujam, 1997; Landwehr, Marwedel, & Dömer, 1994; Narasimhan, 1998; Baruch, 1996; Dias, Akil, Lavarenne, & Sorel, 2000; Gruian, & Kuchcinski, 1998; Casseau, Jégo, & Martin, 2004; Bhattacharyya, Murthy, & Lee, 1999; Schreiber et al, 2000).

Scheduling assigns operations into control steps. A control step is the basic time unit in synchronous systems; it corresponds to a clock cycle.

Allocation is the task of estimating the type and number of resources required to arrive to the RTL description.

Assignment is the task of assigning operations to functional units, from variables to storage units, from data transfers to interconnection units.

The role of architectural synthesis is therefore to find the number and type of material resources required for the implementation of operations (allocation), to define the instants of operations execution (scheduling) and to define the resources attributed to each operation(Heijligers, 1996). These tasks are highly interdependent. In fact, allocation requires information about operations that are executed simultaneously and cannot share the same resource. Conversely, scheduling needs information on the resources allocation to know which operations can perform at the same time. Thus, the case where the scheduling is performed before the allocation, leaves little margin to possible parallelism between operations (Kission, 1996). Otherwise, additional restrictions are imposed on the scheduling step, since each element is bound to a time-limit for execution; the steps executed largely determine the temporal exploitation by the various. In order to have an optimal design, the ideal would be the simultaneous execution of scheduling and allocation. However, because of the complexity of solving such problems (NP-complete problem), several systems perform the scheduling and allocation steps separately(Tseng, & Siewiorek, 1986; Paulin, & Knight, 1989; (Farland, Parker, & Camposano, 1990; Parker, Pizarro, & Mlinar, 1986)Other systems execute them as intertwined tasks. But even when scheduling and allocation are considered separately, they remain NP-complete problems.

SCHEDULING

Scheduling is one of the key steps in the high-level synthesis. Gajski and Al consider it as (Narasimhan, 1998) "perhaps the most important step during structural synthesis". Scheduling is an important issue in the automation of VLSI design. Thus, many HLS researches have been devoted to this problem. In the literature, the scheduling step is almost the most studied at the behavioral synthesis. This is due to the fact that this step probably has the most influence the performance of the generated circuit (Sugar, 2000; Narasimhan, 1998; Nestor, & Krishnamoorthy, 1993). The scheduling is the partitioning of the behavioral description, and partitioning of operations into control steps or cycles (Sugar, 2000; Baruch, 1996). These control cycles correspond to the states of the resulting finite state machine which may include several operations running in parallel. In other words, the scheduling defines an execution date for each operation of the behavioral description, while respecting that at the moment of operation execution.

To explore the design space, a method generally used by most synthesis tools consists of limiting the duration required for the algorithm execution; the duration is generally measured in control steps for architectural models of Synchronous circuits with a single clock, and limiting the number of physical resources.

The architectural compiler then generates an architectural solution that minimizes the second dimension. There are two types of scheduling problems:

- Resource constraint scheduling, i.e. minimizing the execution time while fixing the amount of hardware resources.
- Scheduling under time constraints, i.e. minimizing the circuit surface while limiting the execution time. The equation to be solved in this case is to determine the number of functional units required to perform all operations within the set time limit (Kission, 1996).

Scheduling Mode

According to the results of the scheduling applied at the execution steps. There are three categories of scheduling (Sugar, 2000).

One-Cycle Mode

Each execution step is ordered in a single control cycle. so the operations must be executed in a single clock cycle. In this case, the initial model is very close to the description generated at the register transfer level. However, to synthesize the behavioral specification, restrictions must be imposed on the writing style of the execution steps: prohibition of multi-cycle operations, a run step cannot contain infinite loops and the clock must be long enough to allow the execution of the most complex execution step.

Macro-States Mode (Super-State Mode)

Unlike the previous mode, each execution step is decomposed into several control cycles whose number is fixed during the synthesis process. In this mode, the only restriction on the write style in this mode is the prohibition of infinite loops or dependent on the data resulting from the execution of the same loop. This mode allows a greater architectural exploration with operators with variable executions. However, the fact that the description generated at the RTL level differs from the initial specification makes the verification process more complex.

Mode of Behavioral States

This mode is the most general scheduling mode. Each execution step is performed by a finite state machine whose transitions can be executed in a single clock cycle and no restrictions are made on the writing style of the initial model allowing an architectural exploration more extensive. On the other hand, the verification process, like the mode of the macro-states, poses problems because of the difficulty of establishing the correspondence between the initial descriptions and those generated.

There exist strategies for the scheduling of operations manipulating inputs / outputs. This is the consequence of decomposition the execution steps into several clock cycles which generally involves changes in the behavior of the inputs and outputs of the circuit. Two modes or strategies for I / O processing are known:

1. **Fixed Inputs/Outputs:** This mode preserves the order of I / O operations between them and with respect to the synchronization instructions . In macro mode or behavioral state for a given execution step, reads are ordered in the first cycle and writes are ordered the last one.
2. **Floating Inputs/Outputs:** In this mode, no constraint is maintained. They are ordered with a manner that satisfies the data dependencies constraints while minimizing run time and allowing a wider architectural exploration. However, the behavior of the generated circuit often becomes extremely remote from the initial behavior.

Different Types of Scheduling

The purpose of scheduling is to optimize the number of control steps required to perform a function, taking into account the limits of the available hardware resources and the cycle time. The scheduling algorithm must take into account control constructs such as conditional loops and branches, data dependencies, and constraints on hardware resources. Other constraints may specify cycle time, consumption, etc. Several algorithms are directed to the signal processing circuits, to minimize the cost of the operative part. The input description of these algorithms is generally a data flow graph that most often contains neither a loop nor a procedure call. However, in the current control circuits, the performance of the control part dominates the overall performance of the circuit. The objective of the scheduling algorithms for such circuits is to minimize the surface and the critical path of the controller, while ensuring that the data path complies with predefined constraints. In general, scheduling algorithms at the behavioral synthesis domain can be classified into two major categories: stream-oriented scheduling algorithms and control flow-oriented scheduling algorithms. A data flow-oriented algorithm uses a DFG data stream graph, while a control-flow-oriented algorithm uses a CFG control flow graph as the underlying representation. Thus, the first consists on serializing a parallel description while the latter consist on parallelizing a sequential description. In the literature, the number of control flow-oriented algorithms is lower than stream-oriented algorithms.

RELATED WORK

Scheduling is the partitioning of the behavioral description into sub-graphs, so that each sub-graph is executed in a single control step. A control step corresponds to a Finite State Machine (FSM) (Lee, 2000). Each control step can include several operations to be performed in parallel. This parallelism is conditioned by the resource allocation algorithm, which must be able to allocate the sufficient number of resources for the execution of the parallel operations. Scheduling is an important task in the HLS, because it is during this phase that the number of control cycles is fixed as well as the parallelism of the operations. The different scheduling algorithms can be distinguished according to three criteria(Belhadj, 1994): the interaction between scheduling and allocation, objective function and constraints and the type of scheduling algorithm.

Scheduling and allocation are highly interdependent, as described above. The second criterion concerns the objective function and the constraints. An objective function is a measure that the system tries to maximize or minimize. Constraints are conditions that must be met. Objective functions and constraints relate to the number of physical resources (surface) and time. There are two classes of scheduling problems: scheduling under hardware constraints and scheduling under time constraints. In the first case, since the quantity of material resources has been limited, scheduling attempts to minimize the execution time (minimize control steps) by taking into account the functional units availability. In the second case, since the execution time is bounded, scheduling attempts to minimize the surface of the circuit by taking into account the maximum number of control steps.

Thus the optimal solution is usually defined with tradeoffs between temporal and surface characteristics (resources). However, in some cases, certain constraints are imposed and must be respected. The third criterion concerns the type of scheduling algorithm used. There are two types of algorithms: transformational algorithms and iterative / constructive algorithms. The transformational algorithms start with an initial scheduling. Then transformations are applied to obtain other scheduling, for example, using exhaustive search, branch-and-bound or heuristics (Silc, 1994). These algorithms differ in the selection criterion used to choose the next operation to be scheduled. Given the importance and complexity of the scheduling step, several algorithms have been published. Some of the more well-known techniques of scheduling are based on Integer Linear Programming (ILP) (Hwang, Lee, & Hsu, 1991; Lee, Hsu, & Lin, 1989) The ILP tries to minimize the number of resources and the execution time at a time (scheduling under temporal constraints and surface constraints). However, the execution time of the algorithm grows exponentially with the number of variables and the number of inequalities (Lin, 1997). Thus, although it has been shown to be relatively efficient, ILP techniques are not easily generalizable to control-oriented scheduling. In practice, the ILP approach is applied only to fairly simple problems. Since the ILP method is impractical for complex circuits, heuristic methods that work efficiently at the cost of optimality have been developed(Haynal, 2000). Heuristic scheduling algorithms generally use iterative / constructive techniques. The simplest constructive approaches are the ASAP (As Soon As Possible) scheduling and ALAP (As Late As Possible) scheduling.

The simplest method for scheduling is the manual scheduling (bu the user) as in the Silc system. However, there is a trend towards automatic scheduling. Over the years, researchers have tried to apply different types of solutions to the scheduling problem. Several algorithms are available. These scheduling algorithms are classified, as previously mentioned, into two categories: scheduling flow-oriented scheduling algorithms and flow-oriented scheduling algorithms.

Classification of Scheduling Algorithms

Over time, it has been realized that the scheduling techniques developed in the early stages of high-level synthesis do not in a position to take into account the needs of control-dominated circuits encountered in areas such as Telecommunications. Typically, control-dominated circuits have a complex control flow structure containing loops and conditional statements. For a high quality design, all control structures must be taken into account.

Recently, a large number of approaches and algorithms have been developed in order to try to satisfy the needs of the market. These approaches can be constructive / iterative approaches or transformational approaches. Constructive approaches use gluttonous heuristics and try to minimize the cost but do not necessarily guarantee an optimal solution. Examples that include simple and fast heuristics is scheduling based on lists (Zimmermann & Trautmann, 2017: Kushwaha & Kumar, 2017) or more complex heuristics such as force-oriented scheduling (Paulin, & Knight, 1987). The disadvantage of heuristic greedy is the possibility of being trapped by a local minimum of the cost function not the optima one. Probabilistic iterative approaches, such as simulated annealing, use a small set of transformations that are applied to randomly selected parts in the scheduled graph. A probabilistic acceptance function avoids local minima.

Since these approaches require a good exploration of the scheduling space, they typically exhibit broad execution times. Exact approaches use techniques such as ILP (Integer Linear Programming) (Chang et al.,2016) and guarantee the overall optimum, but they require a long execution times. The ILP can then be used to find the optimal scheduling with respect to a certain cost function. However, the execution time in these approaches depends on the number of variables and constraints. This value can be extremely large for wide problems.

Data Flow-Oriented Scheduling Algorithms

The scheduling algorithms can be classified into time constrained scheduling algorithms and resource constrained scheduling algorithms. For the first scheduling type, the number of resources is minimized for a fixed number of control steps. Whereas for the second, the number of control steps is minimized for a given design cost (number of resources). Time-constrained scheduling is important for designs targeting applications in real-time systems such as signal processing systems. Resource constrained scheduling is important for applications where limitations are imposed on the circuit surface. The objective in this case is to produce a circuit with the highest possible performance while satisfying the imposed hardware constraints (Baruch, 1996). There are also the basic scheduling algorithms: ASAP (As Soon As Possible) and ALAP (As Late As Possible

ASAP and ALAP

The ASAP approach is used by FECET systems of General Electric, early CMUDA, MIMOLA, and Flamel . With the ASAP approach, firstly operations are classified in a list according to their topological order. Then, the operations are placed, one by one, in the rather possible control step. A similar approach is to schedule operations at the latest ALAP. The critical path in the flow graph can be determined by taking the intersection of the ASAP and ALAP scheduling algorithms. The operations that appear in the same control steps, in both cases, belong to the critical paths. The problem with ASAP scheduling, as well as ALAP scheduling, is that critical path operations have no priority. Hence, less critical operations can be scheduled first and thus cause delay. List scheduling solves this problem, using a global criterion for selecting the next operation to be scheduled. For each control step, a list of operations that can be executed during this step is drawn up. Each operation is scheduled if the necessary resources are available; otherwise it is deferred to the next control step. In some cases, this approach is widely used as the branch-and-bound algorithm. In the Slicer system, the priority function is based on mobility, which is the difference between the ASAP and ALAP values of an operation. A compound priority is used in the MAHA system, where critical path operations are scheduled first. Then the other operations are ordered according to the least mobility. The HAL system performs list scheduling with force as a priority. The force between an operation and a certain control step is proportional to the number of operations of the same type that can be scheduled during this step. These values are represented by distribution graphs. An example of algorithms using this global selection criterion is Force Dried Scheduling (FDS) algorithm. The FDS algorithm uses the ASAP and ALAP scheduling algorithms to determine the scope of each operation with respect to the control steps. Then he draws up the distribution graphs. This algorithm tries to minimize the total number of functional units used in the implementation by the uniform distribution of operations of the same type over all available states (Lin, 1997). Until now, only simple structures have been considered, whereas a description of a circuit also contains conditional structures and loops.

Several approaches have been proposed for the scheduling of conditional structures. The Fast As Possible (AFAP) scheduling algorithm one of the best known algorithms. AFAP considers, all possible execution paths, and then orders them separately. The results obtained for the different paths are then combined, by resolving the conflicts between the execution paths. For loop structures, different approaches have been proposed such as pipelining.

ILP Integer Programming

Integer Linear Programming (ILP) is one of the methods used for time constraint scheduling, and is also used for resource constrained scheduling. It allows finding an optimal solution using the Branch-and-Bound algorithm. Integer programming methods describe the scheduling problem accurately, in the form of a system of inequalities. The resolution of this system is an exhaustive search. The search space is defined according to the constraints imposed by the designer.

The ILP formulation grows rapidly with the number of control steps. An additional unit in the number of control steps involves n additional x variables. As a result, the execution time increases rapidly. That's why the ILP approach is used only for small (simple) problems. The elimination of backtracking in the ILP method can considerably reduce the execution time. Heuristic methods have adopted this idea, by scheduling an operation based on a given criterion ((Baruch, 1996).

List Scheduling Algorithm

The scheduling techniques proposed in the literature mainly use heuristics.They obtain fast and quasi optimal solutions to an NP-complete problem. One of the most frequently used heuristics is LS algorithm(Nazar & Murugan, 2016). The list scheduling algorithm is a generalization of the ASAP algorithm with inclusion of hardware constraints]. This algorithm sequentially schedules operations according to a priority list.

The priority list of operations is scheduled according to a priority function. At each iteration, the highest priority operations are scheduled first and the lower priority operations are delayed. In general, the priority list is dynamic, since the scheduling of an operator at a control stage makes his successors ready thus included in the priority list. Existing methods differ essentially by the priority function used to select the operations to be delayed. The principle of a list scheduling algorithm is as follows: at each control step, the operations are stored in a list and ordered by a priority function. Each operation of the list is taken into account.it is scheduled if it respects the constraints, otherwise it is delayed. This process is repeated until the behavioral description is fully ordered. The scheduling by solve both the problem of the scheduling under temporal constraints and under material constraints.

Static List Scheduling Algorithm

The static list scheduling algorithm (Jain, Mujumdar, Sharma, & Wang, 1991) differs from the ordinary list scheduling algorithm in assigning control steps as well as in the priority list. This approach begins with the creation of a large and unique list. The algorithm uses the ASAP and ALAP approaches to obtain the limiting steps to which each operation is assigned. Then all operations are ordered in descending order according to the GCS value as the first key, and the LCS value as the second key. Once the priority list is created, operations are sequentially ordered starting with the last operation in the list. The type of the operation is important. This algorithm has an advantage over list scheduling that a list is statically constructed and it does not dynamically evolve.

Force Directed Scheduling FDS

Force Directed Scheduling (FDS) is a constructive heuristic method. It is a very popular scheduling technique for time-constrained scheduling, which aims to reduce the number of hardware resources by allocating them. The Time Frame interval is included between the ASAP and ALAP execution start time.

The FDS algorithm is capable of producing a suboptimal solution. This disadvantage can be overcome by reordering some operations at each partial scheduling. The iterative reordering method IR (Iterative Reordering), proceeds by reordering one operation at a time. The assignment of an operation to a control step is done taking into account data dependencies and constraints. A random shift is selected, and the operation is blocked in the new position temporarily. The costs of these displacements are calculated, the one that produces a maximum gain (maximum cost reduction) is chosen as this displacement are released (unblocked). Then the procedure is repeated with the new scheduling scheme. The quality of the result produced by this algorithm depends on the initial solution.

Control Flow-Oriented Scheduling Algorithms

The PBS based scheduling algorithm was the first algorithm proposed to satisfy the requirements of the scheduling problem of control-dominated applications. Since then, several scheduling algorithms have been developed for the control, but there are no algorithms that take into account all related questions such as: manipulation of control constructs including loops, dependency exploitation and optimization. The tree-based scheduling algorithm (Huang, Jeang, Hwang, Hsu, & Wang, 1993) has come to reduce, in a way, the problem of path explosion in PBS. This algorithm, too, performs an AFAP algorithm for all paths. The rooted tree used to represent the collection of paths is less compact compared to the case where all paths are considered individually. The tree-based scheduling algorithm takes into consideration data dependencies.Thus, unlike the PBS algorithm, it is not constrained by the initial order of operations. It also takes into account resource constraints, but only those that involve operator delays. In all these approaches oriented to the scheduling of circuits dominated by the control, an important aspect is still ignored which is the loops in the control flow. In all these cases, the loops are neglected at the beginning by the removal of the feedback loops. The Loop Directed Scheduling (LDS) algorithm traverses all control flow paths, including those that contain loopback arcs. However, this algorithm does not take into account time and I / O constraints, and does not allow the concatenation of operations. The Dynamic Loop Scheduling (DLS) algorithm is similar to the LDS algorithm. Another scheduling algorithm called Segment Based Scheduling (SBS) has been developed to fill the boundaries of other algorithms. This algorithm has two major components: an efficient procedure for the flow of control / data flow and an extended procedure for analyzing or locating state boundaries taking into account a variety of constraints and requirements. It avoids the construction of all control paths before scheduling. During scheduling, the control graph is traversed and divided into segments. Each segment represents a collection of paths from one state to another.

PROPOSED APPROACH

As previously mentioned the main idea is to develop an algorithm that optimizes the dynamic power consumed by a system and keeps its same whole latency. That decrease in power is attained through the minimizing of clock frequency of some components whenever it is feasible.

Initially, let's assume a system S= (C1, C2, C3, C4…Cx) which is the input of the algorithm. Where $C_1, C_2, \ldots, C_x$ are the components that run with a main master clock having a frequency value equal to F_0. Each component is characterized by its latency; in terms of cycle numbers. The proposed algorithm contains five steps:

Step 1: Transform the input system into a graph on which the nodes refer to the components and the edges present the dependencies between them.
Step 2: create two virtual nodes C_d and C_f which respectively represent the first and the last node.
Step 3: Find all possible paths from the first node C_d to the last one C_f.
Step 4: Determine the latency of each path and find the critical path. The critical path is the path that has the biggest value of latency.
Step 5: Determine the new latency of each node T_{new}. In fact, $T_{new} = 1/F_j \times T_{old}$, Where T_{old} is the old latency and F_j is the new frequency. $F_j = L_{pi}/L_{pc} \times F_0$, Where L_{pi} is the latency of the path number i and L_{pc} is latency of the critical path.

In some cases, one node (component) will obtain two or more different latencies. If that node doesn't belong to the critical path, the biggest valueof latency will be chosen. In any other circumstances the same latency value will be kept.

Figure 1. Pseudo-code of our algorithm

```
Begin
Build a graph G (n, e) from S (c1, c2, ci...cx)
Create two virtual nodes Cd and Cf
Find all possible paths from the first node Cd to the last one Cf.
Compute the latency of each path Lpi
Compute the critical latency Lpc
For all path p (i) differ from the critical path
      For all nodes n (j)
         Fj= (Lpi/Lpc) × F0   /* -Fj the new frequency and F0 is the pervious one, Lpi is the latency of the path i
         Tnew=1/ Fj × Told                     and Lpc is the latency of the critical path */
      End for
End for
If there exist more than Tnew then
      If that node does not belong to the critical path then
         Take the superior value
      End if
End if
End begin
```

How to Reduce Power

One may still wonder how the proposed algorithm brings a power minimization. For instance, a system is considered with x components $S(C_1,C_2..C_x)$ that are running with a master clock, and have the same frequency F_0; which is the same frequency of the master clock. Then, the dynamic consumed power of that system will be calculated before and after applying the new proposed algorithm on it. The dynamic power named P_d of a system can be written as follows (Lizhe et al., 2013):

$$P_d = A \times C \times V^2 \times F \quad (1)$$

where A is the percentage of active logic gates; that are charged dynamically; C refers to the total capacitance load, V presents the supply voltage, and F is the execution frequency.

As allcomponents are using the master clock, the consumed power of the whole system before scheduling is:

$$P_{dtotal\_before} = \sum A(i) \times C(i) \times V^2(i) \times F_0 \quad (2)$$

$$P_{dtotal\_before} = n\alpha F_0 \quad (3)$$

where $\alpha = A(i) \times C(i) \times V^2(i)$

We suppose that after scheduling, an increase of the latency of at least one component is possible. As mentioned previously one of the existing solutions which manages to increase of latency of a component is to decrease its clock frequency to a lower level that will be named F_j. authors have assumed that the new frequency F_j is equal to μF_0 where $\mu$is a real <1. Based on that, the total consumed power after scheduling becomes:

$$P_{dtotal\_after} = (n-1)\,\alpha F_0 + \alpha\, F_j = (n-1)\,\alpha F_0 + \alpha\mu F_0 = \alpha F_0\,(n-1+\mu) \quad (4)$$

Based on equation (4) and (5), the gain in power equals as follow:

$$G = 100 \times (P_{dtotal\_before} - P_{dtotal\_after}) / P_{dtotal\_before} = 100 \times (1-\mu)/n) \quad (5)$$

Since $\mu < 1$, then $G > 0$.

So given any specific system, if any opportunity of increasing component latency takes place, then the new technique will absolutely provide a gain in dynamic power.

How to Keep the Same Global Latency

Step four of the proposed algorithm is to find the latency of the critical path. The critical path is the longer path that has the longest latency value. So, the global system latency is equal to the critical path latency. Since the components that belong to the critical path will not be changed then the latency of the critical path will not be distorted either. Thus, the global latency will not be modified.

Illustrative Example

To further clarify the proposed algorithm, the system presented in Figure 2 will be applied on it. Let's presume that the frequency of the master clock equals F_0.

First and second steps are presented in Figure 3, where the system is transformed to a graph and the two virtual nodes C_d and C_f where added.

Thus, after applying the third and the fourth step, the possible paths are:

P1: $C_1C_2C_3$ where 8 is its latency.
P2: C_4C_5 where 5 is its latency.
P3: C_1C_5 with latency equal to 7.

After taking the second path P_2, the new latencies of C4 and C5 will be calculated.

$$F_j=(L_{pi}/L_{pc})\times F_0=(5/8)F_0=0.625\ F_0$$

$$T_{new}(C_4)=1/F_j\times T_{old}=(1/0.625)\times 3=4.8$$

$$T_{new}(C_5)=1/F_j\times T_{old}=(1/0.625)\times 2=3.2$$

Taking the third path P3, since C1 belongs to the critical path, its latency will not change. Thus, only the new latency of C5 will be calculated.

$$F_j=(L_{pi}/L_{pc})\times F_0=(7/8)\times F_0=0.875F_0$$

$$T_{new}(C_5)=1/F_j\times T_{old}=(1/0.875)\times 2=2.29$$

Figure 2. The input system

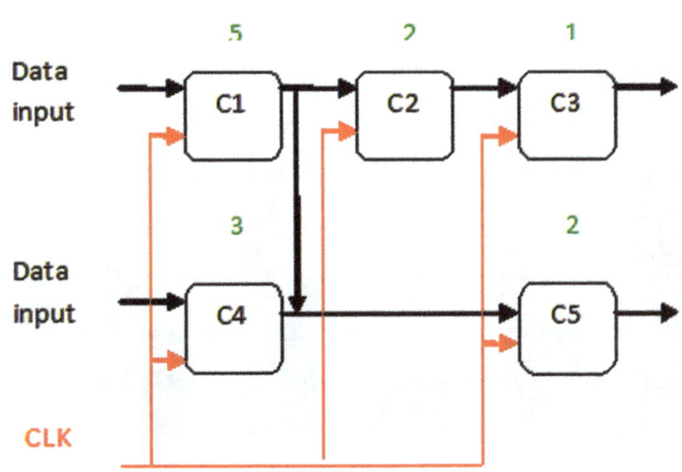

Figure 3. The graph

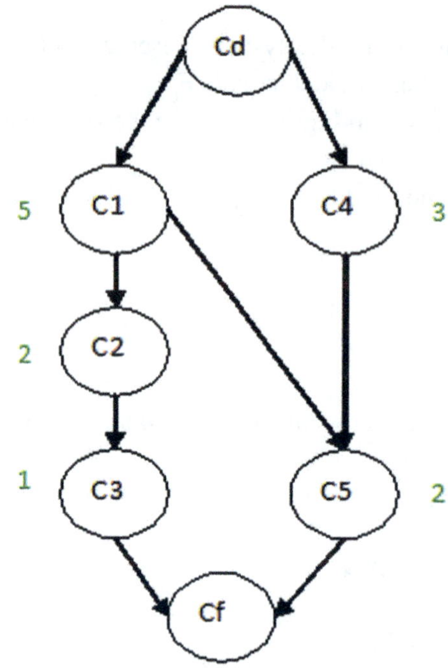

Authors figure out that C5 has two new different frequencies which are 3.2 and 2.29. Hence, the highest one will be chosen which is 3.2.

The Figure 4 summarizes the old and new values of latencies.

From the above results,researchers can conclude that the clock frequency of C_4 and C_5 have been decreased to (0.625) F_0 and (0.625) F_0 respectively. That decrease will be followed by an augmentation in clock period which will provide a simultaneous decrease in power consumption. As a justification, the consumed power before and after scheduling will be calculated.

Figure 4. Frequency values

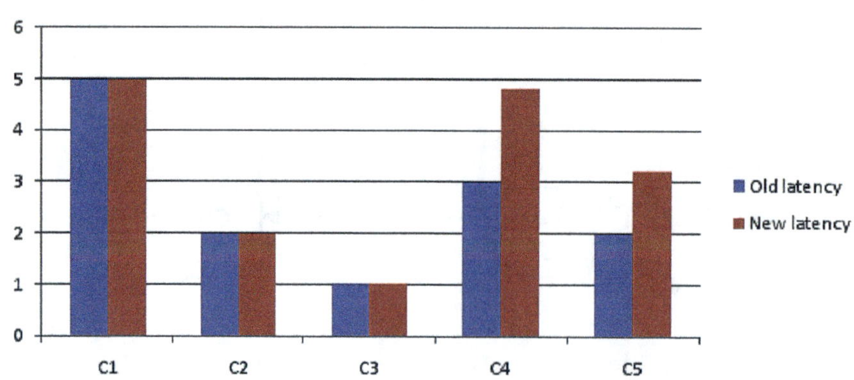

$$P_{dtotal\_before} = n\alpha F_0 = 5\,\alpha F_0 \quad (6)$$

where n is the number of components.

$$P_{dtotal\_after} = (5\text{-}2)\,\alpha F_0 + (0.625)\,F_0 + (0.625)\,F_0 = 4.25\alpha F_0 \quad (7)$$

Based on (6) and (7), authors can conclude that the proposed algorithm offers a gain of 15% in power compared to the classical technique.

As mentioned previously, the global latency is the latency of the critical path. Based on Figure 5, authors notice that after increasing the components latencies, the global latency has been kept.

EXPERIMENT

To prove the efficiency of the proposed algorithm, a comparative study has been illustrated between a system on which an application of the new proposed technique is done and another standard system (without applying the proposed algorithm). In fact, an application which is widely used in the area of image compression and processing has been applied. That application serves to convert the RGB color to HMMD (Hue, Max, Min, and Difference) color. Authors have been interested in the converter block which is consisting of eight blocks as shown Figure 6. The color space HMMD composed of black, white, color and hue. There are four attributes in the color space HMMD: Hue, Max, Min and Diff. First, Max indicates how much black color the image has, giving the flavor of shade or blackness. Second, indicates how much white color the image has, giving the flavor of tint or whiteness. Third, Diff indicates how much gray the image contains and how close to the pure color, giving the flavor of tone or colorfulness. The converter components intended to resolve the following equations:

Figure 5. Paths latencies

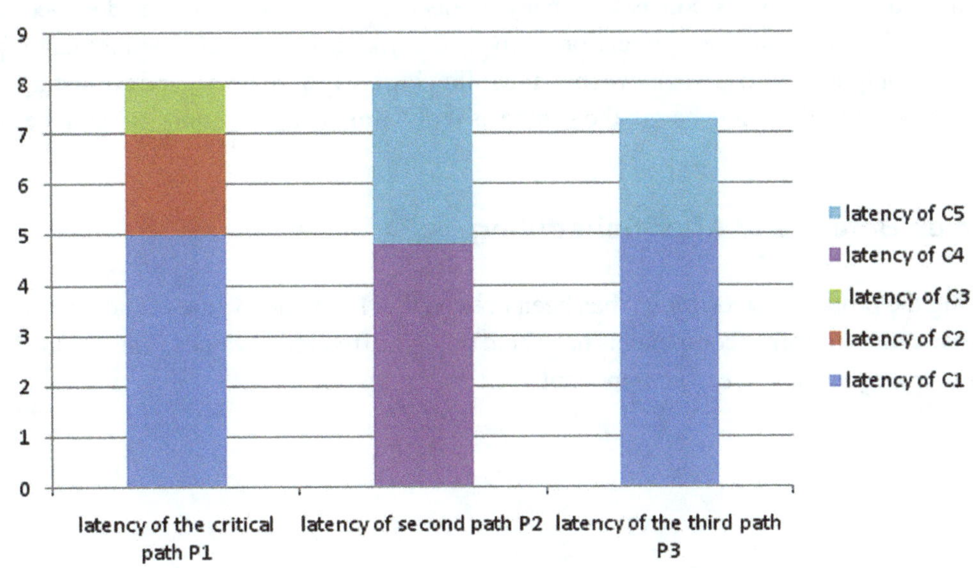

$$\begin{cases} Max = Max\left\{RGB\right\}, \\ Min = Min\left\{RGB\right\}, \text{ (8-10)} \\ Diff = Max - Min. \end{cases}$$

$$\begin{cases} If\left(Max = Min\right); \\ Hue = 0, \end{cases} \tag{11}$$

$$\begin{cases} ElseIf\left(Max = RandG < B\right) \\ Hue = 360 + 60 * \dfrac{\left|G - B\right|}{\left|Max - Min\right|}, \end{cases} \tag{12}$$

$$\begin{cases} ElseIf\left(Max = G\right) \\ Hue = 60 * \dfrac{2 + \left|B - R\right|}{\left|Max - Min\right|}, \end{cases} \tag{13}$$

$$\begin{cases} Else \\ Hue = 60 * \dfrac{4 + \left|R - G\right|}{\left|Max - Min\right|}, \end{cases} \tag{14}$$

Based on Figure 6, this block consists of many inputs such as "Start" which is destined to activate the application, when it is valid the application starts its treatment. The end of the treatment is presented when the "Stop" output is valid. To achieve that, the following tools have been used consisting of ML505 based on Virtex®-5 the electronic board, the synthesis tool Xilinx ISE (version 13.4) and the simulation tool test bench.

Frequencies Before and After Scheduling

After applying the proposed algorithm, it has been noticed that frequency values of adder2 and devider2 have been changed, while other components have held the same frequency values. The Figure 7 illustrates the old and new frequencies of each component.

Figure 6. The target application

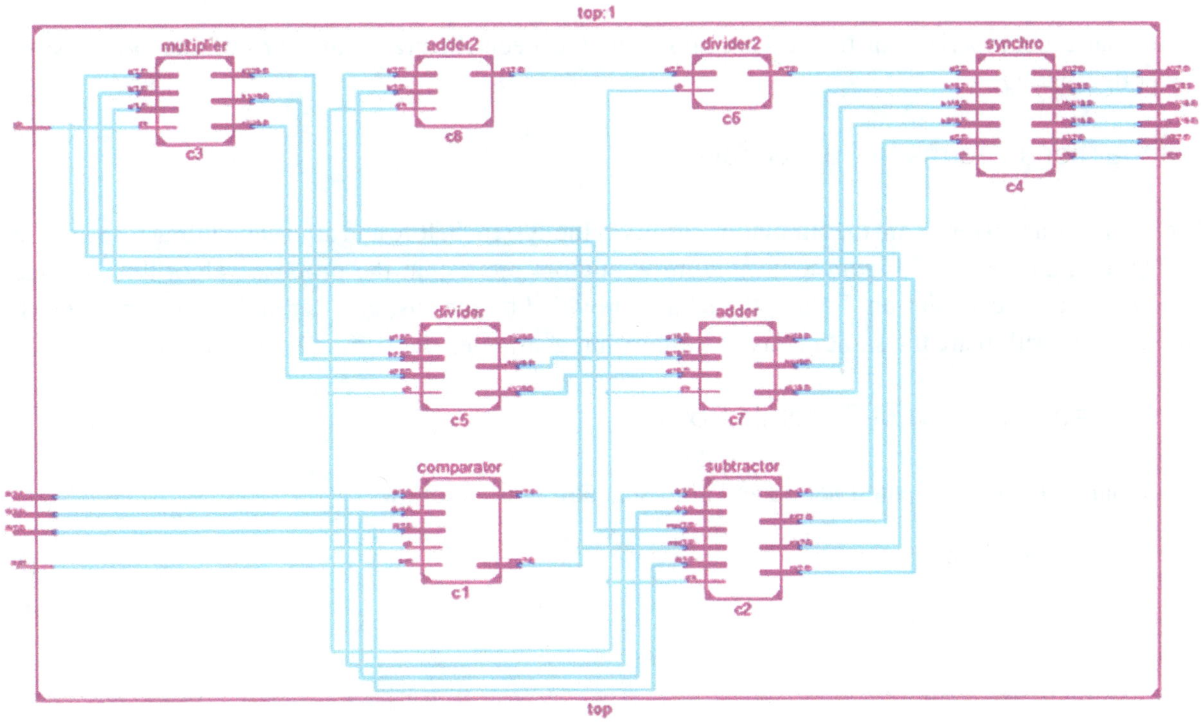

Figure 7. Frequencies of each component after and before scheduling

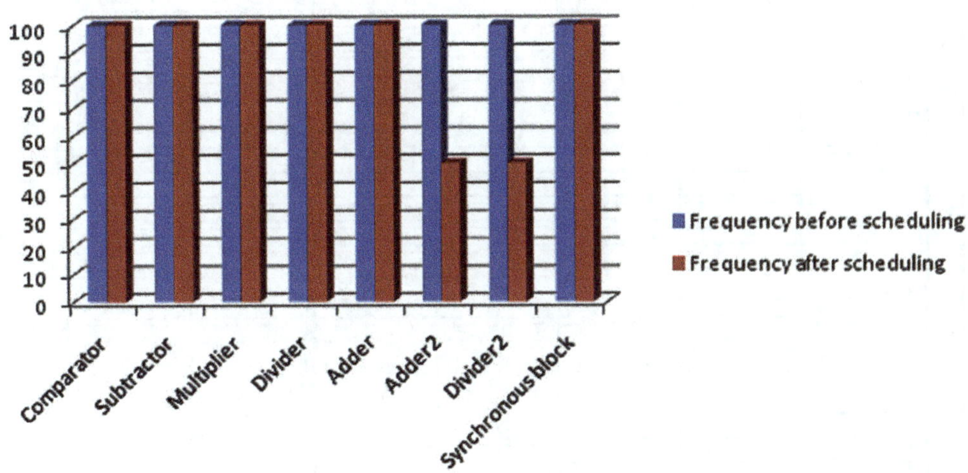

Design Results at Simulation Level

In this part, a comparison of the simulation results has been illustrated in terms of functionality and power dissipating.

Comparison in Terms of Functionality

The Figures 8 and 9 represent the simulation results of the system before and after scheduling respectively.

Through an analys of Figure 8 and Figure 9, authors prove that the proposed algorithm provides accurate and correct solution. In fact, the results provided by the system after and before applying the proposed algorithm are the same. Thus, the algorithm guaranties the correct functionality.

Comparison in Terms of Dynamic Power

The dynamic power dissipated before and after scheduling is as follows:

$$P_{dtotal\_before} = n\alpha F_0 = 8\ \alpha F_0 \quad (15)$$

Figure 8. Before scheduling

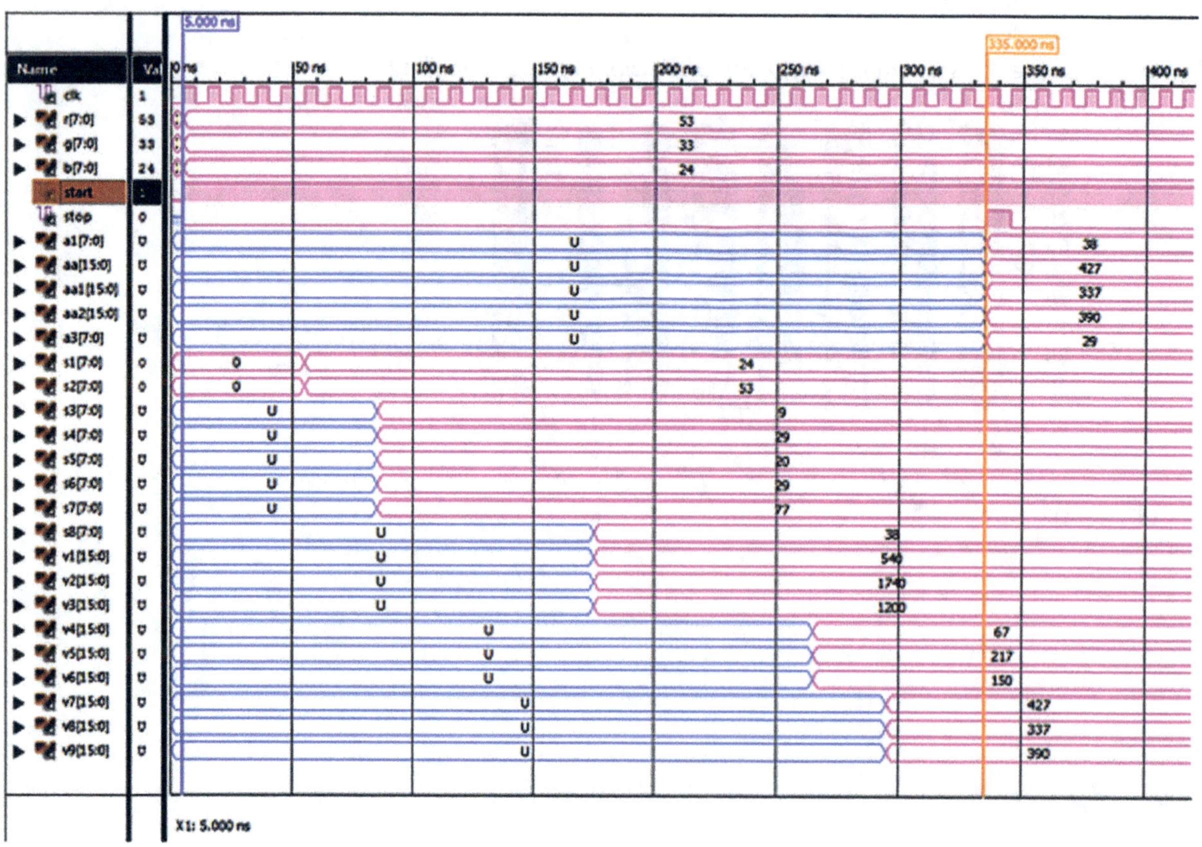

Figure 9. After scheduling

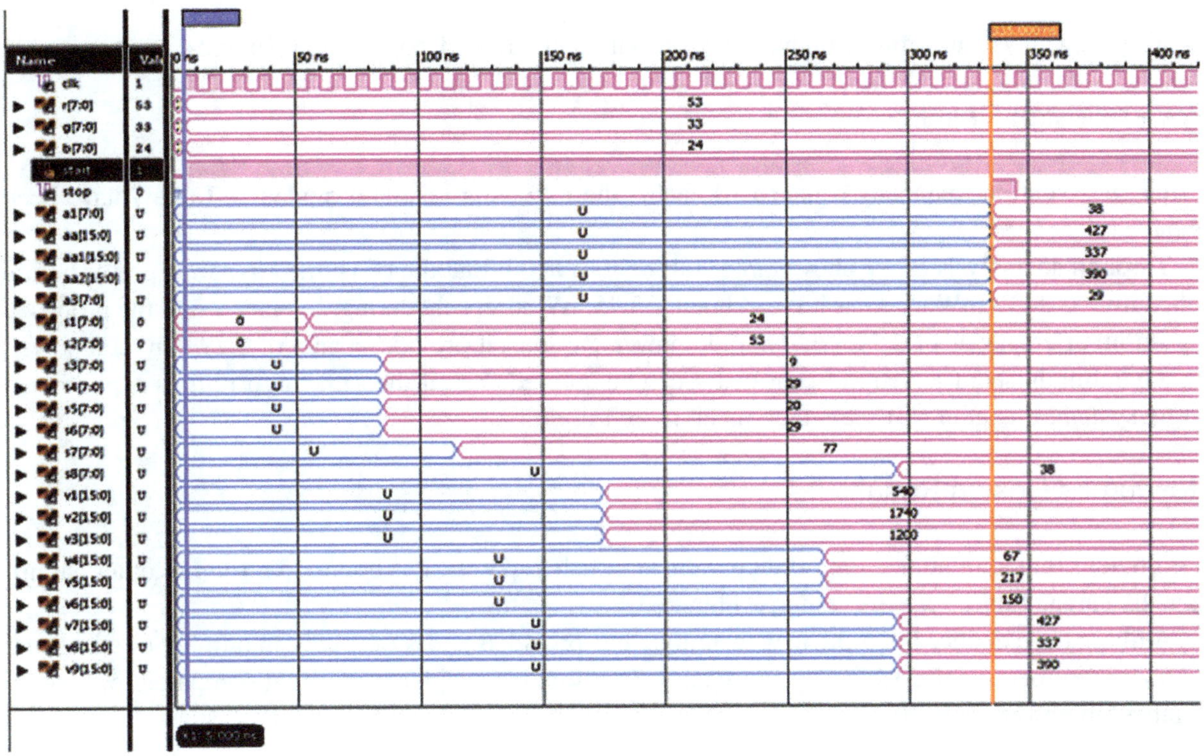

$$P_{dtotal\_after} = (8\text{-}2)\ \alpha F_0 + 0.5\ \alpha F_0 + 0.5\alpha F_0 = 7\ \alpha F_0 \tag{16}$$

The gain in dynamic power equals:

$$G = ((P_{dtotal\_before} - P_{dtotal\_after}) / P_{dtotal\_before}) \times 100$$

$$G = 12.50\% \tag{17}$$

Based on (17), authors can conclude that the proposed algorithm has offered a gain of 12.5% in dynamic power.

Comparison in Terms of Global Latency

It is widely acknowledged that power and time are antagonistic. In fact, existing technologies have managed to augment the processing capacity of the processers at the cost of a considerable simultaneous augmentation in their power and vice versa. In other word, a minimization in power consumption of a system, will lead to an augmentation of its latency. However, according to figure 7 and 8 the system responds in 330ns before and after scheduling. Thus, the new technique has kept the same global latency while minimizing the consumed power.

Design Results at Physical Level

In this part, a physical comparison has been presented in terms of power and resources.

Power Design Results

Using an execution frequency equal to 100 Mhz, the following results provided at Table 1 have been obtained.

Through an analysis of the above table, authors figure out that the consumed dynamic power before scheduling is 34.21mW whereas it is as high as 33.45 mW after scheduling. In other word, the proposed algorithm offers a gain of 2.22% of dynamic power. Besides, it provides a gain of 0.79% in total power. That gain might seem so insignificant that it can be ignored. Yet, if the same algorithm is applied on a larger system, that gain will seem then much more important.

Resources Design Results

It is obvious that to minimize important parameters such as power, it is bound to influence other parameters like area. Table 2 presents the loss obtained.

Authors have deduced that the new technique has provided a loss of 13.5% in number of slices and 4.60% in number of Slice LUT-Flip Flop pairs. This loss is considered minimal compared to the gain obtained in power consumption.

FUTURE RESEARCH DIRECTIONS

Recently, new alternative technology that combines logic elements and memory along with an intellectual property processor core has appeared to remedy the excessive need for better performing systems. This technology called System on Programmable Chip SoPC that allows and facilitates the SW/HW partitioning. In fact, Hardware Software partitioning has become an important phase during the design

Table 1. Power design results

	Before Scheduling	After Scheduling	Gain %
Dynamic Power (mW)	34.21	33.45	2.22
Total power (W)	0.626	0.621	0.79

Table 2. Resources design result

	Before Scheduling	After Scheduling	Loss %
Number of Slices	76	86	13.15
Number of Slice Flip Flops	163	163	0
Number of Slice LUT-Flip Flop pairs	239	250	4.60

flow of SoC. This phase aims to determine which parts of the system should be implemented using the software and which ones should be implemented using the hardware. Such partitioning will ultimately lead to a less consuming system. Thus, the implementation of the above proposed method during the phase of Hw/Sw partitioning all along taking into consideration the notion of tasks parallelism will definitely lead to a better performing system that consumes less power. This method will be the objective of future research.

CONCLUSION

Power management has become a multidimensional problem in many fields such as physics, mechanical engineering, electrical engineering, design automation, high level logic and synthesis, computer architecture, Operating systems, compiler design, and application development. The dilemma is that power problem exponentially arises. For that, it has been addressed along multiple levels ranging from transistors to applications. These levels have been described in this book chapter all along with a description of the different proposed method existing in the literature.

As generally reckoned, a well performed system is a system that consumes less power without influencing the constraint of time. In fact, current technologies have focused on creating faster systems that consume more power or concentrate on optimizing power at the cost of an increase in global latency of a system (execution time). Motivated by the fact, a new algorithm has been proposed that is meant to optimize consumed power without augmenting the global latency of the system. In fact, design results have proved the efficiency of the new technique which should be used by designers in order to obtain less consuming systems.

REFERENCES

Aichouchi, M. (1994). *Etude des liens entre la synthese architecturale et la synthese au niveau transfert de registres* (Doctoral dissertation). Institut National Polytechnique de Grenoble-INPG.

Baruch, Z. (1996). Scheduling algorithms for high-level synthesis. *ACAM Scientific Journal*, *5*(1-2), 48–57.

Belhadj, M. (1994). *Conception d'architectures en utilisant SIGNAL et VHDL* (Doctoral dissertation).

Ben Haj Hassine, S., Jemai, M., & Ouni, B. (2017). Power and Execution Time Optimization through Hardware Software Partitioning Algorithm for Core Based Embedded System. *Journal of Optimization, 2017*.

Berrebi, E. (1997). *Méthodologie pour l'application industrielle de la synthèse comportementale* (Doctoral dissertation). Institut National Polytechnique de Grenoble-INPG.

Bhattacharyya, S. S., Murthy, P. K., & Lee, E. A. (1999). Synthesis of embedded software from synchronous dataflow specifications. *The Journal of VLSI Signal Processing*, *21*(2), 151–166. doi:10.1023/A:1008052406396

Camposano, R. (1991). Path-based scheduling for synthesis. *IEEE Transactions on Computer-Aided Design of Integrated Circuits and Systems, 10*(1), 85–93. doi:10.1109/43.62794

Canis, A., Brown, S. D., & Anderson, J. H. (2014, September). Modulo SDC scheduling with recurrence minimization in high-level synthesis. In *Field Programmable Logic and Applications (FPL), 2014 24th International Conference on* (pp. 1-8). IEEE. 10.1109/FPL.2014.6927490

Canis, A., Choi, J., Aldham, M., Zhang, V., Kammoona, A., Czajkowski, T., ... Anderson, J. H. (2013). LegUp: An open-source high-level synthesis tool for FPGA-based processor/accelerator systems. *ACM Transactions on Embedded Computing Systems, 13*(2), 24. doi:10.1145/2514740

Casseau, E., Jégo, C., & Martin, E. (2004). Synthèse architecturale d'applications temps réel pour technologies submicroniques. *Technique et science informatiques, 23*(1), 35-66.

Chang, J. U., & Lin, C. H. (2016). A Circuit Complexity Optimization ILP Algorithm of High-level Synthesis System for New Multiprocessor Design. The Journal of The Institute of Internet. *Broadcasting and Communication, 16*(3), 137–144. doi:10.7236/JIIBC.2016.16.3.137

De Man, H., Rabaey, J., Six, P., & Claesen, L. (1986). Cathedral-II: A silicon compiler for digital signal processing. *IEEE Design & Test of Computers, 3*(6), 13–25. doi:10.1109/MDT.1986.295047

Dias, A. F., Akil, M., Lavarenne, C., & Sorel, Y. (2000). Vers la synthèse automatique de circuits à partir de graphes algorithmiques factorisés. In Workshop AAA sur l'adéquation algorithme architecture (pp. 165-172). Academic Press.

Ding, H. (1996). *Synthese architecturale interactive et flexible* (Doctoral dissertation). Institut National Polytechnique de Grenoble-INPG.

Ellervee, P. (2000). *High-level synthesis of control and memory intensive applications* (Doctoral dissertation). Institutionen för elektronisk systemkonstruktion.

Fisher, J. A. (1981). Trace scheduling: A technique for global microcode compaction. *IEEE Transactions on Computers, 30*(7), 478–490. doi:10.1109/TC.1981.1675827

Friedman, T. D., & Yang, S. C. (1969). Methods used in an automatic logic design generator (ALERT). *IEEE Transactions on Computers, 100*(7), 593–614. doi:10.1109/T-C.1969.222727

Gajski, D. D., Dutt, N. D., Wu, A. C., & Lin, S. Y. (2012). *High—Level Synthesis: Introduction to Chip and System Design*. Springer Science & Business Media.

Govindarajan, S. (2000). *Algorithms for Design Space Exploration and High-level Synthesis for Multi-FPGA Reconfigurable Computers* (Doctoral dissertation). University of Cincinnati.

Gruian, F., & Kuchcinski, K. (1998). A Constraint Logic Programming Based Approach to High-Level Synthesis for Low Power. Proceedings of Euromicro.

Haynal, S. (2000). *Automata-based symbolic scheduling*. Academic Press.

Heijligers, M. J. M. (1996). *The application of genetic algorithms to high-level synthesis.* Technische Universiteit Eindhoven.

Huang, S. H., Jeang, Y. L., Hwang, C. T., Hsu, Y. C., & Wang, J. F. (1993, July). A tree-based scheduling algorithm for control-dominated circuits. In *Proceedings of the 30th international Design Automation Conference* (pp. 578-582). ACM. 10.1145/157485.165051

Hwang, C. T., Lee, J. H., & Hsu, Y. C. (1991). A formal approach to the scheduling problem in high level synthesis. *IEEE Transactions on Computer-Aided Design of Integrated Circuits and Systems, 10*(4), 464–475. doi:10.1109/43.75629

Jadhav, A. (2016). Resource Scheduling in High Level Synthesis of Digital Filter using Differential Evolution Algorithm. *Journal of Digital Integrated Circuits in Electrical Devices, 1*(1-2).

Jain, R., Mujumdar, A., Sharma, A., & Wang, H. (1991, June). Empirical evaluation of some high-level synthesis scheduling heuristics. In *Proceedings of the 28th ACM/IEEE Design Automation Conference* (pp. 686-689). ACM. 10.1145/127601.127751

Jemai, M., Dimassi, S., Ouni, B., & Mtibaa, A. (2017). A metaheuristic based on the tabu search for hardware-software partitioning. *Turkish Journal of Electrical Engineering and Computer Sciences, 25*(2), 901–912. doi:10.3906/elk-1501-64

Jervan, G. (1998). *Decision Diagram Synthesis from VHDL.* Tallinn Technical University.

Kission, P. (1996). *Exploitation de la hierarchie et de la reutilisation de blocs existants par la synthese de haut niveau* (Doctoral dissertation).

Ku, D. C., & De Micheli, G. (1988). *Hardware C-a language for hardware design (No. CSL-TR-88-362).* Stanford Univ CA Computer Systems Lab. doi:10.21236/ADA207317

Kuchcinski, K. (1998, August). An approach to high-level synthesis using constraint logic programming. In *Euromicro Conference, 1998. Proceedings. 24th* (Vol. 1, pp. 74-82). IEEE. 10.1109/EURMIC.1998.711780

Kushwaha, S., & Kumar, S. (2017). An Investigation of List Heuristic Scheduling Algorithms for Multiprocessor System. *IUP Journal of Computer Sciences, 11*(2), 29.

Landwehr, B., Marwedel, P., & Dömer, R. (1994, September). OSCAR: Optimum simultaneous scheduling, allocation and resource binding based on integer programming. In *Proceedings of the conference on European design automation* (pp. 90-95). IEEE Computer Society Press.

Laurent, B. (1999). *Conception des blocs réutilisables. Réflexion sur la méthodologie* (Doctoral dissertation). Institut National Polytechnique de Grenoble-INPG.

Lee, B. (2000). *Specification and Design of Reactive Systems* (PhD Thesis). University of California, Berkeley, CA.

Li, Z., Lai, K. Y. T., Yu, P. H., Ho, T. Y., Chakrabarty, K., & Lee, C. Y. (2016, June). High-level synthesis for micro-electrode-dot-array digital microfluidic biochips. In *Proceedings of the 53rd Annual Design Automation Conference* (p. 146). ACM. 10.1145/2897937.2898028

Lin, Y. L. (1997). Recent developments in high-level synthesis. *ACM Transactions on Design Automation of Electronic Systems*, *2*(1), 2–21. doi:10.1145/250243.250245

Luo, J., & Jha, N. K. (2000, November). Power-conscious joint scheduling of periodic task graphs and aperiodic tasks in distributed real-time embedded systems. In *Proceedings of the 2000 IEEE/ACM international conference on Computer-aided design* (pp. 357-364). IEEE Press.

McFarland, M. C., Parker, A. C., & Camposano, R. (1990). The high-level synthesis of digital systems. *Proceedings of the IEEE*, *78*(2), 301–318. doi:10.1109/5.52214

Memik, S. O., Kastner, R., Bozorgzadeh, E., & Sarrafzadeh, M. (2005). A scheduling algorithm for optimization and early planning in high-level synthesis. *ACM Transactions on Design Automation of Electronic Systems*, *10*(1), 33–57. doi:10.1145/1044111.1044115

Narasimhan, M. (1998). *Exact scheduling techniques for high level synthesis. Master of Science in Electrical Engineering*. Lousiana State University.

Narasimhan, M., & Ramanujam, J. (1997). *Tighter lower bounds for scheduling problems in high-level synthesis*. Academic Press.

Nazar, F., & Murugan, S. (2016, March). Implementation of JPEG-LS compression algorithm for real time applications. In *Electrical, Electronics, and Optimization Techniques (ICEEOT), International Conference on* (pp. 2772-2774). IEEE. 10.1109/ICEEOT.2016.7755200

Nestor, J. A., & Krishnamoorthy, G. (1993). SALSA: A new approach to scheduling with timing constraints. *IEEE Transactions on Computer-Aided Design of Integrated Circuits and Systems*, *12*(8), 1107–1122. doi:10.1109/43.238604

Note, S., Geurts, W., Catthoor, F., & De Man, H. (1991, June). Cathedral-III: Architecture-driven high-level synthesis for high throughput DSP applications. In *Proceedings of the 28th ACM/IEEE Design Automation Conference* (pp. 597-602). ACM. 10.1145/127601.127739

Parker, A. C., Pizarro, J. T., & Mlinar, M. (1986, July). MAHA: A program for datapath synthesis. In *Proceedings of the 23rd ACM/IEEE Design Automation Conference* (pp. 461-466). IEEE Press.

Paulin, P. G., & Knight, J. P. (1987, October). Force-directed scheduling in automatic data path synthesis. In *Proceedings of the 24th ACM/IEEE Design Automation Conference* (pp. 195-202). ACM. 10.1145/37888.37918

Paulin, P. G., & Knight, J. P. (1989). Algorithms for high-level synthesis. *IEEE Design & Test of Computers*, *6*(6), 18–31. doi:10.1109/54.41671

Potasman, R., Lis, J., Nicolau, A., & Gajski, D. (1991, January). Percolation based synthesis. In *Proceedings of the 27th ACM/IEEE Design Automation Conference* (pp. 444-449). ACM.

Schreiber, R., Aditya, S., Rau, B. R., Kathail, V., Mahlke, S., Abraham, S., & Snider, G. (2000). High-level synthesis of nonprogrammable hardware accelerators. In *Application-Specific Systems, Architectures, and Processors, 2000. Proceedings. IEEE International Conference on* (pp. 113-124). IEEE. 10.1109/ASAP.2000.862383

Silc, J. (1994). Scheduling strategies in high-level synthesis. *Informatica, 18*(1), 71–79.

Sugar, Z. (2000). *Synthèse comportementale basée sur l'ordonnancement* (Doctoral dissertation). Institut National Polytechnique de Grenoble-INPG.

Tchoumatchenko, V. (1998). *Modelisation, architecture et outils de synthese pour additionneurs rapides* (Doctoral dissertation).

Thomas, D., & Moorby, P. (2008). *The Verilog® Hardware Description Language.* Springer Science & Business Media.

Thomas, D. E., Lagnese, E. D., Walker, R. A., Rajan, J. V., Blackburn, R. L., & Nestor, J. A. (1989). *Algorithmic and Register-Transfer Level Synthesis: The System Architect's Workbench: The System Architect's Workbench* (Vol. 85). Springer Science & Business Media.

Timmer, A. H., & Jess, J. A. (1995, March). Exact scheduling strategies based on bipartite graph matching. In *Proceedings of the 1995 European conference on Design and Test* (p. 42). IEEE Computer Society. 10.1109/EDTC.1995.470422

Tseng, C. J., & Siewiorek, D. P. (1986). Automated synthesis of data paths in digital systems. *IEEE Transactions on Computer-Aided Design of Integrated Circuits and Systems, 5*(3), 379–395. doi:10.1109/TCAD.1986.1270207

Wang, L., & Samee, U. (2013, September). Energy-aware parallel task scheduling in a cluster. *Future Generation Computer Systems, 29*(7), 1661–1670. doi:10.1016/j.future.2013.02.010

Zimmermann, A., & Trautmann, N. (2017). A list-scheduling heuristic for the short-term planning of assessment centers. *Journal of Scheduling*, 1–12.

Zimmermann, G. (1979, June). The MIMOLA design system a computer aided digital processor design method. In *Proceedings of the 16th Design Automation Conference* (pp. 53-58). IEEE Press. 10.1109/DAC.1979.1600088

KEY TERMS AND DEFINITIONS

ASIC (Application Specific Integrated Circuit): Is an integrated circuit (IC) manufactured for a particular use, rather than intended for general purpose use.

Dynamic Power: Is the power dissipated due to the switching activity during charging and discharging of the load capacities and due to the short-circuit power caused by a short-circuit current that occurs during changes in the states of the logic gates.

FPGA (Field Programmable Gate Array): Is an integrated circuit designed to be programmed by the user after manufacturing.

Heuristic Algorithm: The term heuristic is used for algorithms that find solutions among many possible ones, but they do not assurance that the best will be found; therefore, they may be considered as approximately algorithms not accurate ones.

HLS (High Level Synthesis): Sometimes referred to C synthesis, behavioral synthesis, algorithmic synthesis, electronic system level (ESL) synthesis, or is an automated design process that interprets an algorithmic description of a desired behavior and creates electronic hardware that implements that behavior.

RTL: (Register Transfer Level): Is a high level description which models a synchronous circuit in terms of the flow of digital signals between hardware registers, and the logical operations performed on those signals.

SoC (System on Chip): Is an integrated circuit that integrates many components of a computer or other electronic systems. It may contain analog, digital, mixed-signal, and often radio frequency functions all on a single substrate.

Chapter 18
An Optimal Photovoltaic Conversion System for Future Smart Grids

Carlo Makdisie
Tishreen University, Syria

Badia Haidar
Tishreen University, Syria

Hassan Haes Alhelou
Tishreen University, Syria

ABSTRACT

Smart grid technology is the key for a reliable and efficient use of distributed energy resources. Amongst all the renewable sources, solar power takes the prominent position due to its availability in abundance. In this chapter, the authors present smart grid infrastructure issues and integrating solar PV-sourced electricity in the smart grid. Smart grid has many features, including reliability, flexibility on network topology, efficiency, sustainability, and market-enabling. The authors select a photovoltaic active power line conditioner as a case study. This line conditioner is a device designed to extract the maximum power of a photovoltaic (PV) system and to compensate the nonlinear and unbalanced loads of the electrical power systems. The performance of the PV conditioner with the neuro-fuzzy control designed has been analyzed through a simulation platform.

INTRODUCTION

For many years the fossil fuel was the major source of energy. But the whole world suffered from the depletion of fossil fuel as well as its bad effects on the atmosphere around the globe. The major bad effect is reflected global warming and changes in the Earth's climate (Partridge, et al. 2017). Due to all the above-mentioned reasons, the researches for other sources of energy begin in the recent years.

DOI: 10.4018/978-1-5225-3935-3.ch018

The renewable energy is the best solution to use a clean energy for the most applications to stop the pollution and have unlimited power sources. Renewable energy resources such as solar energy and wind power can be operated in an optimal mode to provide electricity in a lower price. This chapter focus in providing a comprehensive survey on the methods that can be used for achieving such operation status. Likewise, a mathematical and theoretical models of solar energy systems are presented. The different methods for achieving maximum power point have discussed. Furthermore, all parts and components of PV systems have been illustrated. Moreover, different case studies have been presented for demonstrating the ability of solar energy system for providing an optimal power by using some optimization methods.

The rest of this chapter is organized as follows. Firstly, a brief background and review literature is presented in background section. Then some definitions and the mathematical model of PV systems are presented. The different components of PV system are then illustrated. Some method for design an optimal PV system are demonstrated in PV system design section. After that, a comprehensive review and discussion in the optimal methods are presented. In the next section, the different case studies are presented to show the important of using the optimization methods in smart grids and PV systems. Likewise, the future trends in smart grid and optimization method are illustrated in future research direction section. Final, a conclusion and the references that have been used in this chapter are presented accordingly.

BACKGROUND

Most of the recent researches have tended to use renewable energy sources found in nature such as, solar energy, wind energy, and tidal energy (Alva, et al. 2017). Due to the increasing demands in clean and new energy resources, the solar energy industry is one of the fastest growing forces in the market. In the last decade, there are several major directions for solar technology development (Thornton, 1992; Chaar, et al. 2008; Mohandes, et al. 2009; Al Hanai, et al. 2010). For example, photovoltaic systems directly convert the solar energy into electrical energy while concentrated solar power systems first convert the solar energy into thermal energy and then further convert it into electrical energy through a thermal engine (Chu & Meisen, 2011).

In literatures, many methods have been used of optimal design and operation of PV systems (Nazaripouya, et al. 2015). Evolutionary optimization methods have the major application in solar systems for getting the maximum power operation point (Patcharaprakiti & Premrudeepreechacharn, 2002). Also, different met-heuristic algorithms such as genetic algorithm, PSO, wind driven optimization algorithm have a good performance for designing and operation of PV systems (Injeti & Padma, 2015).

A comprehensive literature reviews and state of arts in these topics could be found in (Ram, et al. 2017; Gordon, J. M. 2013). Due to space consideration, in this chapter, Authors discuss in depth the solar energy system component and the optimization methods used in solar systems to make their operation more effective.

PV Solar Cells Effects and Definition

Definition

The photovoltaic effect refers to the voltage being generated across p-n junctions in a semiconductor, based on the light radiation absorption. The elements involved in this effect are referred to as photovoltaic device. (figure 1a)

Electron-Hole Formation

The solar light consists of photons which are fluxes of particles. On a sunny day about 4.4 x 10^{17} photons hit each squared centimetre of Earth every second. Only the photons with energy in excess of the band gap can be transformed to electric energy by using a solar cell. When these photons hit the semiconductor, they can be absorbed giving enough energy to an electron to move it from the valance band to the conductive band. The photovoltaic energy conversation depends entirely on the numbers of those photons.

A Solar Panel (or) Solar Array

Single Solar Cell

it consists of n-type layer and p-type layer arranged in a form of sandwich referred to as p-n junction. A very known type of this cell is a silicon wafer p-n junction with a large square area. Those Single Solar Cells are present in only small electronic devices like watches and calculators due to its tiny productivity of electrical energy. (Figure 1b)

Solar Panel (or) Solar Array (or) Solar Module

The interconnection of number of solar module allows to form the solar panel (or) solar array with significant power. A solar module format is a number of interconnected solar cells. These interconnected cells planted between two glass plates to protect from the bad weather. Since the capacity area of module is high, more energy can be produced (Maghami, et al. 2017).

Figure 1. a) p-n junction b) single solar cell

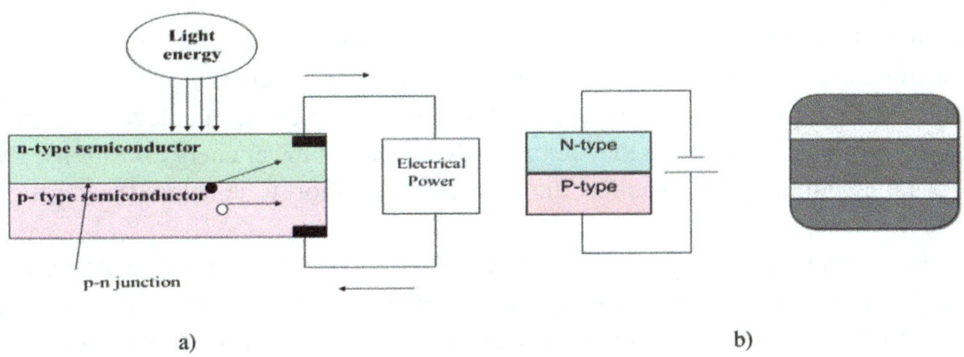

Figure 2. Photovoltaic cells, modules, panels, and array

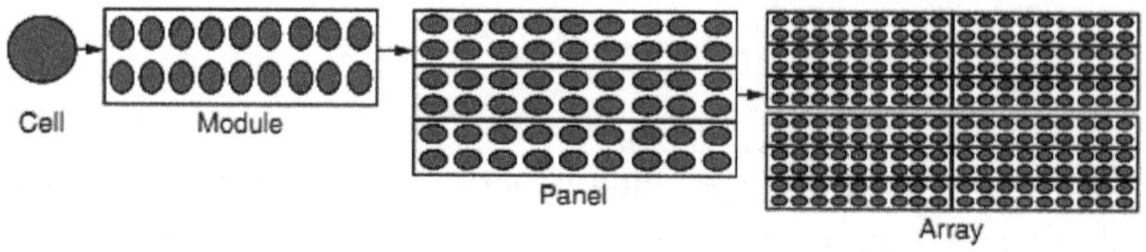

Table 1. Types of solar cell based on the types of crystal used

Type	Efficiency	Description
Monocrystalline silicon cell	(14-17%)	The effectiveness of a cell will be higher due to the fact that it consists of pure silicone in a form of Monocrystalline which is a defect free pure crystal.
polycrystalline silicon cell	(13-15%)	The effectiveness of this cell is lower than the Monocrystalline one because it contains a various crystalline sizes. It's made by solidification process of liquid silicone.
Amorphous silicon cell	(5-7%)	The thickness of layer in this type of cell is less than 1μm. And it's made by placing a silicone film in a substrate, a glass plate for example. It's only used in small devices like watches and small calculators and other low power consumption devices. Because its efficiency it has is much low than the previous two types of cells.

Principle, Construction, and Working of Solar Cell

- **Principle:** The photovoltaic effect is considered the base element of the solar cell. The photo generation of charge carriers in light absorbing materials as a result of absorption of light radiation contribute to the photovoltaic activity.
- **Construction:** the p-n junction in a solar cell is made of p-type semiconductor layer as a base and n-type semiconductor layer as emitter. The layers are sandwiched together to make the p-n junction. The contacts on n-type and p-type semiconductor are made of proper metal elements to make fine electrical connection. The whole surface of the solar cell is covered with anti-reflection coating to minimize the loss of light energy due to reflecting.
- **Working:** when sunlight hits the surface of solar panel, it energies are absorbed in the semiconductors. Causing liberty of some electrons and forming the external DC current. Therefor the DC current could be converted into 240-volt AC current via DC−AC converter for various applications.
- **Mechanism:** First, the sunlight is absorbed by a solar cell in a solar panel. The electrons in the material are liberated to increase in energy because of the absorbed light. And making them move around the material in a free form. Anyway the electrons maintain this level of high energy for only a short period of time. Then they return to their original lower energy positions. Therefor the use of p-n junction is typical to collect the carriers in that short time before their loss of energy.

The p-n junction is made of two regions one is made of a p-type semiconductor material which called the p-type region and the other made of n-type semiconductor material and called the n-type region. While exposed to the light the p-type material electrons gain the energy needed to move to the n-type

Figure 3. How solar cell works

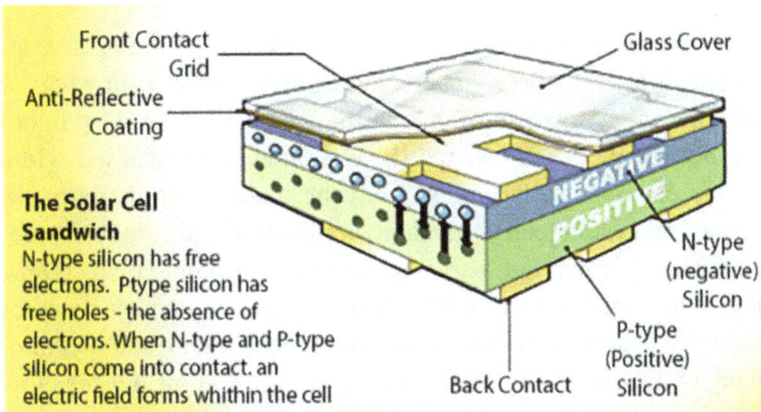

region. After that happens they can't go back to low power position and continue to be in higher energy. In this situation the electrons can be extracted from the solar device to produce a current, or can be kept in the solar device to gain a higher voltage. In the first case the electrons give their energy to anything that's connected to the solar device and gain energy again when they return to it.

PV Solar Cells Applications

1. Water supply can be obtained by solar pumps.
2. Producing power supply domestically for economic applications like refrigeration, lights, washing machines and television.
3. Lighthouses and buoys power supply can be obtained by solar power systems to maintain ocean navigation aid.
4. Producing electric power to fulfill the power needs of radio transceivers and telephone boxes in any telecommunication system.
5. Producing power in near space to fulfill the power needs of satellites and space stations in orbit.
6. Compensation power instruments can be powered by solar cells.
7. Independent and hybrid power supply for various electrical grids.

Table 2. PV Solar cells advantages and disadvantages

Advantages	Disadvantages
1. It's very clean and produce no polluting elements. 2. it's sustainable and long lasting source of energy as it's based on the sunlight (never-ending source of energy). 3. solar cells are silent and they make no noise when extracting energy. 4. the maintenance costs are much cheaper compared to other methods of energy production 5. they have a long life time cycle. 6. Energy could be stored in batteries. 7. it doesn't need fuel, so it has no fuel cost and doesn't require fuel supply.	1. the power production is paused during night time 2. producing the solar panels have a considerable cost. 3. whether and air pollution will have an effect on the power production of the solar panels. 4. they occupy a large area of land in order to extract an efficient power supply. 5. In most applications the solar cells need a maximum power tracking techniques MPPTT. 6. The complexity of control system of converters connected to the solar cells.

SOLAR CELL CHARACTERISTICS

To get additional positive charge transporters (p-type) or an extra flow of negative charge ones (n-type), doping method is adopted. A p-n joint is produced on the border at the point when two layers of various doping are connected. The disengagement of charge transporters released by light are occurred when an interior electrical field is established causing the electrons to be released within the electric filed nearness which make the electrons to be pulled from the p-side towards the n-side (Lana El Chaar, 2007). The components of main solar cell equivalent circuit shown in (Figure 4) are as follows: current source with a diode connected in parallel, parasitic series (R_s; usually small) (Dezso, et al. 2007) and shunt (R_{sh}) resistances (quite large) (Chu & Chen, 2009). R_s is mostly influenced by many factors, for example, the semiconductor material bulk resistance, interconnections, and metallic contacts, while R_{sh} is influenced principally by the p-n joint nonidealities and infection influences close to the joint (Lana El Chaar, 2007). To analyze the PV circuit, a simplified equivalent circuit shown in Figure 5 can be used. Shockley equation is used to get the diode current value:

$$I_D = I_0\left[\exp\left(\frac{qV}{nkT}\right) - 1\right]$$

where I_0 is the reverse saturation current, q is the charge carrier, k is the Boltzmann constant, T is the cell temperature, and n is the ideality factor. As shown in Figure 5, there are two limiting components in the PV module: V_{oc}, the voltage of open-circuit and I_{sc}, current value of short-circuit. Depending on the equivalent circuit and by substituting $V=0$ and $I_{sc}= I_{ph}$, the current I_{sc} value could be calculated, and this value varies proportionally to the cell irradiance value. To conclude V_{OC}, set the cell current $I_L= 0$, hence the open circuit voltage can be expressed by the following equation:

$$V_{OC} = \frac{nkT}{q}\ln\left[\frac{I}{I_0}\right]$$

When the product ($V\times I_{mp}$) value reaches its maximum, then the PV module is characterized by the maximum point. Where V & I_{mp} are the voltage and current values at the maximum power point respectively. The maximum power output value is calculated by using the following equation(Lana El Chaar, 2007):

Figure 4. Solar cell equivalent circuit

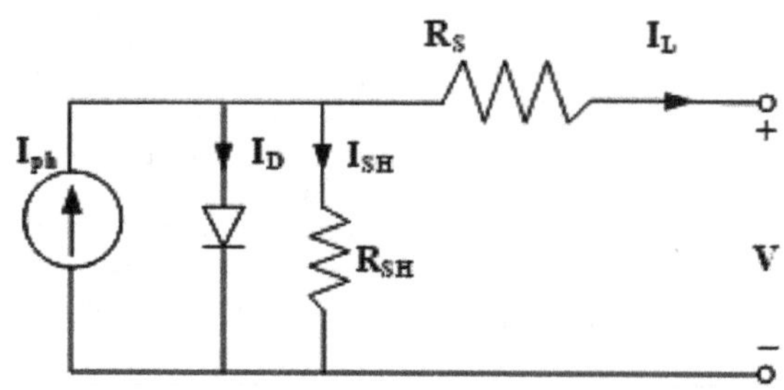

Figure 5. Model for a PV cell

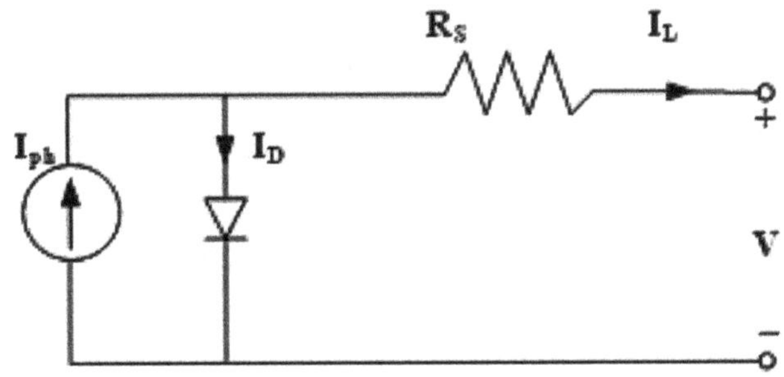

$$\frac{d\left(V \times I\right)}{dt} = 0$$

$$V_{mp} = V_{OC} - \frac{kT}{q}\ln\left[\frac{V_{mp}}{nkT / q} + 1\right]$$

- **Efficiency Calculation of PV Module:** The factor W_p is usually used to rate the PV module, under standard test conditions (STC) the value of this this factor is normally 1 kW/m^2, it defines the PV performance at sunlight of 1000 W/m^2, a cell temperature of 25 °C (77°F), and an air mass (AM) of 1.5. The product ($V_{mp} \times I_{mp}$) is related to the product generated by ($V_{OC} \times I_{SC}$) by a fill factor (FF) that is a measure of the junction quality and series resistance, and it is given by:

$$FF = \frac{V_{mp} \times I_{mp}}{V_{OC} \times I_{SC}}$$

As the value of FF is close to unity, the PV module quality will be higher. As a final point, the value of the PV module efficiency will be defined. This factor is considered to be the most important factor of PV module quality, it is calculated as:

$$\eta = \frac{FF \times V_{OC} \times L_{OC}}{P_{in}}$$

P_{in} represents the incident power depending on the light spectrum incident on the PV cell.

To achieve the desired voltage and current levels, a PV module is formed by using solar cells connected in a combinations of series (*N*s) and parallel (*N*). The PV parameters are then affected as shown below (Bentley, 2004):

$$I_{phtotal} = N_p I_{ph}$$

$$I_{0total} = N_p I_0$$

$$n_{total} = N_s n$$

$$R_{stotal} = \frac{N_s}{N_p} R_s$$

The above mentioned model is presented in Figure 6. To form the photovoltaic module, single solar cells are connected in series to obtain the higher voltage value and then in parallel to obtain the higher current value. Hence forming the photovoltaic array is done by connecting many of these modules to each other. This array is then installed within aluminum or stainless steel frame and wrapped with transparent glass on the front side (Figure 2). The voltage produced by the array depends essentially on the cell design and materials, where the electric current depends mainly on the cell area and the fallen solar irradiance. since the path of the sun varies dramatically over the year in where winter and summer reasons considered to be the two extreme excursions (Lana El Chaar, 2007), The photovoltaic current will fluctuate. The sun elevation angle (θ elevation sun) is expressed in degrees above the horizon. Sun azimuth angle (φ azimuth sun) is expressed in degrees from true north. And finally sun zenith angle (θ zenith sun) equals 90 degrees less than the sun elevation, or

$$\theta_{sun}^{zenith} = 90^\circ - \theta_{sun}^{elevation}$$

The above mentioned angles are explained in Figure 7. In a classic solar cell exposed to the sun the output rises from zero value at sunrise to a maximum one at middle of the day and at the dusk it falls again to the zero. When the sun radiation reaches the earth it varies because of the effect of absorption and disseminating in the atmosphere. Designers of PV system need an accurate estimation for the insolation probable to fall on a arbitrarily inclined surface, therefore need a good assessment of global radiation on a horizontal surface, horizontal direct and diffuse components, in order to estimate the amount of irradiation hitting sloping plane (Lana El Chaar, 2007).

Furthermore, as the temperature of the cell increases the open-circuit voltage decreases and the current of short-circuit slightly increases, causing the cell efficiency to be decreased. To meet the sun at an angle matching the country's latitude, it is very important to mount the PV panels on the roof or near an open area. If convenient, seasonal changing of the module's direction in the direction of the sun is performed manually. As the technology of solar power has pretty high cost, it is essential to run panels at their maximum power circumstances. . Nevertheless, to gather the maximum amount of solar radiation, it is much more effective and really better using a sun tracking technique making the module's surface to track the sun during the day. There are two types of tracking: one axis tracking and double axes tracking. However, double axes tracking offers greater value of the power output. Thus, depending on the geographical location and seasonal climate, yielded energy could be increased by about 20% to 30% (Karimov, et al. 2002; Khalil, et al. 2003; Dankoff, 2009). Even though some state that the low

Figure 6. PV module circuit model

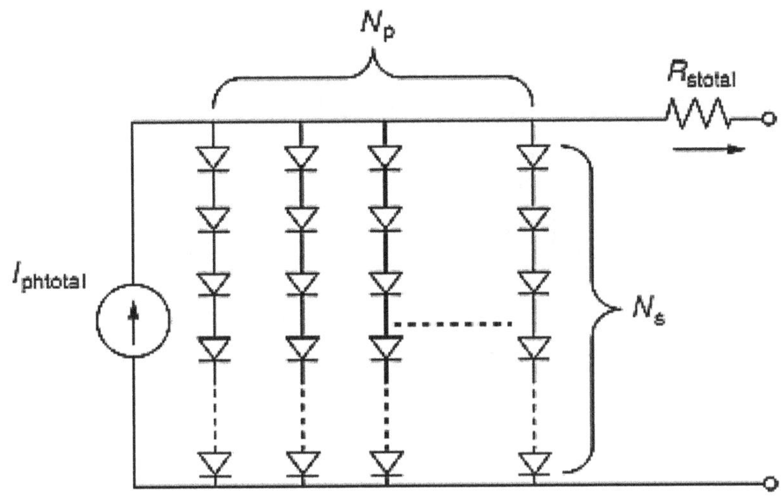

Figure 7. Azimuth, zenith, and elevation angles of a vector pointed toward the sun

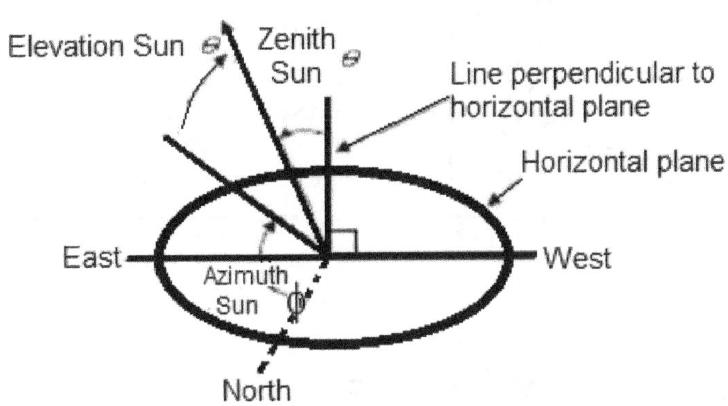

costs of fixed system besides they require nearly no maintenance (Lana El Chaar, 2007), various tracking techniques used to control the PV panels orientation have established their prevalence over fixed systems in terms of transformed power competence. To acquire PV panel maximum power the prevalent temperature and insolation circumstances (Lana El Chaar, 2007), either the operating voltage or current ought to be controlled by a maximum power point tracker (MPPT) that should take the following conditions into consideration (Kroutoulis, et al. 2001):

- The PV system to be operated close to the MPP regardless the changes in atmosphere.
- Having high level of conversion efficiency and low cost.
- The provided output interface should be compatible with the requirements of accumulator charging.

Figure 8. (a) I-V characteristic of a solar cell showing maximum power point (MPP); (b) I-V characteristic of a solar cell showing MPP

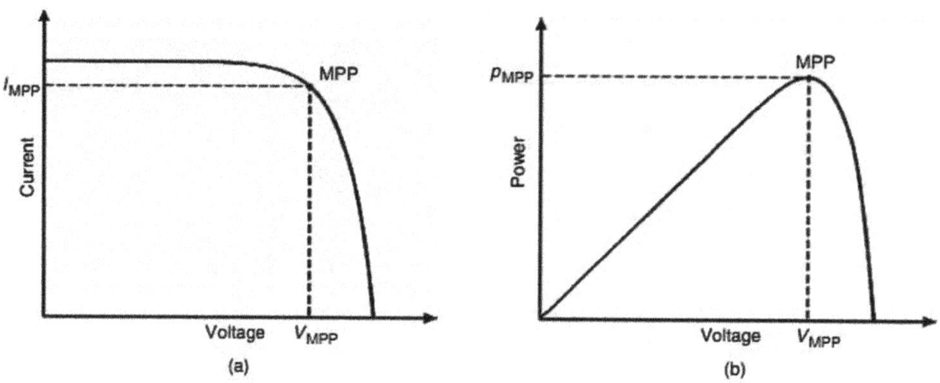

Figure 9. (a) Characteristics of PV panel insolation; (b) Characteristics of PV panel temperature

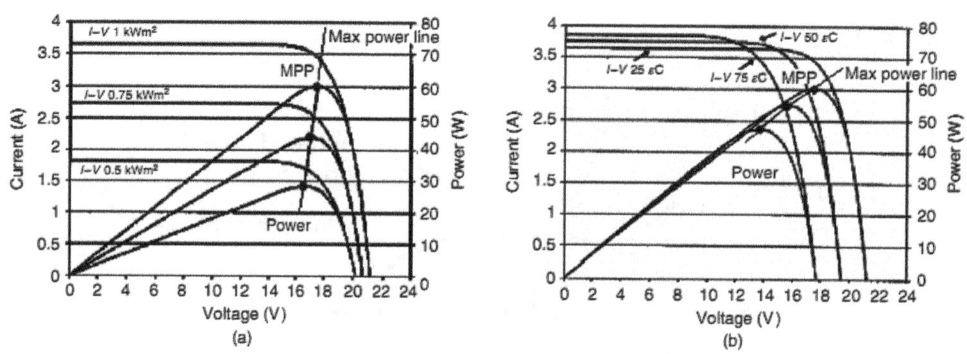

PHOTOVOLTAIC SYSTEM COMPONENTS

Wiring Symbol and Function

The interface between the AC (load) grid and the PV generator is the solar inverter. This inverter is used for converting the solar DC electricity generated by the PV generator into AC electricity with a required parameters of frequency voltage and level of the integrated grid. So, this solar inverter is known as a DC-AC converter. This conversion for Dc to AC has a small losses when inverter's modern control techniques are used. The principle of a grid-connected inverter is shown in Figure 10. Depending on their use, there are two distinguished inverters: grid inverter (grid-connected systems) and stand-alone inverters (stand-alone systems). In stand-alone systems, inverters enable the operation of conventional AC loads. The inverter, in PV systems of grid-connected type, will be linked to the mains electricity grid either directly or via the building's grid. In the case of direct connecting, only the mains grid are supplied by the generated electricity. When the inverter coupled to the building's grid, firstly the building is consumed by the generated solar power and then the mains grid is supplied by any extra power if there any. Figure 11 shows the principle of coupling PV systems with single- and three-phase inverters to the electricity grid.

To achieve delivering the maximum power into the grid, the operation of the inverter must be at the MPP of the PV generator. The MPP tracker, according to the climatic conditions, is used to adjust the inverter to the MPP point. The MPP tracker mainly consists of a DC-DC converter (chopper). This is connected in series with the input of the inverter and adjusts the inverter's input voltage to the voltage of the MPP. Modern solar inverters can be used to do those functions:

- Convert the DC current into AC current so the PV generated current can meet the grid requirements.
- MPP tracing which means the adjustment of the operating point of the Inverter to the MPP of PV generator.
- Signalling and recording operational data
- DC and AC protective devices (e.g. automatic disconnection in the event of a loss of mains or the mains supply falling outside the permitted voltage/frequency window).

Figure 10. Principle of a grid-connected inverter

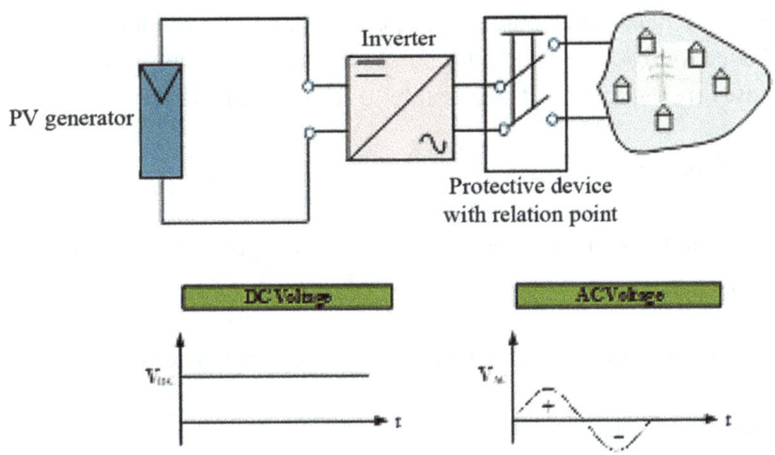

Figure 11. Principle of grid-connected PV systems with single- and three-phase inverters

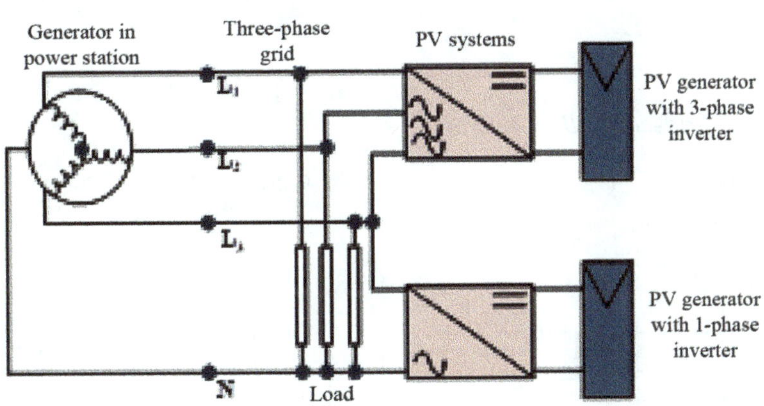

Grid-Connected Inverters

Depending on their base of operation, inverters in grid-connected systems includes two sections, grid-controlled and self-commutated inverters.

Line-Commutated Inverters

The configuration of this inverter is a bridge circuit with thyristor switches. These are still being used today, particularly in high power larger PV systems. For the lower power single-phase inverters (< 5 kW_p), there are now only a few manufacturers who still build inverters based on this principle and replaced by transistor inverters.

The principle of grid-controlled inverters is shown in Figure 12. Each pair of thyristors in the bridge circuit alternately receives a trigger pulse and work together simultaneously, synchronized with the mains frequency. Therefor the commutation circuit is needed to achieve the switch of the turned on thyristors. The produced output currents of theses inverters have a square wave form (square-wave inverters). for reducing (or removing) the harmonic content, compensation equipment and output filters are required. A microprocessor is used to produce the trigger pulses for the switching of used thyristors. By delaying the trigger pulse (phase angle control), it is possible to implement MPP tracking.

Self-Commutated Inverters

A bridge circuit is used in self-commutated inverters. Many types of semiconductor switches can be used in these inverters:

- MOSFET (metal-oxide semiconductor power field effect transistors)
- bipolar transistors
- GTO (gate turn-off thyristors, up to 1 kHz)
- IGBT (insulated gate bipolar transistors).

Getting a sinusoidal output waveforms implies the use of pulse width modulation (PWM). See Figure 13.

Figure 12. Principle of grid-controlled inverters

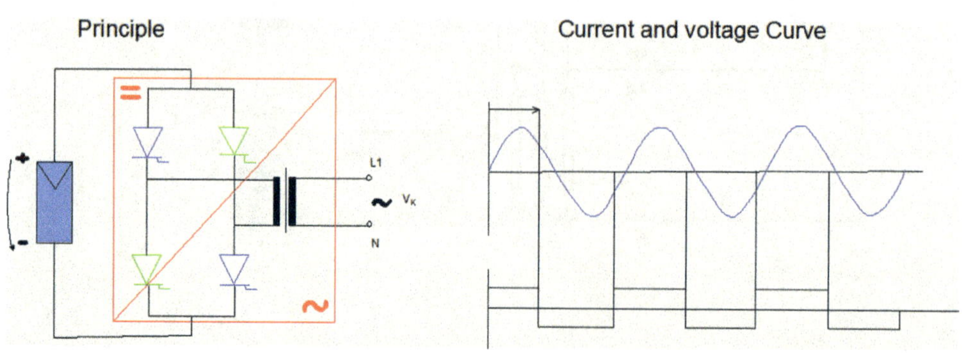

Figure 13. Principle of self-commutated inverters

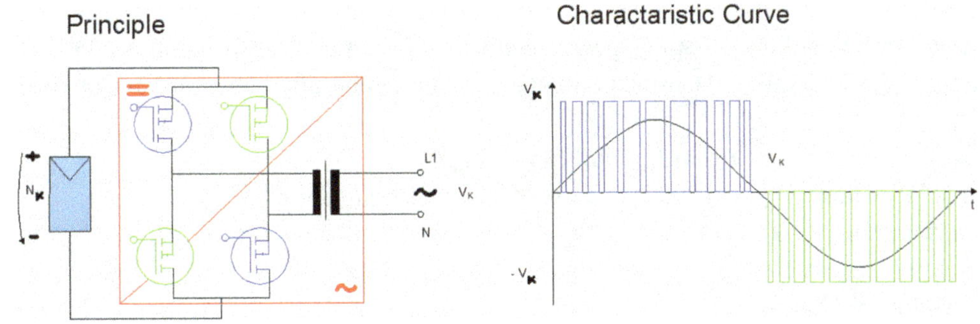

The fast switching of the power-switching devices on and off at a frequency in the region of 10-100 kHz, pulses are formed, the duration and spacing of which are shaped to correspond to the sine wave. Thus, there is good agreement between the power being fed in and the sine wave of the grid, after smoothing with a downstream low pass filter.

Parameters, Characteristics, Curves, and Properties of Inverters

- **Converter Efficiency η_{CON}:** The conversion efficiency expresses the losses that arise when converting DC to AC current. In inverters, these comprise the losses caused by the transformer (in devices that have a transformer), the power switching devices, and by internal consumption controller, recording operating data, etc.

$$\eta_{CON} = \frac{P_{AC} \textit{effective output power (of the fundamental component)}}{P_{DC} \textit{ effective input power}} \tag{12}$$

Overload Behaviour

The important factor for dimensioning when selecting inverters is the overload behaviour of the inverter. Especially in systems without the perfect alignment or systems with fractional shading, it can make sense from an economical and technical point of view to undersize the inverter somewhat. In order to have certainty in planning, it is important to consider the inverter's behaviour under overload. Inverters commonly have three different methods of dealing with overloads:

1. Shift the operating point
2. Power limitation
3. Switch-off.

Inverters using methods1 and 2 can be used variably even with slight undersizing. In method 3, an undersized inverter would react by switching off in the case of an optimum irradiance situation and the accompanying overload.

Recording Operational Data

The data can either be read on a display, performed by LEDs and/or can be sent to a PC. This enables the PV system to be observed and estimated. Data recording generally covers the following values:

- Input side: voltage V_{DC}, current I_{DC}, and power P_{DC}
- Output side: voltage V_{AC}, current I_{AC}, power P_{AC}, and frequency f
- Inverter operating time
- Generated electricity
- Device status and faults.

The number of values recorded is anywhere between one value per minute and five values per day. Operating time and energy volume are listed as daily, weekly, monthly and/or yearly values. The data are either saved directly or sent to a data logger and output on a computer. Data are estimated using suitable software that manufacturers provide for their devices. The storage capacity of commercially available devices is between 28 and 450 days. Some devices demand an external data logger, or a computer takes over this function. Most inverters have a serial or parallel PC interface communication protocols such as an RS-232 or RS-485 interface. The increasing modularity of inverters has led many manufacturers to provide external data loggers with multiple communications interfaces. These allow the data from several inverters to be saved and evaluated centrally (Alhelou, et al. 2015). Automated system monitoring concepts are also increasingly being realized. Fault-signaling contacts enable the output of acoustic alarm signals or visual signals on display units, as well as messages by fax, computer, e-mail, text message or internet.

Figure 14. Module inverter with operational data capture and PC interface

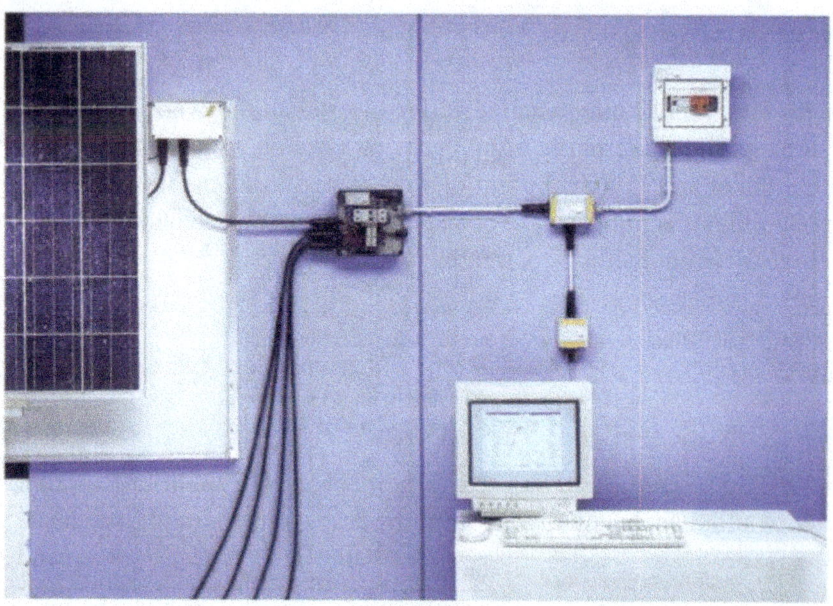

Further Development in Inverter Technology

Multi-String Concept

The use of inverters having high input voltages, particularly string inverters, led to energy losses in some system concepts. In shaded systems, or systems having different generator orientations, there were high energy losses. One way for reducing these losses is known as the multi-string concept. Here the system is designed so that modules having similar irradiance conditions are wired together in one ‘string’. Each string has its own DC-to-DC converter with a separate MPP tracker for operating the respective string at its maximum power. The strings’ MPPs can therefore be different. Via a DC bus, a constant DC voltage is provided via the DC-to-DC converter of the actual inverter unit (Figure 15).

Master-Slave Concept in Low Power Ranges (Team Concept)

The development of the optimizing energy depends on the master-slave concept. It is used for large inverters (over 20 kW), was developed for inverters in low power ranges (up to 5 kW). Several small inverters work together as a master-slave unit. The master device works when irradiance is low. The first slave device switches in, as soon as the power limit of the master device is reached as irradiance increases. As soon as its power limit is reached, the subsequent slave apparatus switches in. This concept allows achieving an optimum efficiency. The characteristic curve of the master-slave unit includes the characteristic curves of the individual inverters, and especially in the low power range has greater efficiency than an individual device with the same overall power. It is possible to split the individual inverters of the master-slave unit to different sub- generators or strings, so that different MPPs of the individual inverters are possible as operating points. One manufacturer also calls this concept the ‘team concept’.

Figure 15. Multi string inverter

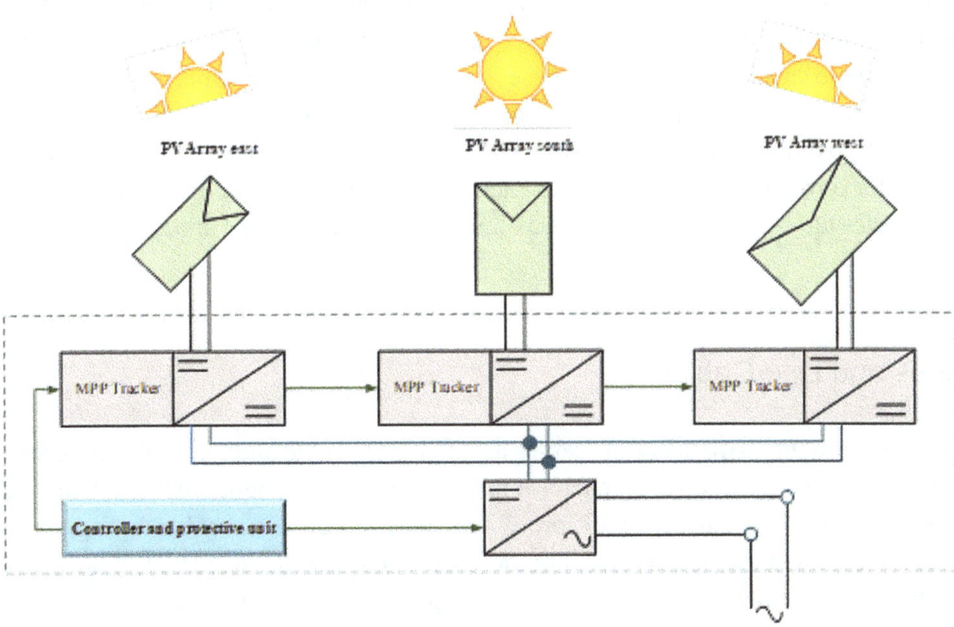

Figure 16. Efficiency characteristic curve of a master-slave unit with three inverters

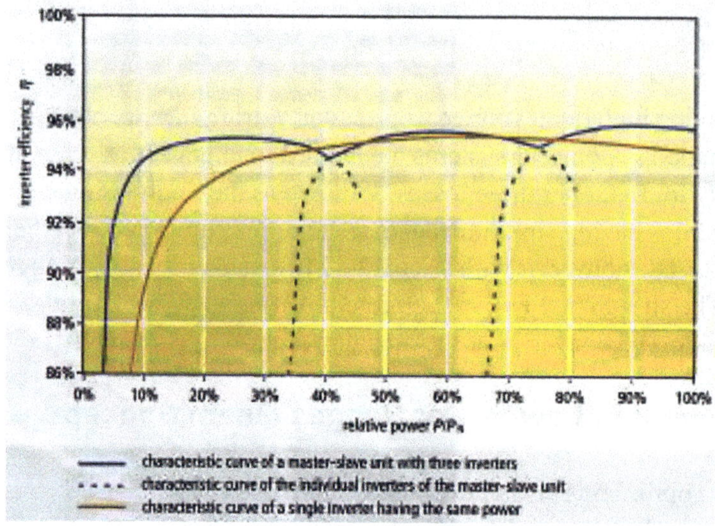

Three-Phase Concept in Low Power Ranges

A three- phase feed is a possible future method for optimizing devices in low power; however, this may not be available to all countries. The anticipated advantages, as well as the high efficiency, are an improvement in grid quality, the simple circuitry, and the resilience and long life of these devices.

Further Developments

Multiple small inverters are joined together on the DC or AC side. This enables several smaller inverter units to be wired together in a modular system to give a demanded higher output. As well as the power components, this modularity also applies to several functions that the devices can be equipped with optionally according to customer necessity. These include, in particular, displays, operating data recording, data logger, communications interfaces, protective functions and so on. In this way, for example, by using communications interfaces, the operating data of multiple inverters can be monitored centrally. Concepts are under development in which, for each string or PV module, a separate MPP tracker and DC-to-DC converter are connected via a DC bus line to a central inverter unit.

Cables

Module Lines and Line Cabling

We can use string cables or termed module cables as AC supply cable and DC main cable and in internal wiring of individual modules in a solar generator and in the generator junction box. This kind of cables is usually used outdoor so the earthing fault-proof and short-circuit-proof cable-laying are assured. We can't place the positive and the negative conductor in the same cable. The cables should be ultraviolet-resistant and weather-resistant, and capable of working with a large temperature range as the cables

Figure 17. Inverter designed with modular technology

Figure 18. Various modules and string cables

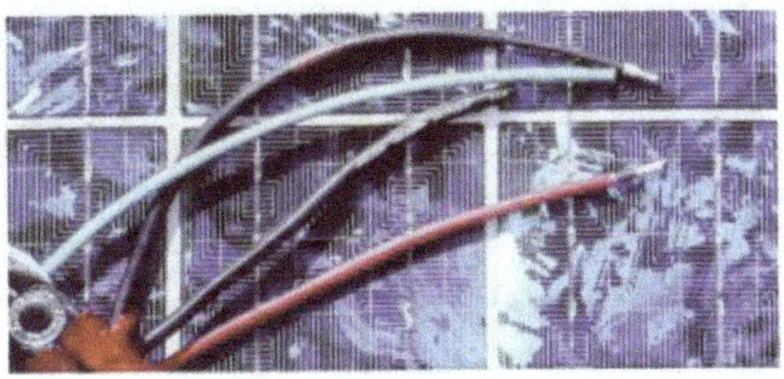

behind the PV array should be rated for 80°c minimum. In the light of this the double insulation single wire cables are considered ideal as they are proven reliable as shown in Table 3.

The module junction boxes enable wires with a cross-section of 1.5-6.0 mm^2 to be clamped. Table 4 lists examples of some module cable types from various manufacturers and their characteristics. The cables are often supplied in the two colors red and black, so enabling a more understandable generator set-up.

Installation Material

Cables should be placed using suitable fixing materials. Installation materials such as cable ties and cable clips should of course also be weather-resistant. The following fixing options can be used: protective pipe, ribbed pipe, cable duct, cable ties, cable clamp clips and cable clips (nailed variety).

Table 3. Requirements for cables

Mechanical Stability	Compression, tension, bending and shear resistance
Weather resistance	UV and ozone resistance for unprotected routing in open air, and resistance for heat and cold(design temperature 70°C on roof, 55°C in loft
Earth-fault-proof & short-circuit-proof installation	Individual line with double insulation

Table 4. Some module cable types and their characteristics

Module Cables for Outdoor Use	Manufacturer	Temperature Range
AEG-Solar module cable 450/750 Vm H07RN-F 1×25 mm^2 (or 4 or 6 mm^2)	AEG	-35°C to +90°C
RADOX 125 SOLARKABEL 600/1000V, 1×25 mm^2 (or 4 or 6 mm^2)	Huber+Suhner	-25°C to +125°C
Siemens solar cable 1800/3000V, 1×25 mm^2 (or 4 or 6 mm^2)	Siemens	-25°C to +90°C
TITANEX 11 450/750 V, H07RN-F 1×25 mm^2 (or 4 or 6 mm^2)	ConCab-Kabel	-35°C to +85°C

DC Switch Disconnector

In most electrical systems of building such as special installations or sites and PV power supply systems, an reachable load switch is required for connecting both of the PV generator and the inverter. The DC main switch should be double-pole and have the ability of load-switching. It must be rated for the maximum open- circuit voltage of the solar generator (at -10°C) and for the maximum generator current (short-circuit current under STC). Usually the DC main switch is placed in the generator junction box. When selecting the switch one should ensure that it can switch the relevant direct current. For reasons of safety, it is preferred to mount it directly before the inverter. The use of touch-proof plug connectors for electric insulation is prevented because of the risk of an electric arc hazard when the direct current connector is disconnected under load.

Figure 19. DC main switch

AC Switch Disconnector

It must be double-pole, obviously classified and lockable in the 'off' position only. In the case of if mounting the inverter in a separate chamber, there is a need for a second isolator nearby the inverter.

Miniature Circuit Breakers (MCBs)

Line circuit breakers are over-current secured devices that after they are triggered they have the ability to be switched back on. Generally an AC isolators are used as automatic circuit breakers. In case of an overload or short circuit happens, these breakers function is to isolate the PV system from the electricity grid.

Earth Leakage Circuit Breakers (Residual Current Devices RCDs)

They are used to measure the difference between the forward flowing and return conductors in the electrical circuit. The RCD isolates the circuit by switching it off within 0.2 seconds, if the value of this difference exceeds 30 mA. If there is an isolation trouble, or a contact between one of the conductor with earth or body, then the RCD will stumble. This is not imperative all the time as most inverters will switch off once there is an earth fault.

Accumulators/Batteries

The solar power generated during the day is for the large part not necessary before sunset, and therefore has to be internally stored. The most common batteries in solar power installations are lead-acid accumulators. These batteries have high efficiency in treating large and small values of charging current and they also have low cost. Storing capability in PV systems are usually between 0.1-100 kWh, even though a little number of systems in the MWh domain have already been applied. Lithium-ion, nickel-metal hydride and nickel-cadmium are other commercially obtainable type of accumulators. Mainly uses for these batteries are in small machines like clocks, radios, torches and laptop computers, but are not use in solar power installations of the magnitudes dealt with here.

Battery Parameters

Parameters that are used to characterize batteries are:

- Voltage
- Capacity
- C-rate
- Battery efficiency
- State of charge and depth of discharge
- Cycle lifetime
- Temperature effects
- Ageing

Figure 20 shows a Ragone-chart, where the particular energy is plotted against the intensity of peak power. Since it is a double logarithmic chart, the comparison of several chart storage techniques with very diverse storage properties in one plot will be possible. Depending on the application, a high energy density and also a fairly high-power density are required in solar energy application s. For instance, because of their very poor energy density, capacitors cannot be used. Of course, the most popular storage technique for short term to medium term storage, is the battery. Batteries have both the correct values of energy and power density to fulfill the daily storage requirement in small and medium-size PV systems.

Charge Controllers

The system voltage of the PV generator must be coincided to the one of the accumulators in stand-alone systems; the usual system voltages are 12, 24 and 48 V. The accumulator voltage value must be less than the charge voltage value. For instance, it can be up to 28.5V when using a 24 V battery. A nominal voltage of 15-18 V could be supplied by using standard modules made of Crystalline with 36-40 solar cells. The nominal voltage must be higher than the batteries' charge voltage so that the MPP voltage at higher temperatures is sufficient to charge the batteries. Furthermore, voltage losses occur through the cables and the string diode, which are usually limited to around 1-2%. At low temperatures, the modules' MPP voltage is approximately 21V and the open-circuit voltage up to 25 V, with the result that the high-voltage disconnection of the battery can be exceeded. For gauging the accumulator voltage value and protecting it from overcharging, the charge controller is used. This is reached by any of the following methods:

- Using MPP charge controller for the purpose of voltage adjusting.
- Short-circuiting the PV generator with a shunt controller
- As it occurs using series controllers, switching out the PV generator when the high voltage disconnect value is surpassed.

Cloudy weather with low amount of irradiance cause breaking down the PV voltage, with the result that the battery discharges via the generator. A reverse-current diode is used to avoid this disadvantage. This is usually integrated in the charge controller.

Charge controllers have to be flexible to ensure an optimized operation of accumulators over a long service life. The value of charge level of the accumulator determines the upper charge and lower discharge

Figure 20. A Ragone chart of different energy storage methods. Capacitors are indicated with "cap"

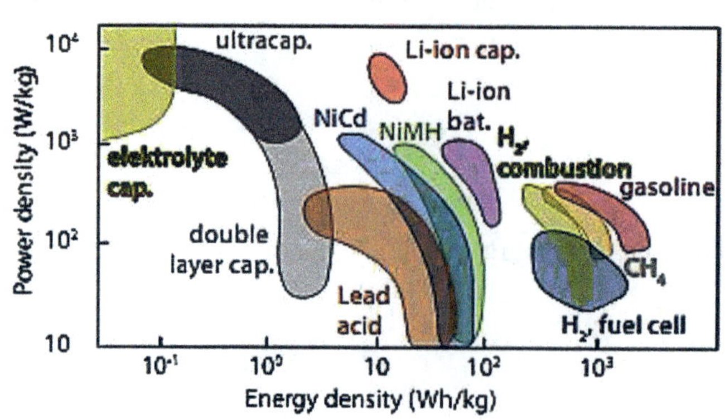

voltages are dependent on. Furthermore, the properties of various battery kinds (wet and gel batteries), the temperature and, rather probably, the battery age all need to be taken into consideration. A temperature sensor connected to the charge controller is used to observe the battery temperature. extending the accumulators lifetime is mainly affected by using high-quality charge controllers which can make a great difference. The main functions of a recent charge controller are:

- Deep discharge protection
- Charge state information.
- Optimum accumulator charging
- Overcharge protection
- Preventing unwanted discharging

Figure 21 shows a charge controller that responds flexibly to the PV generator and accumulator system. The charge controller is able to discover the state of charge of the accumulator and can adjust to the available capacity, the age and the temperature of the accumulator. In addition, using pulse width modulation (PWM) for the charge current ensures the safe charging of the battery.

Series Controllers

Once the maximum value of charge voltage is gotten, a series controller uses power semiconductor or relay to disconnect the module power, and once a defined voltage value drop the controller switches the module back on. This process causes long-lasting forward losses in addition to an fluctuating status of fixed switching procedures nearby the upper charge voltage. To remove these shortcomings, a charge controller using continuous regulation was established as shown in Figure 22.

Shunt Controllers (Parallel Controllers)

As soon as the upper voltage value is attained, the shunt controller continuously minimize the value of module power. However as the module continues generating power, the of the module's power extra amount is utilized as short-circuit current in the module – that is, it is turned into heat producing minor surplus warming. Since charging using this method is safe and fast, it is considered to be an ideal battery chagrin method Figure 23.

Figure 21. Charge controller for outdoor installation with charge level display

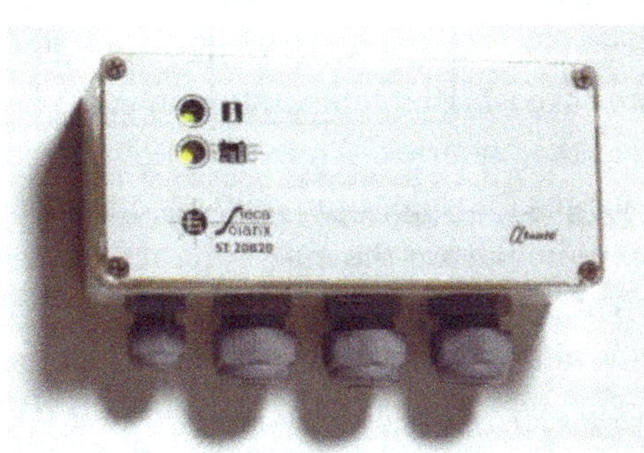

Figure 22. Principle of series charge controller

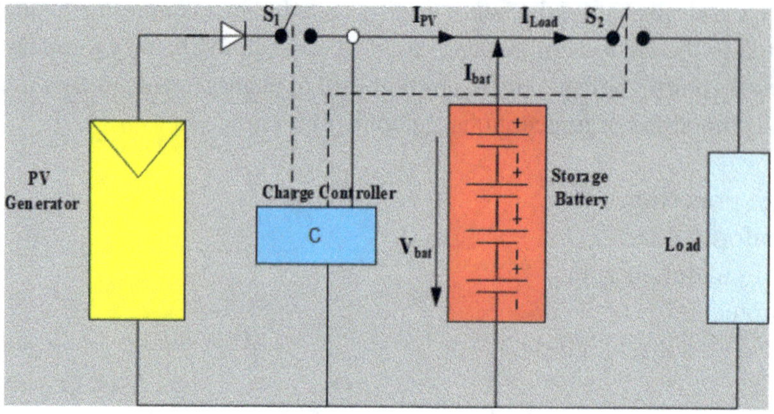

Figure 23. Principle of shunt charge controller

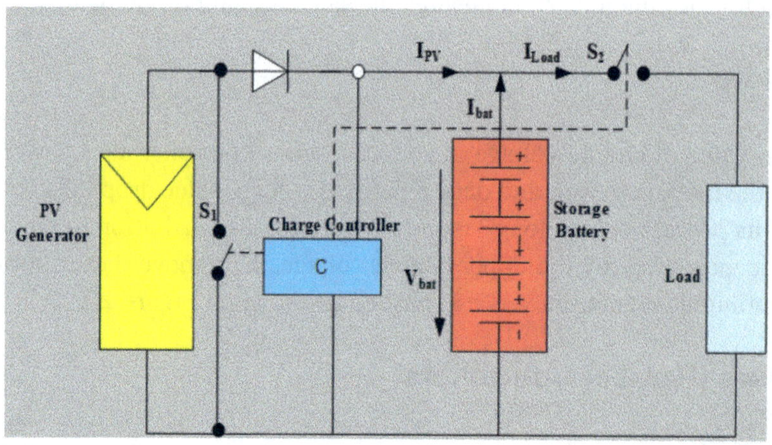

Depletion Protection

A depletion protection with relays is inserted into most charge controllers (S_2 in Figures 22 and 23 and S_1 in Figure 24) to isolate the loads from the battery. It is suitable that the power needs to be switched off if there is a signal, in enough time before reaching the switch-off point. In such cases it is convenient to use a built-in reset button. Normally the charge controller is placed in direct close to the battery, so the temperatures of both devices are very similar. Only in the event that the battery is extremely heavily loaded will it heat up and therefore have a higher temperature than the charge controller. This eventuality can be ruled out by suitable sizing the batteries. An external temperature sensor mounted on the battery can prove useful here too. Charge controllers can only accommodate a limited current, both on the module side and on the load side. A display for the most important electrical parameters such as charge state, battery voltage, charge current and load current is incorporated into many devices. To protect the sensitive power electronics, a convenient fuse is built in. Generally, the same limit values are set for the maximum permitted module current and the maximum permitted load current, which means a common fuse is built in for both currents.

Figure 24. Principle of MPP charge controller

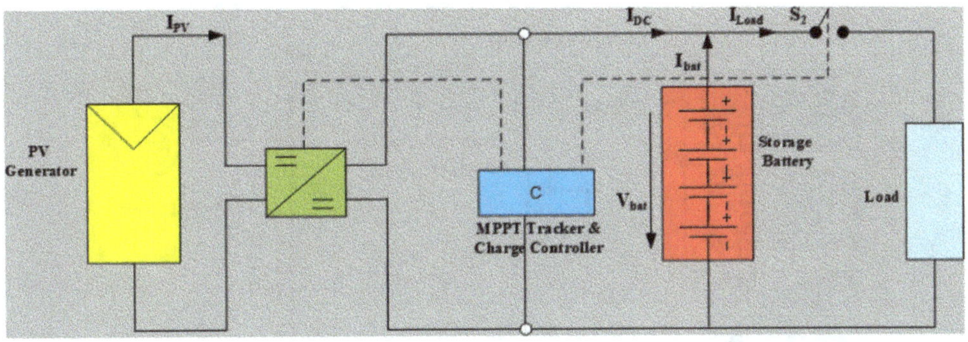

MPP Charge Controller

The series and shunt charge controllers mentioned above do not always make the perfect use of the available solar energy. Then, by means of an MPP tracker the power losses can be avoided. The essential part of MPP charge controller is a regulated DC-to-DC converter. MPP tracker defines the MPP power and carries out regulation, that passes about every 5 minutes along the I-V curve of the PV generator and. Then and there the DC-to-DC converter is set so that it adjusts the PV generator optimum power value to the value of accumulator charge voltage (Alhelou, et al. 2016).

The efficiency of the DC-to-DC converter is around 90-96%. Due to energy reasons, the use of an MPP tracker only makes sense with generator powers of 200 W_p and above. With lower-powered generators, the converter's conversion losses are generally higher than the gain from the controller. Since the circuitry involved is more complicated, the price of the MPP charge controller is somewhat higher. This means that, for reasons of economy, at present MPP charge controllers are often used only with PV generators of 500 W_p and more.

Stand-Alone Inverters

In a PV stand-alone system, the energy storage in the accumulators and the operation of numerous consumers is realized using direct current. To be able to use conventional 230 V AC appliances as well on a DC grid, stand-alone inverters are used (as shown in Figure 26).

Figure 25. MPP charge controller

Figure 26. Stand-alone system with stand-alone inverter

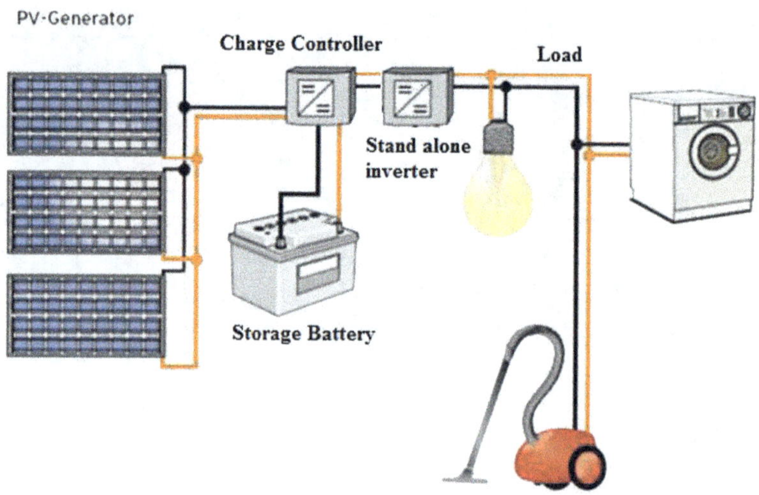

The objective of a stand-alone inverter is to enable the operation of a large range of different kinds of consumers. These vary from harsh erection machinery through local uses to sensitive electronic apparatus used in communications technology. The next demands may be made of a stand-alone inverter:

- Producing an alternating current is as sinusoidal as possible, with stable frequency and voltage
- Very good efficiency of conversion, even in the range of partial load
- Tolerance against battery voltage fluctuations
- Deep discharge protection for the battery
- Switch-on and starting sequences' high overload capability
- On the output side, it must be protected against short circuit damage
- Economical standby state with automatic load detection
- High electromagnetic compatibility (good EMI suppression)
- Small harmonic content
- Overvoltage protection.

Two different inverter models control the market: the sine-wave and the trapezoidal inverter.

Sine-Wave Inverters

The requirements for a stand-alone inverter are best performed by sine-wave inverters. Therefore, the pulse width modulation technique is used. These inverters are appropriate even for running sensitive electronic apparatus. They are better than trapezoidal inverters but have higher cost and more complexity of their circuitry.

Figure 27. Stand-alone inverter with optional charge control function

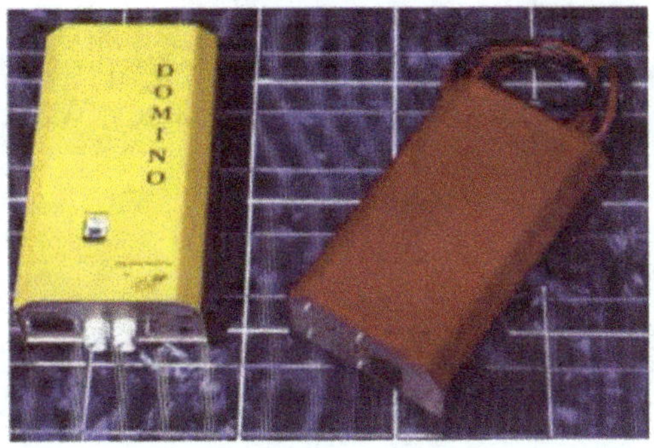

Trapezoidal Inverters

These inverters are cheaper than sine wave inverters and are gradually being pushed out of the market. In these inverters, the DC current is converted into a 50 Hz alternating current with square wave form and stepped up using a transformer to a voltage of 230.

FACTORS AFFECTING PV OUTPUT

PV system produced power amount is very small due to the relation of this power with the strength of sunlight hitting the surface of solar array. The overall PV output are affected by some factors which are discussed below.

Dirt and Dust

As Dirt and dust can stack on the surface of solar module, they prevent some of the sunlight causing output reduction. An annual dust reduction factor of 93% usually used. Sand and dust can also cause corrosion of the PV surface decreasing the output power value to more than 10% which affecting the system's running performance (Thornton, 1992; Chaar, et al. 2008; Mohandes, et al. 2009; Al Hanai, et al. 2010)

Temperature

The output power value of a PV system decreases by increasing the temperature of the module. For modules made of crystalline, the California Energy Commission (CEC) proposed a representative temperature decreasing factor with value of 89% in the mid of spring or in a autumn day under full-light circumstances.

Snow and Clouds

The positive effects of low temperatures on solar panel power production in colder climates can be countered by clouds and snow that decrease solar panel efficiency. While there is little to be done about cloudy skies.

DC–AC Conversion

As shown earlier, because of converting the PV array power back to AC, some power is being wasted, in addition to wiring losses during the conversion process. Common inverters used have peak efficiencies of about 88–90%.

Optimized Methods for Enhancing the Impact of Factors Affecting PV Output

Conventional Methods

As the solar cells tend to produce less voltage in higher temperatures and more of it in lower ones due to their characteristics. Then using the MPPT to change the temperature of the solar cell will make it more efficient. The efficiency of energy conversation will be boosted by decreasing reflection of the incident light. And another way of improving the efficiency is to use a maximum power tracker which change the equivalent load take in the solar array and adjust the working point of the array. The optimum factors give the possibility to achieve greater possibilities out of solar electricity while maintaining the cost at lower rates. Cooling the panels by dousing them with water may enhance power efficiency, though economic efficiency would depend on whether water was inexpensive and plentiful. Numerous steps can be taken to help reducing the snow accumulated on solar panels surfaces. Sloping the units of solar tracker or re-setting up fixed ground array units at a steeper angle can make snow to slide off faster. Using of a roof rake, which looks like a push-broom with a long, angled handle, helps in removing snow from both ground and roof-mounted panels. Sweeping off the snow with a roof rake can also prevent accumulation that might cascade dangerously off the slippery panels (Worcester, et al. 2017).

This study presents a new method to reduce the soiling and accumulation of dust on the surface of PV module. The method can briefly described by a moving cover with a cotton cloth attached firmly to the backside of the cover. At sunset, the module should be covered and the cloth wipes down part of the dust and the opposite happens at the sunrise. This method ensures that the exposure time of the covered module is reduced to less than half the time of the uncovered module, wiping the module surface two times a day, reduce water consumption for cleaning, less complicated than robotic systems and can be designed for a large panel or long string of modules.

Advanced Methods

Separate Photovoltaic Panels Cleaning System consists of a mechanism, which pursues much in the way of a wiper blade on a car, is used to clean the solar panel. This "wiper blade" is adapted with an electrostatic cloth, related to a servo motor and interfaced with and operated by the Basic Stamp 2 (BS2) controller. In addition to this, a dust detecting system is used to set when it is best to clean the panels.

Finally an emergency system is equipped to provide an automatic shut off in the form of a button and a digital thermometer (Gabriele, et al. 2012).

Development of an automatic robotic cleaning system for photovoltaic plants. The aim of this project was to preserve the PVs to generate power at high efficiency. The i7 cleaner robot which was firstly designed to clean homes was studied and new software was developed to be able to effectively work on PV panels. Then the robot structure will be developed so that it will be ready to use. The developed software reduced the cleaning time for a 1x2.20 meter panel from eight minutes to two minutes by setting a path on the surface of the PV panels for the robot, instead of just a semi random movement. Whereas the old software could clean 11 panels in 90 minutes, the new program can clean 37 panels in the same time (Albadawi, 2014).

PV SYSTEM DESIGN

Many steps must be considered to achieve the aim for a solar direct electricity generation system or photovoltaic system is to provide high-quality, reliable, and clean electrical power.

Design Procedures

Realizing the power demand of all loads, the number of used hours per day and the functioning voltage value in assessment of the system's load which is the first step in designing PV system procedures (Sandia Nat. Lab., et al. 1988). The power demand is calculated from the given operating voltage and load ampere-hours for each load. For a stand-alone system, the system voltage is the possible necessary by the biggest load. In case of AC loads control, the voltage of the DC system should be selected to be convenient with the inverter input.

Power-Conditioning Unit (PCU)

PCU choosing plays a major role in determining the performance of the system and its economic cost. This choice of PCU is related to the required output waveform, which depends in turn on the type of used converter, as well as on the filtering techniques used to eliminate unwanted high harmonics.

Several factors must be considered when selecting or designing the converter:

- The power conversion efficiency
- Power factor
- Rated power
- Voltage regulation
- Voltage protection
- Duty assessment, the duration the inverter can provide maximum load
- Frequency requirement
- Input voltage

Battery Sizing

The necessary accumulator storage amount count on the load energy requirement and on weather patterns at the site. The converters output must achieve the low cost and required energy quality (cost-quality effectiveness).

Criteria for a Quality PV System

The criteria for quality PV system are as follows:

- Good control circuit to reduce electrical losses, overcurrent protection, switches, and inverters
- Good charge controller and battery management system, should the system contain batteries
- Providing electrical power and energy implies suitable sizing and orientation

Optimization Based Power Conditioning Unit for Improving the Quality of the System

The electrical distribution grid contains various renewable energy resources, such as wind, solar and others. Due to this, there is an identical increase in power quality problems and power irregularities on the distribution grid. Many methods are used to reduce the irregularities and improve the power quality of the distribution grid.

Super Capacitor Integration

It is useful for handling the output irregularity and for improving the voltage sag and swell reparations, to integrate a super capacitor, because it gives the power conditioner active power capability. Super capacitors have low energy density, high power density, and fast charge/discharge rates, which are all ideal characteristics for meeting high-power low energy events like grid intermittencies, sags/swells. To ensure a stiff DC-link voltage, a bidirectional DC–DC converter is used to integrate the super capacitor into DC-link of the power conditioner. This integration helps in providing active/reactive power support, intermittency smoothing, and sag/swell compensation (Injeti & Padma, 2015).

Compensators and Filters

The producers of photovoltaic system are aiming at removing the disturbance before it comes to the grid in spite of existing numerous technical methods in the distribution grid used to enhance power quality like constant var compensators (Kumar & Kumar, 2013; Akhtar & Shamsudheen, 2012). Therefore, making PV system safe to connect to the grid imply using voltage source converters and filters with their controllers. Injecting a reparations current that cancels harmonics in the line current is the way used by active power filters to cancel out unwanted harmonic currents. By introducing opposite equal harmonic compensating current, shunt active power filters compensate the harmonics of load current. In general, by using four-leg converters, four-wire active power filters have been conceived which has confirmed superior controllability than the classical three-leg four-wire one (Chen, et al. 2008). Figure 28 depicts the considered active power system and its grid interconnection. As it is shown a three-leg four-wire voltage source inverter form this system. In such application kind, the voltage source inverter works as a source of current controlled voltage. An inverter coupled to a dc-link connects the PV system to grid. The DC-capacitor decouples the PV system from the grid and also allows independent control of converters on either side of the dc-link (Alhelou, et al. 2018).

The voltage source inverter is a key element of a photovoltaic system as it interfaces with the renewable energy source to the grid and delivers the generated power. The device developed for compensating non-linear and unbalanced grid loads and extracting PV system maximum power is called the controller or the conditioner which in most cases use neuro-fuzzy control (Vazquez, et al. 2008).

Voltage Source Converters (VSC)

It is another kind of power quality reduction. As it is illustrated in Figure 29, VSC is a power electronic apparatus connected to the system in shunt or in parallel. VSC converts the DC voltage through storage instruments into a set of three phase AC output voltages, and it can produce a voltage in sinusoidal shape with any needed magnitude, frequency and phase angle. It also offers the ability of producing or absorbing reactive power. It is assumed to be in capacitive mode if the value of AC bus terminal voltages is smaller than the value of VSC output voltage . Therefore, the VSC will compensate the reactive power over AC system. An IGBT in anti-parallel mode with a diode is the used power switched (Archana & Kondaiah, 2012).

Figure 28. Active power system topology

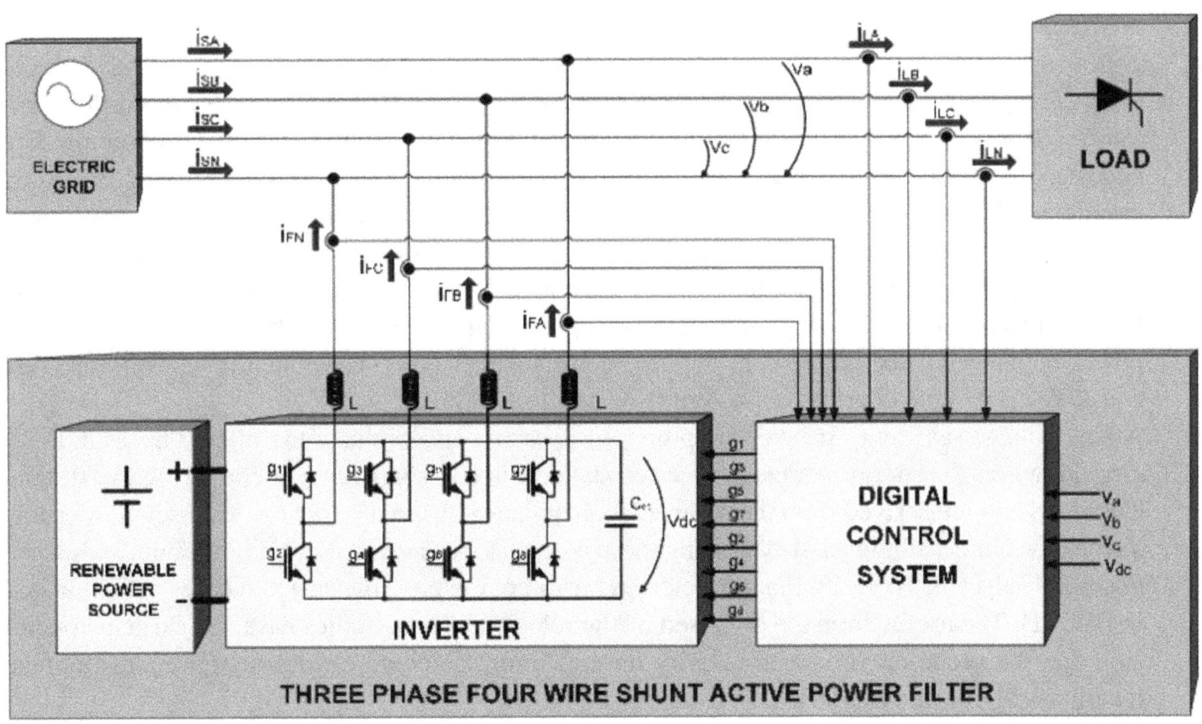

Figure 29. Voltage source converter

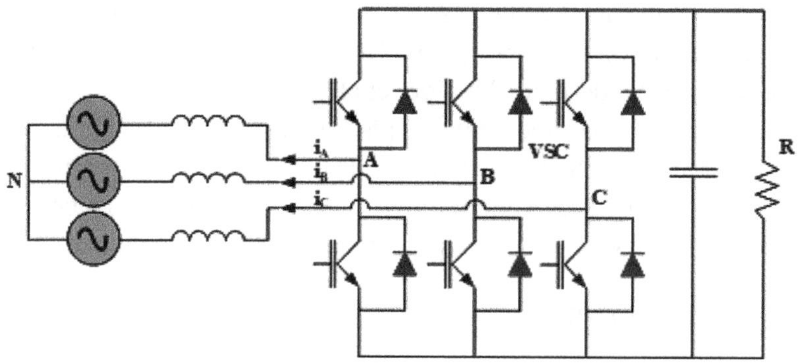

An Optimization Technique for Defining the Optimal Battery Energy Storage Size Used in a Specific PV System Integrated Into Smart Grids and Microgrids

- A control scheme using the optimal sizing of BESS-based PSO with load shedding scheme is suggested in order to enhance the system frequency after islanding happened. Depending on the results, BESS is shown a good performance in controlling load frequency and can be adopted for emergency control objectives when the system load shedding is unavoidable. Results show that the optimal sizing of BESS based PSO can gives better stability than the optimal sizing of BESS-based analytic algorithm. Furthermore, the optimal sizing of BESS-based PSO can also supply the possible voltage to the system much better in comparison with the optimal Sizing of BESS-based analytic algorithm. It is clear that the suggested algorithm provides faster, smoother and more secure system stability than the optimal sizing of BESS-based analytic algorithm from the emergency situation to the normal equilibrium state (Kerdphol, et al. 2014).
- A new approach to attain voltage regulation in distributed power systems in the subsistence of solar energy sources and battery storage systems is suggested. The objective is to find the minimum size of accumulator storage and its corresponding location in the network based on the size and place of the integrated solar generation. The proposed technique formulates the problem by employing the network impedance matrix to get an analytical solution instead of using a recursive algorithm such as power flow. The use of reactive power control to regulate the voltage regulation is not always an optimum solution as in distribution systems *R/X* is large. The minimum size and the best place of battery storage is attained by optimizing the amount of both active and reactive power exchanged by battery storage and its grid tie inverter (GTI) based on the network topology and *R/X* ratios in the distribution system (Nazaripouya, et al. 2015).
- A new voltage sensitivity indexes are proposed to search and evaluate the filtered buses in a microgrid, where the energy storage systems can be installed to share most effectively to the system. These indexes are derived from the reverse Jacobian matrix from the Newton Raphson power flow analysis. After calculating and evaluating these indexes for the studied microgrid, four candidates' buses are selected. To verify that the selected buses are the best, the selected cases are simulated in PSCAD. The results from the proposed methodology and case studies have a consistent conclusion that the proposed voltage sensitivity indexes imply potential candidates for energy storage locations (Montoya, et al. 2014).

OPTIMAL OPERATION OF PV SYSTEM INTEGRATED TO MICROGRID SYSTEMS

Smart Grid Structure

To create a smart grid many technologies must be combined. One part of an interconnected system is smart meters which receive the most attention from the public. A few examples include: advanced communication and control systems, new types of conductors, and various types of distributed electricity generation. Numerous other components are also in the process of design, development and implementation. It will take many years to fully implement the smart grid technology, which is very important technology brings essential possibility for a more flexible and resilient electric power system.

Communication Architecture

Figure 31 illustrates a concept of the smart communication control layers.

Core Smart Grid Technologies

To form a smart grid the distributor will likely look to ensure that the following components (technological stuff) are present.

Active Network Management (ANM)

The (ANM) term stands for the technologies which give the developed network the monitoring and intelligence needed to preform automatic management on functions like network restoration, voltage control and fault levels. The most critical part of the Active Network Management system is the presence of a reliable communication component between the network and the distribution management system along with a suitable software application that control the electric system operations. Using the ANM

Figure 30. Smart grid components

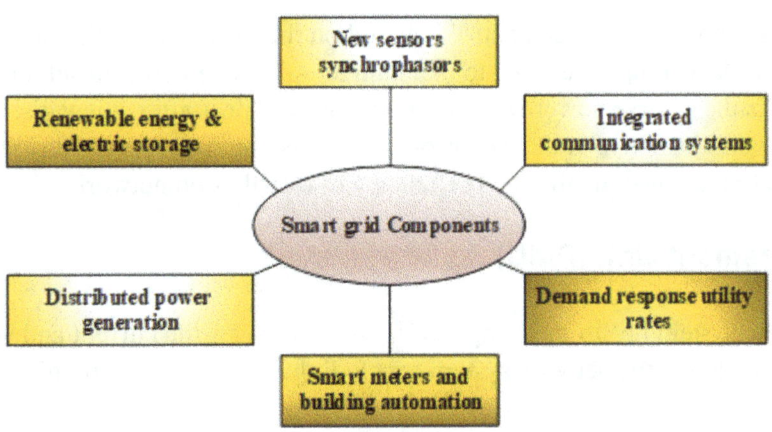

Figure 31. 5 layers of communication in a microgrid

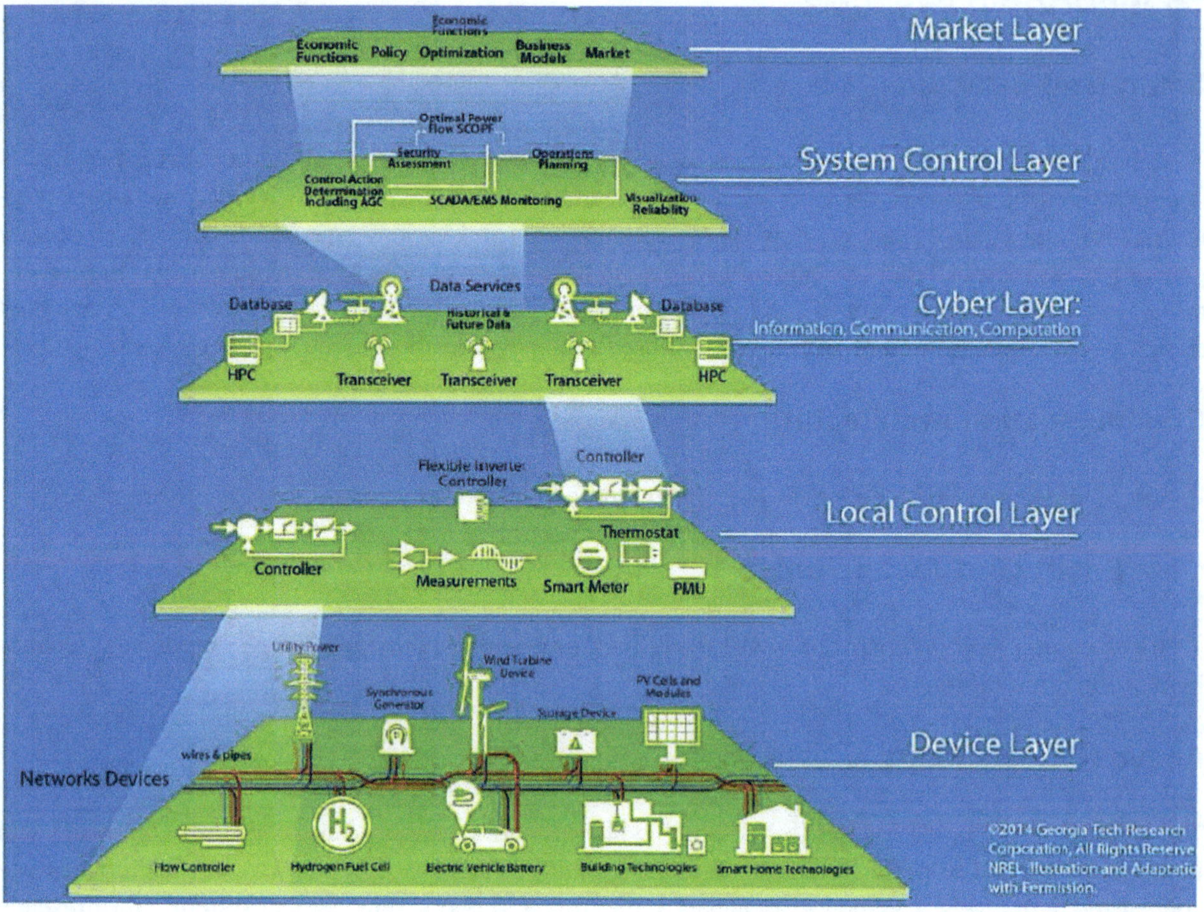

as the optimization method grant the grid distributer the ability to make multiple distributed generators connection as its demonstrated in Figure 32 and it's a very inexpensive way to improve the network.

Automatic Voltage Control

The principle of the automatic voltage control is based on the idea of installing intelligence in the network to keep track of the voltage levels and to maintain its value within the preset limit by automatically adjusting control. As the reserve power flow conditions may appear after a new distributed generator is installed in the network the active voltage control have to get smarter. As conclusion the active voltage control will add up to the power quality and efficiency of distribution network.

Phasor Measurement Unit (PMU)

It's considered as the system health metering device. As it samples the current and voltage many times per second in any location in the network system concluding a near real time monitoring of the behavior

Figure 32. Active network management

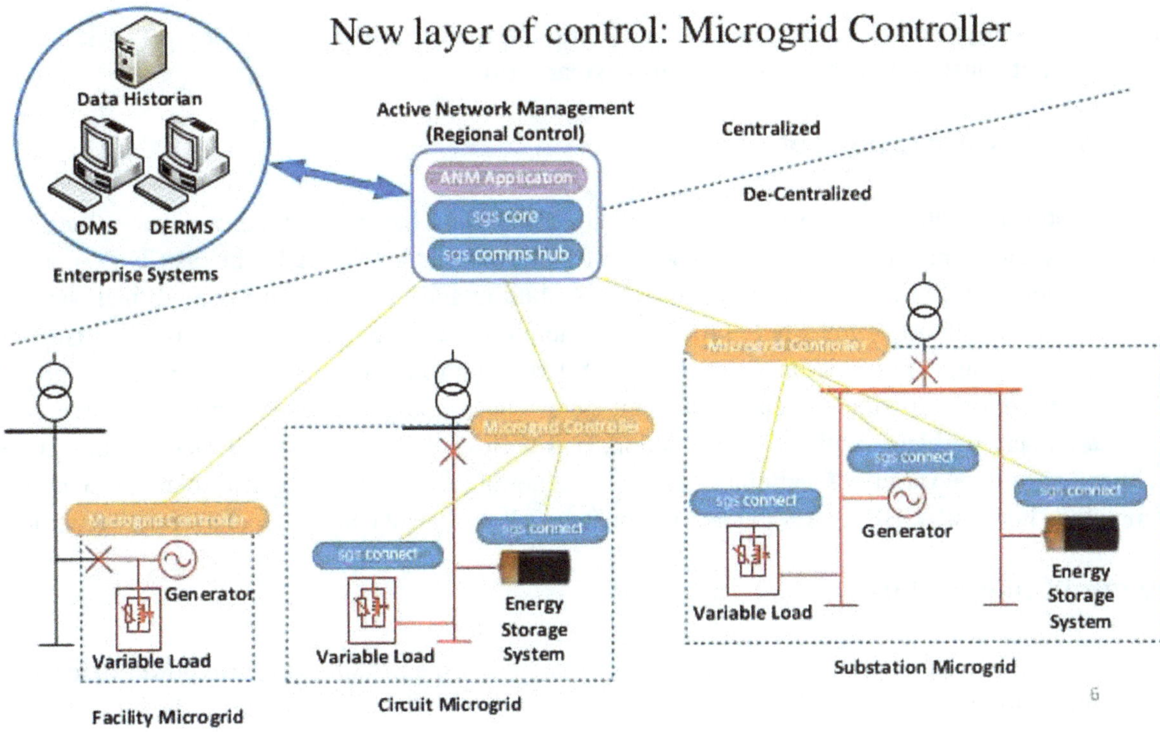

of the power systems. So it provides a much more advantages compared to the traditional Supervisory Control and Data Acquisition.

Dynamic Line Rating (DLR)

The DLR give the network the ability to boost the capacity of any network without expanding the existing infrastructure. As example of the DLR work we can notice the strong windy conditions increase line capacity by offering better cooling. The DLR keep track of weather conditions and line parameters. Then it can mark at any time the capacity of any network section and use this data to help the network to adjust its function to peak performance.

Intelligent Electronic Device (IED)

The Intelligent Electronic Device used in the network are built to provide measurement and recording capabilities of power quality, substation protection and control in one device.

Reactive Power Compensation

It's the ability to increase the available capacity of a network by absorbing or injecting reactivate power to control the voltage.

Peripheral Smart Grid Technologies

In the following part of this work the content is more peripheral to the distributor's network than the past one put it's just as important to form a truly smart grid.

Distributed Generation (DG)

the generator and delivery are usually stationed at central plant but in this type of network the DG is located in stations that are close to the consumers fig33. And that help to make the distribution process more efficient due to the fact that it cuts down distribution and transmission losses. Which saves up to 30% of the total cost of the whole process of generating electricity. The concept of DG brings the distributers both an opportunity and a challenge. If it was controlled in a proper manner it can provide a very useful master distributed generation, so it can encourage more of it, but if it was unharnessed it can make many problems with voltage function, power flow, voltage levels and thermal ratings. DG include a very wide variety of techniques including renewable sources like hydro, solar and wind with complained heat and (CHP) power plants. And on its smallest scale it can even include microgeneration.

Dynamic Demand (DD)

When using this technology the electronic devices that don't require time-specific power demands in the system will take place in balancing the system. System Balancing is defined as the art of keeping the lights on, and it's a task which depends on the transmission system operator (TSO). The TSO makes sure that there's enough electricity in each and every part of the network at any given moment. The TSO operates in the system as reaction to the system frequency which is the voltage of the network to keep it in the acceptable range. The whole process is based on the concept of adjusting the devices duty cycle by altering the time period in which it consumes power as a reaction to the changes in the system frequency in the network. And this type of intervention is immediate and automatic. Giving the TSOs the ability to work as a powerful balancing tool.

Figure 33. Distributed generation

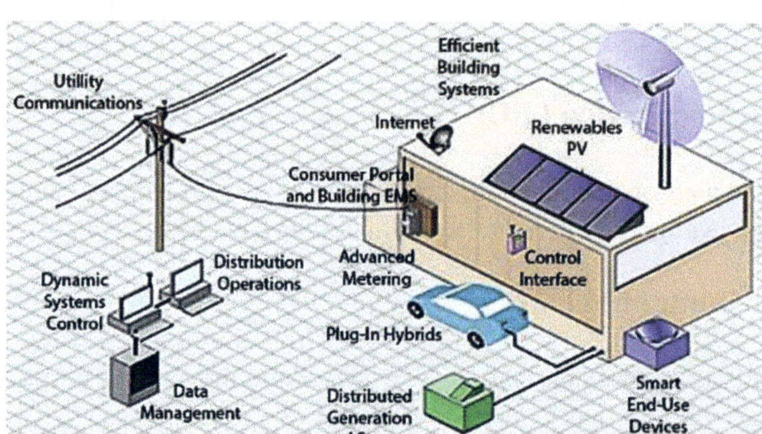

Grid Energy Storage

The difficulty of storing electric energy in large volumes is an age-old matter. In older models of networks this obstacle was overcome by always matching the demand value to the generation one in real-time, so no storage is needed. But in the networks, that depends on the renewable sources overlooking this problem is not possible because they're unpredictable. For example, the generation value exceeds the demand when using wind farm in case of a wind storm in the middle of the night, in other hand the demand value exceeds the generation value during night time when we use solar power generation. This problem needed to be addressed by storing the energy generated to be used when needed. One of the ways used to do this is pump storage which means using the extra power generated to pump water up to a holding tank and then releasing it through turbines to generate electricity in times of need.

One solution used for some time is *pump storage* in which water is pumped up to a holding tank when electricity is available and released through turbines to generate electricity at times of high demand. Problem is, pump storage facilities require a suitable location and are expensive to build, typically in the mountains – away from areas of high demand.

Research is on-going into new grid energy storage technologies including:

- Batteries, which are expensive to produce, costly to maintain and have a limited lifespan.
- Compressed air, which requires similar large-scale facilities as pump storage.
- Flywheels, only good for small scale storage.
- Hydrogen manufactured using off-peak electricity and then combined with oxygen to produce electricity at peak time but with lower efficiency than pumped storage or batteries.
- Superconducting magnetic energy storage (SMES), a means of storing energy in the magnetic field created by a direct current flow in a cryogenically-cooled superconducting coil. It works only for small amounts of energy and is expensive to boot.

Smart Appliances

Smart appliances are your traditional domestic white goods but with some added ICT cleverness. That cleverness comes in different ways – some smart appliances can be controlled remotely by you, the owner, by your smart home or by a third party (an energy services company, for example) so as to run when power is at its cheapest or shut down when peaks occur. Smart appliances can, therefore, help flatten demand by moving consumption from peak periods to off-peak periods – a process known as *peak shaving*.

Smart Houses

The definition of a smart house is a residence that have the ability of optimizing the energy consumption by communicating and coordinating number of smart devices through a smart hub. In this way it can help the owner of the house with many tasks. Not only the owner of the house will benefit from the smart house but it also offers more privileges to the distributors as it will involve them in working with the smart house instead of working with individual smart devices or consumers.

Smart Meters

This component will offer in-depth view of what's happening in the network to the distributor. As they offer full inspecting capabilities down to the end user's devices level instead of just the classical view that used to stop at the level substations of the power flow. And depending on its level of IQ, it can even give the distributor the ability to fully adjust the loads remotely which prove to be a very helpful tool of network management. In countries where metering is the distributer responsibility, setting a smart grid up is the reason for using smart meters.

OPTIMIZATION TECHNIQUES AND ALGORITHMS FOR MPPT AND OPTIMAL POWER POINT OPERATION OF SOLAR SYSTEMS

Reaching maximum efficiency needs the photovoltaic (PV) arrays to be work at their maximum power point (MPP). Therefore, implementing an MPP tracking (MPPT) scheme between the PV system and the load is necessary to obtain maximum power.

Maximum Power Point Tracking Components

Voltage Feedback Control

As the controlled variable is the PV array terminal voltage, voltage feedback controller changes the array terminal voltage and leave out the variation in the temperature and insolation level making the PV array to operate at its MPP (Rashid, 2011).

A simple control system can be implemented if the system does not use an accumulator to keep the bus voltage to a nearly constant level which is represented by using a DC-to-DC converter that transform the voltage level at a photovoltaic cell array to another voltage level at the load.

Feedback of the panel voltage and comparison with a constant reference voltage can be used to continuously set the duty rate of the converter to operate the solar panel at a predefined operating point

Figure 34. (a) Smart home, (b) smart meter

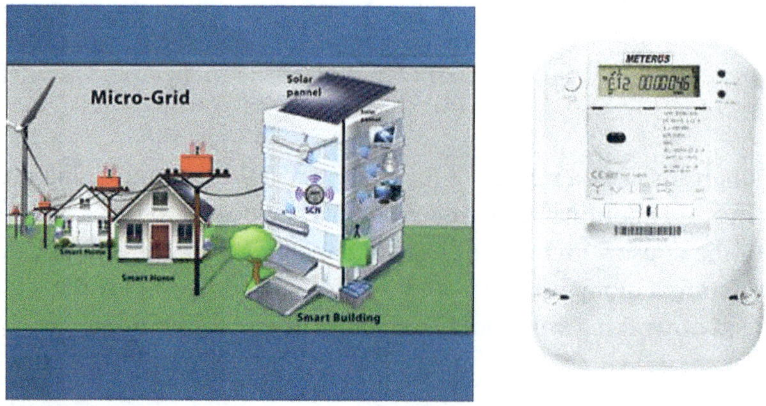

close to the MPP. This method makes it possible to operate a solar array under unknown or changing load conditions and still be able to choose a desirable operating point for the panel.

Power Feedback Control

In this method, the controlled variable is power delivered to the load. The value of dp/dv must be zero in order to reach maximum power. PV array quality doesn't affect this control pattern so far it increases power to the load and not power from the PV array. Fast shadows is considered one of many factors cause trackers to loose MPP momentarily. It is very critical to ensure that the time lost in seeking MPP again, which equates the energy lost while the array is off power point, is very short. On the other hand, to avoid energy loss, the tracker needs to respond quickly to the change of the lightning conditions. Therefore, the controller's ability to fast modify the system to work back at the MPPT is the most important feature.

MPPT Controlling Algorithms

Conventional Algorithms

There are many suggested algorithms to achieve MPPT:

1. **Perturb and Observe (PAO):** This technique is cost-effective and can be easily executed. Continuous tracking and high efficient ability of PV maximum power extracting are the main characteristics of this technique. As it has a slow tracking speed, this method may fail under quickly varying atmospheric conditions. As shown in Figure 35. the movement of the operation point is determined by measuring the derivative of power Δp and the derivative of voltage Δv ($\Delta p / \Delta v$).
2. **Incremental Conductance Technique (ICT):** This technique is very accurate and suitable for fast variation of climate conditions. In this algorithm the MPP reveal is accomplished by comparing the values of di/dv and $-I/V$ till it achieves the voltage operating point at which the incremental conductance value and the source conductance one are identical (figure 36).
3. **Constant Reference (CR):** Very simple, but due to the consideration of the variations of temperature and irradiation values, as well as the difficulty of selection of the optimal point it is not very accurate.

Figure 35. PAO technique

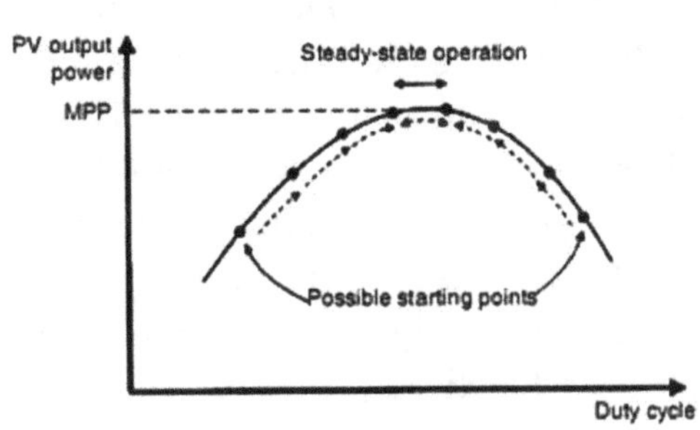

Figure 36. The slope "conductance" of the P-V curve

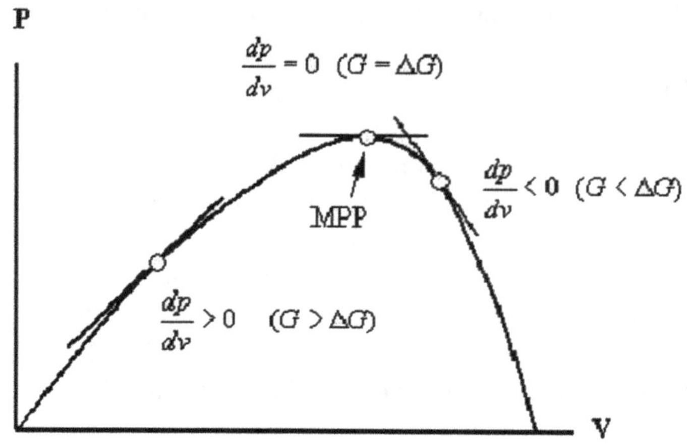

4. **Current Based Maximum Point Tracker (CMPPT):** Easy to carry out, needs additional switch to convert the power periodically among the set of PV, high cost. The main disadvantage connected to periodically adjusting requirements.
5. **Voltage Based Maximum Power Point Tracker (VMPPT):** This method is based on I-V characteristics and needs periodical adjustment.

The last two methods are considered fast and robust methods to evaluate MPP in various conditions of temperature and irradiation.

Evolutionary Algorithms

There are other automatic techniques of MPPT such as: Fibonacci line search, ripple correlation control method, ANN and Fuzzy logic.

Fuzzy Logic Control

Because of its effectiveness in MPP applications, where it improves the microcontroller processing power and reduces costs, Fuzzy logic control has become increasingly popular in MPP applications . The three stages of fuzzy logic controllers are as follows: fuzzification, rule base table seeking, and defuzzification. In fuzzification stage, based on a number of defined membership functions, numerical input variables are transformed into linguistic ones (Hilloowala & Sharaf, 1992). Usually, as the number of used membership functions is great, the controller will work more rightly (Karbakhsh, et al. 2012; Mahmoud, et al. 2000). An error signal, E, and the change in error; ΔE are the fuzzy logic controller input. The user has the ability to freely choose E and ΔE computation method. These signals are then transformed to linguistic variables that describe how the user controls the system. The capability of handling non-linear systems and the no need for a system precise mathematical model are considered the main advantages of fuzzy logic control. The foremost disadvantage of this control is that its efficiency is influenced by user knowledge and capability in selecting the right error calculation and developing a proper rule base table relied on the selected membership functions (Patcharaprakiti & Premrudeepreechacharn, 2002).

Neural Network

The neural network has to be trained to work with the PV array and set for time changing features of the system to ensure accurate MPPT running. As the neural network is a black box technique, so once it trained, it does not need detailed information describing the PV system. The neural network has three layers: input, hidden layer, and output (Hiyama, et al. 1995). The nodes number in each layer is not constant and is user dependent. The input variables are PV array V_{oc} and I_{sc}. Where the network output is generally denoted by one of numerous reference signals, or by the signal of duty cycle used to run power converter. Finally, the hidden layers act to attain the MPP; the performance of the system is greatly dependent on how fully a neural network has been primarily trained (Sayal, 2012).

Fibonacci Line Search

A novel maximum power point tracking method based on Fibonacci search algorithm has been improved and adopted for the MPPT of partial shading or rapidly changing insolation conditions of PV arrays. The line search method with Fibonacci sequence has been employed and modulated to apply for time-variant power–voltage characteristics of PV arrays to realize simple control system to track the real maximum power. This method requires neither the measurement of temperature and insolation level nor prior knowledge about the model of the used PV panel. The proposed method can track the global maximum power point in most cases by doing a wide-range search and more power can be extracted with partial shaded PV modules as compared to the other methods.

Ant Colony Optimization

When the irradiance has a uniform distribution on the PV, many traditional MPPT methods can track the MPP effectively. However, when the PV arrays are partially shaded, multiple MPPs show up, which usually results in the failure of finding the global MPP. In order to overcome this difficulty, a novel ant colony optimization (ACO)- based MPPT scheme for PV systems is adopted. Most of the MPP control schemes are relatively complicated or fail to guarantee the MPP under all shading situations. A new control scheme is also introduced based on the proposed MPPT method. This heuristic algorithm based technique not only ensures the ability to find the global MPP, but also gives a simpler control scheme and lower system cost.

OPTIMIZATION BASED POWER CONDITIONING UNIT FOR IMPROVING THE QUALITY OF THE SYSTEM: CASE STUDY

To extract the maximum power of a photovoltaic (PV) system and to compensate the nonlinear and unbalanced loads of the electrical power systems, a specific device should be used. A Photovoltaic Active Power Line Conditioner is such one and it will be discussed in details in the following section. DC-DC and DC-AC power converters are required in grid-connected PV systems to maximize the generated PV power and to inject an AC current to the network (Vazquez, et al. 2008).

In this chapter, to guarantee the Maximum Power Point Tracking (MPPT) of the PV system associated at DC side and to accomplish the Maximum Power Point (MPP), a neuro-fuzzy system and a fuzzy logic control loop are composed. An active compensation of the non-linear and unbalanced loads connected to the electrical system are embedded in the DC-AC power converter control methodology. Using a simulation platform, the issue of PV conditioner with the nonlinear and unbalanced loads connected to the electrical system will be resolved (Vazquez, et al. 2008).

It is known that the photovoltaic energy systems can have long termed renewable resources. The components of a grid-connected PV power system are as follows: solar modules arrays, a DC-DC power converter, a DC-AC power inverter and a control system. Figure 37 shows the whole diagram of the electrical system, including a load and other electrical sources.

To carry out the MPPT algorithms, methods based on the principle of perturb and observe (P and O) using a DC-DC power converter are used (Esposito, et al. 2008). Although the PAO method is simple, but it sufferers from the local maxima dilemma. As it shows an oscillating behavior. To overcome the previous disadvantages, a fuzzy logic control has been designed to track the MPP. The suitable performance can be achieved using this method which is easier than the previous one. As a neuro-fuzzy system with a plane regression makes the control faster, it is also used in order to obtain the operation point. In order to track the MPP, the system to keep DC-DC power converter duty cycle in a fixed situation, by measuring voltage and current at the converter input point.

The control of DC-AC power inverter of the PV system was designed using an approach based on the modified vectorial theory in order to compensate unbalanced loads of the electrical system, harmonics and reactive power (Akagi, 2005; Macken, et al. 2004). So, by this way of global control, the regulation of the DC-AC power inverter input voltage, the compensation of the non-linear and unbalanced loads of the system and the extraction of the PV system maximum power will be possible. using this approach, imply an improved Power Quality (PQ) of the electrical network (Singh, 2010; Masand, et al. 2008). In order to acquire it, a PV active conditioner system has been used.

PV POWER LINE CONDITIONER

In the following sections, the model representing the photovoltaic active power line conditioner system which consists of: PV array, DC-DC power converter and the inverter, will be clarified.

Figure 37. Basic structure of PV conditioner system

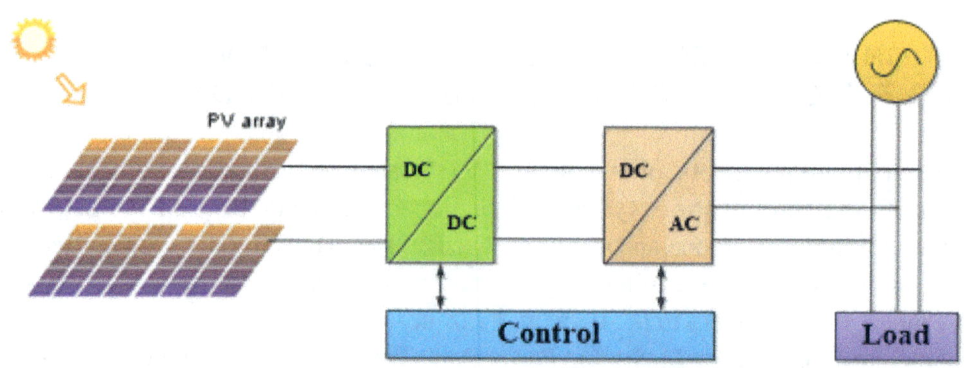

PV Array Model

The equivalent circuit shown in Figure 38 is the model used to represent the solar cell. In this model the current generated by photons is modeled by a current source I [A], the current leakage is represented by means of an antiparallel Diode D_I, a shunt electrical resistance R_{sh}, [Ω], where the ohmic losses is modeled by a series resistance R_s [Ω].

Equation (13) shows the relationship between I, the solar cell output current [A], and V, the solar cell output voltage [V].

$$I = I_l - I_o \left[e^{q\frac{(V+R_S I)}{nKT_K}} - 1 \right] - \frac{V + R_s I}{R_{sh}} \tag{13}$$

I_l is the light generated current [A], I_o is the cell reverse saturation current [A], q is the charge of an electron [C], n is the ideal factor (dimensionless), K is the Boltzmann's constant [J/K] and T_K is the working temperature of the cell [Kelvin].

Therefore, to design a solar module depending on the capacity requirements, the solar cells are connected in series and in parallel. The studied panel (matrix) consists of sixteen modules, each of them with 72 solar cells connected in series and one solar cell in parallel. There are four modules shunt connected and four modules in parallel. Under standard conditions, 1000 W/m^2 and 25 °C, the electrical parameters of the solar system are shown in Table 5.

There is only one operating point for a PV array with a maximum output power. That point changes with the cell temperature and the solar radiation. Table 6 shows the MPPs with different irradiations and a constant cell temperature of 25°C. The PV model has been implemented in Matlab/Simulink.

DC-DC Power Converter

A buck-boost converter is used. The converter's structure is shown in Figure 39. Regulating the solar modules voltage to get the maximum power is done by means of duty cycle control. This can be done by connecting the DC-DC converter at the PV output point.

The duty cycle of the converter is $D = t_{ON} / T$, t_{ON} is the time which the switch is ON and T is the switching period (0<D<1). The polarity of the buck-boost converter output voltage is opposite to the input one. And the output/input ratio can be calculated, as shown in (14).

Figure 38. PV cell equivalent circuit of a

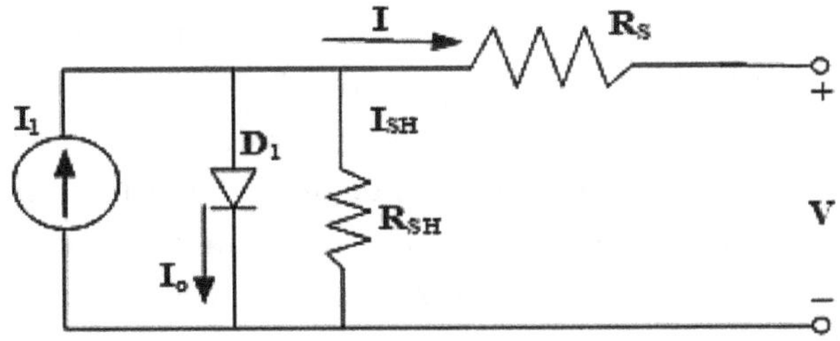

Table 5. Electrical parameters of the solar module

Parameter	Value
Maximum Power (P_{max})	1555 (W)
Maximum Power Voltage (V_{mpp})	102.6 (V)
Maximum Power Current (I_{mpp})	15.16 (A)
Open-Circuit Voltage (V_{OC})	105.6 (V)
Short-Circuit Current (I_{SC})	17.56 (A)

Table 6. MPPs at 25 °C and different irradiances

Radiation (W/m²)	Maximum Power (W)
200	383
400	748
600	1066
800	1332
1000	1555
1200	1733

Figure 39. Buck-boost converter

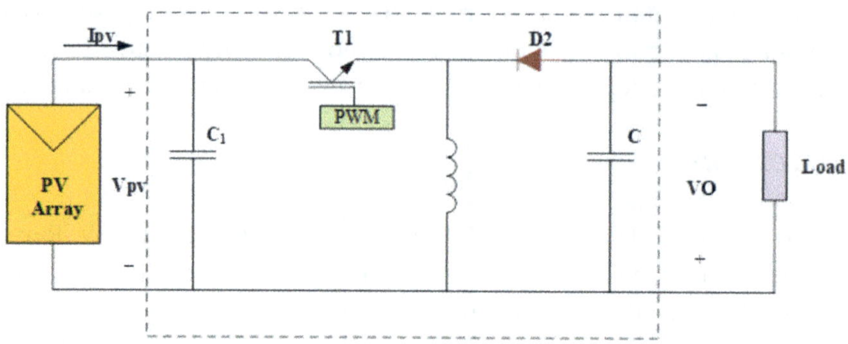

$$\frac{v_o}{V_{PV}} = \frac{D}{1-D} \tag{14}$$

Getting an output voltage higher or lower than the input one is achieved by the ON/OFF switch commutation which also allows energy charge and discharge of the storage elements. The converter non-linear behavior is acquired by the transistor T_1 and the diode D_2. The variable *vo* [V] is the voltage value of the buck boost converter output, I_{PV} is the voltage value of the converter input or the voltage of the solar array output[V], *I* is the output current value of the PV array [A] and i_L is the inductor current value [A]. The load in this case is a resistance. The values of resistance [Ω] *R*, inductor *L* [H], and capacitors C_1 & *C* [F] parameters are constants (Vazquez, et al. 2008).

The maximum energy extraction of the PV modules is guaranteed due to adjusting the input voltage by fixing the reference one using a neuro-fuzzy system which is the goal of our control approach.

By this approach, the peak of the curve is achieved, at the first, in a faster way as the local maximum is avoided by fixing this reference (theoretical) voltage instead of the duty cycle such as in PAO method. Depending on implementing a fuzzy logic control instead of a standard PI control, the optimum voltage will be obtained changing the voltage around the reference (theoretical) voltage. To meet the voltage value that supplies the maximum power at the converter input point, this type of control is needed.

DC-AC Converter

to achieve the connection of the PV system to the electrical network, a DC-AC power inverter is implemented. The goal of using such a converter is to inject the maximum amount of power to the network where in the common connection point it works as an active power line conditioner. So, a global compensation in distributed systems could be accomplished. The key advantage is that the both functions: PV inverter and power conditioner could be achieved by using only one inverter (Vazquez, et al. 2008).

This work is based on using a three-phase inverter which involves a power circuit to follow the reference current determined by a control circuit and to supply the network with the final current. To achieve the needed AC output with a specific series of transistors commutation, the power circuit is designed based on consisting of six (IGBTs) with anti-parallel diodes. To regulate the voltage two capacitors have been embedded. Figure 40 shows the DC-AC structure. The DC-DC power converter output will be linked to the inverter DC side.

To form a pulse-width modulation (PWM) current control, the DC-AC output must be connected to the network through an output reactance. In this work, a hysteresis-band technique has been used.

Figure 40. DC-AC power inverter

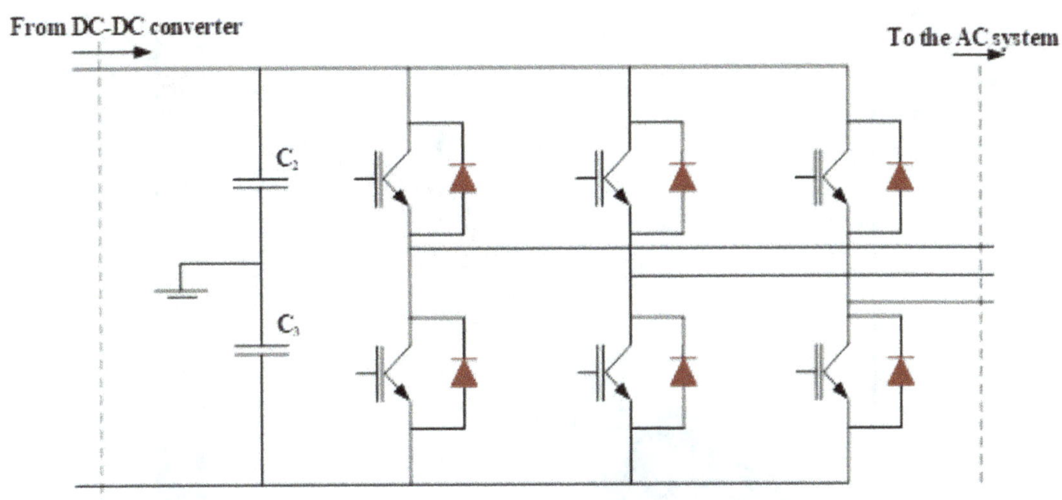

CONTROL STRATEGY OF THE PV ACTIVE POWER LINE CONDITIONER

To attain the MPP, the regulation of the PV array output voltage with the DC-DC power converter is needed. In this work, to gain the reference voltage and to reach the MPP, a neuro-fuzzy system and a fuzzy logic control loop is proposed respectively. Also, the control approach is done to compensate non-linear and unbalanced loads.

Neuro-Fuzzy System and Fuzzy Logic Control

Knowing the optimum voltage value requires a neuro-fuzzy system to be used. This voltage will be used as reference. As this system is multidimensional and it works with non-linear plants, it has been chosen as the adopted optimization method. The data inputs are irradiation and temperature, where the data output is the solar array output voltage that supplies the maximum power value. It is displayed in Figure 41 and results are presented in Figure 42.

In order to control the system using the MPPT algorithm, the required voltage should be reached. This could be done by implementing an adaptive neuro-based fuzzy inference system (ANFIS) which used to produce a trained system with the rules and the membership functions. So, the reference voltage at the input of the DC-DC power converter could be provided by this neuro fuzzy system. The difference between the theoretical voltage value and the value measured at the output of the solar array is considered as the fuzzy control input while the DC-DC power converter duty cycle is the control output.

Thus, the used input fuzzy set is Gaussian curve membership function ("Low" is corresponding to positive error, "Perfect" is for zero and "High" according to negative error) and the output fuzzy set is Triangular membership function ("Less_D", "Hold_D", "More_D" to reduce, remain the same or increase

Figure 41. Reference voltage of the MPP depending on the irradiance and temperature

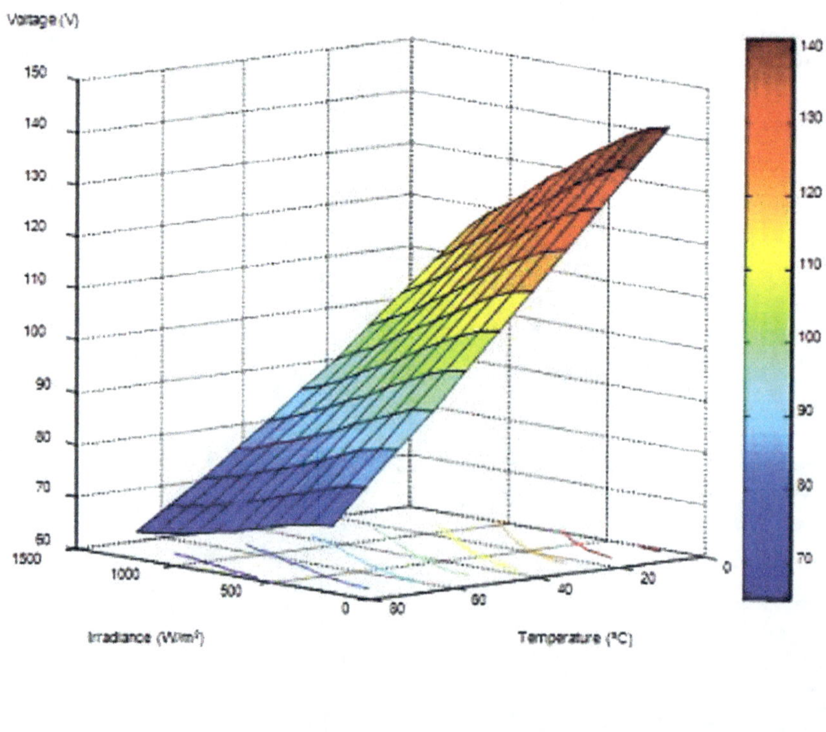

Figure 42. Membership functions; inputs are temperature and irradiation

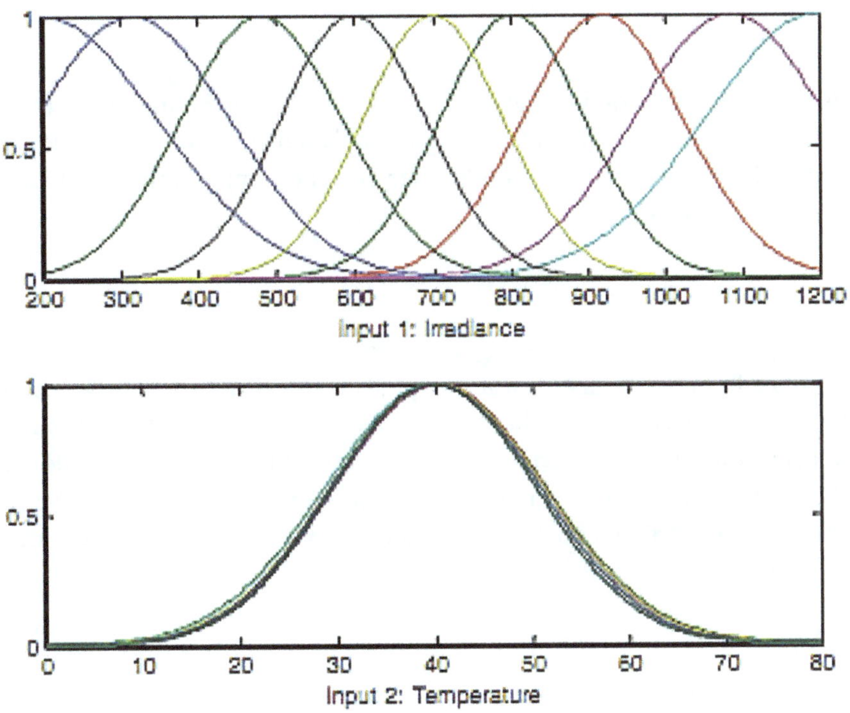

the value of duty cycle). The control rules are as following: if "Low" then "Less_D", if "Perfect" then "Hold_D" and finally if "High" then "More_D".

This produced active power could be introduced to the electrical system as an alternating current, i_{PV} [A], in phase with the voltage of the common coupling point. Since current will be in phase to the major and direct sequence of this voltage.

Inverter Control Strategy

The injection of generated PV active power and the compensation of non-linear and unbalanced loads will be included in the proposed global control approach. In this work, in order to achieve sine and balanced direct sequence network current, sine compensation has been proposed. Henceforth, the approach is presented. For the reason that it is simple, effective and easy to implement with a reduced computation time, the above-mentioned approach has been suggested. The total current of PV conditioner i_T [A] consists of two components. The first one is i_{PV}, that transfers the active power generated by the PV array. The second component is i_C [A], the current required to compensate the electrical loads non-active power (Vazquez, et al. 2008). The calculation of the load active current value i_a [A] is necessary to determine the value of i_C the difference between the load current and its active component is supplied by the shunt active conditioner. The current i_a value depends on the values of both of voltage and active power P [W], as shown in (15).

$$i_a = \frac{P}{V_{1+}^2} v_{1+} = G_1 v_{1+} \tag{15}$$

where G_l [1/Ω] is the equivalent conductance of the load, v_{l+} [V] is the major direct sequence voltage at the point of common connection. By calculating the difference between the load current value and its active component, the value of the PV conditioner current compensation component, i_C, could be computed (Vazquez, et al. 2008).

$$i_C = i_L - G_1 v_{1+} \tag{16}$$

The above-mentioned approach has been accomplished in order to attain the maximum power of PV system connected at inverter DC side. In addition, the value of inverter input voltage need to keep fixed because the capacitor voltage value has to be constant. This is reached by adding to the conductance a factor ⟩ that be influenced by the error between the capacitor voltage and the reference voltage (Vazquez, et al. 2008). A proportional control to guaranty a constant voltage at DC side will by possible by implementing this parameter.

$$i_C = i_L - \left(G_1 \pm \rho\right) v_{1+} \tag{17}$$

The value of PV conditioner total reference current is:

$$i_T = i_{PV} + i_C = i_L - K v_{1+} \tag{18}$$

where K is $G_l - G_l' \pm \rho$, K is an adjustment parameter that can be set by using a PI control. If the hysteresis band is lower than the error between the value of reference current and the value of inverter output one, then the semiconductors state of will be changed to follow the reference.

RESULTS

The suggested control has been executed in Matlab/Simulink using the Simpower System blockset.

Figure 43 presents the block graph of PV Active Power Line Conditioner connected to the electrical system.

Connecting different loads such as a three-phase diode rectifier with inductive impedance in DC side and a non-linear three-phase regulator with inductive load in DC side are used to test the designed control. Performance validation has been done by analyzing the simulation results.

Case 1: PV System With Three Phase Rectifier Under Standard Conditions

Three diode rectifiers are used to establish the unbalanced and non-linear load. The PV conditioner works at 25 °C and 800 W/m^2. Figure 44 shows the power value at the DC-DC converter output point.

Figure 43. Simulink block diagram of the electrical system

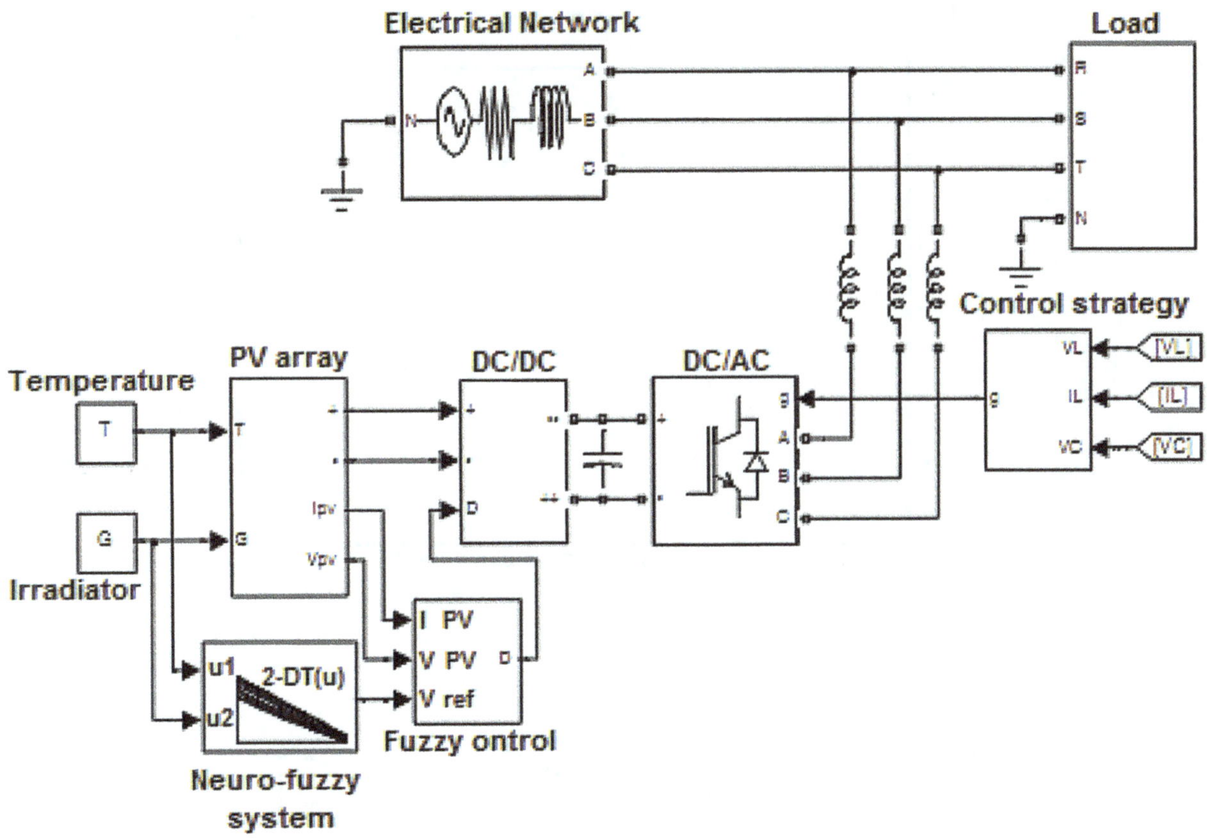

The attained power value is 1256 W, very close to the maximum power value listed in table 04 which is 1332 W, so the performance of the control has a value of 94.29%. Figure 10 displays the power flows between the network, the PV power line conditioner and the grid (load) in order to validate the control of the reactive power compensation (Vazquez, et al. 2008). The PV conditioner produces the reactive power that the load requests. Taking into account that the active power flows from the PV array to the load, all the essential, and the residual to the network.

Figures. 46, 47 and 48 present the load, the PV active conditioner and the source currents, respectively. The studied control allows the extraction of PV array maximum power and the compensation of the electrical power system loads reactive power and harmonic currents.

As it is shown in Figure 47, the PV system current is acting as an active filter. It includes the component comes from the photovoltaic system (active component) and another component without active power supplied by the inverter to act as a shunt active conditioner. Figure 48 shows the source current connected to the PV system.

The value of the source current THD before connecting the PV active conditioner is 14.78% (as load current) while the THD value after connecting the PV power conditioner is 1.88%. Therefore, a source current with sinusoidal shape is reached (Vazquez, et al. 2008).

Figure 44. Output power

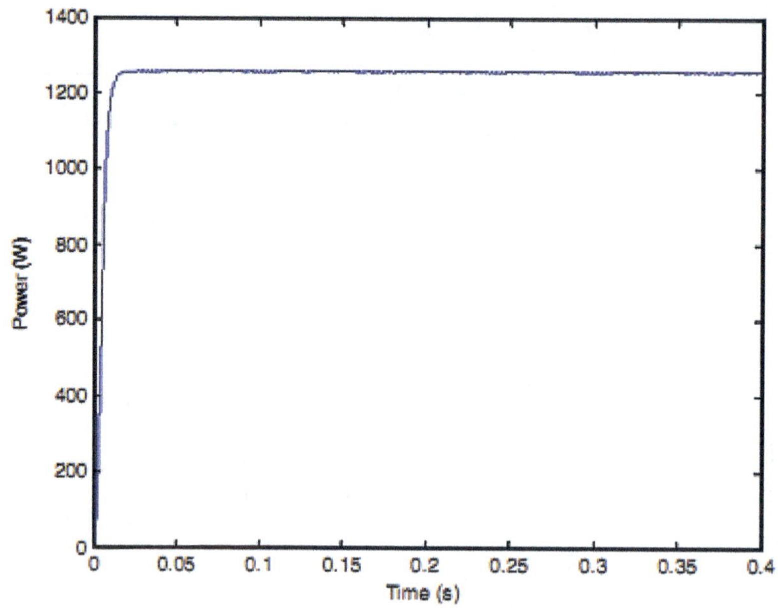

Figure 45. Active and reactive power flows per phase

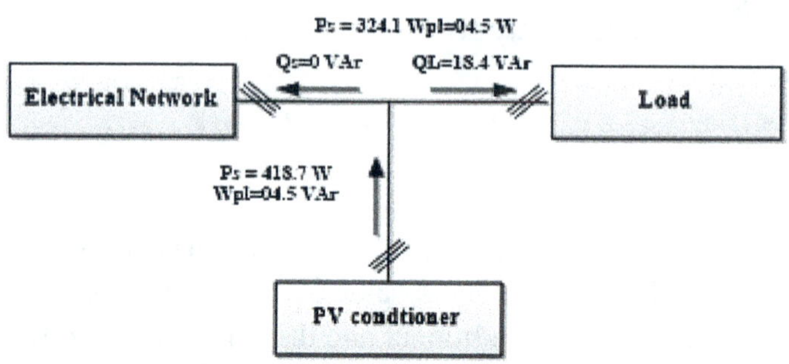

Case 2: PV System With AC Regulator Under Changeable Environmental Conditions

The goal in this case is to check the transient performance of the suggested control where An AC regulator presents the load and the PV conditioner works under various environmental circumstances (Vazquez, et al. 2008). The temperature value is constant 25 °C, but the irradiation change from 1000 W/m^2 to 500 W/m^2 in 0.09 s and to 800 W/m^2 in 0.12 s. The achieved power at the buck-boost converter output point with variable irradiance values is shown in Figure 49.

As it is in Table 4, the maximum power value is always attained with a performance of about 95% where a sufficient performance can be clearly noticed from the power transient response, which is less than a period. Likewise the first case, the reactive power is compensated Figures 50, 51 and 52 show the values of

Figure 46. Unbalanced non sinusoidal load current

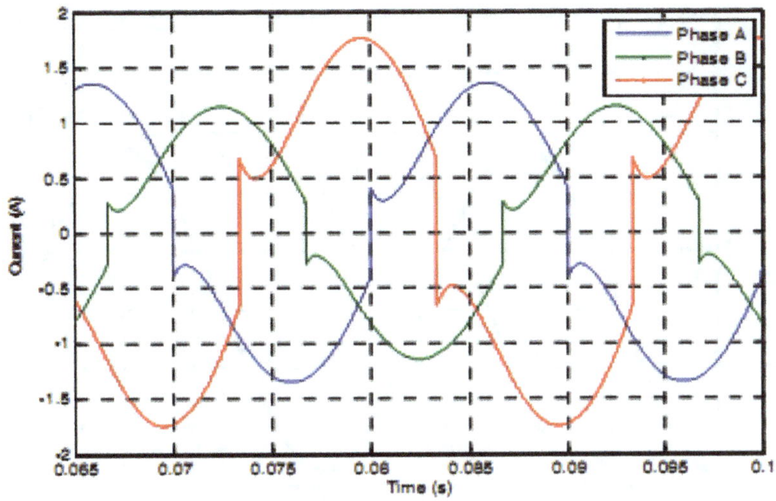

Figure 47. Output current of PV active conditioner

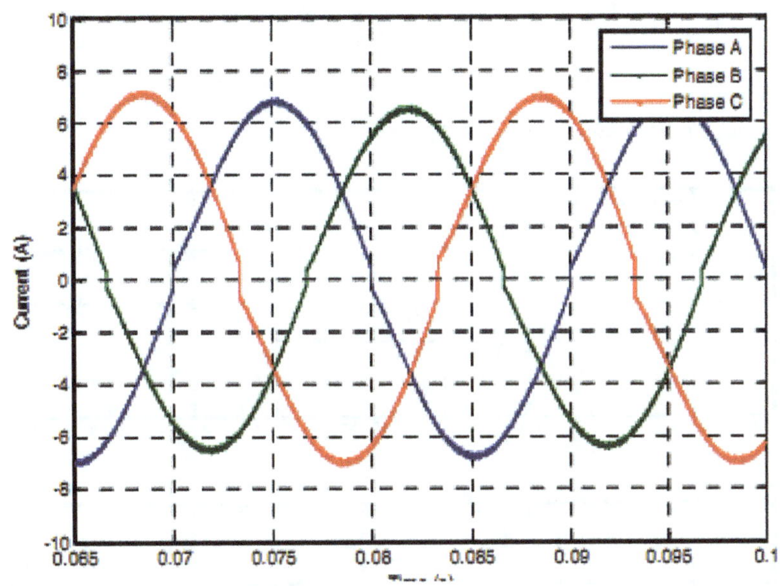

load, the PV power line conditioner and the source currents respectively so as to confirm the harmonic compensation (Vazquez, et al. 2008). From Figure 50 the non-linear and un-balanced behavior of the load current is clearly noticeable.

Figure 51 shows the variations of the current resulted in PV active conditioner. It has two components, the first one that results from the photovoltaic system "active component" and the second provided by the inverter when it works as a shunt active conditioner " non-active component" (Vazquez, et al. 2008). Figure 52 shows the source current when the PV system is connected.

Figure 48. Balanced sinusoidal source current

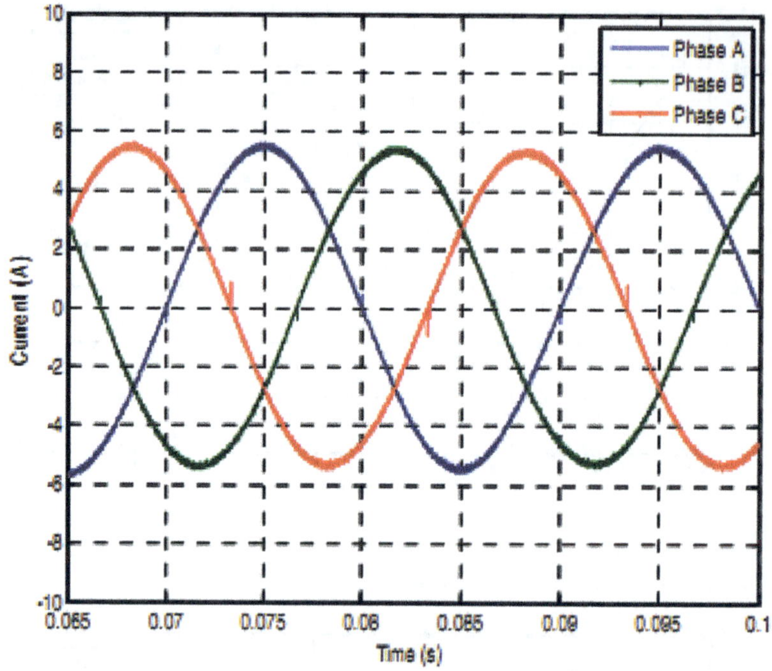

Figure 49. Development of PV power and irradiance

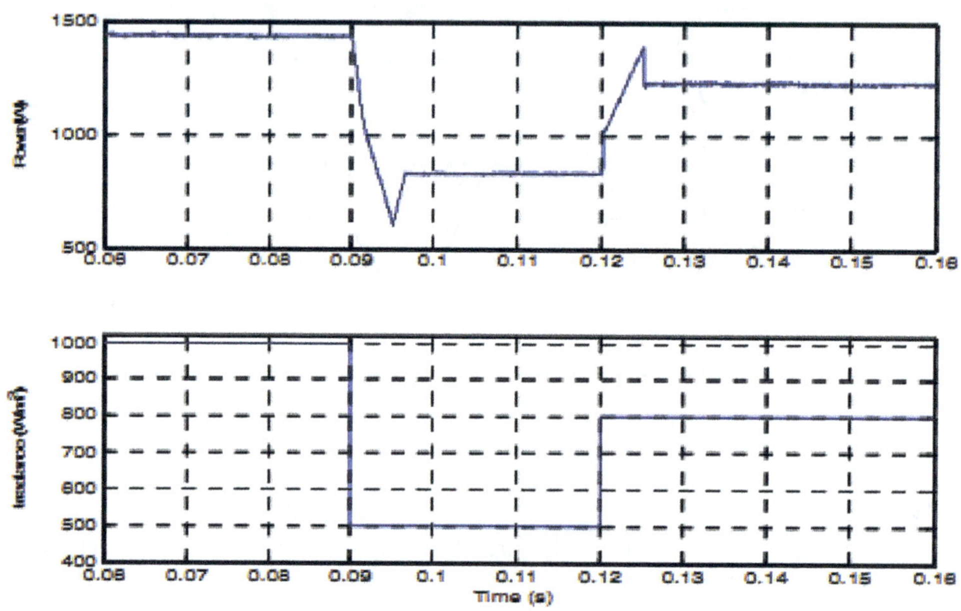

Figure 50. Non-linear and unbalanced load current

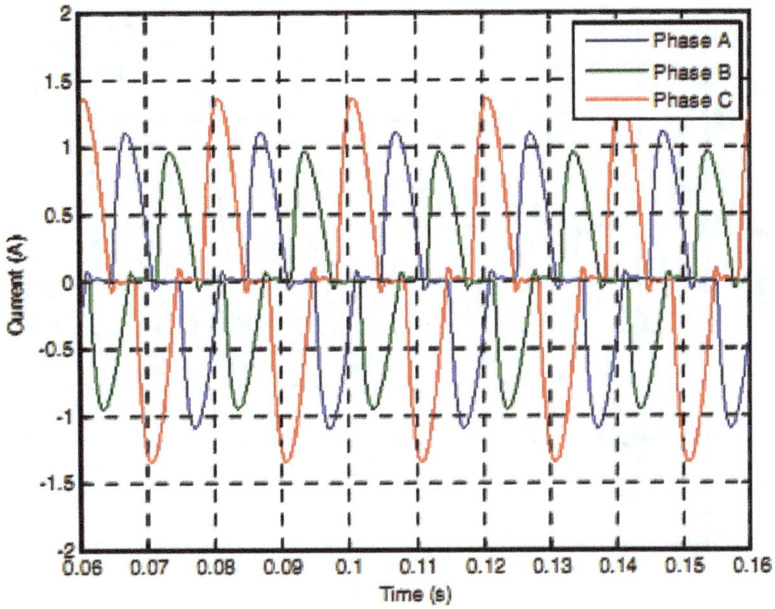

Figure 51. Current of PV power line conditioner

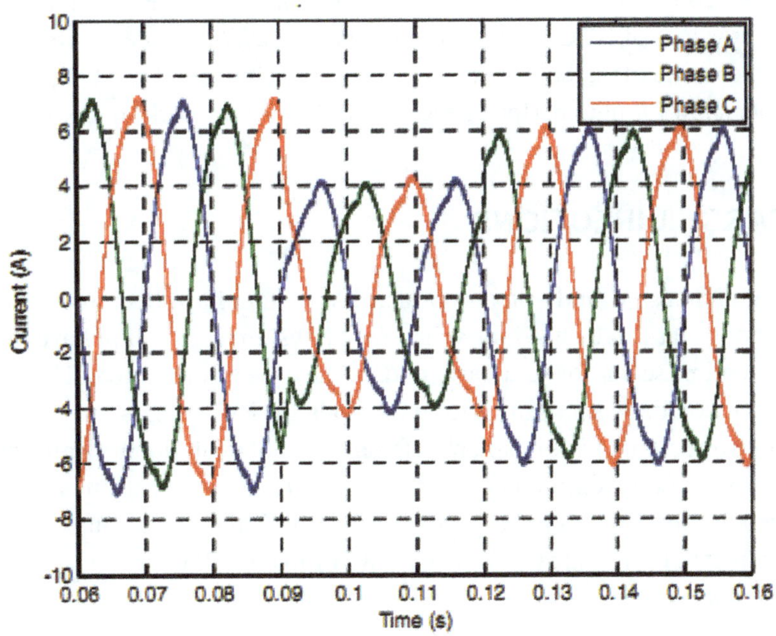

Figure 52. Sinusoidal source current

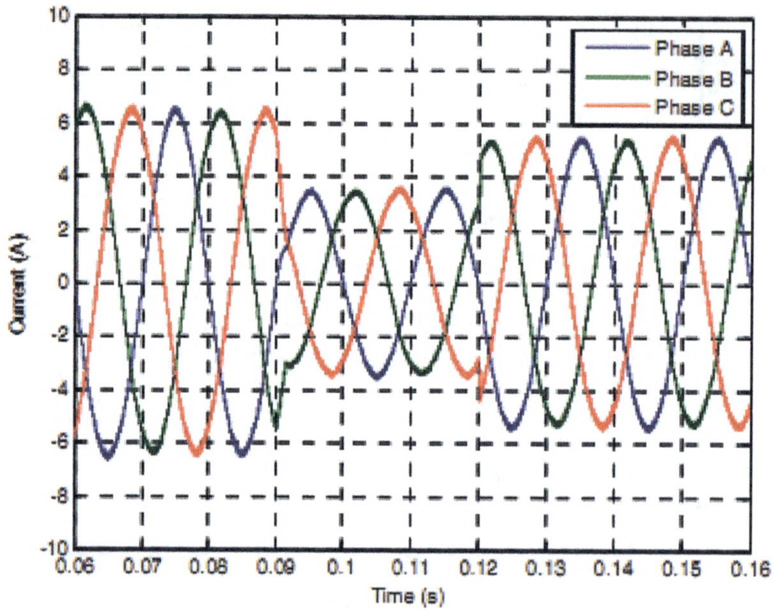

In this case, the value source current THD before connecting the PV conditioner is 47.84% (as load current (Vazquez, et al. 2008)) and the THD value after connecting the PV system is 1.26%, 2.25% and 1.51% with irradiances values of 1000 W/m^2, 500 W/m^2 and 800 W/m^2 respectively, Figure 49. Therefore, with the PV conditioner connected, the source current showed sinusoidal and balanced behavior. The PV power line conditioner shows similar performance even if the system works under low or high irradiation.

FUTURE RESEARCH DIRECTIONS

In this chapter, a comprehensive study on an optimal PV system have been presented. Different types of optimization methods that can be used for solar energy system conversion have also been discussed. Likewise, some simulation scenarios and case studies have been presented. This chapter provided a comprehensive comparison between optimization methods for PV systems.

As a future research direction of this chapter, different types of evolutionary optimization methods can be used for PV systems to get the optimal operation point. Also, a comparison between the well-known algorithm such as genetic algorithm, PSO algorithm, and WDO algorithm can be studied. Furthermore, the researchers are directed to investigate the feasibility of using GPS system to get the optimal operation point in PV systems.

CONCLUSION

In this chapter, a discussion in depth on the topics related to optimal Photovoltaic conversion systems for future smart grids was introduced. The models of solar system components were provided. Then, the conversion system used in solar energy system were illustrated. Likewise, the power electronic devices used to convert the electric power from DC/AC to AC/DC were demonstrated. Furthermore, the optimal methods used in solar energy systems to maximize the delivered solar energy were discussed. Also, a Photovoltaic Active Power Line Conditioner system was introduced. It supplies the system by the PV maximum power and compensates the nonlinear loads that connected to the system. In addition, for the goal of developing the control, a neuro-fuzzy system for calculating the reference value of PV output voltage and a fuzzy logic control loop were used. It attains the maximum power value under variable circumstances. Furthermore, using this method helps in getting better results than results attained from perturb and observe based-methods, since they occasionally can have the problem of getting local maximum instead of global one. The suggested global control approach supplies the PV maximum power value and compensates the components of loads non-active current simultaneously. This method is simple, can easily implemented and needs a very reduced calculations time. The transient response value is lower than half a period.

REFERENCES

Akagi, H. (2005). Active harmonic filters. *Proceedings of the IEEE, 93*(12), 2128–2141. doi:10.1109/JPROC.2005.859603

Akhtar, J., Shamsudheen, P. M., (2012). Power Quality Improvement Using Fuzzy Logic Control Static Var Compensator In Power System Network. *IOSR Journal of Engineering*, *2*(8), 1-8.

Al Hanai, T., Bani Hashim, R., El-Chaar, L., & Lamont, L. (2010). Study of a 900 W, Thin-film, Amorphous Silicon PV System in a Dusty Environment. *International Conference on Renewable Energy: Generation and Applications - ICREGA'10.*

Albadawi, N. (2014). *SCCC701 Engineering Development Project "Development of an automatic robotic cleaning system for photovoltaic plants"*. Available on: www.researcharchive.wintec.ac.nz/3361/1/Final%20reportPDF.pdf

Alhelou, H. H., & Golshan, M. E. H. (2016, May). Hierarchical plug-in EV control based on primary frequency response in interconnected smart grid. In *Electrical Engineering (ICEE), 2016 24th Iranian Conference on* (pp. 561-566). IEEE.

Alhelou, H. H., Golshan, M. E. H., & Askari, J. (in press). Robust Sensor Fault Detection and Isolation Scheme for Interconnected Smart Power Systems in Presence of RER and EVs using Unknown Input Observer. *International Journal of Electrical Power & Energy Systems.*

Alhelou, H. S. H., Golshan, M. E. H., & Fini, M. H. (2015, December). Multi agent electric vehicle control based primary frequency support for future smart micro-grid. In *Smart Grid Conference (SGC)*, (pp. 22-27). IEEE. 10.1109/SGC.2015.7857385

Alva, G., Liu, L., Huang, X., & Fang, G. (2017). Thermal energy storage materials and systems for solar energy applications. *Renewable & Sustainable Energy Reviews, 68*, 693–706. doi:10.1016/j.rser.2016.10.021

Archana, M., & Kondaiah, Y. C. V. (2012). Grid Interfacing Inverter of Renewable Energy Sources to Improve the Power Quality in Distribution System. *International Journal of Advanced Research in Electrical, Electronics and Instrumentation Engineering*, *1*(5), 411–418.

Bentley, R. (2004). Global Oil Depletion–Methodologies and Results. In *Proc. 3rd Int. Workshop on Oil and Gas Depletion*. Australian Association for the Study of Peak Oil and Gas (ASPO).

Chaar, L., Jamaleddine, A., Ajmal, F., & Khanm, H. (2008). Effect of Wind Blown Sand and Dust on PV Arrays Especially in UAE. *Power Systems Conference (PSC)*.

Chen, H.-C., Qiu, J.-C., & Liu, C.-H. (2008). *Dynamic Modelling and Simulation of Renewable Energy Based Hybrid Power Systems*. DRPT.

Chu, C., & Chen, C. (2009). Robust maximum power point tracking method for photovoltaic cells: A sliding mode control approach. *Solar Energy*, *83*(8), 1370–1378. doi:10.1016/j.solener.2009.03.005

Chu, Y., & Meisen, P. (2011). *Review and comparison of different solar energy technologies. Global Energy Network Institute*. San Diego, CA: GENI.

Dankoff, W. (2009). *Glossary of Solar Water Pumping Terms and Related Components*. Available at www.conergy.us/Desktopdefault.aspx/ tabid-332/449 read-3816/

Dezso, S., Remus, T., & Pedro, R. (2007). PV panel model based on datasheet values. *Industrial Electronics, ISIE 2007, IEEE International Symposium*, 2392–2396.

EE362L, Power Electronics, Solar Power, I–V Characteristics, Version October 14, 2005. (n.d.). Retrieved from http://www.ece.utexas.edu/~grady/EE362L Solar.pdf

Enslin, J., Wolf, M., Snyman, D., & Swiegers, W. (1997). Integrated Photovoltaic Maximum Power Point Tracking Converter. *IEEE Transactions on Industrial Electronics*, *44*(6), 769–773. doi:10.1109/41.649937

Esposito, F., Isastia, V., Meo, S., & Piegari, L. (2008). An Improved Perturb and Observe Algorithm for Tracking Maximum Power Points of Photovoltaic Power Systems. International Review on Modelling and Simulations (I.RE.MO.S.), 10-16.

Fini, M. H., Yousefi, G. R., & Alhelou, H. H. (2016). Comparative study on the performance of many-objective and single-objective optimisation algorithms in tuning load frequency controllers of multi-area power systems. *IET Generation, Transmission & Distribution*, *10*(12), 2915–2923. doi:10.1049/iet-gtd.2015.1334

Gabriele, L., Javed, N., & Huailei, Y. (2012). *Autonomous Photovoltaic Panels Cleaning System*. Available on: www.engineering.nyu.edu

Gordon, J. M. (Ed.). (2013). *Solar energy: the state of the art*. Routledge.

Hilloowala, R. M., & Sharaf, A. M. (1992) A Rule-Based Fuzzy Logic Controller for a PWM Inverter in Photo-Voltaic Energy Conversion Scheme. *Conference Record of the 1992 IEEE Industry Applications Society Annual Meeting*, *1*, 762-769. 10.1109/IAS.1992.244319

Hiyama, T., Kouzuma, S., & Imakubo, T. (1995). Identification of Optimal Operating Point of PV Modules Using Neural Network for Real Time Maximum Power Tracking Control. *IEEE Transactions on Energy Conversion, 10*(2), 360–367. doi:10.1109/60.391904

Injeti, S., & Padma, K. (2015). A super capacitor based power conditioning system for power quality improvement. *IRJET, 2*(7).

Karbakhsh, M., Abutorabi, H., & Khazaee, A. (2012). An Enhanced MPPT Fuzzy Control of a Wind Turbine Equipped with Permanent Magnet Synchronous Generator. *2nd International Conference on Computer and Knowledge Engineering (ICCKE)*, 77-82.

Karimov, K., Chattha, J., & Ahmed, M. (2002). Journal of References. *Academy of Sciences of Tajikistan, XLV*(9), 75–83.

Kerdphol, T., Qudaih, Y., & Mitani, Y. (2014). Battery Energy Storage System Size Optimization in Microgrid using Particle Swarm Optimization. 5th IEEE PES Innovative Smart Grid Technologies Europe (ISGT Europe), Istanbul, Turkey. doi:10.1109/ISGTEurope.2014.7028895

Khalil, A., & El-Singaby, M. (2003). Position control of sun tracking system; Circuits and Systems. *Proceedings of the 46th IEEE International Midwest Symposium, 3*, 1134–1137.

Kroutoulis, E., Kalaitzakis, K., & Voulgaris, N. C. (2001). Development of a Microcontroller-Based, Photovoltaic Maximum Power Tracking Control System. *IEEE Transactions on Power Electronics, 16*(1).

Kumar, G., & Kumar, A. (2013). Renewable Energy Interconnection at Distribution Level to improve Power Quality. *International Journal of Engineering Science, 2*(5), 39–48.

Lana El Chaar. (2007). *Photovoltaic System Conversion*. Power Electronics Handbook.

Macken, J., Vanthournout, K., Van den Keybus, J., Deconinck, G., & Belmans, R. (2004). Distributed control of renewable generation units with integrated active filter. *IEEE Transactions on Power Electronics, 19*(5), 1353–1360. doi:10.1109/TPEL.2004.833450

Maghami, M. R., Hizam, H., Gomes, C., Radzi, M. A., Rezadad, M. I., & Hajighorbani, S. (2016). Power loss due to soiling on solar panel: A review. *Renewable & Sustainable Energy Reviews, 59*, 1307–1316. doi:10.1016/j.rser.2016.01.044

Mahmoud, A. M. A., Mashaly, H. M., Kandil, S. A., El Khashab, H., & Nashed, M. N. F. (2000) Fuzzy Logic Implementation for Photovoltaic Maximum Power Tracking. *26th Annual Conference of the IEEE Industrial Electronics Society, 1*, 735-740.

Mahmoud, A. M. A., Mashaly, H. M., Kandil, S. A., El Khashab, H., & Nashed, M. N. F. (2000). Fuzzy Logic Implemen- tation for Photovoltaic Maximum Power Tracking. *Proceedings of 9th IEEE International Workshop on Robot and Human Interactive Communication*, 155-160.

Masand, D., Jain, S., & Agnihotri, G. (2008). Control Strategies for Distribution Static Compensator for Power Quality Improvement. *Journal of the Institution of Electronics and Telecommunication Engineers, 54*(6), 421–428. doi:10.4103/0377-2063.48632

Mohandes, B., El-Chaar, L., & Lamont, L. (2009). Application Study of 500W Photovoltaic (PV) System in the UAE. *Applied Solar Energy Journal*, *45*(4), 242–247. doi:10.3103/S0003701X09040057

Montoya, L. F., Fu, Q., Nasiri, A., Bhavaraju, V., & Yu, D. (2014). *Novel methodology to determine the optimal energy storage location in a microgrid and address power quality and stability*. Eaton White Paper.

Nazaripouya, H., Wang, Y., Chu, P., Pota, H. R., & Gadh, R. (2015, July). Optimal sizing and placement of battery energy storage in distribution system based on solar size for voltage regulation. In Power & Energy Society General Meeting, 2015 IEEE (pp. 1-5). IEEE.

Partridge, T., Thomas, M., Harthorn, B. H., Pidgeon, N., Hasell, A., Stevenson, L., & Enders, C. (2017). Seeing futures now: Emergent US and UK views on shale development, climate change and energy systems. *Global Environmental Change*, *42*, 1–12. doi:10.1016/j.gloenvcha.2016.11.002

Patcharaprakiti, N., & Premrudeepreechacharn, S. (2002). Maximum Power Point Tracking Using Adaptive Fuzzy Logic Control for Grid-Connected Photovoltaic System. *IEEE Power Engineering Society Winter Meeting*, *1*, 372-377.

Ram, J. P., Babu, T. S., & Rajasekar, N. (2017). A comprehensive review on solar PV maximum power point tracking techniques. *Renewable & Sustainable Energy Reviews*, *67*, 826–847. doi:10.1016/j.rser.2016.09.076

Rashid, M. (2011). *Power Electronics Handbook* (3rd ed.). Elsevier.

Sandia National Laboratories. (1988). Stand Alone Photovoltaic Systems: A Handbook of Recommended Design Practices. National Technical Information Service.

Sayal, A. (2012) MPPT Techniques for Photovoltaic System under Uniform Insolation and Partial Shading Conditions. *Students Conference on Engineering and Systems*, 1-6.

Singh, A. (2010). Performance Evaluation of Three Different Configurations of DSTATCOM with Nonlinear Loads. *Journal of the Institution of Electronics and Telecommunication Engineers*, *56*(6), 313–326. doi:10.1080/03772063.2010.10876320

Thornton, P. (1992). The Effect of Sand-Storm on Photovoltaic Array and Components. *Solar Energy Conference*.

Vazquez, J. R., Martin, A. D., & Herrera, R. S. (2013). Neuro-fuzzy Control of a Grid-connected Photovoltaic System with Power Quality Improvement. *Euro. Con. 2013*, 850-856.

KEY TERMS AND DEFINITIONS

Optimization: The action to find the best or the most effective solution of a problem.

Photovoltaic: By using photovoltaic cells, the energy from sunlight can be converted to electricity.

Renewable Energy: This term describes the energy collected from renewable resources such as sunlight, rain, wind, and geothermal heat energies.

Semiconductor: A semiconductor is a substance that can conduct electricity under some conditions but not others, making it a good medium for the control of electrical current.

Smart Grids: A flexible power system with high penetration level of renewable energy resources provided with a great communication infrastructure and information management.

Solar Energy: The energy received to earth from the sun which is usually converted into electrical or thermal energy.

Chapter 19
A Suggestion for Energy Policy Planning System Dynamics

Arzu Eren Şenaras
Uludag Unıversıty, Turkey

ABSTRACT

The system dynamics approach was developed by Jay Forrester from MIT during 1950s to analyze the complex behavior in administration with computer simulation in social sciences. System dynamics is a form of systems approach as a methodology to understand the dynamic behavior of complex systems. The basis of system dynamics is to understand how system structures cause system behavior and system events. System thinking and system dynamics provide computer technology and conceptual and numerical modeling techniques that could help explain the dynamics and forces underlying the complexity and change in business, politics, socio-economic, and environmental systems. System dynamics, policy analysis, and design are used for learning and decision making.

INTRODUCTION

Decision making in energy systems is a complex task at best. Complexity stems primarily from the non-linear nature of energy-economy-environment interactions. Given the global trend of focusing on low carbon economics, energy policy design and assessment has become a dynamic problem. Uncertainties such as the dynamics of energy supply and demand, the price of fossil fuels, various regulatory regimes, the advances and challenges of energy production and consumption-related Technologies, and impact of energy related emissions abound. Therefore, the use of model-based analysis and scenarios in energy policy design and assessment has seen phenomenal growth during the past several decades. The key focus is on development, validation and applications of System Dynamics and agent-based models in service of energy policy design and assessment in the twenty-first century (Qudrat-Ullah, 2013:7). In order to provide useful energy policy design and assessment, therefore, the required modeling approach needs to overcome the above-mentioned limitations. Researchers, especially from system dynamics and agent-based modeling approaches, have found these models capable of addressing these unique challenges. For instance, system dynamics models have been successfully used to study national energy policy evaluation (Qudrat-Ullah, 2013).

DOI: 10.4018/978-1-5225-3935-3.ch019

System dynamics approach was developed by Jay Forrester from MIT during 1950's to analyze especially the complex behavior in administration with computer simulation in social sciences. System dynamics is a form of systems approach as a methodology to understand the dynamic behavior of complex systems. The basis of system dynamics is to understand how system structures cause system behavior and system events (Sezen, 2009: 298).

Jay Forrester initially constructed his first dynamic model upon his meeting with the management of General Electric corporation. Big fluctuations in production, inventory, labor force and profitability were compelling GE management. Despite hard efforts of the management, these fluctuations were mostly associated with external factors. Especially, the fluctuations in the business were related to received orders. Forrester interacted with the management to observe the system operations in other departments. In the first model he developed, he observed that simulations were necessary since the system could not be monitored analytically. He demonstrated that the corporation could experience serious fluctuations due to management policies even when the demand is considered constant with the weekly simulation he ran. Later on, he designed the computer simulation for the same problem. In his later studies, Forrester demonstrated how the feedback control theory could be adapted for complex administration and human systems. He published his initial findings in an article in Harvard Business Review. Later on, he developed this study to write his famous work "Industrial Dynamics" (Lane and Sterman, 2011; Ramage and Shipp, 2009: 100-101).

A SHORT LITERATURE REVIEW

A short literature review concerning about developing energy policy can be listed as below:

Fazeli and Davidsdottir (2015), developed a dynamic building stock model, which captures the effects of building aging on total energy consumption for space heating demand. They studied the behavioral dynamics by focusing on the key processes affecting the implementation of retrofitting measures.

Vilchez et.al.(2015), developed a model integrating discrete choice and accounting frameworks. The developed System Dynamics model is applied to examine alternative policies and to estimate energy use and emissions in each of the markets under various scenarios.

Batinge (2015), developed to create a vivid understanding of the complex feedback loops within the electricity sector. The results present an outlook of the electricity situation in Ghana. Policies discussed include the ideal investment pathways for sustainable electricity supply in the future.

Al-Sarihi et.al. (2015), informed by system dynamics approach that allows providing a systemic account for the impacts of renewable energy development in Oman. Their paper seeks to support the national policy makers in their on-going attempts to integrate RETs in Omani power sector through national planning and policies.

Ebrahimi (2016), defined a general work structure of a small and medium-sized enterprise (SME) which operates newly in the technological innovation and related manufacturing of solar energy technologies, and providing a template for managers in the same firms. Policies of the company are examined which lead to a better performance defined as more profit through serving more customers as a result of faster dealing with inquiries and fulfilling orders. Model is validated based on the structure verification, behavior reproduction, and behavior prediction tests.

Vilchez et al. (2016), developed a model grounded on dynamic econometrics and system dynamics. The model encompasses nine car technologies powered by seven types of fuel: gasoline, diesel, flex-fuel, liquefied petroleum gas, natural gas, hybrid, plug-in hybrid, electric and fuel cell. Model-based policy analysis is performed and the impacts of several types of carmix until 2030 are derived. The model is applied to China, France, Germany, India, Japan and the US.

STRUCTURE OF THE SYSTEM DYNAMICS MODEL

In order to analyze the dynamics of the actual system, system models are investigated. Models are simplified structures designed to examine the system behavior (Karnopp et al., 1990: 4). Models are constructed to better understand the economic performance of alternative decisions and their impact on environmental quality. Models are important tools in obtaining new knowledge (Ruth and Hannon, 2012: 4).

The model must have a clear objective, and this objective should provide a solution to a specific problem. The clear objective is the most important component of a successful model. A clear model, of course, could be inaccurate, too broad and difficult to understand. However, an explicit goal allows the users of the model to ask questions that demonstrate whether solving the problem is beneficial (Sterman, 1991: 5).

According to Forrester (1969), four hierarchical structures must be defined to model the dynamic behavior of a system:

1. A closed boundary around the system
2. Feedback loops as basic structural elements within the closed boundary
3. Level (status) variables indicating the totals of the feedback loops
4. Rate (flow) variables that indicate the actions within the feedback loops
    a. Objective
    b. Observed situations
    c. Components of rate variables
    d. Determination of the difference
    e. Actions based on the difference

To develop the whole systems concept, boundaries in which the interactions forming the system's characteristic behavior occur must be established (Forrester, 1969: 12).

In order to achieve the expected objectives in the dynamic model, the model should include the following characteristics (Erkut, 1983: 42-43):

1. The ability to define the predicted cause and effect relationship
2. A simple mathematical structure
3. Incorporating concepts adequate for the economic and social structure
4. Ability to expand to several variables without exceeding computer application limits
5. Ability to produce and process continuous interactions.

Forrester created a language called system dynamics to define system behavior. This language consists of four components: stocks, flows, decision functions and information flow. As a result, the only thing we need to define the system is to create these four blocks despite the level of complexity of the system. System dynamics language is presented in Figure 1.

Figure 1. System dynamics language
Source: Yamaguchi, 2013.

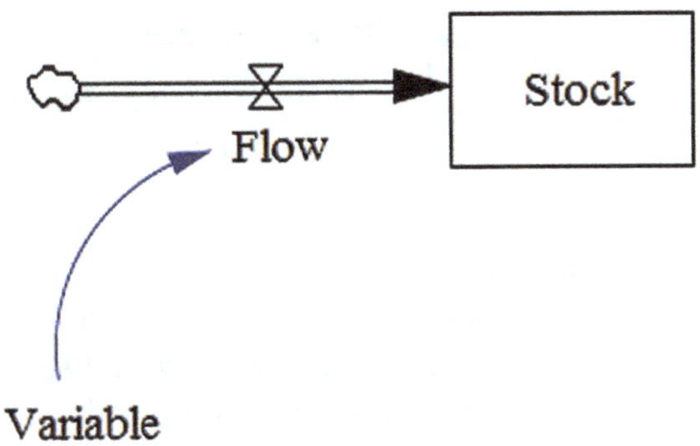

Dynamic analysis deals with time flow. So first, we have to define time. Time could simply be defined as a dimensionless real number. There is a starting point and it continues in the positive direction in the coordinate system. In this case, two different concepts stand out. The first concept is defined as a moment in time and denoted as $\tau = 1, 2, 3, \ldots$ The second concept is the time period and denoted as $t = 1., 2., 3., \ldots$ The unit of time could be expressed in seconds, minutes, hours, weeks, months, years, decades, hundreds of years…, depending on the structure of the examined system. In system dynamics, the two time concepts must be distinguished correctly, because the stocks and flows need to be defined explicitly as τ and t. (Yamaguchi, 2013: 23).

The elements that constitute the dynamic system model could be explained in four groups such as stocks, ratios, decision functions and information flow.

Stocks (Levels) and Flows (Ratios)

While the stock could be defined as accumulation, the flow refers to the rate that stock levels change. Stocks could be clearly defined as inventory of raw materials or finished products. Some are less clear, partly because the everyday language does not identify them as stocks. An example is the cash balance generated by the deposit and withdrawal of funds. Stocks are the accumulations within the system. These accumulations could include concepts such as inventory, business volume, number of employees in different qualifications and experiences, and workforce, etc. Stocks are the present values of the variables that are formed by the accumulated difference between the inflow and outflow (Sezen and Günal, 2009: 306-307; Erkut, 1983: 44).

The most important of the four components that form the foundation of system dynamics is the stocks. System could be defined as the combination of stocks. Stock is the sum of objects that exist at a given time τ. If we express stocks with an x in a specific time τ, stocks could be expressed with $x(\tau)$, when τ is a real number (Yamaguchi, 2013: 23). The stock structure is presented in Figure 2.

A common way utilized to explain the change in stocks is to determine the stock value $\tau = 1, 2, 3, \ldots$ at specific points in time and take into account the amount of change in stocks at the next point in time. The period between τ and $\tau + 1$ is called the unit time interval. The unit could be defined as seconds, minutes, hours, days, weeks, months, years as mentioned above. This period of time is defined as τ. τ's are defined starting from point $\tau = 0$ (the point of origin), (1st time interval, 2nd time interval, 3rd time interval, ...).

Figure 2. Stocks and flow

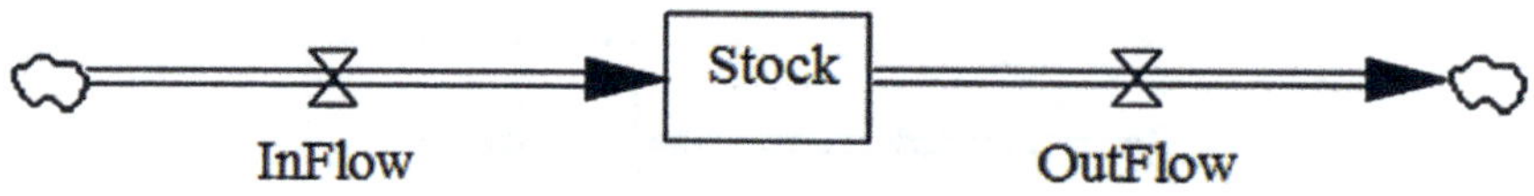

On the other hand, flow is defined as the increase or decrease in the unit time interval in stocks and denoted by f (t). The flow could be defined in each intermittent time period, which is then called the intermittent flow. The flow is related to the amount of time between two points in time (unit time interval), while stocks are defined as the quantity at a certain point in time. In other words, τ is used to define stocks. This corresponds to a point in time. τ is used to define the flows and the tth unit time interval is the time between τ point in time and the next $\tau+1$ point. Any dynamic behavior could be understood functionally with the terms of stocks and rates. Flow is the rate of change in stocks. It refers to the changes in stocks. In the bathtub example, the flow is the water entering the bathtub through the faucet and the water leaving the bathtub via the duct. Births and deaths, revenues and expenses, glucose intake and glucose consumption, and etc. are all examples of flow. In these examples, birth represents the inflow and death represents the outflow (Yamaguchi, 2013, Martin, 1997a, Sterman, 2000, Barlas, 2005b).

As can be seen in Figure 3, the flow is an integral part of stocks. In this context, flow and stock units should overlap physically and qualitatively. We could denote the relation between stocks and flow as follows:

$$x(\tau+1) = x(\tau) + f(t) \quad \tau \text{ ve } t = 0, 1, 2, 3, \ldots \ldots \quad (1)$$

Table 1 demonstrates distinctions between stocks and flow in different disciplines.

To avoid the notation confusion between τ and t, we need to express these two concepts with one of the two. Since t denotes the time interval between τ and $\tau+1$, the stock amount in t^{th} time interval x(t) could be defined as the stock amounts of τ, which is the beginning point of the time interval, and $\tau+1$, which is the end point of the time interval. Accordingly,

Figure 3. Relationship between stocks and flow
Source: Yamaguchi, 2013.

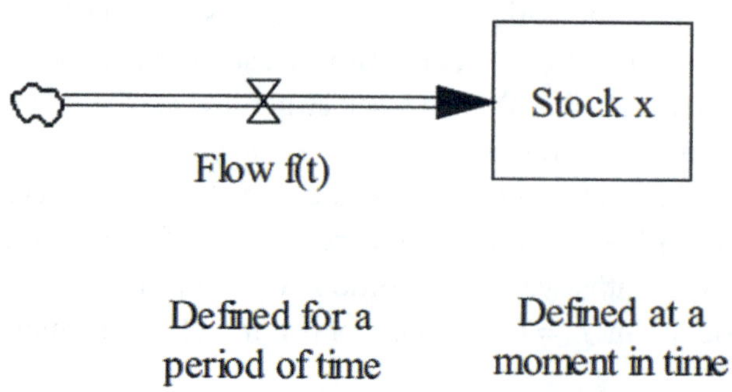

Table 1. Terminology used to distinguish stocks and flow in different disciplines

Field	Stocks	Flows
Mathematics, Physics and Engineering	Integrals, States, State variables and inventories	Derivatives, Rates of change, Flows
Chemistry	Reagents and Reaction products	Reaction rates
Manufacturing	Internal inventories, Inventories	Production rate
Economics	Levels	Ratios
Accounting	Stocks, Balance sheet items	Cash flow
Biology, Psychology, Medicine, Epidemiology	Reservoirs, Prevalence	Prevalence, Mortality rates

Source: Sterman, 2000: 198.

$x(t) = x(\tau)$:Stock at the beginning of the process (2)

or

$x(t) = x(\tau+1)$:Stock at the end of the process (3)

If the stock at the beginning of the process is used, the stock flow equation (1) could be written as follows:

$x(t+1) = x(t) + f(t)$ $t= 0,1,2,3,\ldots\ldots$ (4)

In the equation above, stock $x(t+1)$ value is the value at the beginning of $t+1$ time interval and $f(t)$ is obtained by adding the flow quantity and the stock value.

If the stock equation(3) at the end of the process is accepted, the stock flow equation (5) could be formulated as follows:

$x(t) = x(t-1) + f(t)$ $t=1,2,3,\ldots\ldots$ (5)

With the above mentioned method, two time concepts (a moment in time and a time interval) are combined.

If $f(t)$ is defined as an intermittent time period such as $t = 1,2,3, \ldots$, the equation (4) is called the difference equation and is expressed as (6).

$$x(t) = x(0) + \sum_{i=0}^{t-1} f(i) \quad (6)$$

If the flow is continuous, time units are divided into n sub time intervals and the equation (4) could be written as follows (Yamaguchi, 2013: 25-26):

$$x(t) = x\left(t - \frac{1}{n}\right) + \frac{f\left(t - \frac{1}{n}\right)}{n} \\ = x(t - \Delta t) + f(t - \Delta t)\Delta t \tag{7}$$

For

$$\Delta t \to 0, i\c{c}in, \lim_{\Delta t \to 0} \frac{x(t) - x(t - \Delta t)}{\Delta t} \equiv \frac{dx}{dt} \tag{8}$$

$$\frac{dx}{dt} = f(t) \tag{9}$$

The above formula is the definition of a differential equation. In cases where flow and stocks are constant, the amount of stock at$t$ is found by solving this differential equation.

$$dx = f(t)dt \tag{10}$$

$$x(t) = x(0) + \int_0^t f(u)du \tag{11}$$

The general solution of the differential equation is obtained by the above formula. The only difference between continuous and intermittent flows is the integral in continuous flow and the use of total functions in intermittent flow (Yamaguchi, 2013: 27).

Jay Forrester (1962) explained stocks and flows with the metaphor of the water that accumulates in the bathtub and drained from the bathtub. Stocks are the amount of water that accumulates in the bathtub. The amount of water in the bathtub is the difference of the amount of water entering and leaving the bathtub. Stocks are the accumulation of the flows or the integral of flows. Net flow reflects the rate of the change in stocks (Sterman, 2000: 194).

Integral Equation

$$Stock(t) = \int_{t_0}^{t} \left[(Inflow(s) - Outflow\ (s)\right] ds + Stock\ (t_0) \tag{12}$$

Inflow (s) denotes the value of any flow at any point in time s between the beginning of inflow and the present time t. Similarly, the net stock change rate is derived from this equation and is obtained by subtracting the outflow from inflow as shown below:

Differential Equation

d(Stocks) / dt = Net Change in Stocks= Inflow (t) – Outflow(t) (13)

Classification of Stocks and Flows

Yamaguchi (2013) classified the stocks based on their natural structures as depicted in Table 2.

- **Natural Stocks:** Include stocks that could be found in the natural environment such as world population, number of birds and trees in the forests, and amount of water in the lakes.
- **Capital Stocks:** Are the means of production such as machinery, factories and structures used to produce consumer goods.
- **Products in Process and in Use:** Include goods produces and intermediate products still in manufacturing process.
- **Information:** Is a non-physical inventory and it is stored in various types of media. For example, books, CDs, DVDs, ...
- **Psychological Passions:** Are individuals' emotional stocks. For example, love, happiness, sadness, nervousness. It is stored somewhere in our brain.
- **Indexed Figures:** Are a special form of stored information. It is used to describe nature and human activities. For example, temperature, price, sales figures, etc. (Yamaguchi, 2013: 24).

Forrester (1962) described flows in six categories. These are information, materials, personnel, capital equipment, orders and money and they are explained below in detail:

- **Information Flow:** Creation, control and distribution of information is the most significant task of business management. It might be difficult to model the information properly in the business process. Other flows have a physical representation, and at the same time these are protected quantities. In contrast, information could be shared in several locations at the same time. Experiments demonstrated that the performance of the process is significantly affected by the changes in the connection of information in a business.
- **Material Flow:** It includes stocks and flow rates of physical goods such as raw materials, inventories in the process or finished goods. These include of stocks and flows of physical products in the production and distribution processes. The raw material could be with the stock in process or a completed product.

Table 2. Classification of stocks

Physical Stocks	Non-Physical Stocks
Natural Resources	Information
Capital	Psychological passions
Products in Process and in Use	Indexed figures

Source: Yamaguchi, 2013: 24.

- **Labor Flow:** It generally includes the current staff.
- **Capital Flow:** It includes all necessary equipment for the business and production of goods.
- **Order Flow:** It includes orders for products, requests for recruitment, contracts for new vehicles and locations. The order flow could often be seen as a piece linking the decision to create an order to the action of order fulfillment.
- **Money Flow:** Money denominates cash in this context. Money flow is the transfer of money between different currency stocks (Kirkwood, 1998: 18, Erkut, 1983: 47-48).

Decision Functions and Information Flow

Decision functions are states of the policy that determines how to convert available information in stock into a decision. All decisions produce an action and could be expressed in flow rate.

The decision functions that determine rates are based solely on the information contained in the stocks. Rates are not determined by other rates. This principle applies to all cases (Erkut, 1983: 46).

SYSTEM DYNAMICS MODELING PROCESS

Forrester's system dynamics modeling steps could be listed as follows (Forrester, 1962: 13):

- Identification of the problem.
- Separation (isolation) of the factors that include the observed symptoms.
- Arriving at new decisions through connecting the decisions to actions and actions to information feedback by following cause and effect information feedback cycles
- Formulation of acceptable and adequate (formal) decision-making policies that define the results of these decisions using existing information flows.
- Construction of the mathematical model for the interactions between decision-making policies, sources of information and system components.
- Formulation of system behavior as defined in the model
- Comparison of the results with all adequate data in the existing system
- Review of the model until it represents the actual system
- Discovery of organizational relationships and policy amendments that could improve system behavior through model redesign
- Obtaining the improved performance by implementing the changes in the actual system as stipulated by the model

The first phase of the modeling process in system dynamics is to identify the problem. This phase includes the identification of the problem clearly and the presence of the individual or individuals involved. In the second stage, the system is defined by the construction of influence diagrams, also called causal loop diagrams. In the third step, qualitative analyzes are conducted. The fourth stage involves the development and testing of the simulation model. In the fifth and final stage, the results of the analysis conducted with system dynamics are implemented with policy design. The structure of the system dynamics approach is presented in below.

Step 1: Define the problem. (Who is interested and why)
Step 2: Understand the Problem and Define the System. (Influence Diagrams)
Step 3: Qualitative Analyses
Step 4: Simulation Modeling (Special computer simulation languages), Testing the model.
Step 5: Test and Implement the Policy, Sensitivity test.

a. Discovery of the Model and Policy Design through Simulation/The Ideas It Contains. (Assessment based on the decision), (Objective function)
b. Policy Design with Optimization/Reliable Policies (Source: Coyle, 1996: 11).

After determining the objective of the problem, the system is divided into subsystems. The variables contained in each subsystem are determined. More than necessary number of variables makes it difficult to understand the model. Lack of certain variables could change the meaning of the model altogether (Erkut, 1983: 23-24).

Definition of the Problem

The most important component in successful modeling is a clear objective. Naturally, a model with a clear objective could still be misleading and difficult to understand. However, an explicit objective allows the clients to make inquiries about the relevance of the model for the problem they are addressing (Sterman, 2000: 89).The first step of the identification of the problem is in fact the identification of for whom and why the problem is significant. The correct answer might rarely be found at this step. An attractive feature of system dynamics in administrative sciences is the fact that while one tries to solve the problem, another could re-examine the problem (Coyle, 1996: 10).

Definition of the System and Construction of Influence Diagrams

The second step is the initial phase of system dynamics process. It includes the expression of the system using influence diagrams. Influence diagrams are also called causal loop diagrams as well (Coyle, 1996: 10).

Qualitative Analyses

After creating the influence diagrams, qualitative analyses are conducted. What is meant by qualitative analysis is the examination of influence diagrams closer to better understand the proposed problem. This step is the most important step in system dynamics and carries the modeling towards meaningful results. Sometimes the problem is solved at this step and there is no need to go further. The dashed line represents this case above.

Construction of the Simulation Model

If qualitative analysis does not provide sufficient information to solve the problem, the fourth step is initiated. The influence diagrams constructed in second and third steps could be redrawn with less detail. The simulation model could be designed without flow chart. Simulation models and influence diagrams are two different representations of the same model. While in the latter, the model is indicated by arrows and words, simulation model sutilize equations and computer code. This feature of system dynamics has some powerful practical consequences.

The first is the speed. Equations could be initiated in any order the modeler chooses. In practice, the order is chosen in order to be understood and explained, but there is no need to worry about the order of execution of the equations, because the special programming language takes care of this. In other words, the model could usually be written ten or hundred times faster with this special code when compared to BASIC or a higher level programming language. Thus, the initial results could be obtained within the same day or hour. The second feature is its ease of review and development. The third characteristic is that it is easy to consider and explain the problem as the model is expressed both in influence diagrams and equations. Therefore, it facilitates communication. Furthermore, this stage includes important steps such as testing the model and debugging the errors. These steps are aptly named as the validity of the model. Here, the validity indicates that the model was constructed in a suitable manner for the objective and structured adequately rather than the absolute and real meaning of the term.

Test and Design of the Policies

At this stage, quantitative analysis starts to produce results. At first, the emphasis is on policy design and testing of the system. In step 5A, it is attempted to conceive what are the effects and how these effects influence the system cycles with the help of simulation of potential changes. The emphasis here is to unearth more system behavior.

The step in 5B is conducted to determine the value that the system implied with the objective function attempts to reach. The objective function also exists in other fields of administrative sciences (such as linear programming). System dynamics utilizes these approaches in enhancement of optimization software to produce more robust policies (Coyle, 1996: 10-13).

DELAYS IN SYSTEM DYNAMICS

An important characteristic of various dynamic systems is the presence of delays that control the system, which are much more difficult to comprehend than it might seem at the first glance. Delays are a characteristic of dynamic systems and affect both material and information flow. The most common ones are exponential and pipeline-type delays (Sezen and Günal, 2009: 326).

Delays are inherent in many administrative processes. For example, it takes time to manufacture a product, to hire or lay out workers, to build a new factory, or to set up services (Kirkwood, 1998: 73).

In principle, there are delays in all flow channels. On the other hand, entering a time delay for each flow results in a quite unnecessarily detailed model, since most of these affect system behavior only very little. When introducing delays into the system, the number of delays could be reduced by two types of shortcuts. First is the case where the system delay is so short that its effect could be ignored. Second, the consecutive delays, which are actually separate, are usually combined and expressed as a single delay (Forrester, 1962: 86).

The delays are actually conversion processes that accept the speed of the inflow and transfer it out at the outflow speed. Delays are a special case of levels. Time delays are indicated as packets in the flow channels. By changing the time relation in the inflow, they create the outflow time relation (Forrester: 1962: 87). The structure of delays is presented in Figure 4.

There are some delays in all feedback processes, but they often cannot be recognized. Unrecognized delays could cause instability or collapse, especially when they last longer. For example, a delay of 10 seconds before setting the shower temperature is more difficult than a delay of one or two seconds (Senge, 1994: 75-76).

The water remains cold during the next ten seconds after the hot water tap is turned on. You would continue to turn the faucet, thinking that the action was ineffective. Boiling water at one hundred degrees finally flows through the faucet and then you would turn the faucet on the other direction. After a delay, the water cools down again. This balancing cycle continues until the desired water temperature is obtained. Figure 8 illustrates this process (Senge, 1994: 75-76).

Two characteristics of the delay are significant. The first is the length of time that reflects the average delay. This determines the steady state of the delay. The steady state is the stability between the levels of inflow and outflow rates. In the steady state, the average time multiplied by the flow rate gives the transferred amount during standby. The second characteristic of delay is the transient response. The transient response determines how the outflow time profile changes based on the inflow time profile (Forrester: 1962: 87).

Delays are divided into two categories: material and information delay. Material delays will be discussed first.

Figure 4. The structure of delays
Source: Yamaguchi, 2013:47.

Figure 5. Setting the shower temperature
Source: Sezen and Günal, 2009: 306.

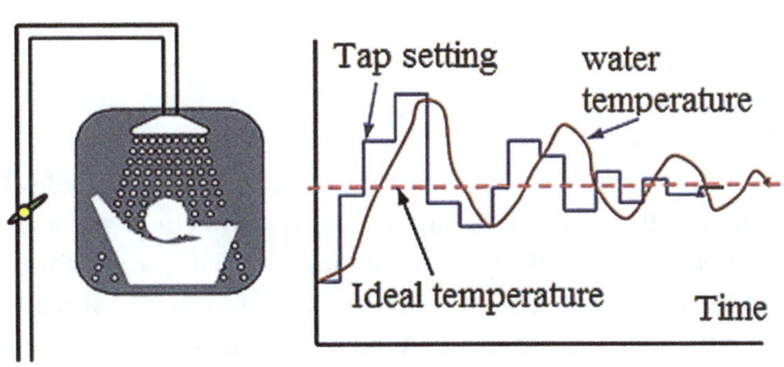

Figure 6. A delay example: Setting the shower temperature
Source: Senge, 1994: 76.

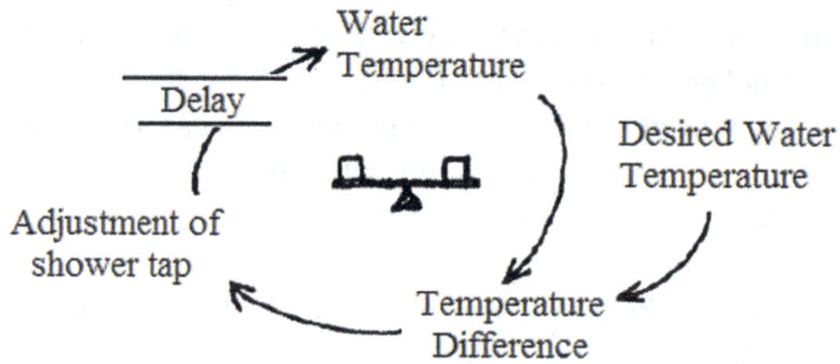

Figure 7. Delay chart
Source: Senge, 1994: 76.

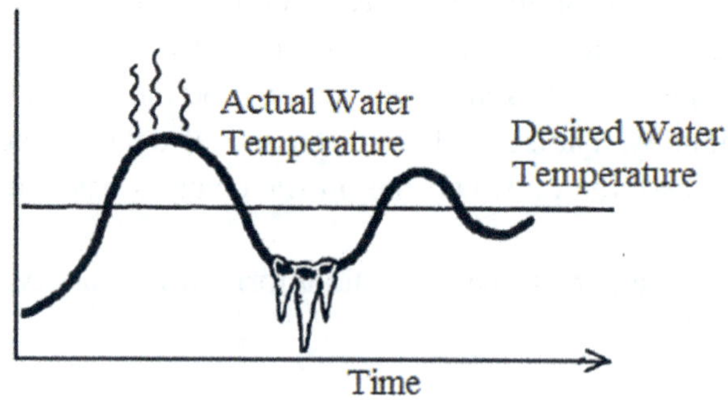

Material Delays

Material delay in its simplest form is the pipeline delay. In this structure, the material moves forward from the beginning to the end of the pipeline without any change during a period of time, similar to a water running through a pipeline (Kirkwood, 1998: 73). First degree material delay is presented in Figure 8.

When there is only one stock, delays are similar to exponential deterioration, which is called a pulse. As an example, 100 units of purchased goods will be transported and stored. 100 units of goods could be purchased and stored in 6 days. The delay time in this example is considered as 6 days. In other words, 1/6 of the goods are shipped as daily output (Yamaguchi, 2013: 47). Material delays could be first degree or higher degrees (Weber, 2010: 13).

Information Delay

Information delays are usually caused by the transfer of information and the delay in the process of taking an action after the information is received. Information delays could occur in vertical hierarchical administrations. For example, there is a delay between the time manager is informed about a sale in the store and the time the same manager places an order for exhausted stocks (Roberts, 2001: 311).

The reason for information delay is the fact that the input information needs to be processed by the human brain to be implemented as an action. This process is related to the perception of the actual state

Figure 8. First degree material delay
Source: Yamaguchi, 2013:48.

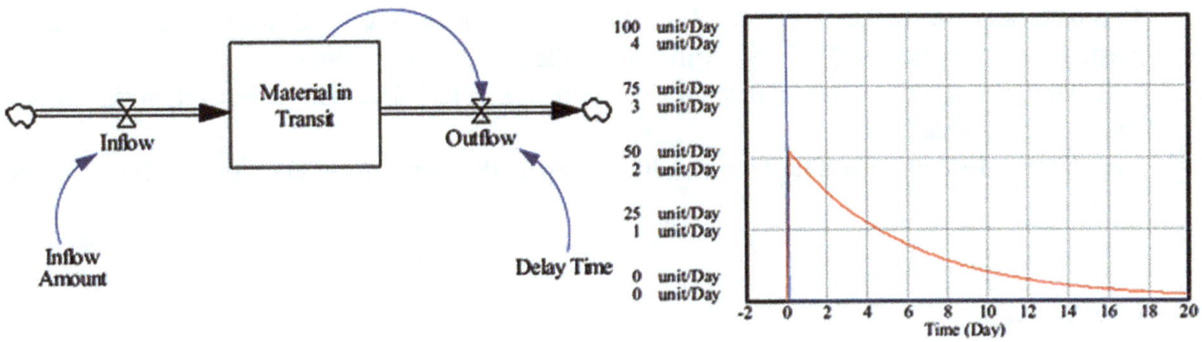

outside our brain accurately by our brain. Structurally, this process is similar to the negative feedback process (Yamaguchi, 2013: 47). Figure 9 demonstrates a first level information delay.

TIME PROCESSING IN SYSTEM DYNAMICS

System dynamics approach usually utilizes the time framing method. The modeler must determine the length of the time frame in advance. One important issue to note here is the accurate selection of the time frame length. If the time frame is extremely long for model behavior, the modeler will miss significant variations, hence, the simulation could provide misleading results. If the time frame is extremely short for model behavior, then model would be examined even at times when a change in the state is not possible, causing excessive collection of data and excessive computer processing time. It is recommended to select a sufficiently short time frame length generally (Sezen, 2009: 320).

System dynamics simulation computes the system status at certain points in time limited by dt, as presented in Figures 10 and 11. Since a constant timeframe such as dt is used in system dynamics, we know when the previous (t-dt) and next calculations (t + dt) will take place (Sezen and Günal, 2009: 320).

Figure 9. First level information delay
Source: Yamaguchi, 2013:47.

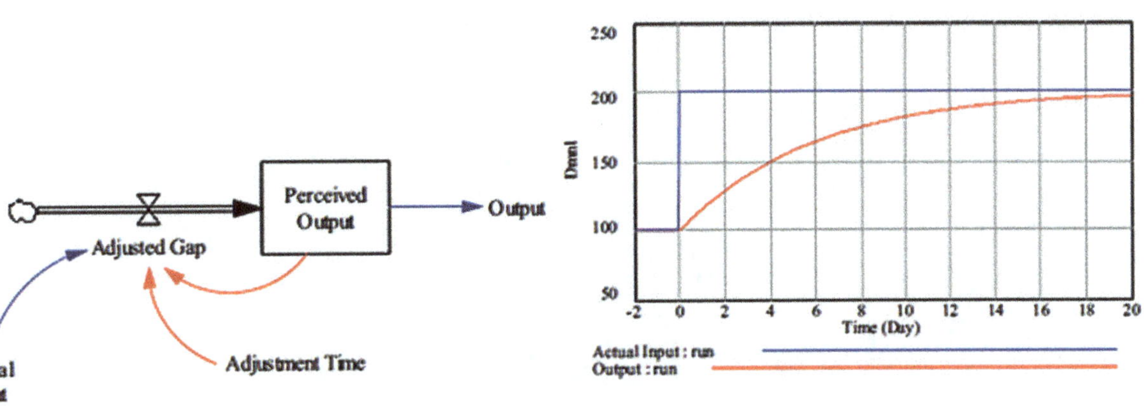

The basic method of time management is as follows:

1. At time t, calculate the new values for levels or stocks using the level and stock equations. For this purpose, use the calculated values at time t-dt or the current flow values.
2. Calculate the values of rates or flows for the next time interval dt. These depend on the current level value information.
3. Step the time further for one time frame (dt) and repeat the same process (Sezen and Günal, 2009: 321).

Figure 10. Timeframe
Source: Flood and Jackson, 1991.

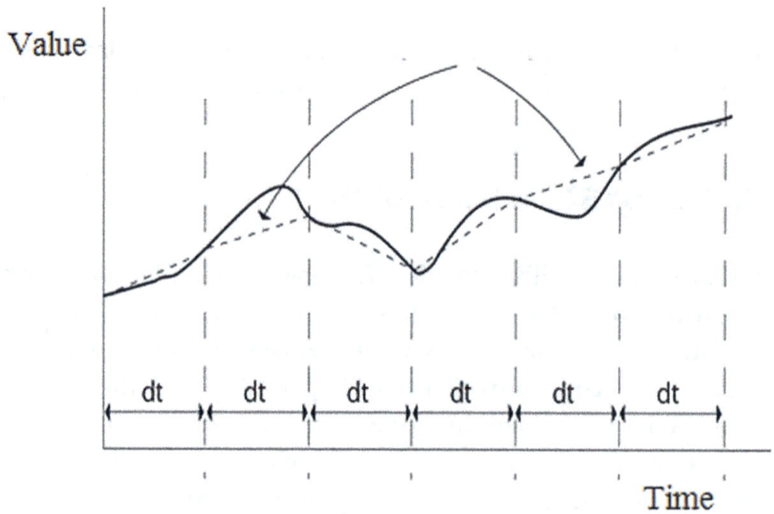

Figure 11. Timeframing in system dynamics
Source: Sezen and Günal, 2009: 320.

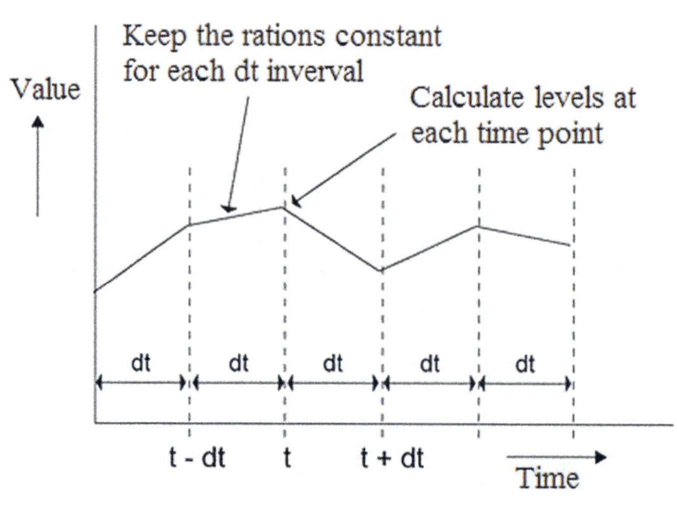

Thus, levels are calculated for the current point in time, and then, the rates are calculated for the next known frame. The rates are assumed to be constant in the frame once they are calculated and are revised only after the next calculation of the levels (Sezen and Günal, 2009: 321).

CLASSIFICATION OF EQUATIONS

Equations can be divided into three classes. These are level equations, rate equations and auxiliary variable equations.

Level Equations

Levels are the changing contents of system stocks (inventory). Even when there is no existing flow in the system, levels are present (Forrester, 1962: 76). A level equation could be expressed as follows (Sezen and Günal, 2009: 322):

$$Level(Current) = Level(Previous) + dt*(Inflow - Outflow) \quad (14)$$

A level equation reflects the accumulation of flow rates that cause increases and decreases in the level content. The new value of the level is calculated by adding to or subtracting from the previous value (Forrester, 1971: 5-7).

Rate Equations

Rate equations define the ratio of flow between the levels of the system. A rate equation is calculated from the existing value of the level. Rates cause changes in levels at the end (Forrester, 1962: 77).

Rate equations demonstrate how the flows in the system are controlled. The rate equation inputs are the system levels and constants, while the output is the control of the flow between the levels (Forrester, 1971: 5-9).

Auxiliary Variable Equations

There are also auxiliary variables in system dynamics models, which are defined as a function of stocks and these are influenced by internal variables and constants and affect the flow variables (Sterman, 2000: 107).

Rate equations would be very complicated if they were only expressed in terms of levels. Rate equations could also be expressed by one or more concepts that are independent of one another. The auxiliary variable equations help to keep the model structure close to the actual system by dividing the rate equations into component equations. As the name implies, the auxiliary variable is insignificant, but only "auxiliary" (Forrester, 1962: 78).

CALCULATION METHODS IN SYSTEM DYNAMICS

In general, the analytical (integral) solution of differential equations is quite difficult or impossible. Usually, numerical approaches are used to obtain solutions (Yamaguchi, 2013: 30). The analytical solution is the real solution. Intermittent approaches are used as an alternative to obtain a solution only when it is difficult to achieve the real solution:

$$x(t) = \sum_{i=0}^{t-1} i \tag{15}$$

The calculations made with this approach are unfortunately far from the real value. Two algorithms are proposed to reduce the difference. These are Euler Cauchy and Runge-Kutta methods.

Euler Cauchy (Continuous Flow *dt→0*)

The first algorithm is the reduction of the intermittent periods of the flow: dt →0, thus, the intermittent flows demonstrate a near-continuous flow (Yamaguchi, 2013: 30).

To examine the effect of this method, let us consider the following equations:

Level= dt * InflowRate –dt* OutflowRate

InflowRate= OutflowRate / A when A is a constant

Figure 12. Effect of the different values of dt in Euler-Cauchy Integral
Source: Sezen and Günal, 2009: 325.

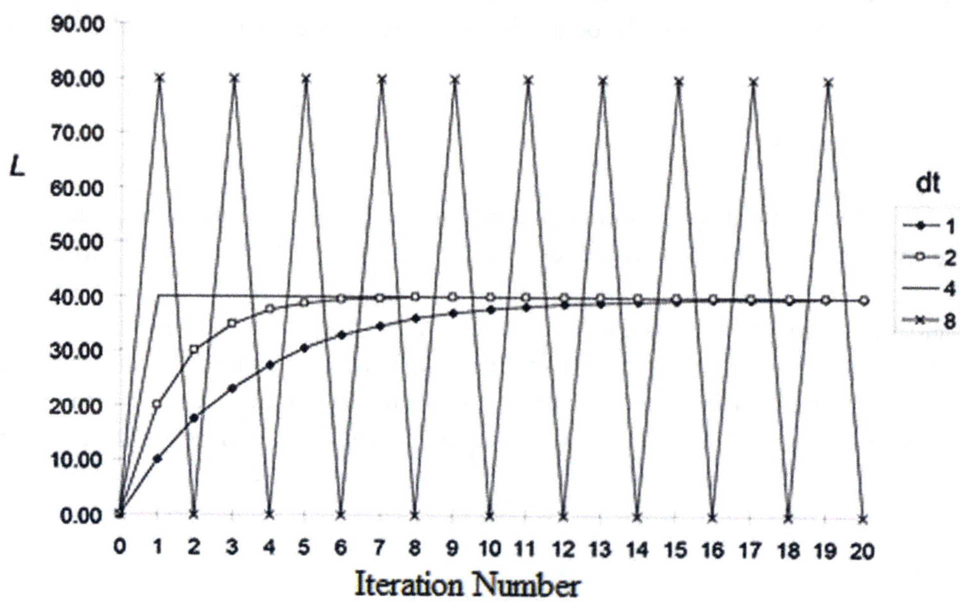

If we write this as a differential equation;

$$\frac{dL}{dt} = I - \frac{L}{A} \tag{16}$$

L and I denotes the level and InflowRate. Euler-Cauchy integral formula for the OutflowRate is as follows:

$$L(n+1) = L(n) + dt\left[I(n).L(n) \,/\, A\right] \tag{17}$$

The Euler-Cauchy numerical integration method is iterative, in this case it utilizes the dT interval (Sezen and Günal, 2009: 325). If we express the Euler method graphically, for each intermittent time period, it is the sum of the areas of the rectangles under the curve. Thinner the rectangles, closer the calculation to the actual result (Yamaguchi, 2013: 30).

$$\frac{dq(t)}{dt} = \lim_{h \to 0} \frac{q(t+h) - q(t)}{h} \tag{18}$$

Figure 13. Euler Cauchy
Source: Zeigler, Praehofer and Kim, 2000: 51.

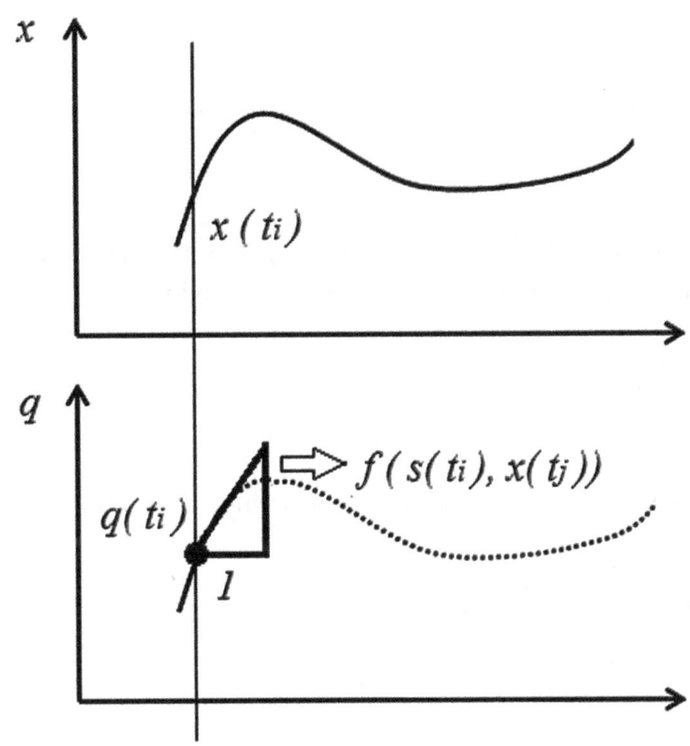

Although it is a simple method, it contains certain drawbacks. On one hand, to obtain consistent results, dt should be as small as possible. However, as step sizes get smaller, more iterations will be needed. This increases the computation time. On the other hand, it is not possible to decrease step sizes indefinitely, because computer processors start to rounding up these values after a certain value. Thus, another algorithm was proposed (Zeigler, Praehofer and Kim, 2000: 52).

The Euler method is still significant, because more advanced algorithms have been derived using the basic ideas provided by this method. The whole of these algorithms is called the Runge-Kutta method (Pegden, 1995: 436).

Runge-Kutta

All Runge-Kutta methods are the solution algorithms for the differential equation given below:

$$x(t+\Delta t) = x(t) + \Phi \Delta t \tag{19}$$

where, Φ is the estimator of dx / dt in time interval Δ t. This estimator is a weighted sum of x (t) values in two or more Δ t interval. Although the Runge-Kutta methods are first-degree derivative assessments, they could also be used with *Taylor series* expansion at higher degrees. The derivative term in the Taylor series expansion that corresponds to this algorithm method is called the level of the method. For example, the Euler method we have described above is the Runge-Kutta method from the first degree. Runge-Kutta methods could be developed at higher degrees to obtain higher accuracy results when compared to the Euler method (Pegden, 1995: 436).

A second degree Runge-Kutta algorithm is as follows. Accordingly, the value at the midpoint of dt is taken as f (t) value.

$$x(t+dt) = x(t) + f\left(t + \frac{dt}{2}\right)dt \tag{20}$$

For instance, for a flow function $f(t) = t^2$ that only depends on time, if we accept the initial stocks as $x_0 = 0$, second degree Runge-Kutta algorithm would be as follows:

$$f\left(t + \frac{dt}{2}\right) = t^2 + tdt + \frac{dt^2}{4} \tag{21}$$

And a fourth degree Runge-Kutta algorithm could be written as follows (Yamaguchi, 2013: 31-33):

$$x(t+dt) = x(t) + \frac{f(t) + 4f\left(t + \frac{dt}{2}\right) + f(t+dt)}{6}dt \tag{22}$$

Table 3. Comparison of second and fourth degree Runge-Kutta methods

t	x(t)	Runge-Kutta 2	x(t)	Runge-Kutta 4
0	0	0,25	0	0,33
1	0,25	2,25	0,33	2,33
2	2,5	6,25	2,66	6,33
3	8,75	12,25	9	12,33
4	21	20,25	21,33	20,33
5	41,25	30,25	41,66	30,33
6	71,5		72	

Source: Yamaguchi, 2013: 33.

VALIDITY OF SYSTEM DYNAMICS MODELS

The computerized simulation model could be used as a powerful tool to test how policies and structure that generate growth and fluctuations in socio-economic systems interact. For the effective use of the model, it needs to reflect the actual system (Forrester, 1977: 541).

Model validity is as controversial topic as it is significant in all model-based methods, especially in system dynamics. The validity of the results of the model-based study depends on the validity of the model. Nevertheless, there is no single well-agreed definition of model validity and model validation (Barlas, 1996: 183).

Significance of a model is measured by how well it serves its objective. The objective of system dynamics models is to help improve the decisions made. The validity of a model should be decided by its adequacy for the unique objective of the model. If a model serves its purpose, it could be justified and is significant. Thus, validity is an abstract concept separated from the objective. A model that is perfect for an objective might be misleading and useless for another objective (Forrester, 1962: 115).

To investigate model validity, two topics should be discussed. First, we cannot expect a model to be completely valid. A model has been constructed for a specific objective and its validity should be tested for this objective. However, a model investigated for other objectives might be unreliable. Second, we must realize that positive proof is impossible. The difficulty begins with selecting the validity criteria. There is no absolute proof, but there is a confidence interval for measuring the link between the model, the actual system and the objective. Concurrently, if two different models are measured, there is no assurance that the model that better reflects the measurement findings is the most efficient model (Forrester, 1975: 162).

The simplest model validity form is called the black box validity. In this method, the model is run under well-defined conditions and findings are obtained. The conditions chosen for running the model should be as similar as possible to the actual system. Then, the simulation model should be rendered compatible with the actual system. Thus, the model results should be the same as the actual system results. This is called the black box validity, because the topics of interest are only the system inputs and outputs. The internal structure of the model is not addressed. In this method, we cannot be sure that the model properly describes the operation of the actual system. In order to make sure of this, we must use a method called white box validity. The idea behind white-box validity is to examine the model's operation in more detail and to verify that the simulated system reflects the operation of the actual system appropriately. This validation method examines the programming logic, rather than investigating the result of the model (Fishwick, 2007: 33-21).

The most important aim of the model validity in system dynamics is verification of the model structure; although the validity of the results produced by the model could be assessed, this is only meaningful when the model structure itself is valid. Thus, verification should be conducted with the model structure first and then the behavior of the model should be verified (Barlas, 1994).

Validity could be examined both structurally and behaviorally. Structural validity could be defined as building trust on the fact that the relationships used in a model represent the actual relationships accurately. Barlas (1994) indicated two types of structural tests: 1-Direct Structural Tests and 2-Structural Behavior Tests. Direct building tests provide model validity only by comparing model equations with available information. In the direct structural test, consistency comparisons are made directly on the model without using a simulation. These tests could be empirical and theoretical. The empirical test involves comparing information obtained from the actual system modeled with equations of the model. On the other hand, the theoretical tests include a comparison of model equations with generalized information on the existing system available in the literature. Both tests are important for direct structural validity. Forrester and Senge (1980) gave the examples of structure and parameter validation test, direct end-value test and unit consistency test for direct structural tests. Parameter validation test is to assess the fixed parameters conceptually and numerically using the actual real information. The direct end-value test involves the evaluation of model equations with extreme values, and is a comparison of the result values with the information about what would occur under similar circumstances in real life. Finally, the unit consistency test analyzes the unit consistency of the model equations (Barlas, 1994, Forrester and Senge, 1980).

The second general category of structural tests is structural behavior tests. These tests assess model validity indirectly by testing certain behavior using behavior patterns created in the model. Indirect structural tests include specialized simulated runs and could provide indirect information about probable errors in model construction. Among these tests, end-value, behavioral sensitivity, limit compliance, and phase-correlation tests could be listed. In a behavioral test for a typical structure, the analyst makes the following proposition: "If the system works in condition K, it should demonstrate behavior D". If the model works under K conditions and yields similar behavior to the expected behavior, it passes this test. End-value behavior tests are the comparison of behavior created by the model by assigning extreme values to selected parameters and the observed or predicted behavior of the actual system with these extreme values. The behavioral sensitivity test aims to determine the parameters that are sensitive to the model and to test whether the actual system has the same high level of sensitivity for these. On the other hand, the adaptive behavior estimation test could be applied where it is possible to find data about the behavior of the adaptive version of the actual system. The model passes this test if it could produce similar adaptive behavior when simulated with similar adaptations. These two tests are designed to evaluate the validity of the model with the model structure. As a result of these tests, if there is sufficient level of confidence about the model structure, we could begin to apply a number of tests designed to measure how accurate the model reproduces the large behavioral patterns observed in the actual system behavior.Here, It is very important to emphasize the sampling estimate (period, frequency, trend wave size...) instead of point estimation. This is a logical consequence of the fact that system dynamics models have a tendency towards long term politics. Behavioral sample tests could be expressed as the multiple test procedure proposed by Barlas (1985, 1989), the general statistical summary proposed by Sterman (1984), and several behavioral sample tests discussed by Forrester and Senge (1980) (Barlas, 1996, Forrester and Senge 1980, Barlas and Saysel, 1996, Kanar, 1995).

CONCLUSION

Modeling energy system for countries is a complex problem due to the presence of multiple decision makers, complexity of consumers' behaviors, feedback processes among module, technological limitations and various kinds of delays. System dynamics is a suitable approach to model such complexities (Ansari and Seifi, 2013).

This chapter aims to explore the system dynamics method for developing energy policy model. System dynamics deal with strategic planning for complex problems. Developing a system dynamic model for designing energy policy will be studied in future studies.

REFERENCES

Al-Sarihia, A., Contestabıleb, M., & Chernia, J. A. (2015). Renewable Energy Policy Evaluation Using A System Dynamics Approach: The Case of Oman. *Proceedings of the 33rd International Conference of the System Dynamics Society.*

Barlas, Y. (1994). Model Validation in System Dynamics. *The 12th International Conference of the System Dynamics Society Stirling.*

Barlas, Y. (1996). Formal Aspects of Model Validity and Validation in System Dynamics. *System Dynamics Review, 12*(3), 183–210. doi:10.1002/(SICI)1099-1727(199623)12:3<183::AID-SDR1033.0.CO;2-4

Barlas Y. (2005b). Dinamik Sistem Yaklaşımı: Modeller, Kurumsal Öğrenme Ve Sorun Çözme - II. *Endüstri Ve Otomasyon Dergisi,* (95).

Batinge, B. (2015). Sustainable Energy Future: A System Dynamics Approach To Solving The Electricity Shortfall in Ghana. *The 33rd International Conference of the System Dynamics Society.*

Coyle, R. G. (1996). *Systems Dynamics Modeling.* London: Chapman & Hall. doi:10.1007/978-1-4899-2935-8

Ebrahimi, M. (2016). Policy Making in a Renewable Energy Manufacturing SME Using System Dynamics. *34th International Conference of the System Dynamics Society.*

Erkut, H. (1983). *Sistem Dinamiğinin Temelleri.* İstanbul: İTÜ Fen Edebiyat Fakültesi Ofset Atölyesi.

Fazeli, R., & Davidsdottir, B. 2015). Energy Modeling of Danish Housing Stock Using System Dynamics. *The 33rd International Conference of the System Dynamics Society.*

Fıshwick, P. A. (2007). *Handbook of Dynamic System Modelling.* Boca Raton, FL: Chapman & Hall / CRC. doi:10.1201/9781420010855

Flood, R. L., & Jackson, M. C. (1991). *Creative Problem Solving: Total Systems Intervention*. Chichester, UK: John Wiley & Sons.

Forrester, J. W. (1962). *Industrial Dynamics*. Waltham, MA: Pegasus Communications.

Forrester, J. W. (1969). *Urban Dynamics*. The MIT Press.

Forrester, J. W. (1971). *Principles Of Systems* (2nd ed.). Cambridge, MA: Wright Allen Press.

Forrester, J. W. (1977). Growth Cycles. *De Economist*, *125*(: 4).

Forrester, J. W., & Senge, P. M. (1980). Tests For Building Confidence In System Dynamic Models. *TIMS Studies in the Management Sciences*, *14*, 209–228.

Kanar, K. (1995). Structure Oriented Behavior Tests. In *Model Validation, Master Of Science*. Industrial Engineering Boğaziçi University.

Karnopp, D. C., Margolis, D. L., & Rosenberg, R. C. (1990). *System Dynamics A Unified Approach* (2nd ed.). John Wiley & Sons, Inc.

Kirkwood, C. W. (1998). *System Dynamics Methods: A Quick Introduction*. College Of Business Arizona State University.

Lane, D. J., & Sterman, J. D. (2011). Profiles in Operations Research: Jay Wright Forrester. In Profiles in Operations Research: Pioneers and Innovators. New York: Springer.

Martın, L. A. (1997a). *Road Map 2: The First Step*. MIT System Dynamics in Education Project.

Pegden, C. D., Shannon, R. E., & Sadowski, R. P. (1995). *Introduction To Simulation Using SIMAN*. New York: McGraw-Hill.

Peter, S., Kleıner, A., Roberts, C., Ross, R., & Smith, B. (1994). The Fifth Discipline Fieldbook. Century.

Qudrat-Ullah, H. (2013). *Energy Policy Modeling in the 21st Century*. New York: Springer. doi:10.1007/978-1-4614-8606-0

Ramage, M., & Ship, K. (2009). *Sytem Thinkers, The Open University Walton Hall*. Milton Keynes, UK: Springer London.

Roberts, N. (2001). *Introduction to Computer Simulation A System Dynamics Modeling Approach*. System Dynamics Group Road Map Series.

Ruth, M., & Hannon, B. (2012). *Modeling Dynamic Economic Systems* (2nd ed.). Springer. doi:10.1007/978-1-4614-2209-9

Sezen, H. K., & Günal, M. (2009). *Yöneylem Araştırmasında Benzetim*. Bursa: Ekin Yayınevi.

Sterman, J. D. (1991). *A Skeptic's Guide to Computer Models*. Cambridge, MA: MIT System Dynamics In Education Project.

Sterman, J. D. (2000). Business Dynamics Systems Thinking And Modelling. In *A Complex World*. New York: Mcgraw-Hill.

Vilchez, J. G., Jochem, P., & Fichtner, W. (2015). Energy Use and Emissions Impacts from Car Technologies Market Scenarios: A Multi-Country System Dynamics Model. *33rd International Conference of the System Dynamics Society.*

Vilchez, J. G., Jochem, P., & Fichtner, W. (2016). Car Technology Market Evolution And Emissions Impacts An Example Of Energy Policy Scenarios Under Uncertainty. *34th International Conference of the System Dynamics Society.*

Weber, L. (2010). *Demographic Change and Economic Growth Simulations on Growth Models.* Springer Verlag Berlin. doi:10.1007/978-3-7908-2590-9

Yamaguchi, K. (2013). *Money and Macroeconomic Dynamics- Accounting System Dynamics Approach.* Awaji Island, Japan: Japan Future Research Center.

Zeigler, B. P., Praehofer, H., & Kim, T. G. (2000). *Theory Of Modelling And Simulation.* London, UK: Academic Press.

KEY TERMS AND DEFINITIONS

Decision Functions: States of the policy that determine how to convert available information in stock into a decision.

Delays: A characteristic of dynamic systems and affect both material and information flow.

Flow: It is defined as the increase or decrease in the unit time interval in stocks and denoted by $f(t)$.

Information Delay: They are usually caused by the transfer of information and the delay in the process of taking an action after the information is received. Information delays could occur in vertical hierarchical administrations.

Information Flow: Creation, control, and distribution of information is the most significant task of business management. It might be difficult to model the information properly in the business process.

Material Delays: In this structure, the material moves forward from the beginning to the end of the pipeline without any change during a period of time, similar to a water running through a pipeline.

Material Flow: It includes stocks and flow rates of physical goods such as raw materials, inventories in the process, or finished goods.

System Dynamic Language: This language consists of four components: stocks, flows, decision functions, and information flow.

Stocks: The present values of the variables that are formed by the accumulated difference between the inflow and outflow.

System Dynamics: The basis of system dynamics is to understand how system structures cause system behavior and system events.

Compilation of References

Abd-Elazim, S. M., & Ali, E. S. (2016). Load frequency controller design via BAT algorithm for nonlinear interconnected power system. *International Journal of Electrical Power & Energy Systems*, *77*, 166–177. doi:10.1016/j.ijepes.2015.11.029

Abdelaziz, A. Y., & Ali, E. S. (2015). Cuckoo Search algorithm based load frequency controller design for nonlinear interconnected power system. *International Journal of Electrical Power & Energy Systems*, *73*, 632–643. doi:10.1016/j.ijepes.2015.05.050

Abdulkadir, M., Yatim, A. H. M., & Yusuf, S. T. (2014). An Improved PSO-Based MPPT Control Strategy for Photovoltaic Systems. *International Journal of Photoenergy*, *2014*, 1–11. doi:10.1155/2014/818232

Abdul-Rahman, K., & Shahidehpour, S. (1993). A fuzzy-based optimal reactive power control. *IEEE Transactions on Power Systems*, *8*(2), 662–670. doi:10.1109/59.260814

Abed, A. M. (1999). WSCC voltage stability criteria, undervoltage load shedding strategy, and reactive power reserve monitoring methodology. In *IEEE Power Engineering Society Summer Meeting* (Vol. 1, pp. 191-197). IEEE. 10.1109/PESS.1999.784345

Abhyakar, A. R., & Khaparde, S. (n.d.). *Introduction to Deregulation in Power Industry*. Retrieved from http://www.nptel.ac.in/courses/108101040/Module 7/L01-Introduction to Deregulation-1.pdf

Abogaleela, M., Elsobki, M., & ElMarsafawy, M. (2012). *A two level optimal DSM load shifting formulation using genetics algorithm, case study: residential load, Power Africa*. IEEE PES. doi:10.1109/PowerAfrica.2012.6498651

Aboushady, H., Dumonteix, Y., Louerat, M. M., & Mehrez, H. (2000). Efficient polyphase decomposition of Comb decimation filters in/spl Sigma//spl Delta/analog-to-digital converters. In *Circuits and Systems, 2000. Proceedings of the 43rd IEEE Midwest Symposium on* (*Vol. 1*, pp. 432-435). IEEE.

Abu-Rub, H., Holtz, J., Rodriguez, J., & Ge Baoming. (2010). Medium- voltage multilevel converters-state of the art, challenges and requirements in industrial applications. *IEEE Transactions on Industrial Electronics*, *57*(8), 2581–2596. doi:10.1109/TIE.2010.2043039

Ackermann, T., Andersson, G., & Soder, L. (2001). Distributed Generation : A Definition. *Electric Power Systems Research*, *57*, 195–204.

Adams, S. F. (2006). Solar thermionic space power technology testing: A historical perspective. In. AIP Conference Proceedings: Vol. 813. *No. 1* (pp. 590–597). AIP.

Adaryani, M. R., & Adaryani, A. K. (2013). Artificial bee colony algorithm for solving multi-objective optimal power flow problem. Electrical Power and Energy Systems, 53, 219-230.

Aditya, S. K., & Das, D. (2000). Load-frequency control of an interconnected hydrothermal power system with new area control error considering battery energy storage facility. *International Journal of Energy Research*, *24*(6), 525–538. doi:10.1002/(SICI)1099-114X(200005)24:6<525::AID-ER6033.0.CO;2-6

Aghatehrani, R., Fan, L., & Kavasseri, R. (2009). Coordinated reactive power control of DFIG rotor and grid sides converters. In IEEE Power & Energy Society General Meeting (pp. 1-6). IEEE. doi:10.1109/PES.2009.5275235

Ahmad & Gupta. (2013). Digital pwm of cascaded multilevel voltage source inverter using fpga. *Engineering and Systems*, 12-14.

Ahmadi, A., & Aldeen, M. (2017). Robust overlapping load frequency output feedback control of multi-area interconnected power systems. *International Journal of Electrical Power & Energy Systems*, *89*, 156–172. doi:10.1016/j.ijepes.2016.12.015

Ahmadi, A., Alinejad-Beromi, Y., & Moradi, M. (2011). Optimal PMU placement for power system observability using binary particle swarm optimization and considering measurement redundancy. *Expert Systems with Applications*, *38*(6), 7263–7269. doi:10.1016/j.eswa.2010.12.025

Aichouchi, M. (1994). *Etude des liens entre la synthese architecturale et la synthese au niveau transfert de registres* (Doctoral dissertation). Institut National Polytechnique de Grenoble-INPG.

Ajami, A., Mokhberdoran, A., & Reza, M. (2013). A new topology of multilevel voltage source inverter to minimize the number of circuit devices and maximize the number of output voltage levels. *Journal of Electrical Engineering & Technology*, *8*(6), 1321–1329. doi:10.5370/JEET.2013.8.6.1328

Ajami, A., & Younesi, M. (2008). Modeling and state feedback controller for current source inverter based STATCOM. *IEEE International Conference on Control, Automation and System*, 2418-2423.

Ajjarapu, V., & Lee, B. (1998). Bibliography on voltage stability. *IEEE Transactions on Power Systems*, *13*(1), 115–125. doi:10.1109/59.651622

Akagi, H. (1994). Trends in active power line conditioners. *IEEE Transactions on Power Electronics*, *9*(9), 263–268. doi:10.1109/63.311258

Akagi, H. (1996). New trends in active filters for power conditioning. *IEEE Transactions on Industry Applications*, *l*(32), 1312–1322. doi:10.1109/28.556633

Akagi, H. (2005). Active harmonic filters. *Proceedings of the IEEE*, *93*(12), 2128–2141. doi:10.1109/JPROC.2005.859603

Akagi, H., Kanazawad, Y., & Nabae, A. (1984). Instantaneous reactive power compensators comprising switching devices without energy storage components. *IEEE Transactions on Industry Applications*, *IA-20*(20), 625–630. doi:10.1109/TIA.1984.4504460

Akhtar, J., Shamsudheen, P. M., (2012). Power Quality Improvement Using Fuzzy Logic Control Static Var Compensator In Power System Network. *IOSR Journal of Engineering*, *2*(8), 1-8.

Al Hanai, T., Bani Hashim, R., El-Chaar, L., & Lamont, L. (2010). Study of a 900 W, Thin-film, Amorphous Silicon PV System in a Dusty Environment. *International Conference on Renewable Energy: Generation and Applications - ICREGA'10*.

Albadawi, N. (2014). *SCCC701 Engineering Development Project "Development of an automatic robotic cleaning system for photovoltaic plants"*. Available on: www.researcharchive.wintec.ac.nz/3361/1/Final%20reportPDF.pdf

Alexandridis, A. T., & Marmidis, G. E. (2010). Modeling wind generators with full-scale frequency converters: Stability and passivity properties. *7th Mediterranean Conference and Exhibition on Power Generation, Transmission, Distribution and Energy Conversion*, 1-5. 10.1049/cp.2010.0870

Alghuwainem, S. M. (1992). Steady-state performance of DC motors supplied from photovoltaic generators with step-up converter. *IEEE Transactions on Energy Conversion*, *7*(2), 267–272. doi:10.1109/60.136220

Al, H., Hassan, H. A. J., Pelov, A., & Nuaymi, L. (2015). Integrating Cellular Networks, Smart Grid, and Renewable Energy: Analysis, Architecture, and Challenges. *IEEE Access: Practical Innovations, Open Solutions*, *3*, 2755–2770.

Alhelou, H. H., & Golshan, M. E. H. (2016, May). Hierarchical plug-in EV control based on primary frequency response in interconnected smart grid. In *Electrical Engineering (ICEE), 2016 24th Iranian Conference on* (pp. 561-566). IEEE.

Alhelou, H. H., Golshan, M. E. H., & Askari, J. (in press). Robust Sensor Fault Detection and Isolation Scheme for Interconnected Smart Power Systems in Presence of RER and EVs using Unknown Input Observer. *International Journal of Electrical Power & Energy Systems.*

Alhelou, H. S. H., Golshan, M. E. H., & Fini, M. H. (2015, December). Multi agent electric vehicle control based primary frequency support for future smart micro-grid. In *Smart Grid Conference (SGC)*, (pp. 22-27). IEEE. 10.1109/SGC.2015.7857385

Ali, I., Thomas, M. S., & Kumar, P. (2013). Distributed resource planning for improved voltage stability of radial distribution system. *International Journal of Advanced Research in Electrical, Electronics and Instrumentation Engineering,* *2*(1), 95-101.

Ali, I., Thomas, M. S., & Kumar, P. (December 2012), Effect of loading pattern on the performance of reconfigured medium size distribution system. *Proc. of IEEE 5th Power India Conference*, 1-6.

Ali, I., Thomas, M. S., & Kumar, P. (2015). Energy efficient reconfiguration for practical load combination in distribution system. *IET Generation, Transmission & Distribution*, *9*(11), 1051–1060.

Ali, I., Thomas, M. S., & Kumar, P. (2015, August). Energy efficient reconfiguration for practical load combination in distribution system. *IET Generation, Transmission & Distribution*, *9*(11), 1051–1060. doi:10.1049/iet-gtd.2014.0915

Aljarah, I., Faris, H., & Mirjalili, S. (2016). Optimizing connection weights in neural networks using the whale optimization algorithm. *Soft Computing*, 1–15.

Al-Mawsawi, S. A., Qader, M. R., & Ali, G. M. (2004). Dynamic Controllers Design for STATCOM. *Iranian Journal of Electrical and Computer Engineering*, *3*(1), 16–22.

Al-Mohammed, A., Abido, M. A., & Mansour, M. (2011). *Optimal PMU placement for power system observability using differential evolution.* Paper presented at the Intelligent Systems Design and Applications (ISDA), 2011 11th International Conference on. 10.1109/ISDA.2011.6121668

Alsac, O., Scott, B. (1974). Optimal load flow with steady state security. *IEEE Trans Power Appar Syst.*, 745–51.

Al-Sarihia, A., Contestabıleb, M., & Chernia, J. A. (2015). Renewable Energy Policy Evaluation Using A System Dynamics Approach: The Case of Oman. *Proceedings of the 33rd International Conference of the System Dynamics Society.*

Alsumait, J. S., Sykulski, J. K., & Al-Othman, A. K. (2010). A hybrid GA–PS–SQP method to solve power system valve-point economic dispatch problems. *Applied Energy*, *87*(5), 1773–1781. doi:10.1016/j.apenergy.2009.10.007

Alva, G., Liu, L., Huang, X., & Fang, G. (2017). Thermal energy storage materials and systems for solar energy applications. *Renewable & Sustainable Energy Reviews*, *68*, 693–706. doi:10.1016/j.rser.2016.10.021

Alves Vitorino, M., & Beltrão de Rossiter Corrêa, M. (2011). An effective induction motor control for photovoltaic pumping. *IEEE Transactions on Industrial Electronics*, *58*(4), 1162–1170. doi:10.1109/TIE.2010.2054053

Anagnostopoulos, K., & Economakos, G. (2005, December). Lowpower design of multipliers using a full-adder isolation technique. In *Electronics, Circuits and Systems, 2005. ICECS 2005. 12th IEEE International Conference on* (pp. 1-4). IEEE. 10.1109/ICECS.2005.4633542

Ananthi, A. C., Ajay, D. P., & Vimal, R. (2014). Adaptive biogeography based predator–prey optimization technique for optimal power flow. *Electrical Power and Energy Systems (Elsevier)*, *62*, 344–352. doi:10.1016/j.ijepes.2014.04.054

Angrist, S. W. (1982). Direct energy conversion (4th ed.). Allyn and Bacon.

Angrist, S. W. (1965). *Direct energy conversion* (4th ed.). Boston: Allyn and Bacon Series in Mechanical Engineering and Applied Mechanics.

Angulo, M., Lezana, P., Kouro, S., Rodrıguez, J., & Wu, B. (2007). Level shifted pwm for cascaded multilevel inverters with even power distribution. *IEEE Power Electronics Specialists Conference*, 2373-2378.

Ansari, A. Q., Singh, B., & Hasan, M. (2015). Algorithm for power angle control to improve power quality in distribution system using unified power quality conditioner. *IET Generation, Transmission & Distribution*, *9*(9), 1439–1439. doi:10.1049/iet-gtd.2014.0734

Ansari, M. M. T., & Velusami, S. (2010). DMLHFLC (Dual mode linguistic hedge fuzzy logic controller) for an isolated wind diesel hybrid power system with BES (battery energy storage) unit. *Energy*, *35*(9), 3827–3837. doi:10.1016/j.energy.2010.05.037

An, W. F., & Tolliday, N. (2010). Cell-based assays for high-throughput screening. *Molecular Biotechnology*, *45*(2), 180–186. doi:10.100712033-010-9251-z PMID:20151227

Appelbaum, J. (1986). Starting and steady-state characteristics of dc-motors powered by solar cell generators. *IEEE Transactions on Energy Conversion*, *EC-1*(1), 17–25. doi:10.1109/TEC.1986.4765665

Archana, M., & Kondaiah, Y. C. V. (2012). Grid Interfacing Inverter of Renewable Energy Sources to Improve the Power Quality in Distribution System. *International Journal of Advanced Research in Electrical, Electronics and Instrumentation Engineering*, *1*(5), 411–418.

Aredes, M., Heumann, K., & Watanabe, E. H. (1998). An universal active power line conditioner. *IEEE Transactions on Power Delivery*, *1*(13), 545–551. doi:10.1109/61.660927

Arora, S., & Kumar, P. (2012). Infrastructural development of energy planning and energy management for smart grid implementation. *Journal of Electrical and Electronics Engineering*, *2*(1), 26-31.

Arroba, P., Moya, J. M., Ayala, J. L., & Buyya, R. (2017). Dynamic Voltage and Frequency Scaling-aware dynamic consolidation of virtual machines for energy efficient cloud data centers. *Concurrency and Computation*, *29*(10), e4067. doi:10.1002/cpe.4067

Arya, Y. (2017). AGC performance enrichment of multi-source hydrothermal gas power systems using new optimized FOFPID controller and redox flow batteries. *Energy*, *127*, 704–715. doi:10.1016/j.energy.2017.03.129

Arya, Y., & Kumar, N. (2016). Optimal AGC with redox flow batteries in multi-area restructured power systems. *Eng Sci Tech. International Journal (Toronto, Ont.)*, *19*(3), 1145–1159.

Ataollah, M. (2014). Symmetric and asymmetric design and implementation of new cascaded multilevel inverter topology. *IEEE Transactions on Power Electronics*, *29*(12), 6712–6724. doi:10.1109/TPEL.2014.2302873

Attia, A. F., Al-Turki, Y. A., & Abusorrah, A. M. (2012). Optimal power flow using adapted genetic algorithm with adjusting population size. *Electric Power Components and Systems*, *40*(11), 1285–1299. doi:10.1080/15325008.2012.689417

Azli, N. A., & Choong, Y. C. (2006). Analysis on the performance of a three-phase cascaded h-bridge multilevel inverter. *First International Power and Energy Conference*, 405-410. 10.1109/PECON.2006.346685

Baetens, R., Jelle, B. P., & Gustavsen, A. (2010). Properties, requirements and possibilities of smart windows for dynamic daylight and solar energy control in buildings: A state-of-the-art review. *Solar Energy Materials and Solar Cells*, *94*(2), 87–105. doi:10.1016/j.solmat.2009.08.021

Bai, P., Li, Q., Zhao, B., & Yang, Y. (2005). Power quality control center and its control method. *Proc.Transmiss. Distrib. Conf. Exhib.*, 1–6.

Bakirtzis, A. G., Biskas, P. N., Zournas, C. E., & Petridis, V. (2002). Optimal power flow by enhanced genetic algorithm. *IEEE Transactions on Power Systems*, *17*(2), 229–236. doi:10.1109/TPWRS.2002.1007886

Balamurugan, C. R., Natarajan, S. P., Bensraj, R., & Shanthi, B. (2016). A review on modulation strategies of multi level inverter. *Indonesian Journal of Electrical Engineering and Computer Science*, *3*(3), 681–705. doi:10.11591/ijeecs.v3.i3.pp681-705

Balasubbareddy, M., Sivanagaraju, S., & Chintalapudi, S. (2015). Multi-objective optimization in the presence of practical constraints using non-dominated sorting hybrid cuckoo search algorithm. *Engineering Science and Technology, an International Journal, 18*, 603-615.

Banaei & Hosseini. (2006). Mitigation of current harmonic using adaptive neural network with active power line conditioner. *Proc. IEEE 5th International Conference on Power Electronics and Motion Control*, 1–5.

Banerjee, N., Raychowdhury, A., Bhunia, S., Mahmoodi, H., & Roy, K. (2005, October). Novel low-overhead operand isolation techniques for low-power datapath synthesis. In *Computer Design: VLSI in Computers and Processors, 2005. ICCD 2005. Proceedings. 2005 IEEE International Conference on* (pp. 206-211). IEEE. 10.1109/ICCD.2005.80

Bansal, R. (2003). Bibliography on the fuzzy set theory applications in power systems (1994-2001). *IEEE Transactions on Power Systems*, *18*(4), 1291–1299. doi:10.1109/TPWRS.2003.818595

Barlas Y. (2005b). Dinamik Sistem Yaklaşımı: Modeller, Kurumsal Öğrenme Ve Sorun Çözme - II. *Endüstri Ve Otomasyon Dergisi,* (95).

Barlas, Y. (1994). Model Validation in System Dynamics. *The 12th International Conference of the System Dynamics Society Stirling.*

Barlas, Y. (1996). Formal Aspects of Model Validity and Validation in System Dynamics. *System Dynamics Review*, *12*(3), 183–210. doi:10.1002/(SICI)1099-1727(199623)12:3<183::AID-SDR1033.0.CO;2-4

Baruch, Z. (1996). Scheduling algorithms for high-level synthesis. *ACAM Scientific Journal*, *5*(1-2), 48–57.

Basetti, V., & Chandel, A. K. (2017). Optimal PMU placement for power system observability using Taguchi binary bat algorithm. *Measurement*, *95*, 8–20. doi:10.1016/j.measurement.2016.09.031

Basu, M., Das, S. P., & Dubey, G. K. (2008). Investigation on the performance of UPQC- for voltage sag mitigation and power quality improvement at a critical load point. *IET Generation, Transmission & Distribution*, *l*(2), 414–423. doi:10.1049/iet-gtd:20060317

Basu, M., Farrell, M., Conlon, F., Gaughan, K., & Coyle, E. (2004). Optimal control strategy of UPQC for minimum operational losses. In *Proc. 39th Int* (pp. 246–250). Univ. Power Eng.

Batinge, B. (2015). Sustainable Energy Future: A System Dynamics Approach To Solving The Electricity Shortfall in Ghana. *The 33rd International Conference of the System Dynamics Society*.

Beck, G., & Povh, D. (2005). *Global Blackouts – Lessons Learned*. POWER-GEN Europe 2005, Milan, Italy.

Becquerel, M. E. (1853). Researches on the electrical conductibility of gases at high temperatures. *Philosophical Magazine Series 4*, *6*(41), 456–457.

Behera, R. K., Dixit, T. V., & Das, S. P. (2006). Analysis of experimental investigation of various carrier-based modulation schemes for three level neutral point clamped inverter-fed induction motor drive. *International Power Electronics, Drives and Energy Systems Conference*, 1-6.

Bei, X., Yoon, Y. J., & Abur, A. (2005). Optimal placement and utilization of phasor measurements for state estimation. *PSERC Publication*, 05-20.

Bejan, A. (1988). *Advanced Engineering Thermodynamics*. New York: Wiley Interscience.

Bejan, A. (2006). *Advanced engineering thermodynamics* (3rd ed.). New York: Willey.

Belhadj, M. (1994). *Conception d'architectures en utilisant SIGNAL et VHDL* (Doctoral dissertation).

Bellaouar, A., & Elmasry, M. (2012). *Low-power digital VLSI design: circuits and systems*. Springer Science & Business Media.

Ben Haj Hassine, S., Jemai, M., & Ouni, B. (2017). Power and Execution Time Optimization through Hardware Software Partitioning Algorithm for Core Based Embedded System. *Journal of Optimization, 2017*.

Bentley, R. (2004). Global Oil Depletion–Methodologies and Results. In *Proc. 3rd Int. Workshop on Oil and Gas Depletion*. Australian Association for the Study of Peak Oil and Gas (ASPO).

Berrebi, E. (1997). *Méthodologie pour l'application industrielle de la synthèse comportementale* (Doctoral dissertation). Institut National Polytechnique de Grenoble-INPG.

Betka, A., & Moussi, A. (2004). Performance optimization of a photovoltaic induction motor pumping system. *Elsevier Journal on Renewable Energy*, *29*(14), 2167–2181. doi:10.1016/j.renene.2004.03.016

Bevrani, H. (2014). *Robust Power System Frequency Control* (2nd ed.). Springer. doi:10.1007/978-3-319-07278-4

Bezerra, P. A., Krismer, F., Andersen, T. M., Kolar, J. W., Sridhar, A., Brunschwiler, T., . . . Wang, N. (2015, November). Modeling and multi-objective optimization of 2.5 D inductor-based Fully Integrated Voltage Regulators for microprocessor applications. In *Power Electronics Conference and 1st Southern Power Electronics Conference (COBEP/SPEC), 2015 IEEE 13th Brazilian* (pp. 1-6). IEEE.

Bhat, S. R., Pittet, A., & Sonde, B. S. (1987). Performance optimization of induction motor-pump system using photovoltaic energy source. *IEEE Transactions on Industry Applications*, *IA-23*(6), 995–1000. doi:10.1109/TIA.1987.4505020

Bhattacharjee, P., & Majumder, A. (2016, December). LECTOR Based Gated Clock Approach to Design Low Power FSM for Serial Adder. In *Nanoelectronic and Information Systems (iNIS), 2016 IEEE International Symposium on* (pp. 250-254). IEEE. 10.1109/iNIS.2016.064

Bhattacharjee, P., Nath, B., & Majumder, A. (2017). LECTOR Based Clock Gating for Low Power Multi-Stage Flip Flop Applications. 대한전자공학회 학술대회, 106-109.

Bhattacharya, S., & Divan, D. (1995). Synchronous frame based controller implementation for a hybrid series active filter system. *Proc. 30th Ind. Appl. Soc. Annu. Meet. Ind. Appl. Conf.*, 2531– 2540. 10.1109/IAS.1995.530625

Bhattacharyya, S. (1999). Chapter. In C. Wu, L. Chen, & J. Chen (Eds.), Recent advances in finite-time thermodynamics (pp. 105–113). New York: Nova Science Publishers.

Bhattacharyya, S. C. (2012). Energy for Sustainable Development Energy access Programmes and Sustainable Development : A Critical Review and Analysis. *Energy for Sustainable Development*, *16*(3), 260–271. doi:10.1016/j.esd.2012.05.002

Bhattacharyya, S. S., Murthy, P. K., & Lee, E. A. (1999). Synthesis of embedded software from synchronous dataflow specifications. *The Journal of VLSI Signal Processing*, *21*(2), 151–166. doi:10.1023/A:1008052406396

Blaabjerg, F., Teodorescu, R., Liserre, M., & Timbus, A. V. (2006, October). Overview of control and grid synchronization for distributed power generation systems. *IEEE Transactions on Industrial Electronics*, *53*(5), 1398–1409. doi:10.1109/TIE.2006.881997

Blazic, B., & Papic, I. (2006). Improved D-StatCom Control for Operation with Unbalanced Currents and Voltages. *IEEE Transactions On Power Delivery*, *21*(1), 225-232.

Bojoi, R., Limongi, L. R., Roiu, D., & Tenconi, A. (2010). Enhanced power quality control strategy for single-phase inverters in distributed generation systems. *Industrial Electronics (ISIE), 2010 IEEE International Symposium on*, 2727-2732.

Borenstein, S., & Bushnell, J. (1998). *An empirical analysis of the potential for market power in California's electricity industry*. Working Paper 6463. NBER Working Paper Series, National Bureau of Economic Research; March. Retrieved from http://www.nber.org/papers/w6463

Borenstein, S., Bushnell, J., & Knittel, C. R. (1999). Market Power in Electricity Markets: Beyond Concentration Measures. *Energy Journal*, *20*(4), 65–88. doi:10.5547/ISSN0195-6574-EJ-Vol20-No4-3

Bortolotti, D., Mamaghanian, H., Bartolini, A., Ashouei, M., Stuijt, J., Atienza, D., ... Benini, L. (2014, August). Approximate compressed sensing: ultra-low power biosignal processing via aggressive voltage scaling on a hybrid memory multi-core processor. In *Proceedings of the 2014 international symposium on Low power electronics and design* (pp. 45-50). ACM. 10.1145/2627369.2627629

Bose, S., Wu, C., Xu, Y., Wierman, A., & Mohsenian-Rad, H. (2015). A Unifying Market Power Measure for Deregulated Transmission-Constrained Electricity Markets. *IEEE Transactions on Power Systems*, *30*(5), 2338–2348. doi:10.1109/TPWRS.2014.2360216

Bouchekaraa, H. R. E. H., Abido, M. A., & Boucherma, M. (2014). Optimal power flow using Teaching-Learning-Based Optimization technique. *Electric Power Systems Research*, *114*, 49–59. doi:10.1016/j.epsr.2014.03.032

Boyle, G. (2004). *Renewable Energy Power for a Sustainable Future* (2nd ed.). Oxford, UK: Oxford University Press.

Brenna, M., Faranda, R., & Tironi, E. (2009). A new proposal for power quality and custom power improvement: OPEN UPQC. *IEEE Transactions on Power Delivery*, *l*(24), 2107–2116. doi:10.1109/TPWRD.2009.2028791

Browne, M. W. (2000). Cross-Validation Methods. *Journal of Mathematical Psychology*, *44*(1), 108–132. doi:10.1006/jmps.1999.1279 PMID:10733860

Bundgaard, E., & Krebs, F. C. (2007). Low band gap polymers for organic photovoltaics. *Solar Energy Materials and Solar Cells*, *91*(11), 954–985. doi:10.1016/j.solmat.2007.01.015

Buragohain, B., Mahanta, P., & Moholkar, V. S. (2010). Biomass gasification for decentralized power generation : The Indian perspective. *Renewable & Sustainable Energy Reviews*, *14*(1), 73–92. doi:10.1016/j.rser.2009.07.034

Butzen, P., Reis, A., Kim, C., & Ribas, R. (2007). Subthreshold leakage modeling and estimation of general CMOS complex gates. *Integrated Circuit and System Design. Power and Timing Modeling, Optimization and Simulation*, 474-484.

Buyuktosunoglu, A., Schuster, S., Brooks, D., Bose, P., Cook, P., & Albonesi, D. (2001). An adaptive issue queue for reduced power at high performance. *Power-Aware Computer Systems*, 25-39.

Cadence document. (n.d.). *"A Practical Guide for Low-power Design": User experience with CPF*. Retrieved from www.cadence.com

Caldognetto, T., Tenti, P., & Brandao, D. I. (2017). Fully-Dispatchable Microgrid: Architecture, Implementation and Experimental Validation. In *IEEE 26th International Symposium on Industrial Electronics (ISIE)* (pp. 93–98). IEEE.

Camposano, R. (1991). Path-based scheduling for synthesis. *IEEE Transactions on Computer-Aided Design of Integrated Circuits and Systems*, *10*(1), 85–93. doi:10.1109/43.62794

Canis, A., Brown, S. D., & Anderson, J. H. (2014, September). Modulo SDC scheduling with recurrence minimization in high-level synthesis. In *Field Programmable Logic and Applications (FPL), 2014 24th International Conference on* (pp. 1-8). IEEE. 10.1109/FPL.2014.6927490

Canis, A., Choi, J., Aldham, M., Zhang, V., Kammoona, A., Czajkowski, T., ... Anderson, J. H. (2013). LegUp: An open-source high-level synthesis tool for FPGA-based processor/accelerator systems. *ACM Transactions on Embedded Computing Systems*, *13*(2), 24. doi:10.1145/2514740

Carrara, G., Gardella, S., Marchesoni, M., Salutarin, R., & Sciutto, G. (1990). A new multilevel PWM method: a theoretical analysis. *Power Electronics Specialists Conference*, 363-371. 10.1109/PESC.1990.131211

Casaro, M. M., & Martins, D. C. (2008, June). Grid-connected PV system: Introduction to behavior matching. In *Power Electronics Specialists Conference, 2008. PESC 2008. IEEE* (pp. 951-956). IEEE. 10.1109/PESC.2008.4592052

Casseau, E., Jégo, C., & Martin, E. (2004). Synthèse architecturale d'applications temps réel pour technologies submicroniques. *Technique et science informatiques, 23*(1), 35-66.

Cengel, Y. A., & Boles, M. A. (2006). *Thermodynamics: An Engineering Approach* (5th ed.). New York: Tata McGraw Hill.

Central Electricity Regulatory Commission. (2010). *Indian Electricity Grid Code*. Retrieved from http://www.cercind.gov.in

Chaar, L., Jamaleddine, A., Ajmal, F., & Khanm, H. (2008). Effect of Wind Blown Sand and Dust on PV Arrays Especially in UAE. *Power Systems Conference (PSC)*.

Chakrabarti, S., Venayagamoorthy, G. K., & Kyriakides, E. (2008). *PMU placement for power system observability using binary particle swarm optimization.* Paper presented at the Power Engineering Conference, 2008. AUPEC'08. Australasian Universities.

Chakrabarti, S., & Kyriakides, E. (2008). Optimal placement of phasor measurement units for power system observability. *IEEE Transactions on Power Systems*, *23*(3), 1433–1440. doi:10.1109/TPWRS.2008.922621

Chakrabarti, S., Kyriakides, E., & Albu, M. (2009). Uncertainty in power system state variables obtained through synchronized measurements. *IEEE Transactions on Instrumentation and Measurement*, *58*(8), 2452–2458. doi:10.1109/TIM.2009.2016387

Chakrabarti, S., Kyriakides, E., & Eliades, D. G. (2009). Placement of synchronized measurements for power system observability. *IEEE Transactions on Power Delivery*, *24*(1), 12–19. doi:10.1109/TPWRD.2008.2008430

Chandrakasan, A. P., Sheng, S., & Brodersen, R. W. (1992). Low-power CMOS digital design. *IEICE Transactions on Electronics*, *75*(4), 371–382.

Chandwani, A., & Kothari, A. (2016). Design, simulation and implementation of Maximum Power Point Tracking (MPPT) for solar based renewable system. *2016 International Conference on Electrical Power and Energy Systems (ICEPES)*, 539-544. 10.1109/ICEPES.2016.7915987

Chang, C. H., Satzoda, R. K., & Sekar, S. (2005, May). A novel multiplexer based truncated array multiplier. In *Circuits and Systems, 2005. ISCAS 2005. IEEE International Symposium on* (pp. 85-88). IEEE.

Chang, J. U., & Lin, C. H. (2016). A Circuit Complexity Optimization ILP Algorithm of High-level Synthesis System for New Multiprocessor Design. The Journal of The Institute of Internet. *Broadcasting and Communication*, *16*(3), 137–144. doi:10.7236/JIIBC.2016.16.3.137

Che, L., & Shahidehpour, M. (2014). DC Microgrids : Economic Operation and Enhancement of Resilience by Hierarchical Control. *IEEE Transactions on Smart Grid*, *5*(5), 2517–2526.

Che, L., Zhang, X., Shahidehpour, M., Alabdulwahab, A., & Abusorrah, A. (2017). Optimal Interconnection Planning of Community Microgrids With Renewable Energy Sources. *IEEE Transactions on Smart Grid*, *8*(3), 1054–1063.

Chen, S., & Joos, G. (2001). A unified series–parallel deadbeat control technique for an active power quality conditioner with full digital implementation. *Proc. IEEE 36th Ind. Appl. Soc. Annu. Meet. Ind. Appl. Conf.*, 172–178.

Chen, A., Hu, L., Chen, L., Deng, Y., & He, X. (2005). A multilevel converter topology with fault-tolerant ability. *IEEE Transactions on Power Electronics*, *20*(2), 405–415. doi:10.1109/TPEL.2004.842983

Chen, G. (2000, August). Actively clamped bi-directional fly back converter. *IEEE Transactions on Industrial Electronics*, *47*(4), 770–779. doi:10.1109/41.857957

Chen, G., Chen, Y., & Smedley, K. M. (2004). Three-phase four-leg active power quality conditioner without references calculation. *Proc. Appl. Power Electron. Conf.*, 587–593.

Cheng, M. Y., & Prayogo, D. (2014). Symbiotic Organisms Search: A new metaheuristic optimization algorithm. *Computers & Structures*, *139*, 98–112. doi:10.1016/j.compstruc.2014.03.007

Cheng, X., & Liang, X. (2012). Entransy loss in thermodynamic processes and its application. *Energy*, *44*(1), 964–972. doi:10.1016/j.energy.2012.04.054

Chen, H.-C., Qiu, J.-C., & Liu, C.-H. (2008). *Dynamic Modelling and Simulation of Renewable Energy Based Hybrid Power Systems*. DRPT.

Chen, J. (1992). Optimization of a solar-driven heat engine. *Journal of Applied Physics*, *72*(8), 3778–3780. doi:10.1063/1.352273

Chen, L., Wu, C., & Sun, F. (1999). Finite time thermodynamic optimization or entropy generation minimization of energy systems. *Journal of Non-Equilibrium Thermodynamics*, *24*(4), 327–359.

Chen, N., Qian, M., Zhu, L., Yao, L., Wu, F., Chen, M., & Wang, N. (2012). A coordinated reactive power and voltage control system for wind farm grid integration. In *IEEE International Conference on Power System Technology* (pp. 1-6). IEEE.

Chen, N., Zhu, L. Z., & Wang, W. (2009). Strategy for reactive power control of wind farm for improving voltage stability in wind power integrated region. *Proceedings of the CSEE*, *10*, 019.

Cherukuri, S. K., & Rayapudi, S. R. (2016). A novel global MPP tracking of photovoltaic system based on whale optimization algorithm. *International Journal of Renewable Energy Development*, *5*(3), 225. doi:10.14710/ijred.5.3.225-232

Chiang, H. C., Ma, T. T., Cheng, Y. H., Chang, J. M., & Chang, W. N. (2010). Design and implementation of a hybrid regenerative power system combining grid-tie and uninterruptible power supply functions. *Renewable Power Generation, IET*, *4*(1), 85–99. doi:10.1049/iet-rpg.2009.0033

China Electric Power Research Institute. (2009). Retrieved from http://www.epri.ac.cn. China.

Chinthamalla, Padmanaban, Karampuri, Jain, Ertas, & Fedak. (2016). A solar PV water pumping solution using a three-level cascaded inverter connected induction motor drive. *Engineering Science and Technology, an International Journal*, *19*(4), 1731-1741.

Chinthamalla, R., Ganesh, K. S., & Jain, S. (2014). *An optimal and efficient PV system using two 2-level cascaded 3-level inverter for Centrifugal pump*. Paper presented at IEEE International Conference on Power Electronics, Drives and Energy Systems (PEDES), Mumbai, India. 10.1109/PEDES.2014.7042060

Cho, I.-K., & Kim, H. (2007). Market power and network constraint in a deregulated electricity market. *Energy Journal*, *28*(2), 1–34. doi:10.5547/ISSN0195-6574-EJ-Vol28-No2-1

Choi, K., Cheng, W. C., & Pedram, M. (2005). Frame-based dynamic voltage and frequency scaling for an MPEG player. *Journal of Low Power Electronics*, *1*(1), 27–43. doi:10.1166/jolpe.2005.005

Choi, K., Lee, W., Soma, R., & Pedram, M. (2004, November). Dynamic voltage and frequency scaling under a precise energy model considering variable and fixed components of the system power dissipation. In *Proceedings of the 2004 IEEE/ACM International conference on Computer-aided design* (pp. 29-34). IEEE Computer Society.

Chong, K. S., Gwee, B. H., & Chang, J. S. (2007). Low energy 16-bit Booth leapfrog array multiplier using dynamic adders. *IET Circuits, Devices & Systems*, *1*(2), 170–174. doi:10.1049/iet-cds:20060109

Chow, T. T. (2010). A review on photovoltaic/thermal hybrid solar technology. *Applied Energy*, *87*(2), 365–379. doi:10.1016/j.apenergy.2009.06.037

Chu, C., & Chen, C. (2009). Robust maximum power point tracking method for photovoltaic cells: A sliding mode control approach. *Solar Energy*, *83*(8), 1370–1378. doi:10.1016/j.solener.2009.03.005

Chung, D.-W., Kim, J.-S., & Sul, S.-K. (1998). Unified voltage modulation technique for real-time three-phase power conversion. *IEEE Transactions on Industry Applications*, *34*(2), 374–380. doi:10.1109/28.663482

Chu, Y., & Meisen, P. (2011). *Review and comparison of different solar energy technologies. Global Energy Network Institute*. San Diego, CA: GENI.

Cohn, N. (1956). Some aspects of tie-line bias control on interconnected power systems. *Transactions of the American Institute of Electrical Engineers*, (Part 3), 75, 1415–1436.

Colak, I., Kabalci, E., & Bayindir, R. (2011). Review of multilevel voltage source inverter topologies and control schemes. *Energy Conversion and Management*, *52*(2), 1114–1128. doi:10.1016/j.enconman.2010.09.006

Colbert, T. (2001). High-Throughput Screening for Induced Point Mutations. *Plant Physiology*, *126*(2), 480–484. doi:10.1104/pp.126.2.480 PMID:11402178

Correa, M., Chakraborty, S., Simoes, G., & Farret, A. (2003). A single phase high frequency AC microgrid with an unified power quality conditioner. *Proc. IEEE Int. Symp. Ind. Electron.*, 956–962.

Correa, J., Farret, F., & Simoes, M. (2005). Application of a modified single phase P-Q Theory in the control of shunt and series active filters in a 400Hz microgrid. *Proc. Power Electron. Spec. Conf.*, 2585–2591.

Corrêa, T. P., Seleme, S. I. Jr, & Silva, S. R. (2012). Efficiency optimization in stand-alone photovoltaic pumping system. *Elsevier Journal on Renewable Energy*, *41*, 220–226. doi:10.1016/j.renene.2011.10.024

Coyle, R. G. (1996). *Systems Dynamics Modeling*. London: Chapman & Hall. doi:10.1007/978-1-4899-2935-8

Crookes, W. (1896). The True Cathode Rays. *The New York Times*.

Currie, M. J., Mapel, J. K., Heidel, T. D., Goffri, S., & Baldo, M. A. (2008). High-efficiency organic solar concentrators for photovoltaics. *Science*, *321*(5886), 226–228. doi:10.1126cience.1158342 PMID:18621664

Daher, S., Schmid, J., & Antunes, F. L. (2008). Multilevel inverter topologies for stand-alone PV systems. *IEEE Transactions on Industrial Electronics*, *55*(7), 2703–2712. doi:10.1109/TIE.2008.922601

Dahidah, M. S. A., & Agelidis, V. G. (2008). Single-carrier sinusoidal PWM equivalent selective harmonic elimination for a five-level voltage source converter. *Electric Power Systems Research*, *78*(11), 1826–1836. doi:10.1016/j.epsr.2008.01.021

Dalali, M., & Karegar, H. K. (2016). Optimal PMU placement for full observability of the power network with maximum redundancy using modified binary cuckoo optimisation algorithm. *IET Generation, Transmission & Distribution*, *10*(11), 2817–2824. doi:10.1049/iet-gtd.2016.0287

Dalton, J. C., & markets. (1997). Assessing the competitiveness of restructured generation service markets. *The Electricity Journal*, *10*(3), 30–39. doi:10.1016/S1040-6190(97)80375-2

Dankoff, W. (2009). *Glossary of Solar Water Pumping Terms and Related Components*. Available at www.conergy.us/Desktopdefault.aspx/ tabid-332/449 read-3816/

Dasgupta, S., Sahoo, S. K., & Panda, S. K. (2011). Single-Phase Inverter Control Techniques for Interfacing Renewable Energy Sources With Microgrid-Part I: Parallel-Connected Inverter Topology With Active and Reactive Power Flow Control Along With Grid Current Shaping. *IEEE Transactions on Power Electronics, 26*(3), 717-731.

Das, M., & Agarwal, V. (2016, May). Design and Analysis of a High-Efficiency DC–DC Converter with Soft Switching Capability for Renewable Energy Applications Requiring High Voltage Gain. *IEEE Transactions on Industrial Electronics*, *63*(5), 2936–2944. doi:10.1109/TIE.2016.2515565

David, A. K., & Wen, F. (2001). Market power in electricity supply. *IEEE Transactions on Energy Conversion*. doi:10.1109/60.969475

Dayananda, B.R., Sekhar, V., & Babu, R.N. (2014). Implementation of Linear Controller of a STATCOM Design in DC Link Voltage. *International Journal of Emerging Technology and Advanced Engineering, 4*(9).

De Almeida, R. G., Castronuovo, E. D., & Peas Lopes, J. (2006). Optimum generation control in wind parks when carrying out system operator requests. *IEEE Transactions on Power Systems*, *21*(2), 718–725. doi:10.1109/TPWRS.2005.861996

De Man, H., Rabaey, J., Six, P., & Claesen, L. (1986). Cathedral-II: A silicon compiler for digital signal processing. *IEEE Design & Test of Computers*, *3*(6), 13–25. doi:10.1109/MDT.1986.295047

DeMeo, E. A., Grant, W., Milligan, M. R., & Schuerger, M. (2005). Wind plant integration. *IEEE Power & Energy Magazine*, *3*(6), 38–46. doi:10.1109/MPAE.2005.1524619

Devassy, S., & Singh, B. (2017). Modified p-q Theory Based Control of Solar PV Integrated UPQC-S. IEEE Transactions on Industry Applications, 1(1) 1-11.

Dezso, S., Remus, T., & Pedro, R. (2007). PV panel model based on datasheet values. *Industrial Electronics, ISIE 2007, IEEE International Symposium*, 2392–2396.

Dhadbanjan, T. (1997). An expert system for alleviation of network overloads. *Electric Power Systems Research*, *40*(2), 143–153. doi:10.1016/S0378-7796(96)01146-7

Dhadbanjan, T., & Parthasarathy, K. (1995). An expert system for voltage control in a power system network. In *International Conference on Energy Management and Power Delivery* (Vol. 1, pp. 364-369). IEEE.

Dhadbanjan, T., Parthasarathy, K., & Prior, D. (1984). Improved algorithm for optimum reactive power allocation. *International Journal of Electrical Power & Energy Systems*, *6*(2), 72–74. doi:10.1016/0142-0615(84)90011-5

Dhadbanjan, T., & Yesuratnam, G. (2006). Comparison of optimum reactive power schedule with different objectives using LP technique. *International Journal of Emerging Electric Power Systems*, *7*(3), 1–31. doi:10.2202/1553-779X.1338

Dhillon, S. S., Lather, J. S., & Marwaha, S. (2016). Multi objective load frequency control using hybrid bacterial foraging and particle swarm optimized PI controller. *International Journal of Electrical Power & Energy Systems*, *79*, 196–209. doi:10.1016/j.ijepes.2016.01.012

Dias, A. F., Akil, M., Lavarenne, C., & Sorel, Y. (2000). Vers la synthèse automatique de circuits à partir de graphes algorithmiques factorisés. In Workshop AAA sur l'adéquation algorithme architecture (pp. 165-172). Academic Press.

Díaz, C. A., Webster, M. D., Villar, J., & Campos, F. A. (2016). Dynamics of Market Power in ERCOT System: A Fundamental CSFE With Network Constraints. *IEEE Transactions on Power Systems*, *31*(2), 861–871. doi:10.1109/TPWRS.2015.2411051

Ding, H. (1996). *Synthese architecturale interactive et flexible* (Doctoral dissertation). Institut National Polytechnique de Grenoble-INPG.

Divya, K., & Rao, P. N. (2006). Models for wind turbine generating systems and their application in load flow studies. *Electric Power Systems Research*, *76*(9), 844–856. doi:10.1016/j.epsr.2005.10.012

Dobson, I., Cutsem, T. V., Vourmas, C., Demarco, C.L., Bramanian, M.V., Ovenbye, T., & Canizares, C.A. (2003). Voltage Stability Assesment: Concepts Practices and Tools. *IEEE Power Engineering Society*.

Dommel, H. W., & Tinney, T. F. (1968). Optimal power flow solutions. *IEEE Transactions on Power Apparatus and Systems*, *87*(5), 1866–1976. doi:10.1109/TPAS.1968.292150

Dorofte, C., Borup, U. & Blaabjerg, F. (2005). A combined two-method MPPT control scheme for grid-connected photovoltaic systems. *Power Electronics and Applications conference*, 10.

Dosluoglu, T., & Gelman, A. (2016). *U.S. Patent No. 9,369,040*. Washington, DC: U.S. Patent and Trademark Office.

Dua, D., Dambhare, S., Gajbhiye, R. K., & Soman, S. (2008). Optimal multistage scheduling of PMU placement: An ILP approach. *IEEE Transactions on Power Delivery*, *23*(4), 1812–1820. doi:10.1109/TPWRD.2008.919046

Dumonteix, Y., & Mehrez, H. (2000). A family of redundant multipliers dedicated to fast computation for signal processing. In *Circuits and Systems, 2000. Proceedings. ISCAS 2000 Geneva. The 2000 IEEE International Symposium on* (Vol. 5, pp. 325-328). IEEE. 10.1109/ISCAS.2000.857430

Du, Z., Tolbert, L. M., Chiasson, J. N., & Ozpineci, B. (2006). A Cascade Multilevel Inverter using a Single DC Source. *Twenty-First Annual IEEE Applied Power Electronics Conference and Exposition*, 426-430. 10.1109/APEC.2006.1620573

E.ON Netz Gmbh. (2016). *Grid Code-High and extra high voltage*. Retrieved from http://www.eon.com

Ebrahimi, M. (2016). Policy Making in a Renewable Energy Manufacturing SME Using System Dynamics. *34th International Conference of the System Dynamics Society*.

Edwin Jose, S., & Titus, S. (2016). A level dependent source concoction multilevel inverter topology with a reduced number of power switches. *Journal of Power Electronics*, *16*(4), 1316–1323. doi:10.6113/JPE.2016.16.4.1316

EE362L, Power Electronics, Solar Power, I–V Characteristics, Version October 14, 2005. (n.d.). Retrieved from http://www.ece.utexas.edu/~grady/EE362L Solar.pdf

Eir Grid. (n.d.). *Eir Grid Plc*. Retrieved from http://www.EirGrid.com

Elfstadius, M., & Gecer, D. (2008). *Method to Detect and Measure Potential Market Power Caused by Transmission Network Congestions on Electricity Markets*. Linköping, Sweden: Department of Electrical Engineering, Linköpings University.

Elgendy, M. A., Zahawi, B., & Atkinson, D. J. (2010). Comparison of Directly Connected and Constant Voltage Controlled Photovoltaic Pumping Systems. *IEEE Transactions on Sustainable Energy*, *1*(3), 184–192. doi:10.1109/TSTE.2010.2052936

Elgerd, O. I., & Fosha, C. E. (1970). Optimum megawatt-frequency control of multiarea electric energy systems. *IEEE Transactions on Power Apparatus and Systems*, *PAS-89*(4), 556–563. doi:10.1109/TPAS.1970.292602

Elgerd, O. I. (1982). *Electic energy systems theory-an introduction*. McGraw-Hill New York.

El-Hawary, M. E. (1998). *Electric power applications of fuzzy systems*. Wiley-IEEE Press. doi:10.1109/9780470544457

El-Keib, A., & Ma, X. (1995). Application of artificial neural networks in voltage stability assessment. *IEEE Transactions on Power Systems*, *10*(4), 1890–1896. doi:10.1109/59.476054

Ellervee, P. (2000). *High-level synthesis of control and memory intensive applications* (Doctoral dissertation). Institutionen för elektronisk systemkonstruktion.

Elnady, A., Goauda, A., & Salama, M. M. A. (2001). Unified power quality conditioner with a novel control algorithm based on wavelet transfor. *Annual IEEE Canadian Conference on Electrical and Computer Engineering*, 1041-1045.

Elnady, A., Goauda, A., & Salama, M. M. A. (2001). Unified power quality conditioner with a novel control algorithm based on wavelet transform. *Proc. Can. Conf. Electr. Comput. Eng.*, 1041–1045. 10.1109/CCECE.2001.933586

Enshaee, A., Hooshmand, R. A., & Fesharaki, F. H. (2012). A new method for optimal placement of phasor measurement units to maintain full network observability under various contingencies. *Electric Power Systems Research*, *89*, 1–10. doi:10.1016/j.epsr.2012.01.020

Enslin, J., Wolf, M., Snyman, D., & Swiegers, W. (1997). Integrated Photovoltaic Maximum Power Point Tracking Converter. *IEEE Transactions on Industrial Electronics*, *44*(6), 769–773. doi:10.1109/41.649937

Erbay, L. B., & Yavuz, H. (1997). Design parameters of a thermionic energy converter. In *Proceedings of the International Conference on Thermodynamics Analysis and Improvement of Energy Systems (TAIES'97)* (pp. 20-24). Academic Press.

Erkut, H. (1983). *Sistem Dinamiğinin Temelleri*. İstanbul: İTÜ Fen Edebiyat Fakültesi Ofset Atölyesi.

Eskander, M. N., & Zaki, A. M. (1997). A maximum efficiency photovoltaic induction motor pump system. *Elsevier Journal on Renewable Energy*, *10*(1), 53–60. doi:10.1016/0960-1481(96)00004-3

Esposito, F., Isastia, V., Meo, S., & Piegari, L. (2008). An Improved Perturb and Observe Algorithm for Tracking Maximum Power Points of Photovoltaic Power Systems. International Review on Modelling and Simulations (I.RE.MO.S.), 10-16.

Estebsari, A., Sadeghzadeh, S. M., & Fotouhi, M. A. (2009). Transmission constraints impacts on monopolistic production capacity under strategic bidding. *2009 International Conference on Electrical Machines and Systems*. 10.1109/ICEMS.2009.5382644

Esther, P. B., & Kumar, K. S. (2016). A survey on residential Demand Side Management architecture, approaches, optimization models and methods. *Renewable & Sustainable Energy Reviews, 59*, 342–351. doi:10.1016/j.rser.2015.12.282

Fan, X., Ellis, C., & Lebeck, A. (2005). The synergy between power-aware memory systems and processor voltage scaling. *Power-Aware Computer Systems*, 151-166.

Farhadi Kangarlu, M., Babaei, E., & Laali, S. (2012). Symmetric multilevel inverter with reduced components based on non-insulated dc voltage sources. *IET Power Electronics, 5*(5), 571–581. doi:10.1049/iet-pel.2011.0263

Farokhnia, N., Khoraminia, R., & Gharehpetian, G. B. (2010). Optimization of PI Controller Gains in Nonlinear Controller of STATCOM Using PSO and GA. *International Conference on Renewable Energies and Power Quality (ICREPQ).*

Fazeli, R., & Davidsdottir, B. 2015). Energy Modeling of Danish Housing Stock Using System Dynamics. *The 33rd International Conference of the System Dynamics Society.*

Federal Energy Regulatory Commission. (2005). Order no. 661 on interconnection for wind farms. Author.

Federal Energy Regulatory Commission. (2005). Principles for efficient and reliable reactive power supply and consumption staff report (Docket No. AD05-1-000). Author.

Federal Energy Regulatory Commission. (2012). Order no. 661a (order on rehearing) on interconnection for wind farms. Author.

Feng, D., Zhong, J., & Gan, D. (2008). Reactive Market Power Analysis Using Must-Run Indices. *IEEE Transactions on Power Systems, 23*(2), 755–765. doi:10.1109/TPWRS.2008.919416

Fini, M. H., Yousefi, G. R., & Alhelou, H. H. (2016). Comparative study on the performance of many-objective and single-objective optimisation algorithms in tuning load frequency controllers of multi-area power systems. *IET Generation, Transmission & Distribution, 10*(12), 2915–2923. doi:10.1049/iet-gtd.2015.1334

Fisher, J. A. (1981). Trace scheduling: A technique for global microcode compaction. *IEEE Transactions on Computers, 30*(7), 478–490. doi:10.1109/TC.1981.1675827

Fıshwick, P. A. (2007). *Handbook of Dynamic System Modelling*. Boca Raton, FL: Chapman & Hall / CRC. doi:10.1201/9781420010855

Fletcher, E. A., & Moen, R. L. (1977). Hydrogen and oxygen from water. *Science, 197*(4308), 1050–1056. doi:10.1126cience.197.4308.1050 PMID:17836063

Flood, R. L., & Jackson, M. C. (1991). *Creative Problem Solving: Total Systems Intervention.* Chichester, UK: John Wiley & Sons.

Fonseca, M., da Costa, E., Bampi, S., & Monteiro, J. (2005, September). Design of a radix-2 m hybrid array multiplier using carry save adder format. In *Proceedings of the 18th annual symposium on Integrated circuits and system design* (pp. 172-177). ACM. 10.1145/1081081.1081126

Forrester, J. W. (1962). *Industrial Dynamics.* Waltham, MA: Pegasus Communications.

Forrester, J. W. (1969). *Urban Dynamics.* The MIT Press.

Forrester, J. W. (1971). *Principles Of Systems* (2nd ed.). Cambridge, MA: Wright Allen Press.

Forrester, J. W. (1977). Growth Cycles. *De Economist, 125*(: 4).

Forrester, J. W., & Senge, P. M. (1980). Tests For Building Confidence In System Dynamic Models. *TIMS Studies in the Management Sciences, 14*, 209–228.

Frank, S., Steponavice, I., & Rebennak, S. (2012). *Optimal power flow: a bibliographic survey I, formulations and deterministic methods*. Springer-Verlag.

Frank, S., Steponavice, I., & Rebennak, S. (2012). *Optimal power flow: a bibliographic survey II, non-deterministic and hybrid methods*. Springer-Verlag.

Freitas, W., Vieira, J. C., Da Suva, L. C. P., Affonso, C. M., & Morelato, A. (2005). Long-term voltage stability of distribution systems with induction generators. In *IEEE Power Engineering Society General Meeting* (pp. 2910-2913). IEEE.

Friedman, T. D., & Yang, S. C. (1969). Methods used in an automatic logic design generator (ALERT). *IEEE Transactions on Computers*, *100*(7), 593–614. doi:10.1109/T-C.1969.222727

Fujino, M., & Moshnyaga, V. G. (2003, May). Dynamic operand transformation for low-power multiplier-accumulator design. In *Circuits and Systems, 2003. ISCAS'03. Proceedings of the 2003 International Symposium on* (*Vol. 5*, pp. V-V). IEEE. 10.1109/ISCAS.2003.1206276

Fujita, H., & Akagi, H. (1998). The Unified Power Quality Conditioner: The Integration of Series- and Shunt-Active Filters. *IEEE Transactions on Power Electronics*, *13*(12), 315–322. doi:10.1109/63.662847

Gabor, D. (1946). Theory of communication. *J IEEE*, *93*(3), 429–457.

Gabriele, L., Javed, N., & Huailei, Y. (2012). *Autonomous Photovoltaic Panels Cleaning System*. Available on: www.engineering.nyu.edu

Gad, M., Shinde, P., & Kulkarni, S. U. (2012). Optimal Location of TCSC by Sensitivity Methods. *International Journal of Computational Engineering Research*, *2*(6), 162–168.

Gaillardon, P. E., Beigne, E., Lesecq, S., & Micheli, G. D. (2015). A survey on low-power techniques with emerging technologies: From devices to systems. *ACM Journal on Emerging Technologies in Computing Systems*, *12*(2), 12. doi:10.1145/2714566

Gajski, D. D., Dutt, N. D., Wu, A. C., & Lin, S. Y. (2012). *High—Level Synthesis: Introduction to Chip and System Design*. Springer Science & Business Media.

Gama, C. A., Ellery, E. H., Azevedo, D. C., & Poute, J. R. R. (1992). Static var Compensator (SVC) versus synchronous compensaters (SC) for inverter Stations. *Compensation- Technical and Economic Aspects in Electronorte studies CIGRE*, 14-103.

Garcia, F. S., Pomilio, J. A., & Spiazzi, G. (2013). Modeling and control design of the interleaved double dual boost converter. *IEEE Transactions on Industrial Electronics*, *60*(8), 3283–3290. doi:10.1109/TIE.2012.2203770

Ghanizadeh, A.J., Mokhtari, G., Abedi, M., Gharehpetian, G.B. (2011). Optimal power flow based on imperialist competitive algorithm. *Int. Rev. Electr. Eng, 6*.

Ghareeb, T., & Mohammed, O. A. (2013). DC Microgrids and Distribution Systems: An Overview. *IEEE Power and Energy Society General Meeting,* 1 - 5.

Ghasemi, M., Ghavidel, S., Ghanbarian, M. M., Reza Massrur, H., & Mahboubi-Moghaddam, E. (2014). Multi-objective optimal power flow using new modified and hybrid modified imperialist competitive algorithms: A comparative study. *Information Sciences (Elsevier)*, *281*, 225–247. doi:10.1016/j.ins.2014.05.040

Ghosh, A., Jindal, K., & Joshi, A. (2004). A unified power quality conditioner for voltage regulation of critical load bus. IEEE Power Engineering Society Meeting, 471-476.

Ghosh, A., Jindal, A. K., & Joshi, A. (2003). Inverter control using output feedback for power compensating devices. *Proc. Convergent Technol. Conf.*, 48–52. 10.1109/TENCON.2003.1273212

Gibson, J. H. (2003). *UVB radiation: Definition and characteristics*. Retrieved from http://uvb. nrel. colostate. edu

Gnanambal, K., & Babulal, C. K. (2012). Maximum loadability limit of power system using hybrid differential evolution with particle swarm optimization. *International Journal of Electrical Power & Energy Systems*, *43*(1), 150–155. doi:10.1016/j.ijepes.2012.04.033

Goldsmid, H. J. (2010). *Introduction to thermoelectricity* (Vol. 121). Heidelberg, Germany: Springer. doi:10.1007/978-3-642-00716-3

Gopakumar, P., Reddy, M. J. B., & Mohanta, D. K. (2015). Pragmatic multi-stage simulated annealing for optimal placement of synchrophasor measurement units in smart power grids. *Frontiers in Energy*, *9*(2), 148–161. doi:10.100711708-015-0344-z

Gopal, M. (2014). *Digital Cntrol and State Variable Methods* (4th ed.). McGrawHill.

Gordon, J. M. (Ed.). (2013). *Solar energy: the state of the art*. Routledge.

Gorripotu, T. S., Sahu, R. K., & Panda, S. (2015). AGC of a multi-area power system under deregulated environment using redox flow batteries and interline power flow controller. *Eng Sci Tech. International Journal (Toronto, Ont.)*, *18*, 555–578.

Gou, B. (2008a). Generalized integer linear programming formulation for optimal PMU placement. *IEEE Transactions on Power Systems*, *23*(3), 1099–1104. doi:10.1109/TPWRS.2008.926475

Gou, B. (2008b). Optimal placement of PMUs by integer linear programming. *IEEE Transactions on Power Systems*, *23*(3), 1525–1526. doi:10.1109/TPWRS.2008.926723

Govindarajan, S. (2000). *Algorithms for Design Space Exploration and High-level Synthesis for Multi-FPGA Reconfigurable Computers* (Doctoral dissertation). University of Cincinnati.

Graditi, G., Ippolito, M., Telaretti, E., & Zizzo, G. (2015, April). An innovative conversion device to the grid interface of combined RES-based generators and electric storage systems. *IEEE Transactions on Industrial Electronics*, *62*(4), 2540–2550. doi:10.1109/TIE.2014.2336620

Greeley, J., Jaramillo, T. F., Bonde, J., Chorkendorff, I. B., & Nørskov, J. K. (2006). Computational high-throughput screening of electrocatalytic materials for hydrogen evolution. *Nature Materials*, *5*(11), 909–913. doi:10.1038/nmat1752 PMID:17041585

Green, M. A. (2003). General temperature dependence of solar cell performance and implications for device modelling. *Progress in Photovoltaics: Research and Applications*, *11*(5), 333–340. doi:10.1002/pip.496

Green, M. A. (2006). *Third generation photovoltaics: Advanced Solar Energy Conversion*. Berlin: Springer.

Green, M. A., Emery, K., Hishikawa, Y., Warta, W., & Dunlop, E. D. (2012). Solar cell efficiency tables (Version 39). *Progress in Photovoltaics: Research and Applications*, *20*(1), 12–20. doi:10.1002/pip.2163

Grid Code. (2004). *Wind power facility technical requirements - Revision 0*. Retrieved from http://www.aeso.ca

Grouz, F., & Sbita, L. A. (2014). Safe and easy methodology for design and sizing of standalone hybrid PV-wind system. *International Conference on Electrical Sciences and Technologies in Maghreb (CISTEM)*, 1-8.

Gruian, F., & Kuchcinski, K. (1998). A Constraint Logic Programming Based Approach to High-Level Synthesis for Low Power. Proceedings of Euromicro.

Guha, D., Roy, P.K., & Banerjee, S. (2016a) Quasi-oppositional differential search algorithm applied to load frequency control. *Engg Science and Technology, an Int. J, 19*, 1635-54.

Guha, D., Roy, P. K., & Banerjee, S. (2016b). Application of backtracking search algorithm in load frequency control of multi-area interconnected power system. *Ain Shams Engg J.* doi:10.1016/j.asej.2016.01.004

Guha, D., Roy, P. K., & Banerjee, S. (2017). Study of differential search algorithm based automatic generation control of an interconnected thermal-thermal system with governor dead-band. *Applied Soft Computing*, *52*, 160–175. doi:10.1016/j.asoc.2016.12.012

Guo, X.-C., Liao, C.-S., & Chu, C.-C. (2016). *Optimal PMU placements under propagation depth constraints by mixed integer linear programming.* Paper presented at the Smart Grid Communications (SmartGridComm), 2016 IEEE International Conference on. 10.1109/SmartGridComm.2016.7778836

Guo, X. Q., & Wu, W. Y. (2010). Improved current regulation of three-phase grid-connected voltage- source inverters for distributed generation systems. *Renewable Power Generation, IET*, *4*(2), 101–115. doi:10.1049/iet-rpg.2009.0044

Gustafsson, O., & Ohlsson, H. (2005, May). A low power decimation filter architecture for high-speed single-bit sigma-delta modulation. In *Circuits and Systems, 2005. ISCAS 2005. IEEE International Symposium on* (pp. 1453-1456). IEEE. 10.1109/ISCAS.2005.1464872

Guthrie, F. (1873). On a relation between heat and static electricity. *The London, Edinburgh and Dublin Philosophical Magazine and Journal of Science*, *46*(306), 257–266.

Hafez, O., & Bhattacharya, K. (2012). Optimal Planning and Design of a Renewable Energy Based Supply System for Microgrids. *Renewable Energy*, *45*, 7–15. doi:10.1016/j.renene.2012.01.087

Hajian, M., Ranjbar, A., Amraee, T., & Shirani, A. (2007). *Optimal placement of phasor measurement units: Particle swarm optimization approach.* Paper presented at the Intelligent Systems Applications to Power Systems, 2007. ISAP 2007. International Conference on. 10.1109/ISAP.2007.4441610

Hajian, M., Ranjbar, A. M., Amraee, T., & Mozafari, B. (2011). Optimal placement of PMUs to maintain network observability using a modified BPSO algorithm. *International Journal of Electrical Power & Energy Systems*, *33*(1), 28–34. doi:10.1016/j.ijepes.2010.08.007

Hammed, A.E., & EI-Sadek, M.Z. (1989). Prevention of transient voltage instabilities due to induction motor loads by static Var Compensators. *IEEE Transactions on Power Systems*, (3).

Hammerstrom, J. (2007). AC Versus DC Distribution Systems: Did we get it Right? *IEEE Power Engineering Society General Meeting,* 1 - 5.

Han, B., Bae, B., Baek, S., & Jang, G. (2006). New configuration of UPQC for medium-voltage application. *IEEE Transactions on Power Delivery*, *l*(21), 1438–1444. doi:10.1109/TPWRD.2005.860235

Harakawa, T., & Tujimoto, T. (2001). A proposal of efficiency improvement with solar power generation system. *Proc. of 27th Annual Conference of IEEE Industrial Electronics Society, 1*, 523-528.

Hasan, Dinesh, & Zafer. (2013). Multimodules of diode clamped multilevel converter: A novel option for high power facts controller. *International Journal of Advance Research in Electrical Electronics and Instrumentation Engineering*, *l*(2), 109–112.

Hasani-Marzooni, M., & Hosseini, S. H. (2013). Short-term market power assessment in a long-term dynamic modeling of capacity investment. *IEEE Transactions on Power Systems*, *28*(2), 626–638. doi:10.1109/TPWRS.2012.2217990

Hasan, M., Ansari, A. Q., & Singh, B. (2015). Parameters estimation of a series VSC and shunt VSC to design a unified power quality conditioner (UPQC). *39th National System Conference*, 1-6. 10.1109/NATSYS.2015.7489111

Hatsopoulos, G. N., & Gyftopoulos, E. P. (1973). Thermionic energy conversion: Vol. 1. Processes and devices. MIT Press.

Hatsopoulos, G. N., & Gyftopoulos, E. P. (1979). Thermionic energy conversion: Vol. 2. Theory, technology, and application (No. DOE/TIC-11035). MIT Press.

Hatsopoulos, G. N., & Gyftopoulos, E. P. (1973). *Thermionic energy conversion* (Vol. 1). Cambridge, MA: MIT Press.

Hatsopoulos, G. N., & Kaye, J. (1958). Analysis and Experimental esults of a Diode Configuration of a Novel Thermo-electron Engine. *Proceedings of the IRE*, *46*(9), 1574–1579. doi:10.1109/JRPROC.1958.286932

Hatsopoulos, G. N., & Kaye, J. (1958). Measured thermal efficiencies of a diode configuration of a thermo electron engine. *Journal of Applied Physics*, *29*(7), 1124–1125. doi:10.1063/1.1723373

Hautier, G., Fischer, C. C., Jain, A., Mueller, T., & Ceder, G. (2010). Finding natures missing ternary oxide compounds using machine learning and density functional theory. *Chemistry of Materials*, *22*(12), 3762–3767. doi:10.1021/cm100795d

Havinga, P. J., & Smit, G. J. (2000). Design techniques for low-power systems. *Journal of Systems Architecture*, *46*(1), 1–21. doi:10.1016/S1383-7621(98)00057-5

Haynal, S. (2000). *Automata-based symbolic scheduling*. Academic Press.

Haynes, T. E., Chu, W. K., Aselage, T. L., & Picraux, S. T. (1986). Initial decomposition of GaAs during rapid thermal annealing. *Applied Physics Letters*, *49*(11), 666–668. doi:10.1063/1.97562

Heijligers, M. J. M. (1996). *The application of genetic algorithms to high-level synthesis*. Technische Universiteit Eindhoven.

Hepbasli, A. (2008). A key review on exergetic analysis and assessment of renewable energy resources for a sustainable future. *Renewable & Sustainable Energy Reviews*, *12*(3), 593–661. doi:10.1016/j.rser.2006.10.001

Hertzberg, R. P., & Pope, A. J. (2000). High-throughput screening: New technology for the 21st century. *Current Opinion in Chemical Biology*, *4*(4), 445–451. doi:10.1016/S1367-5931(00)00110-1 PMID:10959774

Hill, C. A., Such, M. C., Chen, D., Gonzalez, J., & Grady, W. M. (2012). Battery Energy Storage for Enabling Integration of Distributed Solar Power Generation. *IEEE Transactions on Smart Grid*, *3*(2), 850–857. doi:10.1109/TSG.2012.2190113

Hilloowala, R. M., & Sharaf, A. M. (1992) A Rule-Based Fuzzy Logic Controller for a PWM Inverter in Photo-Voltaic Energy Conversion Scheme. *Conference Record of the 1992 IEEE Industry Applications Society Annual Meeting*, *1*, 762-769. 10.1109/IAS.1992.244319

Hingorani, N. G. (1988). High Power Electronics and Flexible AC Transmission System. *IEEE Power Engineering Review*, *8*(7), 3–4. doi:10.1109/MPER.1988.590799

Hingorani, N. G. (1995). Introducing Custom Power. *IEEE Spectrum*, *32*(6), 41–48. doi:10.1109/6.387140

Hingorani, N. G., & Gyugyi, H. (2000). *Understanding FACTS: concepts and technology of flexible AC transmission systems*. New York: IEEE Press, Inc.

Hingorani, N. G., & Gyugyi, L. (2000). *Understanding FACTS Concepts and Technology of Flexible AC Transmission Systems*. New York: IEEE Press.

Hiyama, T., Kouzuma, S., & Imakubo, T. (1995). Identification of Optimal Operating Point of PV Modules Using Neural Network for Real Time Maximum Power Tracking Control. *IEEE Transactions on Energy Conversion*, *10*(2), 360–367. doi:10.1109/60.391904

Ho, B. M. T., Chung, H. S. H., & Lo, W. L. (2004, March). Use of system oscillation to locate the MPP of PV panels. *IEEE Power Electronics Letters*, *2*(1), 1–5. doi:10.1109/LPEL.2004.828443

Homayoun, H., Veidenbaum, A., & Gaudiot, J. L. (2008, October). Adaptive techniques for leakage power management in L2 cache peripheral circuits. In *Computer Design, 2008. ICCD 2008. IEEE International Conference on* (pp. 563-569). IEEE. 10.1109/ICCD.2008.4751917

Homayounzade, M., Keshmiri, M., & Danesh, M. (2010). An Observer-Based State Feedback Controller Design for Robot Manipulators Considering Actuators' Dynamic. *IEEE 2010 15th International Conference on Methods and Models in Automation and Robotics (MMAR)*, 240-248.

HongmeiT.FernandoM. D.KevinE.PeterJ.EduardM. (2012). Retrieved from http://www.nrel.gov/docs/fy12osti/54601.pdf

Horizontal Market Power in Restructured Electricity Markets. (2000). Washington, DC: Office of Economic,Electricity and Natural Gas Analysis, U.S. Department of Energy.

Hornik, K., Stinchcombe, M., & White, H. (1989). Multilayer feedforward networks are universal approximators. *Neural Networks*, *2*(5), 359–366. doi:10.1016/0893-6080(89)90020-8

Hossain, S., Khan, A.R., & Noorm, J.E. (2014). Design of STATCOM for Power System Stability Improvement. *Global Journal of Researches in Engineering: Electrical and Electronics Engineering, 14*.

Hossain, M. J., Pota, H. R., Mahmud, M. A., & Ramos, R. A. (2012). Investigation of the impacts of large-scale wind power penetration on the angle and voltage stability of power systems. *IEEE Systems Journal*, *6*(1), 76–84. doi:10.1109/JSYST.2011.2162991

Houston, J. M. (1959). Theoretical efficiency of the thermionic energy converter. *Journal of Applied Physics*, *30*(4), 481–487. doi:10.1063/1.1702392

Hsiao, Y. Y., Chang, W. C., & Chen, S. L. (2010). A mathematic model of thermoelectric module with applications on waste heat recovery from automobile engine. *Energy*, *35*(3), 1447–1454. doi:10.1016/j.energy.2009.11.030

Hsieh, Y. P., Chen, J. F., Liang, T. J., & Yang, L. S. (2013, April). Novel high step-up DC-DC converter for distributed generation system. *IEEE Transactions on Industrial Electronics*, *60*(4), 1473–1482. doi:10.1109/TIE.2011.2107721

Hsu, C. H., & Feng, W. C. (2004, December). Effective dynamic voltage scaling through CPU-boundedness detection. In PACS (pp. 135-149). Academic Press.

Hsu, C. H., & Kremer, U. (2003). The design, implementation, and evaluation of a compiler algorithm for CPU energy reduction. *ACM SIGPLAN Notices*, *38*(5), 38–48. doi:10.1145/780822.781137

Hua, C., & Shen, C. (1998, May). Study of maximum power tracking techniques and control of DC/DC converters for photovoltaic power system. In *Power Electronics Specialists Conference, 1998. PESC 98 Record. 29th Annual IEEE* (Vol. 1, pp. 86-93). IEEE.

Huang, G.-B., Zhou, H., Ding, X., & Zhang, R. (2012). Extreme learning machine for regression and multiclass classification. *IEEE Transactions on Systems, Man, and Cybernetics. Part B, Cybernetics*, *42*(2), 513–529. doi:10.1109/TSMCB.2011.2168604 PMID:21984515

Huang, G.-B., Zhu, Q., & Siew, C. (2006). Extreme learning machine: Theory and applications. *Neurocomputing*, *70*(1-3), 489–501. doi:10.1016/j.neucom.2005.12.126

Huang, S. H., Jeang, Y. L., Hwang, C. T., Hsu, Y. C., & Wang, J. F. (1993, July). A tree-based scheduling algorithm for control-dominated circuits. In *Proceedings of the 30th international Design Automation Conference* (pp. 578-582). ACM. 10.1145/157485.165051

Huang, Z., & Ercegovac, M. D. (2005). High-performance low-power left-to-right array multiplier design. *IEEE Transactions on Computers*, *54*(3), 272–283. doi:10.1109/TC.2005.51

Humphrey, T. E., O'Dwyer, M. F., & Linke, H. (2005). Power optimization in thermionic devices. *Journal of Physics. D, Applied Physics*, *38*(12), 2051–2054. doi:10.1088/0022-3727/38/12/029

Huneault, M., & Galiana, F. (1991). A survey of the optimal power flow literature. *Power Syst IEEE Trans*, *6*(2), 762–770. doi:10.1109/59.76723

Hurtgen, M., & Maun, J.-C. (2010). Optimal PMU placement using iterated local search. *International Journal of Electrical Power & Energy Systems*, *32*(8), 857–860. doi:10.1016/j.ijepes.2010.01.028

Hwang, C. T., Lee, J. H., & Hsu, Y. C. (1991). A formal approach to the scheduling problem in high level synthesis. *IEEE Transactions on Computer-Aided Design of Integrated Circuits and Systems*, *10*(4), 464–475. doi:10.1109/43.75629

Hwang, E., Vahid, F., & Hsu, Y. C. (1999, January). FSMD functional partitioning for low power. In *Proceedings of the conference on Design, automation and test in Europe* (p. 7). ACM.

Injeti, S., & Padma, K. (2015). A super capacitor based power conditioning system for power quality improvement. *IRJET*, *2*(7).

Ishaque, K., Salam, Z., Amjad, M., & Mekhilef, S. (2012, August). An Improved Particle Swarm Optimization (PSO)–Based MPPT for PV With Reduced Steady-State Oscillation. *IEEE Transactions on Power Electronics*, *27*(8), 3627–3638. doi:10.1109/TPEL.2012.2185713

Islam, M., Inal, O. T., & Luke, J. R. (2006). Development of electron reflection suppression materials for improved thermionic energy converter performance using thin film deposition techniques. *Journal of Applied Physics*, *100*(8), 084903. doi:10.1063/1.2356785

Itoh, K., Kimura, S. I., & Sakata, T. (1999, September). VLSI memory technology: Current status and future trends. In *Solid-State Circuits Conference, 1999. ESSCIRC'99. Proceedings of the 25th European* (pp. 3-10). IEEE.

Ito, T., & Cappelli, M. A. (2012). Optically pumped cesium plasma neutralization of space charge in photon-enhanced thermionic energy converters. *Applied Physics Letters*, *101*(21), 213901. doi:10.1063/1.4767349

ITRS. (2013). *Process Integration, Devices, & Structures (PIDS) Chapter*. Retrieved from http://www.itrs.net/Links/2013ITRS/Home2013.htm

Jadhav, A. (2016). Resource Scheduling in High Level Synthesis of Digital Filter using Differential Evolution Algorithm. *Journal of Digital Integrated Circuits in Electrical Devices*, *1*(1-2).

Jafarian, P., & Bina, M. T. (2010). State-Feedback Current Control of VSI-Based D-STATCOM for Load Compensation. *IEEE International Conference on Environment and Electrical Engineering*, 214-217. 10.1109/EEEIC.2010.5490431

Jain, Karampuri, & Somasekhar. (2016). An integrated control algorithm for a single-stage PV pumping system using an open-end winding induction motor. *IEEE Transactions on Inductrial Electronics, 63*(2), 956-965.

Jain, Ramulu, Sanjeevikumar, Ojo, & Ertas. (2016). Dual MPPT algorithm for dual PV source fed Open-End Winding Induction Motor Drive for pumping application. *Engineering Science and Technology, an International Journal, 19*(4), 1771-1780.

Jain, R., Mujumdar, A., Sharma, A., & Wang, H. (1991, June). Empirical evaluation of some high-level synthesis scheduling heuristics. In *Proceedings of the 28th ACM/IEEE Design Automation Conference* (pp. 686-689). ACM. 10.1145/127601.127751

Jain, S., Thopukara, A. K., Karampuri, R., & Somasekhar, V. T. (2015). A single-stage photovoltaic system for a dual-inverter-fed open-end winding induction motor drive for pumping applications. *IEEE Transactions on Power Electronics, 30*(9), 4809–4818. doi:10.1109/TPEL.2014.2365516

Javadi, A., Hamadi, A., Woodward, L., & Al-Haddad, K. (2016). Experimental Investigation on a Hybrid Series Active Power Compensator to Improve Power Quality of Typical Households. *IEEE Transactions on Industrial Electronics*, 63.

Jemai, M., Dimassi, S., Ouni, B., & Mtibaa, A. (2017). A metaheuristic based on the tabu search for hardware-software partitioning. *Turkish Journal of Electrical Engineering and Computer Sciences, 25*(2), 901–912. doi:10.3906/elk-1501-64

Jensen, K. L. (2007). General formulation of thermal, field, and photoinduced electron emission. *Journal of Applied Physics, 102*(2), 024911. doi:10.1063/1.2752122

Jeong, K. S., Jyung, T. Y., & Baek, Y. S. (2011). Modeling and Dynamic Analysis of STATCOM for Short-Term Voltage Stability Improvement using EMTP/RV. *Journal of International Council on Electrical Engineering, 1*(2), 129–134. doi:10.5370/JICEE.2011.1.2.129

Jervan, G. (1998). *Decision Diagram Synthesis from VHDL.* Tallinn Technical University.

Jeter, S. M. (1981). Maximum conversion efficiency for the utilization of direct solar radiation. *Solar Energy, 26*(3), 231–236. doi:10.1016/0038-092X(81)90207-3

Jewell, W. T. (1989, June). Electric utility experience with solar photovoltaic generation. *IEEE Transactions on Energy Conversion, 4*(2), 166–171. doi:10.1109/60.17907

Jindal, K. A., Ghosh, A., & Joshi, A. (2007). Interline unified power quality conditioner. *IEEE Transactions on Power Delivery, 1*(22), 364–372. doi:10.1109/TPWRD.2006.881581

Jinn, C. W., Kuen, D. W., Hurng, L. J., & Sheng, K. C. (2014). Small capacity Grid connected solar power generation system. *IET Power Electronics, 7*(11), 2717–2725. doi:10.1049/iet-pel.2014.0015

Joorabian, M., & Afsan, E. (2014). Optimal power flow under both normal and contingent operation conditions using the hybrid fuzzy particle swarm optimization and nedler-mead algorithm (HFPSO-NM). *Applied Soft Computing, 14*, 623–633. doi:10.1016/j.asoc.2013.09.015

Jordan, M. I., & Mitchell, T. M. (2015). Machine learning: Trends, perspectives, and prospects. *Science, 349*(6245), 255–260. doi:10.1126cience.aaa8415 PMID:26185243

Joshi, A. S., Dincer, I., & Reddy, B. V. (2009). Thermodynamic assessment of photovoltaic systems. *Solar Energy, 83*(8), 1139–1149. doi:10.1016/j.solener.2009.01.011

Joskow, P. L., & Tirole, J. (2000). Transmission rights and market power on electric power networks. *The Rand Journal of Economics, 31*(3), 450–487. doi:10.2307/2600996

Kadri, A., Marzougui, H., & Bacha, F. (2016). MPPT control methods in wind energy conversion system using DFIG. *4th International Conference on Control Engineering & Information Technology (CEIT)*, 1-6. 10.1109/CEIT.2016.7929115

Kalogirou, S. (1999). Applications of artificial neural networks in energy systems. *Energy Conversion and Management*, *40*(10), 1073–1087. doi:10.1016/S0196-8904(99)00012-6

Kanar, K. (1995). Structure Oriented Behavior Tests. In *Model Validation, Master Of Science*. Industrial Engineering Boğaziçi University.

Kao, J., Narendra, S., & Chandrakasan, A. (2002, November). Subthreshold leakage modeling and reduction techniques. In *Proceedings of the 2002 IEEE/ACM international conference on Computer-aided design* (pp. 141-148). ACM.

Kapoor, A. (2015). *U.S. Patent Application No. 14/973,616*. Washington, DC: US Patent Office.

Karaboga, D. (2005). *An idea based on honey bee swarm for numerical optimization*. Technical Report-TR06, Erciyes University of Engineering, Faculty of Computer Engineering Department.

Karbakhsh, M., Abutorabi, H., & Khazaee, A. (2012). An Enhanced MPPT Fuzzy Control of a Wind Turbine Equipped with Permanent Magnet Synchronous Generator. *2nd International Conference on Computer and Knowledge Engineering* (*ICCKE*), 77-82.

Karimov, K., Chattha, J., & Ahmed, M. (2002). Journal of References. *Academy of Sciences of Tajikistan*, *XLV*(9), 75–83.

Karnopp, D. C., Margolis, D. L., & Rosenberg, R. C. (1990). *System Dynamics A Unified Approach* (2nd ed.). John Wiley & Sons, Inc.

Karthikeyan, S. P., Raglend, I. J., & Kothari, D. P. (2013). A review on market power in deregulated electricity market. *International Journal of Electrical Power & Energy Systems*, *48*, 139–147. doi:10.1016/j.ijepes.2012.11.024

Karthikeyan, S. P., Sahoo, S. K., Roopesh, R., Mathur, S., Raglend, I. J., & Kothari, D. P. (2012). Utilizing FACTS devices in enhancement of market power under congestion-a threat to the deregulated electricity market. *2012 Annual IEEE India Conference (INDICON)*. 10.1109/INDCON.2012.6420632

Kastha, D., & Bose, B. K. (1994). Investigation of Fault Modes of Voltage-Fed Inverter System for Induction Motor Drive. *IEEE Transactions on Industry Applications*, *30*(4), 1028–1038. doi:10.1109/28.297920

Kaveh, A., & Ghazaan, M. I. (2016). Enhanced whale optimization algorithm for sizing optimization of skeletal structures. *Mechanics Based Design of Structures and Machines*, 1–18.

Kawakami, N., Ota, S., Kon, H., Konno, S., Akagi, H., Kobayashi, H., & Okada, N. (2014, November). Development of a 500-kW Modular Multilevel Cascade Converter for Battery Energy Storage Systems. *IEEE Transactions on Industry Applications*, *50*(6), 3902–3910. doi:10.1109/TIA.2014.2313657

Kayikci, M., & Milanovic, J. V. (2007). Reactive power control strategies for dfig-based plants. *IEEE Transactions on Energy Conversion*, *22*(2), 389–396. doi:10.1109/TEC.2006.874215

Kazemi, A., Sarlak, M., & Barkhordary, M. (2006). An Adaptive Noise Canceling Method for Single-Phase Unified Power Quality Conditioner. *I*[st] *IEEE Conference on Industrial Electronics and Applications*, 6.

Kazimierczuk, M. K. (2008). *Pulse-Width Modulated DC-DC Power Converters* (1st ed.). Hoboken, NJ, USA: Wiley. doi:10.1002/9780470694640

Keane, A., Ochoa, L. F., Vittal, E., Dent, C. J., & Harrison, G. P. (2011). Enhanced utilization of voltage control resources with distributed generation. *IEEE Transactions on Power Systems*, *26*(1), 252–260. doi:10.1109/TPWRS.2009.2037635

Kerdphol, T., Qudaih, Y., & Mitani, Y. (2014). Battery Energy Storage System Size Optimization in Microgrid using Particle Swarm Optimization. 5th IEEE PES Innovative Smart Grid Technologies Europe (ISGT Europe), Istanbul, Turkey. doi:10.1109/ISGTEurope.2014.7028895

Ketabi, A., Nosratabadi, S., & Sheibani, M. (2010). *Optimal PMU placement based on mean square error using differential evolution algorithm.* Paper presented at the Power Quality Conference (PQC), 2010 First.

Khadkikar, V., & Chandra, A. (2011). UPQC-S: A Novel Concept of Simultaneous Voltage Sag/Swell and Load Reactive Power Compensations Utilizing Series Inverter of UPQC. IEEE Transactions on Power Electronics, 1(26), 2414-2425.

Khadkikar, V. (2012). Enhancing Electric Power Quality Using UPQC: A comprehensive overview. *IEEE Transactions on Power Electronics, 27*(5), 2284–2297. doi:10.1109/TPEL.2011.2172001

Khadkikar, V., & Chandra, A. (2009). A novel structure for three-phase four wire distribution system utilizing unified power quality conditioner (UPQC). *IEEE Transactions on Industry Applications, 45*(5), 1897–1902. doi:10.1109/TIA.2009.2027147

Khadkikar, V., Chandra, A., & Singh, B. (2011). Digital signal processor implementation and performance evaluation of split capacitor, four-leg and three H-bridge-based three-phase four-wire shunt active filters. *IET Power Electronics, 4*(4), 463–470. doi:10.1049/iet-pel.2010.0198

Khajeh, H., Foroud, A. A., & Firoozi, H. (2016). New ex-ante indices of market power. In *24th Iranian Conference on Electrical Engineering (ICEE).* 10.1109/IranianCEE.2016.7585391

Khajeh, K. G., Bashar, E., Rad, A. M., & Gharehpetian, G. B. (2017). Integrated model considering effects of zero injection buses and conventional measurements on optimal PMU placement. *IEEE Transactions on Smart Grid, 8*(2), 1006–1013.

Khalil, A., & El-Singaby, M. (2003). Position control of sun tracking system; Circuits and Systems. *Proceedings of the 46th IEEE International Midwest Symposium, 3*, 1134–1137.

Khan, M. T., Siddiqui, A. S., Sarwar, M., & Iqbal, F. (2015). Mitigation of Market Power in deregulated power system using Nash Equilibrium. *2015 Annual IEEE India Conference (INDICON)*, 1–5. 10.1109/INDICON.2015.7443645

Khodaei, A., & Shahidehpour, M. (2013). Microgrid-based Co-Optimization of Generation and Transmission Planning in Power Systems. *IEEE Transactions on Power Systems, 28*(2), 1582–1590. doi:10.1109/TPWRS.2012.2224676

Khoor, M., Machmoum, M., & Abdelli, Y. (2005). Using a simple analogical identification method and PID controllers for unified power quality conditioning. *European Conference on Power Electronics and Applications,* 10.

Khoor, M., & Machmoum, M. (2005). Simplified analogical control of a unified power quality conditioner. *Proc. Power Electron. Spec. Conf.*, 2565–2570. 10.1109/PESC.2005.1581994

Khoor, S. M., & Machmoum, M. (2007). A novel single-phase reduced parts on-line UPS with power quality conditioning capability. *Proc. Power Electron. Spec. Conf.*, 1170–1175. 10.1109/PESC.2007.4342158

Khoucha, F., Lagoun, M. S., Kheloui, A., & Benbouzid, M. E. H. (2011). A comparison of symmetrical and asymmetrical three-phase h-bridge ltilevel inverter for dtc induction motor drives. *IEEE Transactions on Energy Conversion, 26*(1), 64–72. doi:10.1109/TEC.2010.2077296

Khounjahan, Banaei, & Farakhor. (2014). A new low cost cascaded transformer multilevel inverter topology using minimum number of components with modified selective harmonic elimination modulation. *Ain Shams Engineering Journal, 6*, 67–73.

Kim, K. K., & Kim, Y. B. (2007, May). Optimal body biasing for minimum leakage power in standby mode. In *Circuits and Systems, 2007. ISCAS 2007. IEEE International Symposium on* (pp. 1161-1164). IEEE. 10.1109/ISCAS.2007.378256

Kim, Kim, Kang, & Hyun. (2003). A simple method for conducting angle calculation in cascaded inverters using step pulse waves. *29th IEEE Annual Conference, Industrial Electronics Society, 3*, 2735-2740.

Kinsey, G. S., Hebert, P., Barbour, K. E., Krut, D. D., Cotal, H. L., & Sherif, R. A. (2008). Concentrator multijunction solar cell characteristics under variable intensity and temperature. *Progress in Photovoltaics: Research and Applications*, *16*(6), 503–508. doi:10.1002/pip.834

Kirkwood, C. W. (1998). *System Dynamics Methods: A Quick Introduction*. College Of Business Arizona State University.

Kisck, D., Navrapescu, V., & Kisck, M. (2007). Single-phase unified power quality conditioner with optimum voltage angle injection for minimum VA requirement. *Proc. IEEE Int. Symp. Ind. Electron.,* 2443–2448.

Kission, P. (1996). *Exploitation de la hierarchie et de la reutilisation de blocs existants par la synthese de haut niveau* (Doctoral dissertation).

Kjaer, S. B., Pedersen, J. K., & Blaabjerg, F. (2005). A review of single-phase grid-connected inverters for photovoltaic modules. *IEEE Transactions on Industry Applications*, *41*(5), 1292–1306. doi:10.1109/TIA.2005.853371

Knittel, C. R., & Metaxoglou, K. (2008). Diagnosing unilateral market power in electricity reserves market. *Journal of Energy Markets*, *1*(1), 65–95. doi:10.21314/JEM.2008.001

Kolhatkar, Y., Errabelli, R., & Das, S. (2005). A sliding mode controller based optimum UPQC with minimum VA loading. *Proc. Power Eng. Soc. Gen. Meet.,* 871–875.

Kolhatkar, Y., & Das, S. (2007). Experimental investigation of a single-phase UPQC with minimum VA loading. *IEEE Transactions on Power Delivery*, (22): 371–380.

Konopinski, R. J., Vijayan, P., & Ajjarapu, V. (2009). Extended reactive capability of dfig wind parks for enhanced system performance. *IEEE Transactions on Power Systems*, *24*(3), 1346–1355. doi:10.1109/TPWRS.2009.2023260

Kotas, T. J. (1995). *The Exergy Method of Thermal Plant Analysis*. Malabar, FL: Krieger Publish Company.

Koutsoukis, N. C., Manousakis, N. M., Georgilakis, P. S., & Korres, G. N. (2013). Numerical observability method for optimal phasor measurement units placement using recursive Tabu search method. *IET Generation, Transmission & Distribution*, *7*(4), 347–356. doi:10.1049/iet-gtd.2012.0377

Kou, X., Corzine, K. A., & Familiant, Y. L. (2004). A unique fault-tolerant design for flying capacitor multilevel inverter. *IEEE Transactions on Power Electronics*, *19*(4), 979–987. doi:10.1109/TPEL.2004.830037

Kraemer, D., Poudel, B., Feng, H. P., Caylor, J. C., Yu, B., Yan, X., ... McEnaney, K. (2011). High-performance flat-panel solar thermoelectric generators with high thermal concentration. *Nature Materials*, *10*(7), 532–538. doi:10.1038/nmat3013 PMID:21532584

Krause, P., Wasynczuk, O., Sudhoff, S., & Pekarek, S. (2013). *Analysis of Electric Machinery and Drive Systems*. John Willey & Sons. doi:10.1002/9781118524336

Kreczanik, P., Venet, P., Hijazi, A., & Clerc, G. (2014, September). Study of supercapacitor aging and lifetime estimation according to voltage, temperature and rms current. *IEEE Transactions on Industrial Electronics*, *61*(9), 4895–4902. doi:10.1109/TIE.2013.2293695

Kreith, F. (1974). *Principles of heat transfer* (3rd ed.). Scranton: International Textbook Company.

Kribus, A., & Mittelman, G. (2008). Potential of polygeneration with solar thermal and photovoltaic systems. *Journal of Solar Energy Engineering*, *130*(1), 011001. doi:10.1115/1.2804618

Krishnamurthy, S. (2009). *Solar and Fuel Cell Circuit Modeling*. Analysis and Integrations with Power Conversion Circuits for Distributed Generation, Electronic Theses and Dissertations.

Krishnan, V., & McCalley, J. D. (2006). Role of Induction Motors in Voltage Instability and coordinated Reactive Power planning. Academic Press.

Kroutoulis, E., Kalaitzakis, K., & Voulgaris, N. C. (2001). Development of a Microcontroller-Based, Photovoltaic Maximum Power Tracking Control System. *IEEE Transactions on Power Electronics*, *16*(1).

Kuchcinski, K. (1998, August). An approach to high-level synthesis using constraint logic programming. In *Euromicro Conference, 1998. Proceedings. 24th* (Vol. 1, pp. 74-82). IEEE. 10.1109/EURMIC.1998.711780

Ku, D. C., & De Micheli, G. (1988). *Hardware C-a language for hardware design (No. CSL-TR-88-362)*. Stanford Univ CA Computer Systems Lab. doi:10.21236/ADA207317

Kulanthaisamy, A., Vairamani, R., Karunamurthi, N. K., & Koodalsamy, C. (2014). A multi-objective PMU placement method considering observability and measurement redundancy using ABC algorithm. *Advances in Electrical and Computer Engineering*, *14*(2), 117–128. doi:10.4316/AECE.2014.02020

Kumar, P., & Singh, S. (2014a). Comprehensive stability analysis of radial distribution system with load growth. *IEEE International Conference IICPE-2014.*

Kumar, P., & Singh, S. (2014a). Comprehensive stability analysis of radial distribution system with load growth. *Proc. IEEE International Conference IICPE-2014.*

Kumar, P., & Singh, S. (2014b). Energy efficiency performance of radial distribution system with reactive power injection. *IEEE International Conference IICPE-2014.*

Kumar, P., & Singh, S. (2014b). Energy efficiency performance of radial distribution system with reactive power injection. *Proc. IEEE International Conference IICPE-2014.*

Kumar, P., & Singh, S. (2014c). Reconfiguration of radial distribution system with static load models for loss minimization. *IEEE International Conference PEDES-2014.*

Kumar, P., & Singh, S. (2014c). Reconfiguration of radial distribution system with static load models for loss minimization. *Proc. IEEE International Conference PEDES-2014.*

Kumar, P., Ali, I., & Thomas, M. S. (2013). Distributed resource planning for improved voltage stability of radial distribution system. *International Journal of Advanced Research in Electrical, Electronics and Instrumentation Engineering*, *2*(1), 95-101.

Kumar, P., Ali, I., & Thomas, M. S. (2015). *Synchronizing solar cell, battery and grid supply for development of smart power system for home*. Presented in IEEE International Conference INDICON-2015 held at Jamia Millia Islamia, New Delhi, India.

Kumar, P., Ali, I., & Thomas, M. S. (2015). Synchronizing solar cell, battery and grid supply for development of smart power system for home. *Proc. IEEE International Conference INDICON.*

Kumar, D., Zare, F., & Ghosh, A. (2017). DC Microgrid Technology: System Architectures, AC Grid Interfaces, Grounding Schemes, Power Quality, Communication Networks, Applications, and Standardizations Aspects. *IEEE Access: Practical Innovations, Open Solutions*, *5*, 12230–12256. doi:10.1109/ACCESS.2017.2705914

Kumar, G., & Kumar, A. (2013). Renewable Energy Interconnection at Distribution Level to improve Power Quality. *International Journal of Engineering Science*, *2*(5), 39–48.

Kumar, I. P., & Kothari, D. P. (2005). Recent philosophies of AGC strategies in power system. *IEEE Transactions on Power Systems*, *20*(1), 345–357.

Kumar, N. J. V., & Ansari, M. M. T. (2015). A new design of dual-mode type-II fuzzy logic load frequency controller for interconnected power systems with parallel AC/DC tie-lines and superconducting magnetic energy storage unit. *Energy, 89*, 118–137. doi:10.1016/j.energy.2015.07.056

Kumar, P., Ali, I., & Thomas, M. S. (2011). Optimal capacitor placement in smart distribution systems to improve its loadability and energy efficiency. *International Journal of Engineering Science and Technology, 3*(8), 271–284.

Kumar, P., Ali, I., & Thomas, M. S. (2015). Synchronizing Solar Cell, Battery and Grid Supply for Development of Smart Power System for Home. In *IEEE International Conference INDICON-2015* (pp. 1–6). IEEE.

Kumar, P., Ali, I., & Thomas, M. S. (2016). Energy efficiency analysis of reconfigured distribution system for practical loads. *Elsevier Journal of Perspectives in Science, 8*, 498–501. doi:10.1016/j.pisc.2016.05.007

Kumar, P., Ali, I., Thomas, M. S., & Singh, S. (2017). Imposing Voltage Security and Network Radiality for Reconfiguration of Distribution System Using Efficient Heuristic and Meta-heuristic approaches. *IET Generation, Transmission & Distribution, 11*(10), 2457–2467.

Kumar, P., Ali, I., Thomas, M. S., & Singh, S. (2017, March). Imposing voltage security and network radiality for reconfiguration of distribution system using efficient heuristic and meta-heuristic approaches. *IET Generation, Transmission & Distribution, 11*(10), 2457–2467. doi:10.1049/iet-gtd.2016.0935

Kumar, P., Jain, G., & Palwalia, D. K. (2015). Genetic Algorithm Based Maximum Power Tracking in Solar Power Generation. *International Conference on Power And Advanced Control engineering (ICPACE)*, 1-6.

Kumar, P., & Singh, S. (2014). Reconfiguration of Radial Distribution System with Static Load Models for Loss Minimization. In *IEEE International Conference PEDES-2014* (pp. 1–5). IEEE.

Kumar, R., & Singh, B. (2016). BLDC Motor-Driven Solar PV Array-Fed Water Pumping System Employing Zeta Converter. *IEEE Transactions on Industry Applications, 52*(3), 2315–2322. doi:10.1109/TIA.2016.2522943

Kumar, R., & Singh, B. (2016). Solar photovoltaic array fed water pump driven by brushless DC motor using Landsman converter. *IET Renewable Power Generation, 10*(4), 474–484. doi:10.1049/iet-rpg.2015.0295

Kumar, V. S. S., Reddy, K. K., & Dhadbanjan, T. (2014). Coordination of reactive power in grid-connected wind farms for voltage stability enhancement. *IEEE Transactions on Power Systems, 29*(5), 2381–2390. doi:10.1109/TPWRS.2014.2300157

Kundur, P., Balu, N. J., & Lauby, M. G. (1994). *Power system stability and control* (Vol. 7). McGraw-Hill New York.

Kuo, C. L., Lin, C. H., Yau, H. T., & Chen, J. L. (2013). Using Self-Synchronization Error Dynamics Formulation Based Controller for Maximum Photovoltaic Power Tracking in Micro- Grid Systems. *IEEE Journal on Emerging and Selected Topics in Circuits and Systems, 3*(3), 459–467. doi:10.1109/JETCAS.2013.2272839

Kuon, I., & Rose, J. (2007). Measuring the gap between FPGAs and ASICs. *IEEE Transactions on Computer-Aided Design of Integrated Circuits and Systems, 26*(2), 203–215. doi:10.1109/TCAD.2006.884574

Kushwaha, S., & Kumar, S. (2017). An Investigation of List Heuristic Scheduling Algorithms for Multiprocessor System. *IUP Journal of Computer Sciences, 11*(2), 29.

Kwan, H. K., Chu, C. Y., & So, L. P. (2009). Model-based H-∞ control of a unified power quality conditioner. *IEEE Transactions on Industrial Electronics, 56*(7), 2493–2504. doi:10.1109/TIE.2009.2020705

Kwan, K. H., So, P. L., & Chu, Y. C. (2007). A harmonic selective unified power quality conditioner using MVR with Kalman filters. *Power Engineering Conference, International, 1* (3), 332-337.

Kwan, K., So, P., & Chu, Y. (2005). Unified power quality conditioner for improving power quality using MVR with Kalman filters. *7th Int. Power Eng. Conf.*, 980–985. 10.1109/IPEC.2005.207050

Kwon, M., & Kwon, B. H. (2009). High step-up active-clamp converter with input-current doubler and output-voltage doubler for fuel cell power systems. *IEEE Transactions on Power Electronics*, *24*(1), 108–115. doi:10.1109/TPEL.2008.2006268

Lahfaoui, B., Zouggar, S., Elhafyani, M. L., & Seddik, M. (2015). Experimental study of P&O MPPT control for wind PMSG turbine. *3rd International Renewable and Sustainable Energy Conference (IRSEC)*, 1-6. 10.1109/IRSEC.2015.7455020

Lai. (1996). Multilevel converters - a new breed of power converter. *IEEE Transactions on Industry Applications*, *32*(3), 509–517. doi:10.1109/28.502161

Lai, L. L. (2001). *Power System Restructuring and Deregulation: Trading, Performance and Information Technology*. John Wiley & Sons, Ltd. doi:10.1002/0470846119

Lai, L. L., Ma, J. T., Yokoyama, R., & Zhao, M. (1997). Improved genetic algorithms for optimal power flow under normal and contingent operation states. *International Journal of Electrical Power & Energy Systems*, *19*(5), 287–292. doi:10.1016/S0142-0615(96)00051-8

Lana El Chaar. (2007). *Photovoltaic System Conversion*. Power Electronics Handbook.

Landsberg, P. T., & Tonge, G. (1980). Thermodynamic energy conversion efficiencies. *Journal of Applied Physics*, *51*(7), R1–R20. doi:10.1063/1.328187

Landwehr, B., Marwedel, P., & Dömer, R. (1994, September). OSCAR: Optimum simultaneous scheduling, allocation and resource binding based on integer programming. In *Proceedings of the conference on European design automation* (pp. 90-95). IEEE Computer Society Press.

Lane, D. J., & Sterman, J. D. (2011). Profiles in Operations Research: Jay Wright Forrester. In Profiles in Operations Research: Pioneers and Innovators. New York: Springer.

Larson, D. L. (1995). Exergy analysis: Essential to effective energy management. *Transactions of the ASAE. American Society of Agricultural Engineers*, *38*(4), 1173–1178. doi:10.13031/2013.27937

Laurent, B. (1999). *Conception des blocs réutilisables. Réflexion sur la méthodologie* (Doctoral dissertation). Institut National Polytechnique de Grenoble-INPG.

Lawan, A.U. (2014). State Feedback Approaches for Designing A Statcom Supplementary Controller for Oscillations Damping. *The International Journal Of Engineering And Science, 3*(7), 27-37.

Lee, B. (2000). *Specification and Design of Reactive Systems* (PhD Thesis). University of California, Berkeley, CA.

Lee, J., Han, B., & Choi, K. (2011). High-efficiency grid-tied power conditioning system for fuel cell power generation. *8th International Conference on Power Electronics - ECCE Asia(ICPE & ECCE)*, 1492-1497.

Lee, J. H., Bae, H., & Cho, B. H. (2006). *Advanced Incremental Conductance MPPT Algorithm with a Variable Step Size*. IEEE Power Electronics and Motion Control. doi:10.1109/EPEPEMC.2006.283227

Lee, J. H., Bargatin, I., Melosh, N. A., & Howe, R. T. (2012). Optimal emitter-collector gap for thermionic energy converters. *Applied Physics Letters*, *100*(17), 173904. doi:10.1063/1.4707379

Lee, J. H., Liang, T. J., & Chen, J. F. (2014, July). Isolated coupled-inductor- integrated ADC–DC converter with nondissipative snubber for solar energy applications. *IEEE Transactions on Industrial Electronics*, *61*(7), 3337–3348. doi:10.1109/TIE.2013.2278517

Lee, J. I., Jeong, Y. H., No, H. C., Hannebauer, R., & Yoo, S. K. (2009). Size effect of nanometer vacuum gap thermionic power conversion device with CsI coated graphite electrodes. *Applied Physics Letters*, *95*(22), 223107. doi:10.1063/1.3266921

Lee, N. E., Kim, W. w., Shin, J. h., & Kim, J. O. (2014). Optimal SOC limit estimation of ESS considering renewable energy sources. *International Conference on Information Science, Electronics and Electrical Engineering*, 1927-1931.

Lee, S., Kim, P., & Choi, S. (2013, July). High step-up soft-switched converters using voltage multiplier cells. *IEEE Transactions on Power Electronics*, *28*(7), 3379–3387. doi:10.1109/TPEL.2012.2227508

Lee, Y. Y., Hur, J., Baldick, R., & Pineda, S. (2011). New Indices of Market Power in Transmission-Constrained Electricity Markets. *IEEE Transactions on Power Systems*, *26*(2), 681–689. doi:10.1109/TPWRS.2010.2060738

Lesieutre, B. C., Rogers, K. M., Overbye, T. J., & Borden, A. R. (2011). A Sensitivity Approach to Detection of Local Market Power Potential. *IEEE Transactions on Power Systems*, *26*(4), 1980–1988. doi:10.1109/TPWRS.2011.2105893

Li, H., Bhunia, S., Chen, Y., Vijaykumar, T. N., & Roy, K. (2003, February). Deterministic clock gating for microprocessor power reduction. In *High-Performance Computer Architecture, 2003. HPCA-9 2003. Proceedings. The Ninth International Symposium on* (pp. 113-122). IEEE.

Liang, T. J., Chen, S. M., Yang, L. S., Chen, J. F., & Ioinovici, A. (2012, April). Ultra- large gain step-up switched-capacitor DC-DC converter with a coupled inductor for alternative sources of energy. *IEEE Transactions on Circuits and Systems*, *59*(4), 864–874. doi:10.1109/TCSI.2011.2169886

Liang, X. (2017). Emerging Power Quality Challenges Due to Integration of Renewable Energy Sources. *IEEE Transactions on Industry Applications*, *53*(2), 855–866.

Li, H., Chen, F., Cheng, K., Zhao, Z., & Yang, D. (2015). Prediction of Zeta Potential of Decomposed Peat via Machine Learning. *Comparative Study of Support Vector Machine and Artificial Neural Networks*, *10*, 6044–6056.

Li, H., Liu, Z., Liu, K., & Zhang, Z. (2017). Predictive Power of Machine Learning for Optimizing Solar Water Heater Performance: The Potential Application of High-Throughput Screening. *International Journal of Photoenergy*, *2017*, 1–10. doi:10.1155/2017/4194251

Li, H., Tang, X., Wang, R., Lin, F., Liu, Z., & Cheng, K. (2016). Comparative Study on Theoretical and Machine Learning Methods for Acquiring Compressed Liquid Densities of 1,1,1,2,3,3,3-Heptafluoropropane (R227ea) via Song and Mason Equation, Support Vector Machine, and Artificial Neural Networks. *Applied Sciences*, *6*(1), 25. doi:10.3390/app6010025

Li, N., & Hedman, K. W. (2015). Economic Assessment of Energy Storage in Systems With High Levels of Renewable Resources. *IEEE Transactions on Sustainable Energy*, *6*(3), 1103–1111.

Lin, T. J., Chen, S. K., Kuo, Y. T., Liu, C. W., & Hsiao, P. C. (2008). Design and implementation of a high-performance and complexity-effective VLIW DSP for multimedia applications. *Journal of Signal Processing Systems for Signal, Image, and Video Technology*, *51*(3), 209–223. doi:10.100711265-007-0061-x

Lin, Y. L. (1997). Recent developments in high-level synthesis. *ACM Transactions on Design Automation of Electronic Systems*, *2*(1), 2–21. doi:10.1145/250243.250245

Liu, F., Zhang, Y., & Duan, S. (2008). Comparison of P&O and hill climbing MPPT methods for grid- connected PV converter. IEEE Industrial Electronics and Applications, 804-807.

Liu, Z., Li, H., & Cao, G. (2017). Quick Estimation Model for the Concentration of Indoor Airborne Culturable Bacteria: An Application of Machine Learning. *International Journal of Environmental Research and Public Health*, *14*(8), 857. doi:10.3390/ijerph14080857 PMID:28758941

Liu, Z., Li, H., Liu, K., Yu, H., & Cheng, K. (2017). Design of high-performance water-in-glass evacuated tube solar water heaters by a high-throughput screening based on machine learning: A combined modeling and experimental study. *Solar Energy*, *142*, 61–67. doi:10.1016/j.solener.2016.12.015

Liu, Z., Li, H., Tang, X., Zhang, X., Lin, F., & Cheng, K. (2016). Extreme learning machine: A new alternative for measuring heat collection rate and heat loss coefficient of water-in-glass evacuated tube solar water heaters. *SpringerPlus*. doi:10.118640064-016-2242-1 PMID:27330892

Liu, Z., Li, H., Zhang, X., Jin, G., & Cheng, K. (2015). Novel method for measuring the heat collection rate and heat loss coefficient of water-in-glass evacuated tube solar water heaters based on artificial neural networks and support vector machine. *Energies*, *8*(8), 8814–8834. doi:10.3390/en8088814

Liu, Z., Liu, K., Li, H., Zhang, X., Jin, G., & Cheng, K. (2015). Artificial Neural Networks-Based Software for Measuring Heat Collection Rate and Heat Loss Coefficient of Water-in-Glass Evacuated Tube Solar Water Heaters. *PLoS One*. doi:10.1371/journal.pone.0143624 PMID:26624613

Li, W., & He, X. (2011, April). Review of non-isolated high-step-up DC/DC converters in photovoltaic grid-connected applications. *IEEE Transactions on Industrial Electronics*, *58*(4), 1239–1250. doi:10.1109/TIE.2010.2049715

Li, Y., & Nejabatkhah, F. J. (2014). Overview of Control, Integration and Energy Management of Microgrids. *Journal of Modern Power Systems and Clean Energy*, *2*(3), 212–222. doi:10.100740565-014-0063-1

Li, Z., Lai, K. Y. T., Yu, P. H., Ho, T. Y., Chakrabarty, K., & Lee, C. Y. (2016, June). High-level synthesis for microelectrode-dot-array digital microfluidic biochips. In *Proceedings of the 53rd Annual Design Automation Conference* (p. 146). ACM. 10.1145/2897937.2898028

Lof, P.-A., Andersson, G., & Hill, D. (1993). Voltage stability indices for stressed power systems. *IEEE Transactions on Power Systems*, *8*(1), 326–335. doi:10.1109/59.221224

Logenthiran, T. (2011). Multi-Agent system for Demand Side Management in smart grid. *2011 IEEE Ninth International Conference on Power Electronics and Drive Systems*. 10.1109/PEDS.2011.6147283

Logenthiran, T., Srinivasan, D., & Shun, T. Z. (2012). Demand side management in smart grid using heuristic optimization. *IEEE Transactions on Smart Grid*, *3*(3), 1244–1252. doi:10.1109/TSG.2012.2195686

Logenthiran, T., Srinivasan, D., & Vanessa, K. W. M. (2014). Demand side management of smart grid: Load shifting and incentives. *Journal of Renewable and Sustainable Energy*, *6*(3), 033136. doi:10.1063/1.4885106

Long, H., Eghlimi, M., & Zhang, Z. (2017, January). Configuration Optimization and Analysis of a Large-Scale PV/Wind System. *IEEE Transactions on Sustainable Energy*, *8*(1), 84–93. doi:10.1109/TSTE.2016.2583469

Lopes, J. A. P., Hatziargyriou, N., Mutale, J., Djapic, P., & Jenkins, N. (2010). Integrating distributed generation into electric power systems : A review of drivers, challenges and opportunities. *Electric Power Systems Research*, *77*(9), 1189–1203. doi:10.1016/j.epsr.2006.08.016

Lopez, O., Alvarez, J., Doval-Gandoy, J., & Freijedo, F. D. (2009). Multilevel multiphase space vector pwm algorithm with switching state redundancy. *IEEE Transactions on Industrial Electronics*, *56*(3), 792–804. doi:10.1109/TIE.2008.2004390

Lu, A., Tang, C., Shuai, Z., Tang, J., Xu, X. Y., & Chen, D. (2009). Fuzzy PI based direct output voltage control for STATCOM. *IEEE Transactions on Industrial Electronics, 56*(7), 2401-2412.

Lukefahr, A., Padmanabha, S., Das, R., Dreslinski, R. Jr, Wenisch, T. F., & Mahlke, S. (2014, August). Heterogeneous microarchitectures trump voltage scaling for low-power cores. In *Proceedings of the 23rd international conference on Parallel architectures and compilation* (pp. 237-250). ACM. 10.1145/2628071.2628078

Luo, J., & Jha, N. K. (2000, November). Power-conscious joint scheduling of periodic task graphs and aperiodic tasks in distributed real-time embedded systems. In *Proceedings of the 2000 IEEE/ACM international conference on Computer-aided design* (pp. 357-364). IEEE Press.

Luo, X., Li, C., Liu, N., Li, R., He, J., & Qiu, T. (2013). The impact of energy spectrum width in the energy selective electron low-temperature thermionic heat engine at maximum power. *Physics Letters. [Part A]*, *377*(25), 1566–1570. doi:10.1016/j.physleta.2013.04.045

Luque, A., & Hegedus, S. (2003). *Handbook of photovoltaic science and engineering*. John Wiley & Sons. doi:10.1002/0470014008

Luque, A., & Martí, A. (1999). Limiting efficiency of coupled thermal and photovoltaic converters. *Solar Energy Materials and Solar Cells*, *58*(2), 147–165. doi:10.1016/S0927-0248(98)00199-8

Lu, X., & Wang, J. (2017, June). A Game Changer: Electrifying Remote Communities by Using Isolated Microgrids. *IEEE Electrification Magazine*, *5*(2), 56–63. doi:10.1109/MELE.2017.2685958

Macken, J., Vanthournout, K., Van den Keybus, J., Deconinck, G., & Belmans, R. (2004). Distributed control of renewable generation units with integrated active filter. *IEEE Transactions on Power Electronics*, *19*(5), 1353–1360. doi:10.1109/TPEL.2004.833450

Maghami, M. R., Hizam, H., Gomes, C., Radzi, M. A., Rezadad, M. I., & Hajighorbani, S. (2016). Power loss due to soiling on solar panel: A review. *Renewable & Sustainable Energy Reviews*, *59*, 1307–1316. doi:10.1016/j.rser.2016.01.044

Mahan, G. D. (1994). Thermionic refrigeration. *Journal of Applied Physics*, *76*(7), 4362–4366. doi:10.1063/1.357324

Mahan, G. D., Sofo, J. O., & Bartkowiak, M. (1998). Multilayer thermionic refrigerator and generator. *Journal of Applied Physics*, *83*(9), 4683–4689. doi:10.1063/1.367255

Mahari, A., & Seyedi, H. (2013). Optimal PMU placement for power system observability using BICA, considering measurement redundancy. *Electric Power Systems Research*, *103*, 78–85. doi:10.1016/j.epsr.2013.04.013

Mahdad, B. (2013). *Power Quality Management Based Security OPF Considering FACTS Using Metaheuristic Optimization Methods*. InTech Education and Publishing KG, Intechweb.org.

Mahdad, B., Srairi, K. (2011). *Understanding Power Quality based FACTS using Interactive Educational GUI Matlab Package*. InTech Education and Publishing KG, Intechweb.org.

Mahdad, B., & Srairi, K. (2012). *Hierarchical adaptive PSO for multi-objective OPF considering emissions based shunt FACTS*. IEEE.

Mahdad, B., & Srairi, K. (2013). Solving multi-objective optimal power flow problem considering wind-STATCOM using differential evolution. *Front. Energy*, *7*(1), 75–89. doi:10.100711708-012-0222-x

Mahdad, B., & Srairi, K. (2014). Multi objective large power system planning under sever loading condition using learning DE-APSO-PS strategy. *Energy Conversion and Mmanagement (Elsevier)*, *87*, 338–350. doi:10.1016/j.enconman.2014.06.090

Mahdad, B., & Srairi, K. (2015). Blackout risk prevention in a smart grid based flexible optimal strategy using Grey Wolf-pattern search algorithms. *Energy Conversion and Management*, *98*, 411–429. doi:10.1016/j.enconman.2015.04.005

Mahdad, B., & Srairi, K. (2015). Security optimal power flow considering loading margin stability using hybrid FFA–PS assisted with brainstorming rules. *Applied Soft Computing*, *35*, 291–309. doi:10.1016/j.asoc.2015.06.037

Mahdad, B., & Srairi, K. (2017). A new interactive sine cosine algorithm for loading margin stability improvement under contingency. *Electrical Engineering*, 1–21.

Mahmoud, A. M. A., Mashaly, H. M., Kandil, S. A., El Khashab, H., & Nashed, M. N. F. (2000) Fuzzy Logic Implementation for Photovoltaic Maximum Power Tracking. *26th Annual Conference of the IEEE Industrial Electronics Society, 1*, 735-740.

Mahmoud, A. M. A., Mashaly, H. M., Kandil, S. A., El Khashab, H., & Nashed, M. N. F. (2000). Fuzzy Logic Implemen- tation for Photovoltaic Maximum Power Tracking. *Proceedings of 9th IEEE International Workshop on Robot and Human Interactive Communication,* 155-160.

Mahmoud, M. S., Saif Ur Rahman, M., & Sunni, F. M. A. L. (2015). Review of microgrid architectures – a system of systems perspective. *IET Renewable Power Generation*, *9*(8), 1064–1078. doi:10.1049/iet-rpg.2014.0171

Maji, T. K., & Acharjee, P. (2017). Multiple Solutions of Optimal PMU Placement Using Exponential Binary PSO Algorithm for Smart Grid Applications. *IEEE Transactions on Industry Applications*, *53*(3), 2550–2559. doi:10.1109/TIA.2017.2666091

Ma, K. W., & Lee, Y. S. (1996, March). An integrated flyback converter for DC uninterruptible power supply. *IEEE Transactions on Power Electronics*, *11*(2), 318–327. doi:10.1109/63.486182

Makhlouf, M., Messai, F., & Benalla, H. (2011). Vectorial command of induction motor pumping system supplied by a photovoltaic generator. *Journal of Electrical Engineering*, *62*(1), 3–10. doi:10.2478/v10187-011-0001-7

Ma, M., Hu, L., Chen, A., & He, X. (2007). Reconfiguration of carrier-based modulation strategy for fault tolerant multilevel inverters. *IEEE Transactions on Power Electronics*, *22*(5), 2050–2060. doi:10.1109/TPEL.2007.904249

Mamandur, K., & Chenoweth, R. (1981). Optimal control of reactive power flow for improvements in voltage profiles and for real power loss minimization. *IEEE Transactions on Power Apparatus and Systems*, (7): 3185–3194. doi:10.1109/TPAS.1981.316646

Mandal, K. K., Mandal, S., Bhattacharya, B., & Chakraborty, N. (2015). Non-convex emission constrained economic dispatch using a new self-adaptive particle swarm optimization technique. *Applied Soft Computing*, *28*, 188–195. doi:10.1016/j.asoc.2014.11.033

Manimekalai, P., Harikumar, R., & Raghavan, S. (2016). SOGI Algorithm-Based Shunt Active Power Filter for Grid Integration of Photovoltaic Systems. *Journal of Circuits, Systems, and Computers*, (25): 20.

Marin, F., Garcia-Lagos, F., Joya, G., & Sandoval, F. (2003). Genetic algorithms for optimal placement of phasor measurement units in electrical networks. *Electronics Letters*, *39*(19), 1403–1405. doi:10.1049/el:20030919

Marques, A. B., Taranto, G. N., & Falc˜ao, D. M. (2005). A knowledge-based system for supervision and control of regional voltage profile and security. *IEEE Transactions on Power Systems*, *20*(1), 400–407. doi:10.1109/TPWRS.2004.831672

Marti, A., & Araújo, G. L. (1996). Limiting efficiencies for photovoltaic energy conversion in multigap systems. *Solar Energy Materials and Solar Cells*, *43*(2), 203–222. doi:10.1016/0927-0248(96)00015-3

Martın, L. A. (1997a). *Road Map 2: The First Step*. MIT System Dynamics in Education Project.

Masand, D., Jain, S., & Agnihotri, G. (2008). Control Strategies for Distribution Static Compensator for Power Quality Improvement. *Journal of the Institution of Electronics and Telecommunication Engineers*, *54*(6), 421–428. doi:10.4103/0377-2063.48632

Masoum, M. A. S., Dehbonei, H., & Fuchs, E. F. (2002). Theoretical and Experimental Analyses of Photovoltaic System with Voltage and Current-Based Maximum Power-Paint Tracking. *IEEE Transactions on Energy Conversion*, *17*(4), 514–522. doi:10.1109/TEC.2002.805205

Mathur, R. M., & Varma, R. K. (2000). *Thyrustor_based FACTS Controllers for Electrical Transmission Systems*. Piscataway, NJ: IEEE press.

Mazlumi, K., & Vahedi, H. (2010). *Optimal placement of PMUs in power systems based on bacterial foraging algorithm.* Paper presented at the Electrical Engineering (ICEE), 2010 18th Iranian Conference on. 10.1109/IRANIANCEE.2010.5506953

McFarland, M. C., Parker, A. C., & Camposano, R. (1990). The high-level synthesis of digital systems. *Proceedings of the IEEE*, *78*(2), 301–318. doi:10.1109/5.52214

Meegahapola, L., Durairaj, S., Flynn, D., & Fox, B. (2011). Coordinated utilisation of wind farm reactive power capability for system loss optimisation. *European Transactions on Electrical Power*, *21*(1), 40–51. doi:10.1002/etep.410

Meillaud, F., Shah, A., Droz, C., Vallat-Sauvain, E., & Miazza, C. (2006). Efficiency limits for single-junction and tandem solar cells. *Solar Energy Materials and Solar Cells*, *90*(18), 2952–2959. doi:10.1016/j.solmat.2006.06.002

Mekhilef, S., Saidur, R., & Safari, A. (2011). A review on solar energy use in industries. *Renewable & Sustainable Energy Reviews*, *15*(4), 1777–1790. doi:10.1016/j.rser.2010.12.018

Melı'in, P. E., Espinoza, J. R., Mu˜noz, J. A., Baier, C. R., & Espinosa, E. E. (2010). Decoupled control of a unified power quality conditioner based on a current source topology for fast AC mains disturbance compensation. *Proc. IEEE Int. Conf. Ind. Technol.*, 730–736. 10.1109/ICIT.2010.5472711

Melosh, N. (2013). *Photon Enhanced Thermionic Emission for Solar Energy Harvesting Final Report to the Global Climate and Energy Project* (No. SLAC-PUB-15455). SLAC National Accelerator Laboratory (SLAC).

Memik, S. O., Kastner, R., Bozorgzadeh, E., & Sarrafzadeh, M. (2005). A scheduling algorithm for optimization and early planning in high-level synthesis. *ACM Transactions on Design Automation of Electronic Systems*, *10*(1), 33–57. doi:10.1145/1044111.1044115

Mendis, N., Muttaqi, K. M., Perera, S., & Uddin, M. N. (2015, March). Remote Area Power Supply System: An Integrated Control Approach Based on Active Power Balance. *IEEE Industry Applications Magazine*, *21*(2), 63–76. doi:10.1109/MIAS.2014.2345817

Meng, J. H., Zhang, X. X., & Wang, X. D. (2014). Multi-objective and multi-parameter optimization of a thermoelectric generator module. *Energy*, *71*, 367–376. doi:10.1016/j.energy.2014.04.082

MermoudA. A.WittmerB. (2014). *PVSYST user's manual.* Retrieved from www.pvsyst.com/images/pdf/PVsyst_Tutorials.pdf

Messenger, R. A., & Ventre, J. (2000). Photovoltaic systems engineering. CRC Press.

Milano, F., Vanfretti, L., & Morataya, J. C. (2008). An open source power system virtual laboratory: The PSAT case and experience. *IEEE Transactions on Education*, *51*(1), 17–23. doi:10.1109/TE.2007.893354

Milosevic, B., & Begovic, M. (2003). Nondominated sorting genetic algorithm for optimal phasor measurement placement. *IEEE Transactions on Power Systems*, *18*(1), 69–75. doi:10.1109/TPWRS.2002.807064

Mirafzal, B. (2014). Survey of fault-tolerance techniques for three-phase voltage source inverters. *IEEE Transactions on Power Electronics*, *61*(10), 5192–5202.

Mirjalili, S., & Lewis, A. (2016). The whale optimization algorithm. *Advances in Engineering Software*, *95*, 51–67. doi:10.1016/j.advengsoft.2016.01.008

Mirjalili, S., Mirjalili, S. M., & Lewis, A. (2014). Grey wolf optimizer. *Advances in Engineering Software*, *69*, 46–61. doi:10.1016/j.advengsoft.2013.12.007

Modiri-Delshad, M., & Abd Rahim, N. (2014). Solving non-convex economic dispatch problem via backtracking search algorithm. *Energy*, *77*, 372–381. doi:10.1016/j.energy.2014.09.009

Moger, T. (2016). *Reactive power planning and operation of power systems with wind farms for voltage stability improvement* (Unpublished doctoral dissertation). Department of Electrical Engineering, Indian Institute of Science, Bangalore, India.

Moger, T., & Dhadbanjan, T. (2015). A novel index for identification of weak nodes for reactive compensation to improve voltage stability. *IET Generation, Transmission & Distribution*, *9*(14), 1826–1834. doi:10.1049/iet-gtd.2015.0054

Moger, T., & Dhadbanjan, T. (2016). Fuzzy logic approach for reactive power coordination in grid connected wind farms to improve steady state voltage stability. *IET Renewable Power Generation*, *11*(2), 351–361. doi:10.1049/iet-rpg.2016.0119

Moger, T., & Dhadbanjan, T. (2017). Impact of different pq models of wind turbine generating units (wtgus) on system voltage performance. *International Journal of Emerging Electric Power Systems*, *18*(4), 1–19. doi:10.1515/ijeeps-2017-0008

Mohammad, Babaei, & Sabahi. (2012). A new single-phase cascade multilevel inverter topology using four-level cells. *International conference in Electrical Engineering*, 348–53.

Mohammad, F. K., & Ebrahim, B. (2013). A generalized cascaded multilevel inverter using series connection of sub-multilevel inverter. *IEEE Transactions on Power Electronics*, *28*(2), 625-636.

Mohammadi-Ivatloo, B. (2009). Optimal placement of PMUs for power system observability using topology based formulated algorithms. *Journal of Applied Sciences (Faisalabad)*, *9*(13), 2463–2468. doi:10.3923/jas.2009.2463.2468

Mohammadi, R. H., Varjani, Y. R., & Mokhtari, H. (2009). Multi-converter unified power-quality conditioning system: MC-UPQC. *IEEE Transactions on Power Delivery*, *l*(24), 1679–1686. doi:10.1109/TPWRD.2009.2016822

Mohandes, B., El-Chaar, L., & Lamont, L. (2009). Application Study of 500W Photovoltaic (PV) System in the UAE. *Applied Solar Energy Journal*, *45*(4), 242–247. doi:10.3103/S0003701X09040057

Mohd Herwan, S., Zuriani, M., Mohd Rusllim, M., & Aliman, O. (2015). Using the gray wolf optimizer for solving optimal reactive power dispatch problem. *Applied Soft Computing*, 286–292.

Mojtaba, G., Ghavidel, S., Gitizadeh, M., & Akbari, E. (2015). An improved teaching–learning-based optimization algorithm using Lévy mutation strategy for non-smooth optimal power flow. *International Journal of Electrical Power & Energy Systems*, *65*, 375–384. doi:10.1016/j.ijepes.2014.10.027

Momoh, J. A., El-Hawary, M. E., & Adapa, R. A. (1999). Review of selected optimal power flow literature to 1993. II. Newton, linear programming and interior point methods. *IEEE Transactions on Power Systems*, *14*(1), 105–111. doi:10.1109/59.744495

Mondal, D., Chakrobarty, A., & Sengupta, A. (2014). Power System small signals stability and Control. Academic Press.

Montoya, L. F., Fu, Q., Nasiri, A., Bhavaraju, V., & Yu, D. (2014). *Novel methodology to determine the optimal energy storage location in a microgrid and address power quality and stability.* Eaton White Paper.

Mukherjee, A., & Mukherjee, V. (2015). Solution of optimal power flow using chaotic krill herd algorithm. *Chaos, Solitons, and Fractals*, *78*, 10–21. doi:10.1016/j.chaos.2015.06.020

Mukhopadhyay, A. K., Kothari, D. P., Chakrabarty, A., & De, A. (2010). *An Introduction to Reactive Power Control and Voltage Stability in Power Transmission System*. PHI.

Muljadi, E. (1997). PV water pumping with a peak-power tracker using a simple six-step square-wave inverter. *IEEE Transactions on Industry Applications*, *33*(3), 714–721. doi:10.1109/28.585862

Murphy, E. L., & Good, R. H. Jr. (1956). Thermionic emission, field emission, and the transition region. *Physical Review*, *102*(6), 1464–1473. doi:10.1103/PhysRev.102.1464

Muthu, S., & Kim, S. M. J. (1997). Steady-state operating characteristics of unified active power filters. *Proc. IEEE Appl. Power Electron. Conf.*, 199–205. 10.1109/APEC.1997.581454

Myrzik, J. M., & Calais, M. (2003, June). String and module integrated inverters for single-phase grid connected photovoltaic systems-a review. In *Power Tech Conference Proceedings, 2003 IEEE Bologna* (Vol. 2). IEEE. 10.1109/PTC.2003.1304589

Nageswararao, B., & Jeyasurya, B. (1998). Fuzzy-expert system for voltage stability monitoring and control. *Electric Power Systems Research*, *47*(3), 215–222. doi:10.1016/S0378-7796(98)00068-6

Naidu, B. R., Panda, G., & Ganesh, R. (2016). Modelling and Hardware-in-Loop validation of a modified controller for intermittent operation of a standalone low voltage DC fuel cell system. *IEEE Region 10 Conference (TENCON)*, 1429-1434.

Naidu, K., Mokhlis, H., Bakar, A. H. A., & Terzija, V. (2017). Performance investigation of ABC algorithm in multi-area power system with multiple interconnected generators. *Applied Soft Computing*, *57*, 436–451. doi:10.1016/j.asoc.2017.03.044

Najafi, E., & Yatim, A. H. M. (2012). Design and implementation of a new multi level inverter topology. *IEEE Transactions on Industrial Electronics*, *59*(11), 4148–4154. doi:10.1109/TIE.2011.2176691

Narasimhan, M., & Ramanujam, J. (1997). *Tighter lower bounds for scheduling problems in high-level synthesis*. Academic Press.

Narasimhan, M. (1998). *Exact scheduling techniques for high level synthesis. Master of Science in Electrical Engineering*. Lousiana State University.

Narges, D., Mehrdad, T. H., & Teimourzadeh, S. (2016). Adaptive group search optimization algorithm for multi-objective optimal power flow problem. *Applied Soft Computing*, *38*, 1012–1024. doi:10.1016/j.asoc.2015.10.057

Nasiri, & Emadi, A. (2003). Different topologies for single-phase unified power quality conditioners. *Proc. 38th Int. Appl. Soc. Annu. Meet. Ind. Appl. Conf.*, 976–981.

Nathalie Julien. (2006). *Caractérisation et modélisation de la consommation sur FPGA, Groupe Lester France*. Nice: Ecofac.

Nathanael, R., Pott, V., Kam, H., Jeon, J., Alon, E., & Liu, T. J. K. (2010). Four-terminal-relay body-biasing schemes for complementary logic circuits. *IEEE Electron Device Letters*, *31*(8), 890–892. doi:10.1109/LED.2010.2050133

National Electricity Amendment. (2007). *Technical Standards for wind and other generators connections rule*. Retrieved from http://www.aemc.gov.au

National Grid Electricity Transmission. (2009). *The Grid Code*. Retrieved from http://www.nationalgrid.com

National Research Council. (2001). *Thermionics Quo Vadis? An Assessment of the DTRA's Advanced Thermionics Research and Development Program*. National Academies Press.

Nazar, F., & Murugan, S. (2016, March). Implementation of JPEG-LS compression algorithm for real time applications. In *Electrical, Electronics, and Optimization Techniques (ICEEOT), International Conference on* (pp. 2772-2774). IEEE. 10.1109/ICEEOT.2016.7755200

Nazari-Heris, M., & Mohammadi-Ivatloo, B. (2015). Application of heuristic algorithms to optimal PMU placement in electric power systems: An updated review. *Renewable & Sustainable Energy Reviews*, *50*, 214–228. doi:10.1016/j.rser.2015.04.152

Nazaripouya, H., Wang, Y., Chu, P., Pota, H. R., & Gadh, R. (2015, July). Optimal sizing and placement of battery energy storage in distribution system based on solar size for voltage regulation. In Power & Energy Society General Meeting, 2015 IEEE (pp. 1-5). IEEE.

Nazir, R., Laksono, H. D., Waldi, E. P., Ekaputra, E., & Coveria, P. (2014). Renewable Energy Sources Optimization : A Micro-Grid Model Design. *Energy Procedia*, *52*, 316–327. doi:10.1016/j.egypro.2014.07.083

Nestor, J. A., & Krishnamoorthy, G. (1993). SALSA: A new approach to scheduling with timing constraints. *IEEE Transactions on Computer-Aided Design of Integrated Circuits and Systems*, *12*(8), 1107–1122. doi:10.1109/43.238604

Niebur, D. (1993). Artificial neural networks for power systems-a literature survey. *International Journal of Engineering Intelligent Systems for Electrical Engineering and Communications*, *1*, 133–158.

Norouzi, A. H., & Sharaf, A. M. (2005). Two control schemes to enhance the dynamic performance of the STATCOM and SSSC. *IEEE Transactions on Power Delivery*, *20*(1), 435–442. doi:10.1109/TPWRD.2004.839725

Note, S., Geurts, W., Catthoor, F., & De Man, H. (1991, June). Cathedral-III: Architecture-driven high-level synthesis for high throughput DSP applications. In *Proceedings of the 28th ACM/IEEE Design Automation Conference* (pp. 597-602). ACM. 10.1145/127601.127739

Nwouh, N. M. (2009). Voltage Stability Improvement using Static Var Compensator in Power Systems. *Leonardo Journal of Sciences*, (14), 167-172.

O'Dwyer, M. F., Lewis, R. A., Zhang, C., & Humphrey, T. E. (2005). Electronic efficiency in nanostructured thermionic and thermoelectric devices. *Physical Review B: Condensed Matter and Materials Physics*, *72*(20), 205330. doi:10.1103/PhysRevB.72.205330

Olivier Rochecouste. (2005). *Architecture et bits significatifs*. Thèse en informatique, Université de Rennes 1, Rennes 1.

Osmulski. (2000). A probabilistic power prediction tool for Xilinx FPGA. *Int. Wksp. on Computer Embedded/Distributed HPC Systems and Applications*, 776 - 783.

Padhan, S., Sahu, R. K., & Panda, S. (2014). Application of Firefly Algorithm for Load Frequency Control of Multi-area Interconnected Power System. *Electric Power Components and Systems*, *42*(13), 1419–1430. doi:10.1080/15325008.2014.933372

Padiyar K. R. (2007). *Facts Controllers in Power Transmission and Distribution*. New Age International (P) Limited.

Palensky, P., & Dietrich, D. (2011, August). Demand Side Management: Demand Response, Intelligent Energy Systems, and Smart Loads. *IEEE Transactions on Industrial Informatics*, *7*(3), 381–388. doi:10.1109/TII.2011.2158841

Pali, B. S., & Vadhera, S. (2016). Renewable Energy Systems for Generating Electric Power : A Review. In *IEEE 1st International Conference on Power Electronics, Intelligent Control and Energy Systems (ICPEICES)* (pp. 1–6). IEEE.

Pal, Y., Swarup, A., & Singh, B. (2010). A comparative analysis of three-phase four-wire UPQC topologies. *Proc. Int. Conf. Power Electron. Drives Energy System*, 1–6. 10.1109/PEDES.2010.5712513

Panagis, P., Stergiopoulos, F., Marabeas, P., & Manias, S. (2008). Comparison of state of the art multilevel inverters. *Electronics Specialists Conference*, 4296-4301. 10.1109/PESC.2008.4592633

Pan, C. T., Chuang, C. F., & Chu, C. C. (2014, September). A novel transformer-less adaptable voltage quadruple DC converter with low switch voltage stress. *IEEE Transactions on Power Electronics*, *29*(9), 4787–4796. doi:10.1109/TPEL.2013.2287020

Panda, A. K., & Patnaik, S. S. (2015). Analysis of cascaded multilevel inverters for active harmonic filtering in distribution networks. *Electrical Power and Energy Systems*, *66*, 216–226. doi:10.1016/j.ijepes.2014.10.034

Panda, S., Mohanty, B., & Hota, P. K. (2013). Hybrid BFOA-PSO algorithm for automatic generation control of linear and nonlinear interconnected power systems. *Applied Soft Computing*, *13*(12), 4718–4730. doi:10.1016/j.asoc.2013.07.021

Parker, A. C., Pizarro, J. T., & Mlinar, M. (1986, July). MAHA: A program for datapath synthesis. In *Proceedings of the 23rd ACM/IEEE Design Automation Conference* (pp. 461-466). IEEE Press.

Partridge, T., Thomas, M., Harthorn, B. H., Pidgeon, N., Hasell, A., Stevenson, L., & Enders, C. (2017). Seeing futures now: Emergent US and UK views on shale development, climate change and energy systems. *Global Environmental Change*, *42*, 1–12. doi:10.1016/j.gloenvcha.2016.11.002

Parvini, S., Siegel, J. B., Stefanopoulou, A. G., & Vahidi, A. (2016, March). Supercapacitor Electrical and Thermal Modeling, Identification, and Validation for a Wide Range of Temperature and Power Applications. *IEEE Transactions on Industrial Electronics*, *63*(3), 1574–1585. doi:10.1109/TIE.2015.2494868

Patcharaprakiti, N., & Premrudeepreechacharn, S. (2002). Maximum Power Point Tracking Using Adaptive Fuzzy Logic Control for Grid-Connected Photovoltaic System. *IEEE Power Engineering Society Winter Meeting*, *1*, 372-377.

Paulin, P. G., & Knight, J. P. (1987, October). Force-directed scheduling in automatic data path synthesis. In *Proceedings of the 24th ACM/IEEE Design Automation Conference* (pp. 195-202). ACM. 10.1145/37888.37918

Paulin, P. G., & Knight, J. P. (1989). Algorithms for high-level synthesis. *IEEE Design & Test of Computers*, *6*(6), 18–31. doi:10.1109/54.41671

Pegden, C. D., Shannon, R. E., & Sadowski, R. P. (1995). *Introduction To Simulation Using SIMAN*. New York: McGraw-Hill.

Peng, C., Sun, H., & Guo, J. (2010). Multi-objective optimal PMU placement using a non-dominated sorting differential evolution algorithm. *International Journal of Electrical Power & Energy Systems*, *32*(8), 886–892. doi:10.1016/j.ijepes.2010.01.024

Peng, H., & Ling, X. (2008). Optimal design approach for the plate-fin heat exchangers using neural networks cooperated with genetic algorithms. *Applied Thermal Engineering*, *28*(5-6), 642–650. doi:10.1016/j.applthermaleng.2007.03.032

Pepermans, G., Driesen, J., Haeseldonckx, D., Belmans, R., & D'haeseleer, W. (2005). Distributed Generation : Definition, Benefits and Issues. *Energy Policy*, *33*(6), 787–798. doi:10.1016/j.enpol.2003.10.004

Petela, R. (1964). Exergy of heat radiation. *ASME J. Heat Transfer*, *86*(2), 187–192. doi:10.1115/1.3687092

Petela, R. (2003). Exergy of undiluted thermal radiation. *Solar Energy*, *74*(6), 469–488. doi:10.1016/S0038-092X(03)00226-3

Peter, S., Kleıner, A., Roberts, C., Ross, R., & Smith, B. (1994). The Fifth Discipline Fieldbook. Century.

Phadke, A. G., & Thorp, J. S. (2008). *Synchronized phasor measurements and their applications*. Springer Science & Business Media. doi:10.1007/978-0-387-76537-2

Phang, J. C. H., Chan, D. S. H., & Phillips, J. R. (1984). Accurate analytical method for the extraction of solar cell model parameters. *Electronics Letters*, *20*(10), 406–408. doi:10.1049/el:19840281

Phatak, D. S., Kahle, S., Kim, H., & Lue, J. (1998, June). Hybrid signed digit representation for low power arithmetic circuits. *Proceedings of the Lower Power Workshop in Conjunction with ISCA*.

Pimentel, D., Herz, M., Glickstein, M., Zimmerman, M., Becker, A. R. K., Evans, J., ... Seidel, T. (2002). Renewable Energy : Current and Potential Issues. *American Institute of Biological Sciences*, *52*(12), 1111–1120.

Pinto, S. J., Panda, G., & Peesapati, R. (2017). An Implementation of Hybrid Control Strategy for Distributed Generation System Interface Using Xilinx System Generator. *IEEE Transactions on Industrial Informatics*.

Ponmurugavel, P. S., & Ghouse, S. M. (2013). Design and Modelling of Fuzzy and Model Predictive Controllers for STATCOM to enhance Transient Stability of Power System. *IACSIT International Journal of Engineering and Technology*, *5*(3), 2609–2619. doi:10.1109/ICPEC.2013.6527635

Potasman, R., Lis, J., Nicolau, A., & Gajski, D. (1991, January). Percolation based synthesis. In *Proceedings of the 27th ACM/IEEE Design Automation Conference* (pp. 444-449). ACM.

Pradeepa, S., Rao, K. U., Deekshit, R., & Shantha, M. S. (2013). State feedback Control of a Voltage Source Inverter-based STATCOM. *IEEE International Conference on Power, Energy and Control (ICPEC)*, 120-123.

Prakash, D., & Lakshminarayana, C. (2016). *Optimal siting of capacitors in radial distribution network using whale optimization algorithm*. Alexandria Engineering Journal.

Prakash, K. R., & Sanju, A. V. (2014). Design and Implementation of Signed, Rounded and Truncated Multipliers using Modified Booth Algorithm for Dsp Systems. *International Journal (Toronto, Ont.)*, *2*(1).

Prieto, J., Salmeron, P., & Herrera, S. R. (2005). A unified power quality conditioner for wide load range: Practical design and experimental results. *Proc. IEEE Russia Power Technol.*, 1–7.

Prudente, M., Pfitscher, L. L., Emmendoerfer, G., Romaneli, E. F., & Gules, R. (2008, March). Voltage multiplier cells applied to non-isolated DC-DC converters. *IEEE Transactions on Power Electronics*, *23*(2), 871–887. doi:10.1109/TPEL.2007.915762

Pudjianto, D., Ahmed, S., & Strbac, G. (2002). Allocation of VAR support using LP and NLP based optimal power flows. *IEE Proceedings. Generation, Transmission and Distribution*, *149*(4), 377–383. doi:10.1049/ip-gtd:20020200

Pyzer-Knapp, E. O., Suh, C., Gómez-Bombarelli, R., Aguilera-Iparraguirre, J., & Aspuru-Guzik, A. (2015). What Is High-Throughput Virtual Screening? A Perspective from Organic Materials Discovery. *Annual Review of Materials Research*, *45*(1), 195–216. doi:10.1146/annurev-matsci-070214-020823

Qinghui, D., & Jun, C. (2009). Improving the efficiency of solar photovoltaic power generation in several important ways. *IET International Technology and Innovation Conference*, 13-15.

Qudrat-Ullah, H. (2013). *Energy Policy Modeling in the 21st Century*. New York: Springer. doi:10.1007/978-1-4614-8606-0

Qu, G., & Potkonjak, M. (2002). Techniques for energy-efficient communication pipeline design. IEEE Transactions on Very Large Scale Integration (VLSI). *Systems*, *10*(5), 542–549.

Qu, M., Yin, H., & Archer, D. H. (2010). A solar thermal cooling and heating system for a building: Experimental and model based performance analysis and design. *Solar Energy*, *84*(2), 166–182. doi:10.1016/j.solener.2009.10.010

Rabaey, J. M., & Pedram, M. (Eds.). (2012). *Low power design methodologies* (Vol. 336). Springer Science & Business Media.

Radziemska, E. (2003). Thermal performance of Si and GaAs based solar cells and modules: A review. *Progress in Energy and Combustion Science*, *29*(5), 407–424. doi:10.1016/S0360-1285(03)00032-7

Rahim, A. H. M. A., & Kandlawala, M. F. (2004). *Robust STATCOM voltage controller design using loop-shaping technique*. Elsevier Electric Power Systems Research.

Rahim, N. A., & Selvaraj, J. (2010). Multistring five-level inverter with novel pwm control scheme for PV application. *IEEE Transactions on Industrial Electronics*, *57*(6), 2111–2123. doi:10.1109/TIE.2009.2034683

Rahman, A., Saikia, L. C., & Sinha, N. (2015). Load frequency control of a hydro-thermal system under deregulated environment using biogeography-based optimised three degree-of-freedom integral-derivative controller. *IET Generation, Transmission & Distribution*, *9*(15), 2284–2293. doi:10.1049/iet-gtd.2015.0317

Rahman, N. H. A., & Zobaa, A. F. (2017). Integrated mutation strategy with modified binary PSO algorithm for optimal PMU placement. *IEEE Transactions on Industrial Informatics*.

Rajderkar, V. P., & Chandrakar, V. K. (2009). Comparison of Series FACTS Devices Via Optimal Location in a Power System for Congestion Management. *2009 Asia-Pacific Power and Energy Engineering Conference*. 10.1109/APPEEC.2009.4918227

Raju, M., Saikia, L. C., & Sinha, N. (2016). Automatic generation control of a multi-area system using ant lion optimizer algorithm based PID plus second order derivative controller. *International Journal of Electrical Power & Energy Systems*, *80*, 52–63. doi:10.1016/j.ijepes.2016.01.037

Ramage, M., & Ship, K. (2009). *Sytem Thinkers, The Open University Walton Hall*. Milton Keynes, UK: Springer London.

Ram, J. P., Babu, T. S., & Rajasekar, N. (2017). A comprehensive review on solar PV maximum power point tracking techniques. *Renewable & Sustainable Energy Reviews*, *67*, 826–847. doi:10.1016/j.rser.2016.09.076

Ramkumar, S., Kamaraj, V., Thamizharasan, S., & Jeevananthan, S. (2012). A new series parallel switched multilevel dc-link inverter topology. *International Journal of Electrical Power & Energy Systems*, *36*(1), 93–99. doi:10.1016/j.ijepes.2011.10.028

Ramulu, C., Praveen Kumar, T., & Jain, S. (2014). *Single stage PV source based Dual Inverter Fed Open-End Winding Induction Motor pump drive*. Paper presented at IEEE Students' Conference on Electrical, Electronics and Computer Science (SCEECS), Bhopal, India. 10.1109/SCEECS.2014.6804459

Rana, A. S., Thomas, M. S., & Senroy, N. (2017). Reliability Evaluation of WAMS using Markov-based Graph Theory Approach. *IET Generation, Transmission & Distribution*, *11*(11), 2930–2937.

Ranjit, R., & Jadhav, H. T. (2015). Optimal power flow solution of power system incorporating stochastic wind power using Gbest guided artificial bee colony algorithm. *International Journal of Electrical Power & Energy Systems*, *64*, 562–578. doi:10.1016/j.ijepes.2014.07.010

Rao, R. V. (2016). JAYA: A simple and new optimization algorithm for solving constrained and unconstrained optimization problems. *Int. J Industrial Eng. Compt.*, *7*, 1–16.

Rao, R. V., Savsani, V. J., & Vakharia, D. P. (2012). Teaching Learning Based Optimization: An optimization method for continuous non-linear large scale problems. *Information Science*, *183*(1), 1–15. doi:10.1016/j.ins.2011.08.006

Rashid, M. (2011). *Power Electronics Handbook* (3rd ed.). Elsevier.

Raygani, S. V., Sharma, R. & Saha, T. K. (2015). PV power output uncertainty in Australia. *IEEE Power & Energy Society General Meeting*, 1-5.

Recioui, A., Bentarzi, H., & Ouadi, A. (2015). *Application of a binary teaching learning-based algorithm to the optimal placement of phasor measurement units Progress in Clean Energy* (Vol. 1, pp. 817–830). Springer.

Reck, K., & Hansen, O. (2014). Thermodynamics of photon-enhanced thermionic emission solar cells. *Applied Physics Letters*, *104*(2), 023902. doi:10.1063/1.4862535

Reddy, S. S., Kumari, M. S., & Sydulu, M. (2010). Congestion management in deregulated power system by optimal choice and allocation of FACTS controllers using multi-objective genetic algorithm. *IEEE PES T&D*, *2010*, 1–7. doi:10.1109/TDC.2010.5484520

Richardeau, F., & Pham, T. T. L. (2013). Reliability calculation of multilevel converters: Theory and applications. *IEEE Transactions on Industrial Electronics*, *60*(10), 4224–4233. doi:10.1109/TIE.2012.2211315

Riesz, F., Dobos, L., & Karanyi, J. (1998). Thermal decomposition of bulk and heteroepitaxial (100) InP surfaces: A combined in situ scanning electron microscopy and mass spectrometric study. *Journal of Vacuum Science & Technology. B, Microelectronics and Nanometer Structures: Processing, Measurement, and Phenomena: An Official Journal of the American Vacuum Society*, *16*(5), 2672–2674.

Roberts, N. (2001). *Introduction to Computer Simulation A System Dynamics Modeling Approach*. System Dynamics Group Road Map Series.

Rocabert, J., Azevedo, G. M. S., Luna, A., Guerrero, J. M., Candela, J. I., & Rodríguez, P. (2011, October). Intelligent Connection Agent for Three-Phase Grid-Connected Microgrids. *IEEE Transactions on Power Electronics*, *26*(10), 2993–3005. doi:10.1109/TPEL.2011.2116126

Rodriguez, Lai, & Peng. (2002). Multilevel inverters: a survey of topologies, controls and applications. *IEEE Transactions on Industrial Electronics*, *49*(4), 724-738.

Rosen, M. A., & Dincer, I. (1999). Thermal storage and exergy analysis: the impact of stratification. *Transactions of the Canadian Society for Mechanical Engineering*, *23*(1B), 173-186.

Rosen, M. A., & Hooper, F. C. (July, 1996). Second Law Analysis of Thermal Energy Storage Systems. In *Proceedings of the First Trabzon International Energy and Environment Symposium* (pp. 361–371). Trabzon, Turkey: Academic Press.

Rosen, M. A. (1999). Second-law analysis: Approaches and implications. *International Journal of Energy Research*, *23*(5), 415–429. doi:10.1002/(SICI)1099-114X(199904)23:5<415::AID-ER4893.0.CO;2-7

Rosen, M. A., Hooper, F. C., & Barbaris, L. N. (1988). Exergy analysis for the evaluation of the performance of closed thermal energy storage systems. *Journal of Solar Energy Engineering*, *110*(4), 255–261. doi:10.1115/1.3268265

Roy, B. S., Sinha, A., & Pradhan, A. (2012). An optimal PMU placement technique for power system observability. *International Journal of Electrical Power & Energy Systems*, *42*(1), 71–77. doi:10.1016/j.ijepes.2012.03.011

Rubilar, I., Espinoza, J., Munoz, J., & Moran, L. (2007). DC link voltage unbalance control in three-phase UPQCs based on NPC topologies. *Proc. 42nd Ind. Appl. Soc. Annu. Meet. Ind. Appl. Conf.*, 597–602.

Ruth, M., & Hannon, B. (2012). *Modeling Dynamic Economic Systems* (2nd ed.). Springer. doi:10.1007/978-1-4614-2209-9

Safari, A., Mawsawi, A., Qader, M. R., & Ali, G. M. (n.d.). Dynamic Controllers Design for STATCOM. *Iranian Journal Of Electrical And Computer Engineering*, *3*(1), 16-22.

Sahin, A. D., Dincer, I., & Rosen, M. A. (2007). Thermodynamic analysis of solar photovoltaic cell systems. *Solar Energy Materials and Solar Cells*, *91*(2), 153–159. doi:10.1016/j.solmat.2006.07.015

Sahoo, A. K., Abhitharan, K. P., Kalaivani, A., & Karthik, T. J. (2015). Feasibility Study of Microgrid Installation in an Educational Institution with Grid Uncertainty. *Procedia Computer Science*, *70*, 550–557. doi:10.1016/j.procs.2015.10.099

Sahu, R. K., Panda, S., & Pradhan, P. C. (2015). Design and analysis of hybrid firefly algorithm-pattern search based fuzzy PID controller for LFC of multi area power systems. *International Journal of Electrical Power & Energy Systems*, *69*, 200–212. doi:10.1016/j.ijepes.2015.01.019

Sahu, R. K., Panda, S., Rout, R. K., & Sahu, D. K. (2016). Teaching learning based optimization algorithm for automatic generation control of linear and nonlinear interconnected power systems. *Applied Soft Computing*, *13*(12), 4718–4730.

Saleh, S. A., Radwan, T. S., & Rahman, M. A. (2007). Real-Time Testing of WPT-Based Protection of Three-Phase VS PWM Inverter-Fed Motor. *IEEE Transactions On Power Delivery, VOL.*, *22*(4), 2108–2115. doi:10.1109/TPWRD.2007.905559

Sandia National Laboratories. (1988). Stand Alone Photovoltaic Systems: A Handbook of Recommended Design Practices. National Technical Information Service.

Sannino, A., Postiglione, G., & Bollen, M. H. J. (2003, October). Feasibility of a DC network for commercial facilities. *IEEE Transactions on Industry Applications*, *39*(5), 1499–1507. doi:10.1109/TIA.2003.816517

Sarhaddi, F., Farahat, S., Ajam, H., & Behzadmehr, A. (2010). Exergetic performance evaluation of a solar photovoltaic (PV) array. *Australian Journal of Basic and Applied Sciences*, *4*(3), 502–519.

Sashidhar, S., & Fernandes, B. G. (2017). A Novel Ferrite SMDS Spoke-Type BLDC Motor for PV Bore-Well Submersible Water Pumps. *IEEE Transactions on Industrial Electronics*, *64*(1), 104–114. doi:10.1109/TIE.2016.2609841

Sayal, A. (2012) MPPT Techniques for Photovoltaic System under Uniform Insolation and Partial Shading Conditions. *Students Conference on Engineering and Systems*, 1-6.

Sayeef, S. (2012). *Solar Intermittency: Australia's Clean Energy Challenge: Characterising the Effect of High Penetration Solar Intermittency on Australian Electricity Networks*. Sydney, Australia: CSIRO.

Scala, M. L., Trovato, M., & Torelli, F. (1996). A neural network-based method for voltage security monitoring. *IEEE Transactions on Power Systems*, *11*(3), 1332–1341. doi:10.1109/59.535674

Schmidlin, C. R. Junior, & Araujo Lima, F. K. (2011, October). Wind Turbine and PMSG Dynamic Modelling in PSIM. *IEEE Latin America Transactions*, *14*(9), 4115–4120. doi:10.1109/TLA.2016.7785941

Schreiber, R., Aditya, S., Rau, B. R., Kathail, V., Mahlke, S., Abraham, S., & Snider, G. (2000). High-level synthesis of nonprogrammable hardware accelerators. In *Application-Specific Systems, Architectures, and Processors, 2000. Proceedings. IEEE International Conference on* (pp. 113-124). IEEE. 10.1109/ASAP.2000.862383

Schultz, O., Glunz, S. W., & Willeke, G. P. (2004). Multicrystalline silicon solar cells exceeding 20% efficiency. *Progress in Photovoltaics: Research and Applications*, *12*(7), 553–558. doi:10.1002/pip.583

Schwede, J. W., Bargatin, I., Riley, D. C., Hardin, B. E., Rosenthal, S. J., Sun, Y., ... Melosh, N. A. (2010). Photon-enhanced thermionic emission for solar concentrator systems. *Nature Materials*, *9*(9), 762–767. doi:10.1038/nmat2814 PMID:20676086

Schwede, J. W., Sarmiento, T., Narasimhan, V. K., Rosenthal, S. J., Riley, D. C., Schmitt, F., & Melosh, N. A. (2013). Photon-enhanced thermionic emission from heterostructures with low interface recombination. *Nature Communications*, *4*, 1576. doi:10.1038/ncomms2577 PMID:23481384

Segev, G., Kribus, A., & Rosenwaks, Y. (2013). High performance isothermal photo-thermionic solar converters. *Solar Energy Materials and Solar Cells*, *113*, 114–123. doi:10.1016/j.solmat.2013.02.002

Segev, G., Rosenwaks, Y., & Kribus, A. (2012). Efficiency of photon enhanced thermionic emission solar converters. *Solar Energy Materials and Solar Cells*, *107*, 125–130. doi:10.1016/j.solmat.2012.08.006

Segev, G., Rosenwaks, Y., & Kribus, A. (2013). Loss mechanisms and back surface field effect in photon enhanced thermionic emission converters. *Journal of Applied Physics*, *114*(4), 044505. doi:10.1063/1.4816256

Segev, G., Rosenwaks, Y., & Kribus, A. (2015). Limit of efficiency for photon-enhanced thermionic emission vs. photovoltaic and thermal conversion. *Solar Energy Materials and Solar Cells*, *140*, 464–476. doi:10.1016/j.solmat.2015.05.001

Seo, Y. H., & Kim, D. W. (2010). A new VLSI architecture of parallel multiplier–accumulator based on Radix-2 modified Booth algorithm. *IEEE Transactions on very large scale integration (vlsi) systems, 18*(2), 201-208.

Serhat, D., Yorukeren, N., & Altas, I. H. (2015). A novel modified hybrid PSOGSA based on fuzzy logic for non-convex economic dispatch problem with valve-point effect. *Electrical Power and Energy Systems*, *64*, 121–135. doi:10.1016/j.ijepes.2014.07.031

Sethy, S.K., & Moharana, J.K. (n.d.). Design, Analysis and Simulation of Linear Model of a STATCOM for Reactive Power Compensation with Variation of DC link Voltage. *International Journal of Engineering and Innovative Technology, 2*(5).

Sezen, H. K., & Günal, M. (2009). *Yöneylem Araştırmasında Benzetim*. Bursa: Ekin Yayınevi.

Shafie-khah, M., Heydarian-Forushani, E., Golshan, M. E. H., Moghaddam, M. P., Sheikh-El-Eslami, M. K., & Catalão, J. P. S. (2015). Strategic Offering for a Price-Maker Wind Power Producer in Oligopoly Markets Considering Demand Response Exchange. *IEEE Transactions on Industrial Informatics*, *11*(6), 1542–1553. doi:10.1109/TII.2015.2472339

Shafie-khah, M., Moghaddam, M. P., & Sheikh-El-Eslami, M. K. (2016). Ex-ante evaluation and optimal mitigation of market power in electricity markets including renewable energy resources. *IET Generation, Transmission & Distribution*, *10*(8), 1842–1852. doi:10.1049/iet-gtd.2015.0981

Shah, A. V., Meier, J., Vallat-Sauvain, E., Wyrsch, N., Kroll, U., Droz, C., & Graf, U. (2003). Material and solar cell research in microcrystalline silicon. *Solar Energy Materials and Solar Cells*, *78*(1), 469–491. doi:10.1016/S0927-0248(02)00448-8

Shahidehpour, M., & Alomoush, M. (2001). *Restructured Electrical Power Systems: Operation, Trading, and Volatility*. Taylor & Francis. Retrieved from https://books.google.co.in/books?id=U7WonDR_b00C

Shankar, G., & Mukherjee, V. (2016). Load frequency control of an autonomous hybrid power system by quasi-oppositional harmony search algorithm. *International Journal of Electrical Power & Energy Systems*, *78*, 715–734. doi:10.1016/j.ijepes.2015.11.091

Shareeq, M. M., Moger, T., & Dhadbanjan, T. (2014). Fuzzy logic approach to curtail number of controllers for voltage control in grid connected wind farms. In *Eighteenth National Power Systems Conference (NPSC)* (pp. 1-6). IEEE. 10.1109/NPSC.2014.7103843

Shashidar, R., Kumar, R. S., MahalingaSwamy, A. M., & Roopa, M. (2017). FPGA Implementation of Low Power Pipelined 32-Bit RISC Processor Using Clock Gating. In *Proceedings of the International Conference on Data Engineering and Communication Technology* (pp. 769-777). Springer Singapore. 10.1007/978-981-10-1678-3_74

Shen, D., & Lehn, P. W. (2002, January). Modeling, Analysis, and Control of a Current Source Inverter-Based STATCOM. *IEEE Transactions on Power Delivery*, *17*(1), 248–253.

Shen, H., Wan, J., Yuan, C., Liu, Y., & Li, G. (2009). Harmonic signal detection algorithm in parallel of UPQC studies based on PSO-FUZZY. *Asia-Pacific Power and Energy Engineering Conference*, 1–4. 10.1109/APPEEC.2009.4918399

Shin, D., Kim, J., & Lee, S. (2001, June). Low-energy intra-task voltage scheduling using static timing analysis. In *Proceedings of the 38th annual Design Automation Conference* (pp. 438-443). ACM. 10.1145/378239.378551

Shiva, C. K., & Mukherjee, V. (2015). A novel quasi-oppositional harmony search algorithm for automatic generation control of power system. *Applied Soft Computing*, *35*, 749–765. doi:10.1016/j.asoc.2015.05.054

Shockley, W., & Queisser, H. J. (1961). Detailed balance limit of efficiency of p-n junction solar cells. *Journal of Applied Physics*, *32*(3), 510–519. doi:10.1063/1.1736034

Shreelakshmi, M. P., Das, M., & Agarwal, V. (2013). High gain, a high efficiency bi-directional DC-DC converter for battery charging applications in stand-alone photovoltaic systems. *IEEE 39th Photovoltaic Specialists Conference (PVSC)*, 2857 – 2861.

Shukla, A., Khare, M., & Shukla, K. N. (2015). Modeling and Simulation of Solar PV Module on MATLAB/Simulink, International Journal of Innovative Research in Science. *Engineering and Technology*, *4*(1), 18516–18527.

Shukla, U. K., & Thampy, A. (2011). Analysis of competition and market power in the whole sale electricity market in India. *Energy Policy*, *2*(39), 2699–2710. doi:10.1016/j.enpol.2011.02.039

Silc, J. (1994). Scheduling strategies in high-level synthesis. *Informatica*, *18*(1), 71–79.

Silveira, G. C., Tofoli, F. L., Bezerra, L. D. S., & Torrico-Bascope, R. P. (2014, December). A non-isolated DC-DC boost converter with high voltage gain and balanced output voltage. *IEEE Transactions on Industrial Electronics*, *61*(12), 6739–6746. doi:10.1109/TIE.2014.2317136

Simoes, M. (1989). A line voltage regulator/conditioner for harmonic-sensitive load isolation. *Proc. IEEE Ind. Appl. Soc. Annu. Meet. Conf.*, 947–951.

Singal, S. K., Varun, & Singh, R. P. (2007). Rural Electrification of a Remote Island by Renewable Energy Sources. *Renewable Energy*, *32*(15), 2491–2501. doi:10.1016/j.renene.2006.12.013

Singh, Singh, Singh, Chandra, & Al-Haddad. (1996). *Optimized performance of solar powered variable speed induction motor drive*. Paper presented at International conference on Power Electronics, Drives and Energy systems for industrial growth, New Delhi, India.

Singh, A. (2010). Performance Evaluation of Three Different Configurations of DSTATCOM with Nonlinear Loads. *Journal of the Institution of Electronics and Telecommunication Engineers*, *56*(6), 313–326. doi:10.1080/03772063.2010.10876320

Singh, B. N., Chandra, A., Al-Haddad, K., & Singh, B. (1998). Fuzzy control algorithm for universal active filter. *Proc. Power Quality Conf.*, 73–80. 10.1109/PQ.1998.710347

Singh, B., Al-Haddad, K., & Chandra, A. (1999). A review of active filters for power quality improvement. *IEEE Transactions on Industrial Electronics*, *46*(5), 960–971. doi:10.1109/41.793345

Singh, B., Mishra, A. K., & Kumar, R. (2016). Solar Powered Water Pumping System Employing Switched Reluctance Motor Drive. *IEEE Transactions on Industry Applications*, *52*(5), 3949–3957. doi:10.1109/TIA.2016.2564945

Singh, P., & Ravindra, N. M. (2012). Temperature dependence of solar cell performance—an analysis. *Solar Energy Materials and Solar Cells*, *101*, 36–45. doi:10.1016/j.solmat.2012.02.019

Sinha, A., Neogi, S., Lahiri, R. N., Chowdhury, S., Chowdhury, S. P., & Chakraborty, N. (2011). Role of Demand Side Management for power distribution utility in India. *2011 IEEE Power and Energy Society General Meeting*. 10.1109/PES.2011.6039352

Sirisukprasert, S., Jih-Sheng Lai, & Tian-Hua Liu. (2002). Optimum harmonic reduction with a wide range of modulation indexes for multilevel converters. *IEEE Transactions on Industrial Electronics*, *49*(4), 875–881. doi:10.1109/TIE.2002.801226

Siwar, B.H.H, Bouraoui, O., & Hatem, S. (2014). Efficient design of Finite State Machine to Minimize Power and Response time. *Wulfenia Journal, 21*(12), 364-383.

Smestad, G. P. (2004). Conversion of heat and light simultaneously using a vacuum photodiode and the thermionic and photoelectric effects. *Solar Energy Materials and Solar Cells*, *82*(1), 227–240. doi:10.1016/j.solmat.2004.01.020

Smestad, G. P., Krebs, F. C., Lampert, C. M., Granqvist, C. G., Chopra, K. L., Mathew, X., & Takakura, H. (2008). Reporting solar cell efficiencies in solar energy materials and solar cells. *Solar Energy Materials and Solar Cells*, *92*(4), 371–373. doi:10.1016/j.solmat.2008.01.003

Smida, M. B., & Sakley, A. (2015). Genetic based algorithm for maximum power point tracking (MPPT) for grid connected PV systems operating under partial shaded conditions. *7th International Conference on Modelling, Identification and Control (ICMIC)*,18-20.

Smith, G. (2014, October). Updates of the ITRS design cost and power models. In *Computer Design (ICCD), 2014 32nd IEEE International Conference on* (pp. 161-165). IEEE. 10.1109/ICCD.2014.6974676

Smith, J. R., Bilbro, G. L., & Nemanich, R. J. (2009). Theory of space charge limited regime of thermionic energy converter with negative electron affinity emitter. *Journal of Vacuum Science & Technology. B, Microelectronics and Nanometer Structures: Processing, Measurement, and Phenomena: An Official Journal of the American Vacuum Society*, *27*(3), 1132–1141.

Sommer, C., & Gerlich, D. W. (2013). Machine learning in cell biology – teaching computers to recognize phenotypes. *Journal of Cell Science*, *126*(24), 5529–5539. doi:10.1242/jcs.123604 PMID:24259662

Somsai, K., Kulworawanichpong, T., & Voraphonpiput, N. (2012). Design of DC Voltage Control for D-STATCOM. *International Scholarly and Scientific Research and Innovation*, *6*, 917–923.

Song, Kang, & Park. (2009). Cascaded multilevel inverter employing three-phase transformers and single DC input. *IEEE Transactions on Industrial Electronics*, *56*(6), 2005-2014.

Song, Y. H., & Johns, A. T. E. (1999). Flexible AC Transmission Systems (FACTS). IEEE Press.

Son, H.-I., Kim, T.-J., Kang, D.-W., & Hyun, D.-S. (2004). Fault Diagnosis. 4nd Neutral Point Voltage Control When The 3-Level Inverter Faults Occur. *35th Annual IEEE Power Electronics Specialists Conference*, 4558-4563.

Spain Grid Code. (2005). *Grid code.* Retrieved from http://www.depeca.uah.es

Spanner, D. C. (1964). *Introduction to thermodynamics*. London: Academic Press.

Specht, D. F. (1991). A general regression neural network. *Neural Networks. IEEE Transactions on*, *2*, 568–576. PMID:18282872

Spitsa, V., Alexandrovitz, A., & Zeheb, E. (2009). Design of a Robust State Feedback Controller for a STATCOM Using a Zero Set Concept. *IEEE Transactions on Power Delivery, 25*(1), 456 – 467.

Srivastava, S. C., Velayutham, A., Agrawal, K. K., & Bakshi, A. S. (2012). Report of the enquiry committee on grid disturbance in northern region on 30th July 2012 and in northern, eastern & north-eastern region on 31st July 2012. Delhi: Academic Press.

Stan, M. R., & Burleson, W. P. (1995). Bus-invert coding for low-power I/O. *IEEE Transactions on very large scale integration (VLSI) systems, 3*(1), 49-58.

Stefan & Fredrik. (1995). *Voltage collapse in power systems* (Licentiate thesis). Chalmers University of Technology.

Stefopoulos, G. K., & Meliopoulos, A. P. (2006). Induction Motor Load Dynamics: Impact on Voltage Recovery Phenomena. *Proceedings of IEEE Transmission and Distribution Conference and Exposition*, 752 – 759. 10.1109/TDC.2006.1668591

Stemmler, H., & Guggenbach, P. (1993). *Configurations of high-power voltage source inverter drives*. Paper presented at European Power Electronic conference, Zurich Switzerland.

Sterman, J. D. (1991). *A Skeptic's Guide to Computer Models*. Cambridge, MA: MIT System Dynamics In Education Project.

Sterman, J. D. (2000). Business Dynamics Systems Thinking And Modelling. In *A Complex World*. New York: Mcgraw-Hill.

Su, G.-J. (2003). *Multilevel DC link inverter.* US Patent 6577087.

Su, C.-T., & Lin, C.-T. (1996). A new fuzzy control approach to voltage profile enhancement for power systems. *IEEE Transactions on Power Systems*, *11*(3), 1654–1659. doi:10.1109/59.535711

Su, C.-T., & Lin, C.-T. (2001). Fuzzy-based voltage/reactive power scheduling for voltage security improvement and loss reduction. *IEEE Transactions on Power Delivery*, *16*(2), 319–323. doi:10.1109/61.915501

Sugar, Z. (2000). *Synthèse comportementale basée sur l'ordonnancement* (Doctoral dissertation). Institut National Polytechnique de Grenoble-INPG.

Sulaiman, D. R. (2008, May). Using clock gating technique for energy reduction in portable computers. In *Computer and Communication Engineering, 2008. ICCCE 2008. International Conference on* (pp. 839-842). IEEE. 10.1109/ICCCE.2008.4580723

Sung-Yeul, P., Jih-Sheng, L., & Woo-Cheol, L. (2010). An easy, simple, and flexible control scheme for a three-phase grid-tie inverter system. In *Energy Conversion Congress and Exposition (ECCE)* (pp. 599–603). IEEE.

Sun, Q., Zhou, J., Guerrero, J. M., & Zhang, H. (2015). Hybrid Three-Phase/Single-Phase Microgrid Architecture With Power Management Capabilities. *IEEE Transactions on Power Electronics*, *30*(10), 5964–5977.

Sun, W., He, Y., & Chang, H. (2015). Forecasting fossil fuel energy consumption for power generation using QHSA-based LSSVM model. *Energies*, *8*(2), 939–959. doi:10.3390/en8020939

Su, S., Wang, Y., Liu, T., Su, G., & Chen, J. (2014). Space charge effects on the maximum efficiency and parametric design of a photon-enhanced thermionic solar cell. *Solar Energy Materials and Solar Cells*, *121*, 137–143. doi:10.1016/j.solmat.2013.11.008

Su, S., Wang, Y., Wang, J., Xu, Z., & Chen, J. (2014). Material optimum choices and parametric design strategies of a photon-enhanced solar cell hybrid system. *Solar Energy Materials and Solar Cells*, *128*, 112–118. doi:10.1016/j.solmat.2014.04.038

Su, S., Zhang, H., Chen, X., Kang, J., & Chen, J. (2013). Parametric optimum design of a photon-enhanced thermionic solar cell. *Solar Energy Materials and Solar Cells*, *117*, 219–224. doi:10.1016/j.solmat.2013.05.001

Susanta, D., Roy, P. K., & Debashis, N. (2015). Optimal location of UPFC controller in transmission network using hybrid chemical reaction optimization algorithm. *International Journal of Electrical Power & Energy Systems*, *64*, 194–211. doi:10.1016/j.ijepes.2014.07.038

Suykens, J. A., & Vandewalle, J. (1999). Least Squares Support Vector Machine Classifiers. *Neural Processing Letters*, *9*(3), 293–300. doi:10.1023/A:1018628609742

Svensson, J. (2001). Synchronization methods for grid connected voltage source converters. *IEE Proceedings Generations, Transmission and Distribution*, *148*(3), 229-235.

Sze, S. M. (1981). *Physics of Semiconductor Devices* (2nd ed.). New York: John Wiley.

Tang, G., Xu, H., & Chen, H. (1993). An Expert System for Voltage/VAR Control of Interconnected Power Systems. In *Proceedings of the Fourth Symposium on Expert Systems Application to Power Systems* (pp. 292-295). Academic Press.

Tang, R., Li, Z., Zhong, H., & Lan, Q. (2006). Assessment of uncertainty in mean heat loss coefficient of all glass evacuated solar collector tube testing. *Energy Conversion and Management*, *47*(1), 60–67. doi:10.1016/j.enconman.2005.03.013

Tan, W., Hao, Y., & Li, D. (2015). Load frequency control in deregulated environments via active disturbance rejection. *International Journal of Electrical Power & Energy Systems*, *66*, 166–177. doi:10.1016/j.ijepes.2014.10.036

Taranto, G. N., Pinto, L. M. V. G., & Pereira, M. V. F. (1992). Representation of FACTS Devices in Power System Economic Dispatch. *IEEE Transactions on Power Systems*, *7*(2), 572–576. doi:10.1109/59.141761

Tavkhelidze, A. N. (2010). Nanostructured electrodes for thermionic and thermotunnel devices. *Journal of Applied Physics*, *108*(4), 044313. doi:10.1063/1.3464256

Taylor Carson, W. (1993). *Modelling of Voltage collapse Including Dynamic Phenomena*. CIGRE Publication.

Taylor Carson, W. (1995). *Criteria and countermeasures for voltage collapse*. CIGRE Publication.

Taylor, C. W. (1994). *Power system voltage stability*. McGraw-Hill Companies.

Tchoumatchenko, V. (1998). *Modelisation, architecture et outils de synthese pour additionneurs rapides* (Doctoral dissertation).

Tersoff, J., Jesson, D. E., & Tang, W. X. (2009). Running droplets of gallium from evaporation of gallium arsenide. *Science*, *324*(5924), 236–238. doi:10.1126cience.1169546 PMID:19359583

Tey, L. H., So, P. L., & Chu, Y. C. (2002). Neural network-controlled unified power quality conditioner for system harmonics compensation. Proc. IEEE/PES Transmiss. Distrib. Conf. Exhib., 1038–1043.

Thamizharasan, S., Baskaran, J., Ramkumar, S., & Jeevananthan, S. (2013). New dual bridge multilevel dc-link inverter topology. *Electrical Power and Energy Systems*, *45*(1), 376–383. doi:10.1016/j.ijepes.2012.09.024

Thomas, D. E., Lagnese, E. D., Walker, R. A., Rajan, J. V., Blackburn, R. L., & Nestor, J. A. (1989). *Algorithmic and Register-Transfer Level Synthesis: The System Architect's Workbench: The System Architect's Workbench* (Vol. 85). Springer Science & Business Media.

Thomas, D., & Moorby, P. (2008). *The Verilog® Hardware Description Language*. Springer Science & Business Media.

Thornton, P. (1992). The Effect of Sand-Storm on Photovoltaic Array and Components. *Solar Energy Conference*.

Thounthong, P., Rael, S., & Davat, B. (2009, March). Analysis of Supercapacitor as Second Source Based on Fuel Cell Power Generation. *IEEE Transactions on Energy Conversion*, *24*(1), 247–255. doi:10.1109/TEC.2008.2003216

Timmer, A. H., & Jess, J. A. (1995, March). Exact scheduling strategies based on bipartite graph matching. In *Proceedings of the 1995 European conference on Design and Test* (p. 42). IEEE Computer Society. 10.1109/EDTC.1995.470422

Tinney, W., Bright, J., Demaree, K., & Hughes, B. (1988). Some deficiencies in optimal power flow. *IEEE Transactions on Power Systems*, *3*(2), 676–683. doi:10.1109/59.192922

Tlusty, J., & Valouch, V. (2004). Effectiveness of unified power quality conditioner for flicker mitigation. *Proc. 4th Int. Power Electron. Motion Control Conf.*, 902–907.

Tomsovic, K. (1992). A fuzzy linear programming approach to the reactive power/voltage control problem. *IEEE Transactions on Power Systems*, *7*(1), 287–293. doi:10.1109/59.141716

Touma, H. J. (2016). Study of the economic dispatch problem on IEEE 30-bus system using whale optimization algorithm. *International Journal of Engineering Technology and Sciences*, *5*(1), 11–18.

Toupchi Khosroshahi, M. (2014). Crisscross cascade multilevel inverter with reduction in number of components. *IET Power Electronics*, *7*(12), 2914–2924. doi:10.1049/iet-pel.2013.0541

Tremblay, O., Dessaint, L. A., & Dekkiche, A. I. (2007). A Generic Battery Model for the Dynamic Simulation of Hybrid Electric Vehicles. *IEEE Vehicle Power and Propulsion Conference*, 284-289. 10.1109/VPPC.2007.4544139

Trichakis, P., Taylor, P. C., Lyons, P. F., & Hair, R. (2008). Predicting the technical impacts of high levels of small-scale embedded generators on low-voltage networks. *IET Renewable Power Generation*, *2*(4), 249–262. doi:10.1049/iet-rpg:20080012

Tripathy, S. C., Balasubramanian, R., & Nair, P. S. C. (1992). Effect of superconducting magnetic energy storage on automatic generation control considering governor dead band and boiler dynamics. *IEEE Transactions on Power Systems*, *7*(3), 1266–1273. doi:10.1109/59.207343

Tsai, H. L., Tu, C. S., & Su, Y. J. (2008, October). Development of generalized photovoltaic model using MATLAB/SIMULINK. In *Proceedings of the world congress on Engineering and computer science* (*Vol. 2008*, pp. 1-6). Academic Press.

Tseng, C. J., & Siewiorek, D. P. (1986). Automated synthesis of data paths in digital systems. *IEEE Transactions on Computer-Aided Design of Integrated Circuits and Systems*, *5*(3), 379–395. doi:10.1109/TCAD.1986.1270207

Tsili, M., & Papathanassiou, S. (2009). A review of grid code technical requirements for wind farms. *IET Renewable Power Generation*, *3*(3), 308–332. doi:10.1049/iet-rpg.2008.0070

Udupa, A. N., Dhadbanjan, T., & Parthasarathy, K. (1999). An expert fuzzy control approach to voltage stability enhancement. *International Journal of Electrical Power & Energy Systems*, *21*(4), 279–287. doi:10.1016/S0142-0615(98)00049-0

Ui-Min Choi, Kyo-Beum Lee, & Blaabjerg, F. (2014). Diagnosis and tolerant strategy of an open-switch fault for T-type three-level inverter systems. *IEEE Transactions on Industry Applications*, *50*(1), 495–508. doi:10.1109/TIA.2013.2269531

Ullah, N. R., & Thiringer, T. (2007). Variable speed wind turbines for power system stability enhancement. *IEEE Transactions on Energy Conversion*, *22*(1), 52–60. doi:10.1109/TEC.2006.889625

Unsal, O. S., & Koren, I. (2003). System-level power-aware design techniques in real-time systems. *Proceedings of the IEEE*, *91*(7), 1055–1069. doi:10.1109/JPROC.2003.814617

US Department of Justice. (2015). Retrieved from https://www.justice.gov/atr/15-concentration-and-market-shares

Van, Moon, & Youn. (2004). Analysis of carrier based PWM methods in relation to common mode voltage for multilevel inverter. *30th Annual Conference of IEEE Industrial Electronics Society*, *2*, 1811 – 1816.

Varpula, A., & Prunnila, M. (2012). Diffusion-emission theory of photon enhanced thermionic emission solar energy harvesters. *Journal of Applied Physics*, *112*(4), 044506. doi:10.1063/1.4747905

Varpula, A., Tappura, K., & Prunnila, M. (2015). Si, GaAs, and InP as cathode materials for photon-enhanced thermionic emission solar cells. *Solar Energy Materials and Solar Cells*, *134*, 351–358. doi:10.1016/j.solmat.2014.12.021

Vassilios & Calais. (1998). Application specific harmonic performance evaluation of multicarrier PWM techniques. *Power Electronics Specialists Conference*, 172-178.

Vazquez, J. R., Martin, A. D., & Herrera, R. S. (2013). Neuro-fuzzy Control of a Grid-connected Photovoltaic System with Power Quality Improvement. *Euro. Con. 2013*, 850-856.

Venugopalan, R., Krishnakumar, N., Sarjila, R., & Rajasekar, N. (2013). Application of Particle Swarm Optimization Technique for the Design of Maximum Power Point Tracking. *Advanced Materials Research*, *768*, 47–56. doi:10.4028/www.scientific.net/AMR.768.47

Verma, A. K., & Ienne, P. (2006, July). Towards the automatic exploration of arithmetic-circuit architectures. In *Proceedings of the 43rd annual Design Automation Conference* (pp. 445-450). ACM. 10.1145/1146909.1147027

Verma, A. K., & Ienne, P. (2007, April). Automatic synthesis of compressor trees: reevaluating large counters. In *Proceedings of the conference on Design, automation and test in Europe* (pp. 443-448). EDA Consortium. 10.1109/DATE.2007.364632

Vilathgamuwa, M., Zhang, H. Y., & Choi, S. S. (1998). Modelling, analysis and control of unified power quality conditioner. In *Proc Harmonics Quality Power*. doi:10.1109/ICHQP.1998.760182

Vilchez, J. G., Jochem, P., & Fichtner, W. (2015). Energy Use and Emissions Impacts from Car Technologies Market Scenarios: A Multi-Country System Dynamics Model. *33rd International Conference of the System Dynamics Society*.

Vilchez, J. G., Jochem, P., & Fichtner, W. (2016). Car Technology Market Evolution And Emissions Impacts An Example Of Energy Policy Scenarios Under Uncertainty. *34th International Conference of the System Dynamics Society*.

Villalva, M. G., Gazoli, J. R., & Filho, E. R. (2009, May). Comprehensive Approach to Modeling and Simulation of Photovoltaic Arrays. *IEEE Transactions on Power Electronics*, *24*(5), 1198–1208. doi:10.1109/TPEL.2009.2013862

Vinothkumar, V. (2013). Fault diagnosis in h-bridge multilevel inverter drive using wavelet transforms. *International Journal of Advanced Research in Electrical Electronics and Instrumentation Engineering*, *2*(4), 1255–1262.

Visakha, K., Dhadbanjan, T., & Jenkins, L. (2004). Application of upfc for system security improvement under normal and network contingencies. *Electric Power Systems Research*, *70*(1), 46–55. doi:10.1016/j.epsr.2003.11.011

Vittal, E., O'Malley, M., & Keane, A. (2009). Impact of wind turbine control strategies on voltage performance. In IEEE Power & Energy Society General Meeting (pp. 1-7). IEEE. doi:10.1109/PES.2009.5275570

Vittal, E., Keane, A., & O'Malley, M. (2008). Varying penetration ratios of wind turbine technologies for voltage and frequency stability. In *IEEE Power and Energy Society General Meeting-Conversion and Delivery of Electrical Energy in the 21st Century* (pp. 1–6). IEEE. doi:10.1109/PES.2008.4596402

Vittal, E., O'Malley, M., & Keane, A. (2010). A steady-state voltage stability analysis of power systems with high penetrations of wind. *IEEE Transactions on Power Systems*, *25*(1), 433–442. doi:10.1109/TPWRS.2009.2031491

Voraphonpiput, N., & Chatratana, S. (2005). STATCOM Analysis and Controller Design for Power System Voltage Regulation. IEEE Transmission and Distribution Conference and Exhibition, 1-6.

Wahler, D., & Reymond, J. L. (2001). High-throughput screening for biocatalysts. *Current Opinion in Biotechnology*, *12*(6), 535–544. doi:10.1016/S0958-1669(01)00260-9 PMID:11849935

Wai, R. J., Duan, R. Y., & Jheng, K. H. (2012, February). High-efficiency bidirectional dc-DC converter with high-voltage gain. *IEEE Transactions on Power Electronics*, *5*(2), 173–184. doi:10.1049/iet-pel.2011.0194

Walker, G. (2001). Evaluating MPPT converter topologies using a MATLAB PV model. *Journal of Electrical and Electronics Engineering (Oradea)*, *21*(1), 49–56.

Wang, F., & Jiang, J. (2011). A novel static frequency converter based on multilevel cascaded h-bridge used for the start-up of synchronous motor in pumped-storage power station. *Energy Conversion and Management*, *52*(5), 2085–2091. doi:10.1016/j.enconman.2010.12.018

Wang, L., & Samee, U. (2013, September). Energy-aware parallel task scheduling in a cluster. *Future Generation Computer Systems*, *29*(7), 1661–1670. doi:10.1016/j.future.2013.02.010

Wang, Y., Dai, C., & Wang, S. (2013). Theoretical analysis of a thermoelectric generator using exhaust gas of vehicles as heat source. *Applied Energy*, *112*, 1171–1180. doi:10.1016/j.apenergy.2013.01.018

Wang, Y., Su, S., Lin, B., & Chen, J. (2013). Parametric design criteria of an irreversible vacuum thermionic generator. *Journal of Applied Physics*, *114*(5), 053502. doi:10.1063/1.4817084

Wang, Y., Su, S., Liu, T., Su, G., & Chen, J. (2015). Performance evaluation and parametric optimum design of an updated thermionic-thermoelectric generator hybrid system. *Energy*, *90*, 1575–1583. doi:10.1016/j.energy.2015.06.120

Wasynczuk, O. (1989). Modeling and dynamic performance of a line-commutated photovoltaic inverter system. *IEEE Transactions on Energy Conversion*, *4*(3), 337–343. doi:10.1109/60.43233

Wasynezuk, O. (1983). Dynamic behavior of a class of photovoltaic power systems. *IEEE Transactions on Power Apparatus and Systems*, *PAS-102*(9), 3031–3037. doi:10.1109/TPAS.1983.318109

Watanabe, E. H., & Aredes, M. (2002). Power quality considerations on shunt/series current and voltage conditioners. *Proc.10th Int. Conf. Harmonics Quality Power*, 595–600. 10.1109/ICHQP.2002.1221503

Watmuff, J. H., Charters, W. W. S., & Proctor, D. (1977). *Solar and wind induced external coefficients for solar collectors. Comples, 2*. Marseille: Rev. Inter. Heliotech.

Weber, L. (2010). *Demographic Change and Economic Growth Simulations on Growth Models*. Springer Verlag Berlin. doi:10.1007/978-3-7908-2590-9

Wenjin, D., & Xiangjie, C. (2009). Transient and Steady-state Analysis for STATCOM Mathematical Model. *2nd International Conference on Power Electronics and Intelligent Transportation System*, 271-275. doi:10.1109/TDC.2005.1546873

Werner, J. H., Kolodinski, S., & Queisser, H. J. (1994). Novel optimization principles and efficiency limits for semiconductor solar cells. *Physical Review Letters*, *72*(24), 3851–3854. doi:10.1103/PhysRevLett.72.3851 PMID:10056313

Wernick, M., Yang, Y., Brankov, J., Yourganov, G., & Strother, S. (2010). Machine learning in medical imaging. *IEEE Signal Processing Magazine*, *27*(4), 25–38. doi:10.1109/MSP.2010.936730 PMID:25382956

Wilson, V. C. (1959). Conversion of heat to electricity by thermionic emission. *Journal of Applied Physics*, *30*(4), 475–481. doi:10.1063/1.1702391

Witulski, A. F. (1995, May). Introduction to modeling of transformers and coupled inductors. *IEEE Transactions on Power Electronics*, *10*(3), 349–357. doi:10.1109/63.388001

Wojtczuk, S., Chiu, P., Zhang, X., Derkacs, D., Harris, C., Pulver, D., & Timmons, M. (2010, June). InGaP/GaAs/InGaAs 41% concentrator cells using bi-facial epigrowth. In *Photovoltaic Specialists Conference (PVSC), 2010 35th IEEE* (pp. 001259-001264). IEEE. 10.1109/PVSC.2010.5614196

Wolak, F. A. (2003). Measuring unilateral market power in wholesale electricity markets: The California market 1998–2000. *The American Economic Review*, *93*(2), 425–430. doi:10.1257/000282803321947461

Wolf, W. (2003). A decade of hardware/software codesign. *Computer*, *36*(4), 38–43. doi:10.1109/MC.2003.1193227

Wong, C. M., Zhan, J. C., Han, D. Y., & Zhao, B. L. (2000). A unified approach for distribution system conditioning: Distribution system unified conditioner. *Proc. Power Eng. Soc. Winter Meet*, 2757–2762.

Wong, K. F. V. (2000). *Thermodynamics for Engineers*. Boca Raton, FL: CRC Press LLC.

Woo, N. L., & Jong, B. P. (2011). Educational Simulator for Particle Swarm Optimization and Economic Dispatch Applications. In MATLAB - A Ubiquitous Tool for the Practical Engineer. InTech Education and Publishing KG.

Wu, Q., Xu, Z., & Østergaard, J. (2010). Grid integration issues for large scale wind power plants (WPPs). In *IEEE Power and Energy Society General Meeting* (pp. 1-6). IEEE.

Wu, C. (1992). Optimal power from a radiating solar-powered thermionic engine. *Energy Conversion and Management*, *33*(4), 279–282. doi:10.1016/0196-8904(92)90119-H

Würfel, P. (2009). *Physics of solar cells: from basic principles to advanced concepts* (2nd ed.). Wiley-VCH.

Xiao, W., Dunford, W. G., & Capel, A. (2004). A novel modelling method for photovoltaic cells. *IEEE 35th Annual Power Electronics Specialist Conference (PESC)*, *3*, 1950–1956.

Xiaohui, Y., Pengtao, W., Yanbin, Y., Yuehua, H., & Xiaopan, Z. (2015). A new quantum inspired chaotic artificial bee colony algorithm for optimal power flow problem. *Energy Conversion and Management*, *100*, 1–9. doi:10.1016/j.enconman.2015.04.051

Xoubi, N. (2015). Viability of a Utility-Scale Grid-Connected Photovoltaic Power Plant in the Middle East. *Journal of Applied Sciences (Faisalabad)*, *15*(11), 1278–1287. doi:10.3923/jas.2015.1278.1287

Xu, B., & Abur, A. (2004). *Observability analysis and measurement placement for systems with PMUs.* Paper presented at the Power Systems Conference and Exposition, 2004. IEEE PES.

Xu, D. &Cheng, K. W. E. (2011). A survey of distributed power system-- AC versus DC distributed power system. *IEEE Power Electronics Systems and Applications (PESA),* 1 - 12.

Xu, Y. (2014). Adaptive PI control of STATCOM for Voltage Regulation. *IEEE Transaction on Power Delivery, 29*(3), 1002-1011.

Xuan, X. C., & Li, D. (2003). Optimization of a combined thermionic–thermoelectric generator. *Journal of Power Sources*, *115*(1), 167–170. doi:10.1016/S0378-7753(02)00620-1

Yadav, R. G., Roy, A., Kharparde, S. A., & Pentayya, P. (2005). India's fast growing power sector from regional development to the growth of a national grid. *IEEE Power & Energy Magazine*, *3*, 39–48. doi:10.1109/MPAE.2005.1458229

Yager, R. R., & Zadeh, L. A. (Eds.). (2012). *An introduction to fuzzy logic applications in intelligent systems* (Vol. 165). Springer Science & Business Media.

Yaghoobi, P., Vahdani Moghaddam, M., & Nojeh, A. (2012). Solar electron source and thermionic solar cell. *AIP Advances*, *2*(4), 042139. doi:10.1063/1.4766942

Yamaguchi, K. (2013). *Money and Macroeconomic Dynamics- Accounting System Dynamics Approach*. Awaji Island, Japan: Japan Future Research Center.

Yang, C., & Smedley, K. (2008, September). Three-Phase Boost-Type Grid-Connected Inverter. *IEEE Transactions on Power Electronics*, *23*(5), 2301–230. doi:10.1109/TPEL.2008.2003025

Yanushkevich, A., Muller, Z., Svec, J., Tlusty, J., & Valouch, V. (2014). Power Quality Enhancement using STATCOM with Energy Storage. *International Conference on Renewable Energies and Power Quality*, 12. 10.24084/repq12.336

Yan, X., & Quantana, V. H. (1999). Improving an interior point based OPF by dynamic adjustments of step sizes and tolerances. *IEEE Transactions on Power Systems*, *14*(2), 709–717. doi:10.1109/59.761902

Yan, Z., Fei, L., Yin, J., & Duan, S. (2008). Study on realizing MPPT by improved Incremental Conductance method with variable step-size. *3rd IEEE Conference on Industrial Electronics and Applications*, 547-550. 10.1109/ICIEA.2008.4582575

Yao, Y., Bustamante, P., & Ramshaw, R. S. (1994). Improvement of induction motor drive systems supplied by photovoltaic arrays with frequency control. *IEEE Transactions on Energy Conversion*, *9*(2), 256–262. doi:10.1109/60.300150

Yazdian-Varjani, A., Perera, S., & Chicharo, J. F. (1998). A centroid-based pwm switching technique for full-bridge inverter applications. *IEEE Transactions on Power Electronics*, *13*(1), 115–124. doi:10.1109/63.654965

Yeap, G. K. (2012). *Practical low power digital VLSI design*. Springer Science & Business Media.

Yishu, Z. (2009). The circuit topology for single-phase Grid connected system and the control technology on converters. *International Conference on Sustainable Power Generation and Supply, SUPERGEN '09*, 1-5. 10.1109/SUPERGEN.2009.5348130

Yi, T., Canbing, L., Yijia, C., Kwang, Y. L., Lijuan, L., Shengwei, T., & Lian, Z. (2015). Improved group search optimization method for optimal power flow problem considering valve-point loading effects. *Neurocomputing*, *148*, 229–239. doi:10.1016/j.neucom.2013.09.065

Yokoyama, R., Niimura, T., & Nakanishi, Y. (1993). A coordinated control of voltage and reactive power by heuristic modeling and approximate reasoning. *IEEE Transactions on Power Systems*, *8*(2), 636–645. doi:10.1109/59.260817

Young, C. M., Chen, M. H., Chang, T. A., Ko, C. C., & Jen, K. K. (2013, February). Cascade Cockcroft–Walton voltage multiplier applied to transformerless high step-up DC-DC converter. *IEEE Transactions on Industrial Electronics*, *60*(2), 523–537. doi:10.1109/TIE.2012.2188255

Zadeh, L. A. (1965). Fuzzy sets. *Information and Control*, *8*(3), 338–353. doi:10.1016/S0019-9958(65)90241-X

Zare, K., Hagh, M. T., & Morsali, J. (2015). Effective oscillation damping of an interconnected multi-source power system with automatic generation control and TCSC. *International Journal of Electrical Power & Energy Systems*, *65*, 220–230. doi:10.1016/j.ijepes.2014.10.009

Zeigler, B. P., Praehofer, H., & Kim, T. G. (2000). *Theory Of Modelling And Simulation*. London, UK: Academic Press.

Zeng, T. (2006). Thermionic-tunneling multilayer nanostructures for power generation. *Applied Physics Letters*, *88*(15), 153104. doi:10.1063/1.2192973

Zhang, H., Liu, J., Huang, X., & Wang, Z. (2007). Design of a new DC link voltage controller for universal power quality controllers. *Proc. Appl. Power Electron. Conf.*, 473–477.

Zhang, X., Zhang, W., Liu, Y., Lv, W., & Wang, Q. (2010). Unified power quality conditioner with model predictive control. *Proc. 5th Int. Conf. Comput. Sci. Educ.*, 1239–1244.

Zhang, Y., Zhang, S., & Chen, J. (2008). The applications of bidirectional full bridge DC–DC isolated converter in UPQC. *Proc. Int. Conf Electr. Mach. Syst.*, 1916–1921.

Zhao, Q., & Lee, F. C. (2003, January). High-efficiency, high step-up DC-DC converters. *IEEE Transactions on Power Electronics*, *18*(1), 65–73. doi:10.1109/TPEL.2002.807188

Zheng, M., & Albicki, A. (1995, October). Low power and high speed multiplication design through mixed number representations. In *Computer Design: VLSI in Computers and Processors, 1995. ICCD'95. Proceedings., 1995 IEEE International Conference on* (pp. 566-570). IEEE.

Zheng, T., Jiao, S., Ding, K., & Lin, L. (2013). A coordinated voltage control strategy of wind farms based on sensitivity method. In IEEE Grenoble PowerTech (POWERTECH) (pp. 1-5). IEEE.

Zhigang, L., Rong, G., Jun, L., & Huang, A. Q. (2011, March). A High-Efficiency PV Module-Integrated DC/DC Converter for PV Energy Harvest in FREEDM Systems. *IEEE Transactions on Power Electronics*, *26*(3), 897–909. doi:10.1109/TPEL.2011.2107581

Zhilei, Y., Lan, X., & Yangguang, Y. (2010). Seamless Transfer of Single-Phase Grid-Interactive Inverters Between Grid-Connected and Stand-Alone Modes. *IEEE Transactions on Power Electronics*, *25*(6), 1597–1603. doi:10.1109/TPEL.2009.2039357

Zhili, T., Xun, L., Jian, C., Yong, K., & Shanxu, D. (2006). A direct control strategy for UPQC in three-phase four-wire system. *Proc. Power Electron. Motion Control Conf.*, 1–5.

Zhong, Ozpineci, Tolbert, & Chiasson. (2009). DC–AC cascaded h-bridge multilevel boost inverter with no inductors for electric hybrid electric vehicle applications. *IEEE Transactions on Industry Applications*, *45*(3), 963-970.

Zhong, J., & Bhattacharya, K. (2002). Toward a competitive market for reactive power. *IEEE Transactions on Power Systems*, *17*(4), 1206–1215. doi:10.1109/TPWRS.2002.805025

Zhou, M., Wan, J. R., Wei, Z. O., & Cui, J. (2006). Control method for power quality compensation based on Levenberg–Marquard optimized BP neural networks. *Proc. IEEE 5th Int. Conf. Power Electron. Motion Control Conf.*, 1–4.

Zhu, L. Z., Chen, N., Wang, W., & Zhu, X. D. (2009). Reactive control strategy for wind farm considering reactive power command. In *International Conference on Sustainable Power Generation and Supply* (pp. 1-5). IEEE. 10.1109/SUPERGEN.2009.5348206

Zimmermann, A., & Trautmann, N. (2017). A list-scheduling heuristic for the short-term planning of assessment centers. *Journal of Scheduling*, 1–12.

Zimmermann, G. (1979, June). The MIMOLA design system a computer aided digital processor design method. In *Proceedings of the 16th Design Automation Conference* (pp. 53-58). IEEE Press. 10.1109/DAC.1979.1600088

Zimmermann, H. J. (2011). *Fuzzy set theory—and its applications*. Springer Science & Business Media.

Zimmirrman, R. D., Murillo-Sanchez, C. E., & Gan, D. (n.d.). *MATPOWER: a Matlab power system simulation package*. Retrieved from http://www.pserc. cornell.edu/matpower/3.2

About the Contributors

Pawan Kumar is the assistant professor in Electrical and Instrumentation Engineering Department at Thapar Institute of Engineering & Technology, Patiala, Punjab, India. He is a Certified Energy Manager cum Auditor (Bureau of Energy Efficiency) with 15 years' (including 7 years of research) academic experience in the Power and Energy fields of Electrical Technology. Dr. Kumar has received his B. Tech from Kurukshetra University in 2001. He has done M. Tech from Indian Institute of Technology, Delhi in 2010 and, PhD from Jamia Millia Islamia (A Central University), New Delhi in 2016. He had excellent independent research training and development while being worked as Coordinator of Energy Elite Club. He has published several research papers in various reputed publishing houses including, IEEE, IET, Elsevier, and AJOL. He has been assumed as Managing Guest Editor for the Special Issue on "Innovative Technologies for Micro-grid and Smart-grid Systems" of Computers and Electrical Engineering, Elsevier Journal. He has also been served as a reviewer for reputed journals and conferences including, Engineering Science and Technology, an International Journal and Electrical Power and Energy Systems, Elsevier, IJITWE IGI Global USA, JCIT IGI Global USA, 12th IEEE INDICON-2015, International Conference on Advances in Computing, Communications and Informatics (ICACCI'17) and 14TH IEEE INDICON-17 held at IIT Roorkee. He has worked as Session Chair and reviewer for ICICS-2016 for the track of power, energy and control. He has also received Best Paper Award during ICICS-2016 International Conference. He has delivered expert lectures on Energy Efficiency in electrical utilities in various institutions. His area of current interests include: Energy efficiency in electrical utilities, optimization in power system's operation and planning, distributed generation, load management and energy auditing, computational intelligence, Innovative technologies for micro-grid and smart-grid systems.

Surjit Singh is currently a faculty member with the Department of Computer Engineering at National Institute of Technology, Kurukshetra, Haryana, India. He has several years of experience in teaching and research. He has submitted Ph.D. in Computer Engineering from NIT Kurukshetra. He has received B.Tech. degree with Honors in Computer Engineering from Kurukshetra University in 2008, and completed his M.Tech. degree with Distinction in Computer Science. He holds State Rank during his diploma course in engineering. He has authored many publications with reputed publishers including ACM, IET, IEEE, Springer, Elsevier, Pearson and IGI Global. He is currently editing a book "Handbook of Research on the IoT, Cloud Computing and Wireless Network Optimization" IGI Global USA. He has already two Books to his credit entitled "Computer Systems and Programming in C" with publisher Pearson Education and "Handbook of Research on Power and Energy System Optimization" with publisher IGI Global USA, also one Book Chapter with IGI Global, USA. He is also serving as a member of the Editorial Advisory Board of Advances in Educational Technologies & Instructional

Design (AETID) Book Series, IGI Global USA. He has served as Guest Editor for the Special Issue of "Journal of Cases on Information Technology (JCIT)" which is listed in WoS/Scopus indexed journal of IGI Global USA. He has chaired Invited Talk during International Conference SmartTech-2017. He is also serving as reviewer and or technical committee member for IEEE and other International Conferences including, NGCT-17, SmartTech-2017, IEEE ICIC 2015 and 12thIEEE INDIACom-18. He is member of the International Advisory Board of WoS/Scopus indexed journals including, International Journal of Hyperconnectivity and Internet of Things (IJHIoT), International Journal of Information and Communication Technology Education (IJICTE), International Journal of Wireless Networks and Broadband Technologies (IJWNBT), Journal of Cases on Information Technology (JCIT), and Reviewer of several journals listed in WoS/Scopus indexed databases. His areas of current interests are Internet of Things, Cloud Computing, Advanced Computer Networks, Wireless Sensor and Ad hoc Networks, and Computational Intelligence.

Ikbal Ali is Associate Professor in the Department of Electrical Engineering, Faculty of Engineering & Technology of Jamia Millia Islamia, New Delhi, India. Dr. Ali has earned the Ph.D. degree for his research work in the field of Substation Automation, where he investigated the communication requirements for substation automation. Dr. Ali's research work has been published in refereed international journals/conferences of repute. He is actively involved in research and development activities in areas of SCADA systems, Power System Automation, Smart Grid, Substation Communication Network, Broadband over Power Line (BPL), IEC61850 based systems, Wide Area Protection, Communication, and Control in Energy Systems, Power and Energy System Applications (generation, transmission, distribution, markets, operations, planning, Energy Management Systems (with applications to building and home automation). Dr. Ali is a senior member in IEEE and an expert lecturer on the topics related to Smart Grid Technologies, Substation Automation Trends and Communication Networks for Substation Automation.

Taha Selim USTUN received B.E. degree in Electrical and Electronics Engineering from Middle East Technical University, Turkey in 2007 and Master of Engineering Science degree from University of Malaya, Malaysia in 2009. He has received Ph.D in Electrical Engineering from Victoria University, Melbourne, Australia. Currently, he is an Assistant Professor in Electrical Engineering, School of Electrical and Compputer Engineering, Carnagie-Mellon University, PA, USA. His research Interests are Power Systems Protection, Communication in Power Networks, Distributed Generation, Microgrids and Smartgrids.

* * *

Subrata Banerjee received his Ph.D in Electrical Engineering from Indian Institute of Technology, Kharagpur in the year 2005. He is presently working as Professor in Electrical Engineering at National Institute of Technology, Durgapur, India. He has published a numbers of research papers in National/ International Journals and Conference Records. His area of interest includes Electromagnetic Levitation, Active Magnetic Bearing, Controller Design, Intelligent Control, Optimization Techniques, Control of Switch-mode converters, non-linear dynamics of converters, load frequency control etc. He is a Fellow of Institute of Engineers (India), life member of Systems Society of India, and senior member of IEEE (USA). He has successfully completed some research and consultancy projects including one major from DST, Govt. of India. His Biographical inclusion is in Marquis Who's Who 2007 and IBC FOREMOST Engineers of world-2008. Prof. Banerjee has received few academic awards including 4 no. of Best papers, TATA RAO Prize in his credit. He has published one Indian Patent in 2014. Prof. Banerjee is acting as regular reviewer of IEEE Transaction on PE, IEEE Transaction on IE, Electrical Power Components and Systems etc.

Siwar Ben Haj Hassine held a Diploma in Computer Engineering in 2013 from the national engineering school of Sousse. Currently, she prepares, in the national engineering School of Monastir, her thesis whose interest includes methodologies development for power optimization of digital circuit.

Ouni Bouraoui is currently an Associate Professor at the National Engineering School of Sousse. He has obtained his Ph.D. entitled 'Synthesis and temporal partitioning for reconfigurable systems' in 2008 from the Faculty of Sciences at Monastir. He is obtained his university habilitation entitled 'Optimization algorithm for reconfigurable architectures' in 2012. Hence, his researches interest cover: models, methods, tools, and architectures for reconfigurable computing; simulation, debugging, synthesis, verification, and test of reconfigurable systems; field programmable gate arrays and other reconfigurable technologies; algorithms implemented on reconfigurable hardware; hardware/software codesign and co-simulation with reconfigurable hardware; and high performance reconfigurable computing.

Ajoy Kumar Chakraborty, Ph.D. (Jadavpur University), is an Associate Professor and was Ex Head, Department of Electrical Engineering, National Institute of Technology (NIT), Agartala. He did his M.Tech in Electrical Engineering from IIT Kharagpur and his B.E.E (Hons) from Jadavpur University. He has to his credits more than 100 numbers of technical articles in national and international journals and conferences including 15 SCI and Scopus indexed journals. Besides teaching experience of more than 23 years at both undergraduate and post- graduate level and industrial experience of 14 years, Dr. Chakraborty has also authored three books. In his teaching tenure he has guided 8 Ph.D. scholars, 12 M.Tech scholars and many other B.Tech students. Currently 8 Ph.D. scholars are working under his guidance. His field of specialization are power system operation and control, power system protection, Network Analysis and Circuit Theory, Soft Computing Techniques and Smart Grid.

Ramulu Chinthamalla received the B.Tech. degree from Gokaraju Rangaraju Institute of Technology, Jawaharlal Nehru Technological University, Hyderabad, Telangana, India in 2001, and the M.Tech. degree in power electronics and drives and Ph.D. from National Institute of Technology, Warangal, India, in 2005 and 2017 respectively. Since 2012, he has been with the Department of Electrical Engineering, National Institute of Technology, Warangal as an Assistant Professor. His research interests include power electronic applications in renewable energy conditioning, multilevel inverters, pulse width-modulation switching strategies, dc and ac drives, etc.

Thukaram Dhadbanjan received the B.E. degree in electrical engineering from Osmania University, Hyderabad, India, in 1974, the M.Tech. degree in integrated power systems from Nagpur University, Nagpur, India, in 1976, and the Ph.D. degree from the Indian Institute of Science, Bangalore, India, in 1986. Since 1976, he has been with the Indian Institute of Science, Bangalore, India, as a Research Fellow and Member of Faculty in various positions, where, since 2002, he has been a Professor with the Department of Electrical Engineering. His research interests include computer-aided power system analysis, reactive power optimization, voltage stability, and AI applications in power systems. He is a senior member of IEEE (USA).

Arzu Eren Şenaras graduated from Uludağ University in 2007. She took her master and Phd degree in the Deparment of Econometrics at Uludag University.

Mohammad Quadeer Fahad has received his B.Tech in Electrical Engineering and M.Tech in Electrical Power System Management both with honors from Jamia Millia Islamia (Central University) in 2014 and 2017 respectively. His areas of interest are Power System Deregulation, Renewable Technology and FACTS devices. Currently he is working in Rural Electrification Corporation (REC), Corporate Office at New Delhi.

Durga Prasad Garapati received the B.Tech. Degree from Pondicherry University, Pondicherry, and M.Tech. from Vellore Institute of Technology (Deemed to be University), Vellore, and completed Ph.D. in Karunya Institute of Technology and Sciences (Declared deemed to be university), Karunya nagar, Coimbatore, TN, India. He is working presently as Associate Professor in Shri Vishnu Engineering College for Women, Bhimavaram, India. He has presented technical papers in various National and International Journals and Conferences. His area of interest includes Power Electronics, Industrial Drives, and Instrumentation.

Dipayan Guha was born in Barasat, North 24 Parganas, West Bengal, India, in 1986. He completed his PhD in Electrical Engineering in the year 2017 from NIT Durgapur, India. He has published 19 research papers in National/International conference records/Journals (out of which four papers are in SCI Journals). He is currently working as an Assistant Professor in Department of Electrical Engineering at Dr. B. C. Roy Engineering College, Durgapur, West Bengal, India. He has total four years experience in teaching. His research interest includes power system operation and control, soft computing, FACTS controllers and nonlinear control system.

Hassan Haes Alhelou was born in Syria. He works at Tishreen University, Faculty of Electrical Power Engineering in Lattakia city, Syria. His scientific and research interests include smart grids, electricity markets, power system stability and control, optimisation, advanced signal processing techniques applied to power systems control and analysis,

Badia Haidar is assistant professor in the Structural Engineering Department, Civil Engineering Faculty Tishreen University. She got her PhD., M.Sc. and B.Sc. from Tishreen University in 1998, 2004, 2011 respectively. She has more than 15 years' experience in teaching engineering. She is currently supervising many M.Sc & PhD students. Her research interests include Optimum Design, Mathematical Modelling, Genetic Algorithm, Programming Languages, Artificial Intelligence and Automatic Control, Using SCADA & Microcontroller in Petroleum Stations, Power Electronics.

Mashhood Hasan received the B.E and M.Tech degree from Jamia Millia Islamia (a central university) New Delhi, India in 2003 and 2009. He has completed his Ph.D during 2012- 2016 from the same university. From March 2004 to Sep 2006 he was a Lecturer in electrical department at Al-Falah School of Engineering and Technology India. Since Sep 2006 to august 2010 worked as Senior Lecturer in electrical department at IEC college of Engineering and Technology, Gr Noida (UP). Presently, working in GALGOTIAS college of Engineering and Technology in Electrical Department, Gr Noida (UP), since august 2010 as a Assistant Professor. His research interests in power quality, power system, FACTS controlling device, and power electronics.

Fahad Iqbal has completed M.Tech in Electrical Power System Management from Department of Electrical Engineering, Jamia Millia Islamia (Central University) New Delhi, India. He is currently pursuing his Ph.D at the same institute. He received B.Tech degree with Honors in Electrical and Electronics engineering from Integral University, Lucknow, India. He was awarded with Silver Medal and Gold Medal for his outstanding performances in Bachelors degree and M.tech degree respectively. His areas of interest are FACTS devices, Power system Management and Power System Deregulation.

Sachin Jain (M' 15, SM' 16) received the B.E. degree from Bhilai Institute of Technology Durg, Pandit Ravishankar Shukla University, Durg, India, in 2000, the M.Tech. degree in integrated power systems from Visvesvaraya National Institute of Technology, Nagpur, India, in 2002, and the Ph.D. degree in electrical engineering from the Indian Institute of Technology Bombay, Mumbai, India, in 2008.From 2007 to 2012, he was a Senior Design Engineer with the R&D Department, Solar Energy Business Group, Schneider Electric, Bangalore, India. He is currently an Assistant Professor with the National Institute of Technology Warangal, Warangal, India. His research interests include power electronic applications in nonconventional energy conditioning, power quality, and distributed generation.

Mehdi Jemai is currently a Contractual teacher at the Faculty of Science of Monastir. He has obtained his PhD in 2016 from the National Engineering School of Monastir. held a Master in Microelectronic in 2011 from the Faculty of Science of Monastir and received his Diploma in Computer Engineering in 2009 from the Higher Institute of Applied Science and Technology of Sousse. Hence, his researches interest cover: models, methods, tools, and architectures for reconfigurable computing; simulation, debugging, synthesis, verification, and test of reconfigurable systems; field programmable gate arrays and other reconfigurable technologies; algorithms implemented on reconfigurable hardware; hardware/ software codesign and cosimulation with reconfigurable hardware; and high performance reconfigurable computing.

Hao Li is a A researcher of energy system optimization and computational modeling.

Ramsha Karampuri received the B.Tech. degree from Srinivas Reddy Institute of Technology, Jawaharlal Nehru Technological University, Hyderabad, India, and the M.Tech. degree from Visvesvaraya National Institute of Technology, Nagpur, India. He is currently working toward the Ph.D. degree in electrical engineering at the National Institute of Technology Warangal, Warangal, India. His research interests include power electronics and drives, application of power electronics to nonconventional energy conditioning, and multiphase drives.

S.C. Kaushik is a former head and currently an Emeritus Professor at Centre for Energy Studies, I.I.T. Delhi. He is a pioneer researcher on Exergy Analysis and Finite time thermodynamics of Energy Systems at National & International level and a leading expert on Alternative Refrigeration and Air Conditioning Technologies. He has made significant contributions in these fields as evident by his above 400 research publications in Journals of repute at national and international levels. Prof. Kaushik has authored several books of the first kind on "Solar Refrigeration & space-conditioning (1989)", "Thermal Control in Passive Solar Buildings" (1989), "Greenhouse Science & Technology" (2006), "Alternative Energy Resources (In Hindi)" (2010), "Energy and Environment Laboratory Experiments" (2011) and "Alternatives in Refrigeration and Air conditioning" (2016). He has supervised and guided 60 Ph.D. Theses and 105 M.Tech Projects. He has contributed to various sponsored/consultancy projects at I.I.T. Delhi in the field of Energy Conservation, Solar thermal devices and Building Space conditioning. His specific contributions on Energy Systems have been recognized by national and international awards to his credit. Recently Prof. Kaushik has been declared as the Top Academic Research Performer Ranked No.1 Researcher in the field of Energy, 2009-2014, DST-Elsevier report-2015.

Mohd Tauseef Khan is currently an Assistant Professor in Rakiya Engineering College Banda. He obtained Bachelor of Engineering (Electrical Engg.) degree from Aligarh Muslim University in 2011, Master of Technology (Electrical Power System management) from Jamia Millia Islamia (Central University) New Delhi, India in 2013 and Ph.D. in Electrical Engineering in February 2018 from Jamia Millia Islamia, New Delhi, India. His research interest includes Power System Deregulation, Power System Management, Optimization techniques, Distributed Generation and FACTS devices.

Vulisi Narendra Kumar received the B.Tech degree in Electrical & Electronics Engineering from Jawaharlal Nehru Technological University Kakinada, India, in 2012. Consequently, he worked as lecturer in A.A.N. M & V.V.R.S.R Polytechnic in Electrical & Electronics Engineering Dept. from 2013 to 2016. He is currently working as M.Tech scholar at Department of Electrical Engineering, National Institute of Technology Meghalaya, India. His research interests are renewable energy technologies, adaptive dynamic energy management & linear/non-linear control of grid-connected DC microgrid, and islanding detection techniques.

Ravita Lamba is a Ph.D. research scholar at Centre for Energy Studies, Indian Institute of Technology Delhi since July 2014. She received her Bachelor of Technology (Electronics Instrumentation and Control Engineering.) degree from Rajasthan Technical University, Kota (Rajasthan) in 2012 and Master of Technology (Renewable Energy) from Indian Institute of Technology Delhi, India in 2014." before She word. She is working in the area of thermodynamic and heat transfer studies of photo thermionic and thermoelectric hybrid energy conversion systems for power generation and space conditioning. She has published several research papers in the journals of international repute. She has also co-authored one book chapter in the book titled "The role of Exergy in Energy and Environment" published by Springer Nature. She has been awarded the prestigious Building Energy Efficiency Higher & Advanced Network (BHAVAN) Fellowship by Indo-US Science and Technology Forum (IUSSTF) and Department of Science and Technology (DST), Government of India to do advanced research on "Building Integrated Solar Photovoltaic-Thermoelectric hybrid systems" at the University of South Florida, USA, for a period of 3 months.

Zhijian Liu is an associate professor of Department of Power Engineering, School of Energy, Power and Mechanical Engineering, North China Electric Power University.

Belkacem Mahdad was born in Biskra, Algeria. He received his B.S degree in Electrical Engineering from Biskra University, Algeria in 1990, and the Magister and Ph.D degrees from Annaba University and Biskra University in 2000 and 2010, respectively. He is an assistant professor at Biskra University. His research interests include power system optimization, FACTS modelling and integration in practical power system, Optimization methods, power system stability and system protection coordination.

Carlo Makdisie is a professor at Power Department, Mechanical and Electrical Engineering Faculty, Tishreen University. He got his Ph.D., from Applied Polytechnical Institute of Kiev, Soviet Union in 1993, and B.Sc. from Tishreen University in 1987. He has more than 24 years' experience in teaching engineering in several countries (Sudan – Jordan – Sultanate of Oman) including many universities and institutes. He is currently supervising number of M.Sc students and has published many technical papers in the referred journals. He has considerable industrial experience. His research interests include smart grid related to modern control and renewable energy engineering, Power Electronics, Using SCADA & Microcontroller in Petroleum Stations, Programmable Logic Controllers PLCs, Artificial Intelligence and Automatic Control, Renewable Power Systems.

Tukaram Moger received B.E. degree in electrical and electronics engineering from Karnatak University, Dharwad, Karnataka, India, in 2001, M.Tech. degree in electrical engineering from Indian Institute of Technology Kanpur (IITK), UP, India, in 2005, and Ph.D. degree in electrical engineering from Indian Institute of Science (IISc) Bengaluru, India, in 2016. He has been associated with academic institutions as a faculty member and currently working as assistant professor in the Department of Electrical and Electronics Engineering, National Institute of Technology Karnataka (NITK), Surathkal, Mangaluru, India. His research interests include grid Integration of renewable energy, solar photovoltaic systems, microgrid, reactive power and voltage control, power system operation and planning, power system deregulation. He is a member of IEEE Eta-Kappa Nu (Mu Xi Chapter of IISc), IEEE (USA), IEEE-PES, IET (UK), CIGRE, Institution of Engineers (India), and life member of Indian Society for Technical Education (ISTE), System Society of India (SSI) and Soft Computing Research Society (SCRS) of India. He also holds Chartered Engineer (India) certificate.

Soumya Mohanty is working as an Assistant Professor with the department of Electrical Engineering, Motilal Nehru National Institute of Technology, Allahabad. His research interests include distributed generation (DG) and smart grid & signal processing applied to Power System Protection and Distributed Generation and allied fields.

Bonu Ramesh Naidu received the B.Tech degree in Electrical & Electronics Engineering from Jawaharlal Nehru Technological University, Kakinada, in 2013 and M.Tech degree in Power & Energy Systems from National Institute of Technology Meghalaya in 2017. He is currently working as Junior Research Fellow in UK India Clean Energy Research Institute at Department of Electrical Engineering, Indian Institute of Technology Kharagpur, India. He received the "Best Poster Award in Engineering Sciences Section" at 99th Indian Science Congress Association Meeting, Bhubaneswar in 2012 and "The Most Innovative Project Award for the year 2016-17" at National Institute of Technology Meghalaya. He also received both "Institute and President Gold Medal" during his M.Tech from National Institute of Technology Meghalaya. His research interests include multi-functional control and energy management in AC/DC microgrid.

Champa Nandi received her B.E. degree in Electrical Engineering from the Tripura University, India in 2004 and M. Tech. in Electrical Engineering from Tripura University, India in 2008. From 2008 to 2010 she served as a lecturer in the Department of Electrical Engineering, NIT, Agartala. She has received her Ph.D in the Dept. of Electrical Engineering at NIT Agartala in May 2016.From 2010 to till date she is serving Department of Electrical Engineering, Tripura University as an Assistant Professor. Her research interest areas are Power system (FACTS Controllers), Control system, Renewable Energy.

Gayadhar Panda received his B. E. degree from the Institute of Engineers, India in 1996. Consequently, he obtained his master's degree from Bengal Engineering College (presently IIEST), India and his PhD degree from Utkal University, India, in 1998 and 2007, respectively. Since January 2013, he has been with the Department of Electrical Engineering, National Institute of Technology Meghalaya, India as a professor. His current research interests include automatic generation control, stability improvements using FACT devices, power quality, power electronic converters, and renewable energy integration.

Tapan Prakash has obtained B. Tech degree in Electrical & Electronics Engineering from West Bengal University of Technology, Kolkata, in the year 2010 and M. Tech degree in Power System Engineering from VSSUT Burla, Odisha, India in the year 2013. He is currently pursuing Ph. D. in Electrical Engineering from National Institute of Technology Raipur, India. His current research interests are power system optimisation, stability and control and Wide area monitoring and its protection and control.

Ankur Singh Rana received B. Tech degree in Electrical and Electronics Engineering from Guru Gobind Singh University, New Delhi, and the M. Tech in Electrical Power System and Management from Jamia Millia Islamia (JMI) in 2010 and 2013 respectively. Currently, He is working as assistant professor in NIT Delhi after submitting Ph.D Thesis for review in Jamia Millia Islamia. His research interests include Wide Area Measurement System, Reliability, SCADA, Power system protection, PMU and Smart Grids.

Provas Kumar Roy was born in 1973 at Mejia, Bankura, West Bengal, India. He received the BE degree in Electrical Engineering from R. E. College, Durgapur, Burdwan, India in 1997; ME degree In Electrical Machine from Jadavpur University, Kolkata, India in 2001 and PhD from NIT Durgapur in 2011. Presently he is working as Professor in the department of Electrical Engineering, Kalyani Government Engineering College, Kalyani, India. He has published more than 120 research papers in international journals and conferences. His field of research interest includes Economic Load Dispatch, Optimal Power flow, FACTS, Unit Commitment, Automatic Generation Control, Power System Stabilizer, Radial distribution system and Evolutionary computing techniques.

Anwar Shahzad Siddiqui obtained his B.Sc. Engg. (Electrical Engineering) and M.Sc. Engg. (Power Systems and Electrical Drives) degrees from AMU, Aligarh, both with Honors in 1992 and 1994 respectively. He earned his Ph.D. degree from Jamia Millia Islamia (Central University) New Delhi, India in 2001. He has been teaching and guiding research in Electrical Engineering for about one and a half decade at AMU, Aligarh; JMI, New Delhi and BITS Pilani – Dubai Campus. His research interests include Power System Management and Applications of Artificial Intelligence Techniques in power System. He has more than 50 research papers published in refereed international and national journals and conference of repute.

Vinay Pratap Singh is an Assistant Professor in Department of Electrical Engineering, National Institute of Technology, Raipur, India. His research interest includes model reduction, optimization techniques and control systems.

Dip V. Thanki is the research scholar in Lovely Professional University, Phagwara. He is doing his thesis on energy management and efficiency in power distribution in LPU campus.

S. K. Tyagi is working as Associate Professor at Centre for Energy Studies, I.I.T. Delhi and earlier he was working as Director/Scientist E at Sardar Swaran Singh National Institute of Renewable Energy, Kapurthala, an autonomous institution of the Ministry of New and Renewable Energy, Government of India. He has also worked as Assistant Professor at SMVDU, Katra, Jammu & Kashmir. Dr. Tyagi has worked also as Invited Scientist at Korea Institute of Energy Research, South Korea and as Post Doctoral Fellow at The Hong Kong Polytechnic University, Hong Kong, Zhejiang University, Hangzhou and Xiamen University, Xiamen, China during 2002–2008. He has made a significant contribution in R&D activities as evident by more than 150 publications in Journals/Conferences of repute. He has guided five Ph.D. Theses and 10 M.Tech. projects.

Jegathesan V. received the B.E. and M.E. degrees from Bharathiar University, Coimbatore, India, in 1999 and 2002 respectively and the Ph.D. degree from Anna University, Chennai, India, in 2010. Currently, he is an Associate Professor in the Department of Electrical and Electronics Engineering at Karunya Institute of Technology and Sciences (Deemed to be University), India. He has presented technical papers in various National and International Conferences in India and Abroad. He has also published papers in National and International peer reviewed Journals. His area of interest includes Electric Circuits and Networks, Power Electronics, Development of heuristic algorithm for Power Electronics applications and Application of Power Electronics to Renewable Energy.

Moorthy Veerasamy is graduated in 1996, Masters in 2002 from Annamalai University, Annamalai Nagar, India and currently pursuing his Ph.D from J.N.T. University(K),Kakinada, India. He is working as Associate Professor in the Department of Electrical Engineering, Vishnu Institute of Technology, Bhimavaram, Andhra Pradesh, India. His research area is Power system operation and control.

Shufali Ashraf Wani received her M. Tech degree from Jamia Millia Islamia, New Delhi, India in the year 2014. She is currently pursuing her Ph.D at the same institute. Her areas of interest are instrumentation, power systems and application of intelligent techniques.

Index

F

G

H

I

J

L

M

O

P

R

S

T

U

V

W

X

Z

www.ingramcontent.com/pod-product-compliance
Lightning Source LLC
LaVergne TN
LVHW060250011026
849161LV00011B/2686